Rubin H. Landau
Manuel J. Páez
Cristian C. Bordeianu

Computational Physics

Related Titles

Paar, H.H.

An Introduction to Advanced Quantum Physics

2010

Print ISBN: 978-0-470-68675-1; also available in electronic formats

Har, J.J., Tamma, K.K.

Advances in Computational Dynamics of Particles, Materials and Structures

2012

Print ISBN: 978-0-470-74980-7; also available in electronic formats ISBN: 978-1-119-96589-3

Cohen-Tannoudji, C., Diu, B., Laloe, F.

Quantum Mechanics

2 Volume Set

1977

Print ISBN: 978-0-471-56952-7; also available in electronic formats

Schattke, W., Díez Muiño, R.

Quantum Monte-Carlo Programming

for Atoms, Molecules, Clusters, and Solids

2013

Print ISBN: 978-3-527-40851-1; also available in electronic formats

Zelevinsky, V.

Quantum Physics 1&2

2 Volume Set

2011

Print ISBN: 978-3-527-41057-6; also available in electronic formats

Rubin H. Landau
Manuel J. Páez
Cristian C. Bordeianu

Computational Physics

Problem Solving with Python

3rd completely revised edition

Verlag GmbH & Co. KGaA

Authors

Rubin H. Landau
Oregon State University
97331 Corvallis OR
United States

Manuel J. Páez
Universad de Antioquia
Departamento Fisica
Medellin
Colombia

Cristian C. Bordeianu
National Military College "Ştefan cal Mare"
Campulung Moldovenesc
Romania

Library of Congress Card No.:
applied for

British Library Cataloguing-in-Publication Data:
A catalogue record for this book is available from the British Library.

Bibliographic information published by the Deutsche Nationalbibliothek
The Deutsche Nationalbibliothek lists this publication in the Deutsche Nationalbibliografie; detailed bibliographic data are available on the Internet at http://dnb.d-nb.de.

Typesetting le-tex publishing services GmbH, Leipzig, Deutschland
Cover Design Formgeber, Mannheim, Deutschland

Print ISBN 978-3-527-41315-7
ePDF ISBN 978-3-527-68466-3
ePub ISBN 978-3-527-68469-4
Mobi ISBN 978-3-527-68467-0

To the memory of Jon Maestri

Contents

Preface

Seventeen years have past since Wiley first published Landau and Páez's *Computational Physics* and twelve years since Cristian Bordeianu joined the collaboration for the second edition. This third edition adheres to the original philosophy that the best way to learn computational physics (CP) is by working on a wide range of projects using the text and the computer as partners. Most projects are still constructed using a computational, scientific problem-solving paradigm:

$$\text{Problem} \rightarrow \text{Theory/Model} \rightarrow \text{Algorithm} \leftrightarrow \text{Visualization} \hookleftarrow \tag{0.1}$$

Our guiding hypothesis remains that CP is a computational science, which means that to understand CP you need to understand some physics, some applied mathematics, and some computer science. What is different in this edition is the choice of Python for sample codes and an increase in the number of topics covered. We now have a survey of CP which is more than enough for a full-year's course.

The use of Python is more than just a change of language, it is taking advantage of the Python ecosystem of base language plus multiple, specialized libraries to provide all computational needs. In addition, we find Python to be the easiest and most accessible language for beginners, while still being excellent for the type of interactive and exploratory computations now popular in scientific research. Furthermore, Python supplemented by the Visual package (VPython) has gained traction in lower division physics teaching, and this may serve as an excellent segue to a Python-based CP course. Nevertheless, the important aspects of computational modeling and thinking transcends any particular computer language, and so having a Python alternative to our previous use of Fortran, C and Java may help promote this view (codes in all languages are available).

As before, we advocate for the use of a compiled or interpreted programming language when learning CP, in contrast to a higher level problem-solving environment like Mathematica or Maple, which we use in daily work. This follows from our experiences that if you want to *understand* how to compute scientifically, then you must look inside a program's black box and get your hands dirty. Otherwise, the algorithms, logic, and the validity of solutions cannot be ascertained, and that is not a good physics. Not surprisingly, we believe all physicists should know how to read programs how to write them as well.

Notwithstanding our beliefs about programming, we appreciate how time-consuming and frustrating debugging programs often is, and especially for beginners. Accordingly, rather than make the learner write all codes from scratch, we have placed a large number of codes within the text and often ask the learner only to run, modify, and extend them. This not only leaves time for exploration and analysis, but also provides experience in the modern work environment in which one must incorporate new developments into the preexisting codes of others. Be that as it may, for this edition we have added problems in which the relevant codes are not in the text (but are available to instructors). This should permit an instructor to decide on the balance of new and second-hand codes with which their students should work.

In addition to the paper version of the text, there is also an eBook of it that incorporates many of the multimodal enhancements possible with modern technologies: video lecture modules, active simulations, editable codes, animations, and sounds. The eBook is available as a Web (HTML5) document appropriate for both PCs or mobile devices. The lecture modules, which can be viewed separately from the eBook, cover most of the topics in the text, are listed in Appendix B, and are available online. They may provide avenues for alternative understanding the text (either as a preview or a review), for an online course, or for a blended course that replaces some lecture time with lab time. This latter approach, which we recommend, provides time for the instructor to assist students more personally with their projects and their learning issues. The studio-produced lectures are truly "modules," with active slides, a dynamic table of context, excellent sound (except maybe for a Bronx accent), and with occasional demonstrations replacing the talking head.

The introductory chapter includes tables listing all of the problems and exercises in the text, their locations in the text, as well as the physics courses in which these problems may be used as computational examples. Although we think it better to have entire courses in CP rather than just examples in the traditional courses, the inclusion of examples may serve as a valuable first step towards modernization.

The entire book has been reedited to improve clarity and useability. New materials have also been added, and this has led to additional and reorganized chapters. Specific additions not found in the second edition include: descriptions of the Python language and its packages, demonstrations of several visualization packages, discussions of algebraic tools, an example on protein folding, a derivation of the Gaussian quadrature rule, searching to obtain the temperature dependence of magnetization, chaotic weather patterns, planetary motion, matrix computing with Numerical Python, expanded and updated discussion of parallel computing including scalability and domain composition, optimized matrix computing with NumPy, GPU computing, CUDA programming, principal components analysis, digital filtering, the fast Fourier transform (FFT), an entire chapter on wavelet analysis and data compression, a variety of predator–prey models, signals of chaos, nonlinear behavior of double pendulum, cellular automata, Perlin noise, ray tracing, Wang–Landau sampling for thermodynamic simulations, finite *element* (in addition to *difference*) solutions of 1D and 2D PDEs, waves on a catenary, finite-difference-time-domain solutions for E&M waves, advection and shock waves in fluids, and a new chapter on fluid dynamics. We hope you enjoy it all!

Redmond, Oregon, June 2014 *RHL*, rubin@science.oregonstate.edu

Acknowledgments

Immature poets imitate;
mature poets steal.
T.S. Elliot

This book and the courses it is based upon could not have been created without continued financial support from the National Science Foundation's CCLI, EPIC, and NPACI programs, as well as support from the Oregon State University. Thank you all and we hope we have done you proud.

Our CP developments have followed the pioneering path paved by the books of Thompson, Gould and Tobochnik, Koonin and Press *et al.*; indubitably, we have borrowed material from them and made it our own with no further thought. We wish to acknowledge valuable contributions by Hans Kowallik, Sally Haerer (video-lecture modules), Paul Fink, Michel Vallières, Joel Wetzel, Oscar A. Restrepo, Jaime Zuluaga, Pavel Snopok, and Henri Jansen. It is our pleasure to acknowledge the invaluable friendship, encouragement, helpful discussions, and experiences we have had with many colleagues and students over the years. We are particularly indebted to Guillermo Avendaño-Franco, Saturo S. Kano, Melanie Johnson, Jon Maestri (deceased), David McIntyre, Shashikant Phatak, Viktor Podolskiy, C.E. Yaguna, Zlatco Dimcovic, and Al Stetz. The new work on principal component analysis resulted from a wonderful collaboration with Jon Wright and Roy Schult in 1997. Our gratitude also goes to the reviewers for their thoughtful and valuable suggestions, and to Bruce Sherwood, who has assisted us in making the Python codes run faster and look better. And finally, Martin Preuss, Nina Stadthaus, Ann Seidel, and Vera Palmer at Wiley-VCH have been a pleasure to work with.

In spite of everyone's best efforts, there are still errors and confusing statements in the book and codes for which we are to blame.

Finally, we extend our gratitude to the wives, Jan and Lucia, whose reliable support and encouragement are lovingly accepted, as always.

1
Introduction

Beginnings are hard.
Chaim Potok

Nothing is more expensive than a start.
Friedrich Nietzsche

This book is really two books. There is a rather traditional paper one with a related Web site, as well as an eBook version containing a variety of digital features best experienced on a computer. Yet even if you are reading from paper, you can still avail yourself of many of digital features, including video-based lecture modules, via the book's Web sites: http://physics.oregonstate.edu/~rubin/Books/CPbook/eBook/Lectures/ and www.wiley.com/WileyCDA.
We start this chapter with a description of how computational physics (CP) fits into physics and into the broader field of computational science. We then describe the subjects we are to cover, and present lists of all the problems in the text and in which area of physics they can be used as computational examples. The chapter finally gets down to business by discussing the Python language, some of the many packages that are available for Python, and some detailed examples of the use of visualization and symbolic manipulation packages.

1.1
Computational Physics and Computational Science

This book presents computational physics (CP) as a subfield of computational science. This implies that CP is a multidisciplinary subject that combines aspects of physics, applied mathematics, and computer science (CS) (Figure 1.1a), with the aim of solving realistic and ever-changing physics problems. Other computational sciences replace physics with their discipline, such as biology, chemistry, engineering, and so on. Although related, computational science is *not* part of computer science. CS studies computing for its own intrinsic interest and develops the hardware and software tools that computational scientists use. Likewise, applied mathematics develops and studies the algorithms that computational scientists use. As much as we also find math and CS interesting for their own sakes,

Computational Physics, 3rd edition. Rubin H. Landau, Manuel J. Páez, Cristian C. Bordeianu.

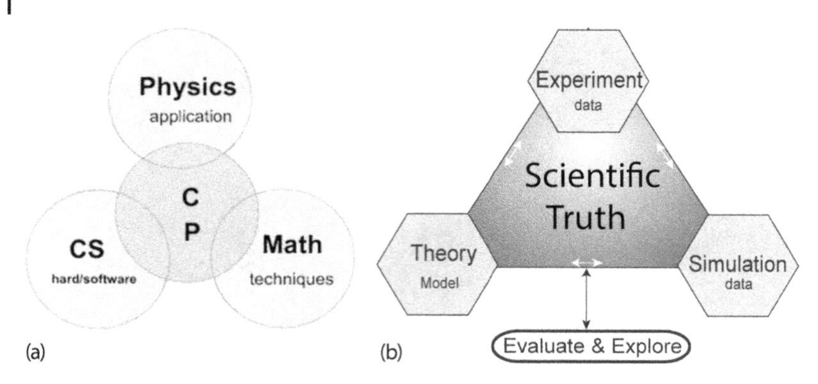

Figure 1.1 (a) A representation of the multidisciplinary nature of computational physics as an overlap of physics, applied mathematics and computer science, and as a bridge among them. (b) Simulation has been added to experiment and theory as a basic approach in the search for scientific truth. Although this book focuses on simulation, we present it as part of the scientific process.

our focus is on helping the reader do better physics for which you need to understand the CS and math well enough to solve your problems correctly, but not to become an expert programmer.

As CP has matured, we have come to realize that it is more than the overlap of physics, computer science, and mathematics. It is also a bridge among them (the central region in Figure 1.1a) containing core elements of it own, such as computational tools and methods. To us, CP's commonality of tools and its problem-solving mindset draws it toward the other computational sciences and away from the subspecialization found in so much of physics. In order to emphasize our computational science focus, to the extent possible, we present the subjects in this book in the form of a *Problem* to solve, with the components that constitute the solution separated according to the scientific problem-solving paradigm (Figure 1.1b). In recent times, this type of problem-solving approach, which can be traced back to the post-World War II research techniques developed at US national laboratories, has been applied to science education where it is called something like *computational scientific thinking*. This is clearly related to what the computer scientists more recently have come to call *Computational Thinking*, but the former is less discipline specific. Our computational scientific thinking is a hands-on, inquiry-based project approach in which there is problem analysis, a theoretical foundation that considers computability and appropriate modeling, algorithmic thinking and development, debugging, and an assessment that leads back to the original problem.

Traditionally, physics utilizes both experimental and theoretical approaches to discover scientific truth. Being able to transform a theory into an algorithm requires significant theoretical insight, detailed physical and mathematical understanding, and a mastery of the art of programming. The actual debugging, testing, and organization of scientific programs are analogous to experimentation, with the numerical simulations of nature being virtual experiments. The synthesis of numbers into generalizations, predictions, and conclusions requires the insight and intuition common to both experimental and theoretical science. In fact, the use of computation and simulation has now become so prevalent and essential a part of the scientific process that many people believe that the scientific paradigm has been extended to include simulation as an additional pillar (Figure 1.1b). Nevertheless, as a science, CP must hold experiment supreme, regardless of the beauty of the mathematics.

1.2
This Book's Subjects

This book starts with a discussion of Python as a computing environment and then discusses some basic computational topics. A simple review of computing hardware is put off until Chapter 10, although it also fits logically at the beginning of a course. We include some physics applications in the first third of this book, by put off most CP until the latter two-thirds of the book.

This text have been written to be accessible to upper division undergraduates, although many graduate students without a CP background might also benefit, even from the more elementary topics. We cover both ordinary and partial differential equation (PDE) applications, as well as problems using linear algebra, for which we recommend the established subroutine libraries. Some intermediate-level analysis tools such as discrete Fourier transforms, wavelet analysis, and singular value/principal component decompositions, often poorly understood by physics students, are also covered (and recommended). We also present various topics in fluid dynamics including shock and soliton physics, which in our experience physics students often do not see otherwise. Some more advanced topics include integral equations for both the bound state and (singular) scattering problem in quantum mechanics, as well as Feynman path integrations.

A traditional way to view the materials in this text is in terms of its use in courses. In our classes (CPUG, 2009), we have used approximately the first third of the text, with its emphasis on computing tools, for a course called *Scientific Computing* that is taken after students have acquired familiarity with some compiled language. Typical topics covered in this one-quarter course are given in Table 1.1, although we have used others as well. The latter two-thirds of the text, with its greater emphasis on physics, has typically been used for a two-quarter (20-week) course in CP. Typical topics covered for each quarter are given in Table 1.2. What with many of the topics being research level, these materials can easily be used for a full year's course or for extended research projects.

The text also uses various symbols and fonts to help clarify the type of material being dealt with. These include:

⊙	Optional material
`Monospace font`	Words as they would appear on a computer screen
Vertical gray line	Note to reader at the beginning of a chapter saying

Table 1.1 Topics for one-quarter (10 Weeks) scientific computing course.

Week	Topics	Chapter	Week	Topics	Chapter
1	OS tools, limits	1, (10)	6	Matrices, N-D search	6
2	Visualization, Errors	1, 3	7	Data fitting	7
3	Monte Carlo,	4, 4	8	ODE oscillations	8
4	Integration, visualization	5, (1)	9	ODE eigenvalues	8
5	Derivatives, searching	5, 7	10	Hardware basics	10

Table 1.2 Topics for two-quarters (20 Weeks) computational physics course.

Computational Physics I			Computational Physics II		
Week	Topics	Chapter	Week	Topics	Chapter
1	Nonlinear ODEs	8, 9	1	Ising model, Metropolis	17
2	Chaotic scattering	9	2	Molecular dynamics	18
3	Fourier analysis, filters	12	3	Project completions	—
4	Wavelet analysis	13	4	Laplace and Poisson PDEs	19
5	Nonlinear maps	14	5	Heat PDE	19
6	Chaotic/double pendulum	15	6	Waves, catenary, friction	21
7	Project completion	15	7	Shocks and solitons	24
8	Fractals, growth	16	8	Fluid dynamics	25
9	Parallel computing, MPI	10, 11	9	Quantum integral equations	26
10	More parallel computing	10, 11	10	Feynman path integration	17

1.3 This Book's Problems

For this book to contribute to a successful learning experience, we assume that the reader will work through what we call the *Problem* at the beginning of each discussion. This entails studying the text, writing, debugging, and running programs, visualizing the results, and then expressing in words what has been performed and what can be concluded. As part of this approach, we suggest that the learner write up a mini lab report for each problem containing sections on

Equations solved	Numerical method	Code listing
Visualization	Discussion	Critique

Although we recognize that programming is a valuable skill for scientists, we also know that it is incredibly exacting and time-consuming. In order to lighten the workload, we provide "bare bones" programs. We recommend that these be used as guides for the reader's own programs, or tested and extended to solve the problem at hand. In any case, they should be understood as part of the text.

While we think it is best to take a course, or several courses, in CP, we recognize that this is not always possible and some instructors may only be able to include some CP examples in their traditional courses. To assist in this latter endeavor, in this section we list the location of each problem distributed throughout the text and the subject area of each problem. Of course this is not really possible with a multidisciplinary subject like CP, and so there is an overlap. The code used in the table for different subjects is: QM = quantum mechanics or modern physics, CM = classical mechanics, NL = nonlinear dynamics, EM = electricity and magnetism, SP = statistical physics, MM = mathematical methods as well as tools, FD = fluid dynamics, CS = computing fundamentals, Th = thermal physics, and BI = biology. As you can see from the tables, there are many problems and exercises, which reflects our view that you learn computing best by doing it, and that many problems cover more than one subject.

Problems and exercises in computational basics

Subject	Section	Subject	Section	Subject	Section
MM, CS	1.6	CS	2.2.2	CS	2.2.2
CS	2.4.3	CS	2.4.5	CS	2.5.2
CS	3.1.2	CS	3.2	CS	3.2.2
CS	3.3	CS	3.3.1	CS	4.2.2
MM, CS	6.6	CS	10.13.1	CS	10.14.1
CS	11.3.1	CS	11.1.2	CS	11.2.1

Problems and exercises in thermal physics and statistical physics

Subject	Section	Subject	Section	Subject	Section
SP, MM	4.3	SP, MM	4.5	QM, SP	4.6
Th, SP	7.4	Th, SP	7.4.1	NL, SP	16.3.3
NL, SP	16.4.1	NL, SP	16.7.1	NL, SP	16.7.1
NL, SP	16.8	NL, SP	16.11	SP, QM	17.4.1
SP, QM	17.4.2	SP, QM	17.6.2	Th, MM	20.2.4
Th, MM	20.3	TH, MM	20.4.2	TH, MM	20.1
TH, MM	17.1	SP	16.2	SP, BI	16.3
SP	16.4	SP, MM	16.5	SP	16.6
SP	16.7				

Problems and exercises in electricity and magnetism

Subject	Section	Subject	Section	Subject	Section
EM, MM	19.6	EM, MM	19.7	EM, MM	19.8
EM, MM	19.9	EM, MM	23.2	EM, MM	23.5
EM, MM	23.5.1	EM, MM	23.6.6	EM, MM	22.7.2
EM, MM	22.10	EM, MM	19.2		

Problems and exercises in quantum mechanics

Subject	Section	Subject	Section	Subject	Section
QM, SP	4.6	QM, MM	7.1	QM, MM	7.2.1
QM, MM	7.3.2	QM, MM	9.1	QM, MM	9.2
QM, MM	9.2.1	QM, MM	9.3	QM	13.6.3
QM, MM	17.7	QM, MM	26.1	QM, MM	26.3
QM, MM	22.1				

Problems and exercises in classical mechanics and nonlinear dynamics

Subject	Section	Subject	Section	Subject	Section
CM, NL	5.16	CM	6.1	CM, NL	8.1
CM, NL	8.7.1	CM, NL	8.8	CM, NL	8.9
CM, NL	8.10	CM, NL	9.4	CM, NL	9.4.3
CM	9.5	CM	9.7	CM	9.7
NL, FD	9.7	CM	9.7	CM, MM	6.6.2
CM, MM	6.6.1	CM, NL	12.1	BI, NL	14.3
CM, MM	6.6.1	BI, NL	14.4	BI, NL	14.5.2
BI, NL	14.5.3	BI, NL	14.10	BI, NL	14.11.1
BI, NL	14.11.4	BI, NL	14.11.5	CM, NL	15.1.3
CM, NL	15.1	NL, BI	14.1	NL, BI	14.9
CM, NL	15.2.2	CM, NL	15.3	CM, NL	15.4
CM, NL	15.5	CM, NL	15.6	CM, NL	15.7
CM, NL	15.7	NL, MM	16.2.1	NL, MM	16.3.3
NL, MM	16.4.1	NL, MM	16.5.3	NL, MM	16.7.1
NL, MM	16.7.1	NL, MM	16.8	NL, MM	16.11
CM, MM	21.2.4	CM, MM	21.3	CM, MM	21.4.3
CM, MM	24.6	CM, MM	21.1	CM, MM	21.5

Problems and exercises in fluid dynamics

Subject	Section	Subject	Section	Subject	Section
NL, FD	9.7	FD, MM	24.3.2	FD, MM	24.5.3
FD, MM	24.5.4	FD, MM	25.1	FD, MM	25.2.3
FD, MM	25.4.4	FD, MM	25.4.5		

Problems and exercises in mathematical methods and computational tools

Subject	Section	Subject	Section	Subject	Section
MM, CS	1.6	MM, SP	4.3	SP, MM	4.3.2
BI, MM	4.4	MM, SP	4.5	MM	5.12.3
MM	5.16	MM	5.17.2	MM	5.5
MM	5.5	QM, MM	7.1	QM, MM	7.2.1
QM, MM	7.3.2	MM, QM	9.1	QM, MM	9.2
QM, MM	9.2.1	QM, MM	9.3	CM, NL	9.4
MM, CS	6.6	CM, MM	6.6.2	CM, MM	6.6.1
MM	7.5.1	MM	7.5.2.1	MM	7.8
MM	7.8.1	MM	7.8.2	MM	12.3
MM	12.5.3	MM	12.7.1	MM	12.11
MM	13.3.1	MM	13.5.2	MM	13.6.3
CM, MM	15.5	NL, MM	16.2.1	NL, MM	16.3.3
NL, MM	16.4.1	NL, MM	16.5.3	NL, MM	16.7.1
NL, MM	16.7.1	NL, MM	16.8	NL, MM	16.11
Th, MM	20.2.4	Th, MM	20.3	TH, MM	20.4.2
EM, MM	19.6	EM, MM	19.7	EM, MM	19.8
EM, MM	19.9	EM, MM	23.5	EM, MM	23.5.1
EM, MM	23.6.6	CM, MM	21.2.4	CM, MM	21.3
CM, MM	21.4.3	QM, MM	22.2.2	QM, MM	22.2.2
QM, MM	22.2.3	EM, MM	22.7.2	EM, MM	22.10
FD, MM	24.3.2	FD, MM	24.5.3	FD, MM	24.5.4
FD, MM	25.2.3	FD, MM	25.4.4	FD, MM	25.4.5
QM, MM	26.2.3	QM, MM	26.2.4	QM, MM	26.4.5
QM, MM	26.4.6	MM, NL	13.1	MM, CM	12.1
MM	12.6	MM	12.8.1	MM	7.5
MM	7.5.2.1	MM	7.6	MM	7.8.2
MM	13.7.2				

Problems and exercises in molecular dynamics and biological applications

Subject	Section	Subject	Section	Subject	Section
BI, MM	4.4	BI, NL	14.3	BI, NL	14.4
BI, NL	14.5.2	BI, NL	14.5.3	BI, NL	14.10
BI, NL	14.11.1	BI, NL	14.11.4	BI, NL	14.11.5
SP, BI	16.3	BI, NL	14.1	BI, NL	14.9
MD, QM	18.3	MD, QM	18.4	MD	18.4
MM, SP	18				

1.4
This Book's Language: The Python Ecosystem

The codes in this edition of *Computational Physics* employ the computer language *Python*. Previous editions have had their examples in Java, Fortran and C, and used post-simulation tools for visualization. Although we have experienced no general agreement in the computational science community as to the best language for scientific computing, this has not stopped many of the users of each language from declaring it to be the best. Even so, we hereby declare that we have found Python to be the best language yet for teaching CP. Python is free, robust (not easily broken), portable (program run without modifications on various devices), universal (available for most every computer system), has a clean syntax that lets students learn the language quickly, has dynamic typing and high-level, built-in data types that enable getting programs to work quickly without having to declare data types or arrays, count matching braces, or use separate visualization programs. Because Python is interpreted, students can learn the language by executing and analyzing individual commands within an interactive shell, or by running the entire program in one fell swoop. Furthermore, Python brings to scientific computing the availability of a myriad of free packages supporting numerical algorithms, state-of the art, or simple, visualizations and specialized toolkits that rival those in Matlab and Mathematica/Maple. And did we mention, all of this is free?

There are literally thousands of Python packages available, but not to worry, we use only a few for numerical and visualization purposes. Because it is essential to be able to run and modify the example codes in this book, we suggest that you spend the time necessary to get Python to function properly on your computer (and then leave notes as to what you did). For learning Python, we recommend the online tutorials (Ptut, 2014; Pguide, 2014; Plearn, 2014), the books by Langtangen Langtangen (2008) and Langtangen (2009), and the *Python Essential Reference* (Beazley, 2009). For general numerical methods, a book by Press *et al.* (1994) is the standard, and most fun to read, while the NIST Digital Library of Mathematical Functions (NIST, 2014) is probably the most convenient.

Python has developed rapidly since its first implementation in December 1989 (History, 2009). Python's combination of language plus packages is now the standard for the explorative and interactive computing that typifies the present-day scientific research. These rapid developments of Python have also led to a succession of new versions, and the inevitable incompatibilities. Most of the codes in this book were written using Python 2, which was released in 2000, and specifically Python 2.6 with the Visual package (also known as "VPython"). However, there have been major changes to the Python development process as well as in features, and this has led to the release of Python 3.0 in December 2008. Unfortunately, some of the changes in Python 3 were not backward compatible with Python 2.6 and 2.7, and so advances in both Python 2 and 3 and their associated packages have been occurring in parallel. (For our codes, the major difference is in the print statement using a parenthesis in 3, which is not hard to correct.) Furthermore, there have been new versions of operating systems and processors from 32- to 64-bit CPUs, and this also has led to the variety of Python versions and associated packages.

To be honest, we have sometimes felt frustrated by these changes and resulting incompatibilities; however, we are intent on not sharing that! While we will describe the packages and distribution briefly, we indicate here that we have adapted to the real world by having both independent Python 2 and 3 implementations exist on our computers. Specifically, our Visual package programs use Python 3.2, while the others use the *Enthought Canopy Distribution Version 1.3.0,* which at present uses *Python 2.7.3.* (The Visual package is not available in Enthought.)

1.4.1 Python Packages (Libraries)

The Python language plus its family of packages comprise a veritable ecosystem for computing. A package or module is a collection of related methods or classes of methods that are assembled together into a subroutine library.[1] Inclusion of the appropriate packages extends the language to meet the specialized needs of various science and engineering disciplines, and lets one obtain state-of-the-art computing for free. In fact, the May/June 2007 and March/April 2011 issues of *Computing in Science and Engineering* (Perez *et al.*, 2010) focus on scientific computing with Python, and we recommend them.

To use a package named `PackageName`, you include in your Python program either an `import PackageName` or a `from PackageName` statement at the beginning of your program. The `import` statement loads the entire package, which is efficient, but may require you to include the package name as a prefix to the method you want. For example,

```
>>> from visual.graph import *        # Import from visual package
>>> y1 = visual.graph.gcurve(color = blue, delta = 3)      # Use of graph
```

1) The Python Package Index (PYPI, 2014), a repository of free Python packages, currently contains more than 40 000 packages!

Here >>> represents the prompt for a Python shell. Some of the typing can be avoided by assigning a symbol to the package name:

```
>>> import visual.graph as p
>>> y1 = p.gcurve(color = blue, delta = 3)
```

There is also a starred version of `from` that copies *all* of the methods of a package (here Matplotlib called pylab) so that you can leave off prefixes:

```
>>> from pylab import *          # Import all pylab methods
>>> plot(x, y, '-', lw=2)        # A pylab method without prefix
```

1.4.2
This Book's Packages

We are about to describe some of the packages that make Python such a rich environment. If you are anxious to get started now, or worry about getting overwhelmed by the Python packages, you may just want to load *VPython* now and move on to the next chapter. You will need some more stuff to do visualizations and matrices, but you can always upgrade your knowledge when you feel more comfortable with Python.

Because all too often you do not know what you do not know, or what you need to know, we list here a few, basic Python packages and what each does. The packages used in the text are underlined and described more fully later.

Boost.Python A C++ library that enables seamless interoperability between C++ and Python, thus extending the lifetime of legacy codes and making use of the speed of C, www.boost.org/doc/libs/1_55_0/libs/python/doc/.

Cython: C Extensions for Python A superset of the Python language that supports calling C functions and intermixing Python and C for legacy purposes and for high performance, http://cython.org/.

f2py: Fortran to Python Interface Generator that provides connection between Python and Fortran languages; great for steering legacy codes, http://cens.ioc.ee/projects/f2py2e/.

IPython: Interactive Python An advanced *shell* (command line interpreter) that extends Python's basic interpreter IDLE. IPython has enhanced interactivity and interactive visualization capabilities that encourage exploratory computing. IPython also has a browser-based notebook like Mathematica that permits embedded code executions, as well as capabilities for parallel computing, http://ipython.org/.

Matplotlib: Mathematics Plotting Library A 2D and 3D graphics library that uses NumPy (Numerical Python), and produces publication quality figures in a variety of hard copy formats, and permits interactive graphics. Similar to MATLAB's plotting (except Matplotlib is free and doesn't need its license renewed yearly). See Section 1.5.3 for examples and discussion, http://matplotlib.sf.net.

Mayavi Interactive and simplified 3D visualization. Also contains TVTK, a wrapper for the more basic Visualization Tool Kit VTK. ("Mayavi" is Sanskrit for magician.) See Section 1.5.6 for examples and discussion, http://mayavi.sf.net.

Mpmath: Multiprecision Floating Point Arithmetic A pure-Python library for multiprecision floating-point arithmetic for transcendental functions, unlimited exponent sizes, complex numbers, interval arithmetic, numerical integration and differentiation, root-finding, linear algebra, and more, https://code.google.com/p/mpmath/.

NumPy: Numerical Python Permits the use of fast, high-level multidimensional arrays in Python, which are used as the basis for many of the numerical procedures in Python libraries (NumPy, 2013; SciPy, 2014) – the successor to both *Numeric* and *Numarray*. Used by Visual and Matplotlib. *SciPy* extends NumPy. See Sections 6.5, 6.5.1, and 11.2 for examples of NumPy array use.

Pandas: Python Data Analysis Library A collection of high-performance, user-friendly data structures and data analysis tools, http://pandas.pydata.org/.

PIL: Python Imaging Library Image processing and graphics routines for various file formats, www.pythonware.com/products/pil/.

Python The Python standard library, http://python.org.

PyVISA Wrappers for the VISA library providing controls for measurement equipment through various busses from within Python programs, http://pyvisa.readthedocs.org/en/latest/.

SciKits: SciPy Toolkits A collection of toolkits that extend SciPy to special disciplines such as audio processing, financial computation, geosciences, time series analysis, computer vision, engineering, machine learning, medical computing, and bioinformatics, https://scikits.appspot.com/.

SciPy: Scientific Python A basic library for mathematics, science, and engineering. (See SciKits for further extensions.) Provides user-friendly and efficient numerical routines for linear algebra, optimization, integration, special functions, signal and image processing, statistics, genetic algorithms, ODE solutions, and others. Uses NumPy's N-dimensional arrays but also extends NumPy. SciPy essentially provides wrapper for many existing libraries in other languages, such as LAPACK (Anderson *et al.*, 2013) and FFT. The SciPy distribution usually includes Python, NumPy, and f2py, http://scipy.org.

Sphinx Python documentation generator for output in various formats, http://sphinx-doc.org/.

SWIG An interface compiler that connects programs written in C and C++ with scripting languages such as Perl, Python, Ruby, and TCL. Useful for extending the lifetime of legacy codes or for making use of the speed of C, http://swig.org.

SyFi: Symbolic Finite Elements Built on top of the symbolic math library GiNaC, SyFi is used in the finite element solution of PDEs. It provides polygonal domains, polynomial spaces and degrees of freedom as symbolic expressions that are easily manipulated, https://pypi.python.org/pypi/SyFi/.

SymPy: Symbolic Python A system for symbolic mathematics using pure Python (no external libraries) to provide a simple computer algebra system that also includes calculus, differential equations, etc. Similar to Maple or Mathematica, with the Sage package being even more complete. Examples in Section 1.7. See also mpmath, http://sympy.org/.

VisIt Distributed, parallel, visualization tool for visualizing data defined on 2D and 3D structured and unstructured meshes, https://wci.llnl.gov/codes/visit/.

Visual (VPython) Python programming language plus the *visual* 3D graphics module, with the *VIDLE* interactive shell replacing Python's standard *IDLE*. Particularly helpful, even for novices, in creating 3D demonstrations and animations for education. We often use Visual for 2D plots of numerical data and animations. Can be installed separately from Canopy, http://vpython.org/.

1.4.3 The Easy Way: Python Distributions (Package Collections)

Although most Python packages are free, there is a true value for both users and vendors to distribute a collection of packages that have been engineered and tuned to work well together, and that can be installed in one fell swoop. (This is similar to what Red Hat and Debian do with Linux.) These distributions can be thought of as complete Python ecosystems assembled for specific purposes, and are highly recommended. Here we mention four with which we are familiar:

Anaconda A free Python distribution including more than 125 packages for science, mathematics, engineering, and data analysis, including Python, NumPy, SciPy, Pandas, IPython, Matplotlib, Numba, Blaze, and Bokeh. Anaconda is self-described as enterprise-ready for large-scale data processing, predictive analytics, and scientific computing, and permits easy switching between Python 2.6, 2.7, and 3.3. As also true for Canopy, Anaconda installs in its own directory and so runs independently from other Python installations on your computer, https://store.continuum.io/cshop/anaconda/.

Enthought Canopy A comprehensive and complete Python analysis environment with easy installation and updates. The commercial distribution includes more than 150 packages, yet is available for free to academic users. In any case, there is an Express version containing more than 50 packages that is free to everyone. The packages include the IPython, NumPy, SciPy, Matplotlib, Mayavi, scikit, SymPy, Chaco, Envisage, and Pandas, /https://www.enthought.com/products/canopy/.

Python XY A free scientific and engineering development collection of packages for numerical computations, data analysis, and data visualization employing the Qt graphical libraries for GUI development and the Spyder interactive scientific development environment, https://code.google.com/p/pythonxy/.

Sage An amazingly complete collection of open-source packages for mathematical computations, both numerically and symbolically using the IPython interface and notebooks. Sage's stated mission is to create a viable, free, open-source alternative to Magma, Maple, Mathematica, and Matlab, www.sagemath.org/.

1.5 Python's Visualization Tools

If I can't picture it, I can't understand it.

Albert Einstein

In the sections to follow we discuss tools to visualize data produced by simulations and measurements. Whereas other books may choose to relegate this discussion to an appendix, or not to include it at all, we believe that visualization is such an integral part of CP, and so useful for your work in the rest of this book, that we have placed it here, right up front. We describe the use of Matplotlib, Visual (VPython), and Mayavi. VPython makes easy 2D plot, solid geometric figures, and animations. Matplotlib makes very nice 3D (surface) plots, while Mayavi can create state-of-the-art visualizations.

Generalities One of the most rewarding aspects of computing is visualizing the results of calculations. While in the past this was performed with 2D plots, in modern times it is a regular practice to use 3D (surface) plots, volume rendering (dicing and slicing), animations, and virtual reality (gaming) tools. These types of visualizations are often breathtakingly beautiful and may provide deep insights into problems by letting us see and "handle" the functions with which we are working. Visualization also assists in the debugging process, the development of physical and mathematical intuition, and the all-around enjoyment of work.

In thinking about ways to view your results, keep in mind that the point of visualization is to make the science clearer and to communicate your work to others. Then it follows that you should make all figures as clear, informative, and self-explanatory as possible, especially if you will be using them in presentations without captions. This means labels for curves and data points, a title, and labels on the axes.[2] After this, you should look at your visualization and ask whether there are better choices of units, ranges of axes, colors, style, and so on, that might get the message across better and provide better insight. And try to remember that those colors which look great on your monitor may turn into uninformative grays when printed. Considering the complexity of human perception and cognition,

2) Although this may not need saying, place the independent variable x along the abscissa (horizontal), and the dependent variable $y = f(x)$ along the ordinate.

there may not be a single best way to visualize a particular data set, and so some trial and error may be necessary to "see" what works best.

Listing 1.1 EasyVisual.py produces two different 2D plot using the Visual package.

```
# EasyVisual.py:          Simple graph object using Visual

from visual.graph import *                        # Import Visual

Plot1 = gcurve(color = color.white)               # gcurve method

for x in arange(0., 8.1, 0.1):                    # x range
    Plot1.plot( pos = (x, 5.*cos(2.*x)*exp(-0.4*x)) ) # Plot pts

graph1 =  gdisplay(width=600, height=450,\
    title='Visual 2D Plot', xtitle='x', ytitle='f(x)',\
     foreground = color.black, background = color.white)

Plot2 = gdots(color = color.black)                # Dots

for x in arange( -5., +5, 0.1 ):
    Plot2.plot(pos = (x, cos(x)))
```

1.5.1
Visual (VPython)'s 2D Plots

As indicated in the description of packets, VPython (Python plus the Visual package) is a simple way to get started with Python and visualizations.[3] The Visual package is useful for creating 3D solids, 2D plots, and animations. For example, in Figure 1.2, we present two plots produced by the program EasyVisual.py in Listing 1.1. Notice that the plotting technique is to create first a plot object, and then to add the points to the object, one by one. (In contrast, Matplotlib creates a vector of points and then plots the entire vector.)

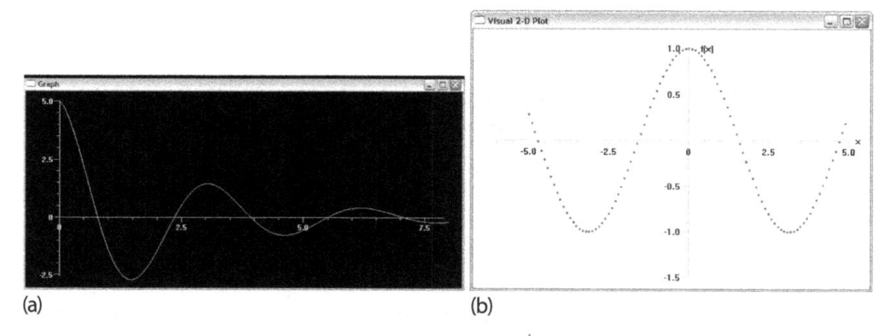

(a) (b)

Figure 1.2 Screen dumps of two x–y plots produced by EasyVisual.py using the Visual package. Plot (a) uses default parameters, while plot (b) uses user-supplied options.

3) Because Visual is not one of the Canopy packages, to run our Visual programs you would need to install the Visual package and the version of Python that runs with it, even if you have Canopy installed. There is no problem doing this because VPython and Canopy go into different folders/directories.

EasyVisual.py is seen to create two plot objects, Plot1 and Plot2, with the plot method used to plot each object. Plot1 uses the gcurve method with no options specified other than the color of the curve (white). We obtain (Figure 1.2a) a connected curve by default, but no labels. In contrast, Plot2 uses the gdisplay method to set the display characteristics for the plot to follow, and then gdots to draw the data as points (Figure 1.2b). You use a gdisplay plotting object before you create a gcurve graph object, in order to set the size, position, and title for the graph window, to specify titles for the x and y axes, and to specify maximum values for each axis. The gdisplay arguments are self-explanatory, with the width and height given in pixels. Because we set Plot2= gdots(), only points are plotted.

Listing 1.2 3GraphVisual.py produces a 2D x–y plot with the Matplotlib and NumPy packages.

```
# 3GraphVisual.py: 3 plots in the same figure, with bars, dots and curve      1

from visual import *                                                           3
from visual.graph import*
                                                                               5
string = "blue: sin^2(x), white: cos^2(x), red: sin(x)*cos(x)"
graph1 = gdisplay(title=string, xtitle='x', ytitle='y')                        7

y1 = gcurve(color=color.yellow, delta=3)          # curve                      9
y2 = gvbars(color=color.white)                     # vertical bars
y3 = gdots(color=color.red, delta=3)               # dots                      11

for x in arange(-5, 5, 0.1):                       # arange for floats         13
    y1.plot( pos=(x, sin(x)*sin(x)) )
    y2.plot( pos=(x, cos(x)*cos(x)/3.) )                                       15
    y3.plot( pos=(x, sin(x)*cos(x)) )
```

Note that the Python codes are listed within shaded boxes with some formatting to improve readability. For example, see Listing 1.2. Note that we have structured the codes so that a line is skipped before major elements like functions, and that indentations indicate structures in Python (where Java and C may use braces).

It is often a good idea to place several plots in the same figure. The program 3GraphVisual.py in Listing 1.2 does that and produces the graph in Figure 1.3a. There are white vertical bars created with gvbars, red dots created with gdots, and a yellow curve created with gcurve (colors appear only as shades of gray in the paper text). Also note in 3GraphVisual.py that we avoid having to include the package name as a prefix to the commands by starting the program with import visual.graph as vg. This both imports Visual's graphing package and assigns the symbol vg to visual.graph.

1.5.1.1 VPython's 3D Objects

Listing 1.3 3Dshapes.py produces a sample of VPython's 3D shapes.

```
# 3Dshapes.py: Some 3D Shapes of VPython
                                                                               2
from visual import *
                                                                               4
graph1 = display(width=500, height=500, title='VPython 3D Shapes',
    range=10)
```

```
sphere(pos=(0,0,0), radius=1, color=color.green)
sphere(pos= (0,1,-3), radius=1.5, color=color.red)
arrow(pos=(3,2,2), axis=(3,1,1), color=color.cyan)
cylinder(pos=(-3,-2,3), axis=(6,-1,5), color=color.yellow)
cone(pos=(-6,-6,0), axis=(-2,1,-0.5), radius=2, color=color.magenta)
helix(pos=(-5,5,-2), axis=(5,0,0), radius=2, thickness=0.4,
     color=color.orange)
ring(pos=(-6,1,0), axis=(1,1,1), radius=2, thickness=0.3,
     color=(0.3,0.4,0.6))
box(pos=(5,-2,2), length=5, width=5, height=0.4, color=(0.4,0.8,0.2))
pyramid(pos=(2,5,2), size=(4,3,2), color=(0.7,0.7,0.2))
ellipsoid(pos=(-1,-7,1), axis=(2,1,3), length=4, height=2, width=5,
     color=(0.1,0.9,0.8))
```

One way to make simulations appear more realistic is to use 3D solid shapes, for example, a sphere for a bouncing ball rather than just a dot. VPython can produce a variety of 3D shapes with one-line commands, as shown in Figure 1.4, and as produced by the code in Listing 1.3. To make the ball bounce, you would need to vary the **position** variable according to some kinematic equations.

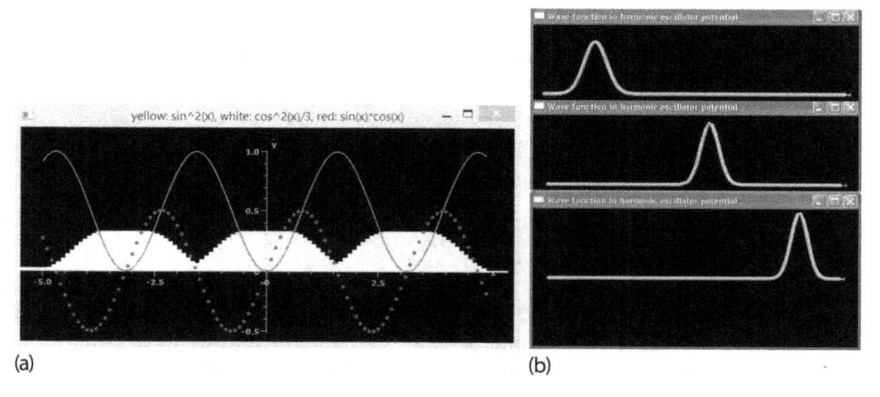

Figure 1.3 (a) Output from the program 3GraphVisual.py that places three different types of 2D plots on one graph using Visual. (b) Three frames from a Visual animation of a quantum mechanical wave packet produced with HarmosAnimate.py.

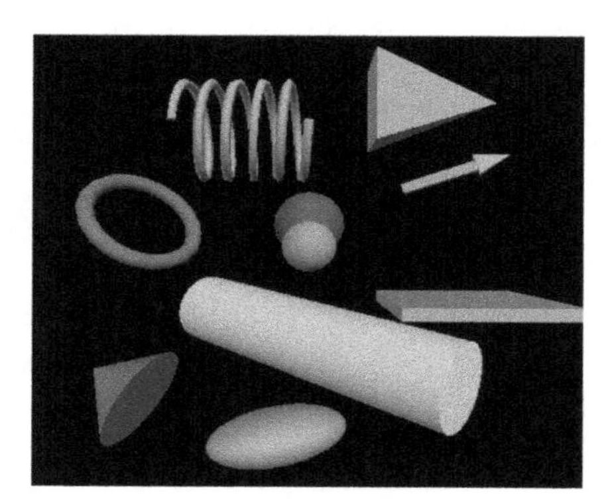

Figure 1.4 Some 3D shapes created with single commands in VPython.

1.5.2
VPython's Animations

Creating animations with Visual is essentially just making the same 2D plot over and over again, with each one at a slightly differing time, and then placing the plots on top of each other. When performed properly, this gives the impression of motion. Several of our sample codes produce animations, for example, HarmosAnimate.py and 3Danimate.py. Three frames produced by HarmosAnimate.py are shown in Figure 1.3b. The major portions of these codes deal with the solution of PDEs, which need not concern us yet. The part which makes the animation is simple:

```
PlotObj= curve(x=xs, color=color.yellow, radius=0.1)
...
while True:                                          # Runs forever
    rate(500)
    psr[1:-1] = ...
    psi[1:-1] = ..
    PlotObj.y = 4*(psr**2 + psi**2)
```

Here PlotObj is a curve that continually gets built from within a while loop and thus appears to be moving. Note that being able to plot points individually without having to store them all in an array for all times keeps the memory demand of the program quite small and leads to fast programs.

Listing 1.4 EasyMatPlot.py produces a, 2D x–y plot using the Matplotlib package (which includes the NumPy package).

```
# EasyMatPlot.py: Simple use of matplotlib's plot command                1

from pylab import *                              # Load Matplotlib     3

Xmin = -5.;   Xmax = +5.;   Npoints= 500                                 5
DelX = (Xmax - Xmin) / Npoints
x = arange(Xmin, Xmax, DelX)                                             7
y =  sin(x) * sin(x*x)                          # function of x array
                                                                         9
print ('arange => x[0], x[1],x[499]=%8.2f %8.2f %8.2f'
      %(x[0],x[1],x[499]))
print ('arange => y[0], y[1],y[499]=%8.2f %8.2f %8.2f'                 11
      %(y[0],y[1],y[499]))
print ("\n Now doing the plotting thing, look for Figure 1 on desktop" )
xlabel('x');        ylabel('f(x)');        title(' f(x) vs x')          13
text(-1.75, 0.75, 'MatPlotLib \n Example')          # Text on plot
plot(x, y, '-', lw=2)                                                    15
grid(True)                                          # Form grid
show()                                                                   17
```

1.5.3
Matplotlib's 2D Plots

Matplotlib is a powerful plotting package that lets you make 2D and 3D graphs, histograms, power spectra, bar charts, error charts, scatter plots, and more, all directly from within your Python program. Matplotlib is free, uses the sophisticated numerics of NumPy and LAPACK (Anderson *et al.*, 2013), and, believe it or not, is easy to use. Specifically, Matplotlib uses the NumPy `array` (vector) object to store the data to be plotted. In Chapter 6, we talk at more length about NumPy arrays, so you may want to go there soon to understand arrays better.

Matplotlib commands are by design similar to the plotting commands of MATLAB, a commercial problem-solving environment that is particularly popular in engineering. As is true for MATLAB, Matplotlib assumes that you have placed the x and y values that you wish to plot into 1D arrays (vectors), and then plots these vectors in one fell swoop. This is in contrast to Visual, which first creates a plot object and then adds points to the object one by one. Because Matplotlib is not part of standard Python, you must import the entire Matplotlib package, or the individual methods you use, into your program. For example, on line 2 of `EasyMatPlot.py` in Listing 1.4 (line numbers are in the dark shading on the right), we import Matplotlib as the `pylab` library:

```
from pylab import *                    # Load Matplotlib
```

Then, on lines 6 and 7 we calculate and input arrays of the x and y values

```
x = arange(Xmin, Xmax, DelX)          # Form x array in range with increment
y =  -sin(x)*cos(x)                   # Form y array as function of x array
```

As you can see, NumPy's `arange` method constructs an array covering "a range" between `Xmax` and `Xmin` in steps of `DelX`. Because the limits are floating-point numbers, so also will be the x_i's. And because `x` is an array, `y = -sin(x)*cos(x)` is automatically one too! The actual plotting is performed on line 14 with a dash "–" used to indicate a line, and `lw` = 2 to set its width. The result is shown in Figure 1.5a with the desired labels and title. The `show()` command produces the graph on your desktop. More commands are given in Table 1.3. We suggest you try out some of the options and types of plots possible.

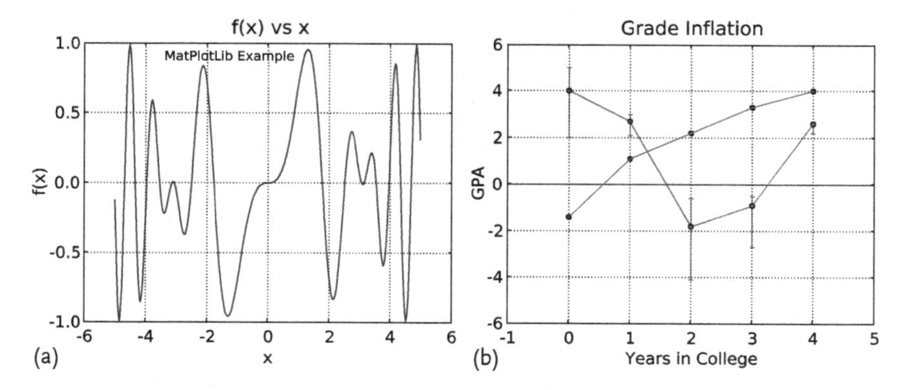

Figure 1.5 Matplotlib plots. (a) Output of EasyMatPlot.py showing a simple, x–y plot. (b) Output from GradesMatPlot.py that places two sets of data points, two curves, and unequal upper and lower error bars, all on one plot.

Table 1.3 Some common Matplotlib commands.

Command	Effect	Command	Effect
plot(x, y, '-', lw=2)	x–y curve, line width 2	myPlot.setYRange(-8., 8.)	Set y range
show()	Show output graph	myPlot.setSize(500, 400)	Size in pixels
xlabel('x')	x-axis label	pyplot.semilogx	Semilog x plot
ylabel('f(x)')	y-axis label	pyplot.semilogy	Semilog y plot
title('f vs x')	Add title	grid(True)	Draw grid
text(x, y, 's')	Add text s at (x, y)	myPlot.setColor(false)	Black and White
myPlot.addPoint (0,x,y,true)	Add (x, y) to set 0 connect	myPlot.setButtons(true)	For zoom button
myPlot.addPoint (1,x,y, false)	Add (x, y) to 1, no connect	myPlot.fillPlot()	Fir ranges to data
pyplot.errorbar	Point + error bar	myPlot.setImpulses(true,0)	Vertical lines, set 0
pyplot.clf()	Clear current figure	pyplot.contour	Contour lines
pyplot.scatter	Scatter plot	pyplot.bar	Bar charts
pyplot.polar	Polar plot	pyplot.gca	For current axis
myPlot.setXRange(-1., 1.)	Set x range	pyplot.acorr	Autocorrelation

Listing 1.5 GradesMatPlot.py produces a, 2D x–y plot using the Matplotlib package.

```
# Grade.py: Using Matplotlib's plot command with multi data sets & curves
                                                                          2
import pylab as p                                          # Matplotlib
from numpy import*                                                        4

p.title('Grade Inflation')                          # Title and labels   6
p.xlabel('Years in College')
p.ylabel('GPA')                                                           8

xa = array([-1, 5])                               # For horizontal line  10
ya = array([0, 0])                                    #  "        "
p.plot(xa, ya)                                   # Draw horizontal line  12

x0 = array([0, 1, 2, 3, 4])                          # Data set 0 points  14
y0 = array([-1.4, +1.1, 2.2, 3.3, 4.0])
p.plot(x0, y0, 'bo')                         # Data set 0 = blue circles  16
p.plot(x0, y0, 'g')                                 # Data set 0 = line
                                                                         18
x1 = arange(0, 5, 1)                                 # Data set 1 points
y1 = array([4.0, 2.7, -1.8, -0.9, 2.6])                                  20
p.plot(x1, y1, 'r')
                                                                         22
errTop = array([1.0, 0.3, 1.2, 0.4, 0.1])        # Asymmetric error bars
errBot = array([2.0, 0.6, 2.3, 1.8, 0.4])                                24
p.errorbar(x1, y1, [errBot, errTop], fmt = 'o')       # Plot error bars
                                                                         26
p.grid(True)                                                # Grid line
p.show()                                        # Create plot on screen  28
```

In Listing 1.5, we give the code GradesMatplot.py, and in Figure 1.5b we show its output. This is not a simple plot. Here we repeat the plot command several times in order to plot several data sets on the same graph, and to plot both the data points and the lines connecting them. On line 3, we import Matplotlib (pylab), and on line 4 we import NumPy, which we need for the `array` command. Because we have imported two packages, we add the `pylab` prefix to the `plot` commands so that Python knows which package to use.

In order to place a horizontal line along $y = 0$, on lines 10 and 11 we create a data set as an array of x values, $-1 \leq x \leq 5$, and a corresponding array of y values, $y_i \equiv 0$. We then plot the horizontal on line 12. Next we place four more curves in the figure. First on lines 14–15, we create data set 0, and then plot the points as blue circles ('bo'), and connect the points with green ('g') lines (the color will be visible on a computer screen, but will appear only as shades of gray in print). On lines 19–21, we create and plot another data set as a red ('r') line. Finally, on lines 23–25, we define unequal lower and upper error bars and place them on the plot. We finish by adding grid lines (line 27) and *showing* the plot on the screen.

Listing 1.6 MatPlot2figs.py produces the two figures shown in Figure 1.6. Each figure contains two plots with one Matplotlib figure.

```
# MatPlot2figs.py: plot of 2 subplots on 1 fig & 2 separate figs
                                                                        2
from pylab import *                                    # Load matplotlib
                                                                        4
Xmin = -5.0;        Xmax = 5.0;         Npoints= 500
DelX= (Xmax-Xmin)/Npoints                                     # Delta x  6
x1 = arange(Xmin, Xmax, DelX)                                 # x1 range
x2 = arange(Xmin, Xmax, DelX/20)                     # Different x2 range  8
y1 = -sin(x1)*cos(x1*x1)                                   # Function 1
y2 = exp(-x2/4.)*sin(x2)                                   # Function 2  10
print("\n Now plotting, look for Figures 1 & 2 on desktop")
#       Figure 1                                                        12
figure(1)
subplot(2,1,1)                                  # 1st subplot in first figure  14
plot(x1, y1, 'r', lw=2)
xlabel('x');        ylabel( 'f(x)' );      title( '-sin(x)*cos(x^2)' )  16
grid(True)                                                  # Form grid
subplot(2,1,2)                                  # 2nd subplot in first figure  18
plot(x2, y2, '-', lw=2)
xlabel('x')                                                # Axes labels  20
ylabel( 'f(x)' )
title( 'exp(-x/4)*sin(x)' )                                             22

#       Figure 2                                                        24
figure(2)
subplot(2,1,1)                                    # 1st subplot in 2nd figure  26
plot(x1, y1*y1, 'r', lw=2)
xlabel('x');        ylabel( 'f(x)' );     title( 'sin^2(x)*cos^2(x^2)' )  28

    # form grid
subplot(2,1,2)                                    # 2nd subplot in 2nd figure
plot(x2, y2*y2, '-', lw=2)                                              30
xlabel('x');        ylabel( 'f(x)' );      title( 'exp(-x/2)*sin^2(x)' )
grid(True)                                                              32

show()                                                     # Show graphs  34
```

Often the science is clearer if there are several curves in one plot, and, several plots in one figures. Matplotlib lets you do this with the `plot` and the `subplot` commands. For example, in `MatPlot2figs.py` in Listing 1.6 and Figure 1.6, we have

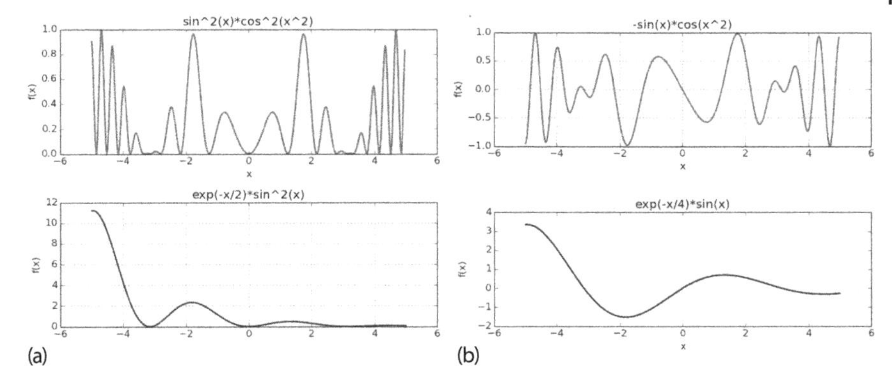

Figure 1.6 (a,b) Columns show two separate outputs, each of two figures, produced by MatPlot2figs.py. (We used the slider button to add some space between the plots.)

placed two curves in one plot, and then output two different figures, each containing two plots. The key here is the repetition of the `subplot` command:

```
figure(1)                # The 1st figure
subplot(2,1,1)           # 2 rows, 1 column, 1st subplot
subplot(2,1,2)           # 2 rows, 1 column, 2nd subplot
```

The listing is self-explanatory, with sections that set the plotting limits, that creates each figure, and then creates the grid.

Listing 1.7 PondMatPlot.py produces the scatter plot and the curve shown in Figure 5.5 in Chapter 5.

```
#   PondMatPlot.py: Monte-Carlo integration via vonNeumann rejection      1

import numpy as np                                                        3
import matplotlib.pyplot as plt
                                                                          5
N = 100
x1 = np.arange(0, 2*np.pi+2*np.pi/N,2*np.pi/N)                            7
fig,ax = plt.subplots()
y1 = x1 * np.sin(x1)**2                          # Integrand              9
ax.plot(x1, y1, 'c', linewidth=4)
ax.set_xlim ((0, 2*np.pi))                                                11
ax.set_ylim((0, 5))
ax.set_xticks([0, np.pi, 2*np.pi])                                        13
ax.set_xticklabels(['0', '$\uppi$','2$\uppi$'])
ax.set_ylabel('$f(x) = x\,\sin^2 x$', fontsize=20)                        15
ax.set_xlabel('x',fontsize=20)
fig.patch.set_visible(False)                                              17
xi=[]; yi=[]; xo=[]; yo=[]
                                                                          19
def fx (x):                                     # Integrand
    return x*np.sin(x)**2                                                 21

j = 0                                           # Inside curve counter    23
Npts = 3000
analyt = np.pi**2                                                         25
xx = 2.* np.pi * np.random.rand(Npts)           # 0 =< x <= 2pi
yy = 5*np.random.rand(Npts)                     # 0 =< y <= 5             27
for i in range(1,Npts):
```

```
        if (yy[i] <= fx(xx[i])):                    # Below curve         29
            if (i <=100): xi.append(xx[i])
            if (i <=100): yi.append(yy[i])                                 31
            j +=1
        else:                                                              33
            if (i <=100): yo.append(yy[i])
            if (i <=100): xo.append(xx[i])                                 35

    boxarea = 2. * np.pi *5.                        # Box area             37
    area = boxarea*j/(Npts-1)                       # Area under curve
    ax.plot(xo,yo,'bo',markersize=3)                                       39
    ax.plot(xi,yi,'ro',markersize=3)
    ax.set_title('Answers:  Analytic = %5.3f, MC = %5.3f'%(analyt,area))   41
plt.show()
```

Scatter Plots Sometimes we need a scatter plot of data points, and maybe even a curve thrown in as well. In Figure 5.5 in Chapter 5, we show such a plot created with the code `PondMapPlot.py` in Listing 1.7. The key statements here are of the form `ax.plot(xo, yo, 'bo', markersize=3)`, which in this case adds a blue point (on screen) of size 3.

1.5.4
Matplotlib's 3D Surface Plots

A 2D plot of the potential $V(r) = 1/r$ vs. r is fine for visualizing the radial dependence of the potential field surrounding a single charge, but if you want to visualize a dipole potential such as $V(x, y) = (B + C(x^2 + y^2)^{-3/2})x$, you need a 3D visualization. We get that by creating a world in which the z dimension (mountain height) is the value of the potential, and the x and y axes define the plane below the mountain. Because the surface we are creating is a 3D object, it is not truly possible to draw it on a flat screen, and so different techniques are used to give the impression of three dimensions to our brains. We do that by rotating the object (by grabbing it with your mouse), shading it, employing parallax, and other tricks.

Listing 1.8 Simple3Dplot.py produces the Matplotlib 3D surface plots in Figure 1.7.

```
# Simple3Dplot.py: matplotlib 3D plot you can rotate and scale via mouse
                                                                            2
import matplotlib.pylab  as p
from mpl_toolkits.mplot3d import Axes3D                                     4

print "Please be patient, I have packages to import & points to plot"       6
delta = 0.1
x = p.arange( -3., 3., delta )                                              8
y = p.arange( -3., 3., delta )
X, Y = p.meshgrid(x, y)                                                     10
Z = p.sin(X) * p.cos(Y)                                 # Surface height
                                                                            12
fig = p.figure()                                        # Create figure
ax = Axes3D(fig)                                        # Plots axes        14
ax.plot_surface(X, Y, Z)                                # Surface
ax.plot_wireframe(X, Y, Z, color = 'r')                 # Add wireframe     16
ax.set_xlabel('X')
ax.set_ylabel('Y')                                                          18
ax.set_zlabel('Z')
                                                                            20
p.show()                                                # Output figure
```

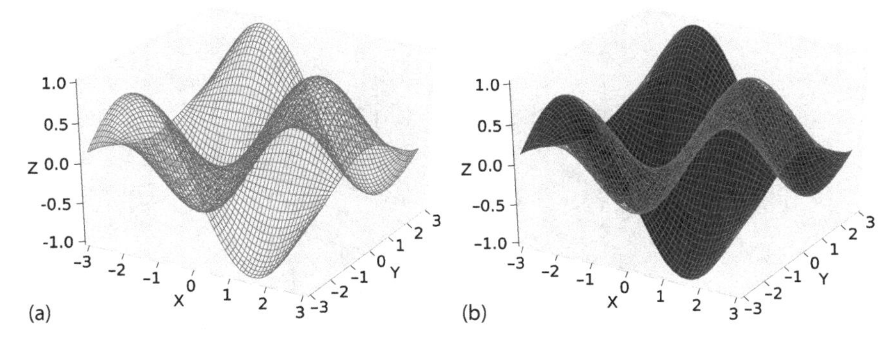

Figure 1.7 (a) A 3D wire frame. (b) A surface plot with wire frame. Both are produced by the program Simple3dplot.py using Matplotlib.

In Figure 1.7a, we show a wire-frame plot and in Figure 1.7b, a surface-plus-wire-frame plot. These are obtained from the program `Simple3Dplot.py` in Listing 1.8. Note that there is an extra import of `Axes3D` from the Matplotlib tool kit needed for 3D plotting. Lines 8 and 9 are the usual creation of x and y arrays of floats using `arange`. Line 11 uses the `meshgrid` method to set up the entire coordinate matrix grid from the x and y coordinate vectors with a vector call, and line 12 constructs the entire Z surface with another vector operation. The remaining of the program is self-explanatory, with `fig` being the plot object, `ax` the 3D axes object, and `plot_wireframe` and `plot_surface` creating wire frame and surface plots, respectively.

Another type of 3D plot is particularly useful when examining data of the form (x_i, y_j, z_k), is a scatter plot into the 3D (x, y, z) volume. In Listing 1.9, we give the program `Scatter3dPlot.py` that created the plot in Figure 1.8. This program, which is taken from the Matplotlib documentation, uses the NumPy random number generator, with the 111 notation being a hand-me-down from MATLAB indicating a $1 \times 1 \times 1$ grid.

Listing 1.9 Scatter3dPlot.py produces a 3D scatter plot using Matplotlib 3D tools.

```
" Scatter3dPlot.py    from matplotlib examples"                          1

import numpy as np                                                       3
from mpl_toolkits.mplot3d import Axes3D
import matplotlib.pyplot as plt                                          5

def randrange(n, vmin, vmax):                                            7
    return (vmax-vmin)*np.random.rand(n) + vmin
                                                                         9
fig = plt.figure()
ax = fig.add_subplot(111, projection='3d')                              11
n = 100
for c, m, zl, zh in [('r', 'o', -50, -25), ('b', '^', -30, -5)]:        13
    xs = randrange(n, 23, 32)
    ys = randrange(n, 0, 100)                                           15
    zs = randrange(n, zl, zh)
    ax.scatter(xs, ys, zs, c=c, marker=m)                               17
```

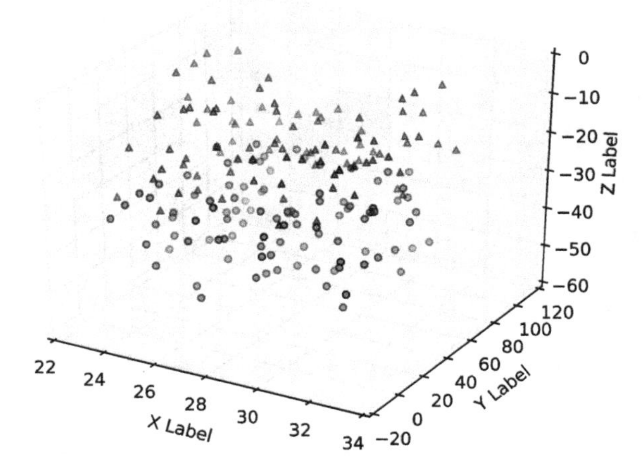

Figure 1.8 A 3D scatter plot produced by the program Scatter3dPlot.py using Matplotlib.

```
ax.set_xlabel('X Label')                                    19
ax.set_ylabel('Y Label')
ax.set_zlabel('Z Label')                                    21

plt.show()                                                  23
```

Finally, the program FourierMatplot.py, written by Oscar Restrepo, performs a Fourier reconstruction of a saw tooth wave, with the number of waves included controlled by the viewer via a slider bar, as shown in Figure 1.9. (We discuss the mathematics of Fourier transforms in Chapter 12.) The slider method is included via the extra lines:

```
from    matplotlib.widgets import Slider
...
snumwaves = Slider(axnumwaves, '# Waves', 1, 20, valinit=T)
...
snumwaves.on_changed(update)
```

1.5.5 Matplotlib's Animations

Matplotlib also can do animations, although not as simply as VPython. The Matplotlib example page shows a number of them. We include some Matplotlib animation codes in the PythonCodes/Visualizations directory, and show a sample code for the heat equation in Listing 1.10. Here too most of the code deals with solving a PDE, which need not interest us yet. The animation is carried out at the bottom of the code.

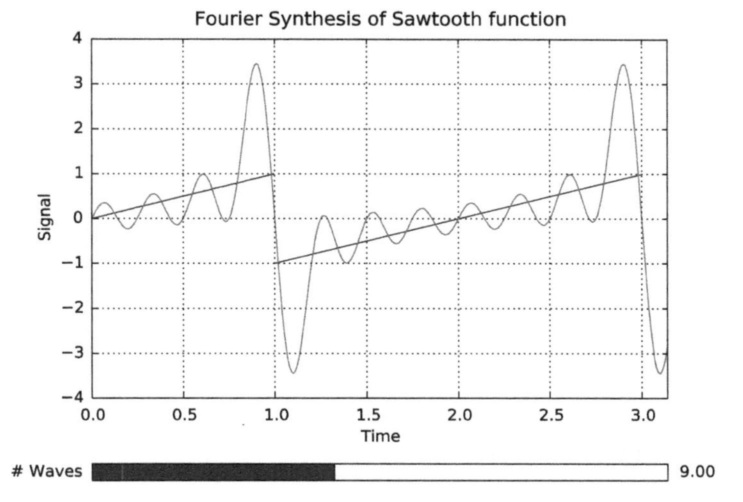

Figure 1.9 A comparison of a saw tooth function to the sum of its Fourier components, with the number of included waves varied interactively by a Matplotlib slider. FourierMatplot.py, which produced this output, was written by Oscar Restrepo.

Listing 1.10 EqHeatAnimateMatPlot.py produces an animation of a cooling bar using Matplotlib.

```
# EqHeat.py Animated heat equation soltn via fine differences

from numpy import *
import numpy as np
import matplotlib.pyplot as plt
import matplotlib.animation as animation

Nx = 101
Dx = 0.01414
Dt = 0.6
KAPPA = 210.                                        # Thermal
    conductivity
SPH = 900.                                          # Specific heat
RHO = 2700.                                         # Density
cons = KAPPA/(SPH*RHO)*Dt/(Dx*Dx);
T = np.zeros( (Nx, 2), float)                       # Temp @ first 2 times
def init():
   for ix in range (1, Nx - 1):                     # Initial temperature
      T[ix, 0] = 100.0;

   T[0, 0] = 0.0                                    # Bar ends T = 0
   T[0, 1] = 0.
   T[Nx - 1, 0] = 0.
   T[Nx - 1, 1] = 0.0
init()
k=range(0,Nx)
fig=plt.figure()                                    # Figure to plot
# select axis; 111: only one plot, x,y, scales given
ax = fig.add_subplot(111, autoscale_on=False, xlim=(-5, 105), ylim=(-5,
    110.0))
ax.grid()                                           # Plot
    grid
plt.ylabel("Temperature")
plt.title("Cooling of a bar")
```

```
line, = ax.plot(k, T[k,0],"r", lw=2)
plt.plot([1,99],[0,0],"r",lw=10)                                           33
plt.text(45,5,'bar',fontsize=20)
                                                                           35
def animate(dum):
    for ix in range (1, Nx - 1):                                           37
        T[ix, 1] = T[ix, 0] +  cons*(T[ix + 1, 0] + T[ix - 1, 0] -
            2.0*T[ix, 0])
    line.set_data(k,T[k,1] )                                               39
    for ix in range (1, Nx - 1):
        T[ix, 0] = T[ix, 1]                      # Row of 100 positions at  41
            t = m
    return line,
                                                                           43
ani = animation.FuncAnimation(fig, animate,1)            # Animation
plt.show()                                                                 45
```

1.5.6
Mayavi's Visualizations Beyond Plotting*

This section on Mayavi is indicated as optional because we do not use it in our sample programs. However, we recommend that, at least, the reader browse through it in order to obtain some ideas about the next level of Python visualization.

Although Matplotlib is excellent for plotting functions vs. one or two of its variables, it is not designed to do the sculpture-like 3D visualizations of functions of three or more variables that are often displayed by supercomputer centers. Mayavi (Sanskrit for "magician") is designed for this next level of visualization. Mayavi is open source, tightly integrated with Python and included in the Canopy distribution.

Mayavi consists of two different packages and two different interfaces to those packages. The package we illustrate here is the set of Matlab- or Mathematica-like commands that operate at a fairly high level of abstraction and works naturally with NumPy arrays. The other package is a set of VTK (Visual Tool Kit) primitives that may be more appropriate for developing your own, research-specific, visualization modules. Even with the high-level package, you have the choice of interacting with Mayavi via scripting from within your Python program (what we demonstrate) or via a stand-alone application that runs separately from your programs.

We will now show a few examples derived from the Enthought tutorial. We start by having Mayavi produce a standard surface plot of $z(x, y) = x^4 + y^4$:

```
import numpy; import Matplotlib; import matplotlib.pyplot
import mayavi;import mayavi.mlab

X, Y = numpy.mgrid[-2:2: 0.1, -2:2: 0.1];                  Z = X**4 + Y**4

mayavi.mlab.surf(Z);                          mayavi.mlab.axes()
mayavi.mlab.outline();                        mlab.show()
```

You see here that we use NumPy's `numpy.mgrid` method to set up the X and Y arrays, and then set up the Z array with a vectorized evaluation of $X^4 + Y^4$. Then we use Mayavi to create the Z surface, to draw the axes and to outline the surface

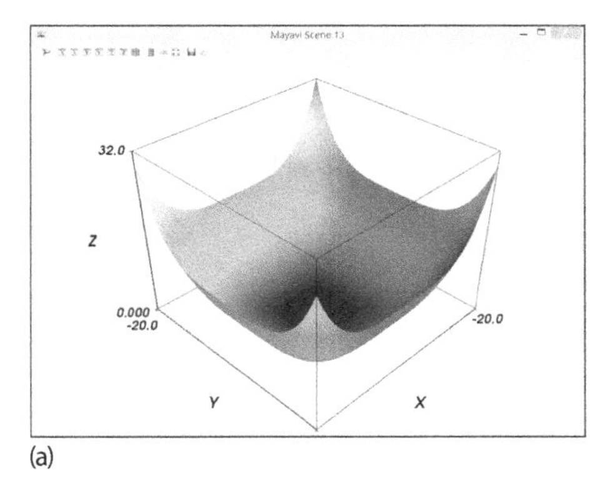

(a)

(b)

Figure 1.10 (a) A Mayavi surface plot of the function $z = x^4 + y^4$ as seen in the screen viewer. (b) A rotatable visualization of a spherical harmonic $Y_l^m(\theta, \phi)$ in which the radial distance represents the value of the function.

with a box. Finally, there is an important call to `mlab.show()` to show the visualization in a display box such as that in Figure 1.10a. This display box is seen (well, if enlarged) to contain a number of (too small) buttons that lets you produce different views and sizes, insert directional arrows, save the file in various formats to disk, edit properties of the visualization, and open the *pipeline* window. The pipeline window shows, and lets the user control, the various stages of a visualization: loading the data into a data source object, transforming the data with filters, and visualizing it with modules.

Now we go beyond the direct plotting of a function's values to the creation of a visualization of a spherical harmonic function $Y_l^m(\theta, \phi)$ that is defined over the surface of a sphere (Figure 1.10b):

```
from numpy import pi, sin, cos, mgrid
from mayavi import mlab
dphi, dtheta = pi/250.0, pi/250.0
[phi,theta] = mgrid[0:pi+dphi*1.5:dphi,0:2*pi+dtheta*1.5:dtheta]
m0 = 4; m1 = 3; m2 = 2; m3 = 3; m4 = 6; m5 = 2; m6 = 6; m7 = 4;
r = sin(m0*phi)**m1 + cos(m2*phi)**m3 + sin(m4*theta)**m5 +
    cos(m6*theta)**m7
x = r*sin(phi)*cos(theta); y = r*cos(phi)              # Function
z = r*sin(phi)*sin(theta)                              # Projections
 # View data
s = mlab.mesh(x, y, z)
mlab.show()
```

Because we do not have four dimensions to use, we take the values of $Y_l^m(\theta, \phi)$ at various grid points and plot those values as the radial distances from the origin for each value of θ and ϕ. The new element here is the statement `s = mlab.mesh(x, y, z)` that produces a mesh throughout 3D space, and then the projection of the radius into its (x, y, z) components.

In the next example, we start with a data set in the form of (x_i, y_j, z_k) values and connect the points with *tubes* of various colors:

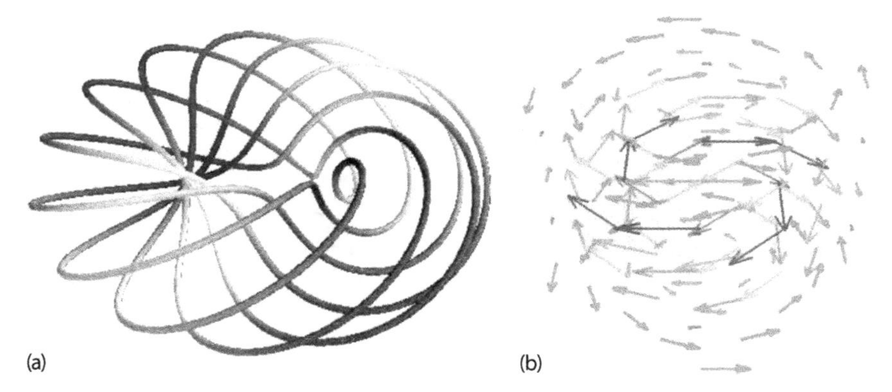

Figure 1.11 (a) A Mayavi visualization in which tubes are used to connect a set of data points. (b) A Mayavi visualization using arrows (glyphs) to represent a vector field.

```
import numpy; import mayavi
from mayavi.mlab import *

n_mer, n_long = 6, 11 ; pi =numpy.pi
dphi   =   pi / 1000.0
phi = numpy.arange(0.0, 2 * pi + 0.5 * dphi, dphi)
mu = phi * n_mer
x = numpy.cos(mu) * (1 + numpy.cos(n_long * mu / n_mer) * 0.5)
y = numpy.sin(mu) * (1 + numpy.cos(n_long * mu / n_mer) * 0.5)
z = numpy.sin(n_long * mu / n_mer) * 0.5

plot3d(x, y, z, numpy.sin(mu), tube_radius=0.025, colormap='Spectral')
mayavi.mlab.show()
```

The new command here is **plot3d**, which is seen in Figure 1.11b to produce rainbow colored ('**Spectral**') tubes connecting the data points. The **arange** command sets up the array of **phi** values, and then the arrays of **x**, **y**, **z**, **mu** and **sin(mu)** values all follow.

A popular style of visualization for vector fields is one in which arrows (*glyphs*) are drawn at various points in space with the directions of the arrows indicating the directions of the field, and with the length of the arrows indicating its strengths. Here we create such a visualization and show its output in Figure 1.11b:

```
import numpy
from mayavi.mlab import *

x, y, z = numpy.mgrid[-2:3, -2:3, -2:3]
r = numpy.sqrt(x ** 2 + y ** 2 + z ** 4)
u = y * numpy.sin(r) / (r + 0.001)
v = -x * numpy.sin(r) / (r + 0.001)
w =   4*numpy.zeros_like(z)

quiver3d(x, y, z, u, v, w, line_width=3, scale_factor =1.5)
show()
```

As before, we use NumPy to set an (x, y, z) grid. Then we set up an array of r values as an intermediate function of (x, y, z), and finally set up arrays of the (u, v, w)

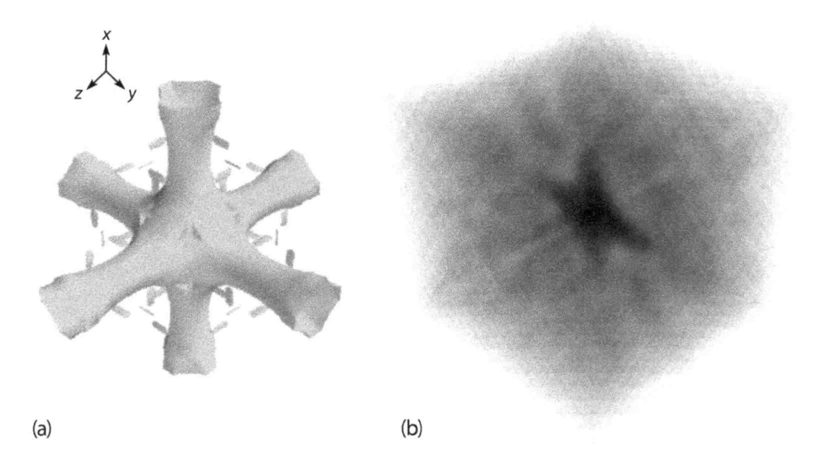

Figure 1.12 (a) A Mayavi visualization of the contours of the scalar field $\phi(x, y, z) = \sin(xyz)/xyz$. (b) A volume rendering of the same scalar field.

components of the vector field as functions of the other arrays. The new command here is `quiver3d` which provides a collection of arrows (cute name).

If the field we wish to visualize is a scalar field, such as $\phi(x, y, z) = \sin(xyz)/xyz$, then the appropriate visualization would be an iso-surface (a 3D contour plot of equal values) throughout a 3D space. We do that with the `contour3d` command:

```
import numpy as np
from mayavi import mlab
x, y, z = np.ogrid[-10:10:20j, -10:10:20j, -10:10:20j]
scalar = np.sin(x*y*z)/(x*y*z)
mlab.contour3d(scalar)
mlab.show()
```

Figure 1.12a shows the output, which is periodic, but not obviously trigonometric.

We now take our visualization of the same scalar field and show how some other Mayavi methods yield different views of the field. First, a volume rendering to produce the nebulous view in Figure 1.12b:

```
import numpy as np
from mayavi import mlab
x, y, z = np.ogrid[-10:10:20j, -10:10:20j, -10:10:20j]
s = np.sin(x*y*z)/(x*y*z)
mlab.pipeline.volume(mlab.pipeline.scalar_field(s))
mlab.show()
```

Next, we take the same field and replace the `mlab.contour3d(s)` command with the pipeline command:

```
mlab.pipeline.volume(mlab.pipeline.scalar_field(s))
```

This produces the nebulous visualization in Figure 1.12b. Next, we produce the visualization in Figure 1.13a by having some planes cut through the scalar field:

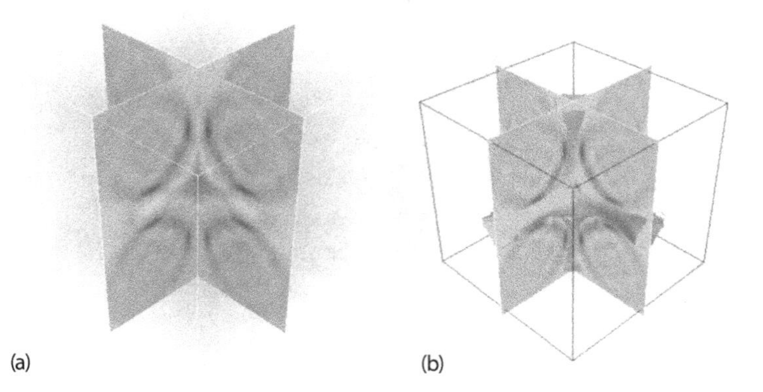

Figure 1.13 (a) A Mayavi visualization of the same scalar field $\phi(x, y, z) = \sin(xyz)/xyz$ using cut planes. (b) A visualization of the same scalar field combining cut planes and contours.

```
import numpy as np
from mayavi import mlab
x, y, z = np.ogrid[-10:10:20j, -10:10:20j, -10:10:20j]
scalar = np.sin(x*y*z)/(x*y*z)
mlab.pipeline.image_plane_widget(mlab.pipeline.scalar_field(scalar),
    plane_orientation='x_axes', slice_index=10,)
mlab.pipeline.image_plane_widget(mlab.pipeline.scalar_field(scalar),
    plane_orientation='y_axes', slice_index=10,)
mlab.outline()
mlab.show()
```

Although we cannot show it, the user can interact with the visualization by moving the cuts and rotating the figures. And to finish, we place both contours and cut planes in the same plot to produce the interesting visualization as shown in Figure 1.13b.

1.6 Plotting Exercises

We encourage you to make your own plots and personalize them by trying out other commands and by including further options in the commands. The Matplotlib documentation is extensive and available on the Web. As an exercise, explore:

1. how to zoom in and zoom out on sections of a plot?
2. how to save your plots to files in various formats?
3. how to print up your graphs?
4. the options available from the pull-down menus?
5. how to increase the space between subplots?
6. and how to rotate and scale the surfaces.

1.7
Python's Algebraic Tools

While this book's focus is on the use of Python for numerical simulations, this is not to discount the importance of computational symbolic manipulations (even though that may be the way we feel). Python actually has (at least) two packages that can be used for symbolic manipulations, and they are quite different. As indicated in Section 1.4.3, the *Sage* package is very much in the same class as Maple and Mathematica, with a notebook graphical interface that lets the user create publication quality text, within which Python programs can be run, or the equations can be manipulated symbolically. Yet Sage is a big and powerful package that goes beyond pure Python by including multiple computer algebra systems as well as visualization tools and more. Using the multiple features of Sage can get to be quite complicated, and, in fact, books have been written and workshops taught on the use of Sage. We refer the interested reader to the online Sage documentation page at www.sagemath.org/help.html.

The *SymPy* package for symbolic manipulations runs very much like any other Python package from within your regular Python shell. It can be downloaded from https://github.com/sympy/sympy/releases, or you can use the Canopy distribution that includes SymPy. Now we give some simple examples of SymPy's use, but you really need to start with the *SymPy Tutorial* http://docs.sympy.org/latest/tutorial/ if you want to use SymPy. (Note, despite the fact that we are working within a Python shell, SymPy has automatically found our LaTeXapplication and used it to format the output.) To start, we will take some derivatives to show that SymPy knows calculus:

```
>>> from sympy import *
>>> x, y = symbols('x y')
>>> y = diff(tan(x),x); y tan²(x) + 1
>>> y = diff(5*x**4 + 7*x**2, x, 1); y                # dy/dx with optional 1
    20x³ + 14x
>>> y = diff(5*x**4+7*x**2, x, 2); y                  # d²y/dx²
    2(30x² + 7)
```

We see here that we must first import methods from SymPy and then use the symbols command to declare the variables x and y as algebraic. The rest is rather obvious, with diff being the derivative operator and the x argument in diff indicating what we are taking the derivative with respect to x. Now let us try expansions:

```
>>> from sympy import *
>>> x, y = symbols('x y')
>>> z = (x + y)**8; z
    (x + y)⁸
>>> expand(z) x⁸ + 8x⁷y + 28x⁶y² + 56x⁵y³ + 70x⁴y⁴ + 56x³y⁵ + 28x²y⁶ + 8xy⁷ + y⁸
```

SymPy knows about infinite series and different expansion points:

```
>>> sin(x).series(x, 0)                               # Usual sinx series about 0
    x − x³/6 + x⁵/120 + O(x⁶)
>>> sin(x).series(x,10)                               # sinx about x= 10
    sin(10) + x cos(10) − x² sin(10)/2 − x³ cos(10)/6 + x⁴ sin(10)/24 + x⁵ cos(10)/120 + O(x⁶)
```

```
>>> z = 1/cos(x); z                              # A division, not an inverse
    1/cos(x)
>>> z.series(x, 0)                               # Expand 1/cos x about x = 0
    1 + x^2/2 + 5x^4/24 + O(x^6)
```

One of the classic difficulties with computer algebra systems is that even if the answer is correct, if it is not simple, then it probably is not useful. And so, SymPy has a **simplify** function as well as a **factor** function (and **collect**, **cancel** and **apart** which we will not illustrate):

```
>>> factor(x**2 -1)
    (x - 1)(x + 1)                                        # A nice answer
>>> factor(x**3 - x**2 + x - 1)
    (x - 1)(x^2 + 1)
>>> simplify((x**3 + x**2 - x - 1)/(x**2 + 2*x + 1))
    x - 1                                                 # Much
          better!
>>> simplify(x**3+3*x**2*y+3*x*y**2+y**3)
    x^3 + 3x^2 y + 3xy^2 + y^3                    # No help!
>>> factor(x**3+3*x**2*y+3*x*y**2+y**3)
    (x + y)^3                                             # Much better!
>>> simplify(1 + tan(x)**2)
    cos(x)^(-2)
>>> simplify(2*tan(x)/(1+tan(x)**2))
    sin(2x)
```

2
Computing Software Basics

This chapter discusses some computing basics starting with computing languages, number representations, and programming tools. The limits and consequences of using floating-point numbers are explored. Related topics dealing with hardware basics are found in Chapter 10.

2.1
Making Computers Obey

The best programs are written so that computing machines can perform them quickly and human beings can understand them clearly. A programmer is ideally an essayist who works with traditional aesthetic and literary forms as well as mathematical concepts, to communicate the way that an algorithm works and to convince a reader that the results will be correct.

Donald E. Knuth

As anthropomorphic as your view of your computer may be, keep in mind that computers always do exactly as they are told. This means that you must tell them exactly everything they have to do. Of course, the programs you run may have such convoluted logic that you may not have the endurance to figure out the details of what you have told the computer to do, but it is always possible in principle. So your first *problem* is to obtain enough understanding so that you feel well enough in control, no matter how illusionary, to figure out what the computer is doing.

Before you tell the computer to obey your orders, you need to understand that life is not simple for computers. The instructions they understand are in a *basic machine language*[1] that tells the hardware to do things like move a number stored in one memory location to another location or to do some simple binary arithmetic. Very few computational scientists talk to computers in a language computers can understand. When writing and running programs, we usually com-

1) The Beginner's All-Purpose Symbolic Instruction Code (BASIC) programming language of the original PCs should not be confused with basic machine language.

Computational Physics, 3rd edition. Rubin H. Landau, Manuel J. Páez, Cristian C. Bordeianu.

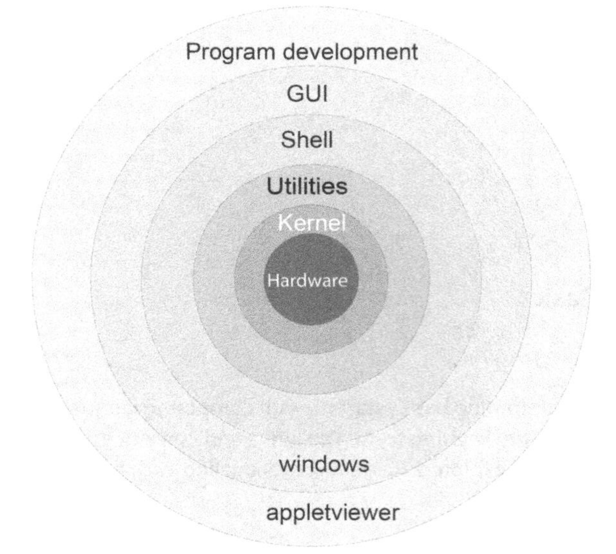

Figure 2.1 A schematic view of a computer's kernel and shells. The hardware is in the center surrounded by increasing higher level software.

municate to the computer through *shells*, in *high-level languages* (Python, Java, Fortran, C), or through *problem-solving environments* (Maple, Mathematica, and Matlab). Eventually, these commands or programs are translated into the basic machine language that the hardware understands.

A *shell* is a *command-line interpreter*, that is, a set of small programs run by a computer that respond to the commands (the names of the programs) that you key in. Usually you open a special window to access the shell, and this window is called a shell as well. It is helpful to think of these shells as the outer layers of the computer's operating system (OS) (Figure 2.1), within which lies a *kernel* of elementary operations. (The user seldom interacts directly with the kernel, except possibly when installing programs or when building an OS from scratch.) It is the job of the shell to run programs, compilers, and utilities that do things like copying files. There can be different types of shells on a single computer or multiple copies of the same shell running at the same time.

Operating systems have names such as *Unix, Linux, DOS, MacOS,* and *MS Windows.* The *operating system* is a group of programs used by the computer to communicate with users and devices, to store and read data, and to execute programs. The OS tells the computer what to do in an elementary way. The OS views you, other devices, and programs as input data for it to process; in many ways, it is the indispensable office manager. While all this may seem complicated, the purpose of the OS is to let the computer do the nitty-gritty work so that you can think higher level thoughts and communicate with the computer in something closer to your normal everyday language.

When you submit a program to your computer in a *high-level language* the computer uses a compiler to process it. The *compiler* is another program that treats your program as a foreign language and uses a built-in dictionary and set of rules to translate it into basic machine language. As you can probably imagine, the final set of instructions is quite detailed and long and the compiler may make several passes through your program to decipher your logic and translate it into a fast code. The translated statements form an *object* or compiled code, and when *linked* together with other needed subprograms, form a load module. A *load module* is a complete set of machine language instructions that can be *loaded* into the computer's memory and read, understood, and followed by the computer.

Languages such as *Fortran* and *C* use compilers to read your entire program and then translate it into basic machine instructions. Languages such as *BASIC* and *Maple* translate each line of your program as it is entered. Compiled languages usually lead to more efficient programs and permit the use of vast subprogram libraries. Interpreted languages give a more immediate response to the user and thereby appear "friendlier." The Python and Java languages are a mix of the two. When you first compile your program, Python interprets it into an intermediate, universal *byte code* which gets stored as a PYC (or PYO) file. This file can be transported to and used on other computers, although not with different versions of Python. Then, when you run your program, Python recompiles the byte code into a machine-specific compiled code, which is faster than interpreting your source code line by line.

2.2 Programming Warmup

Before we go on to serious work, we want to establish that your local computer is working right for you. Assume that calculators have not yet been invented and that you need a program to calculate the area of a circle. Rather than use any specific language, write that program in pseudocode that can be converted to your favorite language later. The first program tells the computer[2]:

```
Calculate area of circle                          # Do this computer!
```

This program cannot really work because it does not tell the computer which circle to consider and what to do with the area. A better program would be

```
read radius                                        # Input
calculate area of circle                           # Numerics
print area                                         # Output
```

The instruction `calculate area of circle` has no meaning in most computer languages, so we need to specify an *algorithm*, that is, a set of rules for the computer to follow:

2) Comments placed in the field to the right are for your information and *not* for the computer to act upon.

```
read radius                                          # Input
PI = 3.141593                                        # Set constant
area = PI * r * r                                    # Algorithm
print area                                           # Output
```

This is a better program, and so let us see how to implement it in Python (other language versions are available online). In Listing 2.1, we give a Python version of our Area program. This is a simple program that outputs to the screen, with its input built into the program.

Listing 2.1 Area.py outputs to the screen, with its input built into the program.

```
# Area.py: Area of a circle , simple program
from math import pi

N = 1
r = 1.
C = 2.* pi* r
A = pi * r**2

print ('Program number =', N, '\n r, C, A = ', r, C, A)

""" Expected OUTPUT
Program number = 1
r, C, A = 1.0 6.283185307179586 3.141592653589793"""
```

2.2.1 Structured and Reproducible Program Design

Programming is a written art that blends elements of science, mathematics, and computer science into a set of instructions that permit a computer to accomplish a desired task. And now, as published scientific results increasingly rely on computation as an essential element, it is increasingly important that the source of your program itself be available to others so that they can reproduce and extend your results. Reproducibility may not be as exciting as a new discovery, but it is an essential ingredient in science (Hinsen, 2013). In addition to the grammar of a computer language, a scientific program should include a number of essential elements to ensure the program's validity and useability. As with other arts, we suggest that until you know better, you follow some simple rules. A good program should

- Give the correct answers.
- Be clear and easy to read, with the action of each part easy to analyze.
- Document itself for the sake of readers and the programmer.
- Be easy to use.
- Be built up out of small programs that can be independently verified.
- Be easy to modify and robust enough to keep giving correct answers after modification and simple debugging.
- Document the data formats used.
- Use trusted libraries.
- Be published or passed on to others to use and to develop further.

One attraction of *object-oriented programming* is that it enforces these rules automatically. An elementary way to make any program clearer is to *structure* it with indentation, skipped lines, and paranetheses placed strategically. This is performed to provide visual clues to the function of the different program parts (the "structures" in structured programming). In fact, Python uses indentations as structure elements as well as for clarity. Although the space limitations of a printed page keep us from inserting as many blank lines as we would prefer, we recommend that you do as we say and not as we do!

In Figure 2.2, we present basic and detailed *flowcharts* that illustrate a possible program for computing projectile motion. A flowchart is not meant to be a detailed description of a program but instead a graphical aid to help visualize its logical flow. As such, it is independent of a specific computer language and is useful for developing and understanding the basic structure of a program. We recommend that you draw a flowchart or (second best) write a pseudocode before you write a program. *Pseudocode* is like a text version of a flowchart that leaves out details and instead focuses on the logic and structures:

```
Store g, Vo, and theta
Calculate R and T
Begin time loop
   Print out "not yet fired" if t < 0
   Print out "grounded" if t > T
   Calculate , print x(t) and y(t)
   Print out error message if x > R, y > H
End time loop    End program
```

2.2.2 Shells, Editors, and Execution

1. To gain some experience with your computer system, use an editor to enter the program **Area.py** that computes the area of a circle (yes, we know you can copy and paste it, but you may need some exercise before getting to work). Then write your file to a disk by saving it in your home (personal) directory (we advise having a separate subdirectory for each chapter). *Note:* Readers familiar with Python may want to enter the program **AreaFormatted.py** instead that uses commands that produce the formatted output.
2. Compile and execute the appropriate version of **Area.py**.
3. Experiment with your program. For example, see what happens if you leave out decimal points in the assignment statement for r, if you assign r equal to a blank, or if you assign a letter to r. Remember, it is unlikely that you will "break" anything on the computer by making a mistake, and it is good to see how the computer responds when under stress.
4. Change the program so that it computes the volume $4/3\,\pi r^3$ of a sphere and prints it out with the proper name. Save the modified program to a file it in your personal directory and give it the name **Vol.py**.
5. Open and execute **Vol.py** and check that your changes are correct by running a number of trial cases. Good input data are $r = 1$ and $r = 10$.

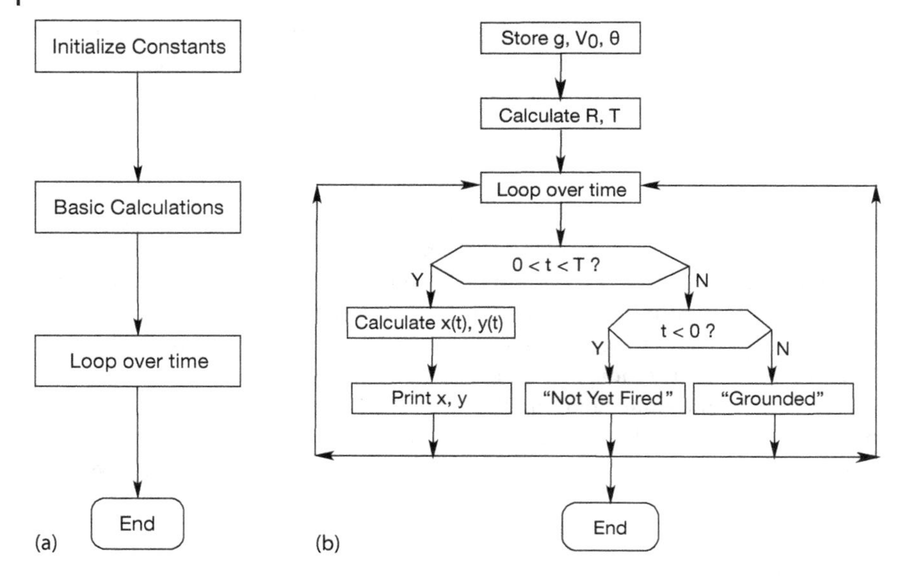

Figure 2.2 A flowchart illustrating a program to compute projectile motion. Plot (a) shows the basic components of the program, and plot (b) shows are some of its details. When writing a program, first map out the basic components, then decide upon the structures, and finally fill in the details. This is called *top-down programming*.

6. Revise `Area.py` so that it takes input from a file name that you have made up, then writes in a different format to another file you have created, and then reads from the latter file.
7. See what happens when the data type used for output does not match the type of data in the file (e.g., floating point numbers are read in as integers).
8. Revise `AreaMod` so that it uses a main method (which does the input and output) and a separate function or method for the calculation. Check that you obtain the same answers as before.

Listing 2.2 AreaFormatted.py does I/O to and from keyboard, as well as from a file. It works with either Python 2 or 3 by switching between *raw_input* and *input*. Note to read from a file using Canopy, you must right click in the Python run window and choose *Change to Editor Directory*.

```
# AreaFormatted: Python 2 or 3 formated output, keyboard input, file input

from numpy import *
from sys import version
if int(version[0])>2:                        # Python 3 uses input, not
    raw_input
    raw_input=input
name = raw_input( 'Key in your name: ')                # raw_input strings
print("Hi ",name)
radius = eval(raw_input('Enter a radius: '))            # For numerical values
print('you entered radius= %8.5f'%radius)                    # formatted output
print('Enter new name and r in file Name.dat')              # raw_input  strings
inpfile = open('Name.dat','r')                       # Read from file Name.dat
```

```
for line in inpfile:
    line = line.split()                          # Splits components of line
    name = line[0]                               # First entry in the list
    print(" Hi  %10s" %(name))                   # print Hi + first entry
    r = float(line[1])                           # convert string to float
    print(" r = %13.5f" %(r))                    # convert to float & print
inpfile.close()
A = math.pi*r**2
print("Done, look in A.dat\n")
outfile = open('A.dat','w')
outfile.write('r=  %13.5f\n'%(r))
outfile.write('A =  %13.5f\n'%(A))
outfile.close()
print('r = %13.5f'%(r), ', A = %13.5f'%(A))                  # Screen output
print('\n Now example of integer input ')
age=int(eval(raw_input ('Now key in your age as an integer:  ')))
print("age: %4d  years old,  you don't look it!\n"%(age))
print("Enter and return a character to finish")
s = raw_input()
```

2.3
Python I/O

The simplest I/O with Python is outputting to the screen with the `print` command (as seen in `Area.py`, Listing 2.1), and inputting from the keyboard with the `input` command (as seen in `AreaFormatted.py`, Listing 2.2). We also see in `AreaFormatted.py` that we can input strings (literal numbers and letters) by either enclosing the string in quotes (single or double), or by using the `raw_input` (Python2) or `input` (Python 3) command without quotes. To print a string with `print`, place the string in quotes. `AreaFormatted.py` also shows how to input both a string and numbers from a file. (But be careful if you are using Canopy: you must right click in the Python run window and choose *Change to Editor Directory* in order to switch the working directory of Canopy's Python shell to your working directory.)

The simplest output prints the value of a float just by giving its name:

```
print 'eps = ', eps              # Output float in default format
```

This uses Python's default format, which tends to vary depending on the precision of the number being printed. As an alternative, you can control the format of your output. For floats, you need to specify two things. First, how many digits (places) after the decimal point are desired, and second, how many spaces overall should be used for the number:

```
print("x=%6.3f,  Pi=%9.6f,  Age=%d \n") % (x, math.pi, age)
print "x=%6.3f, %(x), "Pi=%9.6f," %(math.pi), "Age=%d "%(age)," \n"
x = 12.345,  Pi = 3.141593, Age=39                      # Output from either
```

Here the `%6.3f` formats a float (which is a double in Python) to be printed in fixed-point notation (the `f`) with three places after the decimal point and with six places overall (one place for the decimal point, one for the sign, one for the digit before the decimal point, and three for the decimal). The directive %9.6f produces six digits after the decimal place and nine overall.

To print an integer, we need to specify only the total number of digits (there is no decimal part), and we do that with the %d (d for digits) format. The % symbol in these output formats indicates a conversion from the computer's internal format to that used for output. Notice in Listing 2.2 how we read from the keyboard, as well as from a file, and then output to both the screen and file. Beware that if you do not create the file Name.dat, the program will issue ("throw") an error message of the sort:

```
IOError: [Errno 2] No such file or directory: 'Name.dat'
```

Note that we have also use a \n directive here to indicate a new line. Other directives, some of which are demonstrated in Directives.py in Listing 2.3 (and some of which like backspace may not yet work right) are:

\"	double quote	\0NNN	octal NNN	\\	backslash
\a	alert (bell)	\b	backspace	\c	no more output
\f	form feed	\n	new line	\r	carriage ret
\t	horizontal tab	\v	vertical tab	%%	a single %

Listing 2.3 Directives.py illustrates formatting via directives and escape characters.

```
# Directives.py illustrates escape and formatting  characters
import sys
print("hello \n")
print("\t it's me")                              # tabulator
b = 73
print("decimal 73 as integer b = %d "%(b))       # for integer
print("as octal b = %o"%(b))                     # octal
print("as hexadecimal b = %x "%(b))              # works hexadecimal
print("learn \"Python\" ")                       # use of double quote symbol
print("shows a backslash \\")                    # use of \\
print('use of single \' quotes \' ')             # print single quotes
```

2.4
Computer Number Representations (Theory)

Computers may be powerful, but they are finite. A problem in computer design is how to represent an arbitrary number using a finite amount of memory space and then how to deal with the limitations arising from this representation. As a consequence of computer memories being based on the magnetic or electronic realizations of a spin pointing up or down, the most elementary units of computer memory are the two binary integers (*bits*) 0 and 1. This means that all numbers are stored in memory in the *binary* form, that is, as long strings of 0s and 1s. Accordingly, N bits can store integers in the range $[0, 2^N]$, yet because the sign of the integer is represented by the first bit (a zero bit for positive numbers), the actual range for N-bit integers decreases to $[0, 2^{N-1}]$.

Long strings of 0s and 1s are fine for computers but are awkward for humans. For this reason, binary strings are converted to *octal*, *decimal*, or *hexadecimal* numbers before the results are communicated to people. Octal and hexadecimal numbers are nice because the conversion maintains precision, but not all that nice because our decimal rules of arithmetic do not work for octals and hexadecimals. Converting to decimal numbers makes the numbers easier for us to work with, but unless the original number is a power of 2, the conversion decreases precision.

A description of a particular computer's system or language normally states the *word length*, that is, the number of bits used to store a number. The length is often expressed in *bytes*, (a mouthful of bits) where

$$1\,\text{byte} \equiv 1\,\text{B} \stackrel{\text{def}}{=} 8\,\text{bits}\ .$$

Memory and storage sizes are measured in bytes, kilobytes, megabytes, gigabytes, terabytes, and petabytes (10^{15}). Some care should be taken here by those who chose to compute sizes in detail, because K does not always mean 1000:

$$1\,\text{K} \stackrel{\text{def}}{=} 1\,\text{kB} = 2^{10}\,\text{bytes} = 1024\,\text{bytes}\ . \tag{2.1}$$

This is often (and confusingly) compensated for when memory size is stated in K, for example,

$$512\,\text{K} = 2^{9}\,\text{bytes} = 524\,288\,\text{bytes} \times \frac{1\,\text{K}}{1024\,\text{bytes}}\ .$$

Conveniently, 1 byte is also the amount of memory needed to store a single letter like "a," which adds up to a typical printed page requiring $\sim 3\,\text{kB}$.

The memory chips in some older personal computers used 8-bit words, with modern PCs using 64 bits. This means that the maximum integer was a rather small $2^7 = 128$ (7 because 1 bits is used for the sign). Using 64 bits permits integers in the range $1–2^{63} \simeq 10^{19}$. While at first this may seem like a large range, it really is not when compared to the range of sizes encountered in the physical world. As a case in point, the size of the universe compared to the size of a proton covers a scale of 10^{41}. Trying to store a number larger than the hardware or software was designed for (*overflow*) was common on older machines, but is less so now. An overflow is sometimes accompanied by an informative error message, and sometimes not.

2.4.1
IEEE Floating-Point Numbers

Real numbers are represented on computers in either *fixed-point* or *floating-point* notation. *Fixed-point notation* can be used for numbers with a fixed number of places beyond the decimal point (radix) or for integers. It has the advantages of being able to use *two's complement* arithmetic and being able to store integers exactly.[3] In the fixed-point representation with N bits and with a two's complement format, a number is represented as

$$N_{\text{fix}} = \text{sign} \times \left(\alpha_n 2^n + \alpha_{n-1} 2^{n-1} + \cdots + \alpha_0 2^0 + \cdots + \alpha_{-m} 2^{-m}\right) , \tag{2.2}$$

where $n + m = N - 2$. That is, 1 bit is used to store the sign, with the remaining $(N - 1)$ bits used to store the α_i values (the powers of 2 are understood). The particular values for N, m, and n are machine dependent. Integers are typically 4 bytes (32 bits) in length and in the range

$$-2\,147\,483\,648 \leq \text{4-B integer} \leq 2\,147\,483\,647 . \tag{2.3}$$

An advantage of the representation (2.2) is that you can count on all fixed-point numbers having the same absolute error of 2^{-m-1} (the term left off the right-hand end of (2.2)). The corresponding disadvantage is that *small* numbers (those for which the first string of α values are zeros) have large *relative* errors. Because relative errors in the real world tend to be more important than absolute ones, integers are used mainly for counting purposes and in special applications (like banking).

Most scientific computations use double-precision floating-point numbers with 64 bits = 8 B. The *floating-point representation* of numbers on computers is a binary version of what is commonly known as *scientific* or *engineering notation*. For example, the speed of light $c = +2.997\,924\,58 \times 10^8$ m/s in scientific notation and $+0.299\,792\,458 \times 10^9$ or 0.299 795 498 E09 m/s in engineering notation. In each of these cases, the number in front is called the *mantissa* and contains nine *significant figures*. The power to which 10 is raised is called the *exponent*, with the plus sign in front a reminder that these numbers may be negative.

Floating-point numbers are stored on the computer as a concatenation (juxtaposition) of a sign bit, an exponent, and a mantissa. Because only a finite number of bits are stored, the set of floating-point numbers that the computer can store exactly, *machine numbers* (the hash marks in Figure 2.3), does not cover the entire the set of real numbers. In particular, machine numbers have a maximum and a minimum (the shading in Figure 2.3). If you exceed the maximum, an error condition known as *overflow* occurs; if you fall below the minimum, an error condition known as *underflow* occurs. In the latter case, the software and hardware may be set up so that underflows are set to zero without your even being told. In contrast, overflows usually halt a program's execution.

The actual relation between what is stored in memory and the value of a floating-point number is somewhat indirect, with there being a number of special cases and relations used over the years. In fact, in the past each computer OS and each computer language contained its own standards for floating-point numbers. Different standards meant that the same program running correctly on different

3) The *two's complement* of a binary number is the value obtained by subtracting the number from 2^N for an N-bit representation. Because this system represents negative numbers by the two's complement of the absolute value of the number, additions and subtractions can be made without the need to work with the sign of the number.

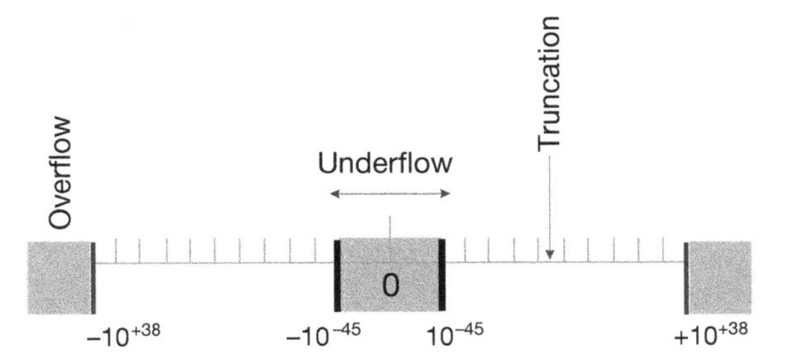

Figure 2.3 The limits of single-precision floating-point numbers and the consequences of exceeding these limits (not to scale). The hash marks represent the values of numbers that can be stored; storing a number in between these values leads to truncation error. The shaded areas correspond to over- and underflow.

computers could give different results. Although the results usually were only slightly different, the user could never be sure if the lack of reproducibility of a test case was as a result of the particular computer being used or to an error in the program's implementation.

In 1987, the Institute of Electrical and Electronics Engineers (IEEE) and the American National Standards Institute (ANSI) adopted the IEEE 754 standard for floating-point arithmetic. When the standard is followed, you can expect the primitive data types to have the precision and ranges given in Table 2.1. In addition, when computers and software adhere to this standard, and most do now, you are guaranteed that your program will produce identical results on different computers. Nevertheless, because the IEEE standard may not produce the most efficient code or the highest accuracy for a particular computer, sometimes you may have to invoke compiler options to demand that the IEEE standard be strictly followed for your test cases. After you know that the code is okay, you may want to run with whatever gives the greatest speed and precision.

Table 2.1 The IEEE 754 Standard for Primitive Data Types.

Name	Type	Bits	Bytes	Range
boolean	Logical	1	$\frac{1}{8}$	true or false
char	String	16	2	'\u0000' ↔ '\uFFFF' (ISO Unicode)
byte	Integer	8	1	$-128 \leftrightarrow +127$
short	Integer	16	2	$-32\,768 \leftrightarrow +32\,767$
int	Integer	32	4	$-2\,147\,483\,648 \leftrightarrow +2\,147\,483\,647$
long	Integer	64	8	$-9\,223\,372\,036\,854\,775\,808 \leftrightarrow 9\,223\,372\,036\,854\,775\,807$
float	Floating	32	4	$\pm 1.401\,298 \times 10^{-45} \leftrightarrow \pm 3.402\,923 \times 10^{38}$
double	Floating	64	8	$\pm 4.940\,656\,458\,412\,465\,44 \times 10^{-324} \leftrightarrow \pm 1.797\,693\,134\,862\,315\,7 \times 10^{308}$

There are actually a number of components in the IEEE standard, and different computer or chip manufacturers may adhere to only some of them. Furthermore, Python as it develops may not follow all standards, but probably will in time. Normally, a floating-point number x is stored as

$$x_{\text{float}} = (-1)^s \times 1.f \times 2^{e\text{-bias}} , \tag{2.4}$$

that is, with separate entities for the sign s, the fractional part of the mantissa f, and the exponential field e. All parts are stored in the binary form and occupy adjacent segments of a single 32-bit word for singles or two adjacent 32-bit words for doubles. The sign s is stored as a single bit, with $s = 0$ or 1 for a positive or a negative sign. Eight bits are used to stored the exponent e, which means that e can be in the range $0 \le e \le 255$. The endpoints, $e = 0$ and $e = 255$, are special cases (Table 2.2). *Normal numbers* have $0 < e < 255$, and with them the convention is to assume that the mantissa's first bit is a 1, so only the fractional part f after the *binary point* is stored. The representations for *subnormal numbers* and for the special cases are given in Table 2.2.

Note that the values ±INF and NaN are not numbers in the mathematical sense, that is, objects that can be manipulated or used in calculations to take limits and such. Rather, they are signals to the computer and to you that something has gone awry and that the calculation should probably stop until you straighten things out. In contrast, the value −0 can be used in a calculation with no harm. Some languages may set unassigned variables to −0 as a hint that they have yet to be assigned, although it is best not to count on that!

Because the uncertainty (error) is only in the mantissa and not in the exponent, the IEEE representations ensure that all normal floating-point numbers have the same relative precision. Because the first bit of a floating point number is assumed to be 1, it does not have to be stored, and computer designers need only recall that there is a *phantom bit* 1 there to obtain an extra bit of precision. During the processing of numbers in a calculation, the first bit of an intermediate result may become zero, but this is changed before the final number is stored. To repeat, for normal cases, the actual mantissa ($1.f$ in binary notation) contains an implied 1 preceding the binary point.

Table 2.2 Representation scheme for normal and abnormal IEEE singles.

Number name	Values of s, e, and f	Value of single
Normal	$0 < e < 255$	$(-1)^s \times 2^{e-127} \times 1.f$
Subnormal	$e = 0,\ f \neq 0$	$(-1)^s \times 2^{-126} \times 0.f$
Signed zero (±0)	$e = 0,\ f = 0$	$(-1)^s \times 0.0$
$+\infty$	$s = 0,\ e = 255,\ f = 0$	+INF
$-\infty$	$s = 1,\ e = 255,\ f = 0$	−INF
Not a number	$s = u,\ e = 255,\ f \neq 0$	NaN

Finally, in order to guarantee that the stored biased exponent e is always positive, a fixed number called the *bias* is added to the actual exponent p before it is stored as the biased exponent e. The actual exponent, which may be negative, is

$$p = e - \text{bias} . \tag{2.5}$$

2.4.1.1 Examples of IEEE Representations

There are two basic IEEE floating-point formats: singles and doubles. *Singles* or *floats* is shorthand for *single-precision floating-point numbers*, and *doubles* is shorthand for *double-precision floating-point numbers*. (In Python, however, all floats are double precision.) Singles occupy 32 bits overall, with 1 bit for the sign, 8 bits for the exponent, and 23 bits for the fractional mantissa (which gives 24-bit precision when the phantom bit is included). Doubles occupy 64 bits overall, with 1 bit for the sign, 10 bits for the exponent, and 53 bits for the fractional mantissa (for 54-bit precision). This means that the exponents and mantissas for doubles are not simply double those of floats, as we see in Table 2.1. (In addition, the IEEE standard also permits *extended precision* that goes beyond doubles, but this is all complicated enough without going into that right now.)

To see the scheme in practice, consider the 32-bit representation (2.4):

	s	e		f	
Bit position	31	30	23	22	0

The sign bit s is in bit position 31, the biased exponent e is in bits 30–23, and the fractional part of the mantissa f is in bits 22–0. Because 8 bits are used to store the exponent e and because $2^8 = 256$, e has the range

$$0 \le e \le 255 . \tag{2.6}$$

The values $e = 0$ and 255 are special cases. With bias $= 127_{10}$, the full exponent

$$p = e_{10} - 127 , \tag{2.7}$$

and, as indicated in Table 2.1, singles have the range

$$-126 \le p \le 127 . \tag{2.8}$$

The mantissa f for singles is stored as the 23 bits in positions 22–0. For *normal numbers*, that is, numbers with $0 < e < 255$, f is the fractional part of the mantissa, and therefore the actual number represented by the 32 bits is

$$\text{Normal floating-point number} = (-1)^s \times 1.f \times 2^{e-127} . \tag{2.9}$$

Subnormal numbers have $e = 0$, $f \neq 0$. For these, f is the entire mantissa, so the actual number represented by these 32 bits is

$$\text{Subnormal numbers} = (-1)^s \times 0.f \times 2^{e-126} . \tag{2.10}$$

The 23 bits $m_{22} - m_0$, which are used to store the mantissa of normal singles, correspond to the representation

$$\text{Mantissa} = 1.f = 1 + m_{22} \times 2^{-1} + m_{21} \times 2^{-2} + \cdots + m_0 \times 2^{-23} , \qquad (2.11)$$

with $0.f$ used for subnormal numbers. The special $e = 0$ representations used to store ± 0 and $\pm\infty$ are given in Table 2.2.

To see how this works in practice (Figure 2.3), the largest positive normal floating-point number possible for a 32-bit machine has the maximum value $e =$ 254 (the value 255 being reserved) and the maximum value for f:

$$\begin{aligned} X_{\max} &= 01111\ 1110\ 1111\ 1111\ 1111\ 1111\ 1111\ 111 \\ &= (0)(1111\ 1110)(1111\ 1111\ 1111\ 1111\ 1111\ 111) , \end{aligned} \qquad (2.12)$$

where we have grouped the bits for clarity. After putting all the pieces together, we obtain the value shown in Table 2.1:

$$\begin{aligned} & s = 0 , \quad e = 1111\ 1110 = 254 , \quad p = e - 127 = 127 , \\ & f = 1.1111\ 1111\ 1111\ 1111\ 1111\ 111 = 1 + 0.5 + 0.25 + \cdots \simeq 2 , \\ & \quad \Rightarrow (-1)^s \times 1.f \times 2^{p=e-127} \simeq 2 \times 2^{127} \simeq 3.4 \times 10^{38} . \end{aligned} \qquad (2.13)$$

Likewise, the smallest positive floating-point number possible is subnormal ($e =$ 0) with a single significant bit in the mantissa:

$$0\ 0000\ 0000\ 0000\ 0000\ 0000\ 0000\ 0000\ 001 . \qquad (2.14)$$

This corresponds to

$$\begin{aligned} & s = 0 , \quad e = 0 , \quad = e - 126 = -126 \\ & f = 0.0000\ 0000\ 0000\ 0000\ 0000\ 001 = 2^{-23} \\ & \quad \Rightarrow (-1)^s \times 0.f \times 2^{p=e-126} = 2^{-149} \simeq 1.4 \times 10^{-45} . \end{aligned} \qquad (2.15)$$

In summary, single-precision (32-bit or 4-byte) numbers have six or seven decimal places of significance and magnitudes in the range

$$1.4 \times 10^{-45} \le \text{single precision} \le 3.4 \times 10^{38} . \qquad (2.16)$$

Doubles are stored as two 32-bit words, for a total of 64 bits (8 B). The sign occupies 1 bit, the exponent e, 11 bits, and the fractional mantissa, 52 bits:

	s	e		f		f(cont.)	
Bit position	63	62	52	51	32	31	0

As we see here, the fields are stored contiguously, with part of the mantissa f stored in separate 32-bit words. The order of these words, and whether the second word with f is the most or least significant part of the mantissa, is machine-dependent. For doubles, the bias is quite a bit larger than for singles,

$$\text{Bias} = 1111111111_2 = 1023_{10} , \qquad (2.17)$$

Table 2.3 Representation scheme for IEEE doubles.

Number name	Values of s, e, and f	Value of double
Normal	$0 < e < 2047$	$(-1)^s \times 2^{e-1023} \times 1.f$
Subnormal	$e = 0,\ f \neq 0$	$(-1)^s \times 2^{-1022} \times 0.f$
Signed zero	$e = 0,\ f = 0$	$(-1)^s \times 0.0$
$+\infty$	$s = 0,\ e = 2047,\ f = 0$	+INF
$-\infty$	$s = 1,\ e = 2047,\ f = 0$	−INF
Not a number	$s = u,\ e = 2047,\ f \neq 0$	NaN

so the actual exponent $p = e - 1023$.

The bit patterns for doubles are given in Table 2.3, with the range and precision given in Table 2.1. To repeat, if you write a program with doubles, then 64 bits (8 bytes) will be used to store your floating-point numbers. Doubles have approximately 16 decimal places of precision (1 part in 2^{52}) and magnitudes in the range

$$4.9 \times 10^{-324} \leq \text{double precision} \leq 1.8 \times 10^{308} . \tag{2.18}$$

If a single-precision number x is larger than 2^{128}, a fault condition known as an *overflow* occurs (Figure 2.3). If x is smaller than 2^{-128}, an underflow occurs. For overflows, the resulting number x_c may end up being a machine-dependent pattern, not a number (NAN), or unpredictable. For underflows, the resulting number x_c is usually set to zero, although this can usually be changed via a compiler option. (Having the computer automatically convert underflows to zero is usually a good path to follow; converting overflows to zero may be the path to disaster.) Because the only difference between the representations of positive and negative numbers on the computer is the sign bit of 1 for negative numbers, the same considerations hold for negative numbers.

In our experience, *serious scientific calculations almost always require at least 64-bit (double-precision) floats.* And if you need double precision in one part of your calculation, you probably need it all over, which means double-precision library routines for methods and functions.

2.4.2
Python and the IEEE 754 Standard

Python is a relatively recent language with changes and extensions occurring as its use spreads and as its features mature. It should be no surprise then that Python does not at present adhere to all aspects, and especially the special cases, of the IEEE 754 standard. Probably the most relevant difference for us is that *Python does not support single (32 bits) precision floating-point numbers.* So when we deal with a data type called a *float* in Python, it is the equivalent of a *double* in the IEEE standard. Because singles are inadequate for most scientific computing, this is not a loss. However be wary, if you switch over to Java or C you should declare your variables as **doubles** and not as **floats**. While Python eliminates single-precision floats, it adds a new data type *complex* for dealing with complex numbers. Complex numbers are stored as pairs of doubles and are quite useful in science.

The details of how closely Python adheres to the IEEE 754 standard depend upon the details of Python's use of the C or Java language to power the Python interpreter. In particular, with the recent 64 bits architectures for CPUs, the range may even be greater than the IEEE standard, and the abnormal numbers (±INF, NaN) may differ. Likewise, the exact conditions for overflows and underflows may also differ. That being the case, the exploratory exercises to follow become all that more interesting because we cannot say that we know what results you should obtain!

2.4.3
Over and Underflow Exercises

1. Consider the 32-bit single-precision floating-point number A:

	s	e	f
Bit position	31	30 23	22 0
Value	0	0000 1110	1010 0000 0000 0000 0000 000

   a) What are the (binary) values for the sign s, the exponent e, and the fractional mantissa f? (*Hint:* $e_{10} = 14$.)
   b) Determine decimal values for the biased exponent e and the true exponent p.
   c) Show that the mantissa of A equals 1.625 000.
   d) Determine the full value of A.
2. Write a program that determines the *underflow* and *overflow* limits (within a factor of 2) for Python on your computer. Here is a sample pseudocode

```
under = 1.
over = 1.
begin do N times
     under = under/2.
     over = over * 2.
     write out: loop number, under, over
end do
```

   You may need to increase N if your initial choice does not lead to underflow and overflow. (Notice that if you want to be more precise regarding the limits of your computer, you may want to multiply and divide by a number smaller than 2.)
   a) Check where under- and overflow occur for double-precision floating-point numbers (floats). Give your answer in decimal.
   b) Check where under- and overflow occur for double-precision floating-point numbers.
   c) Check where under- and overflow occur for integers. *Note:* There is no exponent stored for integers, so the smallest integer corresponds to the most

negative one. To determine the largest and smallest integers, you must observe your program's output as you explicitly pass through the limits. You accomplish this by continually adding and subtracting 1. (Because integer arithmetic uses *two's complement* arithmetic, you should expect some surprises.)

2.4.4
Machine Precision (Model)

A major concern of computational scientists is that the floating-point representation used to store numbers is of limited precision. As we have shown for a 32-bit-word machine, *single-precision numbers are good to 6–7 decimal places, while doubles are good to 15–16 places.* To see how limited precision affects calculations, consider the simple computer addition of two single-precision numbers:

$$7 + 1.0 \times 10^{-7} = ? \tag{2.19}$$

The computer fetches these numbers from memory and stores the bit patterns

$$7 = 0\ 1000\ 0010\ 1110\ 0000\ 0000\ 0000\ 0000\ 000\ , \tag{2.20}$$

$$10^{-7} = 0\ 0110\ 0000\ 1101\ 0110\ 1011\ 1111\ 1001\ 010\ , \tag{2.21}$$

in *working registers* (pieces of fast-responding memory). Because the exponents are different, it would be incorrect to add the mantissas, and so the exponent of the smaller number is made larger while progressively decreasing the mantissa by *shifting bits* to the right (inserting zeros) until both numbers have the same exponent:

$$\begin{aligned} 10^{-7} &= 0\ 01100001\ 0110\ 1011\ 0101\ 1111\ 1100\ 101\ (0) \\ &= 0\ 0110\ 0010\ 0011\ 0101\ 1010\ 1111\ 1110\ 010\ (10) \\ &\dots \\ &= 0\ 1000\ 0010\ 0000\ 0000\ 0000\ 0000\ 0000\ 000\ (0001\ 101\dots 0) \end{aligned} \tag{2.22}$$

$$\Rightarrow 7 + 1.0 \times 10^{-7} = 7\ . \tag{2.23}$$

Because there is no room left to store the last digits, they are lost, and after all this hard work the addition just gives 7 as the answer (truncation error in Figure 2.3). In other words, because a 32-bit computer stores only 6 or 7 decimal places, it effectively ignores any changes beyond the sixth decimal place.

The preceding loss of precision is categorized by defining the *machine precision* ϵ_m as the maximum positive number that, on the computer, can be added to the number stored as 1 without changing that stored 1:

$$1_c + \epsilon_m \stackrel{\text{def}}{=} 1_c\ , \tag{2.24}$$

where the subscript c is a reminder that this is a computer representation of 1. Consequently, an arbitrary number x can be thought of as related to its floating-point representation x_c by

$$x_c = x(1 \pm \epsilon) , \quad |\epsilon| \leq \epsilon_m , \tag{2.25}$$

where the actual value for ϵ is not known. In other words, except for powers of 2 that are represented exactly, we should assume that all single-precision numbers contain an error in the sixth decimal place and that all doubles have an error in the 15th place. And as is always the case with errors, we must assume that we really do not know what the error is, for if we knew, then we would eliminate it! Consequently, the arguments we are about to put forth regarding errors should be considered approximate, but that is to be expected for unknown errors.

2.4.5
Experiment: Your Machine's Precision

Write a program to determine the machine precision ϵ_m of your computer system within a factor of 2. A sample pseudocode is

```
eps = 1.
begin do N times
  eps = eps/2.                                    # Make smaller
  one = 1. + eps                  # Write loop number, one, eps
 end do
```

A Python implementation is given in Listing 2.4, while a more precise one is `ByteLimit.py` on the instructor's guide.

Listing 2.4 Limits.py determines machine precision within a factorof 2.

```python
# Limits.py: determines approximate machine precision

N = 10
eps = 1.0

for i in range(N):
    eps = eps/2
    one_Plus_eps = 1.0 + eps
    print('eps = ', eps, ', one + eps = ', one_Plus_eps)
```

1. Determine experimentally the precision of double-precision floats.
2. Determine experimentally the precision of complex numbers.

To print out a number in the decimal format, the computer must convert from its internal binary representation. This not only takes time, but unless the number is an exact power of 2, leads to a loss of precision. So if you want a truly precise indication of the stored numbers, you should avoid conversion to decimals and instead print them out in octal or hexadecimal format (\0NNN).

2.5 Problem: Summing Series

A classic numerical problem is the summation of a series to evaluate a function. As an example, consider the infinite series for $\sin x$:

$$\sin x = x - \frac{x^3}{3!} + \frac{x^5}{5!} - \frac{x^7}{7!} + \cdots \quad \text{(exact)} . \tag{2.26}$$

Your *problem* is to use just this series to calculate $\sin x$ for $x < 2\pi$ and $x > 2\pi$, with an absolute error in each case of less than 1 part in 10^8. While an infinite series is exact in a mathematical sense, it is not a good algorithm because errors tend to accumulate and because we must stop summing at some point. An algorithm would be the finite sum

$$\sin x \simeq \sum_{n=1}^{N} \frac{(-1)^{n-1} x^{2n-1}}{(2n-1)!} \quad \text{(algorithm)} . \tag{2.27}$$

But how do we decide when to stop summing? (Do not even think of saying, "When the answer agrees with a table or with the built-in library function.") One approach would be to stop summing when the next term is smaller than the precision desired. Clearly then, if x is large, this would require large N as well. In fact, for really large x, one would have to go far out in the series before the terms start decreasing.

2.5.1 Numerical Summation (Method)

Never mind that the algorithm (2.27) indicates that we should calculate $(-1)^{n-1}x^{2n-1}$ and then divide it by $(2n-1)!$ This is not a good way to compute. On the one hand, both $(2n-1)!$ and x^{2n-1} can get very large and cause overflows, despite the fact that their quotient may not be large. On the other hand, powers and factorials are very expensive (time-consuming) to evaluate on the computer. Consequently, a better approach is to use a single multiplication to relate the next term in the series to the previous one:

$$\frac{(-1)^{n-1} x^{2n-1}}{(2n-1)!} = \frac{-x^2}{(2n-1)(2n-2)} \frac{(-1)^{n-2} x^{2n-3}}{(2n-3)!}$$

$$\Rightarrow \quad n\text{th term} = \frac{-x^2}{(2n-1)(2n-2)} \times (n-1)\text{th term} . \tag{2.28}$$

While we want to establish absolute accuracy for $\sin x$, that is not so easy to do. What is easy to do is to assume that the error in the summation is approximately the last term summed (this assumes no round-off error, a subject we talk about in Chapter 3). To obtain a relative error of 1 part in 10^8, we then stop the calculation when

$$\left| \frac{n\text{th term}}{\text{sum}} \right| < 10^{-8} , \tag{2.29}$$

where "term" is the last term kept in the series (2.27) and "sum" is the accumulated sum of all the terms. In general, you are free to pick any tolerance level you desire, although if it is too close to, or smaller than, machine precision, your calculation may not be able to attain it. A pseudocode for performing the summation is

```
term = x, sum = x, eps = 10^(-8)                    # Initialize do
  do term = -term*x*x/(2n+1)/(2*n-2);               # New wrt old
  sum = sum + term                                  # Add term
  while abs(term/sum) > eps                         # Break iteration
end do
```

2.5.2
Implementation and Assessment

1. Write a program that implements this pseudocode for the indicated x values. Present the results as a table with headings `x imax sum |sum- sin(x)|/sin(x)`, where `sin(x)` is the value obtained from the built-in function. The last column here is the relative error in your computation. Modify the code that sums the series in a "good way" (no factorials) to one that calculates the sum in a "bad way" (explicit factorials).
2. Produce a table as above.
3. Start with a tolerance of 10^{-8} as in (2.29).
4. Show that for sufficiently small values of x, your algorithm converges (the changes are smaller than your tolerance level) and that it converges to the correct answer.
5. Compare the number of decimal places of precision obtained with that expected from (2.29).
6. Without using the identity $\sin(x + 2n\pi) = \sin(x)$, show that there is a range of somewhat large values of x for which the algorithm converges, but that it converges to the wrong answer.
7. Show that as you keep increasing x, you will reach a regime where the algorithm does not even converge.
8. Now make use of the identity $\sin(x + 2n\pi) = \sin(x)$ to compute $\sin x$ for large x values where the series otherwise would diverge.
9. Repeat the calculation using the "bad" version of the algorithm (the one that calculates factorials) and compare the answers.
10. Set your tolerance level to a number smaller than machine precision and see how this affects your conclusions.

3
Errors and Uncertainties in Computations

To err is human, to forgive divine.

Alexander Pope

Whether you are careful or not, errors and uncertainties are part of computation. Some errors are the ones that humans inevitably make, but some are introduced by the computer. Computer errors arise because of the limited precision with which computers store numbers or because algorithms or models can fail. Although it stifles creativity to keep thinking "error" when approaching a computation, it certainly is a waste of time, and possibly harmful, to work with results that are meaningless "garbage" because of errors. In this chapter, we examine some of the errors and uncertainties that may occur in computations. Although we do not keep repeating a mantra about watching for errors, the lessons of this chapter apply to all other chapters as well.

3.1
Types of Errors (Theory)

Let us say that you have a program of high complexity. To gauge why errors should be of concern, imagine that the program that has the logical flow

$$\text{start} \rightarrow U_1 \rightarrow U_2 \rightarrow \cdots \rightarrow U_n \rightarrow \text{end}\,, \tag{3.1}$$

where each unit U might be a statement or a step. If each unit has probability p of being correct, then the joint probability P of the whole program being correct is $P = p^n$. Let us say, we have a medium-sized program with $n = 1000$ steps and that the probability of each step being correct is almost 1, $p \simeq 0.9993$. This means that you end up with $P \simeq 1/2$, that is, a final answer that is as likely wrong as right (not a good way to build a bridge). The problem is that, as a scientist, you want a

Computational Physics, 3rd edition. Rubin H. Landau, Manuel J. Páez, Cristian C. Bordeianu.

result that is correct – or at least in which the uncertainty is small and of known size, even if the code executes millions of steps.

Four general types of errors exist to plague your computations:

1. *Blunders or bad theory:* Typographical errors entered with your program or data, running the wrong program or having a fault in your reasoning (theory), using the wrong data file, and so on. (If your blunder count starts increasing, it may be time to go home or take a break.)
2. *Random errors:* Imprecision caused by events such as fluctuations in electronics, cosmic rays, or someone pulling a plug. These may be rare, but you have no control over them and their likelihood increases with running time; while you may have confidence in a 20 s calculation, a week-long calculation may have to be run multiple times to check reproducibility.
3. *Approximation errors:* Imprecision arising from simplifying the mathematics so that a problem can be solved on the computer. They include the replacement of infinite series by finite sums, infinitesimal intervals by finite ones, and variable functions by constants. For example,

$$\begin{aligned}\sin(x) &= \sum_{n=1}^{\infty} \frac{(-1)^{n-1}x^{2n-1}}{(2n-1)!} && \text{(exact)}\\ &\simeq \sum_{n=1}^{N} \frac{(-1)^{n-1}x^{2n-1}}{(2n-1)!} + \mathcal{E}(x,N) && \text{(algorithm)}\,, \end{aligned} \tag{3.2}$$

   where $\mathcal{E}(x, N)$ is the approximation error and $\mathcal{E}$ is the series from $N + 1$ to ∞. Because the approximation error arises from the algorithm we use to approximate the mathematics, it is also called the *algorithmic error*. For every reasonable approximation, the approximation error should decrease as N increases and should vanish in the limit $N \to \infty$. Specifically for (3.2), because the scale for N is set by the value of x, a small approximation error requires $N \gg x$. So if x and N are close in value, the approximation error will be large.
4. *Round-off errors:* Imprecision arising from the finite number of digits used to store floating-point numbers. These "errors" are analogous to the uncertainty in the measurement of a physical quantity encountered in an elementary physics laboratory. The overall round-off error accumulates as the computer handles more numbers, that is, as the number of steps in a computation increases. This may cause some algorithms to become *unstable* with a rapid increase in error. In some cases, round-off error may become the major component in your answer, leading to what computer experts call *garbage*.
   For example, if your computer kept four decimal places, then it will store $1/3$ as 0.3333 and $2/3$ as 0.6667, where the computer has "rounded off" the last digit in $2/3$. Accordingly, if we ask the computer to do as simple a calculation as $2(1/3) - 2/3$, it produces

$$2\left(\frac{1}{3}\right) - \frac{2}{3} = 0.6666 - 0.6667 = -0.0001 \neq 0\,. \tag{3.3}$$

So although the result is small, it is not 0, and if we repeat this type of calculation millions of times, the final answer might not even be small (small garbage begets large garbage).

When considering the precision of calculations, it is good to recall our discussion in Chapter 2 of significant figures and scientific notation given in your early physics or engineering classes. For computational purposes, let us consider how the computer may store the floating-point number

$$a = 11\,223\,344\,556\,677\,889\,900 = 1.122\,334\,455\,667\,788\,99 \times 10^{19} \,. \quad (3.4)$$

Because the exponent is stored separately and is a small number, we may assume that it will be stored in full precision. In contrast, some of the digits of the mantissa may be truncated. In double precision, the mantissa of a will be stored in two words, the *most significant part* representing the decimal 1.122 33, and the *least significant part* 44 556 677. The digits beyond 7 are lost. As we shall see soon, when we perform calculations with words of fixed length, it is inevitable that errors will be introduced (at least) into the least significant parts of the words.

3.1.1
Model for Disaster: Subtractive Cancelation

Calculations employing numbers that are stored only approximately on the computer can only be expected to yield approximate answers. To demonstrate the effect of this type of uncertainty, we model the computer representation x_c of the exact number x as

$$x_c \simeq x(1 + \epsilon_x) \,. \quad (3.5)$$

Here ϵ_x is the relative error in x_c, which we expect to be of a similar magnitude to the machine precision ϵ_m. If we apply this notation to the simple subtraction $a = b - c$, we obtain

$$a = b - c \Rightarrow a_c \simeq b_c - c_c \simeq b(1 + \epsilon_b) - c(1 + \epsilon_c)$$
$$\Rightarrow \frac{a_c}{a} \simeq 1 + \epsilon_b \frac{b}{a} - \frac{c}{a}\epsilon_c \,. \quad (3.6)$$

We see from (3.6) that the resulting error in a is essentially a weighted average of the errors in b and c, with no assurance that the last two terms will cancel. Of special importance here is the observation that the error in the answer a_c increases when we subtract two nearly equal numbers ($b \simeq c$) because then we are subtracting off the most significant parts of both numbers and leaving the error-prone least-significant parts:

$$\frac{a_c}{a} \stackrel{\text{def}}{=} 1 + \epsilon_a \simeq 1 + \frac{b}{a}(\epsilon_b - \epsilon_c) \simeq 1 + \frac{b}{a}\max(|\epsilon_b|, |\epsilon_c|) \,. \quad (3.7)$$

This shows that even if the relative errors in b and c cancel somewhat, they are multiplied by the large number b/a, which can significantly magnify the error. Because we cannot assume any sign for the errors, we must assume the worst (the "max" in (3.7)).

Theorem If you subtract two large numbers and end up with a small one, then the small one is less significant than the large numbers.

We have already seen an example of subtractive cancelation in the power series summation for $\sin x \simeq x - x^3/3! + \cdots$ for large x. A similar effect occurs for $e^{-x} \simeq 1 - x + x^2/2! - x^3/3! + \cdots$ for large x, where the first few terms are large but of alternating sign, leading to an almost total cancelation in order to yield the final small result. (Subtractive cancelation can be eliminated by using the identity $e^{-x} = 1/e^x$, although the round-off error will still remain.)

3.1.2 Subtractive Cancelation Exercises

1. Remember back in high school when you learned that the quadratic equation

$$ax^2 + bx + c = 0 \tag{3.8}$$

has an analytic solution that can be written as either

$$x_{1,2} = \frac{-b \pm \sqrt{b^2 - 4ac}}{2a} \quad \text{or} \quad x'_{1,2} = \frac{-2c}{b \pm \sqrt{b^2 - 4ac}} . \tag{3.9}$$

Inspection of (3.9) indicates that subtractive cancelation (and consequently an increase in error) arises when $b^2 \gg 4ac$ because then the square root and its preceding term nearly cancel for one of the roots.

   a) Write a program that calculates all four solutions for arbitrary values of a, b, and c.
   b) Investigate how errors in your computed answers become large as the subtractive cancelation increases and relate this to the known machine precision. (*Hint*: A good test case utilizes $a = 1, b = 1, c = 10^{-n}, n = 1, 2, 3, \ldots$.)
   c) Extend your program so that it indicates the most precise solutions.

2. As we have seen, subtractive cancelation occurs when summing a series with alternating signs. As another example, consider the finite sum

$$S_N^{(1)} = \sum_{n=1}^{2N} (-1)^n \frac{n}{n+1} . \tag{3.10}$$

If you sum the even and odd values of n separately, you get two sums:

$$S_N^{(2)} = -\sum_{n=1}^{N} \frac{2n-1}{2n} + \sum_{n=1}^{N} \frac{2n}{2n+1} . \tag{3.11}$$

All terms are positive in this form with just a single subtraction at the end of the calculation. Yet even this one subtraction and its resulting cancelation can be avoided by combining the series analytically to obtain

$$S_N^{(3)} = \sum_{n=1}^{N} \frac{1}{2n(2n+1)} . \tag{3.12}$$

Although all three summations $S^{(1)}$, $S^{(2)}$, and $S^{(3)}$ are mathematically equal, they may give different numerical results.

a) Write a double-precision program that calculates $S^{(1)}$, $S^{(2)}$, and $S^{(3)}$.
b) Assume $S^{(3)}$ to be the exact answer. Make a log–log plot of the relative error vs. the number of terms, that is, of $\log_{10}|(S_N^{(1)} - S_N^{(3)})/S_N^{(3)}|$ vs. $\log_{10}(N)$. Start with $N = 1$ and work up to $N = 1\,000\,000$. (Recollect that $\log_{10} x = \ln x/\ln 10$.) The negative of the ordinate in this plot gives an approximate value for the number of significant figures.
c) See whether straight-line behavior for the error occurs in some region of your plot. This indicates that the error is proportional to a power of N.

3. In spite of the power of your trusty computer, calculating the sum of even a simple series may require some thought and care. Consider the two series

$$S^{(\text{up})} = \sum_{n=1}^{N} \frac{1}{n} , \quad S^{(\text{down})} = \sum_{n=N}^{1} \frac{1}{n} . \tag{3.13}$$

Both series are finite as long as N is finite, and when summed analytically both give the same answer. Nonetheless, because of round-off error, the numerical value of $S^{(\text{up})}$ will not be precisely that of $S^{(\text{down})}$.

a) Write a program to calculate $S^{(\text{up})}$ and $S^{(\text{down})}$ as functions of N.
b) Make a log–log plot of $(S^{(\text{up})} - S^{(\text{down})})/(|S^{(\text{up})}| + |S^{(\text{down})}|)$ vs. N.
c) Observe the linear regime on your graph and explain why the downward sum is generally more precise.

3.1.3 Round-off Errors

Let us start by seeing how error arises from a single division of the computer representations of two numbers:

$$\begin{aligned} a = \frac{b}{c} \Rightarrow a_c &= \frac{b_c}{c_c} = \frac{b(1+\epsilon_b)}{c(1+\epsilon_c)} , \\ \Rightarrow \frac{a_c}{a} &= \frac{1+\epsilon_b}{1+\epsilon_c} \simeq (1+\epsilon_b)(1-\epsilon_c) \simeq 1+\epsilon_b-\epsilon_c , \\ \Rightarrow \frac{a_c}{a} &\simeq 1+|\epsilon_b|+|\epsilon_c| . \end{aligned} \tag{3.14}$$

Here we ignore the very small ϵ^2 terms and add errors in absolute value because we cannot assume that we are fortunate enough to have unknown errors canceling each other. Because we add the errors in absolute value, this same rule holds for multiplication. Equation 3.14 is just the basic rule of error propagation from elementary laboratory work: You add the uncertainties in each quantity involved in an analysis to arrive at the overall uncertainty.

We can even generalize this model to estimate the error in the evaluation of a general function $f(x)$, that is, the difference in the value of the function evaluated at x and at x_c:

$$\mathcal{E} = \frac{f(x) - f(x_c)}{f(x)} \simeq \frac{\mathrm{d}f(x)/\mathrm{d}x}{f(x)}(x - x_c) . \tag{3.15}$$

So, for example,

$$f(x) = \sqrt{1+x} , \quad \frac{\mathrm{d}f}{\mathrm{d}x} = \frac{1}{2}\frac{1}{\sqrt{1+x}} = \frac{1}{4} f(x)(x - x_c) \tag{3.16}$$

$$\Rightarrow \mathcal{E} \simeq \frac{1}{2}\sqrt{1+x}(x - x_c) = \frac{x - x_c}{2(1+x)} . \tag{3.17}$$

If we evaluate this expression for $x = \pi/4$ and assume an error in the fourth place of x, we obtain a similar relative error of 1.5×10^{-4} in $\sqrt{1+x}$.

3.1.4
Round-off Error Accumulation

There is a useful model for approximating how round-off error accumulates in a calculation involving a large number of steps. As illustrated in Figure 3.1, we view the error in each step of a calculation as a literal "step" in a *random walk*, that is, a walk for which each step is in a random direction. As we derive and simulate in Chapter 4, the total distance R covered in N steps of length r, is, on average,

$$R \simeq \sqrt{N} r . \tag{3.18}$$

By analogy, the total relative error ϵ_{ro} arising after N calculational steps each with the machine precision error ϵ_{m} is, on average,

$$\epsilon_{\mathrm{ro}} \simeq \sqrt{N} \epsilon_{\mathrm{m}} . \tag{3.19}$$

If the round-off errors in a particular algorithm do not accumulate in a random manner, then a detailed analysis is needed to predict the dependence of the error on the number of steps N. In some cases there may be no cancelation, and the error may increase as $N\epsilon_{\mathrm{m}}$. Even worse, in some recursive algorithms, where the error generation is coherent, such as the upward recursion for spherical Bessel functions, there may be an $N!$ increase in error.

3.2
Error in Bessel Functions (Problem)

Accumulating round-off errors often limits the ability of a program to calculate accurately. Your *problem* is to compute the spherical Bessel and Neumann functions $j_l(x)$ and $n_l(x)$. These function are, respectively, the regular/irregular (nonsingular/singular at the origin) solutions of the differential equation

$$x^2 f(x) + 2x f'(x) + \left[x^2 - l(l+1)\right] f(x) = 0 . \tag{3.20}$$

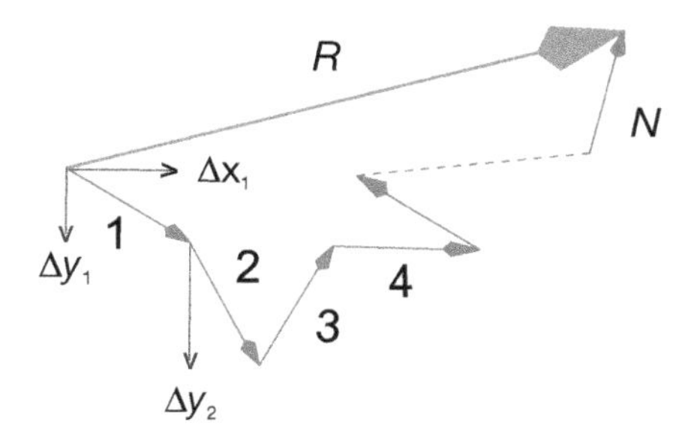

Figure 3.1 A schematic of the N steps in a random walk simulation that end up a distance $R = \sqrt{N}$ from the origin. Notice how the Δx's for each step add vectorially.

Table 3.1 Approximate values for spherical Bessel functions (from maple).

x	$j_3(x)$	$j_5(x)$	$j_8(x)$
0.1	$+9.518\,519\,719 \times 10^{-6}$	$+9.616\,310\,231 \times 10^{-10}$	$+2.901\,200\,102 \times 10^{-16}$
1	$+9.006\,581\,118 \times 10^{-3}$	$+9.256\,115\,862 \times 10^{-05}$	$+2.826\,498\,802 \times 10^{-08}$
10	$-3.949\,584\,498 \times 10^{-2}$	$-5.553\,451\,162 \times 10^{-02}$	$+1.255\,780\,236 \times 10^{-01}$

The spherical Bessel functions are related to the Bessel function of the first kind by $j_l(x) = \sqrt{\pi/2x}J_{n+1/2}(x)$. They occur in many physical problems, such as the expansion of a plane wave into spherical partial waves,

$$e^{ik\cdot r} = \sum_{l=0}^{\infty} i^l(2l+1)j_l(kr)P_l(\cos\theta) . \tag{3.21}$$

Figure 3.2 shows what the first few j_l look like, and Table 3.1 gives some explicit values. For the first two l values, explicit forms are

$$j_0(x) = +\frac{\sin x}{x} , \quad j_1(x) = +\frac{\sin x}{x^2} - \frac{\cos x}{x} , \tag{3.22}$$

$$n_0(x) = -\frac{\cos x}{x} , \quad n_1(x) = -\frac{\cos x}{x^2} - \frac{\sin x}{x} . \tag{3.23}$$

3.2.1
Numerical Recursion (Method)

The classic way to calculate $j_l(x)$ would be by summing its power series for small values of x/l and summing its asymptotic expansion for large x/l values. The approach we adopt is based on the *recursion relations*

$$j_{l+1}(x) = \frac{2l+1}{x} j_l(x) - j_{l-1}(x) , \quad \text{(up)} , \tag{3.24}$$

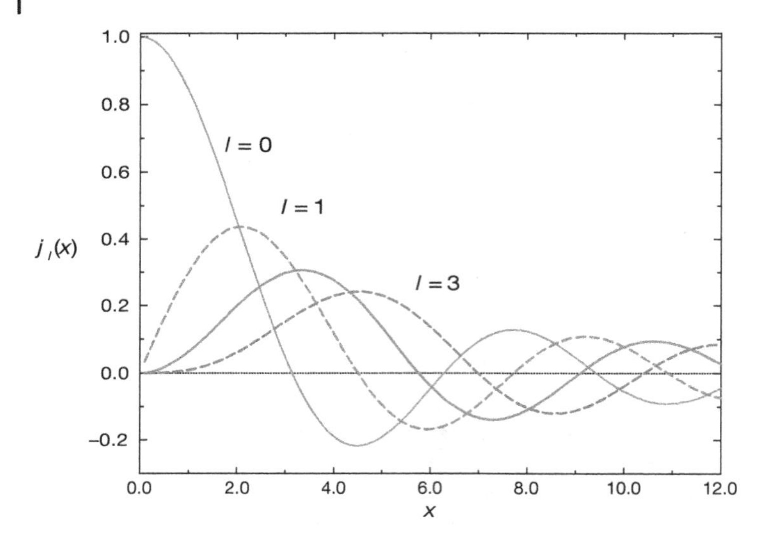

Figure 3.2 The first four spherical Bessel functions $j_l(x)$ as functions of x. Notice that for small x, the values for increasing l become progressively smaller.

$$j_{l-1}(x) = \frac{2l+1}{x} j_l(x) - j_{l+1}(x) , \quad \text{(down)} . \tag{3.25}$$

Equations (3.24) and (3.25) are the same relation, one written for upward recurrence from small to large l values, and the other for downward recurrence from large l to small l. With just a few additions and multiplications, recurrence relations permit rapid, simple computation of the entire set of j_l values for fixed x and all l.

To recur upward in l for fixed x, we start with the known forms for j_0 and j_1 (3.22) and use (3.24). As you will prove for yourself, this upward recurrence usually seems to work at first but then fails. The reason for the failure can be seen from the plots of $j_l(x)$ and $n_l(x)$ vs. x (Figure 3.2). If we start at $x \simeq 2$ and $l = 0$, we will see that as we recur j_l up to larger l values with (3.24), we are essentially taking the difference of two "large" functions to produce a "small" value for j_l. This process suffers from subtractive cancelation and always reduces the precision. As we continue recurring, we take the difference of two small functions with large errors and produce a smaller function with yet a larger error. After a while, we are left with only the round-off error (garbage).

To be more specific, let us call $j_l^{(c)}$ the numerical value we compute as an approximation for $j_l(x)$. Even if we start with pure j_l, after a short while the computer's lack of precision effectively mixes in a bit of $n_l(x)$:

$$j_l^{(c)} = j_l(x) + \epsilon n_l(x) . \tag{3.26}$$

This is inevitable because both j_l and n_l satisfy the same differential equation and, on that account, the same recurrence relation. The admixture of n_l becomes a problem when the numerical value of $n_l(x)$ is much larger than that of $j_l(x)$ because even a minuscule amount of a very large number may be large.

The simple solution to this problem (*Miller's device*) is to use (3.25) for downward recursion of the j_l values starting at a large value $l = L$. This avoids subtractive cancelation by taking small values of $j_{l+1}(x)$ and $j_l(x)$ and producing a larger $j_{l-1}(x)$ by addition. While the error may still behave like a Neumann function, the actual magnitude of the error will *decrease* quickly as we move downward to smaller l values. In fact, if we start iterating downward with arbitrary values for $j_{L+1}^{(c)}$ and $j_L^{(c)}$, after a short while we will arrive at the correct l dependence for this value of x. Although the precise value of $j_0^{(c)}$ so obtained will not be correct because it depends upon the arbitrary values assumed for $j_{L+1}^{(c)}$ and $j_L^{(c)}$, the relative values will be accurate. The absolute values are fixed from the known value (3.22), $j_0(x) = \sin x/x$. Because the recurrence relation is a linear relation between the j_l values, we only need normalize all the computed values via

$$j_l^{\mathrm{N}}(x) = j_l^{\mathrm{c}}(x) \times \frac{j_0^{\mathrm{anal}}(x)}{j_0^{\mathrm{c}}(x)} . \tag{3.27}$$

Accordingly, after you have finished the downward recurrence, you obtain the final answer by normalizing all $j_l^{(c)}$ values based on the known value for j_0.

3.2.2
Implementation and Assessment: Recursion Relations

A program implementing recurrence relations is most easily written using subscripts. If you need to polish up on your skills with subscripts, you may want to study our program **Bessel.py** in Listing 3.1 before writing your own.

1. Write a program that uses both upward and downward recursion to calculate $j_l(x)$ for the first 25 l values for $x = 0.1$, 1, and 10.
2. Tune your program so that at least one method gives "good" values (meaning a relative error $\simeq 10^{-10}$). See Table 3.1 for some sample values.
3. Show the convergence and stability of your results.
4. Compare the upward and downward recursion methods, printing out l, $j_l^{(\mathrm{up})}$, $j_l^{(\mathrm{down})}$, and the relative difference $|j_l^{(\mathrm{up})} - j_l^{(\mathrm{down})}|/(|j_l^{(\mathrm{up})}| + |j_l^{(\mathrm{down})}|)$.
5. The errors in computation depend on x, and for certain values of x, both up and down recursions give similar answers. Explain the reason for this.

Listing 3.1 Bessel.py determines spherical Bessel functions by downward recursion (you should modify this to also work by upward recursion).

```
# Bessel.py
from visual import *
from visual.graph import *

Xmax = 40.
Xmin = 0.25
step = 0.1                                               # Global class variables
order = 10; start = 50              # Plot j_order
graph1 = gdisplay(width = 500, height = 500, title = 'Sperical Bessel, \
```

```
    L = 1 (red), 10',xtitle = 'x', ytitle = 'j(x)',\
    xmin=Xmin,xmax=Xmax,ymin=-0.2,ymax=0.5)
funct1 = gcurve(color=color.red)
funct2 = gcurve(color=color.green)
def down (x, n, m):                          # Method down, recurs downward
    j = zeros( (start  +  2), float)
    j[m  +  1] = j[m] = 1.                           # Start with anything
    for k in range(m, 0,  - 1):
        j[k - 1] = ( (2.*k + 1.)/x)*j[k]  -  j[k + 1]
    scale = (sin(x)/x)/j[0]                  # Scale solution to known j[0]
    return j[n] * scale

for x in arange(Xmin, Xmax, step):
    funct1.plot(pos = (x, down(x, order, start)))

for x in arange(Xmin, Xmax, step):
    funct2.plot(pos = (x, down(x,1,start)))
```

3.3
Experimental Error Investigation

Numerical algorithms play a vital role in computational physics. Your *problem* is to take a general algorithm and decide

1. Does it converge?
2. How precise are the converged results?
3. How expensive (time-consuming) is it?

On first thought you may think, "What a dumb problem! All algorithms converge if enough terms are used, and if you want more precision, then use more terms." Well, some algorithms may be asymptotic expansions that just approximate a function in certain regions of parameter space and converge only up to a point. Yet even if a uniformly convergent power series is used as the algorithm, including more terms will decrease the algorithmic error but will also increase the round-off errors. And because round-off errors eventually diverge to infinity, the best we can hope for is a "best" approximation. Good algorithms are good not only because they are fast but also because they require fewer steps and thus incur less round-off error.

Let us assume that an algorithm takes N steps to find a good answer. As a rule of thumb, the approximation (algorithmic) error decreases rapidly, often as the inverse power of the number of terms used:

$$\epsilon_{\text{app}} \simeq \frac{\alpha}{N^{\beta}} . \tag{3.28}$$

Here α and β are empirical constants that change for different algorithms and may be only approximately constant, and even then only as $N \to \infty$. The fact that the error must fall off for large N is just a statement that the algorithm converges.

In contrast to algorithmic error, round-off error grows slowly and somewhat randomly with N. If the round-off errors in each step of the algorithm are not correlated, then we know from the previous discussion that we can model the accumulation of error as a random walk with step size equal to the machine precision ϵ_m:

$$\epsilon_{ro} \simeq \sqrt{N}\epsilon_m . \tag{3.29}$$

This is the slow growth with N that we expect from the round-off error. The total error in a computation is the sum of the two types of errors:

$$\epsilon_{tot} = \epsilon_{app} + \epsilon_{ro} \tag{3.30}$$

$$\epsilon_{tot} \simeq \frac{\alpha}{N^\beta} + \sqrt{N}\epsilon_m . \tag{3.31}$$

For small N, we expect the first term to be the larger of the two, but as N grows it will be overcome by the growing round-off error.

As an example, in Figure 3.3 we present a log–log plot of the relative error in numerical integration using the Simpson integration rule (Chapter 5). We use the $\log_{10}$ of the relative error because its negative tells us the number of decimal places of precision obtained.[1] Let us assume $\mathcal{A}$ is the exact answer and $A(N)$ the computed answer. If

$$\frac{\mathcal{A} - A(N)}{\mathcal{A}} \simeq 10^{-9} , \quad \text{then} \quad \log_{10}\left|\frac{\mathcal{A} - A(N)}{\mathcal{A}}\right| \simeq -9 . \tag{3.32}$$

We see in Figure 3.3 that the error does show a rapid decrease for small N, consistent with an inverse power law (3.28). In this region, the algorithm is converging. As N keeps increasing, the error starts to look somewhat erratic, with a slow increase on average. In accordance with (3.30), in this region the round-off error has grown larger than the approximation error and will continue to grow for increasing N. Clearly then, the smallest total error will be obtained if we can stop the calculation at the minimum near 10^{-14}, that is, when $\epsilon_{approx} \simeq \epsilon_{ro}$.

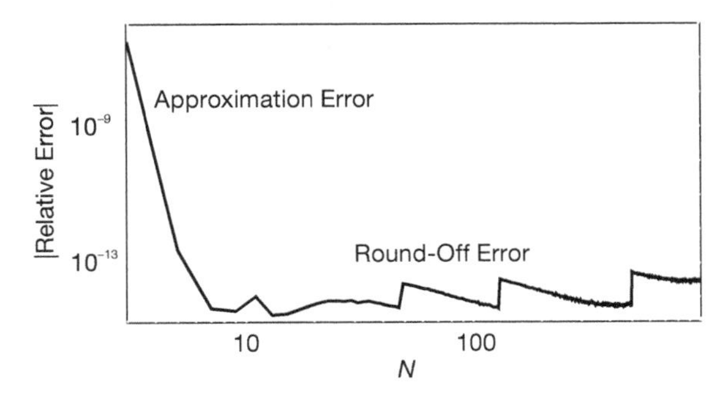

Figure 3.3 A log–log plot of relative error vs. the number of points used for a numerical integration. The ordinate value of $\sim 10^{-14}$ at the minimum indicates that $\sim$ 14 decimal places of precision are obtained before round-off error begins to build up. Notice that while the round-off error does fluctuate indicating a statistical aspect of error accumulation, on average it is increasing but more slowly than did the algorithm's error decrease.

1) Most computer languages use $\ln x = \log_e x$. Yet because $x = a^{\log_a x}$, we have $\log_{10} x = \ln x / \ln 10$.

In realistic calculations, you would not know the exact answer; after all, if you did, then why would you bother with the computation? However, you may know the exact answer for a similar calculation, and you can use that similar calculation to perfect your numerical technique. Alternatively, now that you understand how the total error in a computation behaves, you should be able to look at a table or, better yet, a graph like Figure 3.3, of your answer and deduce the manner in which your algorithm is converging. Specifically, at some point you should see that the mantissa of the answer changes only in the less significant digits, with that place moving further to the right of the decimal point as the calculation executes more steps. Eventually, however, as the number of steps becomes even larger, round-off error leads to a fluctuation in the less significant digits, with a gradual move towards the left on average. It is best to quit the calculation before this occurs.

Based upon this understanding, an approach to obtaining the best approximation is to deduce when your answer behaves like (3.30). To do that, we call $\mathcal{A}$ the exact answer and $A(N)$ the computed answer after N steps. We assume that for large enough values of N, the approximation converges as

$$A(N) \simeq \mathcal{A} + \frac{\alpha}{N^\beta} , \tag{3.33}$$

that is, that the round-off error term in (3.30) is still small. We then run our computer program with $2N$ steps, which should give a better answer, and use that answer to eliminate the unknown $\mathcal{A}$:

$$A(N) - A(2N) \simeq \frac{\alpha}{N^\beta} . \tag{3.34}$$

To see if these assumptions are correct and determine what level of precision is possible for the best choice of N, plot $\log_{10} |[A(N) - A(2N)]/A(2N)|$ vs. $\log_{10} N$, similar to what we have performed in Figure 3.3. If you obtain a rapid straight-line drop off, then you know you are in the region of convergence and can deduce a value for β from the slope. As N gets larger, you should see the graph change from a straight-line decrease to a slow increase as the round-off error begins to dominate. A good place to quit is before this. In any case, now you understand the error in your computation and therefore have a chance to control it.

As an example of how different kinds of errors enter into a computation, we assume that we know the analytic form for the approximation and round-off errors:

$$\epsilon_{\text{app}} \simeq \frac{1}{N^2} , \tag{3.35}$$

$$\epsilon_{\text{ro}} \simeq \sqrt{N} \epsilon_{\text{m}} , \tag{3.36}$$

$$\Rightarrow \epsilon_{\text{tot}} = \epsilon_{\text{approx}} + \epsilon_{\text{ro}} \tag{3.37}$$

$$\simeq \frac{1}{N^2} + \sqrt{N} \epsilon_{\text{m}} . \tag{3.38}$$

The total error is then a minimum when

$$\frac{d\epsilon_{\text{tot}}}{dN} = \frac{-2}{N^3} + \frac{1}{2}\frac{\epsilon_{\text{m}}}{\sqrt{N}} = 0\,, \tag{3.39}$$

$$\Rightarrow N^{5/2} = \frac{4}{\epsilon_{\text{m}}}\,. \tag{3.40}$$

For a double-precision calculation ($\epsilon_{\text{m}} \simeq 10^{-15}$), the minimum total error occurs when

$$N^{5/2} \simeq \frac{4}{10^{-15}} \Rightarrow N \simeq 1099\,, \Rightarrow \epsilon_{\text{tot}} \simeq 4 \times 10^{-6}\,. \tag{3.41}$$

In this case, most of the error is as a result of round-off and is not the approximation error.

Seeing that the total error is mainly the round-off error $\propto \sqrt{N}$, an obvious way to decrease the error is to use a smaller number of steps N. Let us assume, we do this by finding another algorithm that converges more rapidly with N, for example, one with approximation error behaving like

$$\epsilon_{\text{app}} \simeq \frac{2}{N^4}\,. \tag{3.42}$$

The total error is now

$$\epsilon_{\text{tot}} = \epsilon_{\text{ro}} + \epsilon_{\text{app}} \simeq \frac{2}{N^4} + \sqrt{N}\epsilon_{\text{m}}\,. \tag{3.43}$$

The number of points for minimum error is found as before:

$$\frac{d\epsilon_{\text{tot}}}{dN} = 0 \Rightarrow N^{9/2} \Rightarrow N \simeq 67 \Rightarrow \epsilon_{\text{tot}} \simeq 9 \times 10^{-7}\,. \tag{3.44}$$

The error is now smaller by a factor of 4, with only 1/16 as many steps needed. Subtle are the ways of the computer. In this case, the better algorithm is quicker and, by using fewer steps, produces less round-off error.

Exercise Estimate the error now for a double-precision calculation.

3.3.1 Error Assessment

In Section 2.5, we have already discussed the Taylor expansion of $\sin x$:

$$\sin(x) = x - \frac{x^3}{3!} + \frac{x^5}{5!} - \frac{x^7}{7!} + \cdots = \sum_{n=1}^{\infty} \frac{(-1)^{n-1}x^{2n-1}}{(2n-1)!}\,. \tag{3.45}$$

We now extend that discussion with errors in mind. The series (3.45) converges in the mathematical sense for all values of x. Accordingly, a reasonable algorithm to compute the $\sin(x)$ might be

$$\sin(x) \simeq \sum_{n=1}^{N} \frac{(-1)^{n-1}x^{2n-1}}{(2n-1)!}\,. \tag{3.46}$$

1. Write a program that calculates $\sin(x)$ as the finite sum (3.46). (If you already did this in Chapter 2, then you may reuse that program and its results here. But remember, you should not be using factorials in the algorithm.)
2. Calculate your series for $x \leq 1$ and compare it to the built-in function `Math.sin(x)` (you may assume that the built-in function is exact). Stop your summation at an N value for which the next term in the series will be no more than 10^{-7} of the sum up to that point,

$$\frac{|(-1)^N x^{2N+1}|}{(2N-1)!} \leq 10^{-7} \left| \sum_{n=1}^{N} \frac{(-1)^{n-1} x^{2n-1}}{(2n-1)!} \right| . \tag{3.47}$$

3. Examine the terms in the series for $x \simeq 3\pi$ and observe the significant subtractive cancelations that occur when large terms add together to give small answers. (Do not use the identity $\sin(x+2\pi) = \sin x$ to reduce the value of x in the series.) In particular, print out the near-perfect cancelation around $n \simeq x/2$.
4. See if better precision is obtained by using trigonometric identities to keep $0 \leq x \leq \pi$.
5. By progressively increasing x from 1 to 10, and then from 10 to 100, use your program to determine experimentally when the series starts to lose accuracy and when it no longer converges.
6. Make a series of graphs of the error vs. N for different values of x. You should get curves similar to those in Figure 3.4.

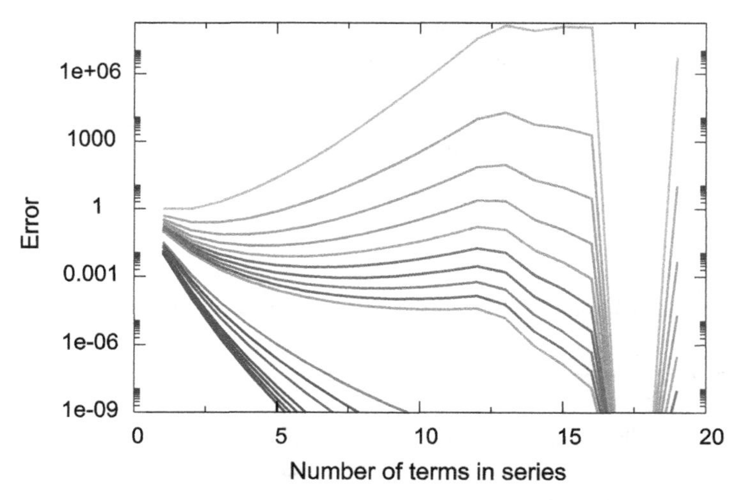

Figure 3.4 The error in the summation of the series for e^{-x} vs. N for various x values. The values of x increase vertically for each curve. Note that a negative initial slope corresponds to decreasing error with N, and that the dip corresponds to a rapid convergence followed by a rapid increase in error. (Courtesy of J. Wiren.)

Because this series summation is such a simple, correlated process, the round-off error does not accumulate randomly as it might for a more complicated computation, and we do not obtain the error behavior (3.33). We will see the predicted error behavior when we examine integration rules in Chapter 5.

4
Monte Carlo: Randomness, Walks, and Decays

This chapter starts with a discussion of how computers generate numbers that appear random, but really are not, and how we can test that. After discussing how computers generate pseudorandom numbers, we explore how these numbers are used to incorporate the element of chance into simulations. We do this first by simulating a random walk and then by simulating the spontaneous decay of an atom or nucleus. In Chapter 5, we show how to use these random numbers to evaluate integrals, and in Chapter 17, we investigate the use of random numbers to simulate thermal processes and the fluctuations of quantum systems.

4.1
Deterministic Randomness

Some people are attracted to computing because of its deterministic nature; it is nice to have a place in one's life where nothing is left to chance. Barring machine errors or undefined variables, you get the same output every time you feed your program the same input. Nevertheless, many computer cycles are used for *Monte Carlo* calculations that at their very core include elements of chance. These are calculations in which random-like numbers generated by the computer are used to *simulate* naturally random processes, such as thermal motion or radioactive decay, or to solve equations on the average. Indeed, much of computational physics' recognition has come about from the ability of computers to solve previously intractable problems using Monte Carlo techniques.

4.2
Random Sequences (Theory)

We define a sequence $r_1, r_2, \dots$ as *random* if there are no correlations among the numbers. Yet being random does not mean that all the numbers in the sequence are equally likely to occur. If all the numbers in a sequence are equally likely to occur, then the sequence is called *uniform*, which does not say anything about being

Computational Physics, 3rd edition. Rubin H. Landau, Manuel J. Páez, Cristian C. Bordeianu.

random. To illustrate, $1, 2, 3, 4, \ldots$ is uniform but probably not random. Furthermore, it is possible to have a sequence of numbers that, in some sense, are random but have very short-range correlations among themselves, for example,

$$r_1, (1 - r_1), r_2, (1 - r_2), r_3, (1 - r_3), \ldots \tag{4.1}$$

have short-range but not long-range correlations.

Mathematically, the likelihood of a number occurring is described by a distribution function $P(r)$, where $P(r)\,\mathrm{d}r$ is the probability of finding r in the interval $[r, r + \mathrm{d}r]$. A *uniform* distribution means that $P(r) =$ a constant. The standard random-number generator on computers generates uniform distributions between 0 and 1. In other words, the standard random-number generator outputs numbers in this interval, each with an equal probability yet each independent of the previous number. As we shall see, numbers can also be more likely to occur in certain regions than other, yet still be random.

By their very nature, computers are deterministic devices and so cannot create a random sequence. Computed random number sequences must contain correlations and in this way cannot be truly random. Although it may be a bit of work, if we know a computed random number r_m and its preceding elements, then it is always possible to figure out r_{m+1}. For this reason, computers are said to generate *pseudorandom numbers* (yet with our incurable laziness we will not bother saying "pseudo" all the time). While more sophisticated generators do a better job at hiding the correlations, experience shows that if you look hard enough or use pseudorandom numbers long enough, you will notice correlations. A primitive alternative to generating random numbers is to read in a table of truly random numbers determined by naturally random processes such as radioactive decay, or to connect the computer to an experimental device that measures random events. These alternatives are not good for production work, but have actually been used as a check in times of doubt.

4.2.1 Random-Number Generation (Algorithm)

The *linear congruent* or *power residue* method is the common way of generating a pseudorandom sequence of numbers $0 \le r_i \le M - 1$ over the interval $[0, M - 1]$. To obtain the next random number r_{i+1}, you multiply the present random number r_i by the constant a, add another constant c, take the *modulus* by M, and then keep just the fractional part (remainder)[1]:

$$r_{i+1} \stackrel{\text{def}}{=} (ar_i + c) \bmod M \tag{4.2}$$

$$= \text{remainder}\left(\frac{ar_i + c}{M}\right) . \tag{4.3}$$

1) You may obtain the same result for the modulus operation by subtracting M until any further subtractions would leave a negative number; what remains is the *remainder*.

The value for r_1 (the *seed*) is frequently supplied by the user, and *mod* is a built-in function on your computer for *remaindering*. In Python, the percent sign % is the modulus operator. This is essentially a bit-shift operation that ends up with the least significant part of the input number and thus counts on the randomness of round-off errors to generate a random sequence.

For example, if $c = 1, a = 4, M = 9$, and you supply $r_1 = 3$, then you obtain the sequence

$$r_1 = 3 , \tag{4.4}$$

$$r_2 = (4 \times 3 + 1) \bmod 9 = 13 \bmod 9 = \text{rem}\frac{13}{9} = 4 , \tag{4.5}$$

$$r_3 = (4 \times 4 + 1) \bmod 9 = 17 \bmod 9 = \text{rem}\frac{17}{9} = 8 , \tag{4.6}$$

$$r_4 = (4 \times 8 + 1) \bmod 9 = 33 \bmod 9 = \text{rem}\frac{33}{9} = 6 , \tag{4.7}$$

$$r_{5-10} = 7, 2, 0, 1, 5, 3 . \tag{4.8}$$

We get a sequence of length $M = 9$, after which the entire sequence repeats. If we want numbers in the range [0, 1], we divide the r's by $M = 9$ to obtain

$$0.333,\ 0.444,\ 0.889,\ 0.667,\ 0.778,\ 0.222,\ 0.000,\ 0.111,\ 0.555,\ 0.333 . \tag{4.9}$$

This is still a sequence of length 9, but is no longer a sequence of integers. If random numbers in the range $[A, B]$ are needed, you only need to *scale*:

$$x_i = A + (B - A)r_i , \quad 0 \leq r_i \leq 1 , \quad \Rightarrow A \leq x_i \leq B . \tag{4.10}$$

As a rule of thumb: *Before using a random-number generator in your programs, you should check its range and that it produces numbers that "look" random.*

Although not a mathematical proof, you should always make a graphical display of your random numbers. Your visual cortex is quite refined at recognizing patterns and will tell you immediately if there is one in your random numbers. For instance, Figure 4.1 shows generated sequences from "good" and "bad" generators, and it is clear which is not random. (Although if you look hard enough at the random points, your mind may well pick out patterns there too.)

The linear congruent method (4.2) produces integers in the range $[0, M-1]$ and therefore becomes completely correlated if a particular integer comes up a second time (the whole cycle then repeats). In order to obtain a longer sequence, a and M should be large numbers but not so large that ar_{i-1} overflows. On a computer using 48-bit integer arithmetic, the built-in random-number generator may use M values as large as $2^{48} \simeq 3 \times 10^{14}$. A 32-bit generator may use $M = 2^{31} \simeq 2 \times 10^9$. If your program uses approximately this many random numbers, you may need to reseed (start the sequence over again with a different initial value) during intermediate steps to avoid the cycle repeating.

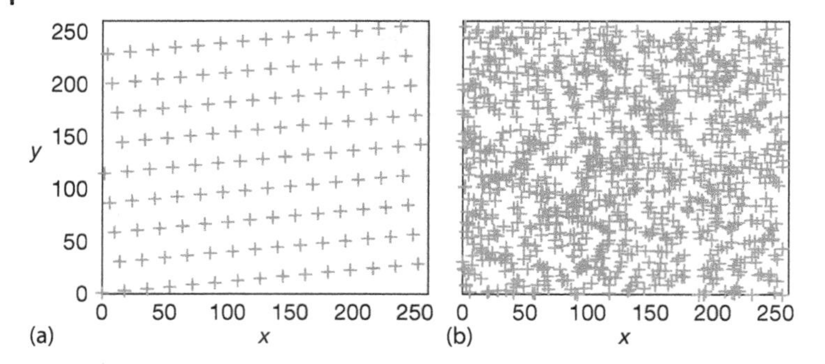

Figure 4.1 (a) A plot of successive random numbers $(x, y) = (r_i, r_{i+1})$ generated with a deliberately "bad" generator. (b) A plot generated with the built in random number generator. While the plot (b) is not the proof that the distribution is random, the plot (a) is a proof enough that the distribution is not random.

Your computer probably has random-number generators that are better than the one you will compute with the power residue method. In Python, we use `random.random()`, the Mersenne Twister generator. We recommend that you use the best one you can find rather than write your own. To initialize a random sequence, you need to plant a seed in it. In Python, the statement `random.seed(None)` seeds the generator with the system time (see `Walk.py` in Listing 4.1). Our old standard, `drand48`, uses:

$$M = 2^{48}, \quad c = B(\text{base16}) = 13\ (\text{base8})\,, \tag{4.11}$$

$$a = 5\text{DEECE66D}(\text{base16}) = 273673163155(\text{base8})\,. \tag{4.12}$$

4.2.2
Implementation: Random Sequences

For scientific work, we recommend using an industrial-strength random-number generator. To see why, here we assess how *bad* a careless application of the power residue method can be.

1. Write a simple program to generate random numbers using the linear congruent method (4.2).
2. For pedagogical purposes, try the unwise choice: $(a, c, M, r_1) = (57, 1, 256, 10)$. Determine the *period*, that is, how many numbers are generated before the sequence repeats.
3. Take the pedagogical sequence of random numbers and look for correlations by observing clustering on a plot of successive pairs $(x_i, y_i) = (r_{2i-1}, r_{2i})$, $i = 1, 2, \ldots$ (Do *not* connect the points with lines.) You may "see" correlations (Figure 4.1), which means that you should not use this sequence for serious work.
4. Make your own version of Figure 4.2; that is, plot r_i vs. i.
5. Test the built-in random-number generator on your computer for correlations by plotting the same pairs as above. (This should be good for serious work.)

6. Test the linear congruent method again with reasonable constants like those in (4.11) and (4.12). Compare the scatterplot you obtain with that of the built-in random-number generator. (This, too, should be good for serious work.)

4.2.3
Assessing Randomness and Uniformity

Because the computer's random numbers are generated according to a definite rule, the numbers in the sequence must be correlated with each other. This can affect a simulation that assumes random events. Therefore, it is wise for you to test a random-number generator to obtain a numerical measure of its uniformity and randomness before you stake your scientific reputation on it. In fact, some tests are simple enough for you to make it a habit to run them simultaneously with your simulation. In the examples to follow, we test for randomness and uniformity.

1. Probably the most obvious, but often neglected, test for randomness and uniformity is just to look at the numbers generated. For example, Table 4.1 presents some output from Python's `random` method. If you just look at these numbers you will know immediately that they all lie between 0 and 1, that they appear to differ from each other, and that there is no obvious pattern (like 0.3333).
2. As we have seen, a quick visual test (Figure 4.2) involves taking this same list and plotting it with r_i as ordinate and i as abscissa. Observe how there appears to be a uniform distribution between 0 and 1 and no particular correlation

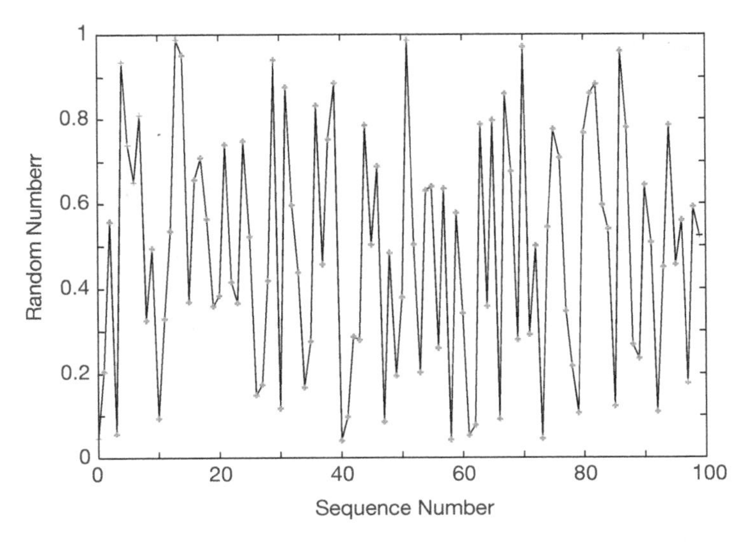

Figure 4.2 A plot of a uniform pseudorandom sequence r_i vs. i. The points are connected to make it easier to follow the order. While this does not prove that a distribution is random, it at least shows the range of values and that there is fluctuation.

Table 4.1 A table of a uniform, pseudo-random sequence r_i generated by Python's random method.

0.046 895 024 385 081 75	0.204 587 796 750 397 95	0.557 190 747 079 725 5	0.056 343 366 735 930 88
0.936 066 864 589 746 7	0.739 939 913 919 486 7	0.650 415 302 989 955 3	0.809 633 370 418 305 7
0.325 121 746 254 331 9	0.494 470 371 018 847 17	0.143 077 126 131 411 28	0.328 581 276 441 882 06
0.535 100 168 558 861 6	0.988 035 439 569 102 3	0.951 809 795 307 395 3	0.368 100 779 256 594 23
0.657 244 381 503 891 1	0.709 076 851 545 567 1	0.563 678 747 459 288 4	0.358 627 737 800 664 9
0.383 369 106 540 338 07	0.740 022 375 602 264 9	0.416 208 338 118 453 5	0.365 803 155 303 808 7
0.748 479 890 046 811 1	0.522 694 331 447 043	0.148 656 282 926 639 13	0.174 188 153 952 713 6
0.418 726 310 120 201 23	0.941 002 689 012 048 8	0.116 704 492 627 128 9	0.875 900 901 278 647 2
0.596 253 540 903 370 3	0.438 238 541 497 494 1	0.166 837 081 276 193	0.275 729 402 460 343 05
0.832 243 048 236 776	0.457 572 427 917 908 75	0.752 028 149 254 081 5	0.886 188 103 177 451 3
0.040 408 674 172 845 55	0.146 901 492 948 813 34	0.286 962 760 984 402 3	0.279 150 544 915 889 53
0.785 441 984 838 243 6	0.502 978 394 047 627	0.688 866 810 791 863	0.085 104 148 559 493 22
0.484 376 438 252 853 26	0.194 793 600 337 003 66	0.379 123 023 471 464 2	0.986 737 138 946 582 1

between points (although your eye and brain will try to recognize some kind of pattern if you look long enough).

3. A simple test of uniformity evaluates the kth moment of a distribution:

$$\langle x^k \rangle = \frac{1}{N} \sum_{i=1}^{N} x_i^k \, . \tag{4.13}$$

If the numbers are distributed *uniformly*, then (4.13) is approximately the moment of the distribution function $P(x)$:

$$\frac{1}{N} \sum_{i=1}^{N} x_i^k \simeq \int_0^1 dx x^k P(x) \simeq \frac{1}{k+1} + O\left(\frac{1}{\sqrt{N}}\right) \, . \tag{4.14}$$

If (4.14) holds for your generator, then you know that the distribution is uniform. If the deviation from (4.14) varies as $1/\sqrt{N}$, then you *also* know that the distribution is random because the $1/\sqrt{N}$ result derives from assuming randomness.

4. Another simple test determines the near-neighbor correlation in your random sequence by taking sums of products for small k:

$$C(k) = \frac{1}{N} \sum_{i=1}^{N} x_i x_{i+k} \, , \quad (k = 1, 2, \dots) \, . \tag{4.15}$$

If your random numbers x_i and x_{i+k} are distributed with the joint probability distribution $P(x_i, x_{i+k}) = 1$ and are independent and uniform, then (4.15) can be approximated as an integral:

$$\frac{1}{N} \sum_{i=1}^{N} x_i x_{i+k} \simeq \int_0^1 dx \int_0^1 dy \, x y P(x, y) = \int_0^1 dy \, x y = \frac{1}{4} \, . \tag{4.16}$$

If (4.16) holds for your random numbers, then you know that they are uniform and independent. If the deviation from (4.16) varies as $1/\sqrt{N}$, then you *also* know that the distribution is random.

5. As we have seen, an effective test for randomness is performed by making a scatterplot of $(x_i = r_{2i}, y_i = r_{2i+1})$ for many i values. If your points have noticeable regularity, the sequence is not random. If the points are random, they should uniformly fill a square with no discernible pattern (a cloud) (as in Figure 4.1b).
6. Test your random-number generator with (4.14) for $k = 1, 3, 7$ and $N = 100, 10\,000, 100\,000$. In each case print out

$$\sqrt{N}\left|\frac{1}{N}\sum_{i=1}^{N} x_i^k - \frac{1}{k+1}\right| \tag{4.17}$$

to check that it is of the order 1.

4.3 Random Walks (Problem)

Consider a perfume molecule released in the front of a classroom. It collides randomly with other molecules in the air and eventually reaches your nose despite the fact that you are hidden in the last row. Your *problem* is to determine how many collisions, on the average, a perfume molecule makes in traveling a distance R. You are given the fact that a molecule travels an average (*root-mean-square*) distance r_{rms} between collisions.

Listing 4.1 Walk.py calls the random-number generator from the random package. Note that a different seed is needed to obtain a different sequence.

```
# Walk.py  Random walk with graph
from visual import *
from visual.graph import *
import random

random.seed(None)                          # Seed generator, None => system clock
jmax = 20
x    = 0.;          y = 0.                                   # Start at origin

graph1 = gdisplay(width=500, height=500, title='Random Walk', xtitle='x',
                  ytitle='y')
pts = gcurve(color = color.yellow)

for i in range(0, jmax + 1):
    pts.plot(pos = (x, y) )                                  # Plot points
    x += (random.random() - 0.5)*2.                          # -1 =< x =< 1
    y += (random.random() - 0.5)*2.                          # -1 =< y =< 1
    pts.plot(pos = (x, y))
    rate(100)
```

4.3.1 Random-Walk Simulation

There are a number of ways to simulate a random walk with (surprise, surprise) different assumptions yielding different physics. We will present the simplest approach for a 2D walk, with a minimum of theory, and end up with a model for *normal diffusion*. The research literature is full of discussions of various versions of this problem. For example, Brownian motion corresponds to the limit in which the individual step lengths approach zero, and with no time delay between steps. Additional refinements include collisions within a moving medium (*abnormal diffusion*), including the velocities of the particles, or even pausing between steps. Models such as these are discussed in Chapter 16, and demonstrated by some of the corresponding applets given online.

In our random-walk simulation (Figure 4.3) an artificial *walker* takes sequential steps with the *direction* of each step *independent* of the direction of the previous step. For our model, we start at the origin and take N steps in the xy plane of *lengths* (not coordinates)

$$(\Delta x_1, \Delta y_1), \quad (\Delta x_2, \Delta y_2), \quad (\Delta x_3, \Delta y_3), \ldots, \quad (\Delta x_N, \Delta y_N). \tag{4.18}$$

Although each step may be in a different direction, the distances along each Cartesian axis just add algebraically. Accordingly, the radial distance R from the starting point after N steps is

$$\begin{aligned} R^2 &= (\Delta x_1 + \Delta x_2 + \cdots + \Delta x_N)^2 + (\Delta y_1 + \Delta y_2 + \cdots + \Delta y_N)^2 \\ &= \Delta x_1^2 + \Delta x_2^2 + \cdots + \Delta x_N^2 + 2\Delta x_1 \Delta x_2 + 2\Delta x_1 \Delta x_3 + 2\Delta x_2 \Delta x_1 + \cdots \\ &\quad + (x \rightarrow y). \end{aligned} \tag{4.19}$$

If the walk is random, the particle is equally likely to travel in any direction at each step. If we take the average of a large number of such random steps, all the cross terms in (4.19) will vanish and we will be left with

$$\begin{aligned} R_{\text{rms}}^2 &\simeq \langle \Delta x_1^2 + \Delta x_2^2 + \cdots + \Delta x_N^2 + \Delta y_1^2 + \Delta y_2^2 + \cdots + \Delta y_N^2 \rangle \\ &= \langle \Delta x_1^2 + \Delta y_1^2 \rangle + \langle \Delta x_2^2 + \Delta y_2^2 \rangle + \cdots \\ &= N \langle r^2 \rangle = N r_{\text{rms}}^2, \\ &\Rightarrow \boxed{R_{\text{rms}} \simeq \sqrt{N} r_{\text{rms}}}, \end{aligned} \tag{4.20}$$

where $r_{\text{rms}} = \sqrt{\langle r^2 \rangle}$ is the *root-mean-square* step size.

To summarize, if the walk is random, then we expect that after a large number of steps the average *vector* distance from the origin will vanish:

$$\langle \boldsymbol{R} \rangle = \langle x \rangle \boldsymbol{i} + \langle y \rangle \boldsymbol{j} \simeq 0. \tag{4.21}$$

Yet $R_{\text{rms}} = \sqrt{\langle R_i^2 \rangle}$ does not vanish. Equation 4.20 indicates that the average *scalar* distance from the origin is $\sqrt{N} r_{\text{rms}}$, where each step is of average length r_{rms}. In

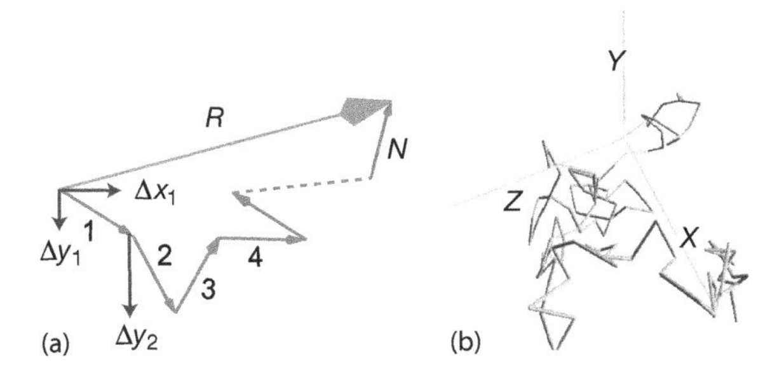

Figure 4.3 (a) A schematic of the N steps in a random walk simulation that end up a distance R from the origin. Note how the Δx's for each step add vectorially. (b) A simulated walk in 3D from *Walk3D.py*.

other words, the vector endpoint will be distributed uniformly in all quadrants, and so the displacement vector averages to zero, but the average length of that vector does not. For large N values, $\sqrt{N}r_{\text{rms}} \ll N r_{\text{rms}}$ (the value if all steps were in one direction on a straight line), but does not vanish. In our experience, practical simulations agree with this theory, but rarely perfectly, with the level of agreement depending upon the details of how the averages are taken and how the randomness is built into each step.

4.3.2
Implementation: Random Walk

The program `Walk.py` in Listing 4.1 is a sample random-walk simulation. It is key element is random values for the x and y components of each step,

```
x += (random.random() - 0.5)*2.                # -1 =< x =< 1
y += (random.random() - 0.5)*2.                # -1 =< y =< 1
```

Here we omit the scaling factor that normalizes each step to length 1. When using your computer to simulate a random walk, you should expect to obtain (4.20) only as the average displacement averaged over many trials, not necessarily as the answer for each trial. You may get different answers depending on just how you take your random steps (Figure 4.4b).

Start at the origin and take a 2D random walk with your computer.

1. To increase the amount of randomness, independently choose random values for $\Delta x'$ and $\Delta y'$ in the range $[-1, 1]$. Then normalize them so that each step is of unit length

$$\Delta x = \frac{1}{L}\Delta x' , \quad \Delta y = \frac{1}{L}\Delta y' , \quad L = \sqrt{\Delta x'^2 + \Delta y'^2} . \tag{4.22}$$

2. Use a plotting program to draw maps of several independent 2D random walks, each of 1000 steps. Using evidence from your simulations, comment on whether these look like what you would expect of a random walk.

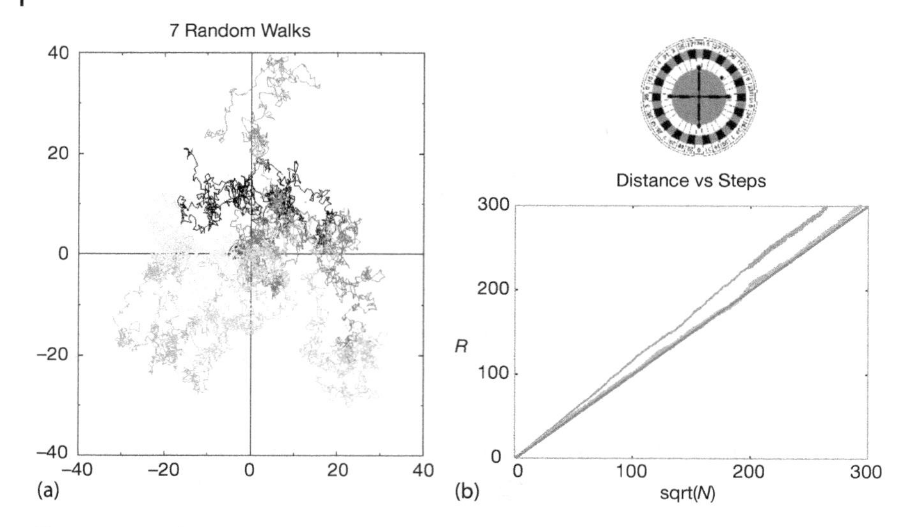

Figure 4.4 (a) The steps taken in seven 2D random walk simulations. (b) The distance covered in two walks of N steps using different schemes for including randomness. The theoretical prediction (4.20) is the straight line.

3. If you have your walker taking N steps in a single trial, then conduct a total number $K \simeq \sqrt{N}$ of trials. Each trial should have N steps and start with a different seed.
4. Calculate the mean square distance R^2 for each trial and then take the average of R^2 for all your K trials:

$$\langle R^2(N) \rangle = \frac{1}{K} \sum_{k=1}^{K} R^2_{(k)}(N) . \tag{4.23}$$

5. Check the validity of the assumptions made in deriving the theoretical result (4.20) by checking how well

$$\frac{\langle \Delta x_i \Delta x_{j \neq i} \rangle}{R^2} \simeq \frac{\langle \Delta x_i \Delta y_j \rangle}{R^2} \simeq 0 . \tag{4.24}$$

   Do your checking for both a single (long) run and for the average over trials.
6. Plot the root-mean-square distance $R_{\text{rms}} = \sqrt{\langle R^2(N) \rangle}$ as a function of $\sqrt{N}$. Values of N should start with a small number, where $R \simeq \sqrt{N}$ is not expected to be accurate, and end at a quite large value, where two or three places of accuracy should be expected on the average.
7. Repeat the preceding and following analysis for a 3D walk as well.

4.4
Extension: Protein Folding and Self-Avoiding Random Walks

A protein is a large biological molecule made up of molecular chains (the residues of amino acids). These chains are formed from *monomers*, that is, molecules that bind chemically with other molecules. More specifically, these chains consist of nonpolar hydrophobic (H) monomers that are repelled by water, and polar (P) monomers that are attracted by water. The actual structure of a protein results from a *folding process* in which random coils of chains rearrange themselves into a configuration of minimum energy. We want to model that process on the computer.

Although molecular dynamics (Chapter 18) may be used to simulate protein folding, it is much slower than Monte Carlo techniques, and even then, it is hard to find the lowest energy states. Here we create a simple Monte Carlo simulation in which you to take a random walk in a 2D square lattice (Yue *et al.*, 2004). At the end of each step, you randomly choose an H or a P monomer and drop it on the lattice, with your choice weighted such that H monomers are more likely than P ones. The walk is restricted such that the only positions available after each step are the three neighboring sites, with the already-occupied sites excluded (this is why this technique is known as a *self-avoiding random walk*).

The goal of the simulation is to find the lowest energy state of an HP sequence of various lengths. These then may be compared to those in nature. Just how best to find such a state is an active research topic (Yue *et al.*, 2004). The energy of a chain is defined as

$$E = -\epsilon f , \tag{4.25}$$

where ϵ is a positive constant and f is the number of H–H neighbor *not* connected directly (P–P and H–P bonds do not count at lowering the energy). So if the neighbor next to an H is another H, it lowers the energy, but if it is a P it does not lower the energy. We show a typical simulation result in Figure 4.5, where a light dot is placed half way between two H (dark-dot) neighbors. Accordingly, for a given length of chain, we expect the natural state(s) of an H–P sequence to be those with the largest possible number f of H–H contacts. That is what we are looking for.

1. Modify the random walk program we have already developed so that it simulates a self-avoiding random walk. The key here is that the walk stops at a corner, or when there are no empty neighboring sites available.
2. Make a random choice as to whether the monomer is an H or a P, with a weighting such that there are more H's than P's.
3. Produce a visualization that shows the positions occupied by the monomers, with the H and P monomers indicated by different color dots. (Our visualization, shown in Figure 4.5, is produced by the program `ProteinFold.py`, available on the Instructor's site.)
4. After the walk ends, record the energy and length of the chain.

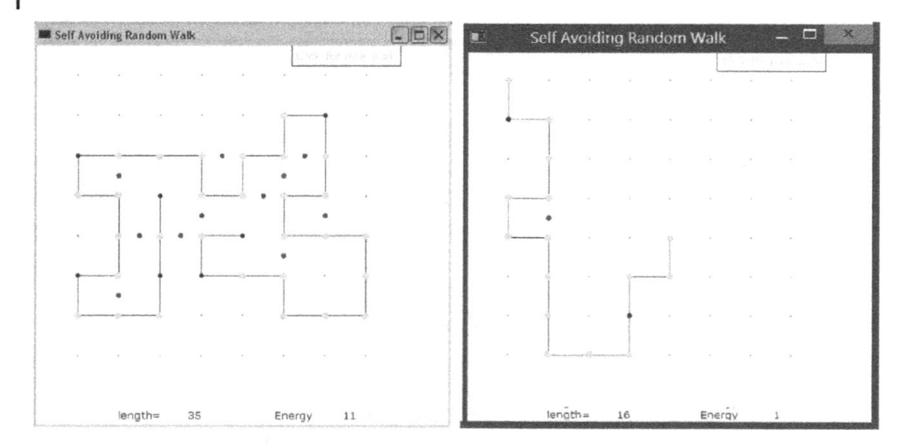

Figure 4.5 Two self-avoiding random walks that simulate protein chains with hydrophobic (H) monomers in light gray, and polar (P) monomers in black. The dark dots indicate regions where two H monomers are not directly connected.

5. Run many folding simulations and save the outputs, categorized by length and energy.
6. Examine the state(s) of the lowest energy for various chain lengths and compare the results to those from molecular dynamic simulations and actual protein structures (available on the Web).
7. Do you think that this simple model has some merit?
8. ⊙ Extend the folding to 3D.

4.5 Spontaneous Decay (Problem)

Your *problem* is to simulate how a small number N of radioactive particles decay.[2)] In particular, you are to determine when radioactive decay looks like exponential decay and when it looks *stochastic* (containing elements of chance). Because the exponential decay law is a large-number approximation to a natural process that always leads to only a small number of nuclei remaining, our simulation should be closer to nature than is the exponential decay law (Figure 4.6). In fact, if you "listen" to the output of the decay simulation code, what you will hear sounds very much like a Geiger counter, an intuitively convincing demonstration of the realism of the simulation.

Spontaneous decay is a natural process in which a particle, with no external stimulation, decays into other particles. Although the probability of decay of any one particle in any time interval is constant, just when it decays is a random event.

2) Spontaneous decay is also discussed in Chapter 7, where we fit an exponential function to a decay spectrum.

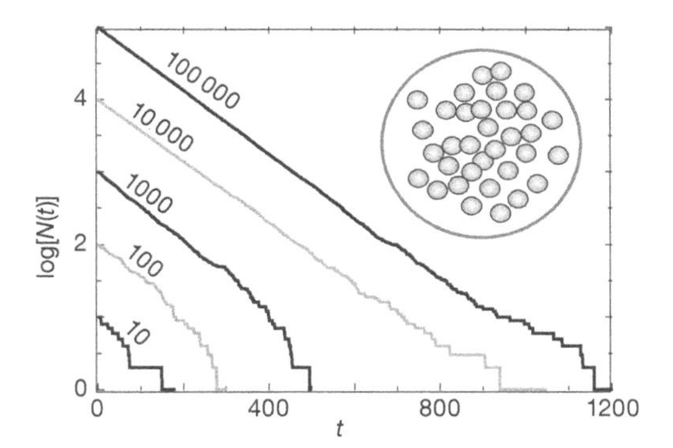

Figure 4.6 *Circle:* A sample containing *N* nuclei, each of which has the same probability of decaying per unit time, *Graphs:* Semilog plots of the number of nuclei vs. time for five simulations with differing initial numbers of nuclei. Exponential decay would be a straight line with bumps, similar to the initial behavior for $N = 100\,000$.

Because the exact moment when any one particle decays is always random, and because one nucleus does not influence another nucleus, the probability of decay is not influenced by how long the particle has been around or whether some other particles have decayed. In other words, the probability $\mathcal{P}$ of any one particle decaying per unit time interval is a constant, yet when that particle decays it is gone forever. Of course, as the total number N of particles decreases with time, so will the number that decay per unit time, but the probability of any one particle decaying in some time interval remains the same as long as that particle exists.

4.5.1 Discrete Decay (Model)

Imagine having a sample containing $N(t)$ radioactive nuclei at time t (Figure 4.6 circle). Let ΔN be the number of particles that decay in some small time interval Δt. We convert the statement "the probability $\mathcal{P}$ of any one particle decaying per unit time is a constant" into the equation

$$\mathcal{P} = \frac{\Delta N(t)/N(t)}{\Delta t} = -\lambda \,, \tag{4.26}$$

$$\Rightarrow \frac{\Delta N(t)}{\Delta t} = -\lambda N(t) \,, \tag{4.27}$$

where the constant λ is called the *decay rate* and the minus sign indicates a decreasing number. Because $N(t)$ decreases in time, the *activity* $\Delta N(t)/\Delta t$ (sometimes called the decay rate) also decreases with time. In addition, because the total activity is proportional to the total number of particles present, it is too stochastic with an exponential-like decay in time. (Actually, because the number of decays $\Delta N(t)$ is proportional to the difference in random numbers, its tends to show even larger statistical fluctuations than does $N(t)$.)

Equation 4.27 is a *finite-difference equation* relating the experimentally quantities $N(t)$, $\Delta N(t)$, and Δt. Although a difference equation cannot be integrated the way a differential equation can, it can be simulated numerically. Because the process is random, we cannot predict a single value for $\Delta N(t)$, although we can predict the average number of decays when observations are made of many identical systems of N decaying particles.

4.5.2
Continuous Decay (Model)

When the number of particles $N \to \infty$ and the observation time interval $\Delta t \to 0$, our difference equation becomes a differential equation, and we obtain the familiar exponential decay law (4.27):

$$\frac{\Delta N(t)}{\Delta t} \to \frac{\mathrm{d}N(t)}{\mathrm{d}t} = -\lambda N(t) . \tag{4.28}$$

This can be integrated to obtain the time dependence of the total number of particles and of the total activity:

$$N(t) = N(0)\mathrm{e}^{-\lambda t} = N(0)\mathrm{e}^{-t/\tau} , \tag{4.29}$$

$$\frac{\mathrm{d}N(t)}{\mathrm{d}t} = -\lambda N(0)\mathrm{e}^{-\lambda t} = \frac{\mathrm{d}N}{\mathrm{d}t}(0)\mathrm{e}^{-\lambda t} . \tag{4.30}$$

In this limit, we can identify the decay rate λ with the inverse lifetime:

$$\lambda = \frac{1}{\tau} . \tag{4.31}$$

We see from its derivation that the exponential decay is a good description of nature for a large number of particles where $\Delta N/N \simeq 0$. However, in nature $N(t)$ can be a small number, and in that case we have a statistical and not a continuous process. The basic law of nature (4.26) is always valid, but as we will see in the simulation, the exponential decay (4.30) becomes less and less accurate as the number of particles gets smaller and smaller.

4.5.3
Decay Simulation with Geiger Counter Sound

A program for simulating radioactive decay is surprisingly simple but not without its subtleties. We increase time in discrete steps of Δt, and for each time interval we count the number of nuclei that have decayed during that Δt. The simulation quits when there are no nuclei left to decay. Such being the case, we have an outer loop over the time steps Δt and an inner loop over the remaining nuclei for each time step. The pseudocode is simple (as is the code):

```
input N, lambda
t=0
while N > 0
 DeltaN = 0
 for i = 1..N
 if (r_i < lambda) DeltaN = DeltaN + 1
 end for
 t = t +1
 N = N - DeltaN
 Output t, DeltaN, N
end while
```

When we pick a value for the decay rate $\lambda = 1/\tau$ to use in our simulation, we are setting the scale for times. If the actual decay rate is $\lambda = 0.3 \times 10^6\ \text{s}^{-1}$ and if we decide to measure times in units of 10^{-6} s, then we will choose random numbers $0 \le r_i \le 1$, which leads to λ values lying someplace near the middle of the range (e.g., $\lambda \simeq 0.3$). Alternatively, we can use a value of $\lambda = 0.3 \times 10^6\ \text{s}^{-1}$ in our simulation and then scale the random numbers to the range $0 \le r_i \le 10^6$. However, unless you plan to compare your simulation to experimental data, you do not have to worry about the scale for time but instead should focus on the physics behind the slopes and relative magnitudes of the graphs.

Listing 4.2 DecaySound.py simulates spontaneous decay in which a decay occurs if a random number is smaller than the decay parameter. The *winsound* package lets us play a beep each time there is a decay, and this leads to the sound of a Geiger counter.

```
# DecaySound.py spontaneous decay simulation

from visual import *
from visual.graph import *
import random
import winsound

lambda1 = 0.005                                              # Decay constant
max = 80.;   time_max = 500;    seed = 68111
number = nloop = max                                         # Initial value
graph1 = gdisplay(title ='Spontaneous Decay',xtitle='Time',\
                  ytitle = 'Number')
decayfunc = gcurve(color = color.green)

for time in arange(0, time_max + 1):                         # Time loop
    for atom in arange(1, number + 1 ):                      # Decay loop
        decay = random.random()
        if (decay < lambda1):
            nloop = nloop - 1                                # A decay
            winsound.Beep(600, 100)                          # Sound beep
    number = nloop
    decayfunc.plot( pos = (time, number) )
    rate(30)
```

Decay.py is our sample simulation of the spontaneous decay. An extension of this program, **DecaySound.py**, in Listing 4.2, adds a beep each time an atom decays (unfortunately this works only with Windows). When we listen to the simulation, it sounds like a Geiger counter, with its randomness and its slowing down with time. This provides some rather convincing evidence of the realism of the simulation.

4.6
Decay Implementation and Visualization

Write a program to simulate the radioactive decay using the simple program in Listing 4.2 as a guide. You should obtain results like those in Figure 4.6.

1. Plot the logarithm of the number left $\ln N(t)$ and the logarithm of the decay rate $\ln \Delta N(t)/\Delta t(= 1)$ vs. time. Note that the simulation measures time in steps of Δt (generation number).
2. Check that you obtain what looks like the exponential decay when you start with large values for $N(0)$, but that the decay displays its stochastic nature for small $N(0)$. (Large $N(0)$ values are also stochastic; they just do not look like it.)
3. Create two plots, one showing that the slopes of $N(t)$ vs. t are *independent* of $N(0)$ and another showing that the slopes are proportional to the value chosen for λ.
4. Create a plot showing that within the expected statistical variations, $\ln N(t)$ and $\ln \Delta N(t)$ are proportional.
5. Explain in your own words how a process that is spontaneous and random at its very heart can lead to the exponential decay.
6. How does your simulation show that the decay is exponential-like and not a power law such as $N = \beta t^{-\alpha}$?

5
Differentiation and Integration

We start this chapter with a short discussion of numerical differentiation, an important but rather simple topic. We derive the forward-difference, central-difference, and extrapolated-difference methods for differentiation. They will be used throughout the book. The majority of this chapter deals with numerical integration, a basic tool of scientific computation. We derive Simpson's rule, the trapezoid rule, and the Gaussian quadrature rule. We discuss Gaussian quadrature (our personal workhorse) in its various forms, and indicate how to map the standard Gauss points to a wide range of intervals. We end the chapter with a discussion of Monte Carlo integration techniques, which are fundamentally different from all the other rules.

5.1
Differentiation

Problem Figure 5.1 shows the trajectory of a projectile with air resistance. The dots indicate the times t at which measurements were made and tabulated. Your *problem* is to determine the velocity dy/dt as a function of time. Note that because there is realistic air resistance present, there is no analytic function to differentiate, only this graph or a table of numbers read from it.

You probably did rather well in your first calculus course and feel competent at taking derivatives. However, you may never have taken derivatives of a table of numbers using the elementary definition

$$\frac{dy(t)}{dt} \stackrel{\text{def}}{=} \lim_{h\to 0} \frac{y(t+h)-y(t)}{h} . \tag{5.1}$$

In fact, even a computer runs into errors with this kind of limit because it is wrought with subtractive cancelation; as h is made smaller, the computer's finite word length causes the numerator to fluctuate between 0 and the machine precision ϵ_m, and as the denominator approaches zero, overflow occurs.

Computational Physics, 3rd edition. Rubin H. Landau, Manuel J. Páez, Cristian C. Bordeianu.

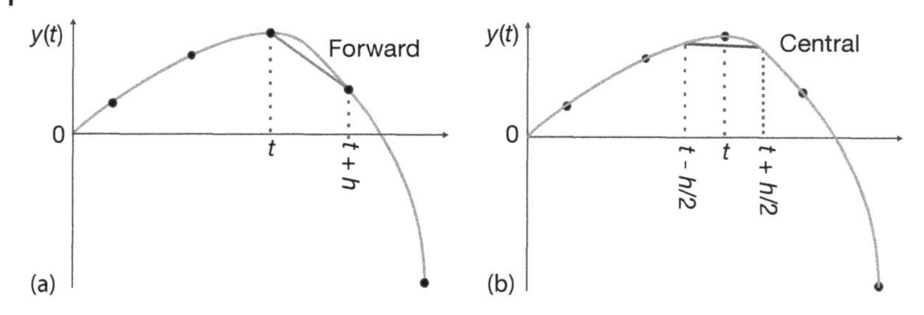

Figure 5.1 A trajectory of a projectile experiencing air resistance. Forward-difference approximation (slanted dashed line) and central-difference approximation (horizontal line) for the numerical first derivative at time t. (A tangent to the curve at t yields the correct derivative.) The central difference is seen to be more accurate.

5.2 Forward Difference (Algorithm)

The most direct method for numerical differentiation starts by expanding a function in a Taylor series to obtain its value at a small step h away:

$$y(t+h) = y(t) + h\frac{\mathrm{d}y(t)}{\mathrm{d}t} + \frac{h^2}{2!}\frac{\mathrm{d}^2y(t)}{\mathrm{d}t^2} + \frac{h^3}{3!}\frac{\mathrm{d}y^3(t)}{\mathrm{d}t^3} + \cdots , \tag{5.2}$$

$$\Rightarrow \qquad \frac{y(t+h) - y(t)}{h} = \frac{\mathrm{d}y(t)}{\mathrm{d}t} + \frac{h}{2!}\frac{\mathrm{d}^2y(t)}{\mathrm{d}t^2} + \frac{h^2}{3!}\frac{\mathrm{d}y^3(t)}{\mathrm{d}t^3} + \cdots . \tag{5.3}$$

If we ignore the h^2 terms in (5.3), we obtain the *forward-difference* derivative algorithm for the derivative (5.2) for $y'(t)$:

$$\left.\frac{\mathrm{d}y(t)}{\mathrm{d}t}\right|_{\mathrm{fd}} \stackrel{\mathrm{def}}{=} \frac{y(t+h) - y(t)}{h} . \tag{5.4}$$

An estimate of the error follows from substituting the Taylor series:

$$\left.\frac{\mathrm{d}y(t)}{\mathrm{d}t}\right|_{\mathrm{fd}} \simeq \frac{\mathrm{d}y(t)}{\mathrm{d}t} - \frac{h}{2}\frac{\mathrm{d}y^2(t)}{\mathrm{d}t^2} + \cdots . \tag{5.5}$$

You can think of this approximation as using two points to represent the function by a straight line in the interval from x to $x+h$ (Figure 5.1a). The approximation (5.4) has an error proportional to h (unless the heavens look down upon you kindly and make y'' vanish). We can make the approximation error smaller by making h smaller, yet precision will be lost through the subtractive cancelation on the left-hand side (LHS) of (5.4) for too small an h.

To see how the forward-difference algorithm works, let $y(t) = a + bt^2$. The exact derivative is $y' = 2bt$, while the computed derivative is

$$\left.\frac{\mathrm{d}y(t)}{\mathrm{d}t}\right|_{\mathrm{fd}} \simeq \frac{y(t+h) - y(t)}{h} = 2bt + bh . \tag{5.6}$$

This clearly becomes a good approximation only for small $h (h \ll 1/b)$.

5.3 Central Difference (Algorithm)

An improved approximation to the derivative starts with the basic definition (5.1) or geometrically as shown in Figure 5.1b. Now, rather than making a single step of h forward, we form a *central difference* by stepping forward half a step and backward half a step:

$$\left.\frac{\mathrm{d}y(t)}{\mathrm{d}t}\right|_{\text{cd}} \equiv D_{\text{cd}}\,y(t) \stackrel{\text{def}}{=} \frac{y(t+h/2)-y(t-h/2)}{h} \,. \tag{5.7}$$

We estimate the error in the central-difference algorithm by substituting the Taylor series for $y(t+h/2)$ and $y(t-h/2)$ into (5.7):

$$\begin{aligned} y\left(t+\frac{h}{2}\right)-y\left(t-\frac{h}{2}\right) &\simeq \left[y(t)+\frac{h}{2}y'(t)+\frac{h^2}{8}y''(t)+\frac{h^3}{48}y(t)+\mathcal{O}(h^4)\right] \\ &\quad -\left[y(t)-\frac{h}{2}y'(t)+\frac{h^2}{8}y''(t)-\frac{h^3}{48}y'''(t)+\mathcal{O}(h^4)\right] \\ &= h\,y'(t)+\frac{h^3}{24}y'''(t)+\mathcal{O}(h^5)\,, \\ \Rightarrow \quad \left.\frac{\mathrm{d}y(t)}{\mathrm{d}t}\right|_{\text{cd}} &\simeq y'(t)+\frac{1}{24}h^2\,y'''(t)+\mathcal{O}(h^4)\,. \end{aligned} \tag{5.8}$$

The important difference between this central-difference algorithm and the forward difference one is that when $y(t-h/2)$ is subtracted from $y(t+h/2)$, all terms containing an even power of h in the two Taylor series cancel. This make the central-difference algorithm accurate to order h^2 (h^3 before division by h), while the forward difference is accurate only to order h. If the $y(t)$ is smooth, that is, if $y'''h^2/24 \ll y''h/2$, then you can expect the central-difference error to be smaller than the one with the central-difference algorithm.

If we now return to our parabola example (5.6), we will see that the central difference gives the exact derivative independent of h:

$$\left.\frac{\mathrm{d}y(t)}{\mathrm{d}t}\right|_{\text{cd}} \simeq \frac{y(t+h/2)-y(t-h/2)}{h} = 2bt\,. \tag{5.9}$$

This is to be expected because the higher derivatives equal zero for a second-order polynomial.

5.4 Extrapolated Difference (Algorithm)

Because a differentiation rule based on keeping a certain number of terms in a Taylor series also provides an expression for the error (the terms not included), we can reduce the theoretical error further by forming a combination of approximations whose summed errors extrapolate to zero. One such algorithm is the central-difference algorithm (5.7) using a half-step back and a half-step forward. A second algorithm is another central-difference approximation, but this time using quarter-steps:

$$\left.\frac{\mathrm{d}y(t, h/2)}{\mathrm{d}t}\right|_{\mathrm{cd}} \stackrel{\mathrm{def}}{=} \frac{y(t + h/4) - y(t - h/4)}{h/2}$$
$$\simeq y'(t) + \frac{h^2}{96}\frac{\mathrm{d}^3 y(t)}{\mathrm{d}t^3} + \cdots \,. \tag{5.10}$$

A combination of the two, called the *extended difference algorithm*, eliminates both the quadratic and linear terms:

$$\left.\frac{\mathrm{d}y(t)}{\mathrm{d}t}\right|_{\mathrm{ed}} \stackrel{\mathrm{def}}{=} \frac{4D_{\mathrm{cd}}\,y(t, h/2) - D_{\mathrm{cd}}\,y(t, h)}{3} \tag{5.11}$$

$$\simeq \frac{\mathrm{d}y(t)}{\mathrm{d}t} - \frac{h^4 y^{(5)}(t)}{4 \times 16 \times 120} + \cdots \,. \tag{5.12}$$

Here (5.11) is the extended-difference algorithm and (5.12) gives its error, with D_{cd} representing the central-difference algorithm. If $h = 0.4$ and $y^{(5)} \simeq 1$, then there will be only one place of the round-off error and the truncation error will be approximately machine precision ϵ_{m}; this is the best you can hope for.

When working with these and similar higher order methods, it is important to remember that while they may work as designed for well-behaved functions, they may fail badly for functions containing noise, as do data from computations or measurements. If noise is evident, it may be better to first smooth the data or fit them with some analytic function using the techniques of Chapter 7 and then differentiate.

5.5
Error Assessment

The approximation errors in numerical differentiation decrease with decreasing step size h. In turn, round-off errors increase with decreasing step size because you have to take more steps and do more calculations. Remember from our discussion in Chapter 3 that the best approximation occurs for an h that makes the total error $\epsilon_{\mathrm{app}} + \epsilon_{\mathrm{ro}}$ a minimum, and that as a rough guide this occurs when $\epsilon_{\mathrm{ro}} \simeq \epsilon_{\mathrm{app}}$.

We have already estimated the approximation error in numerical differentiation rules by making a Taylor series expansion of $y(x + h)$. The approximation error with the forward-difference algorithm (5.4) is $\mathcal{O}(h)$, while that with the central-difference algorithm (5.8) is $\mathcal{O}(h^2)$:

$$\epsilon_{\mathrm{app}}^{\mathrm{fd}} \simeq \frac{y'' h}{2}\,, \quad \epsilon_{\mathrm{app}}^{\mathrm{cd}} \simeq \frac{y''' h^2}{24}\,. \tag{5.13}$$

To obtain a rough estimate of the round-off error, we observe that differentiation essentially subtracts the value of a function at argument x from that of the same function at argument $x + h$ and then divide by h: $y' \simeq [y(t+h) - y(t)]/h$. As h is made continually smaller, we eventually reach the round-off error limit where $y(t+h)$ and $y(t)$ differ by just machine precision ϵ_m:

$$\epsilon_{ro} \simeq \frac{\epsilon_m}{h} . \tag{5.14}$$

Consequently, round-off and approximation errors become equal when

$$\epsilon_{ro} \simeq \epsilon_{app} , \tag{5.15}$$

$$\frac{\epsilon_m}{h} \simeq \epsilon_{app}^{fd} = \frac{y^{(2)} h}{2} , \qquad \frac{\epsilon_m}{h} \simeq \epsilon_{app}^{cd} = \frac{y^{(3)} h^2}{24} , \tag{5.16}$$

$$\Rightarrow \quad h_{fd}^2 = \frac{2\epsilon_m}{y^{(2)}} , \qquad \Rightarrow \quad h_{cd}^3 = \frac{24\epsilon_m}{y^{(3)}} . \tag{5.17}$$

We take $y' \simeq y^{(2)} \simeq y^{(3)}$ (which may be crude in general, although not bad for e^t or $\cos t$) and assume double precision, $\epsilon_m \simeq 10^{-15}$:

$$h_{fd} \simeq 4 \times 10^{-8} , \qquad h_{cd} \simeq 3 \times 10^{-5} , \tag{5.18}$$

$$\Rightarrow \quad \epsilon_{fd} \simeq \frac{\epsilon_m}{h_{fd}} \simeq 3 \times 10^{-8} , \quad \Rightarrow \quad \epsilon_{cd} \simeq \frac{\epsilon_m}{h_{cd}} \simeq 3 \times 10^{-11} . \tag{5.19}$$

This may seem backward because the better algorithm leads to a larger h value. It is not. The ability to use a larger h means that the error in the central-difference method is about 1000 times smaller than the error in the forward-difference method.

The programming for numerical differentiation is simple:

```
FD = (y(t+h) - y(t)) /h;                                    // forward diff
CD = (y(t+h/2) - y(t-h/2)) /h;                              // central diff
ED = (8*(y(t+h/4)-y(t-h/4)) - (y(t+h/2)-y(t-h/2)))/3/h;    //extrap
```

1. Use forward-, central-, and extrapolated-difference algorithms to differentiate the functions $\cos t$ and e^t at $t = 0.1$, 1., and 100.
   a) Print out the derivative and its relative error $\mathcal{E}$ as functions of h. Reduce the step size h until it equals machine precision $h \simeq \epsilon_m$.
   b) Plot $\log_{10} |\mathcal{E}|$ vs. $\log_{10} h$ and check whether the number of decimal places obtained agrees with the estimates in the text.
   c) See if you can identify regions where algorithmic (series truncation) error dominates at large h and the round-off error at small h in your plot. Do the slopes agree with our model's predictions?

5.6 Second Derivatives (Problem)

Let us say that you have measured the position $y(t)$ vs. time for a particle (Figure 5.1). Your *problem* now is to determine the force on the particle. Newton's second law tells us that force and acceleration are linearly related:

$$F = ma = m\frac{\mathrm{d}^2 y}{\mathrm{d}t^2}\,, \tag{5.20}$$

where F is the force, m is the particle's mass, and a is the acceleration. So by determining the derivative $\mathrm{d}^2y/\mathrm{d}t^2$ from the $y(t)$ values, we determine the force.

The concerns we expressed about errors in first derivatives are even more valid for second derivatives where additional subtractions may lead to additional cancelations. Let us look again at the central-difference method:

$$\left.\frac{\mathrm{d}y(t)}{\mathrm{d}t}\right|_{\mathrm{cd}} \simeq \frac{y(t+h/2)-y(t-h/2)}{h}\,. \tag{5.21}$$

This algorithm gives the derivative at t by moving forward and backward from t by $h/2$. We take the second derivative $\mathrm{d}^2y/\mathrm{d}t^2$ to be the central difference of the first derivative:

$$\left.\frac{\mathrm{d}^2 y(t)}{\mathrm{d}t^2}\right|_{\mathrm{cd}} \simeq \frac{y'(t+h/2)-y'(t-h/2)}{h}$$

$$\simeq \frac{[y(t+h)-y(t)]-[y(t)-y(t-h)]}{h^2} \tag{5.22}$$

$$= \frac{y(t+h)+y(t-h)-2y(t)}{h^2}\,. \tag{5.23}$$

As we did for the first derivatives, we determine the second derivative at t by evaluating the function in the region surrounding t. Although the form (5.23) is more compact and requires fewer steps than (5.22), it may increase subtractive cancelation by first storing the "large" number $y(t+h)+y(t-h)$ and then subtracting another large number $2y(t)$ from it. We ask you to explore this difference as an exercise.

5.6.1 Second-Derivative Assessment

Write a program to calculate the second derivative of $\cos t$ using the central-difference algorithms (5.22) and (5.23). Test it over four cycles. Start with $h \simeq \pi/10$ and keep reducing h until you reach machine precision. Is there any noticeable differences between (5.22) and (5.23)?

5.7 Integration

Problem: Integrating a Spectrum An experiment has measured $\mathrm{d}N(t)/\mathrm{d}t$, the number of particles entering a counter per unit time. Your *problem* is to integrate this spectrum to obtain the number of particles $N(1)$ that entered the counter in the first second:

$$N(1) = \int_0^1 \frac{\mathrm{d}N(t)}{\mathrm{d}t}\,\mathrm{d}t\,. \tag{5.24}$$

5.8 Quadrature as Box Counting (Math)

The integration of a function may require some cleverness to do analytically, but is relatively straightforward on a computer. A traditional way to perform numerical integration by hand is to take a piece of a graph paper and count the number of boxes or *quadrilaterals* lying below a curve of the integrand. For this reason, numerical integration is also called *numerical quadrature* even when it becomes more sophisticated than simple box counting.

The Riemann definition of an integral is the limit of the sum over boxes as the width h of the box approaches zero (Figure 5.2):

$$\int_a^b f(x)\,\mathrm{d}x = \lim_{h\to 0}\left[h\sum_{i=1}^{(b-a)/h} f(x_i)\right]\,. \tag{5.25}$$

The numerical integral of a function $f(x)$ is approximated as the equivalent of a finite sum over boxes of height $f(x)$ and width w_i:

$$\int_a^b f(x)\,\mathrm{d}x \simeq \sum_{i=1}^{N} f(x_i)w_i\,, \tag{5.26}$$

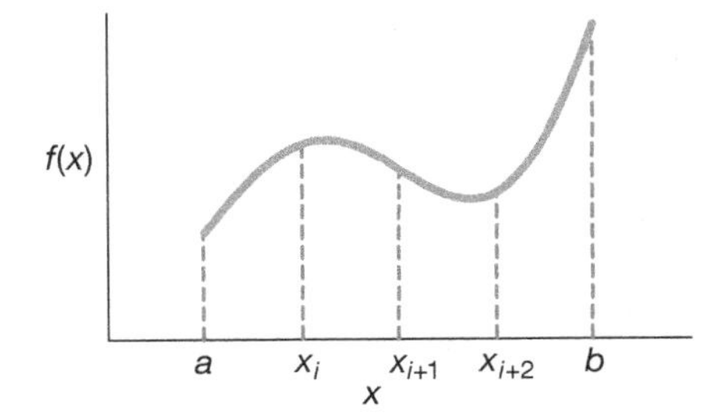

Figure 5.2 The integral $\int_a^b f(x)\,\mathrm{d}x$ is the area under the graph of $f(x)$ from a to b. Here we break up the area into four regions of equal widths h and five integration points.

which is similar to the Riemann definition (5.25) except that there is no limit to an infinitesimal box size. Equation 5.26 is the standard form for all integration algorithms; the function $f(x)$ is evaluated at N points in the interval $[a, b]$, and the function values $f_i \equiv f(x_i)$ are summed with each term in the sum weighted by w_i. While, in general, the sum in (5.26) gives the exact integral only when $N \to \infty$, it may be exact for finite N if the integrand is a polynomial. The different integration algorithms amount to different ways of choosing the points x_i and weights w_i. Generally, the precision increases as N gets larger, with round-off error eventually limiting the increase. Because the "best" integration rule depends on the specific behavior of $f(x)$, there is no universally best approximation. In fact, some of the automated integration schemes found in subroutine libraries switch from one method to another, as well as change the methods for different intervals until they find ones that work well for each interval.

In general, you should not attempt a numerical integration of an integrand that contains a singularity without first removing the singularity by hand.[1] You may be able to do this very simply by breaking the interval down into several subintervals so the singularity is at an endpoint where an integration point is not placed or by a change of variable:

$$\int_{-1}^{1} |x| f(x)\,\mathrm{d}x = \int_{-1}^{0} f(-x)\,\mathrm{d}x + \int_{0}^{1} f(x)\,\mathrm{d}x \,, \tag{5.27}$$

$$\int_{0}^{1} x^{1/3}\,\mathrm{d}x = \int_{0}^{1} 3y^3\,\mathrm{d}y \,, \quad (y \stackrel{\text{def}}{=} x^{1/3}) \,, \tag{5.28}$$

$$\int_{0}^{1} \frac{f(x)\,\mathrm{d}x}{\sqrt{1-x^2}} = 2\int_{0}^{1} \frac{f(1-y^2)\,\mathrm{d}y}{\sqrt{2-y^2}} \,, \quad (y^2 \stackrel{\text{def}}{=} 1-x) \,. \tag{5.29}$$

Likewise, if your integrand has a very slow variation in some region, you can speed up the integration by changing to a variable that compresses that region and places few points there, or divides up the interval and performs several integrations. Conversely, if your integrand has a very rapid variation in some region, you may want to change to variables that expand that region to ensure that no oscillations are missed.

Listing 5.1 TrapMethods.py integrates a function $f(y)$ with the trapezoid rule. Note that the step size h depends upon the size of interval here and that the weights at the ends and middle of the intervals differ.

```
# TrapMethods.py: trapezoid integration, a<x<b, N pts, N-1 intervals

from numpy import *
```

1) In Chapter 26, we show how to remove such a singularity even when the integrand is unknown.

```
def func(x):
    return 5*(sin(8*x))**2*exp(-x*x)-13*cos(3*x)

def trapezoid(A,B,N):
    h = (B - A)/(N - 1)                         # step size
    sum = (func(A)+func(B))/2                   # (1st + last)/2
    for i in range(1, N-1):
       sum += func(A+i*h)
    return h*sum
A = 0.5
B = 2.3
N = 1200
print(trapezoid(A,B,N-1))
```

5.9
Algorithm: Trapezoid Rule

Trapezoid and Simpson integration rules use evenly spaced values of x (Figure 5.3). They use N points $x_i (i = 1, N)$ evenly spaced at a distance h apart throughout the integration region $[a, b]$ and *include the endpoints* in the integration region. This means that there are $(N - 1)$ intervals of length h:

$$h = \frac{b-a}{N-1}\,, \quad x_i = a + (i-1)h\,, \quad i = 1, N\,, \tag{5.30}$$

where we start our counting at $i = 1$. The trapezoid rule takes each integration interval i and constructs a trapezoid of width h in it (Figure 5.3). This approximates $f(x)$ by a straight line in each interval i and uses the average height $(f_i + f_{i+1})/2$ as the value for f. The area of each such trapezoid is

$$\int_{x_i}^{x_i+h} f(x)\,\mathrm{d}x \simeq \frac{h(f_i + f_{i+1})}{2} = \frac{1}{2}hf_i + \frac{1}{2}hf_{i+1}\,. \tag{5.31}$$

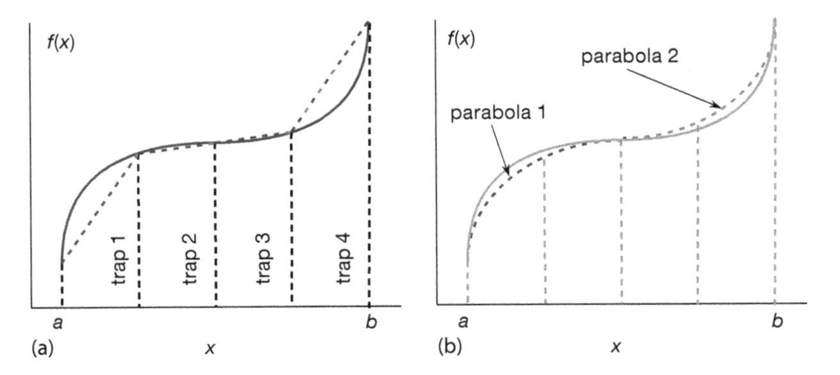

Figure 5.3 Different shapes used to approximate the areas under the curve. (a) Straight-line sections used for the trapezoid rule. (b) Two parabolas used in Simpson's rule.

Table 5.1 Elementary weights for uniform-step integration rules.

Name	Degree	Elementary weights
Trapezoid	1	$(1,1)\frac{h}{2}$
Simpson's	2	$(1,4,1)\frac{h}{3}$
$\frac{3}{8}$	3	$(1,3,3,1)\frac{3}{8}h$
Milne	4	$(14,64,24,64,14)\frac{h}{45}$

In terms of our standard integration formula (5.26), the "rule" in (5.31) is for $N = 2$ points with weights $w_i \equiv \frac{1}{2}$ (Table 5.1).

In order to apply the trapezoid rule to the entire region $[a, b]$, we add the contributions from each subinterval:

$$\int_a^b f(x)\,\mathrm{d}x \simeq \frac{h}{2} f_1 + h f_2 + h f_3 + \cdots + h f_{N-1} + \frac{h}{2} f_N \,. \tag{5.32}$$

You will notice that because the internal points are counted twice (as the end of one interval and as the beginning of the next), they have weights of $h/2 + h/2 = h$, whereas the endpoints are counted just once, and on that account have weights of only $h/2$. In terms of our standard integration rule (5.57), we have

$$w_i = \left\{ \frac{h}{2}, h, \ldots, h, \frac{h}{2} \right\} \quad \text{(Trapezoid rule)} \,. \tag{5.33}$$

In Listing 5.1, we provide a simple implementation of the trapezoid rule.

5.10
Algorithm: Simpson's Rule

Simpson's rule approximates the integrand $f(x)$ by a parabola for each interval (Figure 5.3b):

$$f(x) \simeq \alpha x^2 + \beta x + \gamma \,. \tag{5.34}$$

Again, all intervals equally spaced. The area under the parabola for each interval is

$$\int_{x_i}^{x_i+h} (\alpha x^2 + \beta x + \gamma)\,\mathrm{d}x = \left. \frac{\alpha x^3}{3} + \frac{\beta x^2}{2} + \gamma x \right|_{x_i}^{x_i+h} \,. \tag{5.35}$$

In order to relate the parameters α, β, and γ to the function, we consider an interval from -1 to $+1$, in which case

$$\int_{-1}^{1} (\alpha x^2 + \beta x + \gamma)\,\mathrm{d}x = \frac{2\alpha}{3} + 2\gamma \,. \tag{5.36}$$

But we notice that

$$f(-1) = \alpha - \beta + \gamma\,, \quad f(0) = \gamma\,, \quad f(1) = \alpha + \beta + \gamma\,, \tag{5.37}$$

$$\Rightarrow \alpha = \frac{f(1) + f(-1)}{2} - f(0)\,, \quad \beta = \frac{f(1) - f(-1)}{2}\,, \quad \gamma = f(0)\,. \tag{5.38}$$

In this way, we can express the integral as the weighted sum over the values of the function at three points:

$$\int_{-1}^{1} (\alpha x^2 + \beta x + \gamma)\, \mathrm{d}x = \frac{f(-1)}{3} + \frac{4f(0)}{3} + \frac{f(1)}{3}\,. \tag{5.39}$$

Because three values of the function are needed, we generalize this result to our problem by evaluating the integral over two adjacent intervals, in which case we evaluate the function at the two endpoints and in the middle (Table 5.1):

$$\begin{aligned}\int_{x_i-h}^{x_i+h} f(x)\, \mathrm{d}x &= \int_{x_i}^{x_i+h} f(x)\, \mathrm{d}x + \int_{x_i-h}^{x_i} f(x)\, \mathrm{d}x \\ &\simeq \frac{h}{3} f_{i-1} + \frac{4h}{3} f_i + \frac{h}{3} f_{i+1}\,. \end{aligned} \tag{5.40}$$

Simpson's rule requires the elementary integration to be over *pairs* of intervals, which in turn requires that the *total number of intervals be even or that the number of points N be odd*. In order to apply Simpson's rule to the entire interval, we add up the contributions from each pair of subintervals, counting all but the first and last endpoints twice:

$$\int_a^b f(x)\, \mathrm{d}x \simeq \frac{h}{3} f_1 + \frac{4h}{3} f_2 + \frac{2h}{3} f_3 + \frac{4h}{3} f_4 + \cdots + \frac{4h}{3} f_{N-1} + \frac{h}{3} f_N\,. \tag{5.41}$$

In terms of our standard integration rule (5.26), we have

$$w_i = \left\{ \frac{h}{3}, \frac{4h}{3}, \frac{2h}{3}, \frac{4h}{3}, \ldots, \frac{4h}{3}, \frac{h}{3} \right\} \quad \text{(Simpson's rule)}\,. \tag{5.42}$$

The sum of these weights provides a useful check on your integration:

$$\sum_{i=1}^{N} w_i = (N-1)h\,. \tag{5.43}$$

Remember, the number of points N must be odd for Simpson's rule.

5.11
Integration Error (Assessment)

In general, you should choose an integration rule that gives an accurate answer using the least number of integration points. We obtain a crude estimate of the *approximation* or *algorithmic error* $\mathcal{E}$ for the equal-spacing rules and their relative error ϵ by expanding $f(x)$ in a Taylor series around the midpoint of the integration interval. We then multiply that error by the number of intervals N to estimate the error for the entire region $[a, b]$. For the trapezoid and Simpson rules this yields

$$\mathcal{E}_t = O\left(\frac{[b-a]^3}{N^2}\right) f^{(2)}, \quad \mathcal{E}_s = O\left(\frac{[b-a]^5}{N^4}\right) f^{(4)}, \quad \epsilon_{t,s} = \frac{\mathcal{E}_{t,s}}{f}, \tag{5.44}$$

where ϵ is a measure of the relative error. We see that the third-derivative term in Simpson's rule cancels (much like the central-difference method does in differentiation). Equations 5.44 are illuminating in showing how increasing the sophistication of an integration rule leads to an error that decreases with a higher inverse power of N, yet it is also proportional to higher derivatives of f. Consequently, for small intervals and functions $f(x)$ with well-behaved derivatives, Simpson's rule should converge more rapidly than the trapezoid rule.

To model the round-off error in integration, we assume that after N steps the *relative* round-off error is random and of the form

$$\epsilon_{ro} \simeq \sqrt{N}\epsilon_m, \tag{5.45}$$

where ϵ_m is the machine precision, $\epsilon \sim 10^{-7}$ for single precision and $\epsilon \sim 10^{-15}$ for double precision (the standard for science). Because most scientific computations are performed with doubles, we will assume double precision. We want to determine an N that minimizes the total error, that is, the sum of the approximation and round-off errors:

$$\epsilon_{tot} \simeq \epsilon_{ro} + \epsilon_{app}. \tag{5.46}$$

This occurs, approximately, when the two errors are of equal magnitude, which we approximate even further by assuming that the two errors are equal:

$$\epsilon_{ro} = \epsilon_{app} = \frac{\mathcal{E}_{trap,simp}}{f}. \tag{5.47}$$

To continue the search for optimum N for a general function f, we set the scale of function size and the lengths by assuming

$$\frac{f^{(n)}}{f} \simeq 1, \quad b - a = 1 \quad \Rightarrow \quad h = \frac{1}{N}. \tag{5.48}$$

The estimate (5.47), when applied to the *trapezoid rule*, yields

$$\sqrt{N}\epsilon_m \simeq \frac{f^{(2)}(b-a)^3}{fN^2} = \frac{1}{N^2}, \tag{5.49}$$

$$\Rightarrow \quad N \simeq \frac{1}{(\epsilon_m)^{2/5}} = \left(\frac{1}{10^{-15}}\right)^{2/5} = 10^6 , \tag{5.50}$$

$$\Rightarrow \epsilon_{ro} \simeq \sqrt{N}\epsilon_m = 10^{-12} . \tag{5.51}$$

The estimate (5.47), when applied to *Simpson's rule*, yields

$$\sqrt{N}\epsilon_m = \frac{f^{(4)}(b-a)^5}{fN^4} = \frac{1}{N^4} , \tag{5.52}$$

$$\Rightarrow N = \frac{1}{(\epsilon_m)^{2/9}} = \left(\frac{1}{10^{-15}}\right)^{2/9} = 2154 , \tag{5.53}$$

$$\Rightarrow \epsilon_{ro} \simeq \sqrt{N}\epsilon_m = 5 \times 10^{-14} . \tag{5.54}$$

These results are illuminating in that they show how

- Simpson's rule requires fewer point and has less error than the trapezoid rule.
- It is possible to obtain an error close to machine precision with Simpson's rule (and with other higher order integration algorithms).
- Obtaining the *best* numerical approximation to an integral is not achieved by letting $N \to \infty$ but with a relatively small $N \leq 1000$. Larger N only makes the round-off error dominate.

5.12 Algorithm: Gaussian Quadrature

It is often useful to rewrite the basic integration formula (5.26) with a weighting function $W(x)$ separate from the integrand:

$$\int_a^b f(x)\,dx \equiv \int_a^b W(x)g(x)\,dx \simeq \sum_{i=1}^{N} w_i g(x_i) . \tag{5.55}$$

In the Gaussian quadrature approach to integration, the N points and weights in (5.55) are chosen to make the integration exact if $g(x)$ were a $(2N-1)$-degree polynomial. To obtain this incredible optimization, the points x_i end up having a specific distribution over $[a, b]$. In general, if $g(x)$ is smooth or can be made smooth by factoring out some $W(x)$ (Table 5.2), Gaussian quadrature will produce higher accuracy than the trapezoid and Simpson rules for the same number of points. Sometimes the integrand may not be smooth because it has different behaviors in different regions. In these cases, it makes sense to integrate each region separately and then add the answers together. In fact, some "smart" integration subroutines decide for themselves how many intervals to use and which rule to use in each.

All the rules indicated in Table 5.2 are Gaussian with the general form (5.55). We can see that in one case the weighting function is an exponential, in another a

Table 5.2 Types of Gaussian integration rules.

Integral	Name	Integral	Name
$\int_{-1}^{1} f(y)\,\mathrm{d}y$	Gauss	$\int_{-1}^{1} \frac{F(y)}{\sqrt{1-y^2}}\,\mathrm{d}y$	Gauss–Chebyshev
$\int_{-\infty}^{\infty} \mathrm{e}^{-y^2} F(y)\,\mathrm{d}y$	Gauss–Hermite	$\int_{0}^{\infty} \mathrm{e}^{-y} F(y)\,\mathrm{d}y$	Gauss–Laguerre
$\int_{0}^{\infty} \frac{\mathrm{e}^{-y}}{\sqrt{y}} F(y)\,\mathrm{d}y$	Associated Gauss–Laguerre		

Table 5.3 Points and weights for 4-point Gaussian quadrature (for checking computation).

$\pm y_i$	w_i
0.339 981 043 584 856	0.652 145 154 862 546
0.861 136 311 594 053	0.347 854 845 137 454

Gaussian, and in several an integrable singularity. In contrast to the equally spaced rules, there is never an integration point at the extremes of the intervals, yet the values of the points and weights change as the number of points N changes, and the points are not spaced equally.

The derivation of the Gaussian points will be outlined below, but we point out here that for ordinary Gaussian (Gauss–Legendre) integration, the points y_i turn out to be the N zeros of the Legendre polynomials, with the weights related to the derivatives

$$P_N(y_i) = 0\,, \qquad w_i = \frac{2}{([(1-y_i^2)[P'_N(y_i)]^2]} \,. \tag{5.56}$$

Programs to generate these points and weights are standard in mathematical function libraries, are found in tables such as those in (Abramowitz and Stegun, 1972), or can be computed. The *gauss* program we provide also scales the points to span specified regions. As a check that the program's points are correct, you may want to compare them to the four-point set given in Table 5.3.

5.12.1 Mapping Integration Points

Our standard convention (5.26) for the general interval $[a, b]$ is

$$\int_a^b f(x)\,\mathrm{d}x \simeq \sum_{i=1}^{N} f(x_i) w_i \,. \tag{5.57}$$

With Gaussian points and weights, the y interval $-1 < y_i \le 1$ must be *mapped* onto the x interval $a \le x \le b$. Here are some mappings we have found useful in our work. In all cases, (y_i, w'_i) are the elementary Gaussian points and weights for the interval $[-1, 1]$, and we want to scale to x with various ranges.

1. $[-1, 1] \to [a, b]$ uniformly, $(a + b)/2 =$ *midpoint:*

$$x_i = \frac{b+a}{2} + \frac{b-a}{2} y_i , \quad w_i = \frac{b-a}{2} w'_i , \tag{5.58}$$

$$\Rightarrow \int_a^b f(x)\,\mathrm{d}x = \frac{b-a}{2} \int_{-1}^{1} f[x(y)]\,\mathrm{d}y . \tag{5.59}$$

2. $[0 \to \infty]$, $a =$ *midpoint:*

$$x_i = a\frac{1+y_i}{1-y_i} , \quad w_i = \frac{2a}{(1-y_i)^2} w'_i . \tag{5.60}$$

3. $[-\infty \to \infty]$, *scale set by a:*

$$x_i = a\frac{y_i}{1-y_i^2} , \quad w_i = \frac{a(1+y_i^2)}{\left(1-y_i^2\right)^2} w'_i . \tag{5.61}$$

4. $[a \to \infty]$, $a + 2b =$ *midpoint:*

$$x_i = \frac{a+2b+ay_i}{1-y_i} , \quad w_i = \frac{2(b+a)}{(1-y_i)^2} w'_i . \tag{5.62}$$

5. $[0 \to b]$, $ab/(b+a) =$ *midpoint:*

$$x_i = \frac{ab(1+y_i)}{b+a-(b-a)y_i} , \quad w_i = \frac{2ab^2}{(b+a-(b-a)y_i)^2} w'_i . \tag{5.63}$$

As you can see, even if your integration range extends out to infinity, there will be points at large but not infinite x. As you keep increasing the number of grid points N, the last x_i gets larger but always remains finite.

5.12.2 Gaussian Points Derivation

We want to perform a numerical integration with N integration points:

$$\int_{-1}^{+1} f(x)\,\mathrm{d}x = \sum_{i=1}^{N} w_i f(x_i) , \tag{5.64}$$

where $f(x)$ is a polynomial of degree $2N - 1$ or less. The unique property of Gaussian quadrature is that (5.64) will be exact, as long as we ignore the effect of round-off error. Determining the x_i's and w_i's requires some knowledge of special functions and some cleverness (Hildebrand, 1956). The knowledge needed is the two properties of Legendre polynomials $P_N(x)$ of order N:

1. $P_N(x)$ is orthogonal to every polynomial of order less than N.
2. $P_N(x)$ has N real roots in the interval $-1 \le x \le 1$.

We define now a new polynomial of degree equal to or less than N obtained by dividing the integrand $f(x)$ by the Legendre polynomial $P_N(x)$:

$$q(x) \stackrel{\text{def}}{=} \frac{f(x)}{P_N(x)} , \tag{5.65}$$

$$\Rightarrow \quad f(x) = q(x)P_N(x) + r(x) . \tag{5.66}$$

Here the remainder $r(x)$ is an (unknown) polynomial of degree N or less, which we will not need to determine. If we now substitute (5.66) into (5.64), and use the fact that P_N is orthogonal to every polynomial of degree less than or equal to N, only the second, $r(x)$, term remains

$$\int_{-1}^{+1} f(x)\,\mathrm{d}x = \int_{-1}^{+1} q(x)P_N(x)\,\mathrm{d}x + \int_{-1}^{+1} r(x)\,\mathrm{d}x = \int_{-1}^{+1} r(x)\,\mathrm{d}x . \tag{5.67}$$

Yet because $r(x)$ is a polynomial of degree N or less, we can use a standard N point rule to evaluate the integral exactly (the type of quadrature we did with the Simpson rule).

Now that we know it is possible to integrate a $2N-1$ or less degree polynomial with just N points, we extert some cleverness to determine just what those points will be. We substitute (5.66) into (5.64) and note that

$$\int_{-1}^{+1} f(x)\,\mathrm{d}x = \sum_{i=1}^{N} w_i q(x_i)P_N(x_i) + \sum_{i=1}^{N} w_i r(x_i) = \sum_{i=1}^{N} w_i r(x_i) . \tag{5.68}$$

The cleverness is realizing that if we choose the N integration points to be the zeros or roots of the Legendre polynomial $P_N(x)$, then the first term on the RHS of (5.68) will vanish because $P_N(x_i) = 0$ for each x_i:

$$\int_{-1}^{+1} f(x)\,\mathrm{d}x = \sum_{i=1}^{N} w_i r(x_i) . \tag{5.69}$$

This is our derivation that the N integration points over the interval $(-1, 1)$ are the N zeros of the Legendre polynomial $P_N(x)$. As indicated in (5.56), the weights are related to the derivative of the Legendre polynomials evaluated at the roots of the polynomial. The actual derivation of the weights we leave to Hildebrand (1956).

5.12.3 Integration Error Assessment

1. Write a double-precision program to integrate an arbitrary function numerically using the trapezoid rule, the Simpson rule, and Gaussian quadrature. For

our assumed *problem*, there is an analytic answer with which to compare:

$$\frac{dN(t)}{dt} = e^{-t} \quad \Rightarrow \quad N(1) = \int_0^1 e^{-t}\, dt = 1 - e^{-1} . \tag{5.70}$$

2. Compute the relative error $\epsilon = |(\text{numerical-exact})/\text{exact}|$ in each case. Present your data in the tabular form

N	ϵ_T	ϵ_S	ϵ_G
2	...	...	...
10	...	...	...

with spaces or tabs separating the fields. Try N values of 2, 10, 20, 40, 80, 160, ... (*Hint*: Even numbers may not be the assumption of every rule.)

3. Make a log–log plot of relative error vs. N (Figure 5.4). You should observe that

$$\epsilon \simeq CN^{\alpha} \quad \Rightarrow \quad \log \epsilon = \alpha \log N + \text{constant} . \tag{5.71}$$

This means that a power-law dependence appears as a straight line on a log–log plot, and that if you use $\log_{10}$, then the ordinate on your log–log plot will be the negative of the number of decimal places of precision in your calculation.

4. Use your plot or table to estimate the power-law dependence of the error ϵ on the number of points N, and to determine the number of decimal places of precision in your calculation. Do this for both the trapezoid and Simpson rules and for both the algorithmic and round-off error regimes. (Note that it may be hard to make N large enough to reach the round-off error regime for the trapezoid rule because the approximation error is so large.)

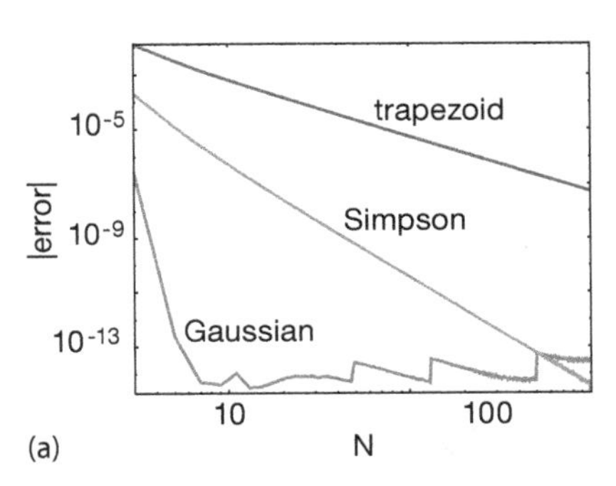

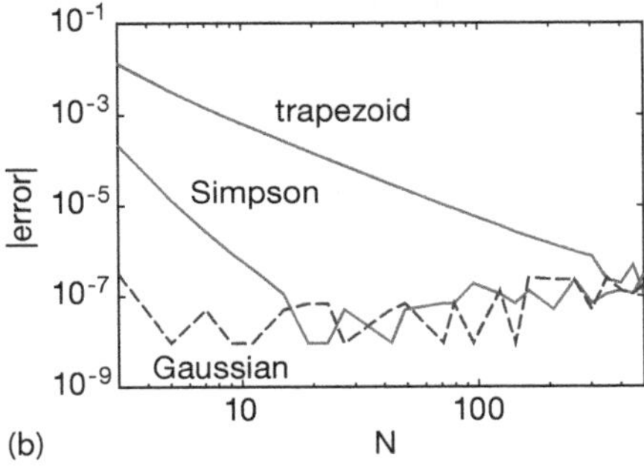

Figure 5.4 Log–log plots of the error in the integration of exponential decay using the trapezoid rule, Simpson's rule, and Gaussian quadrature vs. the number of integration points N. Approximately 15 decimal places of precision are attainable with double precision (a), and seven places with single precision (b). The algorithms are seen to stop converging when round-off error (the fluctuating and increasing part near the bottom) starts to dominate.

In Listing 5.2, we give a sample program that performs an integration with Gaussian points. The method **gauss** generates the points and weights and may be useful in other applications as well.

Listing 5.2 IntegGauss.py integrates the function $f(x)$ via Gaussian quadrature. The points and weights are generated in the method gauss, which will be the same for other applications as well. Note that the level of desired precision is set by the parameter eps, which should be set by the user, as should the value for job, which controls the mapping of the points onto arbitrary intervals (they are generated in $(-1, 1)$).

```
# IntegGauss.py: Gaussian quadrature generator of pts & wts

from numpy import *
from sys import version

max_in = 11                                    # Numb intervals
vmin = 0.; vmax = 1.                           # Int ranges
ME = 2.7182818284590452354E0                   # Euler's const
w = zeros( (2001), float)
x = zeros( (2001), float)

def f(x):                                      # The integrand
    return (exp( - x) )

def gauss(npts, job, a, b, x, w):
    m  = i = j = t = t1 = pp = p1 = p2 = p3 = 0.
    eps = 3.E-14                          # Accuracy: ******ADJUST THIS*******!
    m = int((npts + 1)/2 )
    for i in range(1, m + 1):
        t = cos(math.pi*(float(i) - 0.25)/(float(npts) + 0.5) )
        t1 = 1
        while( (abs(t - t1) ) >= eps):
            p1 = 1. ;  p2 = 0.
            for j in range(1, npts + 1):
                p3 = p2;   p2 = p1
                p1 = ((2.*float(j)-1)*t*p2 - (float(j)-1.)*p3)/(float(j))
            pp = npts*(t*p1 - p2)/(t*t - 1.)
            t1 = t; t = t1  -  p1/pp
        x[i - 1] = - t;    x[npts - i] = t
        w[i - 1] = 2./( (1. - t*t)*pp*pp)
        w[npts - i] = w[i - 1]
    if (job == 0):
        for i in range(0, npts):
            x[i] = x[i]*(b - a)/2. + (b + a)/2.
            w[i] = w[i]*(b - a)/2.
    if (job == 1):
        for i in range(0, npts):
            xi   = x[i]
            x[i] = a*b*(1. + xi) / (b + a - (b - a)*xi)
            w[i] = w[i]*2.*a*b*b/( (b + a - (b-a)*xi)*(b + a - (b-a)*xi))
    if (job == 2):
        for i in range(0, npts):
            xi = x[i]
            x[i] = (b*xi +  b + a + a) / (1. - xi)
            w[i] = w[i]*2.*(a + b)/( (1. - xi)*(1. - xi) )

def gaussint (no, min, max):
    quadra = 0.
    gauss (no, 0, min, max, x, w)                       # Returns pts & wts
    for n in  range(0, no):
        quadra   += f(x[n]) * w[n]                      # Calculate integral
    return (quadra)

for i in range(3, max_in + 1, 2):
```

```
    result = gaussint(i, vmin, vmax)
    print (" i ", i, " err ", abs(result - 1 + 1/ME))
print ("Enter and return any character to quit")
```

5.13 Higher Order Rules (Algorithm)

As in numerical differentiation, we can use the known functional dependence of the error on interval size h to reduce the integration error. For simple rules like the trapezoid and Simpson rules, we have the analytic estimates (5.47), while for others you may have to experiment to determine the h dependence. To illustrate, if $A(h)$ and $A(h/2)$ are the values of the integral determined for intervals h and $h/2$, respectively, we know that the integrals have expansions with a leading error term proportional to h^2,

$$A(h) \simeq \int_a^b f(x)\,\mathrm{d}x + \alpha h^2 + \beta h^4 + \cdots , \tag{5.72}$$

$$A\left(\frac{h}{2}\right) \simeq \int_a^b f(x)\,\mathrm{d}x + \frac{\alpha h^2}{4} + \frac{\beta h^4}{16} + \cdots . \tag{5.73}$$

Consequently, we make the h^2 term vanish by computing the combination

$$\frac{4}{3}A\left(\frac{h}{2}\right) - \frac{1}{3}A(h) \simeq \int_a^b f(x)\,\mathrm{d}x - \frac{\beta h^4}{4} + \cdots . \tag{5.74}$$

Clearly, this particular trick (Romberg's extrapolation) works only if the h^2 term dominates the error and then only if the derivatives of the function are well behaved. An analogous extrapolation can also be made for other algorithms.

In Table 5.1, we gave the weights for several equal interval rules. Whereas the Simpson rule used two intervals, the three-eighths rule uses three intervals, and the Milne[2)] rule uses four intervals. (These are single-interval rules and must be strung together to obtain a rule *extended* over the entire integration range. This means that the points that end one interval and begin the next are weighted twice.) You can easily determine the number of elementary intervals integrated over, and check whether you and we have written the weights right, by summing the weights for any rule. The sum is the integral of $f(x) = 1$ and must equal h times the number of intervals (which in turn equals $b - a$):

$$\sum_{i=1}^{N} w_i = h \times N_{\text{intervals}} = b - a . \tag{5.75}$$

2) There is, not coincidentally, a Milne Computer Center at Oregon State University, although there is no longer a central computer there.

5.14
Monte Carlo Integration by Stone Throwing (Problem)

Imagine yourself as a farmer walking to your furthermost field to add algae-eating fish to a pond having an algae explosion. You get there only to read the instructions and discover that you need to know the area of the pond in order to determine the correct number of the fish to add. Your *problem* is to measure the area of this irregularly shaped pond with just the materials at hand (Gould *et al.*, 2006).

It is hard to believe that Monte Carlo techniques can be used to evaluate integrals. After all, we do not want to gamble on the values! While it is true that other methods are preferable for single and double integrals, it turns out that Monte Carlo techniques are best when the dimensionality of integrations gets large! For our pond problem, we will use a *sampling* technique (Figure 5.5):

1. Walk off a box that completely encloses the pond and remove any pebbles lying on the ground within the box.
2. Measure the lengths of the sides in natural units like *feet*. This tells you the area of the enclosing box A_{box}.
3. Grab a bunch of pebbles, count their number, and then throw them up in the air in random directions.
4. Count the number of splashes in the pond N_{pond} and the number of pebbles lying on the ground within your box N_{box}.
5. Assuming that you threw the pebbles uniformly and randomly, the number of pebbles falling into the pond should be proportional to the area of the pond A_{pond}. You determine that area from the simple ratio

$$\frac{N_{\text{pond}}}{N_{\text{pond}} + N_{\text{box}}} = \frac{A_{\text{pond}}}{A_{\text{box}}} \quad \Rightarrow \quad A_{\text{pond}} = \frac{N_{\text{pond}}}{N_{\text{pond}} + N_{\text{box}}} A_{\text{box}} . \tag{5.76}$$

5.14.1
Stone Throwing Implementation

Use sampling (Figure 5.5) to perform a 2D integration and thereby determine π:

1. Imagine a circular pond enclosed in a square of side $2(r = 1)$.
2. We know the analytic answer that the area of a circle $\oint dA = \pi$.
3. Generate a sequence of random numbers $-1 \le r_i \le +1$.
4. For $i = 1$ to N, pick $(x_i, y_i) = (r_{2i-1}, r_{2i})$.
5. If $x_i^2 + y_i^2 < 1$, let $N_{\text{pond}} = N_{\text{pond}} + 1$; otherwise let $N_{\text{box}} = N_{\text{box}} + 1$.
6. Use (5.76) to calculate the area, and in this way π.
7. Increase N until you get π to three significant figures (we don't ask much – that's only slide-rule accuracy).

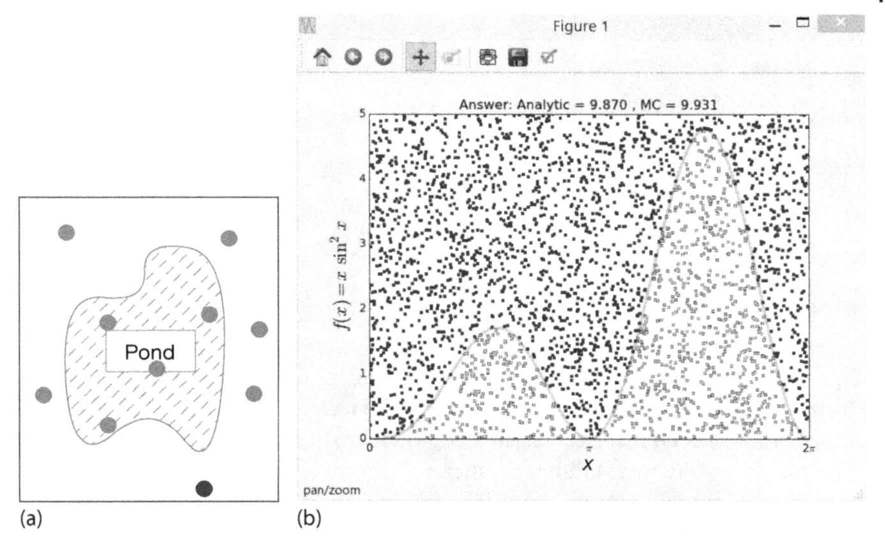

Figure 5.5 (a) Stones into a pond as a technique for measuring its area. The ratio of "hits" to total number of stones thrown equals the ratio of the area of the pond to that of the box. (b) The evaluation of an integral via a Monte Carlo (stone throwing) technique of the ratio of areas.

5.15 Mean Value Integration (Theory and Math)

The standard Monte Carlo technique for integration is based on the *mean value theorem* (presumably familiar from elementary calculus):

$$I = \int_a^b dx\, f(x) = (b-a)\langle f \rangle \,. \tag{5.77}$$

The theorem states the obvious if you think of integrals as areas. The value of the integral of some function $f(x)$ between a and b equals the length of the interval $(b-a)$ times the mean value of the function over that interval $\langle f \rangle$ (Figure 5.6). The Monte Carlo integration algorithm uses random points to evaluate the mean in (5.77). With a sequence $a \le x_i \le b$ of N uniform random numbers, we want to determine the *sample mean* by *sampling* the function $f(x)$ at these points:

$$\langle f \rangle \simeq \frac{1}{N} \sum_{i=1}^{N} f(x_i) \,. \tag{5.78}$$

This gives us the very simple integration rule:

$$\int_a^b dx\, f(x) \simeq (b-a) \frac{1}{N} \sum_{i=1}^{N} f(x_i) = (b-a)\langle f \rangle \,. \tag{5.79}$$

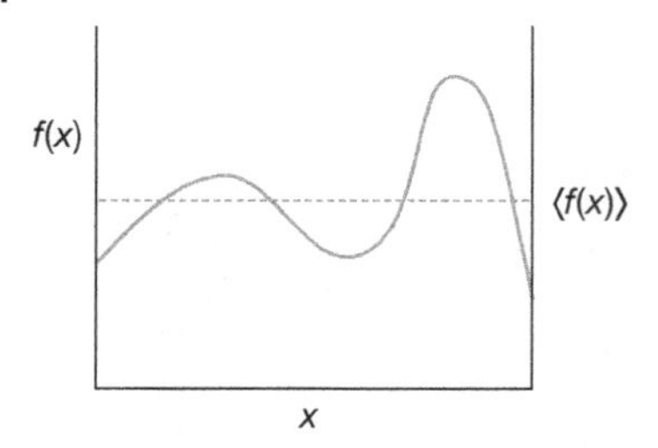

Figure 5.6 The area under the curve $f(x)$ is the same as that under the horizontal line whose height $y = \langle f \rangle$.

Equation 5.79 looks much like our standard algorithm for integration (5.26) with the points x_i chosen randomly and with uniform weights $w_i = (b-a)/N$. Because no attempt has been made to obtain an optimal answer for a given value of N, this does not seem like it would be an efficient means to evaluate integrals; but you must admit that it is simple. If we let the number of samples of $f(x)$ approach infinity $N \to \infty$ or if we keep the number of samples finite and take the average of infinitely many runs, the laws of statistics ensure us that (5.79) will approach the correct answer, at least if there were no round-off errors.

For readers who are familiar with statistics, we remind you that the uncertainty in the value obtained for the integral I after N samples of $f(x)$ is measured by the standard deviation σ_I. If σ_f is the standard deviation of the integrand f in the sampling, then for normal distributions we have

$$\sigma_I \simeq \frac{1}{\sqrt{N}} \sigma_f \ . \tag{5.80}$$

So for large N, the error in the value obtained for the integral decreases as $1/\sqrt{N}$.

5.16 Integration Exercises

1. Here are two integrals that quadrature may find challenging:

$$F_1 = \int_0^{2\pi} \sin(100x)\,\mathrm{d}x, \quad F_2 = \int_0^{2\pi} \sin^x(100x)\,\mathrm{d}x \ . \tag{5.81}$$

   a) Evaluate these integrals using two different integration rules and compare the answers.
   b) Explain why the computer may have trouble with these integrals.
2. The next three problems are examples of how *elliptic integrals* enter into realistic physics problems. It is straightforward to evaluate any integral numerically using the techniques of this chapter, but it may be difficult for you to

know if the answers you obtain are correct. One way to hone your integral-evaluating skills is to compare your answers from quadrature to power series expressions, or to a polynomial approximations of know precision. To help you in this regard, we present here a polynomial approximation for an elliptic integral (Abramowitz and Stegun, 1972):

$$
\begin{aligned}
K(m) &= \int_0^{\pi/2} (1 - m \sin^2 \theta)^{-1/2} \, d\theta \\
&\simeq a_0 + a_1 m_1 + a_2 m_1^2 - \left[b_0 + b_1 m_1 + b_2 m_1^2 \right] \ln m_1 + \epsilon(m) , \\
m_1 &= 1 - m , \quad 0 \le m \le 1 , \quad |\epsilon(m)| \le 3 \times 10^{-5} , \\
& \begin{array}{lll} a_0 = 1.386\,294\,4 & a_1 = 0.111\,972\,3 & a_2 = 0.072\,529\,6 \\ b_0 = 0.5 & b_1 = 0.121\,347\,8 & b_2 = 0.028\,872\,9 \end{array} .
\end{aligned} \tag{5.82}
$$

3. Compute $K(m)$ by evaluating the integral in (5.82) numerically. Tune your integral evaluation until you obtain agreement at the $\le 3 \times 10^{-5}$ level with the polynomial approximation.
4. In Section 15.1.2, we will derive an expression for the period T of a realistic pendulum for which the maximum angle of displacement θ_m is not necessarily small:

$$
T = \frac{T_0}{\pi} \int_0^{\theta_m} \frac{d\theta}{\left[\sin^2(\theta_m/2) - \sin^2(\theta/2) \right]^{1/2}} \tag{5.83}
$$

$$
\simeq T_0 \left[1 + \left(\frac{1}{2} \right)^2 \sin^2 \frac{\theta_m}{2} + \left(\frac{1 \cdot 3}{2 \cdot 4} \right)^2 \sin^4 \frac{\theta_m}{2} + \cdots \right] , \tag{5.84}
$$

where T_0 is the period for small-angle oscillations. The integral in (5.83) can be expressed in terms of an elliptic integral of the first kind. If you think of an elliptic integral as a generalized trigonometric function, then this is a closed-form solution; otherwise, it is an integral needing numerical evaluation.

a) Use numerical quadrature to determine the ratio T/T_0 for five values of θ_m between 0 and π. Show that you have attained at least four places of accuracy by progressively increasing the number of integration points until changes occur only in the fifth place, or beyond.

b) Use the power series (5.84) to determine the ratio T/T_0. Continue summing terms until changes in the sum occur only in the fifth place, or beyond.

c) Plot the values you obtain for T/T_0 vs. θ_m for both the integral and power series solution. Note that any departure from 1 indicates breakdown of the familiar small-angle approximation for the pendulum.

5. In the classic E&M text (Jackson, 1988), there is the problem of an infinite, grounded, thin, plane sheet of conducting material with a hole of radius a cut

in it. The hole contains a conducting disk of slightly smaller radius kept at potential V and separated from the sheet by a thin ring of insulating material. solves for the potential a perpendicular distance z above the *edge* of the disk in terms of an elliptic integral:

$$\Phi(z) = \frac{V}{2}\left(1 - \frac{kz}{\pi a}\int_0^{\pi/2} \frac{d\phi}{\sqrt{1-k^2\sin^2\phi}}\right) , \tag{5.85}$$

where $k = 2a/(z^2 + 4a^2)^{1/2}$.·Use numerical integration to calculate and then plot the potential for $V = 1$, $a = 1$ and values of z in the interval (0.05, 10). Compare to a $1/r$ fall off.

6. Figure 5.7 shows a current loop of radius a carrying a current I. The point P is a distance r from the center of the loop with spherical coordinates (r, θ, ϕ). Jackson (1988) solves for the ϕ component of the vector potential at point P in terms of elliptic integrals:

$$A_\phi(r,\theta) = \frac{\mu_0}{4\pi}\frac{4Ia}{\sqrt{a^2+r^2+2ar\sin\theta}}\left[\frac{(2-k^2)K(k)-2E(k)}{k^2}\right] , \tag{5.86}$$

$$K(k) = \int_0^{\pi/2} \frac{d\phi}{\sqrt{1-k^2\sin^2\phi}} , \qquad E(k) = \int_0^{\pi/2} \sqrt{1-k^2\sin^2\phi}\, d\phi , \tag{5.87}$$

$$k^2 = \frac{4ar\sin\theta}{a^2+r^2+2ar\sin\theta} . \tag{5.88}$$

Here $K(k)$ is a complete elliptic integral of the first kind and $E(k)$ is a complete elliptic integral of the second kind. For $a = 1$, $I = 3$, and $\mu_0/4\pi = 1$, compute and plot

a) $A_\phi(r = 1.1, \theta)$ vs. θ.
b) $A_\phi(r, \theta = \pi/3)$ vs. r.

5.17 Multidimensional Monte Carlo Integration (Problem)

Let us say that we want to calculate some properties of a small atom such as magnesium with 12 electrons. To do that, we need to integrate atomic wave functions over the three coordinates of each of 12 electrons. This amounts to a $3 \times 12 = 36$D integral. If we use 64 points for each integration, this requires about $64^{36} \simeq 10^{65}$ evaluations of the integrand. If the computer were fast and could evaluate the integrand a million times per second, this would take about 10^{59} s, which is significantly longer than the age of the universe ($\sim 10^{17}$ s).

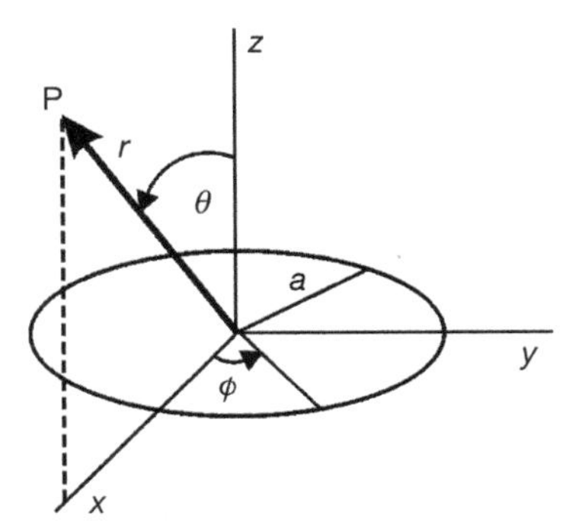

Figure 5.7 A ring of radius a carries a current I. Find vector potential at point P.

Your *problem* is to find a way to perform multidimensional integrations so that you are still alive to savor the results. Specifically, evaluate the 10D integral

$$I = \int_0^1 dx_1 \int_0^1 dx_2 \cdots \int_0^1 dx_{10} \left(x_1 + x_2 + \cdots + x_{10}\right)^2 . \tag{5.89}$$

Check your numerical answer against the analytic one, 155/6.

It is easy to generalize mean value integration to many dimensions by picking random points in a multidimensional space. For example, in 2D:

$$\int_a^b dx \int_c^d dy\, f(x, y) \simeq (b-a)(d-c)\frac{1}{N}\sum_i^N f(\boldsymbol{x}_i) = (b-a)(d-c)\langle f \rangle . \tag{5.90}$$

5.17.1 Multi Dimension Integration Error Assessment

When we perform a multidimensional integration, the relative error in the Monte Carlo technique, being statistical, decreases as $1/\sqrt{N}$. This is valid even if the N points are distributed over D dimensions. In contrast, when we use these same N points to perform a D-dimensional integration as D separate 1D integrals using a rule such as Simpson's, we use N/D points for each integration. For fixed N, this means that the number of points used for each integration decreases as the number of dimensions D increases, and so the error in each integration *increases* with D. Furthermore, the total error will be approximately N times the error in each integral. If you put these trends together and do the analysis for a particular integration rule, you will find that at a value of $D \simeq$ 3–4 the error in Monte Carlo integration is approximately equal to that of conventional schemes. For larger values of D, the Monte Carlo method is always more accurate!

5.17.2
Implementation: 10D Monte Carlo Integration

Use a built-in random-number generator to perform the 10D Monte Carlo integration in (5.89).

1. Conduct 16 trials and take the average as your answer.
2. Try sample sizes of $N = 2, 4, 8, \dots, 8192$.
3. Plot the relative error vs. $1/\sqrt{N}$ and see if linear behavior occurs.
4. What is your estimate for the accuracy of the integration?
5. Show that for a dimension $D \simeq 3$–4, the error in multidimensional Monte Carlo integration is approximately equal to that of conventional schemes, and that for larger values of D, the Monte Carlo method is more accurate.

5.18
Integrating Rapidly Varying Functions (Problem)

It is common in many physical applications to integrate a function with an approximately Gaussian dependence on x. The rapid falloff of the integrand means that our Monte Carlo integration technique would require an incredibly large number of points to have sufficient points where the integrand is large. Your *problem* is to make Monte Carlo integration more efficient for rapidly varying integrands.

5.19
Variance Reduction (Method)

If the function being integrated never differs much from its average value, then the standard Monte Carlo mean value method (5.79) should work well with a large, but manageable, number of points. Yet for a function with a large *variance* (i.e., one that is not “flat”), many of the evaluations of the function may occur for x values at which the function is very small, and thus makes an insignificant contribution to the integral; this is, basically, a waste of time. The method can be improved by mapping the function f into a different function g that has a smaller variance over the interval. We indicate two methods here and refer you to Press *et al.* (1994) and Koonin (1986) for more details.

The first method is a *variance reduction* or *subtraction technique* in which we devise a flatter function on which to apply the Monte Carlo technique. Suppose we construct a function $g(x)$ with the following properties on $[a, b]$:

$$|f(x) - g(x)| \le \epsilon \,, \qquad \int_a^b dx\, g(x) = J \,. \tag{5.91}$$

We now evaluate the integral of the difference $f(x) - g(x)$ and add the result to J to obtain the required integral

$$\int_a^b \mathrm{d}x\, f(x) = \int_a^b \mathrm{d}x [f(x) - g(x)] + J\,. \tag{5.92}$$

If we are clever enough to find a simple $g(x)$ that makes the variance of $f(x) - g(x)$ less than that of $f(x)$, and that we can integrate analytically, we can obtain even more accurate answers in less time.

5.20 Importance Sampling (Method)

A second method for improving Monte Carlo integration is called *importance sampling* because samples the integrand in the most important regions. It derives from the identity

$$I = \int_a^b \mathrm{d}x\; f(x) = \int_a^b \mathrm{d}x\, w(x) \frac{f(x)}{w(x)}\,. \tag{5.93}$$

If we now use a *probability distribution* for our random numbers that includes $x(x)$, the integral can be approximated as

$$I = \left\langle \frac{f}{w} \right\rangle \simeq \frac{1}{N} \sum_{i=1}^{N} \frac{f(x_i)}{w(x_i)}\,. \tag{5.94}$$

The improvement arising from (5.94) is that with a judicious choice of weighting function $w(x) \propto f(x)$, we can make $f(x)/w(x)$ more constant and thus easier to integrate accurately.

5.21 von Neumann Rejection (Method)

A simple and ingenious method for generating random points with a probability distribution $w(x)$ was deduced by von Neumann and is implemented in Listing 5.3. This method is essentially the same as the rejection or sampling method used to guess the area of a pond, only now the pond has been replaced by the weighting function $w(x)$, and the arbitrary box around the lake by the arbitrary constant W_0. Imagine a graph of $w(x)$ vs. x (Figure 5.8). Walk off your box by placing the line $W = W_0$ on the graph, with the only condition being $W_0 \geq w(x)$. We next "throw stones" at this graph and count only those splashes that fall into the $w(x)$ pond. That is, we generate uniform distributions in x and $y \equiv W$ with

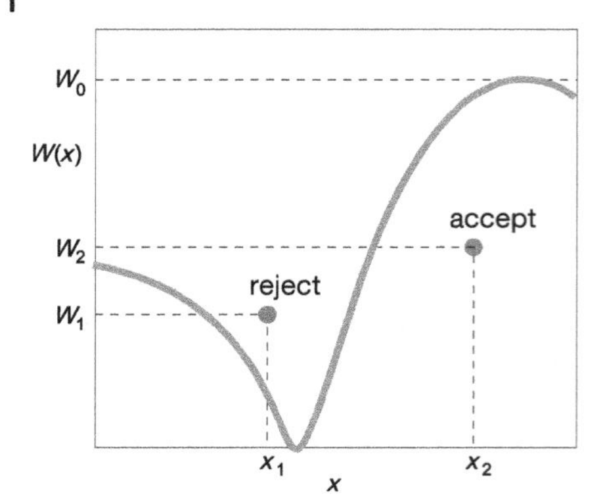

Figure 5.8 The von Neumann rejection technique for generating random points with weight $W(x)$. A random point is accepted if it lies below the curve of $W(x)$ and rejected if it lies above. This generates a random distribution weighted by whatever $W(x)$ function is plotted.

the maximum y value equal to the width of the box W_0:

$$(x_i, W_i) = (r_{2i-1}, W_0 r_{2i}) . \tag{5.95}$$

We then reject all x_i that do not fall into the pond:

$$\text{If } W_i < w(x_i), \text{ accept}, \quad \text{If } W_i > w(x_i), \text{ reject}. \tag{5.96}$$

The x_i values so accepted will have the weighting $w(x)$ (Figure 5.8). The largest acceptance occurs where $w(x)$ is large, in this case for midrange x. In Chapter 17, we apply a variation of the rejection technique known as the *Metropolis algorithm*. This algorithm has now become the cornerstone of computation thermodynamics.

Listing 5.3 vonNeuman.py uses rejection to generate a weighted random distribution.

```
# vonNeuman: Monte-Carlo integration via stone throwing

import random
from visual.graph import *

N        = 100    # points to plot the function
graph    = display(width=500,height=500,title='vonNeumann Rejection Int')
xsinx    = curve(x=list(range(0,N)), color=color.yellow, radius=0.5)
pts      = label(pos=(-60, -60), text='points=', box=0)           # Labels
pts2     = label(pos=(-30, -60), box=0)
inside   = label(pos=(30,-60), text='accepted=', box=0)
inside2  = label(pos=(60,-60), box=0)
arealbl  = label(pos=(-65,60), text='area=', box=0)
arealbl2 = label(pos=(-35,60), box=0)
areanal  = label(pos=(30,60), text='analytical=', box=0)
zero     = label(pos=(-85,-48), text='0', box=0)
five     = label(pos=(-85,50), text='5', box=0)
twopi    = label(pos=(90,-48), text='2pi',box=0)

def fx (x):  return x*sin(x)*sin(x)                               # Integrand

def plotfunc():                                                  # Plot function
```

```
    incr = 2.0*pi/N
    for i in range(0,N):
        xx          = i*incr
        xsinx.x[i]  = ((80.0/pi)*xx-80)
        xsinx.y[i]  = 20*fx(xx)-50
    box             = curve(pos=[(-80,-50), (-80,50), (80,50)
                      ,(80,-50), (-80,-50)], color=color.white)       # box

plotfunc()                                  # Box area = h x w =5*2pi
j              = 0
Npts           = 3001                                  # Pts inside box
analyt         = (pi)**2                             # Analytical integral
areanal.text = 'analytical=%8.5f'%analyt
genpts         = points(size=2)
for i in range(1,Npts):                              # points inside box
    rate(500)                                             # slow process
    x = 2.0*pi*random.random()
    y = 5*random.random()
    xp = x*80.0/pi-80
    yp = 20.0*y-50
    pts2.text = '%4s' %i
    if y <= fx(x):                                           # Below curve
        j += 1
        genpts.append(pos=(xp,yp), color=color.cyan)
        inside2.text='%4s'%j
    else: genpts.append(pos=(xp,yp), color=color.green)
    boxarea = 2.0*pi*5.0
    area = boxarea*j/(Npts-1)
    arealbl2.text = '%8.5f'%area
```

5.21.1 Simple Random Gaussian Distribution

The central limit theorem can be used to deduce a Gaussian distribution via a simple summation. The theorem states, under rather general conditions, that if $\{r_i\}$ is a sequence of mutually independent random numbers, then the sum

$$x_N = \sum_{i=1}^{N} r_i \tag{5.97}$$

is distributed normally. This means that the generated x values have the distribution

$$P_N(x) = \frac{\exp\left[-\frac{(x-\mu)^2}{2\sigma^2}\right]}{\sqrt{2\pi\sigma^2}}, \quad \mu = N\langle r\rangle, \quad \sigma^2 = N(\langle r^2\rangle - \langle r\rangle^2). \tag{5.98}$$

5.22 Nonuniform Assessment ⊙

Use the von Neumann rejection technique to generate a normal distribution of standard deviation 1 and compare it to the simple Gaussian method.

5.22.1 Implementation ⊙

In order for $w(x)$ to be the weighting function for random numbers over $[a,b]$, we want it to have the properties

$$\int_a^b \mathrm{d}x\, w(x) = 1\,, \quad [w(x) > 0]\,, \quad \mathrm{d}\mathcal{P}(x \to x + \mathrm{d}x) = w(x)\,\mathrm{d}x\,, \tag{5.99}$$

where $\mathrm{d}\mathcal{P}$ is the probability of obtaining an x in the range $x \to x + \mathrm{d}x$. For the uniform distribution over $[a,b]$, $w(x) = 1/(b-a)$.

Inverse transform/change of variable method ⊙ Let us consider a change of variables that takes our original integral I (5.93) to the form

$$I = \int_a^b \mathrm{d}x\, f(x) = \int_0^1 \mathrm{d}W \frac{f[x(W)]}{w[x(W)]}\,. \tag{5.100}$$

Our aim is to make this transformation such that there are equal contributions from all parts of the range in W; that is, we want to use a uniform sequence of random numbers for W. To determine the new variable, we start with $u(r)$, the uniform distribution over $[0,1]$,

$$u(r) = \begin{cases} 1, & \text{for } 0 \le r \le 1, \\ 0, & \text{otherwise.} \end{cases} \tag{5.101}$$

We want to find a mapping $r \leftrightarrow x$ or probability function $w(x)$ for which probability is conserved:

$$w(x)\,\mathrm{d}x = u(r)\,\mathrm{d}r, \quad \Rightarrow \quad w(x) = \left|\frac{\mathrm{d}r}{\mathrm{d}x}\right| u(r)\,. \tag{5.102}$$

This means that even though x and r are related by some (possibly) complicated mapping, x is also random with the probability of x lying in $x \to x + \mathrm{d}x$ equal to that of r lying in $r \to r + \mathrm{d}r$.

To find the mapping between x and r (the tricky part), we change variables to $W(x)$ defined by the integral

$$W(x) = \int_{-\infty}^{x} \mathrm{d}x'\, w(x')\,. \tag{5.103}$$

We recognize $W(x)$ as the (incomplete) integral of the probability density $u(r)$ up to some point x. It is another type of distribution function, the integrated probability of finding a random number less than the value x. The function $W(x)$ is on that account called a *cumulative distribution function* and can also be thought of as the area to the left of $r = x$ on the plot of $u(r)$ vs. r. It follows immediately from the definition (5.103) that $W(x)$ has the properties

$$W(-\infty) = 0\,; \quad W(\infty) = 1\,, \tag{5.104}$$

$$\frac{\mathrm{d}W(x)}{\mathrm{d}x} = w(x)\,, \quad \mathrm{d}W(x) = w(x)\,\mathrm{d}x = u(r)\,\mathrm{d}r\,. \tag{5.105}$$

Consequently, $W_i = \{r_i\}$ is a uniform sequence of random numbers, and we just need to invert (5.103) to obtain x values distributed with probability $w(x)$.

The crux of this technique is being able to invert (5.103) to obtain $x = W^{-1}(r)$. Let us look at some analytic examples to get a feel for these steps (numerical inversion is possible and frequent in realistic cases).

Uniform weight function w We start with the familiar uniform distribution

$$w(x) = \begin{cases} \frac{1}{b-a}, & \text{if } a \le x \le b, \\ 0, & \text{otherwise.} \end{cases} \tag{5.106}$$

After following the rules, this leads to

$$W(x) = \int_a^x \mathrm{d}x' \frac{1}{b-a} = \frac{x-a}{b-a} \tag{5.107}$$

$$\Rightarrow \quad x = a + (b-a)W \quad \Rightarrow \quad W^{-1}(r) = a + (b-a)r\,, \tag{5.108}$$

where $W(x)$ is always taken as uniform. In this way, we generate uniform random $0 \le r \le 1$ and uniform random $a \le x \le b$.

Exponential weight We want random points with an exponential distribution:

$$w(x) = \begin{cases} \frac{1}{\lambda}\mathrm{e}^{-x/\lambda}, & \text{for } x > 0, \\ 0, & \text{for } x < 0, \end{cases}$$

$$W(x) = \int_0^x \mathrm{d}x' \frac{1}{\lambda}\mathrm{e}^{-x'/\lambda} = 1 - \mathrm{e}^{-x/\lambda}\,, \tag{5.109}$$

$$\Rightarrow \quad x = -\lambda \ln(1 - W) \equiv -\lambda \ln(1 - r)\,. \tag{5.110}$$

In this way, we generate uniform random r: $[0, 1]$ and obtain $x = -\lambda \ln(1 - r)$ distributed with an exponential probability distribution for $x > 0$. Notice that our prescription (5.93) and (5.94) tells us to use $w(x) = \mathrm{e}^{-x/\lambda}/\lambda$ to remove the exponential-like behavior from an integrand and place it in the weights and scaled points ($0 \le x_i \le \infty$). Because the resulting integrand will vary less, it may be approximated better as a polynomial:

$$\int_0^\infty \mathrm{d}x \mathrm{e}^{-x/\lambda} f(x) \simeq \frac{\lambda}{N} \sum_{i=1}^{N} f(x_i)\,, \quad x_i = -\lambda \ln(1 - r_i)\,. \tag{5.111}$$

Gaussian (normal) distribution We want to generate points with a normal distribution:

$$w(x') = \frac{1}{\sqrt{2\pi}\sigma} e^{-(x'-\overline{x})^2/2\sigma^2} . \tag{5.112}$$

This by itself is rather hard but is made easier by generating uniform distributions in angles and then using trigonometric relations to convert them to a Gaussian distribution. But before doing that, we keep things simple by realizing that we can obtain (5.112) with mean $\overline{x}$ and standard deviation σ by scaling and a translation of a simpler $w(x)$:

$$w(x) = \frac{1}{\sqrt{2\pi}} e^{-x^2/2} , \quad x' = \sigma x + \overline{x} . \tag{5.113}$$

We start by generalizing the statement of probability conservation for two different distributions (5.102) to two dimensions (Press *et al.*, 1994):

$$p(x, y)\,\mathrm{d}x\,\mathrm{d}y = u(r_1, r_2)\,\mathrm{d}r_1\,\mathrm{d}r_2 \tag{5.114}$$

$$\Rightarrow \quad p(x, y) = u(r_1, r_2) \left| \frac{\partial(r_1, r_2)}{\partial(x, y)} \right| . \tag{5.115}$$

We recognize the term in vertical bars as the Jacobian determinant:

$$J = \left| \frac{\partial(r_1, r_2)}{\partial(x, y)} \right| \stackrel{\text{def}}{=} \frac{\partial r_1}{\partial x} \frac{\partial r_2}{\partial y} - \frac{\partial r_2}{\partial x} \frac{\partial r_1}{\partial y} . \tag{5.116}$$

To specialize to a Gaussian distribution, we consider $2\pi r$ as angles obtained from a uniform random distribution r, and x and y as Cartesian coordinates that will have a Gaussian distribution. The two are related by

$$x = \sqrt{-2\ln r_1} \cos 2\pi r_2 , \quad y = \sqrt{-2\ln r_1} \sin 2\pi r_2 . \tag{5.117}$$

The inversion of this mapping produces the Gaussian distribution

$$r_1 = e^{-(x^2+y^2)/2} , \quad r_2 = \frac{1}{2\pi} \tan^{-1} \frac{y}{x} , \quad J = -\frac{e^{-(x^2+y^2)/2}}{2\pi} . \tag{5.118}$$

The solution to our problem is at hand. We use (5.117) with r_1 and r_2 uniform random distributions, and x and y are then Gaussian random distributions centered around $x = 0$.

6 Matrix Computing

This chapter examines various aspects of computing with matrices, and in particular applications of the Python packages. Because these packages are optimized and robust, we strongly recommend their use even in all your small programs (small programs often grow big). The two-mass-on-a-string problem is formulated as a matrix problem, and extends the Newton–Raphson search technique to be discussed in Chapter 7. Although there is some logic in having this chapter after Chapter 7, it is placed here to provide the matrix tools needed for visualizations (which is why Problem 3 preceds 1).

6.1 Problem 3: N–D Newton–Raphson; Two Masses on a String

Problem Two weights $(W_1, W_2) = (10, 20)$ are hung from three pieces of string with lengths $(L_1, L_2, L_3) = (3, 4, 4)$ and a horizontal bar of length $L = 8$ (Figure 6.1). Find the angles assumed by the strings and the tensions exerted by the strings.

In spite of the fact that this is a simple problem requiring no more than first-year physics to formulate, the coupled transcendental equations that result are

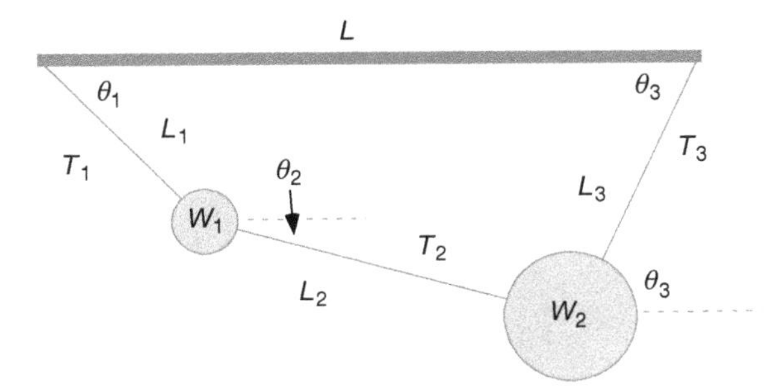

Figure 6.1 Two weights connected by three pieces of string and suspended from a horizontal bar of length L. The lengths are all known, but the angles and the tensions in the strings are to be determined.

Computational Physics, 3rd edition. Rubin H. Landau, Manuel J. Páez, Cristian C. Bordeianu.

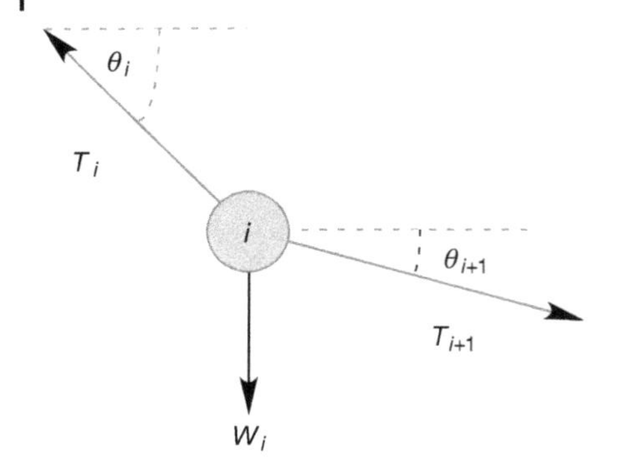

Figure 6.2 A free body diagram for one weight in equilibrium. Balancing the forces in the x and y directions for all weights leads to the equations of static equilibrium.

just about inhumanely painful to solve analytically.[1] However, we will show you how the computer can solve this problem, but even then only by a trial-and-error technique with no guarantee of success. Your *problem* is to test this solution for a variety of weights and lengths and then to extend it to the three-weight problem (not as easy as it may seem). In either case check the physical reasonableness of your solution; the deduced tensions should be positive and similar in magnitude to the weights of the spheres, and the deduced angles should correspond to a physically realizable geometry, as confirmed with a sketch such as Figure 6.2. Some of the exploration you should do is to see at what point your initial guess gets so bad that the computer is unable to find a physical solution.

6.1.1 Theory: Statics

We start with the geometric constraints that the horizontal length of the structure is L and that the strings begin and end at the same height (Figure 6.1):

$$L_1 \cos\theta_1 + L_2 \cos\theta_2 + L_3 \cos\theta_3 = L \,, \tag{6.1}$$

$$L_1 \sin\theta_1 + L_2 \sin\theta_2 - L_3 \sin\theta_3 = 0 \,, \tag{6.2}$$

$$\sin^2\theta_1 + \cos^2\theta_1 = 1 \,, \tag{6.3}$$

$$\sin^2\theta_2 + \cos^2\theta_2 = 1 \,, \tag{6.4}$$

$$\sin^2\theta_3 + \cos^2\theta_3 = 1 \,. \tag{6.5}$$

Observe that the last three equations include trigonometric identities as independent equations because we are treating $\sin\theta$ and $\cos\theta$ as independent variables; this makes the search procedure easier to implement. The basics physics says that because there are no accelerations, the sum of the forces in the horizontal and

1) Almost impossible anyway, as L. Molnar has supplied me with an analytic solution.

vertical directions must equal zero (Figure 6.2):

$$T_1 \sin\theta_1 - T_2 \sin\theta_2 - W_1 = 0 , \tag{6.6}$$

$$T_1 \cos\theta_1 - T_2 \cos\theta_2 = 0 , \tag{6.7}$$

$$T_2 \sin\theta_2 + T_3 \sin\theta_3 - W_2 = 0 , \tag{6.8}$$

$$T_2 \cos\theta_2 - T_3 \cos\theta_3 = 0 . \tag{6.9}$$

Here W_i is the weight of mass i and T_i is the tension in string i. Note that because we do not have a rigid structure, we cannot assume the equilibrium of torques.

6.1.2
Algorithm: Multidimensional Searching

Equations (6.1)–(6.9) are nine simultaneous nonlinear equations. While linear equations can be solved directly, nonlinear equations cannot (Press *et al.*, 1994). You can use the computer to *search* for a solution by guessing, but there is no guarantee of finding one.

Unfortunately, not everything in life is logical, as we need to use a search technique now that will not be covered until Chapter 7. While what we do next is self-explanatory, you may want to look at Chapter 7 now if you are not at all familiar with searching.

We apply to our set the same Newton–Raphson algorithm as used to solve a single equation by renaming the nine unknown angles and tensions as the subscripted variable y_i and placing the variables together as a vector:

$$y = \begin{pmatrix} x_1 \\ x_2 \\ x_3 \\ x_4 \\ x_5 \\ x_6 \\ x_7 \\ x_8 \\ x_9 \end{pmatrix} = \begin{pmatrix} \sin\theta_1 \\ \sin\theta_2 \\ \sin\theta_3 \\ \cos\theta_1 \\ \cos\theta_2 \\ \cos\theta_3 \\ T_1 \\ T_2 \\ T_3 \end{pmatrix} . \tag{6.10}$$

The nine equations to be solved are written in a general form with zeros on the right-hand sides and placed in a vector:

$$f_i(x_1, x_2, \ldots, x_N) = 0 , \quad i = 1, N , \tag{6.11}$$

$$\boldsymbol{f}(\boldsymbol{y}) = \begin{pmatrix} f_1(\boldsymbol{y}) \\ f_2(\boldsymbol{y}) \\ f_3(\boldsymbol{y}) \\ f_4(\boldsymbol{y}) \\ f_5(\boldsymbol{y}) \\ f_6(\boldsymbol{y}) \\ f_7(\boldsymbol{y}) \\ f_8(\boldsymbol{y}) \\ f_9(\boldsymbol{y}) \end{pmatrix} = \begin{pmatrix} 3x_4 + 4x_5 + 4x_6 - 8 \\ 3x_1 + 4x_2 - 4x_3 \\ x_7 x_1 - x_8 x_2 - 10 \\ x_7 x_4 - x_8 x_5 \\ x_8 x_2 + x_9 x_3 - 20 \\ x_8 x_5 - x_9 x_6 \\ x_1^2 + x_4^2 - 1 \\ x_2^2 + x_5^2 - 1 \\ x_3^2 + x_6^2 - 1 \end{pmatrix} = \boldsymbol{0} . \tag{6.12}$$

The solution to these equations requires a set of nine x_i values that make all nine f_i's vanish simultaneously. Although these equations are not very complicated (the physics after all is elementary), the terms quadratic in x make them nonlinear, and this makes it hard or impossible to find an analytic solution. The search algorithm guesses a solution, expands the nonlinear equations into linear form, solves the resulting linear equations, and continues to improve the guesses based on how close the previous one was to making $\boldsymbol{f} = 0$. (We discuss search algorithms using this procedure in Chapter 7.)

Explicitly, let the approximate solution at any one stage be the set x_i and let us assume that there is an (unknown) set of corrections Δx_i for which

$$f_i(x_1 + \Delta x_1, x_2 + \Delta x_2, \dots, x_9 + \Delta x_9) = 0 , \quad i = 1, 9 . \tag{6.13}$$

We solve for the approximate Δx_i's by assuming that our previous solution is close enough to the actual one for two terms in the Taylor series to be accurate:

$$f_i(x_1 + \Delta x_1, \dots, x_9 + \Delta x_9) \simeq f_i(x_1, \dots, x_9) + \sum_{j=1}^{9} \frac{\partial f_i}{\partial x_j} \Delta x_j = 0 \qquad i = 1, \dots, 9 . \tag{6.14}$$

We now have a solvable set of nine linear equations in the nine unknowns Δx_i, which we express as a single matrix equation

$$\begin{gathered} f_1 + \partial f_1/\partial x_1 \Delta x_1 + \partial f_1/\partial x_2 \Delta x_2 + \dots + \partial f_1/\partial x_9 \Delta x_9 = 0 , \\ f_2 + \partial f_2/\partial x_1 \Delta x_1 + \partial f_2/\partial x_2 \Delta x_2 + \dots + \partial f_2/\partial x_9 \Delta x_9 = 0 , \\ \ddots \\ f_9 + \partial f_9/\partial x_1 \Delta x_1 + \partial f_9/\partial x_2 \Delta x_2 + \dots + \partial f_9/\partial x_9 \Delta x_9 = 0 , \end{gathered}$$

$$\begin{pmatrix} f_1 \\ f_2 \\ \ddots \\ f_9 \end{pmatrix} + \begin{pmatrix} \partial f_1/\partial x_1 & \partial f_1/\partial x_2 & \cdots & \partial f_1/\partial x_9 \\ \partial f_2/\partial x_1 & \partial f_2/\partial x_2 & \cdots & \partial f_2/\partial x_9 \\ \ddots & & & \\ \partial f_9/\partial x_1 & \partial f_9/\partial x_2 & \cdots & \partial f_9/\partial x_9 \end{pmatrix} \begin{pmatrix} \Delta x_1 \\ \Delta x_2 \\ \ddots \\ \Delta x_9 \end{pmatrix} = 0 . \tag{6.15}$$

Note now that the derivatives and the f's are all evaluated at known values of the x_i's, so that only the vector of the Δx_i values is unknown. We write this equation in matrix notation as

$$\boldsymbol{f} + F'\boldsymbol{\Delta x} = 0\,, \quad \Rightarrow \quad F'\boldsymbol{\Delta x} = -\boldsymbol{f}\,, \tag{6.16}$$

$$\boldsymbol{\Delta x} = \begin{pmatrix} \Delta x_1 \\ \Delta x_2 \\ \ddots \\ \Delta x_9 \end{pmatrix}, \quad \boldsymbol{f} = \begin{pmatrix} f_1 \\ f_2 \\ \ddots \\ f_9 \end{pmatrix}, \quad F' = \begin{pmatrix} \partial f_1/\partial x_1 & \cdots & \partial f_1/\partial x_9 \\ \partial f_2/\partial x_1 & \cdots & \partial f_2/\partial x_9 \\ & \ddots & \\ \partial f_9/\partial x_1 & \cdots & \partial f_9/\partial x_9 \end{pmatrix}.$$

Here we use bold italic letters to emphasize the vector nature of the columns of f_i and Δx_i values, and denote the matrix of the derivatives F' (it is also sometimes denoted by J because it is the *Jacobian* matrix).

The equation $F'\boldsymbol{\Delta x} = -\boldsymbol{f}$ is in the standard form for the solution of a linear equation (often written as $A\boldsymbol{x} = \boldsymbol{b}$), where $\boldsymbol{\Delta x}$ is the vector of unknowns and $\boldsymbol{b} = -\boldsymbol{f}$. Matrix equations are solved using the techniques of linear algebra, and in the sections to follow we shall show how to do that. In a formal sense, the solution of (6.16) is obtained by multiplying both sides of the equation by the inverse of the F' matrix:

$$\boldsymbol{\Delta x} = -F'^{-1}\boldsymbol{f}\,, \tag{6.17}$$

where the inverse must exist if there is to be a unique solution. Although we are dealing with matrices now, this solution is identical in form to that of the 1D problem, $\Delta x = -(1/f')f$. In fact, one of the reasons, we use formal or abstract notation for matrices is to reveal the simplicity that lies within.

As we indicate for the single-equation Newton–Raphson method in Section 7.3, even in a case such as this where we can deduce analytic expressions for the derivatives $\partial f_i/\partial x_j$, there are $9 \times 9 = 81$ such derivatives for this (small) problem, and entering them all would be both time-consuming and error-prone. In contrast, especially for more complicated problems, it is straightforward to program a forward-difference approximation for the derivatives,

$$\frac{\partial f_i}{\partial x_j} \simeq \frac{f_i(x_j + \delta x_j) - f_i(x_j)}{\delta x_j}\,, \tag{6.18}$$

where each individual x_j is varied independently because these are partial derivatives and δx_j are some arbitrary changes you input. While a central-difference approximation for the derivative would be more accurate, it would also require more evaluations of the f's, and once we find a solution it does not matter how accurate our algorithm for the derivative was.

As also discussed for the 1D Newton–Raphson method (Section 7.3.1), the method can fail if the initial guess is not close enough to the zero of f (here all N of them) for the f's to be approximated as linear. The *backtracking* technique may be applied here as well, in the present case, progressively decreasing the corrections Δx_i until $|f|^2 = |f_1|^2 + |f_2|^2 + \cdots + |f_N|^2$ decreases.

6.2
Why Matrix Computing?

Physical systems are often modeled by systems of simultaneous equations written in matrix form. As the models are made more realistic, the matrices correspondingly become larger, and it becomes more important to use a good linear algebra library. Computers are unusually good with matrix manipulations because those manipulations typically involve the continued repetition of a small number of simple instructions, and algorithms exist to do this quite efficiently. Further speedup may be achieved by *tuning* the codes to the computer's architecture, as discussed in Chapter 11.

Industrial-strength subroutines for matrix computing are found in well-established scientific libraries. These subroutines are usually an order of magnitude or more faster than the elementary methods found in linear algebra texts,[2] are usually designed to minimize the round-off error, and are often "robust," that is, have a high chance of being successful for a broad class of problems. For these reasons, we recommend that you *do not write your own matrix methods* but instead get them from a library. An additional value of library routines is that you can often run the same program either on a desktop machine or on a parallel supercomputer, with matrix routines automatically adapting to the local architecture. The thoughtful reader may be wondering when a matrix is "large" enough to require the use of a library routine. One rule of thumb is "if you have to wait for the answer," and another rule is if the matrices you are using take up a good fraction of your computer's random-access memory (RAM).

6.3
Classes of Matrix Problems (Math)

It helps to remember that the rules of mathematics apply even to the world's most powerful computers. For example, you *should* encounter problems solving equations if you have more unknowns than equations, or if your equations are not linearly independent. But do not fret. While you cannot obtain a unique solution when there are not enough equations, you may still be able to map out a space of allowable solutions. At the other extreme, if you have more equations than unknowns, you have an *overdetermined* problem, which may not have a unique solution. An overdetermined problem is sometimes treated using data fitting techniques in which a solution to a sufficient set of equations is found, tested on the unused equations, and then improved if needed. Not surprisingly, this technique

2) Although we prize the book, *Numerical Recipes* by Press *et al.* (1994), and what it has accomplished, we cannot recommend taking subroutines from it. They are neither optimized nor documented for easy, stand-alone use, whereas the subroutine libraries recommended in this chapter are.

is known as the *linear least-squares method* (as in Chapter 7) because the technique minimizes the disagreement with the equations.

The most basic matrix problem is a system of linear equations:

$$\mathbf{A}\boldsymbol{x} = \boldsymbol{b} \,, \tag{6.19}$$

where $\mathbf{A}$ is a known $N \times N$ matrix, $\boldsymbol{x}$ is an unknown vector of length N, and $\boldsymbol{b}$ is a known vector of length N. The obvious way to solve this equation is to determine the inverse of $\mathbf{A}$ and then form the solution by multiplying both sides of (6.19) by $\mathbf{A}^{-1}$:

$$\boldsymbol{x} = \mathbf{A}^{-1}\boldsymbol{b} \,. \tag{6.20}$$

Both the direct solution of (6.19) and the determination of a matrix's inverse are standards in a matrix subroutine library. A more efficient way to solve (6.19) is by Gaussian elimination or lower–upper (LU) decomposition. This yields the vector $\boldsymbol{x}$ without explicitly calculating $\mathbf{A}^{-1}$. However, sometime you may want the inverse for other purposes, in which case (6.20) is preferred.

If you have to solve the matrix equation

$$\mathbf{A}\boldsymbol{x} = \lambda\boldsymbol{x} \,, \tag{6.21}$$

with $\boldsymbol{x}$ an unknown vector and λ an unknown parameter, then the direct solution (6.20) will not be of much help because the matrix $\boldsymbol{b} = \lambda\boldsymbol{x}$ contains the unknowns λ and $\boldsymbol{x}$. Equation 6.21 is the *eigenvalue problem*. It is harder to solve than (6.19) because solutions exist for only certain, if any, values of λ. To find a solution, we use the identity matrix to rewrite (6.21) as

$$[\mathbf{A} - \lambda I]\boldsymbol{x} = 0 \,. \tag{6.22}$$

We see that multiplication of (6.22) by $[\mathbf{A} - \lambda I]^{-1}$ yields the *trivial solution*

$$\boldsymbol{x} = 0 \quad \text{(trivial solution)} \,. \tag{6.23}$$

While the trivial solution is a bona fide solution, it is nonetheless trivial. A more interesting solution requires the existence of a condition that forbids us from multiplying both sides of (6.22) by $[\mathbf{A} - \lambda I]^{-1}$. That condition is the nonexistence of the inverse, and if you recall that Cramer's rule for the inverse requires division by $\det[\mathbf{A} - \lambda I]$, it is clear that the inverse fails to exist (and in this way eigenvalues *do* exist) when

$$\det[\mathbf{A} - \lambda I] = 0 \,. \tag{6.24}$$

The λ values that satisfy this *secular equation* are the eigenvalues of (6.21). If you are interested in only the eigenvalues for (6.21), you should look for a matrix routine that solves (6.24). To do that, you need a subroutine to calculate the determinant of a matrix, and then a search routine to find the zero of (6.24). Such routines are available in libraries.

The traditional way to solve the eigenvalue problem (6.21) for both eigenvalues and eigenvectors is by *diagonalization*. This is equivalent to successive changes of basis vectors, each change leaving the eigenvalues unchanged while continually decreasing the values of the off-diagonal elements of **A**. The sequence of transformations is equivalent to continually operating on the original equation with the transformation matrix **U** until one is found for which $\mathbf{UAU}^{-1}$ is diagonal:

$$\mathbf{UA}(\mathbf{U}^{-1}\mathbf{U})\boldsymbol{x} = \lambda \mathbf{U}\boldsymbol{x} , \tag{6.25}$$

$$(\mathbf{UAU}^{-1})(\mathbf{U}\boldsymbol{x}) = \lambda \mathbf{U}\boldsymbol{x} , \tag{6.26}$$

$$\mathbf{UAU}^{-1} = \begin{pmatrix} \lambda'_1 & & \cdots & 0 \\ 0 & \lambda'_2 & \cdots & 0 \\ 0 & 0 & \lambda'_3 & \cdots \\ 0 & \cdots & & \lambda'_N \end{pmatrix} . \tag{6.27}$$

The diagonal values of $\mathbf{UAU}^{-1}$ are the eigenvalues with eigenvectors

$$\boldsymbol{x}_i = \mathbf{U}^{-1}\hat{e}_i , \tag{6.28}$$

that is, the eigenvectors are the columns of the matrix $\mathbf{U}^{-1}$. A number of routines of this type are found in subroutine libraries.

6.3.1
Practical Matrix Computing

Many scientific programming bugs arise from the improper use of arrays.[3] This may be as a result of the extensive use of matrices in scientific computing or to the complexity of keeping track of indices and dimensions. In any case, here are some rules of thumb to observe.

Computers are finite Unless you are careful, your matrices may so much memory that your computation will slow down significantly, especially if it starts to use virtual memory. As a case in point, let us say that you store data in a 4D array with each index having a *physical dimension* of 100: `A[100] [100] [100] [100]`. This array of $(100)^4$ 64-byte words occupies $\simeq$1 GB of memory.

Processing time Matrix operations such as inversion require on the order of N^3 steps for a square matrix of dimension N. Therefore, doubling the dimensions of a 2D square matrix (as happens when the number of integration steps is doubled) leads to an *eightfold* increase in the processing time.

Paging Many operating systems have *virtual memory* in which disk space is used when a program runs out of RAM (see Chapter 10) for a discussion of how computers arrange memory). This is a slow process that requires writing a full *page*

3) Even a vector $V(N)$ is called an array, albeit a 1D one.

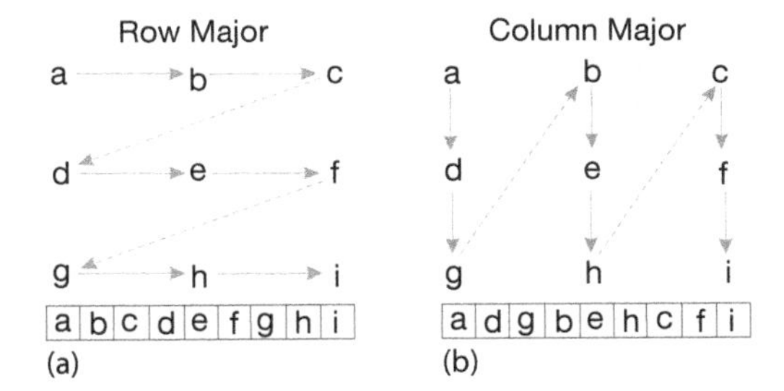

Figure 6.3 (a) Row-major order used for matrix storage in Python, C and Java. (b) Column-major order used for matrix storage in Fortran. The table at the bottom shows how successive matrix elements are actually stored in a linear fashion in memory.

of words to the disk. If your program is near the memory limit at which paging occurs, even a slight increase in a matrix's dimension may lead to an order of magnitude increase in the execution time.

Matrix storage While we think of matrices as multidimensional blocks of stored numbers, the computer stores them as linear strings. For instance, a matrix `a[3,3]` in Python, is stored in *row-major order* (Figure 6.3a):

$$a_{0,0} \quad a_{0,1} \quad a_{0,2} \quad a_{1,0} \quad a_{1,1} \quad a_{1,2} \quad a_{2,0} \quad a_{2,1} \quad a_{2,2} \quad \dots , \tag{6.29}$$

while in Fortran, starting subscripts at 0, it is stored in *column-major order* (Figure 6.3b):

$$a_{0,1} \quad a_{1,0} \quad a_{2,0} \quad a_{0,1} \quad a_{1,1} \quad a_{2,1} \quad a_{0,2} \quad a_{1,2} \; a_{2,2} \quad \dots \tag{6.30}$$

It is important to keep this linear storage scheme in mind in order to write proper code and to permit the mixing of Python and Fortran programs.

When dealing with matrices, you have to balance the clarity of the operations being performed against the efficiency with which the computer performs them. For example, having one matrix with many indices such as `V[L,Nre,Nspin,k,kp,Z,A]` may be neat packaging, but it may require the computer to jump through large blocks of memory to get to the particular values needed (large *strides*) as you vary `k`, `kp`, and `Nre`. The solution would be to have several matrices such as `V1[Nre,Nspin,k,kp,Z,A]`, `V2[Nre,Nspin,k,kp,Z,A]`, and `V3[Nre,Nspin,k,kp,Z,A]`.

Subscript 0 It is standard in Python, C, and Java to have array indices begin with the value 0. While this is now permitted in Fortran, the standard in Fortran and in most mathematical equations has been to start indices at 1. On that account, in addition to the different locations in memory as a result of row-major and column-major ordering, the same matrix element may be referenced differently in the different languages:

Location	Python/C element	Fortran element
Lowest	a [0,0]	a (1,1)
	a [0,1]	a (2,1)
	a [1,0]	a (3,1)
	a [1,1]	a (1,2)
	a [2,0]	a (2,2)
Highest	a [2,1]	a (3,2)

Tests *Always* test a library routine on a small problem whose answer you know (such as the exercises in Section 6.6). Then you will know if you are supplying it with the right arguments and if you have all the links working.

6.4 Python Lists as Arrays

A *list* is Python's built-in sequence of numbers or objects. Although called a "list" it is similar to what other computer languages call an "array." It may be easier for you to think of a Python list as a container that holds a bunch of items in a definite order. (Soon we will describe the higher level **array** data types available with the *NumPy* package.) In this section we review some of Python's native *list* features.

Python interprets a sequence of ordered items, $L = l_0, l_1, \ldots, l_{N-1}$, as a *list* and represents it with a single symbol `L`:

```
>>> L = [1, 2, 3]                    # Create list                    1
>>> L[0]                             # Print element 0 (first)
1                                    # Python output                  3
>>> L                                # Print entire list
[1, 2, 3]                            # Output                         5
>>> L[0]= 5                          # Change element 0
>>> L                                                                 7
[5, 2, 3]
>>> len(L)                           # Length of list                 9
3
>>> for items in L: print items      # for loop over items            11
5
2                                                                     13
3
```

We observe that square brackets with comma separators $[1, 2, 3]$ are used for lists, and that a square bracket is also used to indicate the index for a list item, as in line 2 (`L[0]`). Lists contain sequences of arbitrary objects that are *mutable* or changeable. As we see in line 7 in the `L` command, an entire list can be referenced as a single object, in this case to obtain its printout.

Python also has a built-in type of list known as a *tuple* whose elements are not mutable. Tuples are indicated by round parenthesis (.., .., .), with individual elements still referenced by square brackets:

```
>>> T = (1, 2, 3, 4)                    # Create tuple
>>> T[3]                                # Print element 3
4
>>> T
(1, 2, 3, 4)                            # Print entire tuple
>>> T[0] = 5                            # Attemp to change element 0
Traceback (most recent call last):
 T[0] = 5
TypeError: 'tuple' object does not support item assignment
```

Note that the error message arises when we try to change an element of a tuple.

Most languages require you to specify the size of an array before you can start storing objects in it. In contrast, Python lists are *dynamic*, which means that their sizes adjust as needed. In addition, while a list is essentially one dimensional because it is a sequence, a compound list can be created in Python by having the individual elements themselves as lists:

```
>>> L = [[1,2], [3,4], [5,6]]           # A list of lists
>>> L
[[1, 2], [3, 4], [5, 6]]
>>> L[0]                                # The first element
[1, 2]
```

Python can perform a large number of operations on lists, for example,

Operation	Effect	Operation	Effect
L = [1, 2, 3, 4]	Form list	L1 + L2	Concatenate lists
L[i]	ith element	len(L)	Length of list L
i in L	True if i in L	L[i:j]	Slice from i to j
for i in L	Iteration index	L.append(x)	Append x to end of L
L.count(x)	Number of x's in L	L.index(x)	Location of 1st x in L
L.remove(x)	Remove 1st x in L	L.reverse()	Reverse elements in L
L.sort()	Order elements in L		

6.5 Numerical Python (NumPy) Arrays

Although basic Python does have an `array` data type, it is rather limited and we suggest using NumPy arrays, which will convert Python lists into arrays. Because *we often use NumPy's* `array` *command* to form computer representations of vectors and matrices, you will have to import *NumPy* in your programs (or Visual, which includes NumPy) to use those examples. For instance, here we show the results of running our program `Matrix.py` from a shell:

```
>>> from visual import *                # Load Visual package
>>> vector1 = array([1, 2, 3, 4, 5])    # Fill 1D array
>>> print('vector1 =',vector1)          # Print array (parens if Python 3)
```

```
vector1 = [1 2 3 4 5]                        # Output
>>> vector2 = vector1 + vector1              # Add 2 vectors                 5
>>> print('vector2=',vector2)                # Print vector2
vector2= [ 2 4 6 8 10]                       # Output                        7
>>> vector2 = 3 * vector1                    # Mult array by scalar
>>> print ('3 * vector1 = ', vector2)        # Print vector                  9
3 * vector1 = [ 3 6 9 12 15]                 # Output
>>> matrix1 = array(([0,1],[1,3]))           # An array of arrays            11
>>> print(matrix1)                           # Print matrix1
[[0 1]                                                                       13
 [1 3]]
>>> print ('vector1.shape= ',vector1.shape)                                  15
vector1.shape = (5)
>>> print (matrix1 * matrix1)                # Matrix multiply               17
 [[0 1]
  [1 9]]                                                                     19
```

We see here that we have initialized an array object, added two 1D array objects together, and printed out the result. Likewise we see that multiplying an array by a constant does in fact multiply each element by that constant (line 8). We then construct a "matrix" as a 1D array of two 1D arrays, and when we print it out, we note that it does indeed look like a matrix. However, when we multiply this matrix by itself, the result is not $\begin{bmatrix} 1 & 3 \\ 3 & 10 \end{bmatrix}$ that one normally expects from matrix multiplication. So if you need actual mathematical matrices, then you need to use NumPy!

Now we give some examples of the use of NumPy, but do refer the reader to the NumPy Tutorial (NumPyTut, 2012) and to the articles in *Computing in Science and Engineering* (Perez *et al.*, 2010) for more information. To start, we note that a NumPy array can hold up to 32 dimensions (32 indices), but each element must be of the same type (a *uniform* array). The elements are not restricted to just floating-point numbers or integers, but can be any object, as long as all elements are of this same type. (Compound objects may be useful, for example, for storing parts of data sets.) There are various ways to create arrays, with square brackets [...] used for indexing in all cases. First we start with a Python list (tuples work as well) and create an array from it:

```
>>> from numpy import *                                                      1
>>> a = array([1, 2, 3, 4])            # Array from a list
>>> a                                  # Check with print                    3
array([1, 2, 3, 4])
```

Notice that it is essential to have the square brackets within the round brackets because the square brackets produce the list object while the round brackets indicate a function argument. Note also that because the data in our original list were all integers, the created array is a 32-bit integer data types, which we can check by affixing the `dtype` method:

```
>>> a.dtype
dtype('int32')                                                               2
```

If we had started with floating-point numbers, or a mix of floats and ints, we would have ended up with floating-point arrays:

```
>>> b = array([1.2, 2.3, 3.4])
>>> b                                                                   2
array([ 1.2, 2.3, 3.4])
>>> b.dtype                                                             4
dtype('float64')
```

When describing NumPy arrays, the number of "dimensions," `ndim`, means the number of indices, which, as we said, can be as high as 32. What might be called the "size" or "dimensions" of a matrix in mathematics is called the *shape* of a NumPy array. Furthermore, NumPy does have a `size` method that returns the total number of elements. Because Python's lists and tuples are all one dimensional, if we want an array of a particular shape, we can attain that by affixing the `reshape` method when we create the array. Where Python has a `range` function to generate a sequence of numbers, NumPy has an `arange` function that creates an array, rather than a list. Here we use it and then reshape the 1D array into a 3 × 4 array:

```
>>> import numpy as np                                                  1
>>> np.arange(12)                        # List of 12 ints in 1D array
array([ 0, 1, 2, 3, 4, 5, 6, 7, 8, 9, 10, 11])                          3
>>> np.arange(12).reshape((3,4))         # Create, shape to 3x4 array
array([[ 0, 1, 2, 3],                                                   5
       [ 4, 5, 6, 7],
       [ 8, 9, 10, 11]])                                                7
>>> a = np.arange(12).reshape((3,4))     # Give array a name
>>> a                                                                   9
array([[ 0, 1, 2, 3],
       [ 4, 5, 6, 7],                                                   11
       [ 8, 9, 10, 11]])
>>> a.shape                              # Shape = ?                    13
(3L, 4L)
>>> a.ndim                               # Dimension?                   15
2
>>> a.size                               # Size of a (number of elements)? 17
12
```

Note that here we imported NumPy as the object `np`, and then affixed the `arange` and `reshape` methods to this object. We then checked the shape of `a`, and found it to have three rows and four columns of long integers (Python 3 may just say `ints`). Note too, as we see on line 9, NumPy uses parentheses () to indicate the shape of an array, and so `(3L,4L)` indicates an array with three rows and four columns of long ints.

Now that we have shapes in our minds, we should note that NumPy offers a number of ways to change shapes. For example, we can transpose an array with the `.T` method, or `reshape` into a vector:

```
>>> from numpy import *
>>> a = arange(12).reshape((3,4))        # Give array a name            2
>>> a
array([[ 0, 1, 2, 3],                                                   4
       [ 4, 5, 6, 7],
       [ 8, 9, 10, 11]])                                                6
>>> a.T                                  # Transpose
array([[ 0, 4, 8],                                                      8
```

```
       [ 1, 5, 9],
       [ 2, 6, 10],                                                10
       [ 3, 7, 11]])
>>> b = a.reshape((1,12))                  # Form vector length 12  12
>>> b
array([[ 0, 1, 2, 3, 4, 5, 6, 7, 8, 9, 10, 11]])                   14
```

And again, (1,12) indicates an array with 1 row and 12 columns. Yet another handy way to take a matrix and extract just what you want from it is to use Python's *slice* operator **start:stop:step:** to take a slice out of an array:

```
>>> a
array([[ 0, 1, 2, 3],                                               2
       [ 4, 5, 6, 7],
       [ 8, 9, 10, 11]])                                            4
>>> a[:2, :]                               # First 2 rows
array([[0, 1, 2, 3],                                                6
       [4, 5, 6, 7]])
>>> a[:,1:3]                               # Columns 1-3            8
array([[ 1, 2],
       [ 5, 6],                                                     10
       [ 9, 10]])
```

We note here that Python indices start counting from 0, and so 1:3 means indices 0, 1, 2 (without 3). As we discuss in Chapter 11, slicing can be very useful in speeding up programs by picking out and placing in memory just the specific data elements from a large data set that need to be processed. This avoids the time-consuming jumping through large segments of memory as well as excessive reading from disk.

Finally, we remind you that while all elements in a NumPy array must be of the same data type, that data type can be compound, for example, an array of arrays:

```
>>> from numpy import *                                             1
>>> M = array([ (10, 20), (30,40), (50, 60) ])   # Array of 3 arrays
>>> M                                                               3
array([[10, 20],
       [30, 40],                                                    5
       [50, 60]])
>>> M.shape                                                         7
(3L, 2L)
>>> M.size                                                          9
6
>>> M.dtype                                                         11
dtype('int32')
```

Furthermore, an array can be composed of complex numbers by specifying the **complex** data type as an option on the **array** command. NumPy then uses the j symbol for the imaginary number i:

```
>>> c = array([ [1,complex(2,2)], [complex(3,2),4] ], dtype=complex)
>>> c                                                               2
array([[ 1.+0.j, 2.+2.j],
       [ 3.+2.j, 4.+0.j]])                                          4
```

In the next section, we discuss using true mathematical matrices with NumPy, which is one use of an array object. Here we note that if you wanted the familiar matrix product from two arrays, you would use the dot function, whereas * is used for an element-by-element product:

```
>>> matrix1= array([[0,1], [1,3]])
>>> matrix1                                                              2
array([[0, 1],
       [1, 3]])                                                          4
>>> print (dot(matrix1,matrix1))          # Matrix or dot product
[[ 1 3]                                                                  6
 [ 3 10]]
>>> print (matrix1 * matrix1)             # Element-by-element product   8
[[0 1]
 [1 9]]                                                                  10
```

NumPy is actually optimized to work well with arrays, and in part this is because arrays are handled and processed much as if they were simple, scalar variables.[4] For example, here is another example of *slicing*, a technique that is also used in ordinary Python with lists and tuples, in which two indices separated by a colon indicate a range:

```
from visual import *
stuff = zeros(10, float)
t = arange(4)
stuff[3:7] = sqrt(t+1)
```

Here we start by creating the NumPy array stuff of floats, all of whose 10 elements are initialized to zero. Then we create the array t containing the four elements [0, 1, 2, 3] by assigning four variables uniformly in the range 0–4 (the "a" in arange creates floating-point variables, range creates integers). Next we use a slice to assign [sqrt(0+1), sqrt(1+1), sqrt(2+1), sqrt(3+1)] = [1, 1.414, 1.732, 2] to the middle elements of the stuff array. Note that the NumPy version of the sqrt function, one of many universal functions (*ufunctions*) supported by NumPy[5], has the amazing property of automatically outputting an array whose length is that of its argument, in this case, the array t. In general, major power in NumPy comes from its *broadcasting* operation, an operation in which values are assigned to multiple elements via a single assignment statement. Broadcasting permits Python to *vectorize* array operations, which means that the same operation can be performed on different array elements in parallel (or nearly so). Broadcasting also speeds up processing because array operations occur in C instead of Python, and with a minimum of array copies being made. Here is a simple sample of broadcasting:

```
w = zeros(100, float)
w = 23.7
```

The first line creates the NumPy array w, and the second line "broadcasts" the value 23.7 to all elements in the array. There are many possible array operations

4) We thank Bruce Sherwood for helpful comments on these points.

5) A ufunction is a function that operates on N–D arrays in an element-by-element fashion, supporting array broadcasting, type casting, and several other standard features. In other words, a ufunc is a vectorized wrapper for a function that takes a fixed number of scalar inputs and produces a fixed number of scalar outputs.

in NumPy and various rules pertaining to them; we recommend that the serious user explore the extensive NumPy documentation for additional information.

6.5.1 NumPy's linalg Package

The array objects of NumPy and Visual are not the same as mathematical matrices, although an array can be used to represent a matrix. Fortunately, there is the `LinearAlgebra` package that treats 2D arrays (a 1D array of 1D arrays) as mathematical matrices, and also provides a simple interface to the powerful LAPACK linear algebra library (Anderson *et al.*, 2013). As we keep saying, there is much to be gained in speed and reliability from using these libraries rather than writing your own matrix routines.

Our first example from linear algebra is the standard matrix equation

$$\mathbf{A}\boldsymbol{x} = \boldsymbol{b} \,, \tag{6.31}$$

where we have used an uppercase bold character to represent a matrix and a lowercase bold italic character to represent a 1D matrix (a vector). Equation 6.31 describes a set of linear equations with $\boldsymbol{x}$ an unknown vector and $\mathbf{A}$ a known matrix. Now we take $yy\mathbf{A}$ to be 3×3, $\boldsymbol{b}$ to be 3×1, and let the program figure out that $\boldsymbol{x}$ must be a 3×1 vector.[6] We start by importing all the packages, by inputting a matrix and a vector, and then by printing out $\mathbf{A}$ and $\boldsymbol{x}$:

```
>>> from numpy import *
>>> from numpy.linalg import*                                                2
>>> A = array([ [1,2,3], [22,32,42], [55,66,100] ]) # Array of arrays
>>> print ('A =', A)                                                         4
A = [[ 1 2 3]
     [ 22 32 42]                                                             6
     [ 55 66 100]]
>>> b = array([1,2,3])                                                       8
>>> print ('b =', b)
b = [1 2 3]                                                                  10
```

Because we have the matrices $\mathbf{A}$ and $\boldsymbol{b}$, we can go ahead and solve $\mathbf{A}\boldsymbol{x} = \boldsymbol{b}$ using NumPy's `solve` command, and then test how close $\mathbf{A}\boldsymbol{x} - \boldsymbol{b}$ is to a zero vector:

```
>>> from numpy.linalg import solve
>>> x = solve(A, b)                              # Finds solution            2
>>> print ('x =', x)
x = [ -1.4057971 -0.1884058 0.92753623]          # The solution              4
>>> print ('Residual =', dot(A, x) - b)          # LHS-RHS
                                                                             6
Residual = [4.44089210e-16 0.00000000e+00 -3.55271368e-15]
```

This is really quite impressive. We have solved the entire set of linear equations (by elimination) with just the single command `solve`, performed a matrix multiplication with the single command `dot`, did a matrix subtraction with the usual operator, and are left with a residual essentially equal to machine precision.

6) Do not be bothered by the fact that although we think these vectors as 3×1, they sometimes get printed out as 1×3; think of all the trees that get saved!

Although there are more efficient numerical approaches, a direct way to solve

$$\mathbf{A}\boldsymbol{x} = \boldsymbol{b} \tag{6.32}$$

is to calculate the inverse $\mathbf{A}^{-1}$, and then multiply both sides of the equation by the inverse, yielding

$$\boldsymbol{x} = y\mathbf{A}^{-1}\boldsymbol{b} \tag{6.33}$$

```
>>> from numpy.linalg import inv                                        1
>>> dot(inv(A), A)                               # Test inverse
                                                                        3
array([[ 1.00000000e+00, -1.33226763e-15, -1.77635684e-15],
       [ 8.88178420e-16, 1.00000000e+00, 0.00000000e+00],               5
       [ -4.44089210e-16, 4.44089210e-16, 1.00000000e+00]])
>>> print ('x =', multiply(inv(A), b))                                  7
x = [-1.4057971  -0.1884058  0.92753623]         # Solution
>>> print ('Residual =', dot(A, x) - b)                                 9

Residual = [ 4.44089210e-16 0.00000000e+00 -3.55271368e-15]            11
```

Here we first tested that `inv(A)` is in fact the inverse of `A` by seeing if `A` times `inv(A)` equals the identity matrix. Then we used the inverse to solve the matrix equation directly, and got the same answer as before (some error at the level of machine precision is just fine).

Our second example occurs in the solution for the principal-axes system of a cube, and requires us to find a coordinate system in which the inertia tensor is diagonal. This entails solving the eigenvalue problem

$$\mathbf{I}\boldsymbol{\omega} = \lambda\boldsymbol{\omega} \,, \tag{6.34}$$

where $\mathbf{I}$ is the inertia matrix (tensor), $\boldsymbol{\omega}$ is an unknown eigenvector, and λ is an unknown eigenvalue. The program **Eigen.py** solves for the eigenvalues and vectors, and shows how easy it is to deal with matrices. Here it is in an abbreviated interpretive mode:

```
>>> from numpy import*                                                  1
>>> from numpy.linalg import eig
>>> I = array([[2./3,-1./4], [-1./4,2./3]])                             3
>>> print('I =\n', I)
I =                                                                     5
 [[ 0.66666667 -0.25 ]
  [ -0.25 0.66666667 ]]                                                 7
>>> Es, evectors = eig(A)                       # Solves eigenvalue problem
>>> print('Eigenvalues =', Es, '\n Eigenvector Matrix =\n', evectors)   9
Eigenvalues = [ 0.91666667 0.41666667 ]
 Eigenvector Matrix =                                                   11
 [[ 0.70710678 0.70710678 ]
  [ -0.70710678 0.70710678 ]]                                           13
>>> Vec = array([ evectors[0, 0], evectors[1, 0] ])
>>> LHS = dot(I, Vec)                           # Matrix x vector       15
>>> RHS = Es[0]*Vec                             # Scalar mult
>>> print('LHS - RHS =', LHS-RHS)               # Test for zero         17
LHS - RHS = [ 1.11022302e-16 -1.11022302e-16]
```

Table 6.1 The operators of NumPy and their effects.

Operator	Effect	Operator	Effect
`dot(a, b[,out])`	Dot product arrays	`vdot(a, b)`	Dot product
`inner(a, b)`	Inner product arrays	`outer(a, b)`	Outer product
`tensordot(a, b)`	Tensor dot product	`einsum()`	Einstein sum
`linalg.matrix_power(M, n)`	Matrix to power n	`kron(a, b)`	Kronecker product
`linalg.cholesky(a)`	Cholesky decomp	`linalg.qr(a)`	QR factorization
`linalg.svd(a)`	Singular val decomp	`linalg.eig(a)`	Eigenproblem
`linalg.eigh(a)`	Hermitian eigen	`linalg.eigvals(a)`	General eigen
`linalg.eigvalsh(a)`	Hermitian eigenvals	`linalg.norm(x)`	Matrix norm
`linalg.cond(x)`	Condition number	`linalg.det(a)`	Determinant
`linalg.slogdet(a)`	Sign and log(det)	`trace(a)`	Diagnol sum
`linalg.solve(a, b)`	Solve equation	`linalg.tensorsolve(a, b)`	Solve $ax = b$
`linalg.lstsq(a, b)`	Least-squares solve	`linalg.inv(a)`	Inverse
`linalg.pinv(a)`	Penrose inverse	`linalg.tensorinv(a)`	Inverse N–D array

We see here how, after we set up the array `I` on line 3, we then solve for its eigenvalues and eigenvectors with the single statement `Es, evectors = eig(I)` on line 8. We then extract the first eigenvector on line 14 and use it, along with the first eigenvalue, to check that (6.34) is in fact satisfied to machine precision.

Well, we think by now you have some idea of the use of NumPy. In Table 6.1 we indicate some more of what is available.

6.6 Exercise: Testing Matrix Programs

Before you direct the computer to go off crunching numbers on a million elements of some matrix, it is a good idea to try out your procedures on a small matrix, especially one for which you know the right answer. In this way, it will take you only a short time to realize how hard it is to get the calling procedure perfect. Here are some exercises.

1. Find the numerical inverse of $\mathbf{A} = \begin{pmatrix} +4 & -2 & +1 \\ +3 & +6 & -4 \\ +2 & +1 & +8 \end{pmatrix}$.
   a) As a general check, applicable even if you do not know the analytic answer, check your inverse in both directions; that is, check that $\mathbf{A}\mathbf{A}^{-1} = \mathbf{A}^{-1}\mathbf{A} = I$, and note the number of decimal places to which this is true. This also gives you some idea of the precision of your calculation.

b) Determine the number of decimal places of agreement there is between your numerical inverse and the analytic result:
$\mathbf{A}^{-1} = \frac{1}{263}\begin{pmatrix} +52 & +17 & +2 \\ -32 & +30 & +19 \\ -9 & -8 & +30 \end{pmatrix}$. Is this similar to the error in $\mathbf{AA}^{-1}$?

2. Consider the same matrix $\mathbf{A}$ as before, here being used to describe three simultaneous linear equations, $\mathbf{A}\boldsymbol{x} = \boldsymbol{b}$, or explicitly,

$$\begin{pmatrix} a_{00} & a_{01} & a_{02} \\ a_{10} & a_{11} & a_{12} \\ a_{20} & a_{21} & a_{22} \end{pmatrix} \begin{pmatrix} x_0 \\ x_1 \\ x_2 \end{pmatrix} = \begin{pmatrix} b_0 \\ b_1 \\ b_2 \end{pmatrix} . \tag{6.35}$$

Now the vector $\boldsymbol{b}$ on the RHS is assumed known, and the problem is to solve for the vector $\boldsymbol{x}$. Use an appropriate subroutine to solve these equations for the three different $\boldsymbol{x}$ vectors appropriate to these three different $\boldsymbol{b}$ values on the RHS:

$$\boldsymbol{b}_1 = \begin{pmatrix} +12 \\ -25 \\ +32 \end{pmatrix}, \quad \boldsymbol{b}_2 = \begin{pmatrix} +4 \\ -10 \\ +22 \end{pmatrix}, \quad \boldsymbol{b}_3 = \begin{pmatrix} +20 \\ -30 \\ +40 \end{pmatrix} .$$

The solutions should be

$$\boldsymbol{x}_1 = \begin{pmatrix} +1 \\ -2 \\ +4 \end{pmatrix}, \quad \boldsymbol{x}_2 = \begin{pmatrix} +0.312 \\ -0.038 \\ +2.677 \end{pmatrix}, \quad \boldsymbol{x}_3 = \begin{pmatrix} +2.319 \\ -2.965 \\ +4.790 \end{pmatrix} . \tag{6.36}$$

3. Consider the matrix $\mathbf{A} = \begin{pmatrix} \alpha & \beta \\ -\beta & \alpha \end{pmatrix}$, where you are free to use any values you want for α and β. Use a numerical eigenproblem solver to show that the eigenvalues and eigenvectors are the complex conjugates:

$$\boldsymbol{x}_{1,2} = \begin{pmatrix} +1 \\ \mp i \end{pmatrix}, \quad \lambda_{1,2} = \alpha \mp i\beta . \tag{6.37}$$

4. Use your eigenproblem solver to find the eigenvalues of the matrix

$$\mathbf{A} = \begin{pmatrix} -2 & +2 & -3 \\ +2 & +1 & -6 \\ -1 & -2 & +0 \end{pmatrix} . \tag{6.38}$$

a) Verify that you obtain the eigenvalues $\lambda_1 = 5$, $\lambda_2 = \lambda_3 = -3$. Notice that double roots can cause problems. In particular, there is a uniqueness issue with their eigenvectors because any combination of these eigenvectors is also an eigenvector.

b) Verify that the eigenvector for $\lambda_1 = 5$ is proportional to

$$\boldsymbol{x}_1 = \frac{1}{\sqrt{6}}\begin{pmatrix} -1 \\ -2 \\ +1 \end{pmatrix} . \tag{6.39}$$

c) The eigenvalue -3 corresponds to a double root. This means that the corresponding eigenvectors are degenerate, which in turn means that they are not unique. Two linearly independent ones are

$$\boldsymbol{x}_2 = \frac{1}{\sqrt{5}}\begin{pmatrix} -2 \\ +1 \\ +0 \end{pmatrix} , \quad \boldsymbol{x}_3 = \frac{1}{\sqrt{10}}\begin{pmatrix} 3 \\ 0 \\ 1 \end{pmatrix} . \tag{6.40}$$

In this case, it is not clear what your eigenproblem solver will give for the eigenvectors. Try to find a relationship between your computed eigenvectors with the eigenvalue -3 and these two linearly independent ones.

5. Imagine that your model of some physical system results in $N = 100$ coupled linear equations in N unknowns:

$$\begin{aligned} a_{00} y_0 + a_{01} y_1 + \cdots + a_{0(N-1)} y_{N-1} &= b_0 , \\ a_{10} y_0 + a_{11} y_1 + \cdots + a_{1(N-1)} y_{N-1} &= b_1 , \\ &\cdots \\ a_{(N-1)0} y_0 + a_{(N-1)1} y_1 + \cdots + a_{(N-1)(N-1)} y_{N-1} &= b_{N-1} . \end{aligned}$$

In many cases, the a and b values are known, so your exercise is to solve for all the x values, taking $\boldsymbol{a}$ as the *Hilbert* matrix and $\boldsymbol{b}$ as its first column:

$$[a_{ij}] = \boldsymbol{a} = \left[\frac{1}{i+j-1}\right] = \begin{pmatrix} 1 & \frac{1}{2} & \frac{1}{3} & \frac{1}{4} & \cdots & \frac{1}{100} \\ \frac{1}{2} & \frac{1}{3} & \frac{1}{4} & \frac{1}{5} & \cdots & \frac{1}{101} \\ \ddots & & & & & \\ \frac{1}{100} & \frac{1}{101} & \cdots & & \cdots & \frac{1}{199} \end{pmatrix} ,$$

$$[b_i] = \boldsymbol{b} = \left[\frac{1}{i}\right] = \begin{pmatrix} 1 \\ \frac{1}{2} \\ \frac{1}{3} \\ \ddots \\ \frac{1}{100} \end{pmatrix} .$$

Compare to the analytic solution

$$\begin{pmatrix} y_1 \\ y_2 \\ \ddots \\ y_N \end{pmatrix} = \begin{pmatrix} 1 \\ 0 \\ \ddots \\ 0 \end{pmatrix} . \tag{6.41}$$

6. *Dirac Gamma Matrices:* The Dirac equation extends quantum mechanics to include relativity and spin 1/2. The extension of the Hamiltonian operator for an electron requires it to contain matrices, and those matrices are expressed in terms of $4 \times 4\,\gamma$ matrices that can be represented in terms of the familiar 2×2 Pauli matrices σ_i:

$$\gamma_i = \begin{pmatrix} 0 & \sigma_i \\ -\sigma_i & 0 \end{pmatrix}, \quad i = 1, 2, 3\,, \tag{6.42}$$

$$\sigma_1 = \begin{pmatrix} 0 & 1 \\ 1 & 0 \end{pmatrix}, \quad \sigma_2 = \begin{pmatrix} 0 & -\mathrm{i} \\ \mathrm{i} & 0 \end{pmatrix}, \quad \sigma_3 = \begin{pmatrix} 1 & 0 \\ 0 & -1 \end{pmatrix}. \tag{6.43}$$

Confirm the following properties of the γ matrices:

$$\gamma_2^\dagger = \gamma_2^{-1} = -\gamma_2\,, \tag{6.44}$$

$$\gamma_1\gamma_2 = -\mathrm{i}\begin{pmatrix} \sigma_3 & 0 \\ 0 & \sigma_3 \end{pmatrix}. \tag{6.45}$$

6.6.1
Matrix Solution of the String Problem

In Section 6.1, we set up the solution to our problem of two masses on a string. Now we have the matrix tools needed to solve it. Your *problem* is to check out the physical reasonableness of the solution for a variety of weights and lengths. You should check that the deduced tensions are positive and the deduced angles correspond to a physical geometry (e.g., with a sketch). Because this is a physics-based problem, we know the sine and cosine functions must be less than 1 in magnitude and that the tensions should be similar in magnitude to the weights of the spheres. Our solution is given in `NewtonNDanimate.py` (Listing 6.1), which shows graphically the step-by-step search for a solution.

Listing 6.1 The code **NewtonNDanimate.py** that shows the step-by-step search for solution of the two-mass-on-a-string problem via a Newton–Raphson search.

```
# NewtonNDanimate.py:                    MultiDimension Newton Search

from visual import *
from numpy.linalg import solve
from visual.graph import *

scene = display(x=0,y=0,width=500,height=500,
                title='String and masses configuration')
tempe = curve(x=range(0,500),color=color.black)

n = 9
eps = 1e-3
deriv = zeros( (n, n), float)
f = zeros( (n), float)
```

```
x = array([0.5, 0.5, 0.5, 0.5, 0.5, 0.5, 0.5, 1., 1., 1.])

def plotconfig():
    for obj in scene.objects:
        obj.visible=0                          # Erase previous configuration
    L1 = 3.0
    L2 = 4.0
    L3 = 4.0
    xa = L1*x[3]                      # L1*cos(th1)
    ya = L1*x[0]                      # L1 sin(th1)
    xb = xa+L2*x[4]                   # L1*cos(th1)+L2*cos(th2)
    yb = ya+L2*x[1]                   # L1*sin(th1)+L2*sen(th2)
    xc = xb+L3*x[5]                   # L1*cos(th1)+L2*cos(th2)+L3*cos(th3)
    yc = yb-L3*x[2]                   # L1*sin(th1)+L2*sen(th2)-L3*sin(th3)
    mx = 100.0                        # for linear coordinate transformation
    bx = -500.0                       # from 0=< x =<10
    my = -100.0                       # to     -500 =<x_window=>500
    by = 400.0                        # same transformation for y
    xap = mx*xa+bx                    # to keep aspect ratio
    yap = my*ya+by
    ball1 = sphere(pos=(xap,yap), color=color.cyan,radius=15)
    xbp = mx*xb+bx
    ybp = my*yb+by
    ball2 = sphere(pos=(xbp,ybp), color=color.cyan,radius=25)
    xcp = mx*xc+bx
    ycp = my*yc+by
    x0 = mx*0+bx
    y0 = my*0+by
    line1 = curve(pos=[(x0,y0),(xap,yap)], color=color.yellow,radius=4)
    line2 = curve(pos=[(xap,yap),(xbp,ybp)], color=color.yellow,radius=4)
    line3 = curve(pos=[(xbp,ybp),(xcp,ycp)], color=color.yellow,radius=4)
    topline = curve(pos=[(x0,y0),(xcp,ycp)], color=color.red,radius=4)

def F(x, f):                                          # F function
    f[0] = 3*x[3]  +  4*x[4]  +  4*x[5]  -  8.0
    f[1] = 3*x[0]  +  4*x[1]  -  4*x[2]
    f[2] = x[6]*x[0]  -  x[7]*x[1]  -  10.0
    f[3] = x[6]*x[3]  -  x[7]*x[4]
    f[4] = x[7]*x[1]  +  x[8]*x[2]  -  20.0
    f[5] = x[7]*x[4]  -  x[8]*x[5]
    f[6] = pow(x[0], 2)  +  pow(x[3], 2)  -  1.0
    f[7] = pow(x[1], 2)  +  pow(x[4], 2)  -  1.0
    f[8] = pow(x[2], 2)  +  pow(x[5], 2)  -  1.0

def dFi_dXj(x, deriv, n):                             # Derivatives
    h = 1e-4
    for j in range(0, n):
        temp = x[j]
        x[j] = x[j] + h/2.
        F(x, f)
        for i in range(0, n):  deriv[i, j] = f[i]
        x[j] = temp
    for j in range(0, n):
        temp = x[j]
        x[j] = x[j] - h/2.
        F(x, f)
        for i in range(0, n): deriv[i, j] = (deriv[i, j] - f[i])/h
        x[j] = temp

for it in range(1, 100):
    rate(1)                                  # 1 second between graphs
    F(x, f)
    dFi_dXj(x, deriv, n)
    B = array([[-f[0]], [-f[1]], [-f[2]], [-f[3]], [-f[4]], [-f[5]],\
    [-f[6]], [-f[7]], [-f[8]]])
    sol = solve(deriv, B)
    dx = take(sol, (0, ), 1)                 # First column of sol
```

```
        for i in range(0, n):
            x[i]  = x[i]  +  dx[i]
        plotconfig()
        errX = errF = errXi = 0.0
        for i in range(0, n):
            if ( x[i] !=  0.): errXi = abs(dx[i]/x[i])
            else:   errXi = abs(dx[i])
            if ( errXi > errX): errX = errXi
            if ( abs(f[i]) > errF ):   errF = abs(f[i])
            if ( (errX <=  eps) and (errF <=  eps) ): break

print('Number of iterations = ', it, "\n Final Solution:")
for i in range(0, n):
    print('x[', i, '] =  ', x[i])
```

6.6.2 Explorations

1. See at what point your initial guess for the angles of the strings gets so bad that the computer is unable to find a physical solution.
2. A possible problem with the formalism we have just laid out is that by incorporating the identity $\sin^2\theta_i + \cos^2\theta_i = 1$ into the equations, we may be discarding some information about the sign of $\sin\theta$ or $\cos\theta$. If you look at Figure 6.1, you can observe that for some values of the weights and lengths, θ_2 may turn out to be negative, yet $\cos\theta$ should remain positive. We can build this condition into our equations by replacing f_7–f_9 with f's based on the form

$$f_7 = x_4 - \sqrt{1 - x_1^2}, \quad f_8 = x_5 - \sqrt{1 - x_2^2}, \quad f_9 = x_6 - \sqrt{1 - x_3^2}. \tag{6.46}$$

   See if this makes any difference in the solutions obtained.
3. ⊙ Solve the similar three-mass problem. The approach is the same, but the number of equations is larger.

7
Trial-and-Error Searching and Data Fitting

In this chapter, we add more tools to our computational toolbox. First, we devise ways to find solutions to equations by a trial-and-error search, sometimes using our new-found numerical differentiation tools. Although trial-and-error searching may not sound very precise, it is in fact widely used to solve problems where analytic solutions do not exist or are not practical. We have already looked at one such example in Chapter 6, where we saw how the two-weights-on-a-string problem led to matrix equations. In Chapter 8, we combine trial-and-error searching with the solution of ordinary differential equations to solve the general quantum eigenvalue problem. The second part of this chapter introduces some aspects of data fitting. We examine how to interpolate within a table of numbers and how to do a least-squares fit of a function to data, the latter often requiring a search.

7.1
Problem 1: A Search for Quantum States in a Box

Many computer techniques are well-defined sets of procedures leading to definite outcomes. In contrast, some computational techniques are trial-and-error algorithms in which decisions on what path to follow are made based on the current values of variables, and the program quits only when it thinks it has solved the problem. (We already did some of this when we summed a power series until the terms became small.) Writing this type of program is usually interesting because we must foresee how to have the computer act intelligently in all possible situations, and running them is very much like an experiment in which it is hard to predict what the computer will come up with.

Computational Physics, 3rd edition. Rubin H. Landau, Manuel J. Páez, Cristian C. Bordeianu.

Problem Probably the most standard problem in quantum mechanics[1] is to solve for the energies of a particle of mass m bound within a 1D square well of radius a:

$$V(x) = \begin{cases} -V_0, & \text{for } |x| \le a\,, \\ 0, & \text{for } |x| \ge a\,. \end{cases} \tag{7.1}$$

As shown in quantum mechanics texts (Gottfried, 1966), the energies of the bound states $E = -E_B < 0$ within this well are solutions of the transcendental equations

$$\sqrt{10 - E_B}\tan\left(\sqrt{10 - E_B}\right) = \sqrt{E_B} \quad \text{(even)}\,, \tag{7.2}$$

$$\sqrt{10 - E_B}\,\text{cotan}\left(\sqrt{10 - E_B}\right) = \sqrt{E_B} \quad \text{(odd)}\,, \tag{7.3}$$

where even and odd refer to the symmetry of the wave function. Here we have chosen units such that $\hbar = 1, 2m = 1, a = 1$, and $V_0 = 10$. Your *problem* is to

1. Find several bound-state energies E_B for even wave functions, that is, the solution of (7.2).
2. See if making the potential deeper, say, by changing the 10 to a 20 or a 30, produces a larger number of, or deeper, bound states.

7.2 Algorithm: Trial-and-Error Roots via Bisection

Trial-and-error root finding looks for a value of x for which

$$f(x) \simeq 0\,, \tag{7.4}$$

where the 0 on the right-hand side is conventional (an equation such as $10\sin x = 3x^3$ can easily be written as $10\sin x - 3x^3 = 0$). The search procedure starts with a guessed value for x, substitutes that guess into $f(x)$ (the "trial"), and then sees how far the LHS is from zero (the "error"). The program then changes x based on the error and tries out the new guess in $f(x)$. The procedure continues until $f(x) \simeq 0$ to some desired level of precision, or until the changes in x are insignificant, or if the search seems endless.

The most elementary trial-and-error technique is the *bisection algorithm*. It is reliable but slow. If you know some interval in which $f(x)$ changes sign, then the bisection algorithm will always converge to the root by finding progressively

1) We solve this same problem in Section 9.1 using an approach that is applicable to almost any potential and which also provides the wave functions. The approach of this section works only for the eigen energies of a square well.

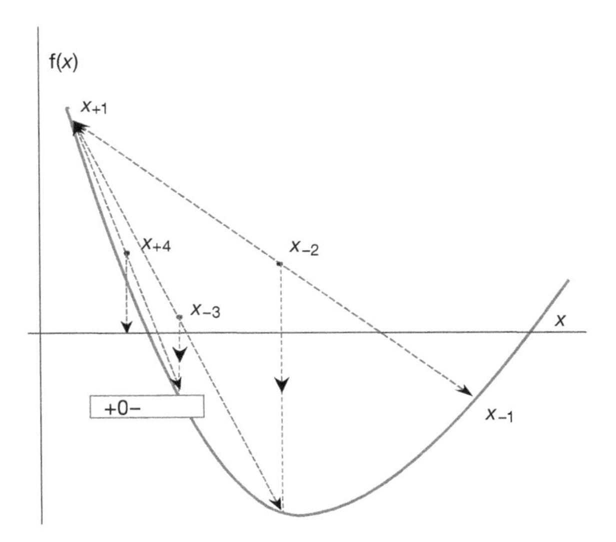

Figure 7.1 A graphical representation of the steps involved in solving for a zero of $f(x)$ using the bisection algorithm. The bisection algorithm takes the midpoint of the interval as the new guess for x, and so each step reduces the interval size by one-half. Four steps are shown for the algorithm.

smaller and smaller intervals within which the zero lies. Other techniques, such as the Newton–Raphson method we describe next, may converge more quickly, but if the initial guess is not close, it may become unstable and fail completely.

The basis of the bisection algorithm is shown in Figure 7.1. We start with two values of x between which we know that a zero occurs. (You can determine these by making a graph or by stepping through different x values and looking for a sign change.) To be specific, let us say that $f(x)$ is negative at x_- and positive at x_+:

$$f(x_-) < 0 , \quad f(x_+) > 0 . \tag{7.5}$$

(Note that it may well be that $x_- > x_+$ if the function changes from positive to negative as x increases.) Thus, we start with the interval $x_+ \le x \le x_-$ within which we know a zero occurs. The algorithm (implemented as in Listing 7.1) then picks the new x as the bisection of the interval and selects as its new interval the half in which the sign change occurs:

```
x = ( xPlus + xMinus ) / 2
if ( f(x) f(xPlus) > 0 ) xPlus = x
else xMinus = x
```

This process continues until the value of $f(x)$ is less than a predefined level of precision or until a predefined (large) number of subdivisions occurs.

Listing 7.1 The **Bisection.py** is a simple implementation of the bisection algorithm for finding a zero of a function, in this case $2\cos x - x$.

```
# Bisection.py                          Find zero via Bisection algorithm

from visual.graph import *

def f(x):                                              # Function = 0?
    return 2*cos(x) - x

def bisection(xminus, xplus, Nmax, eps):           # x+, x-, Nmax, error
    for it in range(0, Nmax):
        x = ( xplus  +  xminus )/2.                         # Mid point
        print(" it ", it, " x  ", x, " f(x) ", f(x))
        if ( f(xplus)*f(x) > 0. ):                   # Root in other half
            xplus = x                                  # Change x+ to x
        else:
            xminus =  x                                # Change x- to x
        if ( abs(f(x) ) < eps ):                            # Converged?
            print("\n Root found with precision eps = ", eps)
            break
        if it == Nmax-1:
            print ("\n Root NOT found after Nmax iterations\n")
    return x

eps = 1e-6                                         # Precision of zero
a = 0.0;            b = 7.0                        # Root in [a,b]
imax = 100                                         # Max no. iterations
root = bisection(a ,b, imax, eps)
print(" Root =", root)
```

The example in Figure 7.1 shows the first interval extending from $x_- = x_{+1}$ to $x_+ = x_{-1}$. We bisect that interval at x, and because $f(x) < 0$ at the midpoint, we set $x_- \equiv x_{-2} = x$ and label it x_{-2} to indicate the second step. We then use $x_{+2} \equiv x_{+1}$ and x_{-2} as the next interval and continue the process. We see that only x_- changes for the first three steps in this example, but for the fourth step x_+ finally changes. The changes then become too small for us to show.

7.2.1
Implementation: Bisection Algorithm

1. The first step in implementing any search algorithm is to get an idea of what your function looks like. For the present problem, you do this by making a plot of $f(E) = \sqrt{10 - E_B}\tan(\sqrt{10 - E_B}) - \sqrt{E_B}$ vs. E_B. Note from your plot some approximate values at which $f(E_B) = 0$. Your program should be able to find more exact values for these zeros.
2. Write a program that implements the bisection algorithm and use it to find some solutions of (7.2).
3. *Warning:* Because the tan function has singularities, you have to be careful. In fact, your graphics program (or Maple) may not function accurately near these singularities. One cure is to use a different but equivalent form of the equation. Show that an equivalent form of (7.2) is

$$\sqrt{E}\cot(\sqrt{10 - E}) - \sqrt{10 - E} = 0 . \tag{7.6}$$

4. Make a second plot of (7.6), which also has singularities but at different places. Choose some approximate locations for zeros from this plot.
5. Evaluate $f(E_B)$ and thus determine directly the precision of your solution.
6. Compare the roots you find with those given by Maple or Mathematica.

7.3 Improved Algorithm: Newton–Raphson Searching

The Newton–Raphson algorithm finds approximate roots of the equation

$$f(x) = 0 \tag{7.7}$$

more quickly than the bisection method. As we see graphically in Figure 7.2, this algorithm is the equivalent of drawing a straight line $f(x) \simeq mx + b$ tangent to the curve at an x value for which $f(x) \simeq 0$ and then using the intercept of the line with the x-axis at $x = -b/m$ as an improved guess for the root. If the "curve" was a straight line, the answer would be exact; otherwise, it is a good approximation if the guess is close enough to the root for $f(x)$ to be nearly linear. The process continues until some set level of precision is reached. If a guess is in a region where $f(x)$ is nearly linear (Figure 7.2), then the convergence is much more rapid than for the bisection algorithm.

The analytic formulation of the Newton–Raphson algorithm starts with an old guess x_0 and expresses a new guess x as a correction Δx to the old guess:

$$x_0 = \text{old guess}\,, \qquad \Delta x = \text{unknown correction} \tag{7.8}$$

$$\Rightarrow \qquad x = x_0 + \Delta x = \text{(unknown) new guess}\,. \tag{7.9}$$

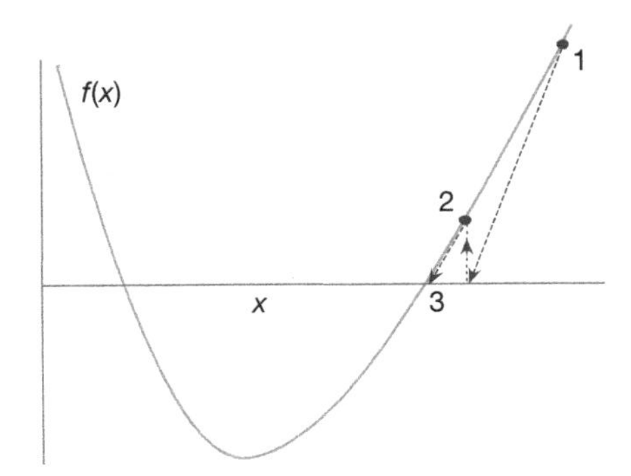

Figure 7.2 A graphical representation of the steps involved in solving for a zero of $f(x)$ using the Newton–Raphson method. The Newton–Raphson method takes the new guess as the zero of the line tangent to $f(x)$ at the old guess. Two guesses are shown.

We next expand the known function $f(x)$ in a Taylor series around x_0 and keep only the linear terms:

$$f(x = x_0 + \Delta x) \simeq f(x_0) + \left.\frac{\mathrm{d}f}{\mathrm{d}x}\right|_{x_0} \Delta x \,. \tag{7.10}$$

We then determine the correction Δx by calculating the point at which this linear approximation to $f(x)$ crosses the x-axis:

$$f(x_0) + \left.\frac{\mathrm{d}f}{\mathrm{d}x}\right|_{x_0} \Delta x = 0 \,, \tag{7.11}$$

$$\Rightarrow \quad \Delta x = -\frac{f(x_0)}{\mathrm{d}f/\mathrm{d}x|_{x_0}} \,. \tag{7.12}$$

The procedure is repeated starting at the improved x until some set level of precision is obtained.

The Newton–Raphson algorithm (7.12) requires evaluation of the derivative $\mathrm{d}f/\mathrm{d}x$ at each value of x_0. In many cases, you may have an analytic expression for the derivative and can build it into the algorithm. However, especially for more complicated problems, it is simpler and less error-prone to use a numerical forward-difference approximation to the derivative[2]:

$$\frac{\mathrm{d}f}{\mathrm{d}x} \simeq \frac{f(x + \delta x) - f(x)}{\delta x} \,, \tag{7.13}$$

where δx is some small change in x that you just chose (different from the Δ used for searching in (7.12)). While a central-difference approximation for the derivative would be more accurate, it would require additional evaluations of the f's, and once you find a zero, it does not matter how you got there. In Listing 7.2, we give a program NewtonCD.py that implement the search with the central difference derivative.

Listing 7.2 NewtonCD.py uses the Newton–Raphson method to search for a zero of the function $f(x)$. A central-difference approximation is used to determine $\mathrm{d}f/\mathrm{d}x$.

```
# NewtonCD.py      Newton Search with central difference

from math import cos

x = 4.;              dx = 3.e-1;            eps = 0.2;                 # Parameters
imax = 100;                                                  # Max no of iterations

def f(x):                                                              # Function
    return 2*cos(x) - x

for it in range(0, imax + 1):
    F = f(x)
    if ( abs(F) <= eps ):                                   # Check for convergence
        print("\n Root found, F =", F, ", tolerance eps = " , eps)
```

2) We discuss numerical differentiation in Chapter 5.

```
        break
    print("Iteration #= ", it , " x = ", x, " f(x) = ", F)
    df = ( f(x + dx/2)  -  f(x - dx/2) )/dx                    # Central diff
    dx = - F/df
    x   += dx                                                   # New guess
```

7.3.1
Newton–Raphson with Backtracking

Two examples of possible problems with the Newton–Raphson algorithm are shown in Figure 7.3. In Figure 7.3a, we see a case where the search takes us to an x value where the function has a local minimum or maximum, that is, where $\mathrm{d}f/\mathrm{d}x = 0$. Because $\Delta x = -f/f'$, this leads to a horizontal tangent (division by zero), and so the next guess is $x = \infty$, from where it is hard to return. When this happens, you need to start your search with a different guess and pray that you do not fall into this trap again. In cases where the correction is very large but maybe not infinite, you may want to try backtracking (described below) and hope that by taking a smaller step you will not get into as much trouble.

In Figure 7.3b, we see a case where a search falls into an infinite loop surrounding the zero without ever getting there. A solution to this problem is called *backtracking*. As the name implies, in cases where the new guess $x_0 + \Delta x$ leads to an increase in the magnitude of the function, $|f(x_0 + \Delta x)|^2 > |f(x_0)|^2$, you can backtrack somewhat and try a smaller guess, say, $x_0 + \Delta x/2$. If the magnitude of f still increases, then you just need to backtrack some more, say, by trying $x_0 + \Delta x/4$ as your next guess, and so forth. Because you know that the tangent line leads to a local decrease in $|f|$, eventually an acceptable small enough step should be found.

The problem in both these cases is that the initial guesses were not close enough to the regions where $f(x)$ is approximately linear. So again, a good plot may help produce a good first guess. Alternatively, you may want to start your search with the bisection algorithm and then switch to the faster Newton–Raphson algorithm when you get closer to the zero.

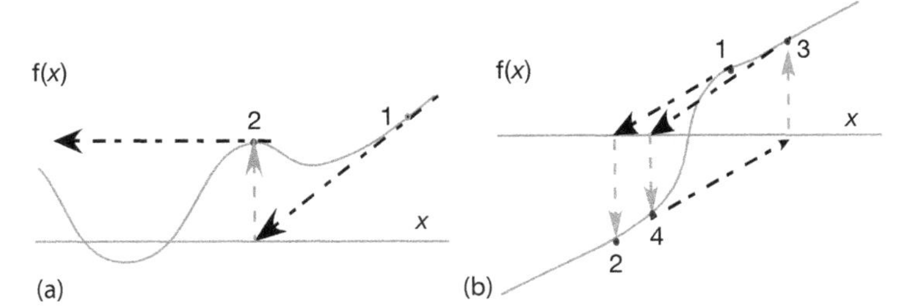

Figure 7.3 Two examples of how the Newton–Raphson algorithm may fail if the initial guess is not in the region where $f(x)$ can be approximated by a straight line. (a) A guess lands at a local minimum/maximum, that is, a place where the derivative vanishes, and so the next guess ends up at $x = \infty$. (b) The search has fallen into an infinite loop. The technique know as "backtracking" could eliminate this problem.

7.3.2
Implementation: Newton–Raphson Algorithm

1. Use the Newton–Raphson algorithm to find some energies E_B that are solutions of (7.2). Compare these solutions with the ones found with the bisection algorithm.
2. Again, notice that the 10 in (7.2) is proportional to the strength of the potential that causes the binding. See if making the potential deeper, say, by changing the 10 to a 20 or a 30, produces more or deeper bound states. (Note that in contrast to the bisection algorithm, your initial guess must be closer to the answer for the Newton–Raphson algorithm to work.)
3. Modify your algorithm to include backtracking and then try it out on some difficult cases.
4. Evaluate $f(E_B)$ and thus determine directly the precision of your solution.

7.4
Problem 2: Temperature Dependence of Magnetization

Problem Determine $M(T)$ the magnetization as a function of temperature for simple magnetic materials.

A collection of N spin-1/2 particles each with the magnetic moment μ is at temperature T. The collection has an external magnetic field B applied to it and comes to equilibrium with N_L particles in the lower energy state (spins aligned with the magnetic field), and with N_U particles in the upper energy state (spins opposed to the magnetic field). The Boltzmann distribution law tells us that the relative probability of a state with energy E is proportional to $\exp(-E/(k_B T))$, where k_B is Boltzmann's constant. For a dipole with moment μ, its energy in a magnetic field is given by the dot product $E = -\boldsymbol{\mu} \cdot \boldsymbol{B}$. Accordingly, spin-up particle have lower energy in a magnetic field than spin-down particles, and thus are more probable.

Applying the Boltzmann distribution to our spin problem, we have that the number of particles in the lower energy level (spin up) is

$$N_L = N \frac{e^{\mu B/(k_B T)}}{e^{\mu B/(k_B T)} + e^{-\mu B/(k_B T)}} , \tag{7.14}$$

while the number of particles in the upper energy level (spin down) is

$$N_U = N \frac{e^{-\mu B/(k_B T)}}{e^{\mu B/(k_B T)} + e^{-\mu B/(k_B T)}} . \tag{7.15}$$

As discussed by (Kittel, 2005), we now assume that the molecular magnetic field $B = \lambda M$ is much larger than the applied magnetic field and so replace B by the molecular field. This permits us to eliminate B from the preceding equations. The *magnetization* $M(T)$ is given by the individual magnetic moment μ times the net number of particles pointing in the direction of the magnetic field:

$$M(T) = \mu \times (N_{\rm L} - N_{\rm U}) \tag{7.16}$$

$$= N\mu \tanh\left(\frac{\lambda\mu M(T)}{k_{\rm B}T}\right) . \tag{7.17}$$

Note that this expression appears to make sense because as the temperature approaches zero, all spins will be aligned along the direction of B and so $M(T = 0) = N\mu$.

Solution via Searching Equation 7.17 relates the magnetization and the temperature. However, it is not really a solution to our problem because M appears on the LHS of the equation as well as within the hyperbolic function on the RHS. Generally, a *transcendental equation* of this sort does not have an analytic solution that would give M as simply a function of the temperature T. But by working backward, we can find a numerical solution. To do that we first express (7.17) in terms of the reduced magnetization m, the reduced temperature t, and the Curie temperature $T_{\rm c}$:

$$m(t) = \tanh\left(\frac{m(t)}{t}\right) , \tag{7.18}$$

$$m(T) = \frac{M(T)}{N\mu} , \quad t = \frac{T}{T_{\rm c}} , \quad T_{\rm c} = \frac{N\mu^2\lambda}{k_{\rm B}} . \tag{7.19}$$

While it is no easier to find an analytic solution to (7.18) than it was to (7.17), the simpler form of (7.18) makes the programming easier as we search for values of t and m that satisfy (7.18).

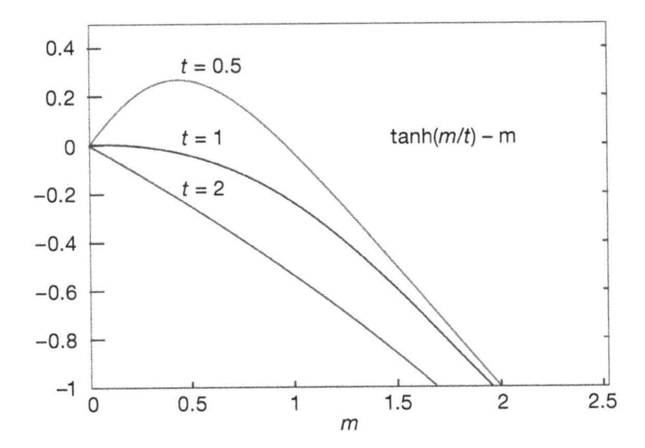

Figure 7.4 A function of the reduced magnetism m at three reduced temperatures t. A zero of this function determines the value of the magnetism at a particular value of t.

One approach to a trial-and-error solution is to define a function

$$f(m,t) = m - \tanh\left(\frac{m(t)}{t}\right), \tag{7.20}$$

and then, for a variety of fixed $t = t_i$ values, search for those m values at which $f(m, t_i) = 0$. (One could just as well fix the value of m to m_j and search for the value of t for which $f(m_j, t) = 0$; once you have a solution, you have a solution.) Each zero so found gives us a single value of $m(t_i)$. A plot or a table of these values for a range of t_i values then provides the best we can do as the desired solution $m(t)$.

Figure 7.4 shows three plots of $f(m, t)$ as a function of the reduced magnetization m, each plot for a different value of the reduced temperature. As you can see, other than the uninteresting solution at $m = 0$, there is only one solution (a zero) near $m = 1$ for $t = 0.5$, and no solution at other temperatures.

7.4.1 Searching Exercise

1. Find the root of (7.20) to six significant figures for $t = 0.5$ using the bisection algorithm.
2. Find the root of (7.20) to six significant figures for $t = 0.5$ using the Newton–Raphson algorithm.
3. Compare the time it takes to find the solutions for the bisection and Newton–Raphson algorithms.
4. Construct a plot of the reduced magnetization $m(t)$ as a function of the reduced temperature t.

7.5 Problem 3: Fitting An Experimental Spectrum

Data fitting is an art worthy of serious study by all scientists (Bevington and Robinson, 2002). In the sections to follow, we just scratch the surface by examining how to interpolate within a table of numbers and how to do a least-squares fit to data. We also show how to go about making a least-squares fit to nonlinear functions using some of the search techniques and subroutine libraries we have already discussed.

Problem The cross sections measured for the resonant scattering of neutrons from a nucleus are given in Table 7.1. Your *problem* is to determine values for the cross sections at energy values lying between those in the table.

You can solve this *problem* in a number of ways. The simplest is to numerically *interpolate* between the values of the experimental $f(E_i)$ given in Table 7.1. This is direct and easy but does not account for there being experimental noise in the data. A more appropriate solution (discussed in Section 7.7) is to find the *best*

Table 7.1 Experimental values for a scattering cross section ($f(E)$ in the theory), each with absolute error $\pm\sigma_i$, as a function of energy (x_i in the theory).

$i =$	1	2	3	4	5	6	7	8	9
E_i (MeV)	0	25	50	75	100	125	150	175	200
$g(E_i)$ (mb)	10.6	16.0	45.0	83.5	52.8	19.9	10.8	8.25	4.7
Error (mb)	9.34	17.9	41.5	85.5	51.5	21.5	10.8	6.29	4.14

fit of a theoretical function to the data. We start with what we believe to be the "correct" theoretical description of the data,

$$f(E) = \frac{f_r}{(E - E_\mathrm{r})^2 + \Gamma^2/4} , \qquad (7.21)$$

where f_r, E_r, and Γ are unknown parameters. We then adjust the parameters to obtain the best fit. This is a best fit in a statistical sense but in fact may not pass through all (or any) of the data points. For an easy, yet effective, introduction to statistical data analysis, we recommend (Bevington and Robinson, 2002).

These two techniques of interpolation and least-squares fitting are powerful tools that let you treat tables of numbers as if they were analytic functions and sometimes let you deduce statistically meaningful constants or conclusions from measurements. In general, you can view data fitting as *global* or *local*. In global fits, a single function in x is used to represent the entire set of numbers in a table like Table 7.1. While it may be spiritually satisfying to find a single function that passes through all the data points, if that function is not the correct function for describing the data, the fit may show nonphysical behavior (such as large oscillations) between the data points. The rule of thumb is that if you must interpolate, keep it local and view global interpolations with a critical eye.

Consider Table 7.1 as ordered data that we wish to interpolate. We call the independent variable x and its tabulated values $x_i (i = 1, 2, \ldots)$, and assume that the dependent variable is the function $g(x)$, with the tabulated values $g_i = g(x_i)$. We assume that $g(x)$ can be approximated as an $(n-1)$-degree polynomial in each interval i:

$$g_i(x) \simeq a_0 + a_1 x + a_2 x^2 + \cdots + a_{n-1} x^{n-1} , \quad (x \simeq x_i) . \qquad (7.22)$$

Because our fit is local, we do not assume that one $g(x)$ can fit all the data in the table but instead use a different polynomial, that is, a different set of a_i values, for each interval. While each polynomial is of low degree, multiple polynomials are needed to span the entire table. If some care is taken, the set of polynomials so obtained will behave well enough to be used in further calculations without introducing much unwanted noise or discontinuities.

The classic interpolation formula was created by Lagrange. He figured out a closed-form expression that directly fits the $(n-1)$-order polynomial (7.22) to n values of the function $g(x)$ evaluated at the points x_i. The formula for each interval is written as the sum of polynomials:

$$g(x) \simeq g_1\lambda_1(x) + g_2\lambda_2(x) + \cdots + g_n\lambda_n(x) , \tag{7.23}$$

$$\lambda_i(x) = \prod_{j(\neq i)=1}^{n} \frac{x - x_j}{x_i - x_j} = \frac{x - x_1}{x_i - x_1}\frac{x - x_2}{x_i - x_2} \cdots \frac{x - x_n}{x_i - x_n} . \tag{7.24}$$

For three points, (7.23) provides a second-degree polynomial, while for eight points it gives a seventh-degree polynomial. For example, assume that we are given the points and function values

$$x_{1-4} = (0, 1, 2, 4) \quad g_{1-4} = (-12, -12, -24, -60) . \tag{7.25}$$

With four points, the Lagrange formula determines a third-order polynomial that reproduces each of the tabulated values:

$$\begin{aligned} g(x) &= \frac{(x-1)(x-2)(x-4)}{(0-1)(0-2)(0-4)}(-12) + \frac{x(x-2)(x-4)}{(1-0)(1-2)(1-4)}(-12) \\ &\quad + \frac{x(x-1)(x-4)}{(2-0)(2-1)(2-4)}(-24) + \frac{x(x-1)(x-2)}{(4-0)(4-1)(4-2)}(-60), \\ \Rightarrow \quad g(x) &= x^3 - 9x^2 + 8x - 12 . \end{aligned} \tag{7.26}$$

As a check, we see that

$$g(4) = 4^3 - 9(4^2) + 32 - 12 = -60 , \quad g(0.5) = -10.125 . \tag{7.27}$$

If the data contain little noise, this polynomial can be used with some confidence within the range of the data, but with risk beyond the range of the data.

Notice that Lagrange interpolation has no restriction that the points x_i be evenly spaced. Usually, the Lagrange fit is made to only a small region of the table with a small value of n, despite the fact that the formula works perfectly well for fitting a high-degree polynomial to the entire table. The difference between the value of the polynomial evaluated at some x and that of the actual function can be shown to be the *remainder*

$$R_n \simeq \frac{(x - x_1)(x - x_2) \cdots (x - x_n)}{n!} g^{(n)}(\zeta) , \tag{7.28}$$

where ζ lies somewhere in the interpolation interval. What significant here is that we see that if significant high derivatives exist in $g(x)$, then it cannot be approximated well by a polynomial. For example, a table of noisy data would have significant high derivatives.

7.5.1 Lagrange Implementation, Assessment

Consider the experimental neutron scattering data in Table 7.1. The expected theoretical functional form that describes these data is (7.21), and our empirical fits to these data are shown in Figure 7.5.

1. Write a subroutine to perform an n-point Lagrange interpolation using (7.23). Treat n as an arbitrary input parameter. (You may also perform this exercise with the spline fits discussed in Section 7.5.2.)
2. Use the Lagrange interpolation formula to fit the entire experimental spectrum with one polynomial. (This means that you must fit all nine data points with an eight-degree polynomial.) Then use this fit to plot the cross section in steps of 5 MeV.
3. Use your graph to deduce the resonance energy E_r (your peak position) and Γ (the full-width at half-maximum). Compare your results with those predicted by a theorist friend, $(E_r, \Gamma) = (78, 55)$ MeV.
4. A more realistic use of Lagrange interpolation is for local interpolation with a small number of points, such as three. Interpolate the preceding cross-sectional data in 5-MeV steps using three-point Lagrange interpolation for each interval. (Note that the end intervals may be special cases.)
5. We deliberately have not discussed *extrapolation* of data because it can lead to serious *systematic* errors; the answer you get may well depend more on the function you assume than on the data you input. Add some adventure to your life and use the programs you have written to extrapolate to values outside Table 7.1. Compare your results to the theoretical Breit–Wigner shape (7.21).

This example shows how easy it is to go wrong with a high-degree-polynomial fit to data with errors. Although the polynomial is guaranteed to pass through all the data points, the representation of the function away from these points can be quite unrealistic. Using a low-order interpolation formula, say, $n = 2$ or 3, in each interval usually eliminates the wild oscillations, but may not have any theoretical justification. If these local fits are matched together, as we discuss in the next section, a rather continuous curve results. Nonetheless, you must recall that if the data contain errors, a curve that actually passes through them may lead you astray. We discuss how to do this properly with least-squares fitting in Section 7.7.

7.5.2
Cubic Spline Interpolation (Method)

If you tried to interpolate the resonant cross section with Lagrange interpolation, then you saw that fitting parabolas (three-point interpolation) within a table may avoid the erroneous and possibly catastrophic deviations of a high-order formula. (A two-point interpolation, which connects the points with straight lines, may not lead you far astray, but it is rarely pleasing to the eye or precise.) A sophisticated variation of an $n = 4$ interpolation, known as *cubic splines*, often leads to surprisingly eye-pleasing fits. In this approach (Figure 7.5), cubic polynomials are fit to the function in each interval, with the additional constraint that the first and second derivatives of the polynomials be continuous from one interval to the next. This continuity of slope and curvature is what makes the spline fit particularly eye-pleasing. It is analogous to what happens when you use the flexible spline

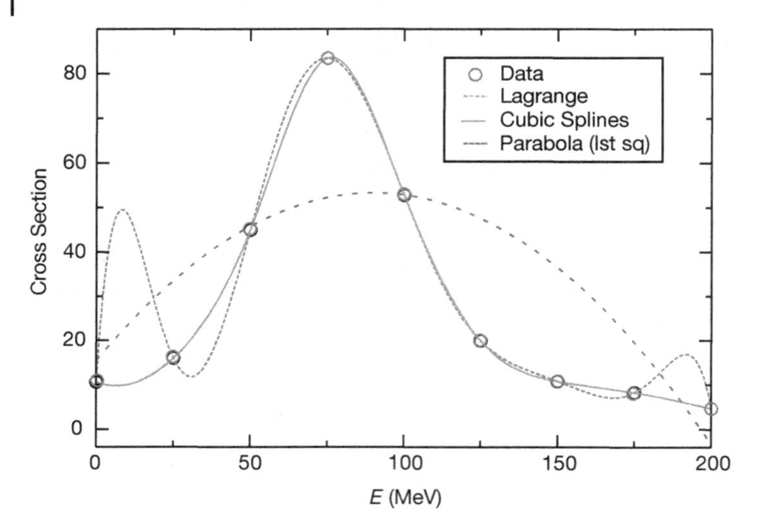

Figure 7.5 Three fits to data. *Dashed:* Lagrange interpolation using an eight-degree polynomial; *Short dashes:* cubic splines fit ; *Long dashed:* Least-squares parabola fit.

drafting tool (a lead wire within a rubber sheath) from which the method draws its name.

The series of cubic polynomials obtained by spline-fitting a table of data can be integrated and differentiated and is guaranteed to have well-behaved derivatives. The existence of meaningful derivatives is an important consideration. As a case in point, if the interpolated function is a potential, you can take the derivative to obtain the force. The complexity of simultaneously matching polynomials and their derivatives over all the interpolation points leads to many simultaneous linear equations to be solved. This makes splines unattractive for hand calculation, yet easy for computers and, not surprisingly, popular in both calculations and computer drawing programs. To illustrate, the smooth solid curve in Figure 7.5 is a spline fit.

The basic approximation of splines is the representation of the function $g(x)$ in the subinterval $[x_i, x_{i+1}]$ with a cubic polynomial:

$$g(x) \simeq g_i(x) , \quad \text{for} \quad x_i \le x \le x_{i+1} , \tag{7.29}$$

$$g_i(x) = g_i + g'_i(x - x_i) + \frac{1}{2}g''_i(x - x_i)^2 + \frac{1}{6}g'''_i(x - x_i)^3 . \tag{7.30}$$

This representation makes it clear that the coefficients in the polynomial equal the values of $g(x)$ and its first, second, and third derivatives at the tabulated points x_i. Derivatives beyond the third vanish for a cubic. The computational chore is to determine these derivatives in terms of the N tabulated g_i values. The matching of g_i at the *nodes* that connect one interval to the next provides the equations

$$g_i(x_{i+1}) = g_{i+1}(x_{i+1}) , \quad i = 1, N - 1 . \tag{7.31}$$

The matching of the first *and* second derivatives at each interval's boundaries provides the equations

$$g'_{i-1}(x_i) = g'_i(x_i) , \quad g''_{i-1}(x_i) = g''_i(x_i) . \tag{7.32}$$

The additional equations needed to determine all constants are obtained by matching the third derivatives at adjacent nodes. Values for the third derivatives are found by approximating them in terms of the second derivatives:

$$g'''_i \simeq \frac{g''_{i+1} - g''_i}{x_{i+1} - x_i} . \tag{7.33}$$

As discussed in Chapter 5, a *central-difference approximation* would be more accurate than a forward-difference approximation, yet (7.33) keeps the equations simpler.

It is straightforward yet complicated to solve for all the parameters in (7.30). We leave that to the references (Thompson, 1992; Press *et al.*, 1994). We can see, however, that matching at the boundaries of the intervals results in only $(N-2)$ linear equations for N unknowns. Further input is required. It usually is taken to be the boundary conditions at the endpoints $a = x_1$ and $b = x_N$, specifically, the second derivatives $g''(a)$ and $g''(b)$. There are several ways to determine these second derivatives:

Natural spline: Set $g''(a) = g''(b) = 0$; that is, permit the function to have a slope at the endpoints but no curvature. This is "natural" because the derivative vanishes for the flexible spline drafting tool (its ends being free).

Input values for g' at the boundaries: The computer uses $g'(a)$ to approximate $g''(a)$. If you do not know the first derivatives, you can calculate them numerically from the table of g_i values.

Input values for g'' at the boundaries: Knowing values is of course better than approximating values, but it requires the user to input information. If the values of g'' are not known, they can be approximated by applying a forward-difference approximation to the tabulated values:

$$g''(x) \simeq \frac{[g(x_3) - g(x_2)]/[x_3 - x_2] - [g(x_2) - g(x_1)]/[x_2 - x_1]}{[x_3 - x_1]/2} . \tag{7.34}$$

7.5.2.1 Cubic Spline Quadrature (Exploration)

A powerful integration scheme is to fit an integrand with splines and then integrate the cubic polynomials analytically. If the integrand $g(x)$ is known only at its tabulated values, then this is about as good an integration scheme as is possible; if you have the ability to calculate the function directly for arbitrary x, Gaussian quadrature may be preferable. We know that the spline fit to g in each interval is the cubic (7.30)

$$g(x) \simeq g_i + g'_i(x - x_i) + \frac{1}{2} g''_i(x - x_i)^2 + \frac{1}{6} g'''_i(x - x_i)^3 . \tag{7.35}$$

It is easy to integrate this to obtain the integral of g for this interval and then to sum over all intervals:

$$\int_{x_i}^{x_{i+1}} g(x)\,\mathrm{d}x \simeq \left(g_i x + \frac{1}{2} g'_i x^2 + \frac{1}{6} g''_i x^3 + \frac{1}{24} g'''_i x^4 \right)\Big|_{x_i}^{x_{i+1}} , \tag{7.36}$$

$$\int_{x_j}^{x_k} g(x)\,\mathrm{d}x = \sum_{i=j}^{k} \left(g_i x + \frac{1}{2} g'_i x_i^2 + \frac{1}{6} g''_i x^3 + \frac{1}{24} g'''_i x^4 \right)\Big|_{x_i}^{x_{i+1}} . \tag{7.37}$$

Making the intervals smaller does not necessarily increase precision, as subtractive cancelations in (7.36) may get large.

Spline Fit of Cross Section (Implementation) Fitting a series of cubics to data is a little complicated to program yourself, so we recommend using a library routine. While we have found quite a few Java-based spline applications available on the Internet, none seemed appropriate for interpreting a simple set of numbers. That being the case, we have adapted the `splint.c` and the `spline.c` functions from (Press *et al.*, 1994) to produce the `SplineInteract.py` program shown in Listing 7.3 (there is also an applet). Your *problem* for this section is to carry out the assessment in Section 7.5.1 using cubic spline interpolation rather than Lagrange interpolation.

7.6 Problem 4: Fitting Exponential Decay

Problem Figure 7.6 presents actual experimental data on the number of decays ΔN of the π meson as a function of time (Stetz *et al.*, 1973). Notice that the time has been "binned" into $\Delta t = 10$ ns intervals and that the smooth curve is the theoretical exponential decay expected for very large numbers of pions (which there is not). Your *problem* is to deduce the lifetime τ of the π meson from these data (the tabulated lifetime of the pion is 2.6×10^{-8} s).

Theory Assume that we start with N_0 particles at time $t = 0$ that can decay to other particles.[3] If we wait a short time Δt, then a small number ΔN of the particles will decay *spontaneously*, that is, with no external influences. This decay is a stochastic process, which means that there is an element of chance involved in just when a decay will occur, and so no two experiments are expected to give exactly the same results. The basic law of nature for spontaneous decay is that the number of decays ΔN in a time interval Δt is proportional to the number of particles $N(t)$ present at that time and to the time interval

$$\Delta N(t) = -\frac{1}{\tau} N(t)\Delta t \Rightarrow \frac{\Delta N(t)}{\Delta t} = -\lambda N(t) . \tag{7.38}$$

3) Spontaneous decay is discussed further and simulated in Section 4.5.

Listing 7.3 SplineInteract.py performs a cubic spline fit to data and permits interactive control. The arrays *x*[] and *y*[] are the data to fit, and the values of the fit at Nfit points are output.

```
# SplineInteract.py Spline fit with slide to control number of points

from visual import *;                    from visual.graph import *;
from visual.graph import gdisplay, gcurve
from visual.controls import slider, controls, toggle

x = array([0., 0.12, 0.25, 0.37, 0.5, 0.62, 0.75, 0.87, 0.99])      # input
y = array([10.6, 16.0, 45.0, 83.5, 52.8, 19.9, 10.8, 8.25, 4.7])
n = 9;   np = 15

# Initialize
y2 = zeros( (n), float); u = zeros( (n), float)
graph1 = gdisplay(x=0,y=0,width=500, height=500,
                 title='Spline Fit', xtitle='x', ytitle='y')
funct1 = gdots(color = color.yellow)
funct2 = gdots(color = color.red)
graph1.visible = 0

def update():                                         # Nfit = 30 = output
    Nfit = int(control.value)
    for i in range(0, n):                             # Spread out points
        funct1.plot(pos = (x[i], y[i]) )
        funct1.plot(pos = (1.01*x[i], 1.01*y[i]) )
        funct1.plot(pos = (.99*x[i], .99*y[i]) )
        yp1 = (y[1]-y[0]) / (x[1]-x[0]) - (y[2]-y[1])/ \
              (x[2]-x[1])+(y[2]-y[0])/(x[2]-x[0])
    ypn = (y[n-1] - y[n-2])/(x[n-1] - x[n-2]) -
        (y[n-2]-y[n-3])/(x[n-2]-x[n-3]) + (y[n-1]-y[n-3])/(x[n-1]-x[n-3])
    if (yp1 > 0.99e30):  y2[0] = 0.;  u[0] = 0.
    else:
        y2[0] = - 0.5
        u[0] = (3./(x[1] - x[0]) )*( (y[1] - y[0])/(x[1] - x[0]) - yp1)
    for i in range(1, n - 1):                              # Decomp loop
        sig = (x[i] - x[i - 1])/(x[i + 1] - x[i - 1])
        p = sig*y2[i - 1] + 2.
        y2[i] = (sig - 1.)/p
        u[i] = (y[i+1]-y[i])/(x[i+1]-x[i]) - (y[i]-y[i-1])/(x[i]-x[i-1])
        u[i] = (6.*u[i]/(x[i + 1] - x[i - 1]) - sig*u[i - 1])/p
    if (ypn > 0.99e30):  qn = un = 0.                      # Test for natural
    else:
        qn = 0.5;
        un = (3/(x[n-1]-x[n-2]))*(ypn - (y[n-1]-y[n-2])/(x[n-1]-x[n-2]))
    y2[n - 1] = (un - qn*u[n - 2])/(qn*y2[n - 2] + 1.)
    for k in range(n - 2, 1, - 1):
        y2[k] = y2[k]*y2[k + 1] + u[k]
    for i in range(1, Nfit + 2):                                # Begin fit
        xout = x[0] + (x[n - 1] - x[0])*(i - 1)/(Nfit)
        klo = 0;     khi = n - 1                          # Bisection algor
        while (khi - klo >1):
            k = (khi + klo) >> 1
            if (x[k] > xout): khi = k
            else: klo = k
        h = x[khi] - x[klo]
        if (x[k] > xout):   khi = k
        else: klo = k
        h = x[khi] - x[klo]
        a = (x[khi] - xout)/h
        b = (xout - x[klo])/h
        yout = a*y[klo] + b*y[khi] +
            ((a*a*a-a)*y2[klo]+(b*b*b-b)*y2[khi])*h*h/6
        funct2.plot(pos = (xout, yout) )
c = controls(x=500,y=0,width=200,height=200)             # Control via slider
```

```
control = slider(pos=(-50,50,0), min = 2, max = 100, action = update)
toggle(pos = (0, 35,  - 5), text1 = "Number of points", height = 0)
control.value = 2
update()

while 1:
    c.interact()
    rate(50)                                          # update < 10/sec
    funct2.visible = 0
```

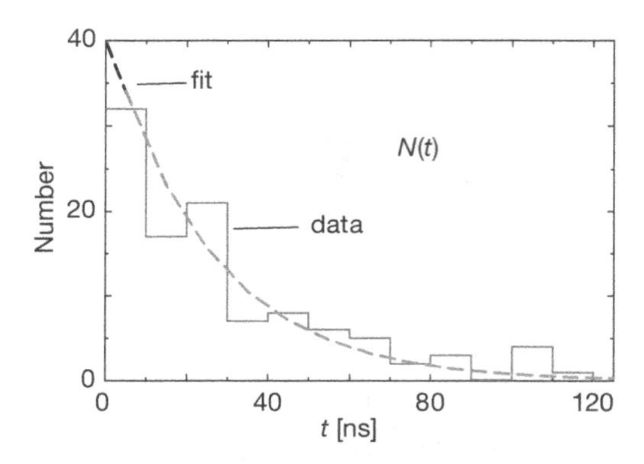

Figure 7.6 A reproduction of the experimental measurement by Stetz *et al.* (1973) giving the number of decays of a π meson as a function of time since its creation. Measurements were made during time intervals (box sizes) of 10-ns width. The dashed curve is the result of a linear least-squares fit to the log $N(t)$.

Here $\tau = 1/\lambda$ is the *lifetime* of the particle, with λ the rate parameter. The actual decay *rate* is given by the second equation in (7.38). If the number of decays ΔN is very small compared to the number of particles N, and if we look at vanishingly small time intervals, then the difference equation (7.38) becomes the differential equation

$$\frac{dN(t)}{dt} \simeq -\lambda N(t) = \frac{1}{\tau} N(t) . \tag{7.39}$$

This differential equation has an exponential solution for the number as well as for the decay rate:

$$N(t) = N_0 e^{-t/\tau} , \quad \frac{dN(t)}{dt} = -\frac{N_0}{\tau} e^{-t/\tau} = \frac{dN(0)}{dt} e^{-t/\tau} . \tag{7.40}$$

Equation 7.40 is the theoretical formula we wish to “fit” to the data in Figure 7.6. The output of such a fit is a “best value” for the lifetime τ.

7.7
Least-Squares Fitting (Theory)

Books have been written and careers have been spent discussing what is meant by a “good fit” to experimental data. We cannot do justice to the subject here and refer the reader to Bevington and Robinson (2002); Press *et al.* (1994); Thompson (1992). However, we will emphasize three points:

1. If the data being fit contain errors, then the "best fit" in a statistical sense should not pass through all the data points.
2. If the theory is not an appropriate one for the data (e.g., the parabola in Figure 7.5), then its best fit to the data may not be a good fit at all. This is good, for this is how we know that the theory is not right.
3. Only for the simplest case of a linear least-squares fit, can we write down a closed-form solution to evaluate and obtain the fit. More realistic problems are usually solved by *trial-and-error* search procedures, sometimes using sophisticated subroutine libraries. However, in Section 7.8.2 we show how to conduct such a nonlinear search using familiar tools.

Imagine that you have measured N_D data values of the independent variable y as a function of the dependent variable x:

$$(x_i, y_i \pm \sigma_i) , \quad i = 1, N_D , \tag{7.41}$$

where $\pm\sigma_i$ is the experimental uncertainty in the ith value of y. (For simplicity we assume that all the errors σ_i occur in the dependent variable, although this is hardly ever true (Thompson, 1992)). For our problem, y is the number of decays as a function of time, and x_i are the times. Our goal is to determine how well a mathematical function $y = g(x)$ (also called a *theory* or a *model*) can describe these data. Additionally, if the theory contains some parameters or constants, our goal can be viewed as determining the best values for these parameters. We assume that the theory function $g(x)$ contains, in addition to the functional dependence on x, an additional dependence upon M_P parameters $\{a_1, a_2, \dots, a_{M_P}\}$. Notice that the parameters $\{a_m\}$ are not variables, in the sense of numbers read from a meter, but rather are parts of the theoretical model, such as the size of a box, the mass of a particle, or the depth of a potential well. For the exponential decay function (7.40), the parameters are the lifetime τ and the initial decay rate $dN(0)/dt$. We include the parameters as

$$g(x) = g(x; \{a_1, a_2, \dots, a_{M_P}\}) = g(x; \{a_m\}) , \tag{7.42}$$

where the a_i's are parameters and x the independent variable. We use the chi-square (χ^2) measure (Bevington and Robinson, 2002) as a gauge of how well a theoretical function g reproduces data:

$$\chi^2 \stackrel{\text{def}}{=} \sum_{i=1}^{N_D} \left(\frac{y_i - g(x_i; \{a_m\})}{\sigma_i} \right)^2 , \tag{7.43}$$

where the sum is over the N_D experimental points $(x_i, y_i \pm \sigma_i)$. The definition (7.43) is such that smaller values of χ^2 are better fits, with $\chi^2 = 0$ occurring if the theoretical curve went through the center of every data point. Notice also that the $1/\sigma_i^2$ weighting means that measurements with larger errors[4] contribute less to χ^2.

Least-squares fitting refers to adjusting the parameters in the theory until a minimum in χ^2 is found, that is, finding a curve that produces the least value for the summed squares of the deviations of the data from the function $g(x)$. In general, this is the best fit possible and the best way to determine the parameters in a theory. The M_P parameters $\{a_m, m = 1, M_P\}$ that make χ^2 an extremum are found by solving the M_P equations:

$$\frac{\partial \chi^2}{\partial a_m} = 0 \,, \quad \Rightarrow \quad \sum_{i=1}^{N_D} \frac{[y_i - g(x_i)]}{\sigma_i^2} \frac{\partial g(x_i)}{\partial a_m} = 0 \,, \quad (m = 1, M_P) \,. \tag{7.44}$$

Often, the function $g(x; \{a_m\})$ has a sufficiently complicated dependence on the a_m values for (7.44) to produce M_P simultaneous nonlinear equations in the a_m values. In these cases, solutions are found by a trial-and-error search through the M_P-dimensional parameter space, as we do in Section 7.8.2. To be safe, when such a search is completed, you need to check that the minimum χ^2 you found is *global* and not *local*. One way to do that is to repeat the search for a whole grid of starting values, and if different minima are found, to pick the one with the lowest χ^2.

7.7.1 Least-Squares Fitting: Theory and Implementation

When the deviations from theory are as a result of random errors and when these errors are described by a Gaussian distribution, there are some useful rules of thumb to remember (Bevington and Robinson, 2002). You know that your fit is good if the value of χ^2 calculated via the definition (7.43) is approximately equal to the number of degrees of freedom $\chi^2 \simeq N_D - M_P$, where N_D is the number of data points and M_P is the number of parameters in the theoretical function. If your χ^2 is much less than $N_D - M_P$, it does not mean that you have a "great" theory or a really precise measurement; instead, you probably have too many parameters or have assigned errors (σ_i values) that are too large. In fact, too small a χ^2 may indicate that you are fitting the random scatter in the data rather than missing approximately one-third of the error bars, as expected if the errors are random. If your χ^2 is significantly greater than $N_D - M_P$, the theory may not be good, you may have significantly underestimated your errors, or you may have errors that are not random.

The M_P simultaneous equations (7.44) can be simplified considerably if the functions $g(x; \{a_m\})$ depend *linearly* on the parameter values a_i, for example,

$$g\left(x; \{a_1, a_2\}\right) = a_1 + a_2 x \,. \tag{7.45}$$

4) If you are not given the errors, you can guess them on the basis of the apparent deviation of the data from a smooth curve, or you can weigh all points equally by setting $\sigma_i \equiv 1$ and continue with the fitting.

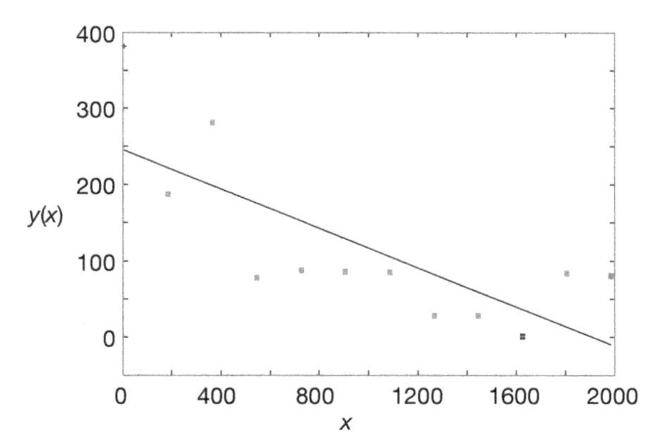

Figure 7.7 A linear least-squares best fit of a straight line to data. The deviation of theory from experiment is greater than would be expected from statistics, which means that a straight line is not a good theory to describe these data.

In this case (also known as *linear regression or straight-line fit*), as shown in Figure 7.7, there are $M_{\text{P}} = 2$ parameters, the slope a_2, and the y intercept a_1. Notice that while there are only two parameters to determine, there still may be an arbitrary number N_{D} of data points to fit. Remember, a unique solution is not possible unless the number of data points is equal to or greater than the number of parameters. For this linear case, there are just two derivatives,

$$\frac{\partial g(x_i)}{\partial a_1} = 1 , \quad \frac{\partial g(x_i)}{\partial a_2} = x_i , \tag{7.46}$$

and after substitution, the χ^2 minimization equations (7.44) can be solved (Press *et al.*, 1994):

$$a_1 = \frac{S_{xx}S_y - S_x S_{xy}}{\Delta} , \quad a_2 = \frac{SS_{xy} - S_x S_y}{\Delta} , \tag{7.47}$$

$$S = \sum_{i=1}^{N_{\text{D}}} \frac{1}{\sigma_i^2} , \quad S_x = \sum_{i=1}^{N_{\text{D}}} \frac{x_i}{\sigma_i^2} , \quad S_y = \sum_{i=1}^{N_{\text{D}}} \frac{y_i}{\sigma_i^2} , \tag{7.48}$$

$$S_{xx} = \sum_{i=1}^{N_{\text{D}}} \frac{x_i^2}{\sigma_i^2} , \quad S_{xy} = \sum_{i=1}^{N_{\text{D}}} \frac{x_i y_i}{\sigma_i^2} , \quad \Delta = SS_{xx} - S_x^2 . \tag{7.49}$$

Statistics also gives you an expression for the *variance* or uncertainty in the deduced parameters:

$$\sigma_{a_1}^2 = \frac{S_{xx}}{\Delta} , \quad \sigma_{a_2}^2 = \frac{S}{\Delta} . \tag{7.50}$$

This is a measure of the uncertainties in the values of the fitted parameters arising from the uncertainties σ_i in the measured y_i values. A measure of the dependence of the parameters on each other is given by the *correlation coefficient*:

$$\rho(a_1, a_2) = \frac{\operatorname{cov}(a_1, a_2)}{\sigma_{a_1}\sigma_{a_2}}, \quad \operatorname{cov}(a_1, a_2) = \frac{-S_x}{\Delta}. \tag{7.51}$$

Here $\operatorname{cov}(a_1, a_2)$ is the *covariance* of a_1 and a_2 and vanishes if a_1 and a_2 are independent. The correlation coefficient $\rho(a_1, a_2)$ lies in the range $-1 \le \rho \le 1$, with a positive ρ indicating that the errors in a_1 and a_2 are likely to have the same sign, and a negative ρ indicating opposite signs.

The preceding analytic solutions for the parameters are of the form found in statistics books but are not optimal for numerical calculations because subtractive cancelation can make the answers unstable. As discussed in Chapter 3, a rearrangement of the equations can decrease this type of error. For example, Thompson (1992) gives improved expressions that measure the data relative to their averages:

$$a_1 = \overline{y} - a_2\overline{x}, \quad a_2 = \frac{S_{xy}}{S_{xx}}, \quad \overline{x} = \frac{1}{N}\sum_{i=1}^{N_d} x_i, \quad \overline{y} = \frac{1}{N}\sum_{i=1}^{N_d} y_i,$$

$$S_{xy} = \sum_{i=1}^{N_d} \frac{(x_i - \overline{x})(y_i - \overline{y})}{\sigma_i^2}, \quad S_{xx} = \sum_{i=1}^{N_d} \frac{(x_i - \overline{x})^2}{\sigma_i^2}. \tag{7.52}$$

In `Fit.py` in Listing 7.4 we give a program that fits a parabola to some data. You can use it as a model for fitting a line to data, although you can use our closed-form expressions for a straight-line fit. In `Fit.py` on the instructor's site, we give a program for fitting to the decay data.

7.8
Exercises: Fitting Exponential Decay, Heat Flow and Hubble's Law

1. Fit the exponential decay law (7.40) to the data in Figure 7.6. This means finding values for τ and $\Delta N(0)/\Delta t$ that provide a best fit to the data, and then judging how good the fit is.
   a) Construct a table of approximate values for $(\Delta N/\Delta t_i, t_i)$, for $i = 1, N_D$ as read from Figure 7.6. Because time was measured in bins, t_i should correspond to the middle of a bin.
   b) Add an estimate of the error σ_i to obtain a table of the form $(\Delta N/\Delta t_i \pm \sigma_i, t_i)$. You can estimate the errors by eye, say, by estimating how much the histogram values appear to fluctuate about a smooth curve, or you can take $\sigma_i \simeq \sqrt{\text{events}}$. (This last approximation is reasonable for large numbers, which this is not.)
   c) In the limit of very large numbers, we would expect a plot of $\ln|\mathrm{d}N/\mathrm{d}t|$ vs. t to be a straight line:

$$\ln\left|\frac{\Delta N(t)}{\Delta t}\right| \simeq \ln\left|\frac{\Delta N_0}{\Delta t}\right| - \frac{1}{\tau}\Delta t. \tag{7.53}$$

This means that if we treat $\ln|\Delta N(t)/\Delta t|$ as the dependent variable and time Δt as the independent variable, we can use our linear-fit results. Plot $\ln|\Delta N/\Delta t|$ vs. Δt.

d) Make a least-squares fit of a straight line to your data and use it to determine the lifetime τ of the π meson. Compare your deduction to the tabulated lifetime of 2.6×10^{-8} s and comment on the difference.
e) Plot your best fit on the same graph as the data and comment on the agreement.
f) Deduce the goodness of fit of your straight line and the approximate error in your deduced lifetime. Do these agree with what your "eye" tells you?
g) Now that you have a fit, look at the data again and estimate what a better value for the errors in the ordinates might be.

2. Table 7.2 gives the temperature T along a metal rod whose ends are kept at a fixed constant temperature. The temperature is a function of the distance x along the rod.
   a) Plot the data in Table 7.3 to verify the appropriateness of a linear relation
   $$T(x) \simeq a + bx \; . \tag{7.54}$$
   b) Because you are not given the errors for each measurement, assume that the least significant figure has been rounded off and so $\sigma \geq 0.05$.
   c) Use that to compute a least-squares straight-line fit to these data.
   d) Plot your best $a + bx$ on the curve with the data.
   e) After fitting the data, compute the variance and compare it to the deviation of your fit from the data. Verify that about one-third of the points miss the σ error band (that is what is expected for a normal distribution of errors).
   f) Use your computed variance to determine the χ^2 of the fit. Comment on the value obtained.
   g) Determine the variances σ_a and σ_b and check whether it makes sense to use them as the errors in the deduced values for a and b.
3. In 1929, Edwin Hubble examined the data relating the radial velocity v of 24 extra galactic nebulae to their distance r from our galaxy (Hubble, 1929). Although there was considerable scatter in the data, he fit them with a straight line:
   $$v = Hr \; , \tag{7.55}$$
   where H is now called the Hubble constant. Table 7.3 contains the distances and velocities used by Hubble.
   a) Plot the data to verify the appropriateness of a linear relation
   $$v(r) \simeq a + Hr \; . \tag{7.56}$$
   b) Because you are not given the errors for each measurement, you may assume that the least significant figure has been rounded off and so $\sigma \geq 1$. Or, you may assume that astronomical measurements are hard to make and that there are at least 10% errors in the data.

Table 7.2 Temperature vs. distance as measured along a metal rod.

x_i (cm)	1.0	2.0	3.0	4.0	5.0	6.0	7.0	8.0	9.0
T_i (C)	14.6	18.5	36.6	30.8	59.2	60.1	62.2	79.4	99.9

Table 7.3 Distance vs. radial velocity for 24 extragalactic nebulae.

Object	r (Mpc)	v (km/s)	Object	r (Mpc)	v (km/s)
	0.032	170	3627	0.9	650
	0.034	290	4826	0.9	150
6822	0.214	−130	5236	0.9	500
598	0.263	−70	1068	1.0	920
221	0.275	−185	5055	1.1	450
224	0.275	−220	7331	1.1	500
5457	0.45	200	4258	1.4	500
4736	0.5	290	4141	1.7	960
5194	0.5	270	4382	2.0	500
4449	0.63	200	4472	2.0	850
4214	0.8	300	4486	2.0	800
3031	0.9	−30	4649	2.0	1090

c) Compute a least-squares straight-line fit to these data.
d) Plot your best $a + Hr$ on the curve with the data.
e) After fitting the data, compute the variance and compare it to the deviation of your fit from the data. Verify that about one-third of the points miss the σ error band (that is what is expected for a normal distribution of errors).
f) Use your computed variance to determine the χ^2 of the fit. Comment on the value obtained.
g) Determine the variances σ_a and σ_b and check whether it makes sense to use them as the errors in the deduced values for a and b.
h) Now that you have a fit, look at the data again and estimate what a better value for the errors in the ordinates might be.

7.8.1 Linear Quadratic Fit

As indicated earlier, as long as the function being fitted depends *linearly* on the unknown parameters a_i, the condition of minimum χ^2 leads to a set of simultaneous linear equations for the a's that can be solved by hand or on the computer using matrix techniques. To illustrate, suppose we want to fit the quadratic poly-

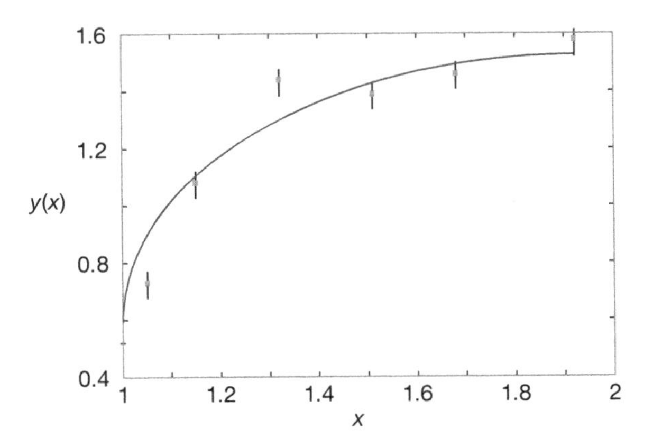

Figure 7.8 A linear least-squares best fit of a parabola to data. Here we see that the fit misses approximately one-third of the points, as expected from the statistics for a good fit.

nomial

$$g(x) = a_1 + a_2 x + a_3 x^2 \tag{7.57}$$

to the experimental measurements $(x_i, y_i, i = 1, N_D)$ (Figure 7.8). Because this $g(x)$ is linear in all the parameters a_i, we can still make a linear fit although x is raised to the second power. (However, if we tried to a fit a function of the form $g(x) = (a_1 + a_2 x)\exp(-a_3 x)$ to the data, then we would not be able to make a linear fit because there is not a linear dependence on a_3.)

The best fit of this quadratic to the data is obtained by applying the minimum χ^2 condition (7.44) for $M_p = 3$ parameters and N_D (still arbitrary) data points. A solution represents the maximum likelihood that the deduced parameters provide a correct description of the data for the theoretical function $g(x)$. Equation 7.44 leads to the three simultaneous equations for a_1, a_2, and a_3:

$$\sum_{i=1}^{N_D} \frac{[y_i - g(x_i)]}{\sigma_i^2} \frac{\partial g(x_i)}{\partial a_1} = 0\,, \qquad \frac{\partial g}{\partial a_1} = 1\,, \tag{7.58}$$

$$\sum_{i=1}^{N_D} \frac{[y_i - g(x_i)]}{\sigma_i^2} \frac{\partial g(x_i)}{\partial a_2} = 0\,, \qquad \frac{\partial g}{\partial a_2} = x\,, \tag{7.59}$$

$$\sum_{i=1}^{N_D} \frac{[y_i - g(x_i)]}{\sigma_i^2} \frac{\partial g(x_i)}{\partial a_3} = 0\,, \qquad \frac{\partial g}{\partial a_3} = x^2\,. \tag{7.60}$$

Note: Because the derivatives are independent of the parameters (the a's), the a dependence arises only from the term in square brackets in the sums, and because that term has only a linear dependence on the a's, these equations are linear in the a's.

Exercise Show that after some rearrangement, (7.58)–(7.60) can be written as

$$S a_1 + S_x a_2 + S_{xx} a_3 = S_y ,$$
$$S_x a_1 + S_{xx} a_2 + S_{xxx} a_3 = S_{xy} ,$$
$$S_{xx} a_1 + S_{xxx} a_2 + S_{xxxx} a_3 = S_{xxy} . \tag{7.61}$$

Here the definitions of the S's are simple extensions of those used in (7.47)–(7.49) and are programmed in `Fit.py` shown in Listing 7.4. After placing the three unknown parameters into a vector $\boldsymbol{x}$ and the known three RHS terms in (7.61) into a vector $\boldsymbol{b}$, these equations assume the matrix form

$$A\boldsymbol{x} = \boldsymbol{b} ,$$
$$A = \begin{bmatrix} S & S_x & S_{xx} \\ S_x & S_{xx} & S_{xxx} \\ S_{xx} & S_{xxx} & S_{xxxx} \end{bmatrix} , \quad \boldsymbol{x} = \begin{bmatrix} a_1 \\ a_2 \\ a_3 \end{bmatrix} , \quad \boldsymbol{b} = \begin{bmatrix} S_y \\ S_{xy} \\ S_{xxy} \end{bmatrix} . \tag{7.62}$$

The solution for the parameter vector $\boldsymbol{a}$ is obtained by solving the matrix equations. Although for 3×3 matrices, we can write out the solution in a closed form, for larger problems the numerical solution requires matrix methods.

Listing 7.4 Fit.py performs a least-squares fit of a parabola to data using the NumPy linalg package to solve the set of linear equations $S\boldsymbol{a} = \boldsymbol{s}$.

```
# Fit.py        Linear least-squares fit; e.g. of matrix computation arrays

import pylab as p
from numpy import*
from numpy.linalg import inv
from numpy.linalg import solve

t = arange(1.0, 2.0, 0.1)                                          # x range curve
x = array([1., 1.1, 1.24, 1.35,  1.451, 1.5, 1.92])              # Given x values
y = array([0.52, 0.8, 0.7, 1.8, 2.9, 2.9, 3.6])                   # Given y values
p.plot(x, y, 'bo' )                                                # Plot data in blue
sig = array([0.1, 0.1, 0.2, 0.3, 0.2, 0.1, 0.1])                  # error bar lenghts
p.errorbar(x,y,sig)                                                 # Plot error bars
p.title('Linear Least Square Fit')                                    # Plot figure
p.xlabel( 'x' )                                                       # Label axes
p.ylabel( 'y' )
p.grid(True)                                                           # plot grid
Nd = 7
A = zeros( (3,3), float )                                             # Initialize
bvec = zeros( (3,1), float )
ss= sx = sxx = sy = sxxx = sxxxx = sxy = sxy = sxxy = 0.

for i in range(0, Nd):
        sig2 = sig[i] * sig[i]
        ss  += 1. / sig2;         sx   += x[i]/sig2;               sy    += y[i]/sig2
        rhl  = x[i] * x[i];       sxx  += rhl/sig2;      sxxy  += rhl * y[i]/sig2
        sxy += x[i]*y[i]/sig2;  sxxx +=rhl*x[i]/sig2;  sxxxx +=rhl*rhl/sig2

A    = array([ [ss,sx,sxx], [sx,sxx,sxxx], [sxx,sxxx,sxxxx] ])
bvec = array([sy, sxy, sxxy])

xvec = multiply(inv(A), bvec)                                        # Invert matrix
Itest = multiply(A, inv(A))                                          # Matrix multiply
```

```
print('\n x vector via inverse')
print(xvec, '\n')
print('A*inverse(A)')
print(Itest, '\n')

xvec = solve(A, bvec)                              # Solve via elimination
print('x Matrix via direct')
print(xvec, 'end= ')
print('FitParabola Final Results\n')
print('y(x) = a0 + a1 x + a2 x^2')                 # Desired fit
print('a0 = ', x[0])
print('a1 = ', x[1])
print('a2 = ', x[2], '\n')
print(' i    xi      yi     yfit    ')
for i in range(0, Nd):
    s = xvec[0] + xvec[1]*x[i] + xvec[2]*x[i]*x[i]
    print(" %d %5.3f  %5.3f  %8.7f \n"  %(i, x[i], y[i], s))
# red line is the fit, red dots the fits at y[i]
curve  = xvec[0] + xvec[1]*t + xvec[2]*t**2
points = xvec[0] + xvec[1]*x + xvec[2]*x**2
p.plot(t, curve,'r', x, points, 'ro')
p.show()
```

Linear Quadratic Fit Assessment

1. Fit the quadratic (7.57) to the following data sets [given as $(x_1, y_1), (x_2, y_2), \ldots$]. In each case, indicate the values found for the a's, the number of degrees of freedom, *and* the value of χ^2.
   a) $(0, 1)$
   b) $(0, 1), (1, 3)$
   c) $(0, 1), (1, 3), (2, 7)$
   d) $(0, 1), (1, 3), (2, 7), (3, 15)$
2. Find a fit to the last set of data to the function

$$y = A e^{-bx^2} . \tag{7.63}$$

Hint: A judicious change of variables will permit you to convert this to a linear fit. Does a minimum χ^2 still have meaning here?

7.8.2 Problem 5: Nonlinear Fit to a Breit–Wigner

Problem Remember how earlier in this chapter we interpolated the values in Table 7.1 in order to obtain the experimental cross section Σ as a function of energy. Although we did not use it, we also gave the theory describing these data, namely, the Breit–Wigner resonance formula (7.21):

$$f(E) = \frac{f_r}{(E - E_r)^2 + \Gamma^2/4} . \tag{7.64}$$

Your *problem* is to determine what values for the parameters E_r, f_r, and Γ in (7.64) provide the best fit to the data in Table 7.1.

Because (7.64) is not a linear function of the parameters (E_r, Σ_0, Γ), the three equations that result from minimizing χ^2 are not linear equations and so cannot be solved by the techniques of *linear* algebra (matrix methods). However, in our study of the masses on a string problem, we showed how to use the Newton–Raphson algorithm to search for solutions of simultaneous nonlinear equations. That technique involved expansion of the equations about the previous guess to obtain a set of linear equations and then solving the linear equations with the matrix libraries. We now use this same combination of fitting, trial-and-error searching, and matrix algebra to conduct a nonlinear least-squares fit of (7.64) to the data in Table 7.1.

Recall that the condition for a best fit is to find values of the M_P parameters a_m in the theory $g(x, a_m)$ that minimize $\chi^2 = \sum_i [(y_i - g_i)/\sigma_i]^2$. This leads to the M_P equations (7.44) to solve

$$\sum_{i=1}^{N_D} \frac{[y_i - g(x_i)]}{\sigma_i^2} \frac{\partial g(x_i)}{\partial a_m} = 0 , \quad (m = 1, M_P) . \tag{7.65}$$

To find the form of these equations appropriate to our problem, we rewrite our theory function (7.64) in the notation of (7.65):

$$a_1 = f_r , \quad a_2 = E_R , \quad a_3 = \Gamma^2/4 , \quad x = E , \tag{7.66}$$

$$\Rightarrow \quad g(x) = \frac{a_1}{(x - a_2)^2 + a_3} . \tag{7.67}$$

The three derivatives required in (7.65) are then

$$\frac{\partial g}{\partial a_1} = \frac{1}{(x - a_2)^2 + a_3} , \quad \frac{\partial g}{\partial a_2} = \frac{-2a_1(x - a_2)}{\left[(x - a_2)^2 + a_3\right]^2} , \quad \frac{\partial g}{\partial a_3} = \frac{-a_1}{\left[(x - a_2)^2 + a_3\right]^2} . \tag{7.68}$$

Substitution of these derivatives into the best-fit condition (7.65) yields three simultaneous equations in a_1, a_2, and a_3 that we need to solve in order to fit the $N_D = 9$ data points (x_i, y_i) in Table 7.1:

$$\sum_{i=1}^{9} \frac{y_i - g(x_i, a)}{(x_i - a_2)^2 + a_3} = 0 ,$$
$$\sum_{i=1}^{9} \frac{y_i - g(x_i, a)}{\left[(x_i - a_2)^2 + a_3\right]^2} = 0 ,$$
$$\sum_{i=1}^{9} \frac{\{y_i - g(x_i, a)\}(x_i - a_2)}{\left[(x_i - a_2)^2 + a_3\right]^2} = 0 . \tag{7.69}$$

Even without the substitution of (7.64) for $g(x, a)$, it is clear that these three equations depend on the a's in a nonlinear fashion. That is okay because in Section 6.1.2 we derived the N-dimensional Newton–Raphson search for the roots of

$$f_i(a_1, a_2, \dots, a_N) = 0 , \quad i = 1, N , \tag{7.70}$$

where we have made the change of variable $y_i \rightarrow a_i$ for the present problem. We use that same formalism here for the $N = 3$ Equation 7.69 by writing them as

$$f_1(a_1, a_2, a_3) = \sum_{i=1}^{9} \frac{y_i - g(x_i, a)}{(x_i - a_2)^2 + a_3} = 0 , \tag{7.71}$$

$$f_2(a_1, a_2, a_3) = \sum_{i=1}^{9} \frac{\{y_i - g(x_i, a)\} (x_i - a_2)}{\left[(x_i - a_2)^2 + a_3\right]^2} = 0 , \tag{7.72}$$

$$f_3(a_1, a_2, a_3) = \sum_{i=1}^{9} \frac{y_i - g(x_i, a)}{\left[(x_i - a_2)^2 + a_3\right]^2} = 0 . \tag{7.73}$$

Because $f_r \equiv a_1$ is the peak value of the cross section, $E_R \equiv a_2$ is the energy at which the peak occurs, and $\Gamma = 2\sqrt{a_3}$ is the full width of the peak at half-maximum, good guesses for the a's can be extracted from a graph of the data. To obtain the nine derivatives of the three f's with respect to the three unknown a's, we use two nested loops over i and j, along with the forward-difference approximation for the derivative

$$\frac{\partial f_i}{\partial a_j} \simeq \frac{f_i(a_j + \Delta a_j) - f_i(a_j)}{\Delta a_j} , \tag{7.74}$$

where Δa_j corresponds to a small, say $\leq 1\%$, change in the parameter value.

Nonlinear Fit Implementation Use the Newton–Raphson algorithm as outlined in Section 7.8.2 to conduct a nonlinear search for the best-fit parameters of the Breit–Wigner theory (7.64) to the data in Table 7.1. Compare the deduced values of (f_r, E_R, Γ) to that obtained by inspection of the graph.

8
Solving Differential Equations: Nonlinear Oscillations

Part of the joy of having computational tools at your disposal is that it is easy to solve almost every differential equation. Consequently, while most traditional (read "analytic") treatments of oscillations are limited to the small displacements about equilibrium where the restoring forces are linear, we eliminate those restrictions here and explore some interesting *nonlinear* physics. First, we look at oscillators that may be harmonic for certain parameter values, but then become anharmonic. We start with simple systems that have analytic solutions, use them to test various differential-equation solvers, and then include time-dependent forces to investigate nonlinear resonances and beating.[1]

8.1
Free Nonlinear Oscillations

Problem In Figure 8.1, we show a mass m attached to a spring that exerts a restoring force toward the origin, as well as a hand that exerts a time-dependent external force on the mass. We are told that the restoring force exerted by the spring is nonharmonic, that is, not simply proportional to displacement from equilibrium, but we are not given details as to how this is nonharmonic. Your *problem* is to solve for the motion of the mass as a function of time. You may assume that the motion is constrained to one dimension.

8.2
Nonlinear Oscillators (Models)

This is a problem in classical mechanics for which Newton's second law provides us with the equation of motion

$$F_k(x) + F_{\text{ext}}(x, t) = m\frac{\mathrm{d}^2x}{\mathrm{d}t^2}\,, \tag{8.1}$$

1) In Chapter 15, we make a related study of the realistic pendulum and its chaotic behavior. Some special properties of nonlinear equations are also discussed in Chapter 24.

Computational Physics, 3rd edition. Rubin H. Landau, Manuel J. Páez, Cristian C. Bordeianu.

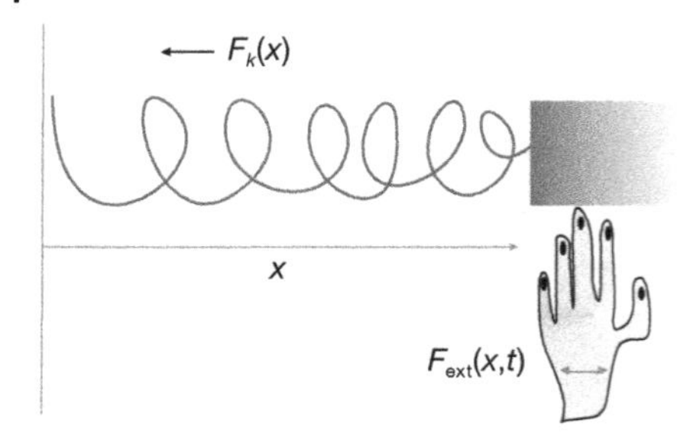

Figure 8.1 A mass m (the block) attached to a spring with restoring force $F_k(x)$ driven by an external time-dependent driving force (the hand).

where $F_k(x)$ is the restoring force exerted by the spring and $F_{\text{ext}}(x, t)$ is the external force. Equation 8.1 is the differential equation we must solve for arbitrary forces. Because we are not told just how the spring departs from being linear, we are free to try out some different models. As our first model, we wish to try a potential that is a harmonic oscillator for small displacements x and also contains a perturbation that introduces a nonlinear term to the force for large x values:

$$V(x) \simeq \frac{1}{2}kx^2\left(1 - \frac{2}{3}\alpha x\right) , \tag{8.2}$$

$$\Rightarrow \quad F_k(x) = -\frac{\mathrm{d}V(x)}{\mathrm{d}x} = -kx(1 - \alpha x) \tag{8.3}$$

$$\Rightarrow \quad m\frac{\mathrm{d}^2x}{\mathrm{d}t^2} = -kx(1 - \alpha x) , \tag{8.4}$$

where we have omitted the time-dependent external force. Equation 8.4 is the second-order ODE we need to solve. If $\alpha x \ll 1$, we should have essentially harmonic motion, but as $x \to 1/\alpha$ the anharmonic effects would be large.

We can understand the basic physics of this model by looking at the curves as shown in Figure 8.2a. As long as $x < 1/\alpha$, there will be a *restoring force* and the motion will be periodic (repeated exactly and indefinitely in time), even although it is harmonic only for small-amplitude oscillations. Yet, if the amplitude of oscillation is large, there will be an asymmetry in the motion to the right and left of the equilibrium position. And if $x > 1/\alpha$, the force will become repulsive and the mass will be pushed away from the origin.

As a second model of a nonlinear oscillator, we assume that the spring's potential function is proportional to some arbitrary *even* power p of the displacement x from equilibrium:

$$V(x) = \frac{1}{p}kx^p , \quad (p \text{ even}) . \tag{8.5}$$

We require an even p to ensure that the force,

$$F_k(x) = -\frac{\mathrm{d}V(x)}{\mathrm{d}x} = -kx^{p-1} , \tag{8.6}$$

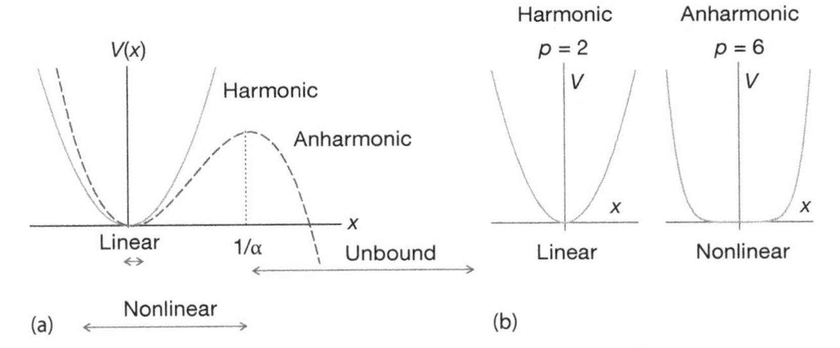

Figure 8.2 (a) The potentials of an harmonic oscillator (solid curve) and of an anharmonic oscillator (dashed curve). If the amplitude becomes too large for the anharmonic oscillator, the motion becomes unbound. (b) The shapes of the potential energy function $V(x) \propto |x|^p$ for $p = 2$ and $p = 6$. The "linear" and "nonlinear" labels refer to the restoring force derived from these potentials.

contains an odd power of p, which guarantees that it is a *restoring* force for positive and negative x values. We display some characteristics of this potential in Figure 8.2b. We see that $p = 2$ is the harmonic oscillator and that $p = 6$ is nearly a square well with the mass moving almost freely until it hits the wall at $x \simeq \pm 1$. Regardless of the p value, the motion will be periodic, but it will be harmonic only for $p = 2$. Newton's law (8.1) gives the second-order ODE we need to solve

$$m\frac{\mathrm{d}^2 x}{\mathrm{d}t^2} = F_{\mathrm{ext}}(x, t) - kx^{p-1} \ . \tag{8.7}$$

8.3 Types of Differential Equations (Math)

The background material in this section is presented to avoid confusion about semantics. The well-versed reader may want to skim or skip it.

Order A general form for a *first-order* differential equation is

$$\frac{\mathrm{d}y}{\mathrm{d}t} = f(t, y) \ , \tag{8.8}$$

where the "order" refers to the degree of the derivative on the LHS. The derivative or force function $f(t, y)$ on the RHS is arbitrary. For instance, even if $f(t, y)$ is a nasty function of y and t such as

$$\frac{\mathrm{d}y}{\mathrm{d}t} = -3t^2 y + t^9 + y^7 \ , \tag{8.9}$$

this is still first order in the derivative. A general form for a *second-order* differential equation is

$$\frac{d^2y}{dt^2} + \lambda\frac{dy}{dt} = f\left(t, \frac{dy}{dt}, y\right) . \tag{8.10}$$

The derivative function f on the RHS is arbitrary and may involve any power of the first derivative as well. To illustrate,

$$\frac{d^2y}{dt^2} + \lambda\frac{dy}{dt} = -3t^2\left(\frac{dy}{dt}\right)^4 + t^9 y(t) \tag{8.11}$$

is a second-order differential equation, as is Newton's law (8.1).

In the differential equations (8.8) and (8.10), the time t is the *independent* variable and the position y is the *dependent* variable. This means that we are free to vary the time at which we want a solution, but not the value of the position y at that time. Note that we often use the symbol y or Y for the dependent variable but that this is just a symbol. In some applications, we use y to describe a position that is an independent variable instead of t.

Ordinary and partial Differential equations such as (8.1) and (8.8) are *ordinary* differential equations because they contain only *one* independent variable, in these cases t. In contrast, an equation such as the Schrödinger equation

$$i\hbar\frac{\partial\psi(\boldsymbol{x},t)}{\partial t} = -\frac{\hbar^2}{2m}\left[\frac{\partial^2\psi}{\partial x^2} + \frac{\partial^2\psi}{\partial y^2} + \frac{\partial^2\psi}{\partial z^2}\right] + V(\boldsymbol{x})\psi(\boldsymbol{x},t) \tag{8.12}$$

contains several independent variables, and this makes it a *partial differential equation* (PDE). The partial derivative symbol ∂ is used to indicate that the dependent variable ψ depends simultaneously on several independent variables. In the early parts of this book, we limit ourselves to ordinary differential equations. In Chapters 19–25, we examine a variety of PDEs.

Linear and nonlinear Part of the liberation of computational science is that we are no longer limited to solving *linear equations*. A *linear* equation is one in which only the first power of y or of $d^n y/d^n t$ appears; a *nonlinear* equation may contain higher powers. For example,

$$\frac{dy}{dt} = g^3(t)y(t) \quad \text{(linear)}, \qquad \frac{dy}{dt} = \lambda y(t) - \lambda^2 y^2(t) \quad \text{(nonlinear)} . \tag{8.13}$$

An important property of linear equations is the *law of linear superposition* that lets us add solutions together to form new ones. As a case in point, if $A(t)$ and $B(t)$ are solutions of the linear equation in (8.13), then

$$y(t) = \alpha A(t) + \beta B(t) \tag{8.14}$$

is also a solution for arbitrary values of the constants α and β. In contrast, even if we were clever enough to guess that the solution of the nonlinear equation in (8.13) is

$$y(t) = \frac{a}{1 + be^{-\lambda t}} \tag{8.15}$$

(which you can verify by substitution), things would be amiss if we tried to obtain a more general solution by adding together two such solutions:

$$y_1(t) = \frac{a}{1 + be^{-\lambda t}} + \frac{a'}{1 + b'e^{-\lambda t}} \tag{8.16}$$

(which you can verify by substitution).

Initial and boundary conditions The general solution of a first-order differential equation always contains one arbitrary constant. The general solution of a second-order differential equation contains two such constants, and so forth. For any specific problem, these constants are fixed by the *initial conditions*. For a first-order equation, the sole initial condition may be the position $y(t)$ at some time. For a second-order equation, the two initial conditions may be the position and velocity at some time. Regardless of how powerful the hardware and software that you utilize, mathematics remains valid, and so you must know the initial conditions in order to solve the problem uniquely.

In addition to the initial conditions, it is possible to further restrict the solutions of differential equations. One such way is by *boundary conditions* that constrain the solution to have fixed values at the boundaries of the solution space. Problems of this sort are called *eigenvalue problems*, and they are so demanding that solutions do not always exist, and even when they do exist, a trial-and-error search may be required to find them. In Chapter 9, we discuss how to extend the techniques of the present unit to boundary-value problems.

8.4 Dynamic Form for ODEs (Theory)

A standard form for ODEs, which has found proven to be useful in both numerical analysis (Press *et al.*, 1994) and classical dynamics (Scheck, 1994; Tabor, 1989; José and Salatan, 1998), is to express ODEs of *any order* as N simultaneous first-order ODEs in the N unknowns $y^{(0)} - y^{(N-1)}$:

$$\frac{dy^{(0)}}{dt} = f^{(0)}(t, y^{(i)}) \, , \tag{8.17}$$

$$\frac{dy^{(1)}}{dt} = f^{(1)}(t, y^{(i)}) \, ,$$

$$\ddots \tag{8.18}$$

$$\frac{dy^{(N-1)}}{dt} = f^{(N-1)}(t, y^{(i)}) \, , \tag{8.19}$$

where a $y^{(i)}$ dependence in f is allowed, but not any dependence on the derivatives $\mathrm{d}y^{(i)}/\mathrm{d}t$. These equations can be expressed more succinctly by use of the N-dimensional vectors (indicated here in boldface italic) $\boldsymbol{y}$ and $\boldsymbol{f}$:

$$\frac{\mathrm{d}\boldsymbol{y}(t)}{\mathrm{d}t} = \boldsymbol{f}(t, \boldsymbol{y}),$$

$$\boldsymbol{y} = \begin{pmatrix} y^{(0)}(t) \\ y^{(1)}(t) \\ \ddots \\ y^{(N-1)}(t) \end{pmatrix}, \quad \boldsymbol{f} = \begin{pmatrix} f^{(0)}(t, \boldsymbol{y}) \\ f^{(1)}(t, \boldsymbol{y}) \\ \ddots \\ f^{(N-1)}(t, \boldsymbol{y}) \end{pmatrix}. \tag{8.20}$$

The utility of such compact notation is that we can study the properties of the ODEs, as well as develop algorithms to solve them, by dealing with the single equation (8.20) without having to worry about the individual components. To see how this works, let us convert Newton's law

$$\frac{\mathrm{d}^2x}{\mathrm{d}t^2} = \frac{1}{m}F\left(t, x, \frac{\mathrm{d}x}{\mathrm{d}t}\right) \tag{8.21}$$

to this standard form. The rule is that the RHS may *not* contain any explicit derivatives, although individual components of $y^{(i)}$ may represent derivatives. To pull this off, we define the position x as the first dependent variable $y^{(0)}$, and the velocity $\mathrm{d}x/\mathrm{d}t$ as the second dependent variable $y^{(1)}$:

$$y^{(0)}(t) \stackrel{\text{def}}{=} x(t), \quad y^{(1)}(t) \stackrel{\text{def}}{=} \frac{\mathrm{d}x}{\mathrm{d}t} = \frac{\mathrm{d}y^{(0)}(t)}{\mathrm{d}t}. \tag{8.22}$$

The second-order ODE (8.21) now becomes two simultaneous first-order ODEs:

$$\frac{\mathrm{d}y^{(0)}}{\mathrm{d}t} = y^{(1)}(t), \quad \frac{\mathrm{d}y^{(1)}}{\mathrm{d}t} = \frac{1}{m}F(t, y^{(0)}, y^{(1)}). \tag{8.23}$$

This expresses the acceleration (the second derivative in (8.21)) as the first derivative of the velocity $[y^{(1)}]$. These equations are now in the standard form (8.20), with the derivative or force function $\boldsymbol{f}$ having the two components

$$f^{(0)} = y^{(1)}(t), \quad f^{(1)} = \frac{1}{m}F(t, y^{(0)}, y^{(1)}), \tag{8.24}$$

where F may be an explicit function of time as well as of position and velocity.

To be even more specific, applying these definitions to our spring problem (8.7), we obtain the coupled first-order equations

$$\frac{\mathrm{d}y^{(0)}}{\mathrm{d}t} = y^{(1)}(t), \quad \frac{\mathrm{d}y^{(1)}}{\mathrm{d}t} = \frac{1}{m}\left[F_{\text{ext}}(x, t) - k y^{(0)}(t)^{p-1}\right], \tag{8.25}$$

where $y^{(0)}(t)$ is the position of the mass at time t and $y^{(1)}(t)$ is its velocity. In the standard form, the components of the force function and the initial conditions are

$$\begin{aligned} f^{(0)}(t, \boldsymbol{y}) &= y^{(1)}(t), & f^{(1)}(t, \boldsymbol{y}) &= \frac{1}{m}\left[F_{\text{ext}}(x, t) - k(y^{(0)})^{p-1}\right], \\ y^{(0)}(0) &= x_0, & y^{(1)}(0) &= v_0. \end{aligned} \tag{8.26}$$

8.5 ODE Algorithms

The classic way to solve an ODE is to start with the known initial value of the dependent variable, $y_0 \equiv y(t=0)$, and then use the derivative function $f(t, y)$ to advance the initial value one small step h forward in time to obtain $y(t=h) \equiv y_1$. Once you can do that, you can solve the ODE for all t values by just continuing stepping to larger times, one small h at a time (Figure 8.3).[2] Error is always a concern when integrating differential equations because derivatives require small differences, and small differences are prone to subtractive cancelations and round-off error accumulation. In addition, because our stepping procedure for solving the differential equation is a continuous extrapolation of the initial conditions, with each step building on a previous extrapolation, this is somewhat like a castle built on sand; in contrast to interpolation, there are no tabulated values on which to anchor your solution.

It is simplest if the time steps used throughout the integration remain constant in size, and that is mostly what we shall do. Industrial-strength algorithms, such as the one we discuss in Section 8.6, adapt the step size by making h larger in regions where y varies slowly (this speeds up the integration and cuts down on the round-off error) and making h smaller in regions where y varies rapidly.

8.5.1 Euler's Rule

Euler's rule (Figure 8.4) is a simple algorithm for integrating the differential equation (8.8) by one step and is just the forward-difference algorithm for the derivative:

$$\frac{\mathrm{d}\boldsymbol{y}(t)}{\mathrm{d}t} \simeq \frac{\boldsymbol{y}(t_{n+1}) - \boldsymbol{y}(t_n)}{h} = \boldsymbol{f}(t, \boldsymbol{y}) \,, \tag{8.27}$$

$$\Rightarrow \quad \boldsymbol{y}_{n+1} \simeq \boldsymbol{y}_n + h\boldsymbol{f}(t_n, \boldsymbol{y}_n) \,, \tag{8.28}$$

where $\boldsymbol{y}_n \stackrel{\text{def}}{=} \boldsymbol{y}(t_n)$ is the value of $\boldsymbol{y}$ at time t_n. We know from our discussion of differentiation that the error in the forward-difference algorithm is $\mathcal{O}(h^2)$, and so then is the error in Euler's rule.

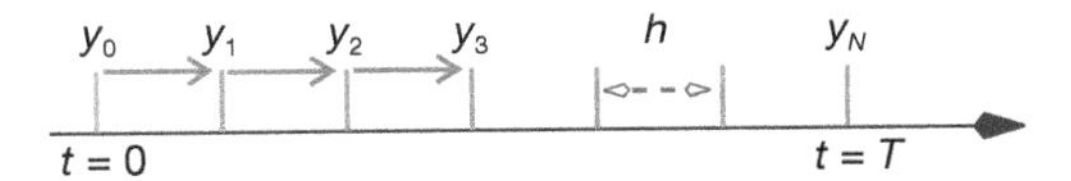

Figure 8.3 A sequence of uniform steps of length h taken in solving a differential equation. The solution starts at time $t = 0$ and is integrated in steps of h until $t = T$.

2) To avoid confusion, notice that $y^{(n)}$ is the nth component of the $\boldsymbol{y}$ vector, while $\boldsymbol{y}_n$ is the value of $\boldsymbol{y}$ after n time steps. Yes, there is a price to pay for elegance in notation.

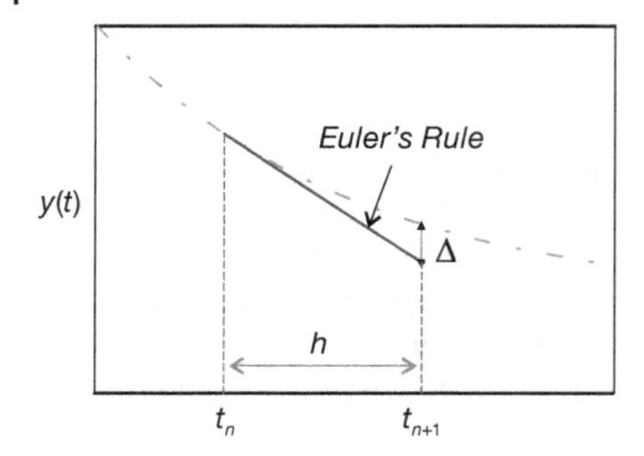

Figure 8.4 Euler's algorithm for integration of a differential equation one step forward in time. This linear extrapolation with the slope evaluated at the initial point is seen to lead to an error Δ.

To indicate the simplicity of this algorithm, we apply it to our oscillator problem (8.4) for the first time step:

$$y_1^{(0)} = x_0 + v_0 h \,, \quad y_1^{(1)} = v_0 + h\frac{1}{m}\left[F_{\text{ext}}(t=0) + F_k(t=0)\right] \,. \tag{8.29}$$

Compare these to the projectile equations familiar from first-year physics,

$$x = x_0 + v_0 h + \frac{1}{2}ah^2 \,, \quad v = v_0 + ah \,, \tag{8.30}$$

and we see that with Euler's rule the acceleration does not contribute to the distance covered (no h^2 term), yet it does contribute to the velocity (and so will contribute belatedly to the distance in the next time step). This is clearly a simple algorithm that requires very small h values to obtain precision. Yet using small values for h increases the number of steps and the accumulation of the round-off error, which may lead to instability.[3] Whereas we do not recommend Euler's algorithm for general use, it is commonly used to start off a more precise algorithm.

8.6 Runge–Kutta Rule

Although no single algorithm is good for solving all ODEs, the fourth-order Runge–Kutta algorithm, `rk4`, or its extension with adaptive step size, `rk45`, has proven to be robust and capable of industrial-strength work. In spite of `rk4` being our recommended standard method, we derive the simpler `rk2` here, and just state the result for `rk4`.

3) Instability is often a problem when you integrate a $y(t)$ that decreases as the integration proceeds, analogous to upward recursion of spherical Bessel functions. In this case, and if you have a linear ODE, you are best off integrating *inward* from large times to small times and then scaling the answer to agree with the initial conditions.

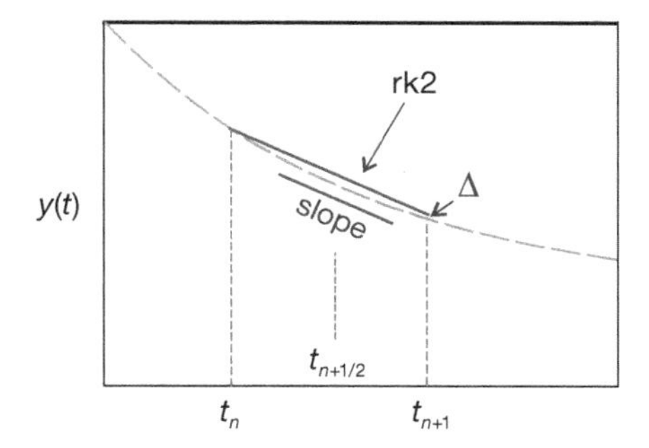

Figure 8.5 The rk2 algorithm for integration of a differential equation uses a slope (bold line segment) evaluated at the interval's midpoint, and is seen to lead to a smaller error than Euler's algorithm in Figure 8.4.

The Runge–Kutta algorithms for integrating a differential equation are based upon the formal (exact) integral of our differential equation:

$$\frac{\mathrm{d}y}{\mathrm{d}t} = f(t, y) \quad \Rightarrow \quad y(t) = \int f(t, y)\,\mathrm{d}t \tag{8.31}$$

$$\Rightarrow \quad y_{n+1} = y_n + \int_{t_n}^{t_{n+1}} f(t, y)\,\mathrm{d}t\,. \tag{8.32}$$

To derive the second-order Runge–Kutta algorithm `rk2` (Figure 8.5 and `rk2.py`), we expand $f(t, y)$ in a Taylor series about the *midpoint* of the integration interval and retain two terms:

$$f(t, y) \simeq f(t_{n+1/2}, y_{n+1/2}) + (t - t_{n+1/2})\frac{\mathrm{d}f}{\mathrm{d}t}(t_{n+1/2}) + \mathcal{O}(h^2)\,. \tag{8.33}$$

Because $(t - t_{n+1/2})$ raised to any odd power is equally positive and negative over the interval $t_n \le t \le t_{n+1}$, the integral of the $(t - t_{n+1/2})$ term in (8.32) vanishes and we obtain our algorithm:

$$\int_{t_n}^{t_{n+1}} f(t, y)\,\mathrm{d}t \simeq f(t_{n+1/2}, y_{n+1/2})h + \mathcal{O}(h^3)\,, \tag{8.34}$$

$$\Rightarrow \quad y_{n+1} \simeq y_n + h f(t_{n+1/2}, y_{n+1/2}) + \mathcal{O}(h^3) \quad (\texttt{rk2})\,. \tag{8.35}$$

We see that while `rk2` contains the same number of terms as Euler's rule, it obtains a higher level of precision by taking advantage of the cancelation of the $\mathcal{O}(h)$ terms. Yet the price for improved precision is having to evaluate the derivative function and the solution y at the middle of the time interval, $t = t_n + h/2$. And there is the rub, for we do not know the value of $y_{n+1/2}$ and cannot use this algorithm to determine it. The way out of this quandary is to use Euler's algorithm to approximate $y_{n+1/2}$:

$$y_{n+1/2} \simeq y_n + \frac{1}{2}h\frac{dy}{dt} = y_n + \frac{1}{2}hf(t_n, y_n). \tag{8.36}$$

Putting the pieces all together gives the complete `rk2` algorithm:

$$\boldsymbol{y}_{n+1} \simeq \boldsymbol{y}_n + \boldsymbol{k}_2, \quad \text{(rk2)} \tag{8.37}$$

$$\boldsymbol{k}_2 = h\boldsymbol{f}\left(t_n + \frac{h}{2}, \boldsymbol{y}_n + \frac{\boldsymbol{k}_1}{2}\right), \quad \boldsymbol{k}_1 = h\boldsymbol{f}(t_n, \boldsymbol{y}_n), \tag{8.38}$$

where we use boldface italic to indicate the vector nature of y and f. We see that the known derivative function $\boldsymbol{f}$ is evaluated at the ends and the midpoint of the interval, but that only the (known) initial value of the dependent variable $\boldsymbol{y}$ is required. This makes the algorithm self-starting.

As an example of the use of `rk2`, we apply it to our spring problem:

$$y_1^{(0)} = y_0^{(0)} + hf^{(0)}\left(\frac{h}{2}, y_0^{(0)} + k_1\right) \simeq x_0 + h\left[v_0 + \frac{h}{2}F_k(0)\right],$$

$$y_1^{(1)} = y_0^{(1)} + hf^{(1)}\left[\frac{h}{2}, y_0 + \frac{h}{2}f(0, y_0)\right]$$

$$\simeq v_0 + \frac{h}{m}\left[F_{\text{ext}}\left(\frac{h}{2}\right) + F_k\left(y_0^{(1)} + \frac{k_1}{2}\right)\right].$$

These equations say that the position $y^{(0)}$ changes because of the initial velocity and force, while the velocity $y^{(1)}$ changes because of the external force at $t = h/2$ and the internal force at two intermediate positions. We see that the position $y^{(0)}$ now has an h^2 time dependence, which at last brings us up to the level of first-year physics.

The fourth-order Runge–Kutta method `rk4` (Listing 8.1) obtains $\mathcal{O}(h^4)$ precision by approximating y as a Taylor series up to h^2 (a parabola) at the midpoint of the interval, which again leads to cancellation of lower order error. `rk4` provides an excellent balance of power, precision, and programming simplicity. There are now four gradient (k) terms to evaluate, with four subroutine calls. This provides an improved approximation to $f(t, y)$ near the midpoint. Although `rk4` is computationally more expensive than the Euler method, its precision is much better, and some time is saved by using larger values for the step size h. Explicitly, `rk4` requires the evaluation of four intermediate slopes, and these are approximated with the Euler algorithm to (Press *et al.*, 1994):

$$\boldsymbol{y}_{n+1} = \boldsymbol{y}_n + \frac{1}{6}(\boldsymbol{k}_1 + 2\boldsymbol{k}_2 + 2\boldsymbol{k}_3 + \boldsymbol{k}_4), \tag{8.39}$$

$$\boldsymbol{k}_1 = h\boldsymbol{f}(t_n, \boldsymbol{y}_n), \qquad \boldsymbol{k}_2 = h\boldsymbol{f}\left(t_n + \frac{h}{2}, \boldsymbol{y}_n + \frac{\boldsymbol{k}_1}{2}\right),$$

$$\boldsymbol{k}_3 = h\boldsymbol{f}\left(t_n + \frac{h}{2}, \boldsymbol{y}_n + \frac{\boldsymbol{k}_2}{2}\right), \quad \boldsymbol{k}_4 = h\boldsymbol{f}(t_n + h, \boldsymbol{y}_n + \boldsymbol{k}_3).$$

Listing 8.1 **rk4.py** solves an ODE with the RHS given by the method $f()$ using a fourth-order Runge–Kutta algorithm. Note that to avoid introducing bugs, the method $f()$, which you will need to change for each problem, is kept separate from the algorithm.

```
# rk4.py 4th order Runge Kutta

from visual.graph import *

#    Initialization
a = 0.
b = 10.
n = 100
ydumb = zeros((2), float);      y = zeros((2), float)
fReturn = zeros((2), float);    k1 = zeros((2), float)
k2 = zeros((2), float);         k3 = zeros((2), float)
k4 = zeros((2), float)
y[0] = 3.;     y[1] = -5.
t = a;         h = (b-a)/n;

def f( t, y):                                            # Force function
    fReturn[0] = y[1]
    fReturn[1] = -100.*y[0]-2.*y[1] + 10.*sin(3.*t)
    return fReturn

graph1 = gdisplay(x=0,y=0, width = 400, height = 400, title = 'RK4',
                  xtitle = 't', ytitle =
                       'Y[0]',xmin=0,xmax=10,ymin=-2,ymax=3)
funct1 = gcurve(color = color.yellow)
graph2 = gdisplay(x=400,y=0, width = 400, height = 400, title = 'RK4',
                  xtitle = 't', ytitle =
                       'Y[1]',xmin=0,xmax=10,ymin=-25,ymax=18)
funct2 = gcurve(color = color.red)

def rk4(t,h,n):
    k1 = [0]*(n)
    k2 = [0]*(n)
    k3 = [0]*(n)
    k4 = [0]*(n)
    fR = [0]*(n)
    ydumb = [0]*(n)
    fR = f(t, y)                                         # Returns RHS's
    for i in range(0, n):
        k1[i] = h*fR[i]
    for i in range(0, n):
        ydumb[i] = y[i] + k1[i]/2.
    k2 = h*f(t+h/2., ydumb)
    for i in range(0, n):
        ydumb[i] = y[i] + k2[i]/2.
    k3 = h*f(t+h/2., ydumb)
    for i in range(0, n):
        ydumb[i] = y[i] + k3[i]
    k4 = h*f(t+h, ydumb)
    for i in range(0, 2):
        y[i] = y[i] + (k1[i] + 2.*(k2[i] + k3[i]) + k4[i])/6.
    return y

while (t < b):                                           # Time loop
    if ((t + h) > b):
        h = b - t                                        # Last step
    y = rk4(t,h,2)
    t = t + h
    rate(30)
    funct1.plot(pos = (t, y[0]) )
    funct2.plot(pos = (t, y[1]) )
```

A variation of `rk4`, known as the Runge–Kutta–Fehlberg method (Mathews, 2002), or `rk45`, varies the step size while doing the integration with the hope of obtaining better precision and maybe better speed. Our implementation, `rk45.py`, is given in Listing 8.2. It automatically doubles the step size and tests to see how an estimate of the error changes. If the error is still within acceptable bounds, the algorithm will continue to use the larger step size and thus speed up the computation; if the error is too large, the algorithm will decrease the step size until an acceptable error is found. As a consequence of the extra information obtained in the testing, the algorithm does obtains $\mathcal{O}(h^5)$ precision, but sometimes at the expense of extra computing time. Whether that extra time is recovered by being able to use a larger step size depends upon the application.

Listing 8.2 rk45.py solves an ODE with the RHS given by the method $f()$ using a fourth-order Runge–Kutta algorithm with adaptive step size.

```
# rk45.py            Adaptive step size Runge Kutta

from visual.graph import *

a = 0.; b = 10.                                  # Error tolerance, endpoints
Tol = 1.0E-8
ydumb = zeros( (2), float)                       # Initialize
y = zeros( (2), float)
fReturn = zeros( (2), float)
err = zeros( (2), float)
k1 = zeros( (2), float)
k2 = zeros( (2), float)
k3 = zeros( (2), float)
k4 = zeros( (2), float)
k5 = zeros( (2), float)
k6 = zeros( (2), float)
n = 20
y[0] = 1. ;    y[1] = 0.

h = (b - a)/n;    t = a;    j = 0
hmin = h/64;    hmax = h*64                      # Min and max step sizes
flops = 0;    Eexact = 0. ;    error = 0.
sum = 0.

def f( t, y, fReturn ):                          # Force function
    fReturn[0] = y[1]
    fReturn[1] =    - 6.*pow(y[0], 5.)

graph1 = gdisplay( width = 600, height = 600, title = 'RK 45',
                   xtitle = 't', ytitle = 'Y[0]')
funct1 = gcurve(color = color.blue)
graph2 = gdisplay( width = 500, height = 500, title = 'RK45',
                   xtitle = 't', ytitle = 'Y[1]')
funct2 = gcurve(color = color.red)
funct1.plot(pos = (t, y[0]) )
funct2.plot(pos = (t, y[1]) )

while (t < b):                                   # Loop over time
    funct1.plot(pos = (t, y[0]) )
    funct2.plot(pos = (t, y[1]) )
    if ( (t  +  h) > b ):
        h = b  -  t                              # Last step
    f(t, y, fReturn)                             # Evaluate f, return in fReturn
    k1[0] = h*fReturn[0];           k1[1] = h*fReturn[1]
    for i in range(0, 2):
```

```
        ydumb[i] = y[i]  +  k1[i]/4
    f(t  +  h/4, ydumb, fReturn)
    k2[0] = h*fReturn[0];        k2[1] = h*fReturn[1]
    for i in range(0, 2):
        ydumb[i] = y[i] + 3*k1[i]/32  +  9*k2[i]/32
    f(t  +  3*h/8, ydumb, fReturn)
    k3[0] = h*fReturn[0];  k3[1] = h*fReturn[1]
    for i in range(0, 2):
        ydumb[i] = y[i]  +  1932*k1[i]/2197 - 7200*k2[i]/2197.  +
             7296*k3[i]/2197
    f(t  +  12*h/13, ydumb, fReturn)
    k4[0] = h*fReturn[0]; k4[1] = h*fReturn[1]
    for i in range(0, 2):
        ydumb[i] = y[i] + 439*k1[i]/216  - 8*k2[i] +  3680*k3[i]/513  -
             845*k4[i]/4104
    f(t  +  h, ydumb, fReturn)
    k5[0] = h*fReturn[0]; k5[1] = h*fReturn[1]
    for i in range(0, 2):
        ydumb[i] = y[i]  - 8*k1[i]/27  +  2*k2[i] - 3544*k3[i]/2565  +
             1859*k4[i]/4104  - 11*k5[i]/40
    f(t  +  h/2, ydumb, fReturn)
    k6[0] = h*fReturn[0]; k6[1] = h*fReturn[1];
    for i in range(0, 2):
        err[i] = abs( k1[i]/360  -  128*k3[i]/4275  -  2197*k4[i]/75240
             +  k5[i]/50.  + 2*k6[i]/55)
    if ( err[0] < Tol or err[1] < Tol or h <=  2*hmin ):  # Accept step
        for i in range(0, 2):
            y[i] = y[i]  +  25*k1[i]/216.  +  1408*k3[i]/2565.  +
                2197*k4[i]/4104.  -  k5[i]/5.
        t = t  +  h
        j = j  +  1
    if ( err[0] == 0 or err[1] == 0 ):
        s = 0                                      # Trap division by 0
    else:
        s = 0.84*pow(Tol*h/err[0], 0.25)           # Reduce step
    if ( s  <  0.75 and h > 2*hmin ):
        h /=  2.                                   # Increase step
    else:
        if ( s > 1.5 and 2* h  <  hmax ):
            h *=  2.
    flops = flops  + 1
    E = pow(y[0], 6.)  +  0.5*y[1]*y[1]
    Eexact = 1.
    error = abs( (E - Eexact)/Eexact)
    sum   +=  error
print(" <error>=  ", sum/flops , ", flops = ", flops)
```

8.7 Adams–Bashforth–Moulton Predictor–Corrector Rule

Another approach for obtaining high precision in an ODE algorithm uses the solution from previous steps y_{n-2} and y_{n-1}, in addition to y_n, to predict y_{n+1}. (The Euler and `rk` methods use just one previous step.) Many of these types of methods tend to be like a Newton's search method; we start with a guess or *prediction* for the next step and then use an algorithm such as `rk4` to check on the prediction and thereby obtain a *correction*. As with `rk45`, one can use the correction as a measure of the error and then adjust the step size to obtain improved precision (Press *et al.*, 1994). For those readers who may want to explore such methods, `ABM.py` in Listing 8.3 gives our implementation of the *Adams–Bashforth–Moulton* predictor–corrector scheme.

$$y_{n+1} = y_n + \frac{1}{6}(k_0 + 2k_1 + 2k_2 + k_3) ,$$

$$k_0 = hf(t_n, y_n) , \qquad k_1 = hf\left(t_n + \frac{h}{2}, y_n + \frac{k_1}{2}\right) ,$$

$$k_2 = hf\left(t_n + \frac{h}{2}, y_n + \frac{k_2}{2}\right) , \quad k_3 = hf(t_n + h, y_n + k_3) .$$

Listing 8.3 ABM.py solves an ODE with the RHS given by the method $f()$ using the ABM predictor–corrector algorithm.

```
# ABM.py:     Adams BM method to integrate ODE
# Solves y' = (t - y)/2,     with y[0] = 1 over [0, 3]

from visual.graph import *

numgr = gdisplay(x=0, y=0, width=600, height=300, xmin=0.0, xmax = 3.0,
            title="Numerical Solution", xtitle='t', ytitle='y', ymax=2.,
                ymin=0.9)
numsol = gcurve(color=color.yellow, display = numgr)
exactgr = gdisplay(x=0, y=300, width=600, height=300, title="Exact
     solution",
            xtitle='t', ytitle='y', xmax=3.0, xmin=0.0, ymax=2.0,
                 ymin=0.9)

exsol =  gcurve(color = color.cyan, display = exactgr)
n = 24                                                 # N steps > 3
A = 0; B = 3.
t =[0]*500;        y =[0]*500;       yy=[0]*4

def f(t, y):                                          # RHS F function
    return  (t - y)/2.0

def rk4(t, yy, h1):
    for i in range(0, 3):
        t  = h1 * i
        k0 = h1 * f(t, y[i])
        k1 = h1 * f(t + h1/2., yy[i] + k0/2.)
        k2 = h1 * f(t + h1/2., yy[i] + k1/2.)
        k3 = h1 * f(t + h1, yy[i] + k2 )
        yy[i + 1] = yy[i]  +  (1./6.) * (k0  +  2.*k1  +  2.*k2 + k3)
        print(i,yy[ i])
    return yy[3]

def ABM(a,b,N):
# Compute 3 additional starting values using rk
    h = (b-a) / N                              # step
    t[0] = a;      y[0] = 1.00;     F0 = f(t[0], y[0])
    for k in range(1, 4):
       t[k] = a  +  k * h
    y[1]  = rk4(t[1], y, h)                            # 1st step
    y[2]  = rk4(t[2], y, h)                            # 2nd step
    y[3] = rk4(t[3], y, h)                             # 3rd step
    F1 = f(t[1], y[1])
    F2 = f(t[2], y[2])
    F3 = f(t[3], y[3])
    h2 = h/24.

    for k in range(3, N):                              # Predictor
```

```
        p = y[k]  +  h2*(-9.*F0  +  37.*F1 - 59.*F2 + 55.*F3)
        t[k + 1] = a + h*(k+1)                          # Next abscissa
        F4 = f(t[k+1], p)
        y[k+1] = y[k] + h2*(F1-5.*F2 + 19.*F3 + 9.*F4)   # Corrector
        F0 = F1                                         # Update values
        F1 = F2
        F2 = F3
        F3 = f(t[k + 1], y[k + 1])
    return t,y

print("   k       t       Y numerical       Y exact")
t, y = ABM(A,B,n)
for k in range(0, n+1):
    print (" %3d  %5.3f  %12.11f  %12.11f "
        %(k,t[k],y[k],(3.*exp(-t[k]/2.)-2.+t[k])))
    numsol.plot(pos = (t[k], y[k]) )
    exsol.plot(pos = (t[k], 3.*exp(-t[k]/2.) -2. + t[k]))
```

8.7.1 Assessment: rk2 vs. rk4 vs. rk45

While you are free to do as you please, unless you are very careful, we recommend that you *not* write your own rk4 or rk45 methods. You will be using this code for some high-precision work, and unless you get every fraction and method call just right, your code may appear to work well but still not give all the precision that you should be obtaining, and therefore we give you rk4, and rk45 codes to use. However, we do recommend that you write your own rk2, as doing so will make it clearer as to how the Runge–Kutta methods work, but without all the pain and danger.

1. Write your own rk2 method. Design your method for a general ODE; this means making the derivative function $f(t, x)$ a separate method.
2. Use your rk2 solver in a program that solves the equation of motion (8.7) or (8.25). Plot both the position $x(t)$ and velocity $\mathrm{d}x/\mathrm{d}t$ as functions of time.
3. Once your ODE solver is running, do a number of things to check that it is working well and that you know what h values to use.
   a) Adjust the parameters in your potential so that it corresponds to a pure harmonic oscillator (set $p = 2$ or $\alpha = 0$). For an oscillator initially at rest, we have an analytic result with which to compare:

$$x(t) = A\sin(\omega_0 t), \quad v = \omega_0 A\cos(\omega_0 t), \quad \omega_0 = \sqrt{k/m}. \tag{8.40}$$

   b) Pick values of k and m such that the period $T = 2\pi/\omega$ is a nice number with which to work (something like $T = 1$).
   c) Start with a step size $h \simeq T/5$ and make h smaller until the solution looks smooth, has a period that remains constant for a large number of cycles, and agrees with the analytic result. As a general rule of thumb, we suggest that you start with $h \simeq T/100$, where T is a characteristic time for the problem at hand. You should start with a large h so that you can see a bad solution turn good.

d) Make sure that you have exactly the same initial conditions for the analytic and numerical solutions (zero displacement, nonzero velocity) and then plot the two together. It is good if you cannot tell them apart, yet that only ensures that there are approximately two places of agreement.

e) Try different initial velocities and verify that a *harmonic* oscillator is *isochronous*, that is, that its period does *not* change as the amplitude varies.

4. Now that you know you can get a good solution of an ODE with `rk2`, compare the solutions obtained with the `rk2`, `rk4`, and `rk45` solvers.
5. Make a table of comparisons similar to Table 8.1. There we compare `rk4` and `rk45` for the two equations

$$2yy'' + y^2 - y'^2 = 0 , \tag{8.41}$$

$$y'' + 6y^5 = 0 , \tag{8.42}$$

with initial conditions $([y(0), y'(0)] = (1, 1)$. Although nonlinear, (8.41) does have the analytic solution $y(t) = 1 + \sin t$[4]. Equation 8.42 corresponds to our standard potential (8.5), with $p = 6$. Although we have not tuned `rk45`, Table 8.2 shows that by setting its tolerance parameter to a small enough number, `rk45` will obtain better precision than `rk4` (Figure 8.6), but that it requires ~ 10 times more floating-point operations and takes ~ 5 times longer. Yet for (8.41), we obtained increased precision in less time.

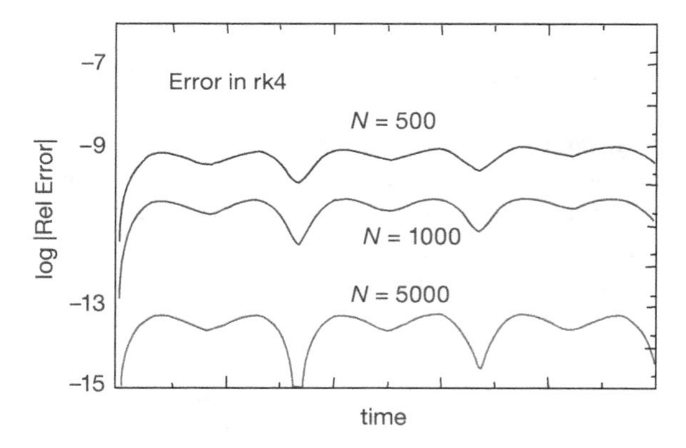

Figure 8.6 The logarithm of the relative error in the solution of an ODE obtained with rk4 using a differing number N of time steps over a fixed time interval. The logarithm approximately equals the negative of the number of places of precision. Increasing the number of steps used for a fixed interval is seen to lead to smaller error.

4) Be warned, the **rk** procedures may be inaccurate for this equation if integrated exactly through the point $y(t) = 0$ because then terms in the equation proportional to y vanish and this leaves $y'^2 = 0$, which is problematic. A different algorithm may be better there.

Table 8.1 Comparison of ODE solvers for different equations.

Eq. No.	Method	Initial h	No. of flops	Time (ms)	Relative error
(8.41)	rk4	0.01	1000	5.2	2.2×10^{-8}
	rk45	1.00	72	1.5	1.8×10^{-8}
(8.42)	rk4	0.01	227	8.9	1.8×10^{-8}
	rk45	0.1	3143	36.7	5.7×10^{-11}

8.8 Solution for Nonlinear Oscillations (Assessment)

Use your rk4 program to study anharmonic oscillations by trying powers in the range $p = 2$–12, or anharmonic strengths in the range $0 \leq \alpha x \leq 2$. Do *not* include any explicit time-dependent forces yet. Note that for large values of p, the forces and accelerations get large near the turning points, and so you may need a smaller step size h than that used for the harmonic oscillator.

1. Check that the solution remains periodic with constant amplitude and period for a given initial condition regardless of how nonlinear you make the force. In particular, check that the maximum speed occurs at $x = 0$ and that the velocity $v = 0$ at maximum x's, the latter being a consequence of energy conservation.
2. Verify that nonharmonic oscillators are *nonisochronous*, that is, vibrations with different amplitudes have different periods (Figure 8.7).
3. Explain why the shapes of the oscillations change for different p's or α's.
4. Devise an algorithm to determine the period T of the oscillation by recording times at which the mass passes through the origin. Note that because the mo-

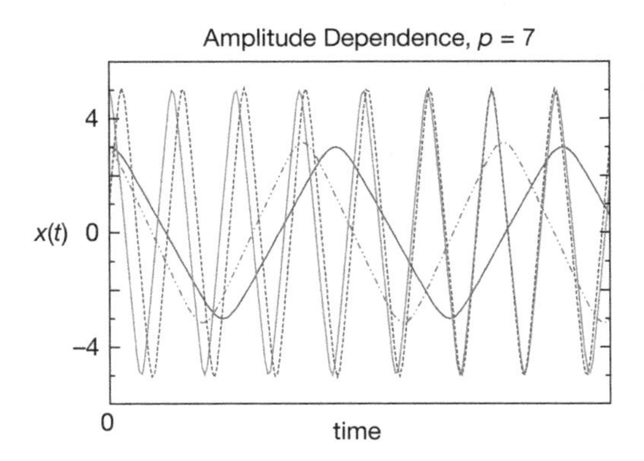

Figure 8.7 The position vs. time for oscillations within the potential $V \propto x^7$ for four different initial amplitudes. Each is seen to have a different period.

tion may be asymmetric, you must record at least three times to deduce the period.

5. Construct a graph of the deduced period as a function of initial amplitude.
6. Verify that the motion is oscillatory, but not harmonic, as the energy approaches $k/6\alpha^2$ or for $p > 6$.
7. Verify that for the anharmonic oscillator with $E = k/6\alpha^2$, the motion changes from oscillatory to translational. See how close you can get to the *separatrix* where a single oscillation takes an infinite time. (There is no separatrix for the power-law potential.)

8.8.1 Precision Assessment: Energy Conservation

We have not explicitly built energy conservation into our ODE solvers. Nonetheless, unless you have explicitly included a frictional force, it follows mathematically from the form of the equations of motion that energy must be a constant for all values of p or α. That being the case, the constancy of energy is a demanding test of the numerics.

1. Plot the potential energy $\text{PE}(t) = V[x(t)]$, the kinetic energy $\text{KE}(t) = mv^2(t)/2$, and the total energy $E(t) = \text{KE}(t) + \text{PE}(t)$, for 50 periods. Comment on the correlation between $\text{PE}(t)$ and $\text{KE}(t)$ and how it depends on the potential's parameters.
2. Check the long-term *stability* of your solution by plotting

$$-\log_{10}\left|\frac{E(t) - E(t=0)}{E(t=0)}\right| \simeq \text{number of places of precision} \tag{8.43}$$

   for a large number of periods (Figure 8.6). Because $E(t)$ should be independent of time, the numerator is the absolute error in your solution, and when divided by $E(0)$, becomes the relative error (approximately 10^{-11}). If you cannot achieve 11 or more places, then you need to decrease the value of h or debug.
3. Because a particle bound by a large-p oscillator is essentially "free" most of the time, you should observe that the average of its kinetic energy over time exceeds its average potential energy. This is actually the physics behind the Virial theorem for a power-law potential (Marion and Thornton, 2003):

$$\langle \text{KE} \rangle = \frac{p}{2}\langle \text{PE} \rangle \,. \tag{8.44}$$

   Verify that your solution satisfies the Virial theorem. (Those readers who have worked the perturbed oscillator problem can use this relation to deduce an effective p value, which should be between 2 and 3.)

8.9 Extensions: Nonlinear Resonances, Beats, Friction

Problem So far our oscillations have been rather simple. We have ignored friction and have assumed that there are no external forces (hands) influencing the system's natural oscillations. Determine

1. How the oscillations change when friction is included?
2. How the resonances and beats of nonlinear oscillators differ from those of linear oscillators?
3. How introducing friction affects resonances?

8.9.1 Friction (Model)

The world is full of friction, and not all of it is bad. For while friction makes it harder to pedal a bike through the wind, it also lets you walk on ice, and generally adds stability to dynamical systems. The simplest models for frictional force are called *static*, *kinetic*, and *viscous* friction:

$$F_f^{(\text{static})} \leq -\mu_s N\,, \quad F_f^{(\text{kinetic})} = -\mu_k N \frac{v}{|v|}\,, \quad F_f^{(\text{viscous})} = -bv\,. \tag{8.45}$$

Here N is the *normal force* on the object under consideration, μ and b are parameters, and v is the velocity. This model for static friction is appropriate for objects at rest, while the model for kinetic friction is appropriate for an object sliding on a dry surface. If the surface is lubricated, or if the object is moving through a viscous medium, then a frictional force dependent on velocity is a better model.[5)]

1. Extend your harmonic oscillator code to include the three types of friction in (8.45) and observe how the motion differs for each.
2. *Hint:* For the simulation with static plus kinetic friction, each time the oscillator has $v = 0$ you need to check that the restoring force exceeds the static force of friction. If not, the oscillation must end at that instant. Check that your simulation terminates at nonzero x values.
3. For your simulations with viscous friction, investigate the qualitative changes that occur for increasing b values:

Underdamped:	$b < 2m\omega_0$	Oscillate within decaying envelope
Critically:	$b = 2m\omega_0$	Nonoscillatory, finite decay time
Overdamped:	$b > 2m\omega_0$	Nonoscillatory, infinite decay time

5) The effect of air resistance on projectile motion is studied in Section 9.6.

8.9.2
Resonances and Beats: Model, Implementation

Stable physical systems will oscillate if displaced slightly from their rest positions. The frequency ω_0 with which such a system executes small oscillations about its rest positions is called its *natural frequency*. If an external sinusoidal force is applied to this system, and if the frequency of the external force equals the natural frequency ω_0, then a *resonance* may occur in which the oscillator absorbs energy from the external force and the amplitude of oscillation increases with time. If the oscillator and the driving force remain in phase over time, the amplitude of oscillation will increase continuously unless there is some mechanism, such as friction or nonlinearities, to limit the growth.

If the frequency of the driving force is close to, but not exactly equal to, the natural frequency of the oscillator, then a related phenomena, known as *beating*, may occur. In this situation there is interference between the natural oscillation, which is independent of the driving force, and the oscillation resulting from the external force. If the frequency of the external driving force is very close to the natural frequency, then the resulting motion

$$x \simeq x_0 \sin \omega t + x_0 \sin \omega_0 t = \left(2x_0 \cos \frac{\omega - \omega_0}{2} t\right) \sin \frac{\omega + \omega_0}{2} t \,, \tag{8.46}$$

resembles the natural oscillation of the system at the average frequency $(\omega + \omega_0)/2$, yet with an amplitude $2x_0 \cos((\omega - \omega_0)/2)t$ that varies slowly with the *beat frequency* $(\omega - \omega_0)/2$.

8.10
Extension: Time-Dependent Forces

To extend our simulation to include an external force,

$$F_{\text{ext}}(t) = F_0 \sin \omega t \,, \tag{8.47}$$

we need to include a time dependence in the force function $\boldsymbol{f}(t, \boldsymbol{y})$ of our ODE solver.

1. Add the sinusoidal time-dependent external force (8.47) to the space-dependent restoring force in your program (do not include friction yet).
2. Start with a very large value for the magnitude of the driving force F_0. This should lead to *mode locking* (the 500-pound-gorilla effect), where the system is overwhelmed by the driving force and, after the transients die out, the system oscillates in phase with the driver regardless of the driver's frequency.
3. Now lower F_0 until it is close to the magnitude of the natural restoring force of the system. You need to have this near equality for beating to occur.
4. Verify that for the harmonic oscillator, the beat frequency, that is, the number of variations in intensity per unit time, equals the frequency difference $(\omega - \omega_0)/2\pi$ in cycles per second, where $\omega \simeq \omega_0$.

5. Once you have a value for F_0 matched well with your system, make a series of runs in which you progressively increase the frequency of the driving force in the range $\omega_0/10 \leq \omega \leq 10\omega_0$.
6. Make of plot of the maximum amplitude of oscillation vs. the driver's ω.
7. Explore what happens when you make a nonlinear system resonate. If the nonlinear system is close to being harmonic, you should get beating in place of the blowup that occurs for the linear system. Beating occurs because the natural frequency changes as the amplitude increases, and thus the natural and forced oscillations fall out of phase. Yet once out of phase, the external force stops feeding energy into the system, and so the amplitude decreases, and with the decrease in amplitude, the frequency of the oscillator returns to its natural frequency, the driver and oscillator get back in phase, and the entire cycle repeats.
8. Investigate now how the inclusion of viscous friction modifies the curve of amplitude vs. driver frequency. You should find that friction broadens the curve.
9. Explain how the character of the resonance changes as the exponent p in the potential $V(x) = k|x|^p/p$ is made larger and larger. At large p, the mass effectively "hits" the wall and falls out of phase with the driver, and so the driver is less effective at pumping energy into the system.

9
ODE Applications: Eigenvalues, Scattering, and Projectiles

Now that we know how to solve ODEs numerically, we use our newfound skills in some different ways. First, we combine our ODE solver with a search algorithm to solve the quantum eigenvalue problem. Then we solve some of the simultaneous ODEs that arise in the scattering problem, and explore classical chaotic scattering. Finally, we look upward to balls falling out of the sky and planets that do not.

9.1
Problem: Quantum Eigenvalues in Arbitrary Potential

Quantum mechanics describes phenomena that occur on atomic or subatomic scales (an elementary particle is subatomic). It is a statistical theory in which the probability that a particle is located in a region $\mathrm{d}x$ around the point x is $\mathcal{P} = |\psi(x)|^2\,\mathrm{d}x$, where $\psi(x)$ is called the *wave function*. If a particle of definite energy E moving in one dimension experiences a potential $V(x)$, its wave function is determined by an ordinary differential equation (partial differential equation for more than one dimension) known as the time-independent Schrödinger equation[1]:

$$\frac{-\hbar^2}{2m}\frac{\mathrm{d}^2\psi(x)}{\mathrm{d}x^2} + V(x)\psi(x) = E\psi(x)\,. \tag{9.1}$$

Although we say we are solving for the energy E, in practice we solve for the wave vector κ, where the two are related for bound states by

$$\kappa^2 = -\frac{2m}{\hbar^2}E = \frac{2m}{\hbar^2}|E|\,. \tag{9.2}$$

The Schrödinger equation now takes the form

$$\frac{\mathrm{d}^2\psi(x)}{\mathrm{d}x^2} - \frac{2m}{\hbar^2}V(x)\psi(x) = \kappa^2\psi(x)\,. \tag{9.3}$$

1) The time-dependent equation requires the solution of a partial differential equation, as discussed in Chapter 22.

Computational Physics, 3rd edition. Rubin H. Landau, Manuel J. Páez, Cristian C. Bordeianu.

When our problem tells us that the particle is bound, this means that it is confined to some finite region of space, which implies that $\psi(x)$ is normalizeable. The only way for that to happen is if $\psi(x)$ decay exponentially as $x \to \pm\infty$ (where the potential vanishes):

$$\psi(x) \to \begin{cases} e^{-\kappa x}, & \text{for } x \to +\infty \,, \\ e^{+\kappa x}, & \text{for } x \to -\infty \,. \end{cases} \tag{9.4}$$

In summary, although it is straightforward to solve the ODE (9.1) with the techniques we have learned so far, we must also require that the solution $\psi(x)$ simultaneously satisfies the boundary conditions (9.4). This extra condition turns the ODE problem into an *eigenvalue problem* that has solutions (*eigenvalues*) for only certain values of the energy E or equivalently κ. The ground-state energy corresponds to the smallest (most negative) eigenvalue. The ground-state wave function (eigenfunction), which we must determine in order to find its energy, must be nodeless and even (symmetric) about $x = 0$. The excited states have higher (less negative) energies and wave functions that may be odd (antisymmetric).

9.1.1
Model: Nucleon in a Box

The numerical methods we describe are capable of handling the most realistic potential shapes. Yet to make a connection with the standard textbook case and to permit some analytic checking, we will use a simple model in which the potential $V(x)$ in (9.1) is a finite square well (Figure 9.1):

$$V(x) = \begin{cases} -V_0 = -83\,\text{MeV}, & \text{for } |x| \le a = 2\,\text{fm} \,, \\ 0, & \text{for } |x| > a = 2\,\text{fm} \,, \end{cases} \tag{9.5}$$

where values of 83 MeV for the depth and 2 fm for the radius are typical for nuclei (these are the units in which we solve the problem). With this potential, the Schrödinger equation (9.3) becomes

$$\frac{d^2\psi(x)}{dx^2} + \left(\frac{2m}{\hbar^2}V_0 - \kappa^2\right)\psi(x) = 0 \,, \quad \text{for } |x| \le a \,, \tag{9.6}$$

$$\frac{d^2\psi(x)}{dx^2} - \kappa^2\psi(x) = 0 \,, \quad \text{for } |x| > a \,. \tag{9.7}$$

To evaluate the ratio of constants here, we insert c^2, the speed of light squared, into both the numerator and the denominator and then these familiar values (Landau, 1996):

$$\frac{2m}{\hbar^2} = \frac{2mc^2}{(\hbar c)^2} \simeq \frac{2 \times 940\,\text{MeV}}{(197.32\,\text{MeV fm})^2} = 0.0483\,\text{MeV}^{-1}\,\text{fm}^{-2} \,. \tag{9.8}$$

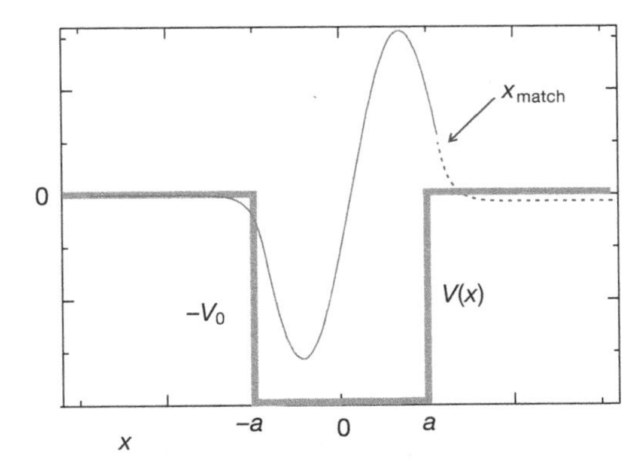

Figure 9.1 Computed wave function and the square-well potential (bold lines). The wave function computed by integration in from the left is matched to the one computed by integration in from the right (dashed curve) at a point near the right edge of the well. Note how the wave function decays rapidly outside the well.

9.2
Dual Algorithms: Eigenvalues via ODE Solver + Search

The solution of the eigenvalue problem combines the numerical solution of the ordinary differential equation (9.3) with a trial-and-error search for a wave function that satisfies the boundary conditions (9.4). This is carried out in several steps:[2)]

1. Start on the very far *left* at $x = -X_{\rm max} \simeq -\infty$, where $X_{\rm max} \gg a$. Because the potential $V = 0$ in this region, the analytic solution here is $e^{\pm\kappa x}$. Accordingly, assume that the wave function there satisfies the left-hand boundary condition:

$$\psi_{\rm L}(x = -X_{\rm max}) = e^{+\kappa x} = e^{-\kappa X_{\rm max}} . \tag{9.9}$$

2. Use your favorite ODE solver to step $\psi_{\rm L}(x)$ in toward the origin (to the right) from $x = -X_{\rm max}$ until you reach the *matching radius* $x_{\rm match}$. The exact value of this matching radius is not important, and our final solution should be independent of it. In Figure 9.1, we show a sample solution with $x_{\rm match} = +a$; that is, we match at the right edge of the potential well. In Figure 9.2 we see some guesses that do not match.
3. Start at the extreme *right*, that is, at $x = +X_{\rm max} \simeq +\infty$, with a wave function that satisfies the right-hand boundary condition:

$$\psi_{\rm R}(x = +\kappa X_{\rm max}) = e^{-\kappa x} = e^{-\kappa X_{\rm max}} . \tag{9.10}$$

2) The procedure outlined here is for a general potential that falls off gradually. For a square well with sharp cutoffs, the analytic solution is valid right up to the walls, and we could start integrating inwards from there in this special case. In contrast, if we were working with a Coulomb potential, its very slow falloff would does not match onto a simple exponential, even at infinity (Landau, 1996).

4. Use your `rk4` ODE solver to step $\psi_R(x)$ in toward the origin (to the left) from $x = +X_{max}$ until you reach the *matching radius* x_{match}
5. In order for probability and current to be continuous at $x = x_{match}$, $\psi(x)$ and $\psi'(x)$ must be continuous there. Requiring the ratio $\psi'(x)/\psi(x)$, called the *logarithmic derivative*, to be continuous there encapsulates both continuity conditions into a single condition and is independent of ψ's normalization.
6. Although we do not know ahead of time which E or κ values are eigenvalues, we still need a starting value for the energy in order to use our ODE solver. Such being the case, we start the solution with a guess for the energy. A good guess for ground-state energy would be a value somewhat up from that at the bottom of the well, $E > -V_0$.
7. Because it is unlikely that any guess will be correct, the left- and right-wave functions will not quite match at $x = x_{match}$ (Figure 9.2). This is okay because we can use the amount of mismatch to improve the next guess. We measure how well the right and left wave functions match by calculating the difference in logarithmic derivatives:

$$\Delta(E, x) = \left. \frac{\psi'_L(x)/\psi_L(x) - \psi'_R(x)/\psi_R(x)}{\psi'_L(x)/\psi_L(x) + \psi'_R(x)/\psi_R(x)} \right|_{x=x_{match}}, \tag{9.11}$$

where the denominator is used to avoid overly large or small numbers. Next, we try a different energy, note how much $\Delta(E)$ has changed, and use this to deduce an intelligent guess at the next energy. The search continues until the left and right ψ'/ψ match within some set tolerance that depends on the precision in energy desired.

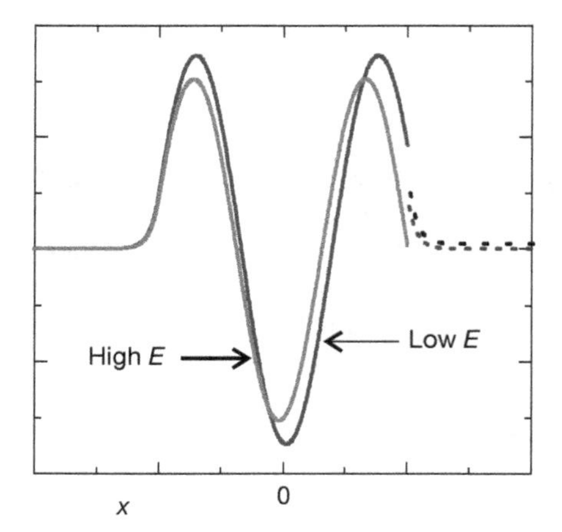

Figure 9.2 Two guesses for the energy that are either too low or too high to be an eigenvalue. We see that the low-E guess does not oscillate enough to match onto a dying exponential, and that the high-E guess oscillates too much.

9.2.1 Numerov Algorithm for Schrödinger ODE ⊙

We generally recommend the fourth-order Runge–Kutta method for solving ODEs, and its combination with a search routine for solving the eigenvalue problem. In this section we present the Numerov method, an algorithm that is specialized for ODEs not containing any first derivatives (such as our Schrödinger equation). While this algorithm is not as general as rk4, it is of $\mathcal{O}(h^6)$ and thus speeds up the calculation by providing additional precision.

We start by rewriting the Schrödinger equation (9.3) in the generic form

$$\frac{\mathrm{d}^2\psi}{\mathrm{d}x^2} + k^2(x)\psi = 0\,, \quad k^2(x) = \frac{2m}{\hbar^2}\begin{cases} E + V_0, & \text{for } |x| < a\,, \\ E, & \text{for } |x| > a\,, \end{cases} \tag{9.12}$$

where $k^2 = -\kappa^2$ for bound states. Observe that although (9.12) is specialized to a square well, other potentials would have a function $V(x)$ in place of $-V_0$. The trick in the Numerov method is to get extra precision in the second derivative by taking advantage of there being no first derivative $\mathrm{d}\psi/\mathrm{d}x$ in (9.12). We start with the Taylor expansions of the wave functions

$$\psi(x+h) \simeq \psi(x) + h\psi^{(1)}(x) + \frac{h^2}{2}\psi^{(2)}(x) + \frac{h^3}{3!}\psi^{(3)}(x) + \frac{h^4}{4!}\psi^{(4)}(x) + \cdots \tag{9.13}$$

$$\psi(x-h) \simeq \psi(x) - h\psi^{(1)}(x) + \frac{h^2}{2}\psi^{(2)}(x) - \frac{h^3}{3!}\psi^{(3)}(x) + \frac{h^4}{4!}\psi^{(4)}(x) + \cdots\,, \tag{9.14}$$

where $\psi^{(n)}$ signifies the nth derivative $\mathrm{d}^n\psi/\mathrm{d}x^n$. Because the expansion of $\psi(x-h)$ has odd powers of h appearing with negative signs, all odd powers cancel when we add $\psi(x+h)$ and $\psi(x-h)$ together:

$$\psi(x+h) + \psi(x-h) \simeq 2\psi(x) + h^2\psi^{(2)}(x) + \frac{h^4}{12}\psi^{(4)}(x) + \mathcal{O}(h^6)\,, \tag{9.15}$$

$$\Rightarrow \quad \psi^{(2)}(x) \simeq \frac{\psi(x+h) + \psi(x-h) - 2\psi(x)}{h^2} - \frac{h^2}{12}\psi^{(4)}(x) + \mathcal{O}(h^4)\,. \tag{9.16}$$

To obtain an algorithm for the second derivative, we eliminate the fourth-derivative term by applying the operator $1 + (h^2/12)(\mathrm{d}^2/\mathrm{d}x^2)$ to the Schrödinger equation (9.12):

$$\psi^{(2)}(x) + \frac{h^2}{12}\psi^{(4)}(x) + k^2(x)\psi + \frac{h^2}{12}\frac{\mathrm{d}^2}{\mathrm{d}x^2}[k^2(x)\psi^{(4)}(x)] = 0\,. \tag{9.17}$$

We eliminate the $\psi^{(4)}$ terms by substituting the derived expression for the $\psi^{(2)}$:

$$\frac{\psi(x+h)+\psi(x-h)-2\psi(x)}{h^2}+k^2(x)\psi(x)+\frac{h^2}{12}\frac{\mathrm{d}^2}{\mathrm{d}x^2}[k^2(x)\psi(x)]\simeq 0\,. \tag{9.18}$$

Now we use a central-difference approximation for the second derivative of $k^2(x)\psi(x)$:

$$h^2\frac{\mathrm{d}^2[k^2(x)\psi(x)]}{\mathrm{d}x^2}\simeq[(k^2\psi)_{x+h}-(k^2\psi)_x]+[(k^2\psi)_{x-h}-(k^2\psi)_x]\,. \tag{9.19}$$

After this substitution, we obtain the Numerov algorithm:

$$\psi(x+h)\simeq\frac{2\left[1-\frac{5}{12}h^2k^2(x)\right]\psi(x)-\left[1+\frac{h^2}{12}k^2(x-h)\right]\psi(x-h)}{\frac{1+h^2k^2(x+h)}{12}}\,. \tag{9.20}$$

We see that the Numerov algorithm uses the values of ψ at the two previous steps x and $x-h$ to move ψ forward to $x+h$. To step backward in x, we need only to reverse the sign of h. Our implementation of this algorithm, Numerov.py, is given in Listing 9.1.

Listing 9.1 QuantumNumerov.py solves the 1D time-independent Schrödinger equation for bound-state energies using a Numerov method (rk4 also works, as we show in Listing 9.2).

```
# QuantumNumerov: Quantum BS via Numerov ODE solver + search

from visual import *
from visual.graph import *

psigr  = display(x=0, y=0, width=600,height=300,title='R & L Wave Funcs')
psi    = curve(x=list(range(0,1000)), display=psigr, color=color.yellow)
psi2gr = display(x=0,y=300,width=600,height=200,title='Wave func^2')
psio   = curve(x=list(range(0,1000)),color=color.magenta, display=psi2gr)
energr = display(x=0, y=500, width=600,height=200,title='Potential & E')
poten  = curve(x=list(range(0,1000)), color=color.cyan, display=energr)
autoen = curve(x=list(range(0,1000)), display=energr)

dl     = 1e-6                         # very small interval to stop bisection
ul     = zeros([1501], float)
ur     = zeros([1501], float)
k2l    = zeros([1501], float)                                # k**2 left wavefunc
k2r    = zeros([1501], float)
n      = 1501
m      = 5                                                   # plot every 5 points
imax   = 100
xl0    = -1000;    xr0    =  1000                           # leftmost, rightmost x
h      = 1.0*(xr0-xl0)/(n-1.)
amin   = -0.001; amax   = -0.00085                                   # root limits
e      = amin                                                    # Initial E guess
de     = 0.01
ul[0]  = 0.0; ul[1] = 0.00001; ur[0] = 0.0; ur[1] = 0.00001
im     = 500                                                         # match point
nl     = im+2; nr = n-im+1                                         # left, right wv
istep  = 0
```

```
def V(x):                                                    # Square well
    if (abs(x)<=500):       v = -0.001
    else:                   v = 0
    return v

def setk2():                                                  #  k2
    for i in range(0,n):
        xl = xl0+i*h
        xr = xr0-i*h
        k2l[i] = e-V(xl)
        k2r[i] = e-V(xr)

def numerov (n,h,k2,u):                                  # Numerov algorithm
    b=(h**2)/12.0
    for i in range(1, n-1):
      u[i+1] = (2*u[i]*(1-5*b*k2[i]) -(1.+b*k2[i-1])*u[i-1])/(1+b*k2[i+1])

setk2()
numerov (nl, h, k2l, ul)                                         # Left psi
numerov (nr, h, k2r, ur)                                        # Right psi
fact= ur[nr-2]/ul[im]                                               # Scale
for i  in range (0,nl): ul[i] = fact*ul[i]
f0 = (ur[nr-1]+ul[nl-1]-ur[nr-3]-ul[nl-3])/(2*h*ur[nr-2])    #  Log deriv

def normalize():
    asum = 0
    for i in range( 0,n):
        if i > im :
           ul[i] = ur[n-i-1]
           asum = asum+ul[i]* ul[i]
    asum         = sqrt(h*asum);
    elabel       = label(pos=(700, 500), text='e=', box=0,display=psigr)
    elabel.text  = 'e=%10.8f' %e
    ilabel       = label(pos=(700,400),text='istep=',box=0,display=psigr)
    ilabel.text  = 'istep=%4s' %istep
    poten.pos    = [(-1500,200),(-1000,200),(-1000,-200),
                                (0,-200),(0,200),(1000,200)]
    autoen.pos = [(-1000,e*400000.0+200),(0,e*400000.0+200)]
    label(pos=(-1150,-240), text='0.001', box=0, display=energr)
    label(pos=(-1000,300),  text='0',     box=0, display=energr)
    label(pos=(-900,180),   text='-500',  box=0, display=energr)
    label(pos=(-100,180),   text='500',   box=0, display=energr)
    label(pos=(-500,180),   text='0',     box=0, display=energr)
    label(pos=(900,120),    text='r',     box=0, display=energr)
    j=0
    for i in range(0,n,m):
        xl          = xl0 + i*h
        ul[i]       = ul[i]/asum                      # wave function normalized
        psi.x[j]    = xl - 500                                       # plot psi
        psi.y[j]    = 10000.0*ul[i]               # vertical line for match of wvfs
        line        = curve(pos=[(-830,-500),(-830,500)],
                      color=color.red, display=psigr)
        psio.x[j]   = xl-500                                         # plot psi
        psio.y[j]   = 1.0e5*ul[i]**2
        j +=1

while abs(de) > dl and istep < imax :                     # bisection algorithm
    rate(2)                                                  # Slow animation
    e1 = e
    e  = (amin+amax)/2
    for i in range(0,n):
        k2l[i] = k2l[i] + e-e1
        k2r[i] = k2r[i] + e-e1
    im = 500;
    nl = im+2
    nr = n-im+1;
    numerov (nl,h,k2l,ul)                                    # New wavefuntions
```

```
numerov (nr,h,k2r,ur)
fact = ur[nr-2]/ul[im]
for i in range(0,nl):   ul[i] = fact*ul[i]
f1 = (ur[nr-1]+ul[nl-1]-ur[nr-3]-ul[nl-3])/(2*h*ur[nr-2]) # Log deriv
rate(2)
if f0*f1 < 0:                                    # Bisection localize root
    amax = e
    de = amax - amin
else:
     amin = e
     de = amax - amin
     f0 = f1
normalize()
istep = istep + 1
```

9.2.2
Implementation: Eigenvalues via ODE Solver + Bisection Algorithm

1. Combine your bisection algorithm search program with your `rk4` or Numerov ODE solver program to create an eigenvalue solver. Start with a step size $h = 0.04$.
2. Write a function that calculates the matching function $\Delta(E, x)$ as a function of energy and matching radius. This subroutine will be called by the bisection algorithm program to search for the energy at which $\Delta(E, x = 2)$ (9.11) vanishes.
3. As a first guess, take $E \simeq 65$ MeV.
4. Search until $\Delta(E, x)$ changes in only the fourth decimal place. We do this in the code `QuantumEigen.py` shown in Listing 9.2.
5. Print out the value of the energy for each iteration. This will give you a feel as to how well the procedure converges, as well as a measure of the precision obtained. Try different values for the tolerance until you are confident that you are obtaining three good decimal places in the energy.
6. Build in a limit to the number of energy iterations you permit, and print out a warning if the iteration scheme fails.
7. As we have shown, plot the wave function and potential on the same graph (you will have to scale one plot to get both of them to fit).
8. Deduce, by counting the number of nodes in the wave function, whether the solution found is a ground state (no nodes) or an excited state (with nodes) and whether the solution is even or odd about the origin (the ground state must be even).
9. Include in your version of Figure 9.1 a horizontal line within the potential indicating the energy of the ground state relative to the potential's depth.
10. Increase the value of the initial energy guess and search for excited states. Make sure to examine the wave function for each state found to establish that it is continuous, and to count the number of nodes to see if you have missed a state.
11. Add each new state found as another horizontal bar within the potential.
12. Verify that you have solved the *problem*, that is, the spacing between levels is on the order of MeV for a nucleon bound in a several-fermi potential well.

Listing 9.2 QuantumEigen.py solves the 1D time-independent Schrödinger equation for bound-state energies using the rk4 algorithm.

```
# QuantumNumerov: Quantum BS via Numerov ODE solver + search

# QuantumEigen.py:                    Finds E and psi via rk4 + bisection

# mass/((hbar*c)**2)= 940MeV/(197.33MeV-fm)**2  =0.4829, well width=20.0 fm
# well depth 10 MeV, Wave function not normalized.

from visual import *

psigr = display(x=0,y=0,width=600,height=300, title='R & L Wavefunc')
Lwf = curve(x=list(range(502)),color=color.red)
Rwf = curve(x=list(range(997)),color=color.yellow)
eps       = 1E-3                                              # Precision
n_steps   = 501
E         = -17.0                                             # E guess
h         = 0.04
count_max = 100
Emax      = 1.1*E                                             # E limits
Emin      = E/1.1

def f(x, y, F,E):
    F[0] = y[1]
    F[1] = -(0.4829)*(E-V(x))*y[0]

def V(x):
    if (abs(x) < 10.):  return (-16.0)                        # Well depth
    else:               return (0.)

def rk4(t, y,h,Neqs,E):
    F  = zeros((Neqs),float)
    ydumb    = zeros((Neqs),float)
    k1 = zeros((Neqs),float)
    k2 = zeros((Neqs),float)
    k3 = zeros((Neqs),float)
    k4 = zeros((Neqs),float)
    f(t, y, F,E)
    for i in range(0,Neqs):
        k1[i] = h*F[i]
        ydumb[i] = y[i] + k1[i]/2.
    f(t + h/2., ydumb, F,E)
    for i in range(0,Neqs):
        k2[i] = h*F[i]
        ydumb[i] = y[i] + k2[i]/2.
    f(t + h/2., ydumb, F,E)
    for i in range(0,Neqs):
        k3[i]=  h*F[i]
        ydumb[i] = y[i] + k3[i]
    f(t + h, ydumb, F,E);
    for i in range(0,Neqs):
        k4[i]=h*F[i]
        y[i]=y[i]+(k1[i]+2*(k2[i]+k3[i])+k4[i])/6.0

def diff(E, h):
    y = zeros((2),float)
    i_match = n_steps//3                                      # Matching radius
    nL = i_match + 1
    y[0] = 1.E-15;                                            # Initial left wf
    y[1] = y[0]*sqrt(-E*0.4829)
    for ix in range(0,nL + 1):
        x = h * (ix  -n_steps/2)
        rk4(x, y, h, 2, E)
    left = y[1]/y[0]                                          # Log  derivative
    y[0] = 1.E-15;                      # slope for even;  reverse for odd
```

```
    y[1] = -y[0]*sqrt(-E*0.4829)                             # Initialize R wf
    for ix in range( n_steps ,nL+1,-1):
        x = h*(ix+1-n_steps /2)
        rk4(x, y, -h, 2, E)
    right = y[1]/y[0]                                        # Log derivative
    return( (left - right)/(left + right) )

def plot(E, h):                                  # Repeat integrations for plot
    x = 0.
    n_steps = 1501                                       # # integration steps
    y = zeros((2) ,float)
    yL = zeros((2,505),float)
    i_match = 500                                            # Matching point
    nL = i_match + 1;
    y[0] = 1.E-40                                           # Initial left wf
    y[1] = -sqrt(-E*0.4829) *y[0]
    for ix in range(0,nL+1):
        yL[0][ix] = y[0]
        yL[1][ix] = y[1]
        x = h * (ix -n_steps/2)
        rk4(x, y, h, 2, E)
    y[0] = -1.E-15                          # - slope: even;  reverse for odd
    y[1] = -sqrt(-E*0.4829)*y[0]
    j=0
    for ix in range(n_steps -1,nL + 2,-1):               # right wave function
        x = h * (ix + 1 -n_steps/2)                              # Integrate in
        rk4(x, y, -h, 2, E)
        Rwf.x[j] = 2.*(ix + 1 -n_steps/2)-500.0
        Rwf.y[j] = y[0]*35e-9 +200
        j +=1
    x = x-h
    normL = y[0]/yL[0][nL]
    j=0
    # Renormalize L wf & derivative
    for ix in range(0,nL+1):
        x = h * (ix-n_steps/2 + 1)
        y[0] = yL[0][ix]*normL
        y[1] = yL[1][ix]*normL
        Lwf.x[j] = 2.*(ix  -n_steps/2+1)-500.0
        Lwf.y[j] = y[0]*35e-9+200                             # Factor for scale
        j +=1
for count in range(0,count_max+1):
    rate(1)                                          # Slow rate to show changes
    # Iteration loop
    E = (Emax + Emin)/2.                                        # Divide E range
    Diff = diff(E, h)
    if (diff(Emax, h)*Diff > 0):   Emax = E              # Bisection algorithm
    else:                          Emin = E
    if ( abs(Diff) < eps ):        break
    if count >3:                                # First iterates too irregular
        rate(4)
        plot(E, h)
    elabel      = label(pos=(700, 400), text='E=', box=0)
    elabel.text = 'E=%13.10f' %E
    ilabel      = label(pos=(700, 600), text='istep=', box=0)
    ilabel.text = 'istep=%4s' %count
elabel      = label(pos=(700, 400), text='E=', box=0)        # Last iteration
elabel.text = 'E=%13.10f' %E
ilabel      = label(pos=(700, 600), text='istep=', box=0)
ilabel.text = 'istep=%4s' %count

print("Final eigenvalue E = ",E)
print("iterations , max = ",count)
```

9.3 Explorations

1. Check to see how well your search procedure works by using arbitrary values for the starting energy. For example, because no bound-state energies can lie below the bottom of the well, try $E \geq -V_0$, as well as some arbitrary fractions of V_0. In every case, examine the resulting wave function and check that it is both symmetric and continuous.
2. Increase the depth of your potential progressively until you find several bound states. Look at the wave function in each case and correlate the number of nodes in the wave function with the position of the bound state in the well.
3. Explore how a bound-state energy changes as you change the depth V_0 of the well. In particular, as you keep decreasing the depth, watch the eigenenergy move closer to $E = 0$ and see if you can find the potential depth at which the bound state has $E \simeq 0$.
4. For a fixed well depth V_0, explore how the energy of a bound state changes as the well radius a is varied. Larger radius should give increased binding.
5. ⊙ Conduct some explorations in which you discover different combinations of (V_0, a) that give the same ground-state energies (discrete ambiguities). The existence of several different combinations means that a knowledge of ground-state energy is not enough to determine a unique depth of the well.
6. Modify the procedures to solve for the eigenvalues and eigenfunctions for odd wave functions.
7. Solve for the wave function of a linear potential:

$$V(x) = -V_0 \begin{cases} |x|, & \text{for } |x| < a\,, \\ 0, & \text{for } |x| > a\,. \end{cases} \tag{9.21}$$

   There is less potential here than for a square well, so you may expect smaller binding energies and a less confined wave function. (For this potential, there are no analytic results with which to compare.)
8. Compare the results obtained, and the time the computer took to get them, using both the Numerov and `rk4` methods.
9. *Newton–Raphson extension:* Extend the eigenvalue search by using the Newton–Raphson method in place of the bisection algorithm. Determine how much faster it is.

9.4 Problem: Classical Chaotic Scattering

Problem One expects the classical scattering of a projectile from a barrier to be a continuous process. Yet it has been observed in experiments conducted on pinball machines (Figure 9.3) that for certain conditions the projectile undergoes multiple internal scatterings and ends up with a final trajectory that is apparently unrelated

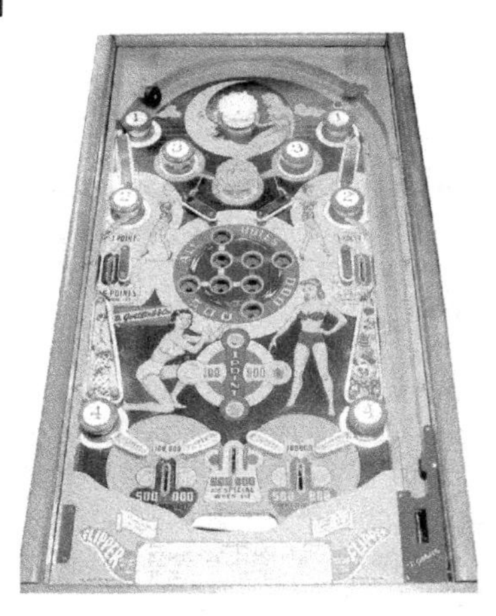

Figure 9.3 A photograph of a pinball machine in which multiple scatterings occur from the bumpers.

to the initial one. Your *problem* is to determine if this process can be modeled as scattering from a static potential, or if there must be active mechanisms built into the pinball machines that cause chaotic scattering.

Although this problem is easy to solve on the computer, the results have some chaotic features that are surprising (chaos is discussed further in Chapter 14). In fact, the applet `Disper2e.html` *(created by Jaime Zuluaga) that simulates this problem continues to be a source of wonderment for readers as well as authors.*

9.4.1 Model and Theory

Our model for balls bouncing off the electrically activated bumpers in pinball machines is a point particle scattering from the stationary 2D potential (Blehel *et al.*, 1990)

$$V(x, y) = \pm x^2 y^2 e^{-(x^2+y^2)} . \tag{9.22}$$

As seen in Figure 9.4, this potential has four circularly symmetric peaks in the xy plane. The two signs correspond to repulsive and attractive potentials, respectively (the pinball machine contains only repulsive interactions). Because there are four peaks in this potential, we suspect that it may be possible to have multiple scatterings in which the projectile bounces back and forth between the peaks, somewhat as in a pinball machine.

The *theory* for this problem is classical dynamics. Visualize a scattering experiment in which a projectile starting out at an infinite distance from a target with a definite velocity v and an impact parameter b (Figure 9.4) is incident on a target. After interacting with the target and moving a large distance from it, the scattered particle is observed at the scattering angle θ. Because the potential cannot recoil and thereby carry off energy, the speed of the projectile does not change,

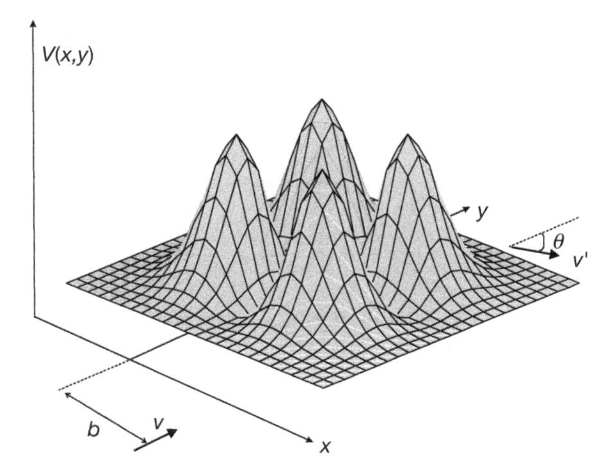

Figure 9.4 Scattering from the potential $V(x,y) = x^2y^2e^{-(x^2+y^2)}$ which in some ways models a pinball machine. The incident velocity v is in the y direction, and the impact parameter (x value) is b. The velocity of the scattered particle is v' and its scattering angle is θ.

but its direction does. An experiment typically measures the number of particles scattered and then converts this to a function, the differential cross section $\sigma(\theta)$, which is independent of the details of the experimental apparatus:

$$\sigma(\theta) = \lim_{\Delta\Omega,\Delta A \to 0} \frac{N_{\text{scatt}}(\theta)/\Delta\Omega}{N_{\text{in}}/\Delta A_{\text{in}}} . \tag{9.23}$$

Here $N_{\text{scatt}}(\theta)$ is the number of particles per unit time scattered into the detector at angle θ subtending a solid angle $\Delta\Omega$, N_{in} is the number of particle per unit time incident on the target of cross-sectional area ΔA_{in}, and the limit in (9.23) is for infinitesimal detector and area sizes.

The definition (9.23) for the cross section is the one that experimentalists use to convert their measurements to a function that can be calculated by theory. We, as theorists, solve for the trajectory of a particle scattered from the potential (9.22) and from that deduce the scattering angle θ. Once we have the scattering angle, we predict the differential cross section from the dependence of the scattering angle upon the classical impact parameter b (Marion and Thornton, 2003):

$$\sigma(\theta) = \left|\frac{\mathrm{d}\theta}{\mathrm{d}b}\right| \frac{b}{\sin\theta(b)} . \tag{9.24}$$

The surprise you should find in the simulation is that for certain potential parameters, $\mathrm{d}\theta/\mathrm{d}b$ can get to be very large or discontinuous, and this, accordingly, leads to large and discontinuous cross sections.

The dynamical equations to solve are just the vector form of Newton's law for the simultaneous x and y motions in the potential (9.22):

$$\begin{aligned} \boldsymbol{F} &= m\boldsymbol{a} \\ -\frac{\partial V}{\partial x}\hat{\imath} - \frac{\partial V}{\partial y}\hat{\jmath} &= m\frac{\mathrm{d}^2\boldsymbol{x}}{\mathrm{d}t^2}, \\ \mp 2xy\mathrm{e}^{-(x^2+y^2)}\left[y(1-x^2)\hat{\imath} + x(1-y^2)\hat{\jmath}\right] &= m\frac{\mathrm{d}^2x}{\mathrm{d}t^2}\hat{\imath} + m\frac{\mathrm{d}^2y}{\mathrm{d}t^2}\hat{\jmath}. \end{aligned} \tag{9.25}$$

The equations for the x and y motions are simultaneous second-order ODEs:

$$m\frac{\mathrm{d}^2x}{\mathrm{d}t^2} = \mp 2y^2x(1-x^2)\mathrm{e}^{-(x^2+y^2)}, \tag{9.26}$$

$$m\frac{\mathrm{d}^2y}{\mathrm{d}t^2} = \mp 2x^2y(1-y^2)\mathrm{e}^{-(x^2+y^2)}. \tag{9.27}$$

Because the force vanishes at the peaks in Figure 9.4, these equations tell us that the peaks are at $x = \pm 1$ and $y = \pm 1$. Substituting these values into the potential (9.22) yields $V_{max} = \pm\mathrm{e}^{-2}$, which sets the energy scale for the problem.

9.4.2 Implementation

Although (9.26) and (9.27) are simultaneous second-order ODEs, we can still use our standard `rk4` ODE solver on them after expressing them in the standard form, only now the arrays will be 4D rather than the previous 2D:

$$\frac{\mathrm{d}\boldsymbol{y}(t)}{\mathrm{d}t} = \boldsymbol{f}(t, \boldsymbol{y}), \tag{9.28}$$

$$y^{(0)} \stackrel{\text{def}}{=} x(t), \quad y^{(1)} \stackrel{\text{def}}{=} y(t), \tag{9.29}$$

$$y^{(2)} \stackrel{\text{def}}{=} \frac{\mathrm{d}x}{\mathrm{d}t}, \quad y^{(3)} \stackrel{\text{def}}{=} \frac{\mathrm{d}y}{\mathrm{d}t}, \tag{9.30}$$

where the order in which the $y^{(i)}$s are assigned is arbitrary. With these definitions and equations (9.26) and (9.27), we can assign values for the force function:

$$f^{(0)} = y^{(2)}, \quad f^{(1)} = y^{(3)}, \tag{9.31}$$

$$f^{(2)} = \frac{\mp 1}{m}2y^2x(1-x^2)\mathrm{e}^{-(x^2+y^2)} \tag{9.32}$$

$$= \frac{\mp 1}{m}2y^{(1)^2}y^{(0)}(1-y^{(0)^2})\mathrm{e}^{-(y^{(0)^2}+y^{(1)^2})}, \tag{9.33}$$

$$f^{(3)} = \frac{\mp 1}{m}2x^2y(1-y^2)\mathrm{e}^{-(x^2+y^2)} \tag{9.34}$$

$$= \frac{\mp 1}{m}2y^{(0)^2}y^{(1)}(1-y^{(1)^2})\mathrm{e}^{-(y^{(0)^2}+y^{(1)^2})}. \tag{9.35}$$

To deduce the scattering angle from our simulation, we need to examine the trajectory of the scattered particle at a very large separation from the target. To approximate that, we wait until the scattered particle no longer feels the potential (say $|\mathrm{PE}|/\mathrm{KE} \leq 10^{-10}$) and call this infinity. The scattering angle is then deduced from the components of velocity,

$$\theta = \tan^{-1}\left(\frac{v_y}{v_x}\right) = \texttt{math.atan2(y, x)} \,. \tag{9.36}$$

Here `atan2` is a function that computes the arctangent in the correct quadrant without requiring any explicit divisions (that can blow up).

9.4.3
Assessment

1. Apply the `rk4` method to solve the simultaneous second-order ODEs (9.26) and (9.27) with a 4D force function.
2. The *initial conditions* are (a) an incident particle with only a y component of velocity and (b) an impact parameter b (the initial x value). You do not need to vary the initial y, but it should be large enough such that $\mathrm{PE}/\mathrm{KE} \leq 10^{-10}$, which means that the $\mathrm{KE} \simeq E$.
3. Good parameters to use are $m = 0.5$, $v_y(0) = 0.5$, $v_x(0) = 0.0$, $\Delta b = 0.05$, $-1 \leq b \leq 1$. You may want to lower the energy and use a finer step size once you have found regions of rapid variation in the cross section.
4. Plot a number of trajectories $[x(t), y(t)]$ that show usual and unusual behaviors. In particular, plot those for which backward scattering occurs, and consequently for which there is much multiple scattering.
5. Plot a number of phase-space trajectories $[x(t), \dot{x}(t)]$ and $[y(t), \dot{y}(t)]$. How do these differ from those of bound states?
6. Determine the scattering angle $\theta =$ `atan2(Vx,Vy)` by determining the velocity of the scattered particle after it has left the interaction region, that is, $\mathrm{PE}/\mathrm{KE} \leq 10^{-10}$.
7. Identify which characteristics of a trajectory lead to discontinuities in $\mathrm{d}\theta/\mathrm{d}b$ and thus $\sigma(\theta)$.
8. Run the simulations for both attractive and repulsive potentials and for a range of energies less than and greater than $V_{\max} = \exp(-2)$.
9. *Time delay:* Another way to find unusual behavior in scattering is to compute the *time delay* $T(b)$ as a function of the impact parameter b. The time delay is the increase in the time taken by a particle to travel through the interaction region. Look for highly oscillatory regions in the semilog plot of $T(b)$, and once you find some, repeat the simulation at a finer scale by setting $b \simeq b/10$ (the structures are fractals, see Chapter 16).

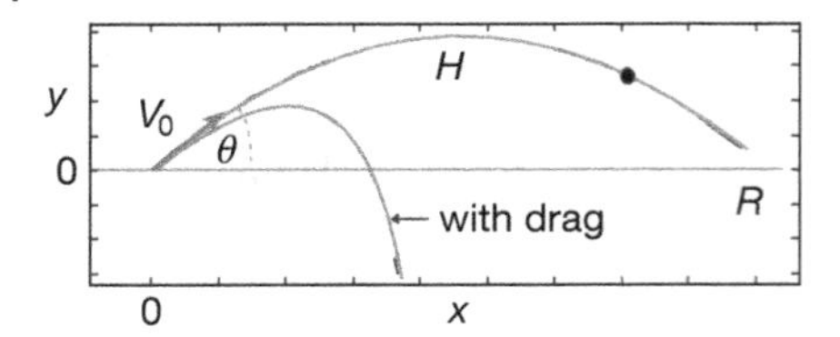

Figure 9.5 The trajectories of a projectile fired with initial velocity V_0 in the θ direction. The lower curve includes air resistance.

9.5 Problem: Balls Falling Out of the Sky

Golf and baseball players claim that balls appear to fall straight down out of the sky at the end of their trajectories (the solid curve in Figure 9.5). Your *problem* is to determine whether there is a simple physics explanation for this effect or whether it is "all in the mind's eye." And while you are wondering why things fall out of the sky, see if you can use your new-found numerical tools to explain why planets do not fall out of the sky.

9.6 Theory: Projectile Motion with Drag

Figure 9.5 shows the initial velocity V_0 and inclination θ for a projectile launched from the origin. If we ignore air resistance, the projectile has only the force of gravity acting on it and therefore has a constant acceleration $g = 9.8\,\mathrm{m/s^2}$ in the negative y direction. The analytic solutions to the equations of motion are

$$x(t) = V_{0x}t\,, \quad y(t) = V_{0y}t - \frac{1}{2}gt^2\,, \tag{9.37}$$

$$v_x(t) = V_{0x}\,, \quad v_y(t) = V_{0y} - gt\,, \tag{9.38}$$

where $(V_{0x}, V_{0y}) = V_0(\cos\theta, \sin\theta)$. Solving for t as a function of x and substituting it into the $y(t)$ equation show that the trajectory is a parabola:

$$y = \frac{V_{0y}}{V_{0x}}x - \frac{g}{2V_{0x}^2}x^2\,. \tag{9.39}$$

Likewise, it is easy to show (solid curve in Figure 9.5) that without friction the range $R = 2V_0^2 \sin\theta\cos\theta/g$ and the maximum height $H = \frac{1}{2}V_0^2\sin^2\theta/g$.

The parabola of frictionless motion is symmetric about its midpoint and so does not describe a ball dropping out of the sky. Maybe air resistance will change that? The basic physics is Newton's second law in two dimensions for a frictional force $\boldsymbol{F}^{(\mathrm{f})}$ opposing motion, and a vertical gravitational force $-mg\hat{\boldsymbol{e}}_y$:

$$\boldsymbol{F}^{(\mathrm{f})} - mg\hat{\boldsymbol{e}}_y = m\frac{\mathrm{d}^2\boldsymbol{x}(t)}{\mathrm{d}t^2}\,, \tag{9.40}$$

$$\Rightarrow \quad F_x^{(\mathrm{f})} = m\frac{\mathrm{d}^2 x}{\mathrm{d}t^2}\,, \quad F_y^{(\mathrm{f})} - mg = m\frac{\mathrm{d}^2 y}{\mathrm{d}t^2}\,, \tag{9.41}$$

where the boldface italic symbols indicate vector quantities.

The frictional force $\boldsymbol{F}^{(\mathrm{f})}$ is not a basic force of nature but rather a simple model of a complicated phenomenon. We know that friction always opposes motion, which means it is in the direction opposite to velocity. One model assumes that the frictional force is proportional to a power n of the projectile's speed (Marion and Thornton, 2003; Warburton and Wang, 2004):

$$\boldsymbol{F}^{(\mathrm{f})} = -km|v|^n \frac{\boldsymbol{v}}{|v|}\,, \tag{9.42}$$

where the $-\boldsymbol{v}/|v|$ factor ensures that the frictional force is always in a direction opposite that of the velocity. Physical measurements indicate that the power n is noninteger and varies with velocity, and so a more accurate model would be a numerical one that uses the empirical velocity dependence $n(v)$. With a constant power law for friction, the equations of motion are

$$\frac{\mathrm{d}^2 x}{\mathrm{d}t^2} = -kv_x^n \frac{v_x}{|v|}\,, \quad \frac{\mathrm{d}^2 y}{\mathrm{d}t^2} = -g - kv_y^n \frac{v_y}{|v|}\,, \quad |v| = \sqrt{v_x^2 + v_y^2}\,. \tag{9.43}$$

We shall consider three values for n, each of which represents a different model for the air resistance: (1) $n = 1$ for low velocities; (2) $n = 3/2$, for medium velocities; and (3) $n = 2$ for high velocities.

9.6.1
Simultaneous Second-Order ODEs

Although (9.43) are simultaneous second-order ODEs, we can still use our regular ODE solver on them after expressing them in the standard form

$$\frac{\mathrm{d}\boldsymbol{y}}{\mathrm{d}t} = \boldsymbol{f}(t, \boldsymbol{y}) \quad \text{(standard form)}\,, \tag{9.44}$$

$$y^{(0)} = x(t)\,, \quad y^{(1)} = \frac{\mathrm{d}x}{\mathrm{d}t}\,, \quad y^{(2)} = y(t)\,, \quad y^{(3)} = \frac{\mathrm{d}y}{\mathrm{d}t}\,. \tag{9.45}$$

We express the equations of motion in terms of $\boldsymbol{y}$ to obtain the standard form:

$$\frac{\mathrm{d}y^{(0)}}{\mathrm{d}t} = y^{(1)}\,, \quad \frac{\mathrm{d}y^{(1)}}{\mathrm{d}t} = \frac{1}{m}F_x^{(\mathrm{f})}(\boldsymbol{y})\,, \tag{9.46}$$

$$\frac{\mathrm{d}y^{(2)}}{\mathrm{d}t} = y^{(3)}\,, \quad \frac{\mathrm{d}y^{(3)}}{\mathrm{d}t} = \frac{1}{m}F_y^{(\mathrm{f})}(\boldsymbol{y}) - g\,. \tag{9.47}$$

And now we just read off the components of the force function $\boldsymbol{f}(t, \boldsymbol{y})$:

$$f^{(0)} = y^{(1)}\,, \quad f^{(1)} = \frac{1}{m}F_x^{(\mathrm{f})}\,, \quad f^{(2)} = y^{(3)}\,, \quad f^{(3)} = \frac{1}{m}F_y^{(\mathrm{f})} - g\,. \tag{9.48}$$

Our implementation, `ProjectiveAir.py`, is given in Listing 9.3.

Listing 9.3 **ProjectileAir.py** solves for projectile motion with air resistance as well as analytically for the frictionless case.

```
# ProjectileAir.py: Numerical solution for projectile with drag

from visual import *
from visual.graph import *

v0 = 22.;     angle = 34.;     g = 9.8;     kf = 0.8;     N = 5
v0x = v0*cos(angle*pi/180.);     v0y = v0*sin(angle*pi/180.)
T = 2.*v0y/g;     H = v0y*v0y/2./g;     R = 2.*v0x*v0y/g
graph1 = gdisplay( title='Projectile with & without Drag',
        xtitle='x', ytitle='y', xmax=R, xmin=-R/20.,ymax=8,ymin=-6.0)
funct = gcurve(color=color.red)
funct1 = gcurve(color=color.yellow)
print('No Drag T =',T,', H =',H,', R =',R)

def plotNumeric(k):
  vx = v0*cos(angle*pi/180.)
  vy = v0*sin(angle*pi/180.)
  x = 0.0
  y = 0.0
  dt =  vy/g/N/2.
  print("\n          With Friction   ")
  print("          x                 y")
  for i in range(N):
     rate(30)
     vx = vx - k*vx*dt
     vy = vy - g*dt - k*vy*dt
     x = x + vx*dt
     y = y + vy*dt
     funct.plot(pos=(x,y))
     print(" %13.10f  %13.10f "%(x,y))

def plotAnalytic():
     v0x = v0*cos(angle*pi/180.)
     v0y = v0*sin(angle*pi/180.)
     dt = 2.*v0y/g/N
     print("\n          No Friction   ")
     print("          x                 y")
     for i in range(N):
        rate(30)
        t = i*dt
        x = v0x*t
        y = v0y*t -g*t*t/2.
        funct1.plot(pos=(x,y))
        print(" %13.10f  %13.10f"%(x,y))

plotNumeric(kf)
plotAnalytic()
```

9.6.2 Assessment

1. Modify your `rk4` program so that it solves the simultaneous ODEs for projectile motion (9.43) with friction ($n = 1$).
2. Check that you obtain graphs similar to those in Figure 9.5.
3. The model (9.42) with $n = 1$ is okay for low velocities. Now modify your program to handle $n = 3/2$ (medium-velocity friction) and $n = 2$ (high-velocity friction). Adjust the value of k for the latter two cases such that the initial force of friction kv_0^n is the same for all three cases.
4. What is your conclusion about balls falling out of the sky?

9.7 Exercises: 2- and 3-Body Planet Orbits and Chaotic Weather

Planets via Two of Newton's Laws Newton's explanation of the motion of the planets in terms of a universal law of gravitation is one of the greatest achievements of science. He was able to prove that planets traveled in elliptical orbits with the sun at one vertex, and then go on to predict the periods of the motion. All Newton needed to postulate was that the force between a planet of mass m and the sun of mass M is

$$F^{(g)} = -\frac{GmM}{r^2} , \tag{9.49}$$

where r is the planet–CM distance, G is the universal gravitational constant, and the attractive force lies along the line connecting the planet and the sun (Figure 9.6a). Seeing that he had to invent calculus to do it, the hard part for Newton was solving the resulting differential equations. In contrast, the numerical solution is straightforward because even for planets the equation of motion is still

$$\boldsymbol{F} = m\boldsymbol{a} = m\frac{\mathrm{d}^2\boldsymbol{x}}{\mathrm{d}t^2} , \tag{9.50}$$

with the force (9.49) having Cartesian components (Figure 9.6)

$$F_x = F^{(g)} \cos\theta = F^{(g)}\frac{x}{r} = F^{(g)}\frac{x}{\sqrt{x^2+y^2}} , \tag{9.51}$$

$$F_y = F^{(g)} \sin\theta = F^{(g)}\frac{y}{r} = F^{(g)}\frac{y}{\sqrt{x^2+y^2}} . \tag{9.52}$$

The equation of motion (9.50) becomes two simultaneous second-order ODEs:

$$\frac{\mathrm{d}^2 x}{\mathrm{d}t^2} = -GM\frac{x}{(x^2+y^2)^{3/2}} , \quad \frac{\mathrm{d}^2 y}{\mathrm{d}t^2} = -GM\frac{y}{(x^2+y^2)^{3/2}} . \tag{9.53}$$

1. Assume units such that $GM = 1$ and the initial conditions

$$x(0) = 0.5 , \quad y(0) = 0 , \quad v_x(0) = 0.0 , \quad v_y(0) = 1.63 . \tag{9.54}$$

2. Modify your ODE solver program to solve (9.53).
3. Establish that you use small enough time steps so that the orbits are closed and fall upon themselves.
4. Experiment with the initial conditions until you find the ones that produce a circular orbit (a special case of an ellipse).
5. Once you have obtained good precision, note the effect of progressively increasing the initial velocity until the orbits open up and the planets become projectiles.
6. Using the same initial conditions that produced the ellipse, investigate the effect of the power in (9.49) being $1/r^{2+\alpha}$ with $\alpha \neq 0$. Even for small α you should find that the ellipses now rotate or precess (Figure 9.6). (A small value for α is predicted by general relativity.)

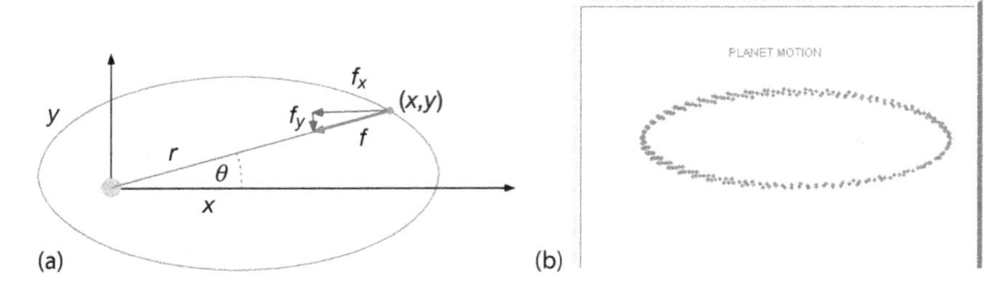

Figure 9.6 (a) The gravitational force on a planet at distance r from the sun. The x and y components of the force are indicated. (b) Output from the applet `Planet` showing the precession of a planet's orbit when the gravitational force $\propto 1/r^4$ (successive orbits do not lie on top of each other).

Three-Body Problem: The Discovery of Neptune The planet Uranus was discovered in 1781 by William Herschel and found to have an orbital period of approximately 84 years. By the year 1846, Uranus had just about completed a full orbit around the sun since its discovery, but did not seem to be following precisely the positions predicted by Newton's law of gravity. However, theoretical calculations indicated that if there was a yet-to-be-discovered planet lying about 50% further away from the sun than Uranus, then its perturbation on the orbit of Uranus could explain the disagreement with Newton's law. The planet Neptune was thus discovered theoretically and confirmed experimentally. (If Pluto is discarded as just a dwarf planet, then Neptune is the most distant planet in the solar system.)

Assume that the orbits of Neptune and Uranus are circular and coplanar (as shown in Figure 9.7), and that the initial angular positions with respect to the x-axis are as follows:

	Mass ($\times 10^{-5}$ **Solar Masses**)	**Distance** **(AU)**	**Orbital period** **(Years)**	**Angular position** **(in 1690)**
Uranus	4.366 244	19.1914	84.0110	∼ 205.640
Neptune	5.151 389	30.0611	164.7901	∼ 288.380

Using these data and rk4, find the variation in angular position of Uranus with respect to the Sun as a result of the influence of Neptune during one complete orbit of Neptune. Consider only the forces of the Sun and Neptune on Uranus. In the astronomical units, $M_s = 1$ and $G = 4\pi^2$.

You can do this calculation following the procedure outlined above in which the problems is reduced to simultaneous ODEs for the x and y Cartesian coordinates, and the components of the forces along x and y are computed. Another approach that you may want to try, computes the explicit values of the derivatives used in

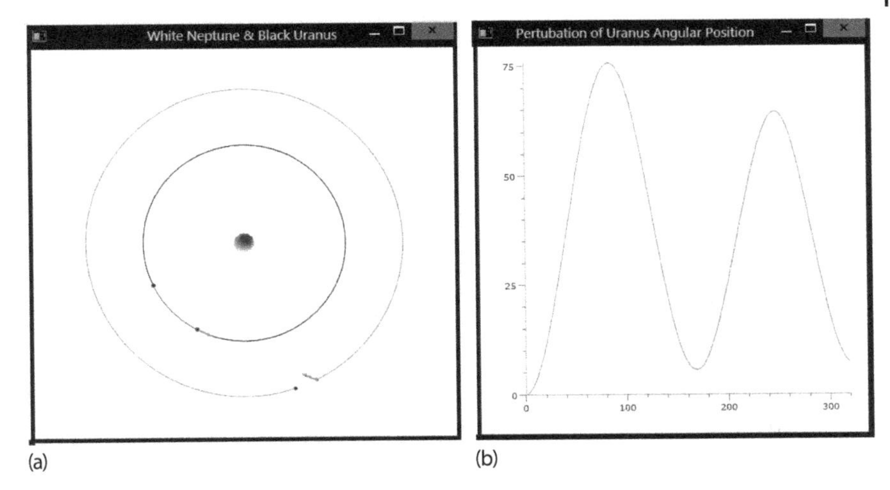

Figure 9.7 A snapshot from the animated output of the code `UranusNeptune.py` (on Instructor's disk) showing: (a) the orbits of Uranus (inner circle) and of Neptune (outer circle) with the Sun in the center. The arrows indicate the Uranus–Neptune force that causes a perturbation in the orbits. (b) The perturbation in the angular position of Uranus as a result of the presence of Neptune.

the rk4 method (http://spiff.rit.edu/richmond/nbody/OrbitRungeKutta4_fixed.pdf):

$$\boldsymbol{k}_{1v} = \frac{\boldsymbol{F}_T(\boldsymbol{r}_{su}, \boldsymbol{r}_{nu})}{m_u}, \quad \boldsymbol{k}_{1r} = \boldsymbol{v}_u, \tag{9.55}$$

$$\boldsymbol{k}_{2v} = \frac{\boldsymbol{F}_T(\boldsymbol{r}_{su} + \boldsymbol{k}_{1r}\frac{\mathrm{d}t}{2}, \boldsymbol{r}_{nu})}{m_u}, \quad \boldsymbol{k}_{2r} = \boldsymbol{v}_u + \boldsymbol{k}_{2v}\frac{\mathrm{d}t}{2}, \tag{9.56}$$

$$\boldsymbol{k}_{3v} = \frac{\boldsymbol{F}_T(\boldsymbol{r}_{su} + \boldsymbol{k}_{2r}\frac{\mathrm{d}t}{2}, \boldsymbol{r}_{nu})}{m_u}, \quad \boldsymbol{k}_{3r} = \boldsymbol{v}_u + \boldsymbol{k}_{3v}\frac{\mathrm{d}t}{2}, \tag{9.57}$$

$$\boldsymbol{k}_{4v} = \frac{\boldsymbol{F}_T(\boldsymbol{r}_{su} + \boldsymbol{k}_{3r}\,\mathrm{d}t, \boldsymbol{r}_{nu})}{m_u}, \qquad \boldsymbol{k}_{4r} = \boldsymbol{v}_u + \boldsymbol{k}_{4v}\,\mathrm{d}t, \tag{9.58}$$

$$\boldsymbol{v}_u = \boldsymbol{v}_u + (\boldsymbol{k}_{1v} + 2\boldsymbol{k}_{2v} + 2\boldsymbol{k}_{3v} + \boldsymbol{k}_{4v})\frac{\mathrm{d}t}{6}, \tag{9.59}$$

$$\boldsymbol{r} = \boldsymbol{r} + (\boldsymbol{k}_{1r} + 2\boldsymbol{k}_{2r} + 2\boldsymbol{k}_{3r} + \boldsymbol{k}_{4r})\frac{\mathrm{d}t}{6}, \tag{9.60}$$

with a similar set of $\boldsymbol{k}$'s defined for Neptune because the force is equal and opposite.

To help you get started, here is a listing of some of the constants used in our program:

```
G = 4*pi*pi                    # AU, Msun=1
mu = 4.366244e-5               # Uranus mass
M = 1.0                        # Sun mass
mn = 5.151389e-5               # Neptune mass
du = 19.1914                   # Uranus Sun distance
dn = 30.0611                   # Neptune sun distance
Tur = 84.0110                  # Uranus Period
Tnp = 164.7901                 # Neptune Period
omeur = 2*pi/Tur               # Uranus angular velocity
omennp = 2*pi/Tnp              # Neptune angular velocity
omreal = omeur
urvel = 2*pi*du/Tur            # Uranus orbital velocity UA/yr
npvel = 2*pi*dn/Tnp            # Neptune orbital velocity UA/yr
radur = (205.64)*pi/180.       # in radians
urx = du*cos(radur)            # init x uranus in 1690
ury = du*sin(radur)            # init y uranus in 1690
urvelx = urvel*sin(radur)
urvely = -urvel*cos(radur)
radnp = (288.38)*pi/180.       # Neptune angular pos.
```

10
High-Performance Hardware and Parallel Computers

This chapter discusses a number of topics associated with high-performance computing (HPC) and parallel computing. Although this may sound like something only specialists should be reading, using history as a guide, present HPC hardware and software will be desktop machines in less than a decade, and so you may as well learn these things now. We start with a discussion of a high-performance computer's memory and central processor design, and then examine various general aspects of parallel computing. Chapter 11 goes on to discuss some practical programming aspects of HPC and parallel computing. HPC is a broad subject, and our presentation is brief and given from a practitioner's point of view. The text (Quinn, 2004) surveys parallel computing and message passing interface from a computer science point of view. Other references on parallel computing include van de Velde (1994); Fox (1994), and Pancake (1996).

10.1
High-Performance Computers

By definition, supercomputers are the fastest and most powerful computers available, and at present, the term refers to machines with hundreds of thousands of processors. They are the superstars of the high-performance class of computers. Personal computers (PCs) small enough in size and cost to be used by an individual, yet powerful enough for advanced scientific and engineering applications, can also be high-performance computers. We define *high-performance computers* as machines with a good balance among the following major elements:

- Multistaged (pipelined) functional units,
- Multiple central processing units (CPUs),
- Multiple cores,
- Fast central registers,
- Very large, fast memories,
- Very fast communication among functional units,
- Vector, video, or array processors,
- Software that integrates the above effectively and efficiently.

As the simplest example, it makes little sense to have a CPU of incredibly high speed coupled to a memory system and software that cannot keep up with it.

10.2
Memory Hierarchy

An idealized model of computer architecture is a CPU sequentially executing a stream of instructions and reading from a continuous block of memory. To illustrate, in Figure 10.1 we have a vector `A[ ]` and an array `M[.. , .. ]` loaded in memory and about to be processed. The real world is more complicated than this. First, arrays are not stored in 2D blocks, but rather in the linear order. For instance, in Python, Java, and C it is in *row-major* order:

$$M(0,0)\,M(0,1)\,M(0,2)\,M(1,0)\,M(1,1)\,M(1,2)\,M(2,0)\,M(2,1)\,M(2,2)\;. \tag{10.1}$$

In Fortran, it is in *column-major* order:

$$M(1,1)\,M(2,1)\,M(3,1)\,M(1,2)\,M(2,2)\,M(3,2)\,M(1,3)\,M(2,3)\,M(3,3)\;. \tag{10.2}$$

Second, as illustrated in Figures 10.2 and 10.3, the values for the matrix elements may not even be in the same physical place. Some may be in RAM, some on the disk, some in cache, and some in the CPU. To give these words more meaning, in Figures 10.3 and 10.2 we show simple models of the memory architecture of a high-performance computer. This hierarchical arrangement arises from an effort to balance speed and cost, with fast, expensive memory supplemented by slow, less expensive memory. The memory architecture may include the following elements:

CPU Central processing unit, the fastest part of the computer. The CPU consists of a number of very high-speed memory units called *registers* containing the *instructions* sent to the hardware to do things like fetch, store, and operate on data. There are usually separate registers for instructions, addresses, and *operands* (current data). In many cases, the CPU also contains some specialized parts for accelerating the processing of floating-point numbers.

Cache A small, very fast bit of memory that holds instructions, addresses, and data in their passage between the very fast CPU registers and the slower RAM (also called a high-speed buffer). This is seen in the next level down the pyramid in Figure 10.2. The main memory is also called *dynamic RAM* (DRAM), while the cache is called *static RAM* (SRAM). If the cache is used properly, it can greatly reduce the time that the CPU waits for data to be fetched from memory.

Cache lines The data transferred to and from the cache or CPU are grouped into cache or data lines. The time it takes to bring data from memory into the cache is called *latency*.

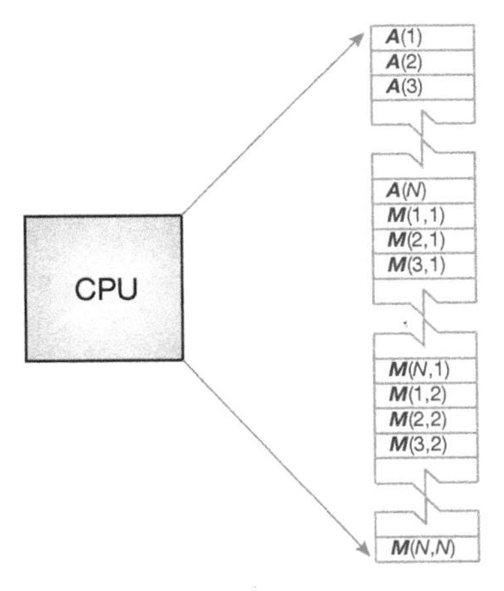

Figure 10.1 The logical arrangement of the CPU and memory showing a Fortran array $A(N)$ and matrix $\mathbf{M}(N, N)$ loaded into memory.

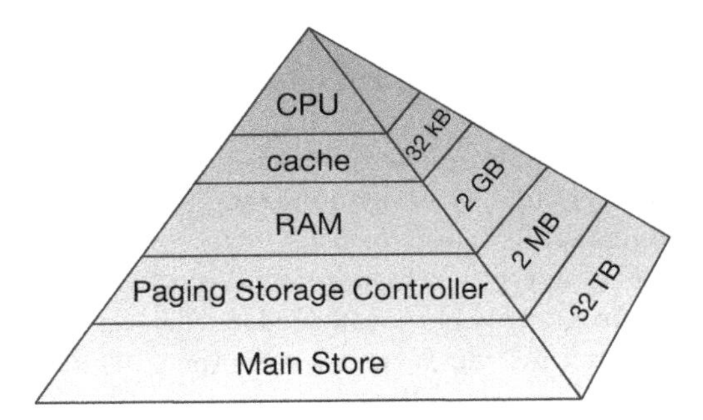

Figure 10.2 Typical memory hierarchy for a single-processor, high-performance computer (B = bytes, k, M, G, T = kilo, mega, giga, tera).

RAM Random-access or central memory is in the middle of the memory hierarchy in Figure 10.2. RAM is fast because its addresses can be accessed directly in random order, and because no mechanical devices are needed to read it.

Pages Central memory is organized into pages, which are blocks of memory of fixed length. The operating system labels and organizes its memory pages much like we do the pages of a book; they are numbered and kept track of with a *table of contents*. Typical page sizes range from 4 to 16 kB, but on supercomputers they may be in the MB range.

Hard disk Finally, at the bottom of the memory pyramid is permanent storage on magnetic disks or optical devices. Although disks are very slow compared to RAM, they can store vast amounts of data and sometimes compensate for their slower speeds by using a cache of their own, the *paging storage controller*.

Virtual memory True to its name, this is a part of memory you will not find in our figures because it is *virtual*. It acts like RAM but resides on the disk.

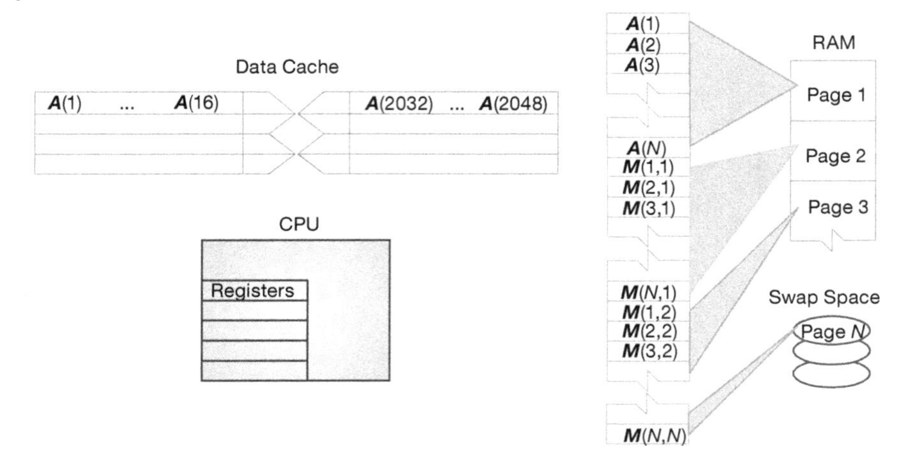

Figure 10.3 The elements of a computer's memory architecture in the process of handling matrix storage.

When we speak of "fast" and "slow" memory we are using a time scale set by the clock in the CPU. To be specific, if your computer has a clock speed or cycle time of 1 ns, this means that it could perform a billion operations per second, if it could get its hands on the needed data quickly enough (typically, more than 10 cycles are needed to execute a single instruction). While it usually takes 1 cycle to transfer data from the cache to the CPU, the other types of memories are much slower. Consequently, you can speedup your program by having all needed data available for the CPU when it tries to execute your instructions; otherwise the CPU may drop your computation and go on to other chores while your data gets transferred from lower memory (we talk more about this soon in the discussion of pipelining or cache reuse). Compilers try to do this for you, but their success is affected by your programming style.

As shown in Figure 10.3, virtual memory permits your program to use more pages of memory than can physically fit into RAM at one time. A combination of operating system and hardware *maps* this virtual memory into pages with typical lengths of 4–16 kB. Pages not currently in use are stored in the slower memory on the hard disk and brought into fast memory only when needed. The separate memory location for this switching is known as *swap space* (Figure 10.4a).

Observe that when an application accesses the memory location for `M[i,j]`, the number of the page of memory holding this address is determined by the computer, and the location of `M[i,j]` within this page is also determined. A *page fault* occurs if the needed page resides on disk rather than in RAM. In this case, the entire page must be read into memory while the least recently used page in RAM is swapped onto the disk. Thanks to virtual memory, it is possible to run programs on small computers that otherwise would require larger machines (or extensive reprogramming). The price you pay for virtual memory is an order-of-magnitude slowdown of your program's speed when virtual memory is actually invoked. But

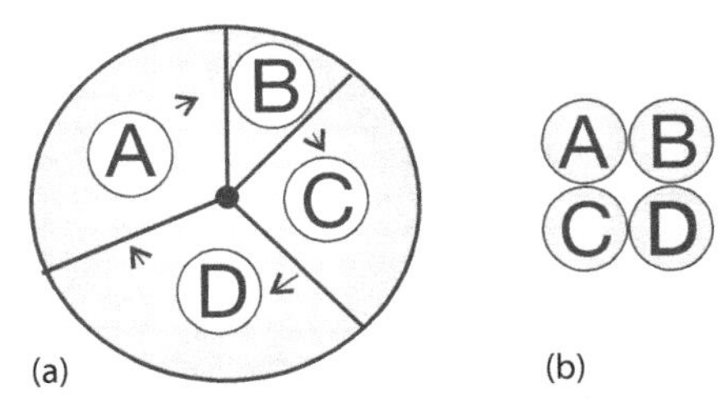

Figure 10.4 (a) Multitasking of four programs in memory at one time. On a SISD computer the programs are executed in round robin order. (b) Four programs in the four separate memories of a MIMD computer.

this may be cheap compared to the time you would have to spend to rewrite your program so it fits into RAM, or the money you would have to spend to buy enough RAM for your problem.

Virtual memory also allows *multitasking*, the simultaneous loading into memory of more programs than can physically fit into RAM (Figure 10.4b). Although the ensuing switching among applications uses computing cycles, by avoiding long waits while an application is loaded into memory, multitasking increases total throughout and permits an improved computing environment for users. For example, it is multitasking that permits a windowing system, such as Linux, Apple OS, or Windows, to provide us with multiple windows. Although each window application uses a fair amount of memory, only the single application currently receiving input must actually reside in memory; the rest are *paged out* to disk. This explains why you may notice a slight delay when switching to an idle window; the pages for the now-active program are being placed into RAM, and the least used application still in memory is simultaneously being paged out.

10.3 The Central Processing Unit

How does the CPU get to be so fast? Often, it utilizes *prefetching* and *pipelining*; that is, it has the ability to prepare for the next instruction before the current one has finished. It is like an assembly line or a bucket brigade in which the person filling the buckets at one end of the line does not wait for each bucket to arrive at the other end before filling another bucket. In the same way, a processor fetches, reads, and decodes an instruction while another instruction is executing. Consequently, despite the fact that it may take more than one cycle to perform some operations, it is possible for data to be entering and leaving the CPU on each cycle. To illustrate, Table 10.1 indicates how the operation $c = (a + b)/(d \times f)$ is handled. Here the pipelined arithmetic units A1 and A2 are simultaneously doing their jobs of fetching and operating on operands, yet arithmetic unit A3 must wait for the first two units to complete their tasks before it has something to do (during which time the other two sit idle).

Table 10.1 Computation of $c = (a + b)/(d \times f)$.

Arithmetic Unit	Step 1	Step 2	Step 3	Step 4
A1	Fetch a	Fetch b	Add	—
A2	Fetch d	Fetch f	Multiply	—
A3	—	—	—	Divide

10.4
CPU Design: Reduced Instruction Set Processors

Reduced instruction set computer (RISC) architecture (also called *superscalar*) is a design philosophy for CPUs developed for high-performance computers and now used broadly. It increases the arithmetic speed of the CPU by decreasing the number of instructions the CPU must follow. To understand RISC, we contrast it with CISC (*complex instruction set computer*) architecture. In the late 1970s, processor designers began to take advantage of *very-large-scale integration* (VLSI), which allowed the placement of hundreds of thousands of elements on a single CPU chip. Much of the space on these early chips was dedicated to *microcode* programs written by chip designers and containing machine language instructions that set the operating characteristics of the computer. There were more than 1000 instructions available, and many were similar to higher level programming languages like *Pascal* and *Forth*. The price paid for the large number of complex instructions was slow speed, with a typical instruction taking more than 10 clock cycles. Furthermore, a 1975 study by Alexander and Wortman of the *XLP* compiler of the IBM System/360 showed that about 30 low-level instructions accounted for 99% of the use with only 10 of these instructions accounting for 80% of the use.

The RISC philosophy is to have just a small number of instructions available at the chip level, but to have the regular programmer's high-level language, such as Fortran or C, translate them into efficient machine instructions for a particular computer's architecture. This simpler scheme is cheaper to design and produce, lets the processor run faster, and uses the space saved on the chip by cutting down on microcode to increase arithmetic power. Specifically, RISC increases the number of internal CPU registers, thus, making it possible to obtain longer pipelines (cache) for the data flow, a significantly lower probability of memory conflict, and some instruction-level parallelism.

The theory behind this philosophy for RISC design is the simple equation describing the execution time of a program:

$$\text{CPU time} = \text{number of instructions} \times \text{cycles/instruction} \times \text{cycle time} \,. \tag{10.3}$$

Here "CPU time" is the time required by a program, "number of instructions" is the total number of machine-level instructions the program requires (sometimes called the *path length*), "cycles/instruction" is the number of CPU clock cycles each instruction requires, and "cycle time" is the actual time it takes for one CPU cycle. After thinking about (10.3), we can understand the CISC philosophy that tries to reduce CPU time by reducing the number of instructions, as well as the RISC philosophy, which tries to reduce the CPU time by reducing cycles/instruction (preferably to 1). For RISC to achieve an increase in performance requires a greater decrease in cycle time and cycles/instruction than is the increase in the number of instructions.

In summary, the elements of RISC are the following:

Single-cycle execution, for most machine-level instructions.
Small instruction set, of less than 100 instructions.
Register-based instructions, operating on values in registers, with memory access confined to loading from and storing to registers.
Many registers, usually more than 32.
Pipelining, concurrent preparation of several instructions that are then executed successively.
High-level compilers, to improve performance.

10.5 CPU Design: Multiple-Core Processors

The present time is seeing a rapid increase in the inclusion of multicore (up to 128) chips as the computational engine of computers, and we expect that number to keep rising. As seen in Figure 10.5, a dual-core chip has two CPUs in one integrated circuit with a shared interconnect and a shared level-2 cache. This type of configuration with two or more identical processors connected to a single shared main memory is called *symmetric multiprocessing*, or SMP.

Although multicore chips were originally designed for game playing and single precision, they are finding use in scientific computing as new tools, algorithms, and programming methods are employed. These chips attain more integrated speed with less heat and more energy efficiency than single-core chips, whose heat generation limits them to clock speeds of less than 4 GHz. In contrast to multiple single-core chips, multicore chips use fewer transistors per CPU and are thus simpler to make and cooler to run.

Parallelism is built into a multicore chip because each core can run a different task. However, because the cores usually share the same communication channel and level-2 cache, there is the possibility of a communication bottleneck if both CPUs use the bus at the same time. Usually the user need not worry about this, but the writers of compilers and software must. Modern compilers automatically make use of the multiple cores, with MPI even treating each core as a separate processor.

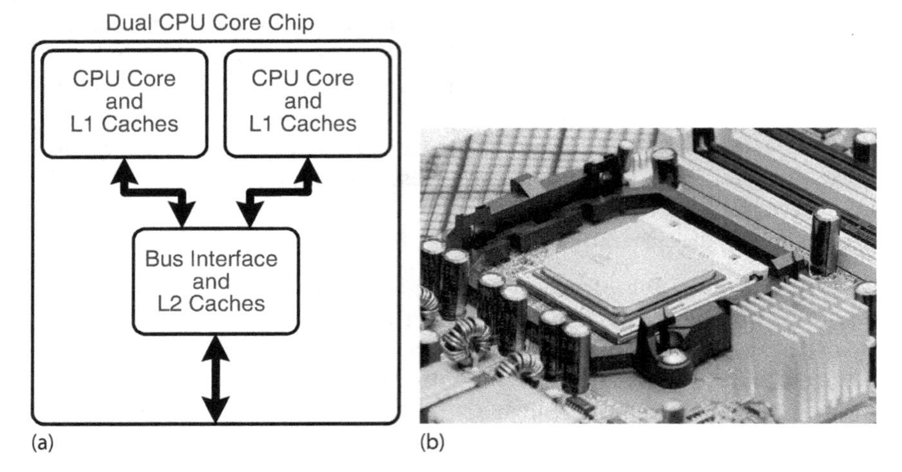

Figure 10.5 (a) A generic view of the Intel core-2 dual-core processor, with CPU-local level-1 caches and a shared, on-die level-2 cache (courtesy of D. Schmitz). (b) The AMD Athlon 64 X2 3600 dual-core CPU (Wikimedia Commons).

10.6
CPU Design: Vector Processors

Often the most demanding part of a scientific computation involves matrix operations. On a classic (von Neumann) scalar computer, the addition of two vectors of physical length 99 to form a third, ultimately requires 99 sequential additions (Table 10.2). There is actually much behind-the-scenes work here. For each element i there is the *fetch* of $a(i)$ from its location in memory, the *fetch* of $b(i)$ from its location in memory, the *addition* of the numerical values of these two elements in a CPU register, and the *storage* in memory of the sum in $c(i)$. This fetching uses up time and is wasteful in the sense that the computer is being told again and again to do the same thing.

When we speak of a computer doing *vector processing,* we mean that there are hardware components that perform mathematical operations on entire rows or columns of matrices as opposed to individual elements. (This hardware can also handle single-subscripted matrices, that is, mathematical vectors.) In the vector processing of $[A] + [B] = [C]$, the successive fetching of and addition of the elements A and B are grouped together and overlaid, and $Z \simeq 64–256$ elements (the *section size*) are processed with one command, as seen in Table 10.3. Depending on the array size, this method may speedup the processing of vectors by a factor of approximately 10. If all Z elements were truly processed in the same step, then the speedup would be $\sim 64–256$.

Vector processing probably had its heyday during the time when computer manufacturers produced large mainframe computers designed for the scientific and military communities. These computers had proprietary hardware and software and were often so expensive that only corporate or military laboratories

Table 10.2 Computation of matrix $[C] = [A] + [B]$-

Step 1	Step 2	...	Step 99
$c(1) = a(1) + b(1)$	$c(2) = a(2) + b(2)$	...	$c(99) = a(99) + b(99)$

Table 10.3 Vector processing of matrix $[A] + [B] = [C]$.

Step 1	Step 2	...	Step Z
$c(1) = a(1) + b(1)$			
	$c(2) = a(2) + b(2)$		
		...	
			$c(Z) = a(Z) + b(Z)$

could afford them. While the Unix and then PC revolutions have nearly eliminated these large vector machines, some do exist, as well as PCs that use vector processing in their video cards. Who is to say what the future holds in store?

10.7 Introduction to Parallel Computing

There is a little question that advances in the hardware for parallel computing are impressive. Unfortunately, the software that accompanies the hardware often seems stuck in the 1960s. In our view, message passing and GPU programming have too many details for application scientists to worry about and (unfortunately) requires coding at an elementary level reminiscent of the early days of computing. However, the increasing occurrence of clusters in which the nodes are symmetric multiprocessors has led to the development of sophisticated compilers that follow simpler programming models; for example, *partitioned global address space* compilers such as *CoArray Fortran*, *Unified Parallel C*, and *Titanium*. In these approaches, the programmer views a global array of data and then manipulates these data as if they were contiguous. Of course, the data really are distributed, but the software takes care of that outside the programmer's view. Although such a program may make use of processors less efficiently than would a hand-coded program, it is a lot easier than redesigning your program. Whether it is worth *your* time to make a program more efficient depends on the problem at hand, the number of times the program will be run, and the resources available for the task. In any case, if each node of the computer has a number of processors with a shared memory and there are a number of nodes, then some type of a hybrid programming model will be needed.

10.8
Parallel Semantics (Theory)

We saw earlier that many of the tasks undertaken by a high-performance computer are run in parallel by making use of internal structures such as pipelined and segmented CPUs, hierarchical memory, and separate I/O processors. While these tasks are run "in parallel," the modern use of *parallel computing* or *parallelism* denotes applying multiple processors to a single problem (Quinn, 2004). It is a computing environment in which some number of CPUs are running asynchronously and communicating with each other in order to exchange intermediate results and coordinate their activities.

For an instance, consider the matrix multiplication:

$$[B] = [A][B] . \tag{10.4}$$

Mathematically, this equation makes no sense unless $[A]$ equals the identity matrix $[I]$. However, it does make sense as an algorithm that produces new value of B on the LHS in terms of old values of B on the RHS:

$$[B^{\text{new}}] = [A][B^{\text{old}}] \tag{10.5}$$

$$\Rightarrow \quad B^{\text{new}}_{i,j} = \sum_{k=1}^{N} A_{i,k} B^{\text{old}}_{k,j} . \tag{10.6}$$

Because the computation of $B^{\text{new}}_{i,j}$ for specific values of i and j is independent of the computation of all the other values of $B^{\text{new}}_{i,j}$, each $B^{\text{new}}_{i,j}$ can be computed in parallel, or each row or column of $[B^{\text{new}}]$ can be computed in parallel. If B were not a matrix, then we could just calculate $B = AB$ with no further ado. However, if we try to perform the calculation using just matrix elements of $[B]$ by replacing the old values with the new values as they are computed, then we must somehow establish that the $B_{k,j}$ on the RHS of (10.6) are the values of $[B]$ that existed *before* the matrix multiplication.

This is an example of *data dependency*, in which the data elements used in the computation depend on the order in which they are used. A way to account for this dependency is to use a temporary matrix for $[B^{\text{new}}]$, and then to assign $[B]$ to the temporary matrix after all multiplications are complete:

$$[\text{Temp}] = [A][B] , \tag{10.7}$$

$$[B] = [\text{Temp}] . \tag{10.8}$$

In contrast, the matrix multiplication $[C] = [A][B]$ is a *data parallel* operation in which the data can be used in any order. So already we see the importance of communication, synchronization, and understanding of the mathematics behind an algorithm for parallel computation.

The processors in a parallel computer are placed at the *nodes* of a communication network. Each node may contain one CPU or a small number of CPUs, and the communication network may be internal to or external to the computer. One way of categorizing parallel computers is by the approach they utilize in handling instructions and data. From this viewpoint there are three types of machines:

Single instruction, single data (SISD) These are the classic (von Neumann) serial computers executing a single instruction on a single data stream before the next instruction and next data stream are encountered.

Single instruction, multiple data (SIMD) Here instructions are processed from a single stream, but the instructions act concurrently on multiple data elements. Generally, the nodes are simple and relatively slow but are large in number.

Multiple instructions, multiple data (MIMD) In this category, each processor runs independently of the others with independent instructions and data. These are the types of machines that utilize *message-passing* packages, such as MPI, to communicate among processors. They may be a collection of PCs linked via a network, or more integrated machines with thousands of processors on internal boards, such as the Blue Gene computer described in Section 10.15. These computers, which do not have a shared memory space, are also called *multicomputers*. Although these types of computers are some of the most difficult to program, their low cost and effectiveness for certain classes of problems have led to their being the dominant type of parallel computer at present.

The running of independent programs on a parallel computer is similar to the multitasking feature used by Unix and PCs. In multitasking (Figure 10.4a), several independent programs reside in the computer's memory simultaneously and share the processing time in a round robin or priority order. On a SISD computer, only one program runs at a single time, but if other programs are in memory, then it does not take long to switch to them. In multiprocessing (Figure 10.4b), these jobs may all run at the same time, either in different parts of memory or in the memory of different computers. Clearly, multiprocessing becomes complicated if separate processors are operating on different parts of the *same* program because then synchronization and load balance (keeping all the processors equally busy) are concerns.

In addition to instructions and data streams, another way of categorizing parallel computation is by *granularity*. A *grain* is defined as a measure of the computational work to be performed, more specifically, the ratio of computation work to communication work.

Coarse-grain parallel Separate programs running on separate computer systems with the systems coupled via a conventional communication network. An illustration is six Linux PCs sharing the same files across a network but with a different central memory system for each PC. Each computer can be operating on a different, independent part of one problem at the same time.

Medium-grain parallel Several processors executing (possibly different) programs simultaneously while accessing a common memory. The processors are usually placed on a common *bus* (communication channel) and communicate with each other through the memory system. Medium-grain programs have different, independent, *parallel subroutines* running on different processors. Because the compilers are seldom smart enough to figure out which parts of the program to run where, the user must include the multitasking routines in the program.[1)]

Fine-grain parallel As the granularity decreases and the number of nodes increases, there is an increased requirement for fast communication among the nodes. For this reason, fine-grain systems tend to be custom-designed machines. The communication may be via a central bus or via shared memory for a small number of nodes, or through some form of high-speed network for massively parallel machines. In the latter case, the user typically divides the work via certain coding constructs, and the compiler just compiles the program. The program then runs concurrently on a user-specified number of nodes. For example, different `for` loops of a program may be run on different nodes.

10.9 Distributed Memory Programming

An approach to concurrent processing that, because it is built from commodity PCs, has gained dominant acceptance for coarse- and medium-grain systems is *distributed memory*. In it, each processor has its own memory and the processors exchange data among themselves through a high-speed switch and network. The data exchanged or *passed* among processors have encoded *to* and *from* addresses and are called *messages*. The *clusters* of PCs or workstations that constitute a *Beowulf*[2)] are examples of distributed memory computers (Figure 10.6). The unifying characteristic of a cluster is the integration of highly replicated compute and communication components into a single system, with each node still able to operate independently. In a Beowulf cluster, the components are commodity ones designed for a general market, as are the communication network and its high-speed switch (special interconnects are used by major commercial manufacturers, but they do not come cheaply). *Note*: A group of computers connected by a network may also be called a cluster, but unless they are designed for parallel processing, with the same type of processor used repeatedly and with only a limited number of processors (the *front end*) onto which users may log in, they are not usually called a Beowulf.

1) Some experts define our medium grain as coarse grain yet this distinction changes with time.

2) Presumably there is an analogy between the heroic exploits of the son of Ecgtheow and the nephew of Hygelac in the 1000 C.E. poem *Beowulf* and the adventures of us common folk assembling parallel computers from common elements that have surpassed the performance of major corporations and their proprietary, multimillion-dollar supercomputers.

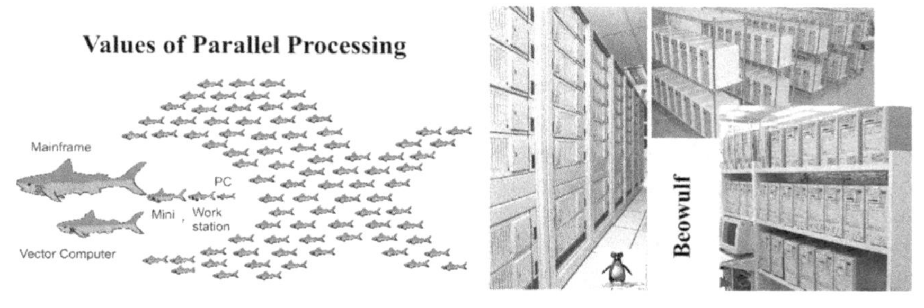

Figure 10.6 Two views of parallel computing (courtesy of Yuefan Deng).

The literature contains frequent arguments concerning the differences among clusters, commodity clusters, Beowulfs, constellations, massively parallel systems, and so forth (Dongarra *et al.*, 2005). Although, we recognize that there are major differences between the clusters on the top 500 list of computers and the ones that a university researcher may set up in his or her lab, we will not distinguish these fine points in the introductory materials we present here.

For a message-passing program to be successful, the data must be divided among nodes so that, at least for a while, each node has all the data it needs to run an independent subtask. When a program begins execution, data are sent to all the nodes. When all the nodes have completed their subtasks, they exchange data again in order for each node to have a complete new set of data to perform the next subtask. This repeated cycle of data exchange followed by processing continues until the full task is completed. Message-passing MIMD programs are also *single-program, multiple-data* programs, which means that the programmer writes a single program that is executed on all the nodes. Often a separate host program, which starts the programs on the nodes, reads the input files and organizes the output.

10.10 Parallel Performance

Imagine a cafeteria line in which all the servers appear to be working hard and fast yet the ketchup dispenser has some relish partially blocking its output and so everyone in line must wait for the ketchup lovers up front to ruin their food before moving on. This is an example of the slowest step in a complex process determining the overall rate. An analogous situation holds for parallel processing, where the ketchup dispenser may be a relatively small part of the program that can be executed only as a series of serial steps. Because the computation cannot advance until these serial steps are completed, this small part of the program may end up being the bottleneck of the program.

As we soon will demonstrate, the speedup of a program will not be significant unless you can get ∼90% of it to run in parallel, and even then most of the speedup will probably be obtained with only a small number of processors. This means that you need to have a computationally intense problem to make parallelization worthwhile, and that is one of the reasons why some proponents of parallel computers with thousands of processors suggest that you should not apply the new machines to old problems but rather look for new problems that are both big enough and well-suited for massively parallel processing to make the effort worthwhile.

The equation describing the effect on speedup of the balance between serial and parallel parts of a program is known as Amdahl's law (Amdahl, 1967; Quinn, 2004). Let

$$p = \text{number of CPUs}\,, \quad T_1 = \text{time to run on 1 CPU}\,,$$
$$T_p = \text{time to run on } p \text{ CPUs}\,. \tag{10.9}$$

The maximum speedup S_p attainable with parallel processing is thus

$$S_p = \frac{T_1}{T_p} \to p\,. \tag{10.10}$$

In practice, this limit is never met for a number of reasons: some of the program is serial, data and memory conflicts occur, communication and synchronization of the processors take time, and it is rare to attain a perfect load balance among all the processors. For the moment, we ignore these complications and concentrate on how the *serial* part of the code affects the speedup. Let f be the fraction of the program that potentially may run on multiple processors. The fraction $1 - f$ of the code that cannot be run in parallel must be run via serial processing and thus takes time:

$$T_s = (1 - f)T_1 \quad \text{(serial time)}\,. \tag{10.11}$$

The time T_p spent on the p parallel processors is related to T_s by

$$T_p = f\frac{T_1}{p}\,. \tag{10.12}$$

That being so, the maximum speedup as a function of f and the number of processors is

$$S_p = \frac{T_1}{T_s + T_p} = \frac{1}{1 - f + f/p} \quad \text{(Amdahl's law)}\,. \tag{10.13}$$

Some theoretical speedups are shown in Figure 10.7 for different numbers of processors p. Clearly the speedup will not be significant enough to be worth the trouble unless most of the code is run in parallel (this is where the 90% of the in-parallel figure comes from). Even an infinite number of processors cannot increase the speed of running the serial parts of the code, and so it runs at one processor speed. In practice, this means many problems are limited to a small number of processors, and that only 10–20% of the computer's peak performance is often all that is obtained for realistic applications.

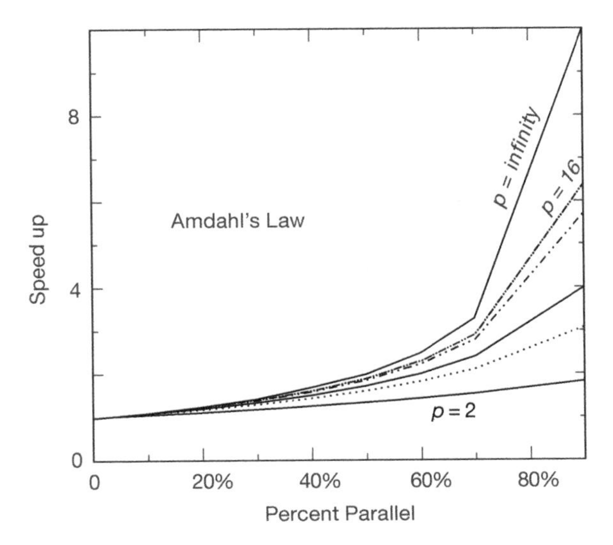

Figure 10.7 The theoretical maximum speedup of a program as a function of the fraction of the program that potentially may be run in parallel. The different curves correspond to different numbers of processors.

10.10.1 Communication Overhead

As discouraging as Amdahl's law may seem, it actually *overestimates* speedup because it ignores the *overhead* for parallel computation. Here we look at communication overhead. Assume a completely parallel code so that its speedup is

$$S_p = \frac{T_1}{T_p} = \frac{T_1}{T_1/p} = p\,. \tag{10.14}$$

The denominator is based on the assumption that it takes no time for the processors to communicate. However, in reality it takes a finite time, called *latency*, to get data out of memory and into the cache or onto the communication network. In addition, a communication channel also has a finite *bandwidth*, that is, a maximum rate at which data can be transferred, and this too will increase the *communication time* as large amounts of data are transferred. When we include communication time T_c, the speedup decreases to

$$S_p \simeq \frac{T_1}{T_1/p + T_c} < p \quad \text{(with communication time)}\,. \tag{10.15}$$

For the speedup to be unaffected by communication time, we need to have

$$\frac{T_1}{p} \gg T_c \quad \Rightarrow \quad p \ll \frac{T_1}{T_c}\,. \tag{10.16}$$

This means that as you keep increasing the number of processors p, at some point the time spent on computation T_1/p must equal the time T_c needed for communication, and adding more processors leads to greater execution time as the processors wait around more to communicate. This is another limit, then, on the maximum number of processors that may be used on any one problem, as well as on the effectiveness of increasing processor speed without a commensurate increase in communication speed.

The continual and dramatic increase in the number of processors being used in computations is leading to a changing view as to how to judge the speed of an algorithm. Specifically, the slowest step in a process is usually the rate-determining step, yet with the increasing availability of CPU power, the slowest step is more often the access to or communication among processors. Such being the case, while the number of computational steps is still important for determining an algorithm's speed, the number and amount of memory access and interprocessor communication must also be mixed into the formula. This is currently an active area of research in algorithm development.

10.11 Parallelization Strategies

A typical organization of a program containing both serial and parallel tasks is given in Table 10.4. The user organizes the work into units called *tasks*, with each task assigning work (*threads*) to a processor. The main task controls the overall execution as well as the subtasks that run independent parts of the program (called *parallel subroutines*, *slaves*, *guests*, or *subtasks*). These parallel subroutines can be distinctive subprograms, multiple copies of the same subprogram, or even Python `for` loops.

It is the programmer's responsibility to establish that the breakup of a code into parallel subroutines is mathematically and scientifically valid and is an equivalent formulation of the original program. As a case in point, if the most intensive part of a program is the evaluation of a large Hamiltonian matrix, you may want to evaluate each row on a different processor. Consequently, the key to parallel programming is to identify the parts of the program that may benefit from parallel execution. To do that the programmer should understand the program's data structures (discussed soon), know in what order the steps in the computation must be performed, and know how to coordinate the results generated by different processors.

The programmer helps speedup the execution by keeping many processors simultaneously busy and by avoiding storage conflicts among different parallel subprograms. You do this *load balancing* by dividing your program into subtasks of approximately equal numerical intensity that will run simultaneously on different processors. The rule of thumb is to make the task with the largest granularity (workload) dominant by forcing it to execute first and to keep all the processors busy by having the number of tasks an integer multiple of the number of processors. This is not always possible.

Table 10.4 A typical organization of a program containing both serial and parallel tasks.

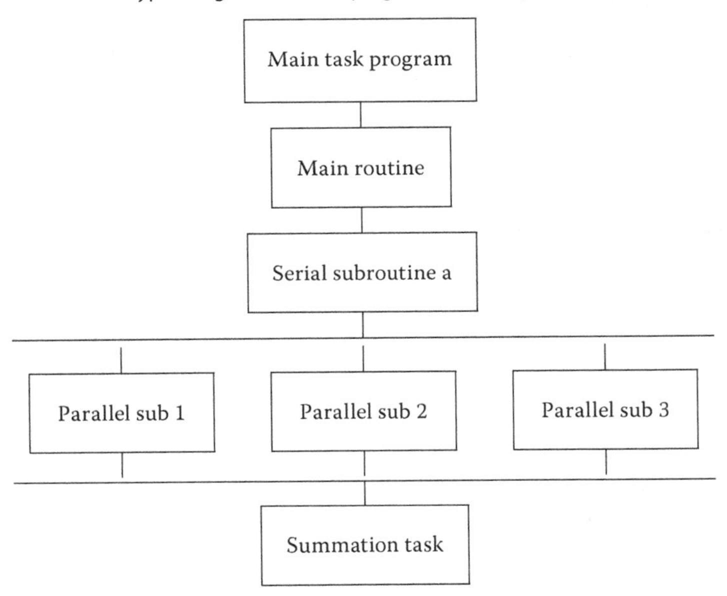

The individual parallel threads can have *shared* or *local* data. The shared data may be used by all the machines, while the local data are private to only one thread. To avoid storage conflicts, design your program so that parallel subtasks use data that are independent of the data in the main task and in other parallel tasks. This means that these data should not be modified *or even examined* by different tasks simultaneously. In organizing these multiple tasks, reduce communication *overhead costs* by limiting communication and synchronization. These costs tend to be high for fine-grain programming where much coordination is necessary. However, *do not* eliminate communications that are necessary to ensure the scientific or mathematical validity of the results; bad science can do harm!

10.12 Practical Aspects of MIMD Message Passing

It makes sense to run only the most numerically intensive codes on parallel machines. Frequently, these are very large programs assembled over a number of years or decades by a number of people. It should come as no surprise, then, that the programming languages for parallel machines are primarily Fortran, which now has explicit structures for the compiler to parallelize, and C. (In the past, we have not obtained good speedup with Java and MPI, yet *FastMPJ* and *MPJ Express* have fixed the problems.)

Effective parallel programming becomes more challenging as the number of processors increases. Computer scientists suggest that it is best *not* to attempt to modify a serial code but instead rewrite one from scratch using algorithms and subroutine libraries best suited to parallel architecture. However, this may involve months or years of work, and surveys find that ~70% of computational scientists revise existing codes instead (Pancake, 1996).

Most parallel computations at present are performed on multiple instruction, multiple-data computers via message passing using MPI. Next we outline some practical concerns based on user experience (Dongarra *et al.*, 2005; Pancake, 1996).

Parallelism carries a price tag There is a steep learning curve requiring intensive effort. Failures may occur for a variety of reasons, especially because parallel environments tend to change often and get "locked up" by a programming error. In addition, with multiple computers and multiple operating systems involved, the familiar techniques for debugging may not be effective.

Preconditions for parallelism If your program is run thousands of times between changes, with execution time in days, and you must significantly increase the resolution of the output or study more complex systems, then parallelism is worth considering. Otherwise, and to the extent of the difference, parallelizing a code may not be worth the time investment.

The problem affects parallelism You must analyze your problem in terms of how and when data are used, how much computation is required for each use, and the type of problem architecture.

Perfectly parallel This is when the same application is run simultaneously on different data sets, with the calculation for each data set independent (e.g., running multiple versions of a Monte Carlo simulation, each with different seeds, or analyzing data from independent detectors). In this case, it would be straightforward to parallelize with a respectable performance to be expected.

Fully synchronous The same operation applied in parallel to multiple parts of the same data set, with some waiting necessary (e.g., determining positions and velocities of particles simultaneously in a molecular dynamics simulation). Significant effort is required, and unless you balance the computational intensity, the speedup may not be worth the effort.

Loosely synchronous Different processors do small pieces of the computation but with intermittent data sharing (e.g., diffusion of groundwater from one location to another). In this case, it would be difficult to parallelize and probably not worth the effort.

Pipeline parallel Data from earlier steps processed by later steps, with some overlapping of processing possible (e.g., processing data into images and then into animations). Much work may be involved, and unless you balance the computational intensity, the speedup may not be worth the effort.

10.12.1 High-Level View of Message Passing

Although it is true that parallel computing programs may become very complicated, the basic ideas are quite simple. All you need is a regular programming language like Python, C, or Fortran, plus four communication statements:[3]

- `send`: One processor sends a message to the network.
- `receive`: One processor receives a message from the network.
- `myid`: An integer that uniquely identifies each processor.
- `numnodes`: An integer giving the total number of nodes in the system.

Once you have made the decision to run your program on a computer cluster, you will have to learn the specifics of a message-passing system such as MPI. Here we give a broader view. When you write a message-passing program, you intersperse calls to the message-passing library with your regular Python, Fortran, or C program. The basic steps are as follows:

1. Submit your job from the command line or a job control system.
2. Have your job start additional processes.
3. Have these processes exchange data and coordinate their activities.
4. Collect these data and have the processes stop themselves.

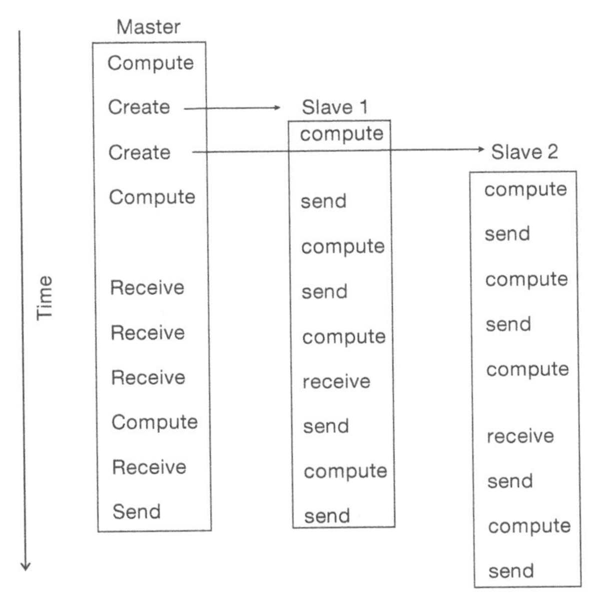

Figure 10.8 A master process and two slave processes passing messages. Notice how in this not-well-designed program there are more sends than receives, and consequently the results may depend upon the order of execution, or the program may even lock up.

3) Personal communication, Yuefan Deng.

We show this graphically in Figure 10.8 where at the top we see a *master* process create two *slave* processes and then assign work for them to do (arrows). The processes then communicate with each other via message passing, output their data to files, and finally terminate.

What can go wrong Figure 10.8 also illustrates some of the difficulties:

- The programmer is responsible for getting the processes to cooperate and for dividing the work correctly.
- The programmer is responsible for ensuring that the processes have the correct data to process and that the data are distributed equitably.
- The commands are at a lower level than those of a compiled language, and this introduces more details for you to worry about.
- Because multiple computers and multiple operating systems are involved, the user may not receive or understand the error messages produced.
- It is possible for messages to be sent or received not in the planned order.
- A *race condition* may occur in which the program results depend upon the specific ordering of messages. There is no guarantee that slave 1 will get its work performed before slave 2, even though slave 1 may have started working earlier (Figure 10.8).
- Note in Figure 10.8 how different processors must wait for signals from other processors; this is clearly a waste of time and has potential for deadlock.
- Processes may *deadlock*, that is, wait for a message that never arrives.

10.12.2
Message Passing Example and Exercise

Start with a simple serial program you have already written that is a good candidate for parallelization. Specifically, one that steps through parameter space in order to generate its results is a good candidate because you can have parallel tasks working on different regions of parameter space. Alternatively, a Monte Carlo calculation that repeats the same step many times is also a good candidate because you can run copies of the program on different processors, and then add up the results at the end. For example, Listing 10.1 is a serial calculation of π by Monte Carlo integration in the C language:

Listing 10.1 Serial calculation of π by Monte Carlo integration.

```
// pi.c: *Monte-Carlo integration to determine pi

#include <stdio.h>
#include <stdlib.h>

// if you don't have drand48 uncomment the following two lines
//    #define drand48 1.0/RAND_MAX*rand
//    #define srand48 srand

#define max 1000                            // number of stones to be thrown
#define seed 68111                              // seed for number generator
```

```
main()  {

  int i, pi = 0;
  double x, y, area;
  FILE *output;                                    // save data in pond.dat
  output = fopen("pond.dat","w");
  srand48(seed);                                // seed the number generator
  for (i = 1; i<= max; i++) {
    x = drand48()*2-1;                             // creates floats between
    y = drand48()*2-1;                                             // 1 and -1
    if ((x*x + y*y)<1) pi++;                         // stone hit the pond
    area = 4*(double)pi/i;                               // calculate area
    fprintf(output, "%i\t%f\n", i, area);
  }
  printf("data stored in pond.dat\n");
  fclose(output);
}
```

Modify your serial program so that different processors are used and perform independently, and then have all their results combined. For example, Listing 10.2 is a parallel version of pi.c that uses the message passing interface (MPI). You may want to concentrate on the arithmetic commands and not the MPI ones at this point.

Listing 10.2 MPI.c: Parallel calculation of π by Monte Carlo integration using MPI.

```
   /* MPI.c uses a monte carlo method to compute PI  by Stone Throwing */
   /* Based on http://www.dartmouth.edu/~rc/classes/soft_dev/mpi.html   */
   /* Note: if the sprng library is not available, you may use rnd */

#include <stdlib.h>
#include <stdio.h>
#include <math.h>
#include <string.h>
#include <stdio.h>
#include <sprng.h>
#include <mpi.h>
#define USE_MPI
#define SEED 35791246

main(int argc, char *argv[])
{
   int niter=0;
   double x,y;
   int i,j,count=0,mycount; /* # of points in the 1st quadrant of unit
        circle */
   double z;
   double pi;
   int myid,numprocs,proc;
   MPI_Status status;
   int master =0;
   int tag = 123;
   int *stream_id;                /* stream id generated by SPRNGS */

   MPI_Init(&argc,&argv);
   MPI_Comm_size(MPI_COMM_WORLD,&numprocs);
   MPI_Comm_rank (MPI_COMM_WORLD,&myid);

   if (argc <=1) {
     fprintf(stderr,"Usage: monte_pi_mpi number_of_iterations\n");
     MPI_Finalize();
     exit(-1);
   }
```

```
    sscanf(argv[1],"%d",&niter); /* 1st argument is the number of
        iterations*/

    /* initialize random numbers */
    stream_id = init_sprng(myid,numprocs,SEED,SPRNG_DEFAULT);
    mycount=0;
    for ( i=0; i<niter; i++) {
        x = (double)sprng(stream_id);
        y = (double)sprng(stream_id);
        z = x*x+y*y;
        if (z<=1) mycount++;
    }
    if (myid ==0) { /* if I am the master process gather results from
        others */
        count = mycount;
        for (proc=1; proc<numprocs; proc++) {
            MPI_Recv(&mycount,1,MPI_REAL,proc,tag,MPI_COMM_WORLD,&status);
            count +=mycount;
        }
        pi=(double)count/(niter*numprocs)*4;
        printf("\n # of trials= %d , estimate of pi is %g
            \n",niter*numprocs,pi);
    }
    else { /* for all the slave processes send results to the master */
        printf("Processor %d sending results= %d to master
            process\n",myid,mycount);
        MPI_Send(&mycount,1,MPI_REAL,master,tag,MPI_COMM_WORLD);
    }

    MPI_Finalize();              /* let MPI finish up */
}
```

Although this small a problem is not worth your efforts in order to obtain a shorter run time, it is worth investing your time to gain some experience in parallel computing.

10.13 Scalability

A common discussion at HPC and Supercomputing conferences of the past heard application scientists get up, after hearing about the latest machine with what seemed like an incredible number of processors, and ask "But how can I use such a machine on my problem, which takes hours to run, but is not trivially parallel like your example?" The response from the computer scientist was often something like "You just need to think up some new problems that are more appropriate to the machines being built. Why use a supercomputer for a problem you can solve on a modern laptop?" It seems that these anecdotal exchanges have now been incorporated into the fabric of parallel computing under the title of *scalability*. In the most general sense, *scalability is defined as the ability to handle more work as the size of the computer or application grows.*

As we have already indicated, the primary challenge of parallel computing is deciding how best to break up a problem into individual pieces that can each be computed separately. In an ideal world, a problem would *scale* in a linear fashion, that is, the program would speedup by a factor of N when it runs on a machine having N nodes. (Of course, as $N \to \infty$ the proportionality cannot hold because communication time must at some point dominate). In present day terminology, this type of scaling is called *strong scaling*, and refers to a situation in which the *problem size remains fixed* while the number of number of nodes (the *machine scale*) increases. Clearly then, the goal when solving a problem that scales strongly is to decrease the amount of time it takes to solve the problem by using a more powerful computer. These are typically CPU-bound problems and are the hardest ones to yield something close to a linear speedup.

In contrast to strong scaling in which the problem size remains fixed, in *weak scaling* we have applications of the type our CS colleagues referred to earlier; namely, ones in which we make the problem bigger and bigger as the number of processors increases. So here, we would have linear or perfect scaling if we could increase the size of the problem solved in proportion to the number N of nodes.

To illustrate the difference between strong and weak scaling, consider Figure 10.9 (based on a lecture by Thomas Sterling). We see that for an application that scales perfectly strongly, the work carried out on each node decreases as the scale of the machine increases, which of course means that the time it takes to complete the problem decreases linearly. In contrast, we see that for an application that scales perfectly weakly, the work carried out by each node remains the same as the scale of the machine increases, which means that we are solving progressively larger problems in the same time as it takes to solve smaller ones on a smaller machine.

The concepts of weak and strong scaling are ideals that tend not to be achieved in practice, with real world applications being a mix of the two. Furthermore, it is the combination of application and computer architecture that determines the type of scaling that occurs. For example, shared memory systems and distributed-

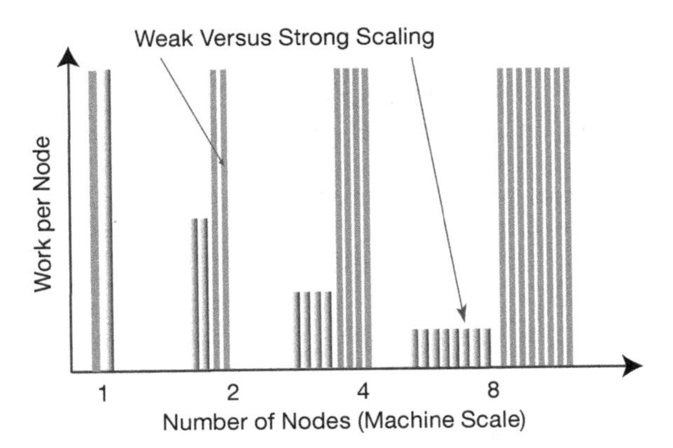

Figure 10.9 A graphical representation of weak vs. strong scaling. Weak scaling keeps each node doing the same amount of work as the problem is made bigger. Strong scaling has each node doing less work (running for less time) as the number of nodes is made bigger.

memory, message passing systems scale differently. Furthermore, a *data parallel* application (one in which each node can work on its own separate data set) will by its very nature scale weakly.

Before we go on and set you working on some examples of scaling, we should introduce a note of caution. Realistic applications tend to have various levels of complexity and so it may not be obvious just how to measure the increase in "size" of a problem. As an instance, it is known that the solution of a set of N linear equations via Gaussian elimination requires $\mathcal{O}(N^3)$ floating-point operations (flops). This means that doubling the number of equations does not make the "problem" twice as large, but rather eight times as large! Likewise, if we are solving partial differential equations on a 3D spatial grid and a 1D time grid, then the problem size would scale like N^4. In this case, doubling the problem size would mean increasing N by only $2^{1/4} \simeq 1.19$.

10.13.1 Scalability Exercises

We have given above, and included in the Codes directory, a serial code pi.c that computes $\pi/4$ by Monte Carlo integration of a quarter of a unit circle. We have also given the code MPIpi.c that computes π by the same algorithm using MPI to compute the algorithm in parallel. Your exercise is to see how well this application scales. You can modify the codes we have given, or you can write your own.

1. Determine the CPU time required to calculate π with the serial calculation using 1000 iterations (stone throws). Make sure that this is the actual run time and does not include any system time. (You can get this type of information, depending upon the operating system, by inserting timer calls in your program.)
2. Get the MPI code running for the same number (1000) of iterations.
3. First do some running that constitutes a *strong scaling test*. This means keeping the problem size constant, or in other words, keeping $N_{\text{iter}} = 1000$. Start by running the MPI code with only one processor doing the numerical computation. A comparison of this to the serial calculation gives you some idea of the overhead associated with MPI.
4. Again keeping $N_{\text{iter}} = 1000$, run the MPI code for 2, 4, 8, and 16 computing nodes. In any case, make sure to go up to enough nodes so that the system no longer scales. Record the run time for each number of nodes.
5. Make a plot of run time vs. number of nodes from the data you have collected.
6. Strong scalability here would yield a straight line graph. Comment on your results.
7. Now do some running that constitutes a *weak scaling test*. This means increasing the problem size simultaneously with the number of nodes being used. In the present case, increasing the number of iterations, N_{iter}.

8. Run the MPI code for 2, 4, 8, and 16 computing nodes, with proportionally larger values for N_{iter} in each case (2000, 4000, 8000, 16 000, etc.). In any case, make sure to go up to enough nodes so that the system no longer scales.
9. Record the run time for each number of nodes and make a plot of the run time vs. number of computing nodes.
10. Weak scaling would imply that the run time remains constant as the problem size and the number of compute nodes increase in proportion. Comment on your results.
11. Is this problem more appropriate for weak or strong scaling?

10.14 Data Parallelism and Domain Decomposition

As you have probably realized by this point, there are two basic, but quite different, approaches to creating a program that runs in parallel. In *task parallelism*, you decompose your program by tasks, with differing tasks assigned to different processors, and with great care taken to maintain *load balance*, that is, to keep all processors equally busy. Clearly you must understand the internal workings of your program in order to do this, and you must also have made an accurate *profile* of your program so that you know how much time is being spent in various parts.

In *data parallelism*, you decompose your program based on the data being created or acted upon, with differing data spaces (*domains*) assigned to different processors. In data parallelism, there often must be data shared at the boundaries of the data spaces, and therefore synchronization among the data spaces. Data parallelism is the most common approach and is well suited to message-passing machines in which each node has its own private data space, although this may lead to a large amount of data transfer at times.

When planning how to decompose global data into subspaces suitable for parallel processing, it is important to divide the data into contiguous blocks in order to minimize the time spent on moving data through the different stages of memory (page faults). Some compilers and operating systems help you in this regard by exploiting *spatial locality*, that is, by assuming that if you are using a data element at one location in data space, then it is likely that you may use some nearby ones as well, and so they too are made readily available. Some compilers and operating systems also exploit *temporal locality*, that is, by assuming that if you are using a data element at one time, then there is an increased probability that you may use it again in the near future, and so it too is kept handy. You can help optimize your programs by taking advantage of these localities while programming.

As an example of *domain decomposition*, consider the solution of a partial differential equation by a finite difference method. It is known from classical electrodynamics that the electric potential $U(\boldsymbol{x})$ in a charge-free region of 2D space satisfies *Laplace's equation* (fully discussed in Section 19.4):

$$\frac{\partial^2 U(x,y)}{\partial x^2} + \frac{\partial^2 U(x,y)}{\partial y^2} = 0 . \tag{10.17}$$

We see that the potential depends simultaneously on x and y, which is what makes it a partial differential equation. The electric charges, which are the source of the field, enter indirectly by specifying the potential values on some boundaries or charged objects.

As shown in Figure 10.10, we look for a solution on a lattice of (x, y) values separated by finite difference Δ in each dimension and specified by discrete locations on the lattice:

$$x = x_0 + i\Delta , \quad y = y_0 + j\Delta , \quad i, j = 0, \dots, N_{\text{max}-1} . \tag{10.18}$$

When the finite difference expressions for the derivatives are substituted into (10.17), and the equation is rearranged, we obtain the finite-difference algorithm for the solution of Laplace's equation:

$$U_{i,j} = \frac{1}{4}[U_{i+1,j} + U_{i-1,j} + U_{i,j+1} + U_{i,j-1}] . \tag{10.19}$$

This equation says that when we have a proper solution, it will be the average of the potential at the four nearest neighbors (Figure 10.10). As an algorithm, (10.19) does not provide a direct solution to Laplace's equation but rather must be repeated many times to converge upon the solution. We start with an initial guess for the potential, improve it by sweeping through all space, taking the average over nearest neighbors at each node, and keep repeating the process until the solution no longer changes to some level of precision or until failure is evident. When converged, the initial guess is said to have *relaxed* into the solution.

In Listing 10.3, we have a serial code `laplace.c` that solves the Laplace equation in two dimensions for a straight wire kept at 100 V in a grounded box, using the relaxation algorithm (10.19). There are five basic elements of the code:

1. Initialize the potential values on the lattice.
2. Provide an initial guess for the potential, in this case $U = 0$ except for the wire at 100 V.
3. Maintain the boundary values and the source term values of the potential at all times.
4. Iterate the solution until convergence ((10.19) being satisfied to some level of precision) is obtained.
5. Output the potential in a form appropriate for 3D plotting.

As you can see, the code is a simple pedagogical example with its essential structure being the array `p[40][40]` representing a (small) regular lattice. Industrial strength applications might use much larger lattices as well as adaptive meshes and hierarchical multigrid techniques.

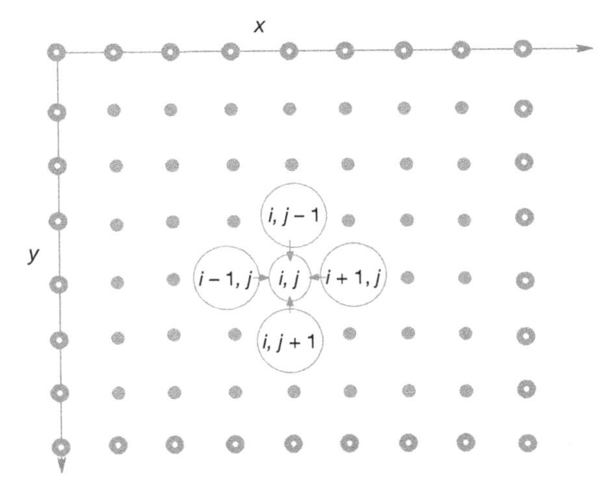

Figure 10.10 A representation of the lattice in a 2D rectangular space upon which Laplace's equation is solved using a finite difference approach. The lattice sites with white centers correspond to the boundary of the physical system, upon which boundary conditions must be imposed for a unique solution. The large circles in the middle of the lattice represent the algorithm used to solve Laplace's equation in which the potential at the point $(x, y) = (i, j)\Delta$ is set to the average of the potential values at the four nearest-neighbor points.

When thinking of parallelizing this program, we note an algorithm being applied to a space of data points, in which case we can divide the domain into subspaces and assign each subspace to a different processor. This is domain decomposition or data parallelism. In essence, we have divided a large boundary-value problem into an equivalent set of smaller boundary-value problems that eventually get fit back together. Often extra storage is allocated on each processor to hold the data values that get communicated from neighboring processors. These storage locations are referred to as *ghost cells, ghost rows, ghost columns, halo cells,* or *overlap areas.*

Two points are essential in domain decomposition: (i) Load balancing is critical, and is obtained here by having each domain contain the same number of lattice points. (ii) Communication among the processors should be minimized because this is a slow process. Clearly the processors must communicate to agree on the potential values at the domain boundaries, except for those boundaries on the edge of the box that remain grounded at 0 V. But because there are many more lattice sites that need computing than there are boundaries, communication should not slow down the computation severely for large lattices.

To see an example of how this is carried out, the serial code `poisson_1d.c` solves Laplace's equation in 1D, and `poisson_parallel_1d.c` solves the same 1D equation in parallel (codes courtesy of Michel Vallieres). This code uses an accelerated version of the iteration algorithm using the parameter Ω, a separate method for domain decomposition, as well as ghost cells to communicate the boundaries.

Listing 10.3 laplace.c Serial solution of Laplace's equation using a finite difference technique.

```
   /* laplace.c:  Solve Laplace equation with finite differences        */

#include <stdio.h>
#define max 40                                /* number of grid points */
main()
{
   double x, p[max][max];
   int i, j, iter, y;
   FILE *output;          /* save data in laplace.dat */
   output = fopen("laplace.dat","w");
   for(i=0; i<max; i++)                      /* clear the array  */
   {     for (j=0; j<max; j++) p[i][j] = 0;}
   for(i=0; i<max; i++) p[i][0] = 100.0;           /* p[i][0] = 100 V */
   for(iter=0; iter<1000; iter++)                   /* iterations */
   {     for(i=1; i<(max-1); i++)                    /* x-direction */
      { for(j=1; j<(max-1); j++)                    /* y-direction */
         {  p[i][j] = 0.25*(p[i+1][j]+p[i-1][j]+p[i][j+1]+p[i][j-1]);    }
      }
   }
   for (i=0; i<max ; i++)              /* write data gnuplot 3D format */
   {     for (j=0; j<max; j++)
      { fprintf(output, "%f\n",p[i][j]);     }
      fprintf(output, "\n");      /* empty line for gnuplot */
   }
   printf("data stored in laplace.dat\n");
   fclose(output);
}
```

10.14.1 Domain Decomposition Exercises

1. Get the serial version of either laplace.c or laplace.f running.
2. Increase the lattice size to 1000 and determine the CPU time required for convergence to six places. Make sure that this is the actual run time and does not include any system time. (You can get this type of information, depending upon the operating system, by inserting timer calls in your program.)
3. Decompose the domain into four subdomains and get an MPI version of the code running on four compute nodes. [Recall, we give an example of how to do this in the Codes directory with the serial code poisson_1d.c and its MPI implementation, poisson_parallel_1d.c (courtesy of Michel Vallières).]
4. Convert the serial code to three dimensions. This makes the application more realistic, but also more complicated. Describe the changes you have had to make.
5. Decompose the 3D domain into four subdomains and get an MPI version of the code running on four compute nodes. This can be quite a bit more complicated than the 2D problem.
6. Conduct a weak scaling test for the 2D or 3D code.
7. Conduct a strong scaling test for the 2D or 3D code.

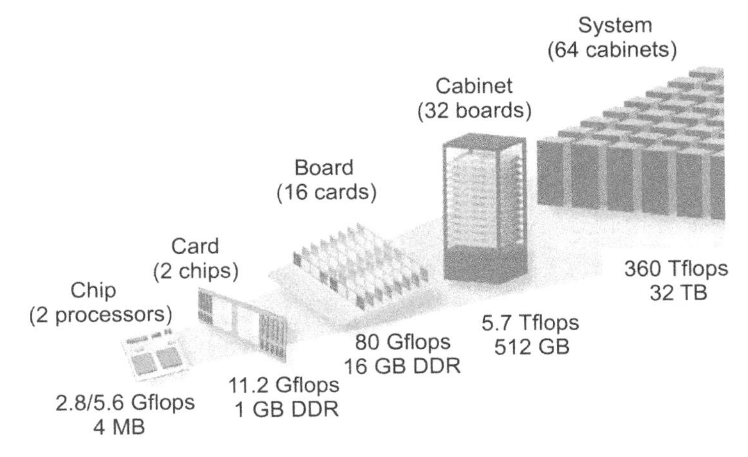

Figure 10.11 The building blocks of Blue Gene (from Gara *et al.* (2005)).

10.15 Example: The IBM Blue Gene Supercomputers

Whatever figures we give to describe the latest supercomputer will be obsolete by the time you read them. Nevertheless, for the sake of completeness, and to set the present scale, we do it anyway. At the time of this writing, one of the fastest computers is the IBM Blue Gene/Q member of the Blue Gene series. In its largest version, its 96 racks contains 98 304 compute nodes with 1.6 million processor cores and 1.6 PB of memory (Gara *et al.*, 2005). In June 2012, it reached a peak speed of 20.1 PFLOPS.

The name Blue Gene reflects the computer's origin in gene research, although Blue Genes are now general-purpose supercomputers. In many ways, these are computers built by committee, with compromises made in order to balance cost, cooling, computing speed, use of existing technologies, communication speed, and so forth. As a case in point, the compute chip has 18 cores, with 16 for computing, 1 for assisting the operating system with communication, and 1 as a redundant spare in case one of the others was damaged. Having communication on the chip reflects the importance of communication for distributed-memory computing (there are both on- and off-chip distributed memories). And while the CPU is fast with 204.8 GFLOPS at 1.6 GHz, there are faster ones made, but they would generate so much heat that it would not be possible to obtain the extreme scalability up to 98 304 compute nodes. So with the high-efficiency figure of 2.1 GFLOPS/watt, Blue Gene is considered a "green" computer.

We look more closely now at one of the original Blue Genes, for which we were able to obtain illuminating figures (Gara *et al.*, 2005). Consider the building-block view in Figure 10.11. We see multiple cores on a chip, multiple chips on a card, multiple cards on a board, multiple boards in a cabinet, and multiple cabinets in an installation. Each processor runs a Linux operating system (imagine what the

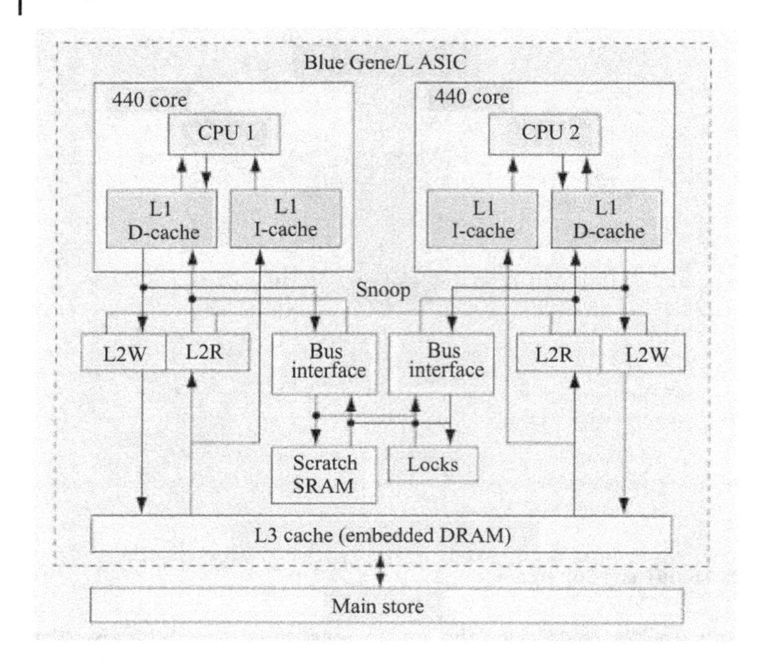

Figure 10.12 The single-node memory system (as presented by Gara *et al.* (2005)).

cost in both time and money would be for Windows!) and utilizes the hardware by running a distributed memory MPI with C, C++, and Fortran90 compilers.

Blue Gene has three separate communication networks (Figure 10.13). At the heart of the network is a 3D torus that connects all the nodes, for example, Figure 10.13 a shows a sample torus of $2 \times 2 \times 2$ nodes. The links are made by special link chips that also compute; they provide both direct neighbor–neighbor communications and cut-through communication across the network. The result of this sophisticated communications network is that there are approximately the same effective bandwidth and latencies (response times) between all nodes. However, the bandwidth may be affected by interference among messages, with the actual latency also depending on the number of *hops* to get from one node to another.[4] A rate of 1.4 Gb/s is obtained for node-to-node communication. The collective network in Figure 10.13b is used for collective communications among processors, such as a *broadcast*. Finally, the control and gigabit ethernet network (Figure 10.13c) is used for I/O to communicate with the switch (the hardware communication center) and with ethernet devices.

The computing heart of Blue Gene is its integrated circuit and the associated memory system (Figure 10.12). This is essentially an entire computer system on a chip, with the exact specifications depending upon the model, and changing with time.

4) The number of hops is the number of devices a given data packet passes through.

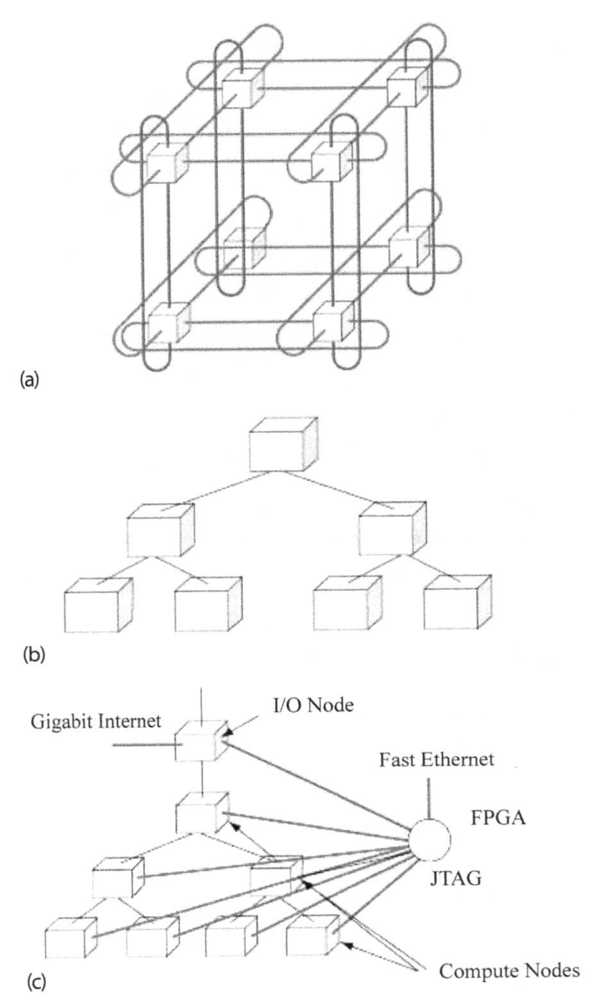

Figure 10.13 (a) A 3D torus connecting $2 \times 2 \times 2$ compute nodes. (b) The global collective memory system. (c) The control and Gb-ethernet memory system (from Gara *et al.* (2005)).

10.16 Exascale Computing via Multinode-Multicore GPUs

The current architecture of top-end supercomputers (Figure 10.14) uses a very large numbers of nodes, with each node containing a chip set that includes multiple cores as well as a graphical processing unit (GPU) attached to the chip set[5]. In the near future, we expect to see laptop computers capable of teraflops (10^{12} floating-point operations per second), deskside computers capable of petaflops,

5) GPUs and their programming are discussed in Chapter 11.

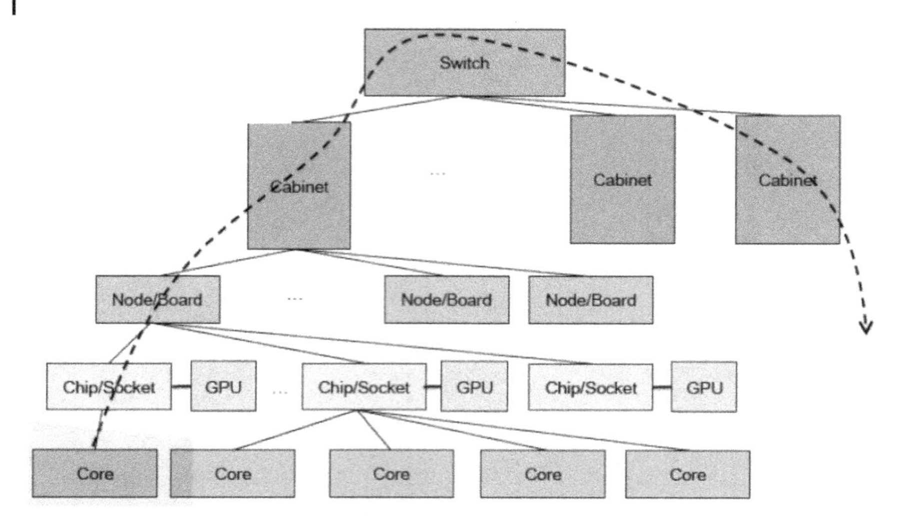

Figure 10.14 A schematic of an exascale computer in which, in addition to each chip having multiple cores, a graphical processing unit is attached to each chip (adapted from Dongarra (2011)).

and supercomputers at the exascale, in terms of both flops and memory, probably with millions of nodes.

Look again at the schematic in Figure 10.14. As in Blue Gene, there really are large numbers of chip boards and large numbers of cabinets. Here we show just one node and one cabinet and not the full number of cores. The dashed line in Figure 10.14 represents communications, and it is seen to permeate all components of the computer. Indeed, communications have become such an essential part of modern supercomputers, which may contain 100s of 1000s of CPUs, that the network interface "card" may be directly on the chip board. Because a computer of this sort contains shared memory at the node level and distributed memory at the cabinet or higher levels, programming for the requisite data transfer among the multiple elements is a fundamental challenge, with significant new investments likely to occur (Dongarra *et al.*, 2014).

11
Applied HPC: Optimization, Tuning, and GPU Programming

> More computing sins are committed in the name of efficiency (without necessarily achieving it) than for any other single reason – including blind stupidity.
>
> *W.A. Wulf*

> We should forget about small efficiencies, say about 97% of the time: premature optimization is the root of all evil.
>
> *Donald Knuth*

> The best is the enemy of the good.
>
> *Voltaire*

This chapter continues the discussion of high-performance computing (HPC) began in the previous chapter. Here we lead the reader through exercises that demonstrates some techniques for writing programs that are optimized for HPC hardware, and how the effects of optimization may be different in different computer languages. We then move on to tips for programming on a multicore computer. We conclude with an example of programming in compute unified device architecture (CUDA) on a graphics processing card, a subject beyond the general scope of this text, but one of high current interest.

11.1
General Program Optimization

Rule 1 Do not do it.
Rule 2 (for experts only): Do not do it yet.
Rule 3 Do not optimize as you go: Write your program without regard to possible optimizations, concentrating instead on making sure that the code is clean, correct, and understandable. If it is too big or too slow when you have finished, then you can consider optimizing it.

Computational Physics, 3rd edition. Rubin H. Landau, Manuel J. Páez, Cristian C. Bordeianu.

Rule 4 Remember the 80/20 rule: In many fields you can get 80% of the result with 20% of the effort (also called the 90/10 rule – it depends on who you talk to). Whenever you're about to optimize code, use profiling to find out where that 80% of execution time is going, so you know where to concentrate your effort.

Rule 5 Always run "before" and "after" benchmarks: How else will you know that your optimizations actually made a difference? If your optimized code turns out to be only slightly faster or smaller than the original version, undo your changes and go back to the original, clear code.

Rule 6 Use the right algorithms and data structures: Do not use an $\mathcal{O}(n^2)$ DFT algorithm to do a Fourier transform of a 1000 elements when there is an $\mathcal{O}(n \log n)$ FFT available. Similarly, do not store a thousand items in an array that requires an $\mathcal{O}(n)$ search when you could use an $\mathcal{O}(\log n)$ binary tree or an $\mathcal{O}(1)$ hash table (Hardwich, 1996).

The type of optimization often associated with *high-performance* or *numerically intensive* computing is one in which sections of a program are rewritten and reorganized in order to increase the program's speed. The overall value of doing this, especially as computers have become so fast and so available, is often a subject of controversy between computer scientists and computational scientists. Both camps agree that using the optimization options of compilers is a good idea. Yet on the one hand, CS tends to think that optimization is best left to the compilers, while computational scientists, who tend to run large codes with large amounts of data in order to solve real-world problems, often believe that you cannot rely on the compiler to do all the optimization.

11.1.1 Programming for Virtual Memory (Method)

While memory paging makes little appear big, you pay a price because your program's run time increases with each page fault. If your program does not fit into RAM all at once, it will be slowed down significantly. If virtual memory is shared among multiple programs that run simultaneously, they all cannot have the entire RAM at once, and so there will be memory access *conflicts* which will decrease performance. The basic rules for programming for virtual memory are:

1. Do not waste your time worrying about reducing the amount of memory used (the *working set size*) unless your program is large. In that case, take a global view of your entire program and optimize those parts that contain the largest arrays.
2. Avoid page faults by organizing your programs to successively perform their calculations on subsets of data, each fitting completely into RAM.
3. Avoid simultaneous calculations in the same program to avoid competition for memory and consequent page faults. Complete each major calculation before starting another.

4. Group data elements close together in memory blocks if they are going to be used together in calculations.

11.1.2 Optimization Exercises

Many of the optimization techniques developed for Fortran and C are also relevant for Python applications. Yet while Python is a good language for scientific programming and is as universal and portable as Java, at present Python code runs slower than Fortran, C or even Java codes. In part, this is a consequence of the Fortran and C compilers having been around longer and thereby having been better refined to get the most out of a computer's hardware, and in part this is also a consequence of Python not being designed for speed. Because modern computers are so fast, whether a program takes 1s or 3s usually does not matter much, especially in comparison to the hours or days of *your* time that it might take to modify a program for increased speed. However, you may want to convert the code to C (whose command structure is similar to that of Python) if you are running a computation that takes hours or days to complete and will be doing it many times.

Especially when asked to, compilers may look at your entire code as a single entity and rewrite it in an attempt to make it run faster. In particular, Fortran and C compilers often speed up programs by being careful in how they load arrays into memory. They also are careful in keeping the cache lines full so as not to keep the CPU waiting or having it move on to some other chore. That being said, in our experience compilers still cannot optimize a program as well as can a skilled and careful programmer who understands the order and structure of the code.

There is no fundamental reason why a program written in Java or Python cannot be compiled to produce a highly efficient code, and indeed such compilers are being developed and becoming available. However, such code is optimized for a particular computer architecture and so is not portable. In contrast, the byte code (`.class` file in Java and `.pyc` file in Python) produced by the compiler is designed to be interpreted or recompiled by the Java or Python *Virtual Machine* (just another program). When you change from Unix to Windows, for example, the Virtual Machine program changes, but the byte code is the same. This is the essence of portability.

In order to improve the performance of Java and Python, many computers and browsers run *Just-in-Time* (JIT) compilers. If a JIT is present, the Virtual Machine feeds your byte code `Prog.class` or `Prog.pyc` to the JIT so that it can be recompiled into native code explicitly tailored to the machine you are using. Although there is an extra step involved here, the total time it takes to run your program is usually 10–30 times faster with a JIT as compared to line-by-line interpretation. Because the JIT is an integral part of the Virtual Machine on each operating system, this usually happens automatically.

In the experiments below, you will investigate techniques to optimize both Fortran and Java or Python programs, and to compare the speeds of both languages for the same computation. If you run your program on a variety of machines you should also be able to compare the speed of one computer to that of another. Note that a knowledge of Fortran or C is not required for these exercises; if you keep an open mind you should be able to look at the code and figure out what changes may be needed.

11.1.2.1 Good and Bad Virtual Memory Use (Experiment)

To see the effect of using virtual memory, convert these simple pseudocode examples (Listings 11.1 and 11.2) into actual code in your favorite language, and then run them on your computer. Use a command such as `time`, `time.clock()` or `timeit` to measure the time used for each example. These examples call functions `force12` and `force21`. You should write these functions and make them have significant memory requirements.

Listing 11.1 BAD program, too simultaneous.

```
for j = 1, n; {   for i = 1, n; {
     f12(i,j) = force12(pion(i), pion(j))      // Fill f12
     f21(i,j) = force21(pion(i), pion(j))      // Fill f21
     ftot = f12(i,j) + f21(i,j) }}             // Fill ftot
```

You see (Listing 11.1) that each iteration of the `for` loop requires the data and code for all the functions as well as access to all the elements of the matrices and arrays. The working set size of this calculation is the sum of the sizes of the arrays `f12(N,N)`, `f21(N,N)`, and `pion(N)` plus the sums of the sizes of the functions `force12` and `force21`.

A better way to perform the same calculation is to break it into separate components (Listing 11.2):

Listing 11.2 GOOD program, separate loops.

```
for j = 1, n;
       { for i = 1, n;   f12(i,j) = force12(pion(i), pion(j)) }
for j = 1, n;
       { for i = 1, n;   f21(i,j) = force21(pion(i), pion(j)) }
for j = 1, n;
       { for i = 1, n;   ftot = f12(i,j) + f21(i,j) }
```

Here, the separate calculations are independent and the *working set size*, that is, the amount of memory used, is reduced. However, you do pay the additional overhead costs associated with creating extra `for` loops. Because the working set size of the first `for` loop is the sum of the sizes of the arrays `f12(N, N)` and `pion(N)`, and of the function `force12`, we have approximately half the previous size. The size of the last `for` loop is the sum of the sizes for the two arrays. The working set size of the entire program is larger than of the working set sizes for the different `for` loops.

As an example of the need to group data elements close together in memory or common blocks if they are going to be used together in calculations, consider the following code (Listing 11.3):

Listing 11.3 BAD Program, discontinuous memory.

```
Common zed, ylt(9), part(9), zpart1(50000), zpart2(50000), med2(9)
  for j = 1, n;
  ylt(j) = zed * part(j)/med2(9)           // Discontinuous variables
```

Here the variables **zed**, **ylt**, and **part** are used in the same calculations and are adjacent in memory because the programmer grouped them together in **Common** (global variables). Later, when the programmer realized that the array **med2** was needed, it was tacked onto the end of **Common**. All the data comprising the variables **zed**, **ylt**, and **part** fit onto one page, but the **med2** variable is on a different page because the large array **zpart2(50000)** separates it from the other variables. In fact, the system may be forced to make the entire 4 kB page available in order to fetch the 72 B of data in **med2**. While it is difficult for a Fortran or C programmer to establish the placement of variables within page boundaries, you will improve your chances by grouping data elements together (Listing 11.4):

Listing 11.4 GOOD program, continuous memory

```
Common zed, ylt(9), part(9), med2(9), zpart1(50000), zpart2(50000)
        for j = 1, n;
            ylt(j) = zed*part(j)/med2(J)          // Continuous
```

11.2 Optimized Matrix Programming with NumPy

In Chapter 6, we demonstrated several ways of handling matrices with Python. In particular, we recommended using the **array** structure and the NumPy package. In this section, we extend that discussion somewhat by demonstrating two ways in which NumPy may speed up a program. The first is by using NumPy arrays rather than Python ones, and the second is by using Python slicing to reduce stride.

Listing 11.5 TuneNumPy.py compares the time it takes to evaluate a function of each element of an array using a for loop, as well as using a vectorized call with NumPy. To see the effect of fluctuations, the comparison is repeated three times.

```
# TuneNumpy.py: Comparison of NumPy op versus for loop

from datetime import datetime
import numpy as np

def f(x):                    # A function requiring some computation
    return x**2-3*x + 4
x = np.arange(1e5)           # An array of 100,000 integers

for j in range(0, 3):   # Repeat comparison three time

    t1 = datetime.now()
    y = [f(i) for i in x]           # The for loop
    t2 = datetime.now()
    print (' For for loop,          t2-t1 =', t2-t1)
```

```
t1 = datetime.now()
y = f(x)                            # Vectorized evaluation
t2 = datetime.now()
print (' For vector function, t2-t1 =', t2-t1)
```

A powerful feature of NumPy is its high-level *vectorization*. This is the process in which the single call of a function operates not on a variable but on an array object as a whole. In the latter case, NumPy automatically *broadcasts* the operation across all elements of the array with efficient use of memory. As we shall see, the resulting speed up can be more than an order of magnitude! While this may sound complicated, it really is quite simple because NumPy does this for you automatically.

For example, in Listing 11.5 we present the code `TuneNumPy.py`. It compares the speed of a calculation using a *for* loop to evaluate a function for each of 100 000 elements in an array, vs. the speed using NumPy's vectorized evaluation of that function for an array object (Perez *et al.*, 2010). And to see the effects of fluctuations as a result of things like background processes, we repeat the comparison three times. We obtained the following results:

```
For for loop,          t2-t1 = 0:00:00.384000
For vector function,   t2-t1 = 0:00:00.009000
For for loop,          t2-t1 = 0:00:00.383000
For vector function,   t2-t1 = 0:00:00.009000
For for loop,          t2-t1 = 0:00:00.387000
For vector function,   t2-t1 = 0:00:00.008000
```

Though a simple calculation, these results show that vectorization speeds the calculation up by a factor of nearly 50; really!

Now, recall from Chapter 10 our discussion of *stride* (the amount of memory skipped in order to get to the next element needed in a calculation). It is important to have your program minimize stride in order not waste time jumping through memory, as well as not to load unneeded data into memory. For example, for a (1000, 1000) array, there is a passage of 1 word to get to the next column, but of 1000 words to get to the next row. Clearly better to do a column-by-column calculation than a row-by-row one.

We start our example by setting up a 3×3 array of integers using NumPy's `arange` to create a 1D array. We then reshape it into a 3×3 array and determine the strides of the matrix for rows and columns:

```
>>> from numpy import *
>>> A = arange(0,90,10)                                          2
>>> A
array([ 0, 10, 20, 30, 40, 50, 60, 70, 80])                      4
>>> A = A.reshape((3,3))
>>> A                                                            6
array([[ 0, 10, 20],
       [30, 40, 50],                                             8
       [60, 70, 80]])
>>> A.strides                                                    10
(12, 4)
```

Line 11 tells us that it takes 12 bytes (three values) to get to the same position in the next row, but only 4 bytes (one value) to get to the same position in the next column. Clearly cheaper to go from column to column. Now, we show you an easy way to do that.

Recall Python's *slice* operator that extracts just the desired part of a list (like taking a "slice" through the center of a jelly doughnut):

```
ListName[StartIndex:StopBeforeIndex:Step] .
```

The convention here is that if no argument is given, then the slice starts at 0 and stops at the end of the list. For example:

```
>>> A = arange(0,90,10).reshape((3,3))                                   1
>>> A
array([[ 0, 10, 20],                                                     3
       [30, 40, 50],
       [60, 70, 80]])                                                    5
>>> A[:2,:]                     # First two rows (start at 2, go to end)
array([[ 0, 10, 20],                                                     7
       [30, 40, 50]])
>>> A[:,1:3]                    # Columns 1-3 (start at 1, end at 4)     9
array([[10, 20],
       [40, 50],                                                         11
       [70, 80]])
>>> A[::2,:]                    # Every second row                       13
array([[ 0, 10, 20],
       [60, 70, 80]])                                                    15
```

Once sliced, Python does not have to jump through memory to work with these elements, or even set up separate arrays for them. This is called *view-based indexing*, with the indexed notation returning a new array object that *points* to the address of the original data rather than store the values of the new array (think "pointers" in C). This does lead to improved speed, but you must remember that if you alter the new array, you are also altering the original array to which it points (think "pointers" in C).

For instance, you can optimize a finite difference calculation of forward- and central-difference derivatives quite elegantly (Perez *et al.*, 2010):

```
>>> x = arange(0,20,2)                                                               1
>>> x
array([ 0,  2,  4,  6,  8, 10, 12, 14, 16, 18])                                      3
>>> y = x**2
>>> y                                                                                5
array([  0,   4,  16,  36,  64, 100, 144, 196, 256, 324], dtype=int32)
>>> dy_dx = ((y[1:]-y[:1])/(x[1:]-x[:-1]))          # Forward difference             7
>>> dy_dx
array([   2.,    8.,   18.,   32.,   50.,   72.,    98.,   128.,   162.])           9
>>> dy_dx_c = ((y[2:]-y[:-2])/(x[2:]-x[:-2]))       # Central difference
>>> dy_dx_c                                                                          11
array([  4.,   8.,  12.,  16.,  20.,  24.,  28.,  32.])
```

(We note that the values of the derivatives look quite different, yet the forward difference is evaluated at the start of the interval and the central difference at the center.)

11.2.1 NumPy Optimization Exercises

1. We have just demonstrated how NumPy's vectorized function evaluation can speed up a calculation by a factor of 50 via broadcast of an operation over an array rather than performing that operation on each individual element in the array. Determine the speedup for the matrix multiplication [**A**][**B**] where the matrices are at least 10^5 in size and contain floating-point numbers. Compare the direct multiplication to application of the elementary rule for each element:

$$[\mathbf{AB}]_{ij} = \sum_k a_{ik} b_{kj} . \tag{11.1}$$

2. We have just demonstrated how Python's slice operator can be used to reduce the stride of a calculation of derivatives. Determine the speedup obtained in evaluating the forward-difference and central-difference derivatives over an array of at least 10^5 floating-point numbers using stripping to reduce stride. Compare to the same calculation without any stripping.

11.3 Empirical Performance of Hardware

In this section, you conduct an experiment in which you run a complete program in several languages and on as many computers as are available to you. In this way, you will explore how a computer's architecture and software affect a program's performance.

The first step in optimization is to try asking the compiler to optimize your program. You control how completely the compiler tries to do this by adding *optimization options* to the compile command. For example, in Fortran (where this works better than in Python):

```
> f90 -O tune.f90
```

Here -O turns on optimization (O is the capital letter "oh," not zero). The actual optimization that is turned on differs from compiler to compiler. Fortran and C compilers have a bevy of such options and directives that let you truly customize the resulting compiled code. Sometimes optimization options make the code run faster, sometimes not, and sometimes the faster-running code gives the wrong answers (but does so quickly).

Because computational scientists may spend a good fraction of their time running compiled codes, the compiler options tend to become quite specialized. As a case in point, most compilers provide a number of levels of optimization for the compiler to attempt (there are no guarantees with these things). Although the speedup obtained depends upon the details of the program, higher levels may give greater speedup. However, we have had the experience of higher levels of optimization sometimes giving wrong answers (presumably this may be related to our programs not following the rules of grammar perfectly).

Some typical Fortran compiler options include the following:

-O Use the default optimization level (-O3)
-O1 Provide minimum statement-level optimizations
-O2 Enable basic block-level optimizations
-O3 Add loop unrolling and global optimizations
-O4 Add automatic inlining of routines from the same source file
-O5 Attempt aggressive optimizations (with profile feedback).

The *gnu compilers* gcc, g77, and g90 accept -O options as well as

-malign-double Align doubles on 64-bit boundaries
-ffloat-store For codes using IEEE-854 extended precision
-fforce-mem, -fforce-addr Improves loop optimization
-fno-inline Do not compile statement functions inline
-nffast-math Try non-IEEE handling of floats
-funsafe-math-optimizations Speeds up float operations; incorrect result possible
-fno-trapping-math Assume no floating-point traps generated
-fstrength-reduce Makes some loops faster
-frerun-cse-after-loop
-fexpensive-optimizations
-fdelayed-branch
-fschedule-insns
-fschedule-insns2
-fcaller-saves
-funroll-loops Unrolls iterative DO loops
-funroll-all-loops Unrolls DO WHILE loops

11.3.1 Racing Python vs. Fortran/C

The various versions of the program tune that are given in the Codes/HPC directory solve the matrix eigenvalue problem

$$\mathbf{H}c = Ec \tag{11.2}$$

for the eigenvalues E and eigenvectors c of a Hamiltonian matrix $\mathbf{H}$. Here the individual Hamiltonian matrix elements are assigned the values

$$\mathbf{H}_{i,j} = \begin{cases} i, & \text{for } i = j\,, \\ 0.3^{|i-j|}, & \text{for } i \neq j\,, \end{cases} \tag{11.3}$$

$$= \begin{bmatrix} 1 & 0.3 & 0.14 & 0.027 & \dots \\ 0.3 & 2 & 0.3 & 0.9 & \dots \\ 0.14 & 0.3 & 3 & 0.3 & \dots \\ \ddots & & & & \end{bmatrix} . \tag{11.4}$$

Listing 11.6 tune.f90 is meant to be numerically intensive enough to show the results of various types of optimizations, but you may have to increase sizes in it to make it more intensive. The program solves the eigenvalue problem iteratively for a nearly diagonal Hamiltonian matrix using a variation of the power method.

```
!    tune.f90: matrix algebra program to be tuned for performance

Program  tune

  parameter (ldim = 2050)
  Implicit Double precision (a - h, o - z)
  dimension ham(ldim, ldim), coef(ldim), sigma(ldim)
                                          ! set up H and starting vector
  Do i = 1, ldim
    Do j = 1, ldim
      If ( abs(j - i)  >  10) then
        ham(j, i) = 0.
      else
        ham(j, i) = 0.3**Abs(j - i)
      EndIf
    End Do
    ham(i, i) = i
    coef(i) = 0.
  End Do
  coef(1) = 1.
                                                        ! start iterating
  err = 1.
  iter = 0
 20   If (iter< 15 .and. err >1.e-6) then
  iter = iter + 1
                                     ! compute current energy & normalize
  ener = 0.
  ovlp = 0.
  Do     i = 1, ldim
    ovlp = ovlp + coef(i)*coef(i)
    sigma(i) = 0.
    Do     j = 1, ldim
      sigma(i) = sigma(i) + coef(j)*ham(j, i)
    End Do
    ener = ener + coef(i)*sigma(i)
  End Do
  ener = ener/ovlp
  Do     I = 1, ldim
    coef(i) = coef(i)/Sqrt(ovlp)
    sigma(i) = sigma(i)/Sqrt(ovlp)
  End Do
                                        ! compute update and error norm
  err = 0.
  Do   i = 1, ldim
    If (i == 1) goto 22
    step = (sigma(i) - ener*coef(i))/(ener - ham(i, i))
    coef(i) = coef(i) + step
    err = err + step**2
 22 Continue
 23 End Do
  err = sqrt(err)
    write(*, '(1x, i2, 7f10.5)') iter, ener, err, coef(1)
```

```
      goto 20
    Endif
  Stop
End Program tune
```

Because the Hamiltonian is almost diagonal, the eigenvalues should be close to the values of the diagonal elements, and the eigenvectors should be close to a set of N-dimensional unit vectors. For example, let us say that $\mathbf{H}$ has dimensions of $N \times N$ with $N = 2000$. The number of elements in the matrix is then $2000 \times 2000 = 4\,000\,000$, and so it will take 4 million $\times$ 8 B = 32 MB to store this many double precision numbers. Because modern PCs have 4 GB or more of RAM, this small a matrix should not have memory issues. Accordingly, *determine the size of RAM on your computer and increase the dimension of the* $\mathbf{H}$ *matrix until it surpasses that size.* (On Windows, this will be indicated as one of the "Properties" of "Computer" or in the information about "System" in the Control Panel.)

We find the solution to (11.2) via a variation of the *power* or *Davidson method.* We start with an arbitrary first guess for the eigenvector $\boldsymbol{c}$ and use it to calculate the energy corresponding to this eigenvector,[1)]

$$\boldsymbol{c}_0 \simeq \begin{pmatrix} 1 \\ 0 \\ \vdots \\ 0 \end{pmatrix} , \quad E \simeq \frac{\boldsymbol{c}_0^\dagger \mathbf{H} \boldsymbol{c}_0}{\boldsymbol{c}_0^\dagger \boldsymbol{c}_0} , \tag{11.5}$$

where $\boldsymbol{c}_0^\dagger$ is the row vector adjoint of $\boldsymbol{c}_0$. Because $\mathbf{H}$ is nearly diagonal with diagonal elements that increase as we move along the diagonal, this guess should be close to the eigenvector with the smallest eigenvalue. The heart of the algorithm is the guess that an improved eigenvector has the kth component

$$\boldsymbol{c}_1|_k \simeq \boldsymbol{c}_0|_k + \frac{[\mathbf{H} - E\mathbf{I}]\boldsymbol{c}_0|_k}{E - \mathbf{H}_{k,k}} , \tag{11.6}$$

where k ranges over the length of the eigenvector. If repeated, this method converges to the eigenvector with the smallest eigenvalue. It will be the smallest eigenvalue because it gets the largest weight (smallest denominator) in (11.6) each time. For the present case, six places of precision in the eigenvalue are usually obtained after 11 iterations. Here are the steps to follow:

1. Vary the value of `err` in `tune` that controls precision and note how it affects the number of iterations required.
2. Try some variations on the initial guess for the eigenvector (11.6) and see if you can get the algorithm to converge to some of the other eigenvalues.
3. Keep a table of your execution times vs. technique.
4. Compile and execute `tune.f90` and record the run time (Listing 11.6). On Unix systems, the compiled program will be placed in the file `a.out`. From a Unix

1) Note that the codes refer to the eigenvector c_0 as `coef`.

shell, the compilation, timing, and execution can all be carried out with the commands

```
> F90 tune.f90        # Fortran compilation
>  cc --lm tune.c     # C compilation, (or gcc instead of cc)
>  time a.out          # Execution
```

Here the compiled Fortran program is given the (default) name of a.out, and the time command gives you the execution (user) time and system time in seconds to execute a.out.

5. As indicated in Section 11.3, you can ask the compiler to produce a version of your program optimized for speed by including the appropriate compiler option:

```
> f90 -O tune.f90
```

Execute and time the optimized code, checking that it still gives the same answer, and note any speedup in your journal.

6. Try out optimization options up to the highest levels and note the run time and accuracy obtained. Usually -O3 is pretty good, especially for as simple a program as tune with only a main method. With only one program unit, we would not expect -O4 or -O5 to be an improvement over -O3. However, we do expect -O3, with its loop unrolling, to be an improvement over -O2.
7. The program tune4 does some *loop unrolling* (we will explore that soon). To see the best we can do with Fortran, record the time for the most optimized version of tune4.f95.
8. The program Tune.py in Listing 11.7 is the Python equivalent of the Fortran program tune.f90.
9. To get an idea of what Tune.py does (and give you a feel for how hard life is for the poor computer), assume ldim =2 and work through one iteration of Tune *by hand*. Assume that the iteration loop has converged, follow the code to completion, and write down the values assigned to the variables.
10. Compile and execute Tune.py. You do not have to issue the time command because we built a timer into the Python program. Check that you still get the same answer as you did with Fortran and note how much longer it takes with Python. You might be surprised how much slower Python is than Fortran.
11. We now want to perform a little experiment in which we see what happens to performance as we fill up the computer's memory. In order for this experiment to be reliable, it is best for you *not* to be sharing the computer with any other users. On Unix systems, the who -a command shows you the other users (we leave it up to you to figure out how to negotiate with them).
12. To get some idea of what aspect of our little program is making it so slow, compile and run Tune.py for the series of matrix sizes ldim = 10, 100, 250, 500, 750, 1025, 2500, and 3000. You may get an error message that Python is out of memory at 3000. This is because you have not turned on the use of virtual memory.

Listing 11.7 Tune.py is meant to be numerically intensive enough to show the results of various types of optimizations, but you may need to increase sizes to make it more intensive. The program solves the eigenvalue problem iteratively for a nearly diagonal Hamiltonian matrix using a variation of the power method.

```
# Tune.py Basic tuning program showing memory allocation

import datetime;   from numpy import zeros;   from math import (sqrt,
    pow)

Ldim = 251;             iter = 0;                step = 0.
diag  = zeros((Ldim, Ldim), float);              coef = zeros( (Ldim), float)
sigma = zeros((Ldim), float);                    ham = zeros( (Ldim, Ldim),
    float)
t0 = datetime.datetime.now()                                    # Initialize
    time
for i in range(1, Ldim):                                        # Set up
    Hamiltonian
    for j in range(1, Ldim):
        if (abs(j - i) >10): ham[j, i] = 0.
        else : ham[j, i] = pow(0.3, abs(j - i) )
    ham[i,i] = i ;                 coef[i] = 0.;
coef[1] = 1.;                      err = 1.;               iter = 0 ;
print("iter   ener        err  ")
while (iter  < 15 and err > 1.e-6):              # Compute current E &
    normalize
    iter = iter + 1; ener = 0. ; ovlp = 0.;
    for i in range(1, Ldim):
        ovlp = ovlp + coef[i]*coef[i]
        sigma[i] = 0.
        for j in range(1, Ldim): sigma[i] = sigma[i] +
            coef[j]*ham[j][i]
        ener = ener + coef[i]*sigma[i]
    ener = ener/ovlp
    for i in range(1, Ldim):
        coef[i] = coef[i]/sqrt(ovlp)
        sigma[i] = sigma[i]/sqrt(ovlp)
    err = 0.;
    for i in range(2, Ldim):                                          #
        Update
        step = (sigma[i]  -  ener*coef[i])/(ener - ham[i, i])
        coef[i] = coef[i] + step
        err = err +  step*step
    err = sqrt(err)
    print" %2d  %9.7f  %9.7f "%(iter, ener, err)
delta_t = datetime.datetime.now()  -  t0                        # Elapsed
    time
print" time = ", delta_t
```

13. Make a graph of run time vs. matrix size. It should be similar to Figure 11.1, although if there is more than one user on your computer while you run, you may get erratic results. Note that as our matrix becomes larger than $\sim 1000 \times 1000$ in size, the curve sharply increases in slope with execution time, in our case increasing like the *third* power of the dimension. Because the number of elements to compute increases as the *second* power of the dimension, something else is happening here. It is a good guess that the additional slowdown is as a result of page faults in accessing memory. In particular, accessing 2D arrays, with their elements scattered all through memory, can be very slow.

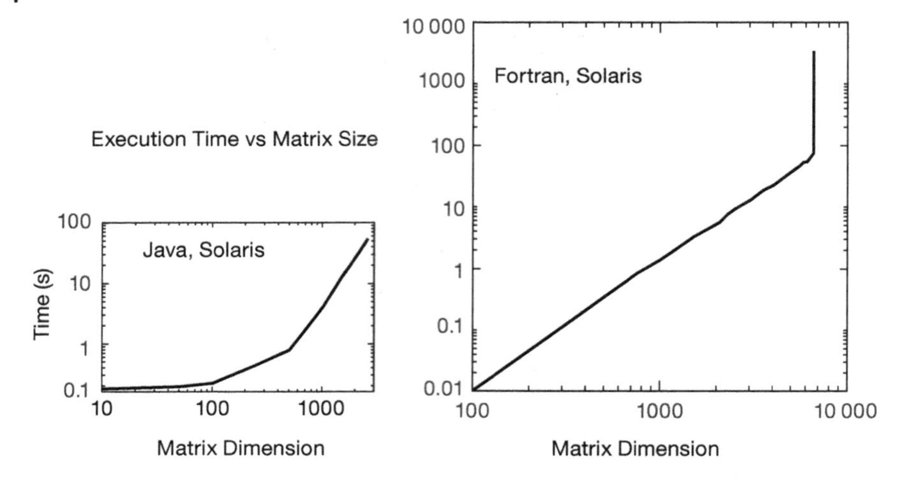

Figure 11.1 Running time vs. dimension for an eigenvalue search using Tune.java and tune.f90.

Listing 11.8 Tune4.py does some loop unrolling by explicitly writing out two steps of a for loop (steps of 2). This results in better memory access and faster execution.

```
# Tune4.py  Model tuning program

import datetime
from numpy import zeros
from math import (sqrt ,pow, pi)
from sys import version
if int(version[0]) >2: #raw_input deprecated in Python 3
    raw_input=input
Ldim = 200;                 iter1 = 0;                  step = 0.
ham = zeros( (Ldim, Ldim), float);                      diag = zeros( (Ldim), float)
coef = zeros( (Ldim), float);                           sigma = zeros( (Ldim), float)
t0 = datetime.datetime.now()                                      # Initialize time

for i in range(1, Ldim):                                        # Set up Hamiltonian
        for j in range(1, Ldim):
            if abs(j - i) >10: ham[j, i] = 0.
            else : ham[j, i] = pow(0.3, abs(j - i) )
for i in range(1, Ldim):
    ham[i, i] = i
    coef[i] = 0.
    diag[i] = ham[i, i]
coef[1] = 1.;         err = 1.;             iter = 0 ;
print("iter          ener              err  ")

while (iter1  < 15 and err > 1.e-6):                    # Compute E & normalize
    iter1 = iter1 + 1
    ener = 0.
    ovlp1 = 0.
    ovlp2 = 0.
    for i in range(1, Ldim - 1, 2):
        ovlp1 = ovlp1 + coef[i]*coef[i]
        ovlp2 = ovlp2 + coef[i + 1]*coef[i + 1]
        t1 = 0.
        t2 = 0.
        for j in range(1, Ldim):
            t1 = t1 + coef[j]*ham[j, i]
            t2 = t2 + coef[j]*ham[j, i + 1]
```

```
            sigma[i] = t1
            sigma[i + 1] = t2
            ener = ener + coef[i]*t1 + coef[i + 1]*t2
    ovlp = ovlp1 + ovlp2
    ener = ener/ovlp
    fact = 1./sqrt(ovlp)
    coef[1] = fact*coef[1]
    err = 0.                                        # Update & error norm
    for i in range(2, Ldim):
        t = fact*coef[i]
        u = fact*sigma[i] - ener*t
        step = u/(ener  -  diag[i])
        coef[i] = t + step
        err = err +  step*step
    err = sqrt(err)
    print (" %2d  %15.13f  %15.13f "%(iter1 , ener, err))
delta_t = datetime.datetime.now()  -  t0                # Elapsed time
print (" time = ", delta_t)
```

Listing 11.9 tune4.f95 does some loop unrolling by explicitly writing out two steps of a Do loop (steps of 2). This results in better memory access and faster execution.

```
!  tune4.f95: matrix algebra with RISC tuning                              !

     Program  tune4
     PARAMETER (ldim = 2050)
     Implicit Double Precision (a-h,o-z)
     Dimension ham(ldim,ldim),coef(ldim),sigma(ldim),diag(ldim)
!       set up Hamiltonian and starting vector
     Do  i = 1,ldim
        Do  j = 1,ldim
           If( Abs(j-i) > 10) Then
              ham(j,i) = 0.0
           Else
              ham(j,i) = 0.3**Abs(j-i)
           EndIf
        End Do
     End Do

!       start iterating towards the solution
     Do  i = 1,ldim
        ham(i,i) = i
        coef(i)  = 0.0
        diag(i)  = ham(i,i)
     End Do
     coef(1) = 1.0
     err = 1.0
     iter = 0
20   If(iter <15 .and. err >1.0e-6) Then
        iter = iter+1
     ener = 0.0
     ovlp1 = 0.0
     ovlp2 = 0.0
     Do     i = 1,ldim-1,2
        ovlp1 = ovlp1+coef(i)*coef(i)
        ovlp2 = ovlp2+coef(i+1)*coef(i+1)
        t1    = 0.0
        t2    = 0.0
        Do     j = 1,ldim
           t1 = t1 + coef(j)*ham(j,i)
           t2 = t2 + coef(j)*ham(j,i+1)
        End Do
     sigma(i)   = t1
     sigma(i+1) = t2
```

```
ener          = ener + coef(i)*t1 + coef(i)*t2
End Do
ovlp = ovlp1 + ovlp2
ener = ener/ovlp
fact  = 1.0/Sqrt(ovlp)
coef(1)  =  fact*coef(1)
err  =  0.0
Do     i = 2,ldim
   t          =    fact*coef(i)
   u          =    fact*sigma(i) - ener*t
   step       =    u/(ener - diag(i))
   coef(i)    =    t + step
   err        =    err + step*step
End Do
err = Sqrt(err)
Write(*,'(1x,i2,7f10.5)') iter,ener,err,coef(1)
GoTo 20
EndIf
Stop
End  Program  tune4
```

14. Repeat the previous experiment with `tune.f90` that gauges the effect of increasing the `ham` matrix size, only now do it for `ldim` = 10, 100, 250, 500, 1025, 3000, 4000, 6000, ... You should get a graph like ours. Although our implementation of Fortran has automatic virtual memory, its use will be exceedingly slow, especially for this problem (possibly a 50-fold increase in time!). So if you submit your program and you get nothing on the screen (though you can hear the disk spin or see it flash busy), then you are probably in the virtual memory regime. If you can, let the program run for one or two iterations, kill it, and then scale your run time to the time it would have taken for a full computation.
15. To test our hypothesis that the access of the elements in our 2D array `ham[i,j]` is slowing down the program, we have modified `Tune.py` into `Tune4.py` in Listing 11.8 (and similar modification with the Fortran versions).
16. Look at `Tune4` and note where the nested `for` loop over `i` and `j` now takes step of $\Delta i = 2$ rather the unit steps in `Tune.py`. If things work as expected, the better memory access of `Tune4.py` should cut the run time nearly in half. Compile and execute `Tune4.py`. Record the answer in your table.
17. In order to cut the number of calls to the 2D array in half, we employed a technique known as *loop unrolling* in which we explicitly wrote out some of the lines of code that otherwise would be executed implicitly as the `for` loop went through all the values for its counters. This is not as clear a piece of code as before, but it evidently permits the compiler to produce a faster executable. To check that `Tune` and `Tune4` actually do the same thing, assume `ldim = 4` and run through one iteration of `Tune4` *by hand*. Hand in your manual trial.

11.4 Programming for the Data Cache (Method)

Data caches are small, very fast memory banks used as temporary storage between the ultrafast CPU registers and the fast main memory. They have grown in impor-

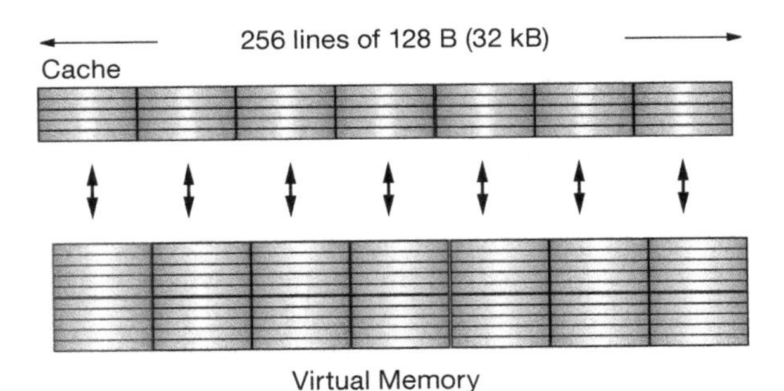

Figure 11.2 The cache manager's view of RAM. Each 128-B cache line is read into one of four lines in cache.

tance as high-performance computers have become more prevalent. For systems that use a data cache, this may well be the single most important programming consideration; continually referencing data that are not in the cache (*cache misses*) may lead to an order-of-magnitude increase in CPU time.

As indicated in Figures 10.3 and 11.2, the data cache holds a copy of some of the data in memory. The basics are the same for all caches, but the sizes are manufacturer dependent. When the CPU tries to address a memory location, the *cache manager* checks to see if the data are in the cache. If they are not, the manager reads the data from memory into the cache, and then the CPU deals with the data directly in the cache. The cache manager's view of RAM is shown in Figure 11.2.

When considering how a matrix operation uses memory, it is important to consider the *stride* of that operation, that is, the number of array elements that are stepped through as the operation repeats. For instance, summing the diagonal elements of a matrix to form the trace

$$\mathrm{Tr}\,\mathbf{A} = \sum_{i=1}^{N} a(i,i) \tag{11.7}$$

involves a large stride because the diagonal elements are stored far apart for large N. However, the sum

$$c(i) = x(i) + x(i+1) \tag{11.8}$$

has stride 1 because adjacent elements of x are involved. Following is the basic rule in programming for a cache:

- Keep the stride low, preferably at 1, which in practice means:
  - Vary the leftmost index first on Fortran arrays.
  - Vary the rightmost index first on Python and C arrays.

11.4.1
Exercise 1: Cache Misses

We have said a number of times that your program will be slowed down if the data it needs are in virtual memory and not in RAM. Likewise, your program will also be slowed down if the data required by the CPU are not in the cache. For high-performance computing, you should write programs that keep as much of the data being processed as possible in the cache. To do this, you should recall that Fortran matrices are stored in successive memory locations with the row index varying most rapidly (column-major order), while Python and C matrices are stored in successive memory locations with the column index varying most rapidly (row-major order). While it is difficult to isolate the effects of the cache from other elements of the computer's architecture, you should now estimate its importance by comparing the time it takes to step through the matrix elements row by row to the time it takes to step through the matrix elements column by column.

Run on machines available to you a version of each of the two simple codes given in Listings 11.10 and 11.11. Check that although each has the same number of arithmetic operations, one takes significantly more time to execute because it makes large jumps through memory, with some of the memory locations addressed not yet read into the cache:

Listing 11.10 Sequential column and row references.

```
for j = 1, 999999;
    x(j) = m(1,j)                          // Sequential column reference
```

Listing 11.11 Sequential column and row references.

```
for j = 1, 999999;
    x(j) = m(j,1)                          // Sequential row reference
```

11.4.2
Exercise 2: Cache Flow

Below in Listings 11.12 and 11.13, we give two simple code fragments that you should place into full programs in whatever computer language you are using. Test the importance of cache flow on your machine by comparing the time it takes to run these two programs. Run for increasing column size `idim` and compare the times for loop A vs. those for loop B. A computer with very small caches may be most sensitive to stride.

Listing 11.12 Loop A: GOOD f90 (min stride), BAD Python/C (max stride).

```
Dimension Vec(idim,jdim)                        // Stride 1 fetch (f90)
      for j = 1, jdim; { for i=1, idim;        Ans = Ans +
           Vec(i,j)*Vec(i,j)}
```

Listing 11.13 Loop B: BAD f90 (max stride), GOOD Python/C (min stride).

```
Dimension Vec(idim, jdim)                       // Stride jdim fetch (f90)
      for i = 1, idim; {for j=1, jdim;   Ans = Ans + Vec(i,j)*Vec(i,j)}
```

Loop A steps through the matrix Vec in column order. Loop B steps through in row order. By changing the size of the columns (the leftmost Python index), we change the step size (*stride*) taken through memory. Both loops take us through all the elements of the matrix, but the stride is different. By increasing the stride in any language, we use fewer elements already present in the cache, require additional swapping and loading of the cache, and thereby slow down the whole process.

11.4.3 Exercise 3: Large-Matrix Multiplication

As you increase the dimensions of the arrays in your program, memory use increases geometrically, and at some point you should be concerned about efficient memory use. The penultimate example of memory usage is large-matrix multiplication:

$$[\mathbf{C}] = [\mathbf{A}] \times [\mathbf{B}] , \tag{11.9}$$

$$c_{ij} = \sum_{k=1}^{N} a_{ik} \times b_{kj} . \tag{11.10}$$

Listing 11.14 BAD f90 (max stride), GOOD Python/C (min stride).

```
for i = 1, N; {                                          // Row
    for j = 1, N; {                                      // Column
         c(i,j) = 0.0                                 // Initialize
    for k = 1, N; {
         c(i,j) = c(i,j) + a(i,k)*b(k,j) }}}          // Accumulate
```

This involves all the concerns with different kinds of memory. The natural way to code (11.9) follows from the definition of matrix multiplication (11.10), that is, as a sum over a row of **A** times a column of **B**. Try out the two codes in Listings 11.14 and 11.15 on your computer. In Fortran, the first code uses matrix **B** with stride 1, but matrix **C** with stride N. This is corrected in the second code by performing the initialization in another loop. In Python and C, the problems are reversed. On one of our machines, we found a factor of 100 difference in CPU times despite the fact that the number of operations is the same!

Listing 11.15 GOOD f90 (min stride), BAD Python/C (max stride).

```
for j = 1, N; {                              // Initialization
    for  i = 1, N; {
         c(i,j) = 0.0 }
    for k = 1, N; {
         for  i = 1, N; {c(i,j) = c(i,j) + a(i,k)*b(k,j) }}}
```

11.5
Graphical Processing Units for High Performance Computing

In Section 10.16, we discussed how the trend toward exascale computing appears to be one using multinode-multicore-GPU computers, as illustrated in Figure 10.14. The GPU component in this figure extends a supercomputer's architecture beyond that of computers such as IBM Blue Gene. The GPUs in these future supercomputers are electronic devices designed to accelerate the creation of visual images. A GPU's efficiency arises from its ability to create many different parts of an image in parallel, an important ability because there are millions of pixels to display simultaneously. Indeed, these units can process 100s of millions of polygons in a second. Because GPUs are designed to assist the video processing on commodity devices such as personal computers, game machines, and mobile phones, they have become inexpensive, high performance, parallel computers in their own right. Yet because GPUs are designed to assist in video processing, their architecture and their programming are different from that of the general purpose CPUs usually used for scientific applications, and it takes some work to use them for scientific computing.

Programming of GPUs requires specialized tools specific to the GPU architecture being used, and while we do discuss them in Section 11.6.2, their low-level details places it beyond the normal level of this book. What is often called "compute unified device architecture (CUDA) programming" refers to programming for the architecture developed by the Nvidia corporation, and is probably the most popular approach to GPU programming at present (Zeller, 2008). However, some of the complications are being reduced via extensions and wrappers developed for popular programming languages such as C, Fortran, Python, Java, and Perl. However, the general principles involved are just an extension of those used already discussed, and after we have you work through some examples of those general principles, in Section 11.6 we give some practical tips for programming multinode-multicore-GPU computers.

In Chapter 10, we discussed some high-level aspects regarding parallel computing. In this section, we discuss some practical tips for multicore, GPU Programming. In the next section, we provide some actual examples of GPU programming using Nvidia's *CUDA* language, with access to it from within Python programs via the use of *PyCUDA* (Klöckner, 2014). But do not think this is nothing more than adding in another package to Python. That section is marked as optional because using CUDA goes beyond the usual level of this text, and because, in our experience, it can be difficult for a regular scientist to get CUDA running on their personal computer without help. As is too often the case with parallel computing, one does have to get one's hands dirty with lower level commands. Our presentation should give the reader some general understanding of CUDA. To read more on the subject, we recommend the CUDA tutorial (Zeller, 2008), as well as (Kirk and Wen-Mei, 2013; Sanders and Kandrot, 2011; Faber, 2010).

11.5.1 The GPU Card

GPUs are hardware components that are designed to accelerate the storage, manipulation, and display of visual images. GPUs tend to have more core chips and more arithmetic data processing units than do multicore central processing units (CPUs), and so are essentially fast, but limited, parallel computers. Because GPUs have become commodity items on most PCs (think gaming), they hold the promise of being a high powered, yet inexpensive, parallel computing environment. At present an increasing number of scientific applications are being adapted to GPUs, with the CPU being used for the sequential part of a program and the GPU for the parallel part.

The most popular GPU programming language is CUDA. CUDA is both a platform and a programming model created by Nvidia for parallel computing on their GPU. The model defines:

1. threads, clocks, and grids,
2. a memory system with registers, local, shared and global memories, and
3. an execution environment with scheduling of threads and blocks of threads.

Figure 11.3 shows a schematic of a typical Nvidia card containing four streaming multiprocessors (SMs), each with eight streaming processors (SPs), two special function units (SFUs), and 16 kB of shared memory. The SFUs are used for the evaluation of the, otherwise, time-consuming transcendental functions such as sin, cosine, reciprocal, and square root. Each group of dual SMs form a texture processing cluster (TPC) that is used for pixel shading or general-purpose graphical processing. Also represented in the figure is the transfer of data among the three types of memory structures on the GPU and those on the host (Figure 11.3a). Having memory on the chip is much faster than accessing remote memory structures.

11.6 Practical Tips for Multicore and GPU Programming ⊙

We have already described some basic elements of exascale computing in Section 10.16. Some practical tips for programming multinode-multicore-GPU computers follow along the same lines as we have been discussing, but with an even

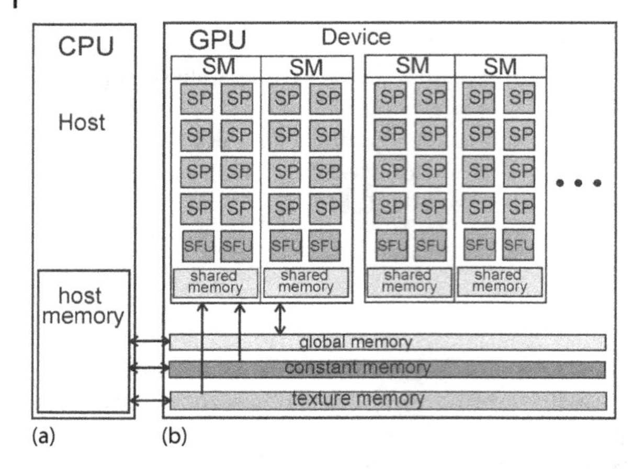

Figure 11.3 A schematic of a GPU (b) and its interactions with a CPU (a). The GPU is seen to have streaming multiprocessors, special function units and several levels of memory. Also shown is the communication between memory levels and with the host CPU.

greater emphasis on minimizing communication costs.[2] Contrary to the traditional view on optimization, this means that the "faster" of two algorithms may be the one that takes more steps, but requires less communications. Because the effort in programming the GPU directly can be quite high, many application programmers prefer to let compiler extensions and wrappers deal with the GPU. But if you must, here is how.

Exascale computers and computing are expected to be *"disruptive technologies"* in that they lead to drastic changes from previous models for computers and computing. This is in contrast to the more evolving technology of continually increasing the clock speed of the CPU, which was the practice until the power consumption and associated heat production imposed a roadblock. Accordingly, we should expect that software and algorithms will be changing (and we will have to rewrite our applications), much as it did when supercomputers changed from large vector machines with proprietary CPUs to cluster computers using commodity CPUs and message passing. Here are some of the major points to consider:

Exacascale data movement is expensive The time for a floating-point operation and for a data transfer can be similar, although if the transfer is not local, as Figures 10.13 and 10.14 show happens quite often, then communication will be the rate-limiting step.

Exacascale flops are cheap and abundant GPUs and local multicore chips provide many, very fast flops for us to use, and we can expect even more of these elements to be included in future computers. So do not worry about flops as much as communication.

Synchronization-reducing algorithms are essential Having many processors stop in order to synchronize with each other, while essential in ensuring that

2) Much of the material in this section comes from talks by John Dongarra (Dongarra, 2011).

the proper calculation is being performed, can slow down processing to a halt (literally). It is better to find or derive an algorithm that reduces the need for synchronization.

Break the fork-join model This model for parallel computing takes a queue of incoming jobs, divides them into subjobs for service on a variety of servers, and then combines them to produce the final result. Clearly, this type of model can lead to completed subjobs on various parts of the computer that must wait for other subjobs to complete before recombination. A big waste of resources.

Communication-reducing algorithms As already discussed, it is best to use methods that lessen the need for communication among processors, even if more flops are needed.

Use mixed precision methods Most GPUs do not have native double precision floating-point numbers (or even full single precision) and correspondingly slow down by a factor-of-two or more when made to perform double precision calculation, or when made to move double-precision data. The preferred approach then is to use single precision. One way to do this is to utilize the perturbation theory in which the calculation focuses on the small (single precision) correction to a known, or separately calculated, large (double precision) basic solution. The rule-of-thumb then is to use the lowest precision required to achieve the required accuracy.

Push for and use autotuning The computer manufactures have advanced the hardware to incredible speeds, but have not produced concordant advances in the software that scientists need to utilize in order to make good use of these machines. It often takes years for people to rewrite a large program for these new machines, and that is a tremendous investment that not many scientists can afford. We need smarter software to deal with such complicated machines, and tools that permit us to optimize experimentally our programs for these machines.

Fault resilient algorithms Computers containing millions or billions of components do make mistakes at times. It does not make sense to have to start a calculation over, or hold a calculation up, when some minor failure such as a bit flip occurs. The algorithms should be able to recover from these types of failures.

Reproducibility of results Science is at its heart the search for scientific truth, and there should be only one answer when solving a problem whose solution is mathematically unique. However, approximations in the name of speed are sometimes made and this can lead to results whose exact reproducibility cannot be guaranteed. (Of course exact reproducibility is not to be expected for Monte Carlo calculations involving chance.)

Data layout is critical As we discussed with Figures 10.3, 11.2, and 11.4, much of HPC deals with matrices and their arrangements in memory. With parallel computing, we must arrange the data into tiles such that each data tile is contiguous in memory. Then, the tiles can be processed in a fine-grained computation. As we have seen in the exercises, the best size for the tiles depends upon the size of the caches being used, and these are generally small.

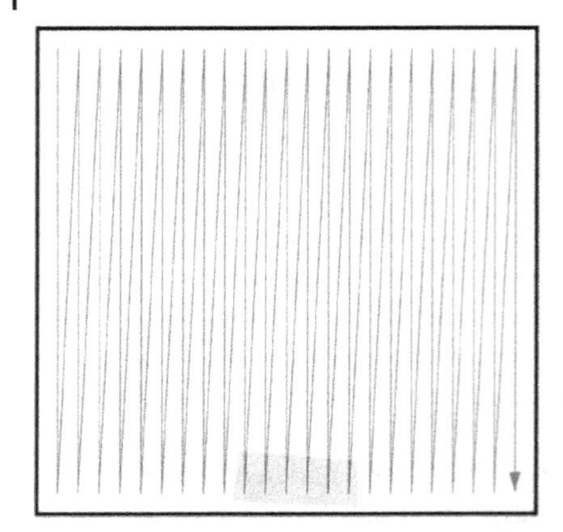

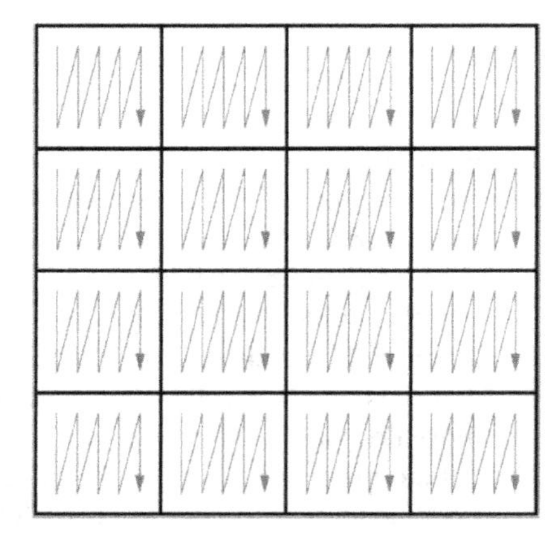

Figure 11.4 A schematic of how the contiguous elements of a matrix must be tiled for parallel processing (from Dongarra, 2011).

11.6.1 CUDA Memory Usage

CUDA commands are extensions of the C programming language and are used to control the interaction between the GPU's components. CUDA supports several data types such as dim2, dim3, and 2D textures, for 2D, 3D, and 2D shaded objects, respectively. PyCUDA (Klöckner, 2014), in turn, provides access to CUDA from Python programs as well as providing some additional extensions of C. Key concepts when programming with CUDA are following:

Kernel A part or parts of a program that are executed in parallel by the GPU using threads. The kernel is called from the host and executed on the *device* (another term for the GPU).

Thread The basic element of data to be processed on the GPU. Sometimes defined as an execution of a kernel with a given index, or the abstraction of a function call acting on data. The numbers of threads that are able to run concurrently on a GPU depends upon the specific hardware being used.

Index All threads see the same kernel, but each kernel is executed with a different parameter or subscript, called an index.

As a working example, consider the summation of the arrays `a` and `b` by a kernel with the result assigned to the array `c`:

$$c_i = a_i + b_i \,. \tag{11.11}$$

If these were calculated using a sequential programming model, then we might use a `for` loop to sum over the index `i`. However, this is not necessary in CUDA where each thread sees the same kernel, but there is a different ID assigned to each thread. Accordingly, one thread associated with an ID would see

$$c_0 = a_0 + b_0 \,, \tag{11.12}$$

while another thread assigned to a different ID might see

$$c_1 = a_1 + b_1 , \qquad (11.13)$$

and so forth. In this way, the summation is performed in parallel by the GPU. You can think of this model as analogous to an orchestra in which all violins plays the same part with the conductor setting the timing (the clock cycles in the computer). Each violinist has the same sheet music (the kernel) yet plays his/her own violin (the threads).

As we have already seen in Figure 11.3, CUDA's GPU architecture contains scalable arrays of multithreaded SMs. Each SM contains eight SPs, also called CUDA cores, and these are the parts of the GPU that run the kernels. The SPs are execution cores like CPUs, but simpler. They typically can perform more operations in a clock cycle than the CPU, and being dedicated just to data processing, constitute the arithmetic logic units of the GPU. The speed of dedicated SMs, along with their working in parallel, leads to the significant processing power of GPUs, even when compared to a multicore CPU computer.

In the CUDA approach, the threads are assigned in *blocks* that can accommodate from one to thousands of threads. For example, the CUDA GeForce GT 540M can handle 1024 threads per block, and has 96 cores. As we have said, each thread runs the same program, which make this a single program multiple data computer. The threads of a block are partitioned into *warps* (set of lengthwise yarns in weaving), with each warp containing 32 threads. The blocks can be 1D, 2D, or 3D, and are organized into *grids*. The grids, in turn, can be 1D or 2D.

When a grid is created or *launched*, the blocks are assigned to the SMs in an arbitrary order, with each block partitioned into warps, and each thread going to a different streaming processor. If more blocks are assigned to an SM than it can process at once, the excess blocks are scheduled for a later time. For example, the GT200 architecture can process 8 blocks or 1024 threads per SM, and contain 30 SMs. This means that the $8 \times 30 = 240$ CUDA cores (SPs) can process $30 \times 1024 = 30\,720$ threads in parallel. Compare that to a 16 core CPU!

11.6.2 CUDA Programming ⊙

The installation of CUDA development tools follows a number of steps:

1. Establish that the video card on your system has a GPU with Nvidia CUDA. Just having an Nvidia card might not do because not all Nvidia cards have a CUDA architecture.
2. Verify that your operating system can support CUDA and PyCuda (Linux seems to be the preferred system).
3. On Linux, verify that your system has gcc installed. On Windows, you will need the Microsoft Visual Studio compiler.
4. Download the Nvidia CUDA Toolkit.
5. Install the Nvidia CUDA Toolkit.

6. Test that the installed software runs and can communicate with the CPU.
7. Establish that your Python environment contains PyCuda. If not, install it.

Let us now return to our problem of summing two arrays and storing the result, a[i]+b[i]=c[i]. The steps needed to solve this problem using a host (CPU) and a device (GPU) are:

1. Declare single precision arrays in the host and initialize the arrays.
2. Reserve space in the device memory to accommodate the arrays.
3. Transfer the arrays from the host to the device.
4. Define a kernel on the host to sum the arrays, with the execution of the kernel to be performed on the device.
5. Perform the sum on the device, transfer the result to the host and print it out.
6. Finally, free up the memory space on the device.

Listing 11.16 The PyCUDA program **SumArraysCuda.py** uses the GPU to do the array sum $a + b = c$.

```
# SumArraysCuda.py: sums arrays a + b = c                                    1

import pycuda.autoinit                                                       3
import pycuda.driver as drv
import numpy                                                                 5
from pycuda.compiler import SourceModule
                                                                             7
                              # The kernel in C
mod = SourceModule("""                                                       9
__global__ void sum_ab(float *c, float *a, float *b)
{ const int i = threadIdx.x;                                                 11
  c[i] = a[i] + b[i]; }
                            """)                                             13
sum_ab = mod.get_function("sum_ab")
N = 32                                                                       15
a = numpy.array(range(N)).astype(numpy.float32)
b = numpy.array(range(N)).astype(numpy.float32)                             17
for i in range (0, N):
  a[i] = 2.0*i                                                               19
  b[i] = 1.0*i
c = numpy.zeros_like(a)                  # intialize c                       21
a_dev = drv.mem_alloc(a.nbytes)          # reserve memory in device
b_dev = drv.mem_alloc(b.nbytes)                                              23
c_dev = drv.mem_alloc(c.nbytes)
drv.memcpy_htod(a_dev, a)                  # copy a to device                25
drv.memcpy_htod(b_dev, b)                  # copy b to device
sum_ab( c_dev, a_dev, b_dev, block=(32,1,1), grid=(1,1))                    27
print("a" + \n + a + \n + "b"+ \n + b)
drv.memcpy_dtoh(c,c_dev)                 # copy c from device                29
print("c" +\n + c)
```

Listing 11.16 presents a PyCUDA program SummArraysCuda.py that performs our array summation on the GPU. The program is seen to start on line 3 with the import pycuda.autoinit command that initializes CUDA to accept the kernel. Line 4, import pycuda.driver as drv, is needed for Pycuda to be able to find the available GPU. Line 7, from pycuda.compiler import SourceModule prepares the Nvidia kernel compiler nvcc. The SourceModule, which nvcc compiles, is given in CUDA C on lines 10–14.

As indicated before, even though we are summing indexed arrays, there is no explicit for loop running over an index i. This is because CUDA and the hardware know about arrays and so take care of the indices automatically. This works by having all threads seeing the same kernel, and also by having each thread process a different value of i. Specifically, notice the command const int i = threadIdx.x on line 11 in the source module. The suffix .x indicates that we have a one dimensional collections of threads. If we had a 2D or 3D collection, you would also need to include threadIdx.y and threadIdx.z commands. Also notice in this same SourceModule the prefix __global__ on line 11. This indicates that the execution of the following summation is to be spread among the processors on the device.

In contrast to Python, where arrays are double precision floating-point numbers, in CUDA arrays are single precision. Accordingly, on lines 17 and 18 of Listing 11.16, we specify the type of arrays with the command

```
a = numpy.array(range(N)).astype(numpy.float32) .
```

The prefix numpy indicates where the array data type is defined, while the suffix indicates that the array is single precision (32 bits). Next on lines 22–24 are the drv.mem_alloc(a.nbytes) statements needed to reserve memory on the device for the arrays. With a already defined, a.nbytes translates into the number of bytes in the array. After that, the drv.memcp_htod() commands are used to copy a and b to the device. The sum is performed with the sum_ab() command, and then the result is sent back to the host for printing.

Several of these steps can be compressed by using some PyCUDA-specific commands. Specifically, the memory allocation, the performance of the summation and the transmission of the results on lines 22–28 can be replaced by the statements

```
sum_ab(
        drv.Out(c), drv.In(a), drv.In(b),
        block=(N,1,1), grid=(1,1))    # a, b sent to device, c to host
```

with the rest of the program remaining the same. The statement grid=(1,1) describes the dimension of the grid in blocks as 1 in x and 1 in y, which is the same as saying that the grid is composed of one block. The block(N,1,1) command indicates that there are N (= 32) threads, with the 1,1 indicating that the threads are only in the x direction.

Listing 11.17 The PyCUDA program **SumArraysCuda2.py** uses the GPU to do the array sum $a + b = c$, using four blocks with a (4,1) grid.

```
# SumArraysCuda2.py: sums arrays a + b = c using a different block          1
    structure
import pycuda.autoinit
import pycuda.driver as drv                                                  3
import numpy
from pycuda.compiler import SourceModule                                     5

                        # kernel in C language                               7
mod = SourceModule("""
__global__ void sum_ab(float *c, float *a, float *b)                         9
```

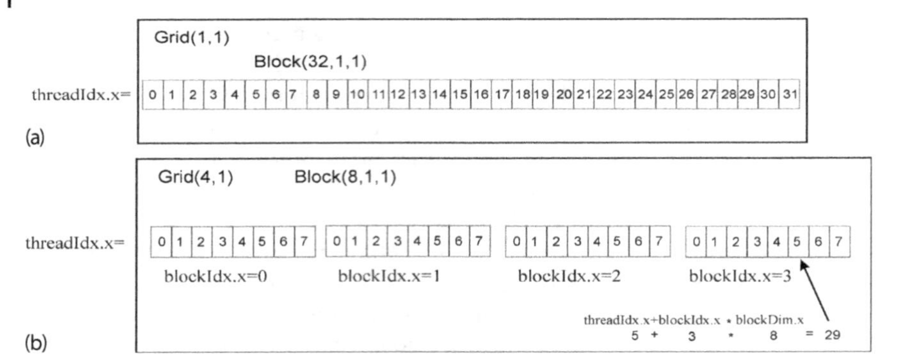

Figure 11.5 (a) A 1×1 grid with a 1D block containing 32 threads, and the corresponding threadIdx.x that goes from 0 to 31. (b) A 1D 4×1 grid, that has 4 blocks with 8 threads each.

```
{ const int i = threadIdx.x+blockDim.x*blockIdx.x;
  c[i] = a[i] + b[i];
}
""")

sum_ab = mod.get_function("sum_ab")
N=32
a = numpy.array(range(N)).astype(numpy.float32)
b = numpy.array(range(N)).astype(numpy.float32)
for i in range (0,N):
  a[i] = 2.0*i                          # assign a
  b[i] = 1.0*i                          # assign b
c = numpy.zeros_like(a)                 # sum on device
sum_ab( drv.Out(c), drv.In(a), drv.In(b), block=(8,1,1), grid=(4,1) )
print("a" + \n + a + \n+ "b" + \n + "c" + \n + c)
```

Our example is so simple that we have organized it into just a single 1D block of 32 threads in a 1×1 grid (Figure 11.5a). However, in more complicated problems, it often makes sense to distribute the threads over a number of different block, with each block performing a different operation. For example, the program `SumArraysCuda2.py` in Listing 11.17 performs the same sum with the four-block structure illustrated in Figure 11.5b. The difference in the program here is the statement on line 23, `block=(8,1,1), grid=(4,1)` that specifies a grid of four blocks in 1D x, with each 1D block having eight threads. To use this structure of the program, on line 10 we have the command `const int i = threadIdx.x+blockDim.x*blockIdx.x;`. Here `blockDim.x` is the dimension of each block in threads (8 in this case numbered 0, 1,…,7), with `blockIdx.x` indicating the block numbering. The threads in each block have `treadIdx.x` from 0 to 7 and the corresponding IDs of the blocks are `blockIdx.x` = 0, 1, 2, and 3.

12
Fourier Analysis: Signals and Filters

We start this chapter with a discussion of Fourier series and Fourier transforms, the standard tools for decomposing periodic and nonperiodic motions, respectively. We find that, as implemented for numerical computation, both the series and the transform become the same discrete Fourier transform (DFT) algorithm, which has a beautiful simplicity in its realization as a program. We then show how Fourier tools can be used to reduce noise in measured or simulated signals. We end the chapter with a discussion of the fast Fourier transform (FFT), a technique so efficient that it permits nearly instantaneous evaluations of DFTs on various devices.

12.1
Fourier Analysis of Nonlinear Oscillations

Consider a particle oscillating either in the nonharmonic potential of (8.5):

$$V(x) = \frac{1}{p}k|x|^p \,, \quad p \neq 2 \,, \tag{12.1}$$

or in the perturbed harmonic oscillator potential (8.2),

$$V(x) = \frac{1}{2}kx^2\left(1 - \frac{2}{3}\alpha x\right) \,. \tag{12.2}$$

While free oscillations in these potentials are always periodic, they are not sinusoidal. Your *problem* is to take the solution of one of these nonlinear oscillators and expand it in a Fourier basis:

$$x(t) = A_0 \sin(\omega t + \phi_0) \,. \tag{12.3}$$

For example, if your oscillator is sufficiently nonlinear to behave like the sawtooth function (Figure 12.1a), then the Fourier spectrum you obtain should be similar to that shown in Figure 12.1b.

In general, when we undertake such a spectral analysis we want to analyze the steady-state behavior of a system. This means that we have waited for the initial

Computational Physics, 3rd edition. Rubin H. Landau, Manuel J. Páez, Cristian C. Bordeianu.

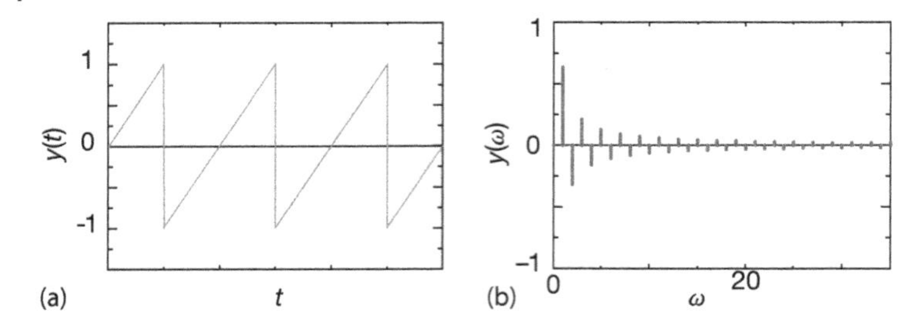

Figure 12.1 (a) A sawtooth function that repeats infinitely in time. (b) The Fourier spectrum of frequencies contained in this function (natural units). Also see Figure 1.9 in which we show the result of summing a finite number of terms of this series.

transient behavior to die out. It is easy to identify just what the initial transient is for linear systems, but may be less so for nonlinear systems in which the "steady state" jumps among a number of configurations. In the latter case, we would have different Fourier spectra at different times.

12.2 Fourier Series (Math)

Part of our interest in nonlinear oscillations arises from their lack of study in traditional physics courses where linear oscillations, despite the fact that they are just a first approximation, are most often studied. If the force on a particle is always toward its equilibrium position (a restoring force), then the resulting motion will be *periodic*, but not necessarily *harmonic*. A good example is the motion in a highly anharmonic potential with $p \simeq 10$ in (12.1) that produces an $x(t)$ looking like a series of pyramids; this motion is periodic but not harmonic.

In a sense, our approach is the inverse of the traditional one in which the *fundamental* oscillation is determined analytically and the higher frequency *overtones* are determined by perturbation theory (Landau and Lifshitz, 1976). We start with the full (numerical) periodic solution and then decompose it into what may be called *harmonics*. When we speak of fundamentals, overtones, and harmonics, we speak of solutions to the linear *boundary-value problem*, for example, of waves on a plucked violin string. In the latter case, and when given the correct conditions (enough musical skill), it is possible to excite individual harmonics or sums of them in the series

$$y(t) = b_0 \sin \omega_0 t + b_1 \sin 2\omega_0 t + \cdots \ . \tag{12.4}$$

Anharmonic oscillators vibrate at a single frequency (which may vary with amplitude) but not with a sinusoidal waveform. Although it is mathematically proper to expand nonlinear oscillations in a Fourier series, this does not imply that the individual harmonics can be excited (played).

You may recall from classical mechanics that the general solution for a vibrating system can be expressed as the sum of the *normal modes* of that system. These expansions are possible only if we have *linear operators* and, subsequently, the *principle of superposition*: If $y_1(t)$ and $y_2(t)$ are solutions of some linear equation, then $\alpha_1 y_1(t) + \alpha_2 y_2(t)$ is also a solution. The principle of linear superposition does not hold when we solve nonlinear problems. Nevertheless, it is always possible to expand a *periodic* solution of a *nonlinear* problem in terms of trigonometric functions with frequencies that are integer multiples of the true frequency of the nonlinear oscillator.[1] This is a consequence of *Fourier's theorem* being applicable to any single-valued periodic function with only a finite number of discontinuities. We assume we know the period T, that is,

$$y(t + T) = y(t) . \tag{12.5}$$

This tells us the *true frequency* ω:

$$\omega \equiv \omega_1 = \frac{2\pi}{T} . \tag{12.6}$$

A periodic function (usually designated as the *signal*) can be expanded as a series of harmonic functions with frequencies that are multiples of the true frequency:

$$y(t) = \frac{a_0}{2} + \sum_{n=1}^{\infty} (a_n \cos n\omega t + b_n \sin n\omega t) . \tag{12.7}$$

This equation represents the signal $y(t)$ as the simultaneous sum of pure tones of frequency $n\omega$. The coefficients a_n and b_n measure the amount of $\cos n\omega t$ and $\sin n\omega t$ present in $y(t)$, respectively. The intensity or power at each frequency is proportional to $a_n^2 + b_n^2$.

The Fourier series (12.7) is a "best fit" in the least-squares sense of Chapter 7, because it minimizes $\sum_i [y(t_i) - y_i]^2$, where i denotes different measurements of the signal. This means that the series converges to the *average* behavior of the function, but misses the function at discontinuities (at which points it converges to the mean) or at sharp corners (where it overshoots). A general function $y(t)$ may contain an infinite number of Fourier components, although low-accuracy reproduction is usually possible with a small number of harmonics.

The coefficients a_n and b_n in (12.7) are determined by the standard techniques for orthogonal function expansion. To find the coefficients, multiply both sides of (12.7) by $\cos n\omega t$ or $\sin n\omega t$, integrate over one period, and project a single a_n or b_n:

$$\begin{pmatrix} a_n \\ b_n \end{pmatrix} = \frac{2}{T} \int_0^T \mathrm{d}t \begin{pmatrix} \cos n\omega t \\ \sin n\omega t \end{pmatrix} y(t) , \quad \omega \stackrel{\text{def}}{=} \frac{2\pi}{T} . \tag{12.8}$$

1) We remind the reader that every periodic system by definition has a period T and consequently a true frequency ω. Nonetheless, this does not imply that the system behaves like $\sin \omega t$. Only harmonic oscillators do that.

As seen in the b_n coefficients (Figure 12.1b), these coefficients usually decrease in magnitude as the frequency increases, and can enter with a negative sign, the negative sign indicating the relative phase.

Awareness of the *symmetry* of the function $y(t)$ may eliminate the need to evaluate all the expansion coefficients. For example,

- a_0 is twice the average value of y:

$$a_0 = 2\langle y(t)\rangle \ . \tag{12.9}$$

- For an *odd function*, that is, one for which $y(-t) = -y(t)$, all a_n coefficients $\equiv 0$, and only half of the integration range is needed to determine b_n:

$$b_n = \frac{4}{T}\int_0^{T/2} dt\ y(t)\sin n\omega t \ . \tag{12.10}$$

  However, if there is no input signal for $t < 0$, we do not have a truly odd function, and so small values of a_n may occur.
- For an *even function*, that is, one for which $y(-t) = y(t)$, all b_n coefficient $\equiv 0$, and only half the integration range is needed to determine a_n:

$$a_n = \frac{4}{T}\int_0^{T/2} dt y(t)\cos n\omega t \ . \tag{12.11}$$

12.2.1 Examples: Sawtooth and Half-Wave Functions

The sawtooth function (Figure 12.1) is described mathematically as

$$y(t) = \begin{cases} \frac{t}{T/2}, & \text{for } 0 \le t \le \frac{T}{2} \ , \\ \frac{t-T}{T/2}, & \text{for } \frac{T}{2} \le t \le T \ . \end{cases} \tag{12.12}$$

It is clearly periodic, nonharmonic, and discontinuous. Yet it is also odd and so can be represented more simply by shifting the signal to the left:

$$y(t) = \frac{t}{T/2} \ , \quad -\frac{T}{2} \le t \le \frac{T}{2} \ . \tag{12.13}$$

Although the general shape of this function can be reproduced with only a few terms of the Fourier components, many components are needed to reproduce the sharp corners. Because the function is odd, the Fourier series is a sine series and (12.8) determines the b_n values:

$$b_n = \frac{2}{T}\int_{-T/2}^{+T/2} dt \sin n\omega t \frac{t}{T/2} = \frac{2}{n\pi}(-1)^{n+1} \ , \tag{12.14}$$

$$\Rightarrow y(t) = \frac{2}{\pi}\left[\sin\omega t - \frac{1}{2}\sin 2\omega t + \frac{1}{3}\sin 3\omega t - \cdots\right] . \tag{12.15}$$

The half-wave function

$$y(t) = \begin{cases} \sin\omega t, & \text{for } 0 < t < \frac{T}{2} , \\ 0, & \text{for } \frac{T}{2} < t < T , \end{cases} \tag{12.16}$$

is periodic, nonharmonic (the upper half of a sine wave), and continuous, but with discontinuous derivatives. Because it lacks the sharp corners of the sawtooth function, it is easier to reproduce with a finite Fourier series. Equation 12.8 determines

$$a_n = \begin{cases} \frac{-2}{\pi(n^2-1)}, & n \text{ even or } 0 , \\ 0, & n \text{ odd} , \end{cases} \qquad b_n = \begin{cases} \frac{1}{2}, & n = 1 , \\ 0, & n \neq 1 , \end{cases}$$

$$\Rightarrow \quad y(t) = \frac{1}{2}\sin\omega t + \frac{1}{\pi} - \frac{2}{3\pi}\cos 2\omega t - \frac{2}{15\pi}\cos 4\omega t + \ldots \tag{12.17}$$

12.3
Exercise: Summation of Fourier Series

Hint: The program **FourierMatplot.py** written by Oscar Restrepo performs a Fourier analysis of a sawtooth function and produces the visualization shown in Figure 1.9b. You may want to use this program to help with this exercise.

1. *Sawtooth function:* Sum the Fourier series for the *sawtooth function* up to order $n = 2, 4, 10, 20$, and plot the results over two periods.
   a) Check that in each case, the series gives the mean value of the function *at* the points of discontinuity.
   b) Check that in each case the series *overshoots* by about 9% the value of the function on either side of the discontinuity (the *Gibbs phenomenon*).
2. *Half-wave function:* Sum the Fourier series for the *half-wave function* up to order $n = 2, 4, 10, 50$ and plot the results over two periods. (The series converges quite well, doesn't it?)

12.4
Fourier Transforms (Theory)

Although a Fourier *series* is the right tool for approximating or analyzing periodic functions, the Fourier *transform* or *integral* is the right tool for nonperiodic functions. We convert from series to transform by imagining a system described by a continuum of "fundamental" frequencies. We thereby deal with *wave packets* containing continuous rather than discrete frequencies.[2] While the difference

2) We follow convention and consider time t the function's variable and frequency ω the transform's variable. Nonetheless, these can be reversed or other variables such as position x and wave vector $\boldsymbol{k}$ may also be used.

between series and transform methods may appear clear mathematically, when we approximate the Fourier integral as a finite sum, the two become equivalent.

By analogy with (12.7), we now imagine our function or signal $y(t)$ expressed in terms of a continuous series of harmonics (*inverse Fourier transform*):

$$y(t) = \int_{-\infty}^{+\infty} d\omega Y(\omega) \frac{e^{i\omega t}}{\sqrt{2\pi}} , \tag{12.18}$$

where for compactness we use a complex exponential function.[3] The expansion amplitude $Y(\omega)$ is analogous to the Fourier coefficients (a_n, b_n), and is called the *Fourier transform* of $y(t)$. The integral (12.18) is the inverse transform because it converts the transform to the signal. The *Fourier transform* converts the signal $y(t)$ to its transform $Y(\omega)$:

$$Y(\omega) = \int_{-\infty}^{+\infty} dt \frac{e^{-i\omega t}}{\sqrt{2\pi}} y(t) . \tag{12.19}$$

The $1/\sqrt{2\pi}$ factor in both these integrals is a common normalization in quantum mechanics, but maybe not in engineering where only a single $1/2\pi$ factor is used. Likewise, the signs in the exponents are also conventions that do not matter as long as you maintain consistency.

If $y(t)$ is the measured response of a system (signal) as a function of time, then $Y(\omega)$ is the *spectral function* that measures the amount of frequency ω present in the signal. In many cases, it turns out that $Y(\omega)$ is a complex function with both positive and negative values, and with powers-of-ten variation in magnitude. Accordingly, it is customary to eliminate some of the complexity of $Y(\omega)$ by making a semilog plot of the squared modulus $|Y(\omega)|^2$ vs. ω. This is called a *power spectrum* and provides an immediate view of the amount of power or strength in each component.

If the Fourier transform and its inverse are consistent with each other, we should be able to substitute (12.18) into (12.19) and obtain an identity:

$$Y(\omega) = \int_{-\infty}^{+\infty} dt \frac{e^{-i\omega t}}{\sqrt{2\pi}} \int_{-\infty}^{+\infty} d\omega' \frac{e^{i\omega' t}}{\sqrt{2\pi}} Y(\omega') \tag{12.20}$$

$$= \int_{-\infty}^{+\infty} d\omega' \left\{ \int_{-\infty}^{+\infty} dt \frac{e^{i(\omega'-\omega)t}}{2\pi} \right\} Y(\omega') . \tag{12.21}$$

3) Recall that $\exp(i\omega t) = \cos \omega t + i \sin \omega t$, and with the law of linear superposition this means that the real part of y gives the cosine series, and the imaginary part the sine series.

For this to be an identity, the term in braces must be the *Dirac delta function*:

$$\int_{-\infty}^{+\infty} dt e^{i(\omega'-\omega)t} = 2\pi\delta(\omega' - \omega) . \tag{12.22}$$

While the delta function is one of the most common and useful functions in theoretical physics, it is not well behaved in a mathematical sense and misbehaves terribly in a computational sense. While it is possible to create numerical approximations to $\delta(\omega' - \omega)$, they may well be borderline pathological. It is certainly better for you to do the delta function part of an integration analytically and leave the nonsingular leftovers to the computer.

12.5 The Discrete Fourier Transform

If $y(t)$ or $Y(\omega)$ is known analytically or numerically, integrals (12.18) and (12.19) can be evaluated using the integration techniques studied earlier. In practice, the signal $y(t)$ is measured at just a finite number N of times t, and these are all we have as input to approximate the transform. The resultant *discrete Fourier transform* is an approximation both because the signal is not known for all times, and because we must integrate numerically (Briggs and Henson, 1995). Once we have a discrete set of (approximate) transform values, they can be used to reconstruct the signal for any value of the time. In this way, the DFT can be thought of as a technique for interpolating, compressing, and extrapolating the signal.

We assume that the signal $y(t)$ is sampled at $(N + 1)$ discrete times (N time intervals), with a constant spacing $\Delta t = h$ between times:

$$y_k \stackrel{\text{def}}{=} y(t_k) , \quad k = 0, 1, 2, \dots , N , \tag{12.23}$$

$$t_k \stackrel{\text{def}}{=} kh , \quad h = \Delta t . \tag{12.24}$$

In other words, we measure the signal $y(t)$ once every hth of a second for a total time of T. This correspondingly define the signal's period T and the *sampling rate s*:

$$T \stackrel{\text{def}}{=} Nh , \quad s = \frac{N}{T} = \frac{1}{h} . \tag{12.25}$$

Regardless of the true periodicity of the signal, when we choose a period T over which to sample the signal, the mathematics will inevitably produce a $y(t)$ that is periodic with period T,

$$y(t + T) = y(t) . \tag{12.26}$$

We recognize this periodicity, and ensure that there are only N independent measurements used in the transform, by defining the first and last y's to be equal:

$$y_0 = y_N . \tag{12.27}$$

If we are analyzing a truly periodic function, then the N points should span one complete period, but not more. This guarantees their independence. Unless we make further assumptions, the N independent data $y(t_k)$ can determine no more than N independent transform values $Y(\omega_k)$, $k = 0, \dots, N$.

The time interval T (which should be the period for periodic functions) is the largest time over which we measure the variation of $y(t)$. Consequently, it determines the lowest frequency contained in our Fourier representation of $y(t)$,

$$\omega_1 = \frac{2\pi}{T} . \tag{12.28}$$

The full range of frequencies in the spectrum ω_n are determined by the number of samples taken, and by the total sampling time $T = Nh$ as

$$\omega_n = n\omega_1 = n\frac{2\pi}{Nh} , \qquad n = 0, 1, \dots, N . \tag{12.29}$$

Here $\omega_0 = 0$ corresponds to the zero-frequency or DC component of the transform, that is, the part of the signal that does not oscillate.

The discrete Fourier transform (DFT) algorithm follows from two approximations. First we evaluate the integral in (12.19) from time 0 to time T, over which the signal is measured, and not from $-\infty$ to $+\infty$. Second, the trapezoid rule is used for the integration[4]:

$$Y(\omega_n) \stackrel{\text{def}}{=} \int_{-\infty}^{+\infty} dt \frac{e^{-i\omega_n t}}{\sqrt{2\pi}} y(t) \simeq \int_0^T dt \frac{e^{-i\omega_n t}}{\sqrt{2\pi}} y(t) , \tag{12.30}$$

$$\simeq \sum_{k=1}^{N} h y(t_k) \frac{e^{-i\omega_n t_k}}{\sqrt{2\pi}} = h \sum_{k=1}^{N} y_k \frac{e^{-2\pi i k n/N}}{\sqrt{2\pi}} . \tag{12.31}$$

To keep the final notation more symmetric, the step size h is factored from the transform Y and a discrete function Y_n is defined as

$$\boxed{Y_n \stackrel{\text{def}}{=} \frac{1}{h} Y(\omega_n) = \sum_{k=1}^{N} y_k \frac{e^{-2\pi i k n/N}}{\sqrt{2\pi}} , \qquad n = 0, 1 \dots, N .} \tag{12.32}$$

With this same care in accounting, and with $d\omega \to 2\pi/Nh$, we invert the Y_n's:

$$y(t) \stackrel{\text{def}}{=} \int_{-\infty}^{+\infty} d\omega \frac{e^{i\omega t}}{\sqrt{2\pi}} Y(\omega) \tag{12.33}$$

$$\Rightarrow \quad y(t) \simeq \sum_{n=1}^{N} \frac{2\pi}{Nh} \frac{e^{i\omega_n t}}{\sqrt{2\pi}} Y(\omega_n) . \tag{12.34}$$

4) The alert reader may be wondering what has happened to the $h/2$ with which the trapezoid rule weights the initial and final points. Actually, they are there, but because we have set $y_0 \equiv y_N$, two $h/2$ terms have been added to produce one h term.

Once we know the N values of the transform, we can use (12.34) to evaluate $y(t)$ for any time t. There is nothing illegal about evaluating Y_n and y_k for arbitrarily large values of n and k, yet there is also nothing to be gained either. Because the trigonometric functions are periodic, we just get the old answers:

$$y(t_{k+N}) = y((k+N)h) = y(t_k) , \tag{12.35}$$

$$Y(\omega_{n+N}) = Y((n+N)\omega_1) = Y(\omega_n) . \tag{12.36}$$

Another way of stating this is to observe that none of the equations change if we replace $\omega_n t$ by $\omega_n t + 2\pi n$. There are still just N independent output numbers for N independent inputs, and so the transform and the reconstituted signal are periodic.

We see from (12.29) that the larger we make the time $T = Nh$ over which we sample the function, the smaller will be the frequency steps or resolution.[5] Accordingly, if you want a smooth frequency spectrum, you need to have a small frequency step $2\pi/T$, which means a longer observation time T. While the best approach would be to measure the input signal for all times, in practice a measured signal $y(t)$ is often extended in time ("padded") by adding zeros for times beyond the last measured signal, which thereby increases the value of T artificially. Although this does not add new information to the analysis, it does build in the experimentalist's view that the signal has no existence, or no meaning, at times after the measurements are stopped.

While periodicity is expected for a Fourier *series*, it is somewhat surprising for Fourier a *integral*, which have been touted as the right tool for nonperiodic functions. Clearly, if we input values of the signal for longer lengths of time, then the inherent period becomes longer, and if the repeat period T is very long, it may be of little consequence for times short compared to the period. If $y(t)$ is actually periodic with period Nh, then the DFT is an excellent way of obtaining Fourier series. If the input function is not periodic, then the DFT can be a bad approximation near the endpoints of the time interval (after which the function will repeat) or, correspondingly, for the lowest frequencies.

The DFT and its inverse can be written in a concise and insightful way, and be evaluated efficiently, by introducing a complex variable Z for the exponential and then raising Z to various powers:

$$\boxed{y_k = \frac{\sqrt{2\pi}}{N}\sum_{n=1}^{N} Z^{-nk} Y_n , \quad Z = e^{-2\pi i/N} ,} \tag{12.37}$$

$$\boxed{Y_n = \frac{1}{\sqrt{2\pi}}\sum_{k=1}^{N} Z^{nk} y_k , \quad Z^{nk} \equiv [(Z)^n]^k .} \tag{12.38}$$

With this formulation, the computer needs to compute only powers of Z. We give our DFT code in Listing 12.1. If your preference is to avoid complex numbers,

5) See also Section 12.5.1 where we discuss the related phenomenon of aliasing.

we can rewrite (12.37) in terms of separate real and imaginary parts by applying Euler's theorem with $\theta \stackrel{\text{def}}{=} 2\pi/N$:

$$Z = e^{-i\theta}, \quad \Rightarrow Z^{\pm nk} = e^{\mp ink\theta} = \cos nk\theta \mp i \sin nk\theta, \tag{12.39}$$

$$\Rightarrow \quad Y_n = \frac{1}{\sqrt{2\pi}} \sum_{k=1}^{N} \left[\cos(nk\theta) \text{Re } y_k + \sin(nk\theta) \text{ Im } y_k \right.$$
$$\left. + i(\cos(nk\theta) \text{ Im } y_k - \sin(nk\theta) \text{Re } y_k) \right], \tag{12.40}$$

$$y_k = \frac{\sqrt{2\pi}}{N} \sum_{n=1}^{N} \left[\cos(nk\theta) \text{Re } Y_n - \sin(nk\theta) \text{Im } Y_n \right.$$
$$\left. + i(\cos(nk\theta) \text{Im } Y_n + \sin(nk\theta) \text{Re } Y_n) \right]. \tag{12.41}$$

Readers new to DFTs are often surprised when they apply these equations to practical situations and end up with transforms Y having imaginary parts, despite the fact that the signal y is real. Equation 12.40 should make it clear that a real signal ($\text{Im } y_k \equiv 0$) will yield an imaginary transform unless $\sum_{k=1}^{N} \sin(nk\theta) \text{Re } y_k = 0$. This occurs only if $y(t)$ is an *even* function over $-\infty \le t \le +\infty$ *and* we integrate exactly. Because neither condition holds, the DFTs of real, even functions may have small imaginary parts. This is not as a result of an error in programming, and in fact is a good measure of the approximation error in the entire procedure.

The computation time for a DFT can be reduced even further by using *fast Fourier transform* algorithm, as discussed in Section 12.9. An examination of (12.37) shows that the DFT is evaluated as a matrix multiplication of a vector of length N containing the Z values by a vector of length N of y value. The time for this DFT scales like N^2, while the time for the FFT algorithm scales as $N \log_2 N$. Although this may not seem like much of a difference, for $N = 10^{2-3}$, the difference of 10^{3-5} is the difference between a minute and a week. For this reason, it is the FFT is often used for online spectrum analysis.

Listing 12.1 DFTcomplex.py uses the built-in complex numbers of Python to compute the DFT for the signal in method f(signal).

```
# DFTcomplex.py: Discrete Fourier Transform with built in complex from
     visual import * from
visual.graph import * import cmath # complex math

signgr = gdisplay(x=0, y=0, width=600, height=250, title ='Signal',\
            xtitle='x', ytitle = 'signal', xmax = 2.*math.pi, xmin = 0,\
        ymax = 30, ymin = 0)
sigfig = gcurve(color=color.yellow, display=signgr)
imagr = gdisplay(x=0,y=250,width=600,height=250,title ='Imag Fourier TF',
          xtitle = 'x',ytitle='TF.Imag',xmax=10.,xmin=-1,ymax=100,ymin=-0.2)
impart = gvbars(delta = 0.05, color = color.red, display = imagr)

N = 50;                       Np = N
signal = zeros( (N+1), float )
twopi  = 2.*pi;              sq2pi = 1./sqrt(twopi);           h = twopi/N
dftz   = zeros( (Np), complex )                          # Complex elements
```

```
def f(signal):                                              # Signal
    step = twopi/N;          x = 0.
    for i in range(0, N+1):
        signal[i] = 30*cos(x*x*x*x)
        sigfig.plot(pos = (x, signal[i]))                   # Plot
        x += step

def fourier(dftz):                                          # DFT
    for n in range(0, Np):
       zsum = complex(0.0, 0.0)
       for  k in range(0, N):
           zexpo = complex(0, twopi*k*n/N)           # Complex exponent
           zsum += signal[k]*exp(-zexpo)
       dftz[n] = zsum * sq2pi
       if dftz[n].imag != 0:
           impart.plot(pos = (n, dftz[n].imag) )            # Plot

f(signal);          fourier(dftz)               # Call signal, transform
```

12.5.1 Aliasing (Assessment)

The sampling of a signal by DFT for only a finite number of times (large Δt) limits the accuracy of the deduced high-frequency components present in the signal. Clearly, good information about very high frequencies requires sampling the signal with small time steps so that all the wiggles can be included. While a poor deduction of the high-frequency components may be tolerable if all we care about are the low-frequency components, the inaccurate high-frequency components remain present in the signal and may contaminate the low-frequency components that we deduce. This effect is called *aliasing* and is the cause of the Moiré pattern distortion in digital images.

As an example, consider Figure 12.2 showing the two functions $\sin(\pi t/2)$ and $\sin(2\pi t)$ for $0 \le t \le 8$, with their points of overlap in bold. If we were unfortunate enough to sample a signal containing these functions at the times $t = 0, 2, 4, 6, 8$, then we would measure $y \equiv 0$ and assume that there was no signal at all. However, if we were unfortunate enough to measure the signal at the filled dots in Figure 12.2 where $\sin(\pi t/2) = \sin(2\pi t)$, specifically, $t = 0, 12/10, 4/3, \ldots$, then our Fourier analysis would completely miss the high-frequency component. In DFT jargon, we would say that the high-frequency component has been *aliased* by the low-frequency component. In other cases, some high-frequency values may be included in our sampling of the signal, but our sampling rate may not be high enough to include enough of them to separate the high-frequency component properly. In this case some high-frequency signals would be included spuriously as part of the low-frequency spectrum, and this would lead to spurious low-frequency oscillations when the signal is synthesized from its Fourier components.

More precisely, aliasing occurs when a signal containing frequency f is sampled at a rate of $s = N/T$ measurements per unit time, with $s \le f/2$. In this case, the frequencies f and $f - 2s$ yield the same DFT, and we would not be able to

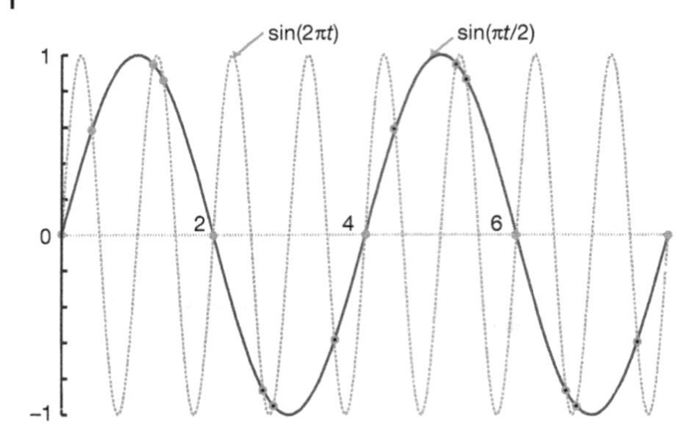

Figure 12.2 A plot of the functions $\sin(\pi t/2)$ and $\sin(2\pi t)$. If the sampling rate is not high enough, these signals may appear indistinguishable in a Fourier decomposition. If the sample rate is too low and if both signals are present in a sample, the deduced low-frequency component may be contaminated by the higher frequency component signal.

determine that there are two frequencies present. That being the case, to avoid aliasing we want no frequencies $f > s/2$ to be present in our input signal. This is known as the *Nyquist criterion*. In practice, some applications avoid the effects of aliasing by filtering out the high frequencies from the signal and then analyzing only the remaining low-frequency part. (The low-frequency *sinc filter* discussed in Section 12.8.1 is often used for this purpose.) Although filtering eliminates some high-frequency information, it lessens the distortion of the low-frequency components, and so may lead to improved reproduction of the signal.

If accurate values for the high frequencies are required, then we will need to increase the sampling rate s by increasing the number N of samples taken within the fixed sampling time $T = Nh$. By keeping the sampling time constant and increasing the number of samples taken, we make the time step h smaller and we pick up the higher frequencies. By increasing the number N of frequencies and that you compute, you move the previous higher frequency components in closer to the middle of the spectrum, and thus away from the error-prone ends.

If we increase the total time sampling time $T = Nh$ and keep h the same, then the sampling rate $s = N/T = 1/h$ remains the same. Because $\omega_1 = 2\pi/T$, this makes ω_1 smaller, which means we have more low frequencies recorded and a smoother frequency spectrum. And as we said, this is often carried out, after the fact, by padding the end of the data set with zeros.

Listing 12.2 DFTreal.py computes the discrete Fourier transform for the signal in method f(signal) using real numbers.

```
# DFTreal.py: Discrete Fourier Transform using real numbers

from visual.graph import *

signgr = gdisplay(x=0,y=0,width=600,height=250, \
        title='Signal y(t)= 3 cos(wt)+2 cos(3wt)+ cos(5wt) ',\
   xtitle='x', ytitle='signal',xmax=2.*math.pi,xmin=0,ymax=7,ymin=-7)
```

```
sigfig = gcurve(color=color.yellow, display=signgr)
imagr = gdisplay(x=0,y=250,width=600,height=250,\
      title='Fourier transform imaginary part',xtitle='x',\
        ytitle='Transf.Imag',xmax=10.0,xmin=-1,ymax=20,ymin=-25)
impart = gvbars(delta=0.05,color=color.red, display=imagr)

N = 200
Np = N
signal = zeros((N+1),float)
twopi = 2.*pi
sq2pi = 1./sqrt(twopi)
h = twopi/N
dftimag = zeros((Np),float)                          # Im. transform

def f(signal):
    step = twopi/N
    t= 0.
    for  i in range(0,N+1):
        signal[i] = 3*sin(t*t*t)
        sigfig.plot(pos=(t, signal[i]))
        t += step

def fourier(dftimag):                                          # DFT
    for n in range(0,Np):
      imag = 0.
      for  k in range(0, N):
        imag += signal[k]* sin((twopi*k*n)/N)
      dftimag[n] = -imag*sq2pi                        # Im transform
      if dftimag[n] !=0:
          impart.plot(pos=(n, dftimag[n]))

f(signal)
fourier(dftimag)
```

12.5.2
Fourier Series DFT (Example)

For simplicity, let us consider the Fourier cosine series:

$$y(t) = \sum_{n=0}^{\infty} a_n \cos(n\omega t) , \quad a_k = \frac{2}{T} \int_0^T \mathrm{d}t \cos(k\omega t) y(t) . \tag{12.42}$$

Here $T \stackrel{\text{def}}{=} 2\pi/\omega$ is the actual period of the system (not necessarily the period of the simple harmonic motion occurring for a small amplitude). We assume that the function $y(t)$ is sampled for a discrete set of times

$$y(t = t_k) \equiv y_k , \quad k = 0, 1, \ldots, N . \tag{12.43}$$

Because we are analyzing a periodic function, we retain the conventions used in the DFT and require the function to repeat itself with period $T = Nh$; that is, we assume that the amplitude is the same at the first and last points:

$$y_0 = y_N . \tag{12.44}$$

This means that there are only N independent values of y being used as input. For these N independent y_k values, we can determine uniquely only N expansion coefficients a_k. If we use the trapezoid rule to approximate the integration in (12.42), we determine the N independent Fourier components as

$$a_n \simeq \frac{2h}{T} \sum_{k=1}^{N} \cos(n\omega t_k) y(t_k) = \frac{2}{N} \sum_{k=1}^{N} \cos\left(\frac{2\pi nk}{N}\right) y_k, \quad n = 0, \ldots, N. \tag{12.45}$$

Because we can determine only N Fourier components from N independent $y(t)$ values, our Fourier series for the $y(t)$ must be in terms of only these components:

$$y(t) \simeq \sum_{n=0}^{N} a_n \cos(n\omega t) = \sum_{n=0}^{N} a_n \cos\left(\frac{2\pi nt}{Nh}\right). \tag{12.46}$$

In summary, we sample the function $y(t)$ at N times, $t_1, \ldots, t_N$. We see that all the values of y sampled contribute to each a_k. Consequently, if we increase N in order to determine more coefficients, we must recompute all the a_n values. In the wavelet analysis in Chapter 13, the theory is reformulated so that additional samplings determine higher spectral components without affecting lower ones.

12.5.3 Assessments

Simple analytic input It is always good to do simple checks before examining more complex problems, even if you are using a package's Fourier tool.

1. Sample the even signal

$$y(t) = 3\cos(\omega t) + 2\cos(3\omega t) + \cos(5\omega t). \tag{12.47}$$

   a) Decompose this into its components.
   b) Check that the components are essentially real and in the ratio 3 : 2 : 1 (or 9 : 4 : 1 for the power spectrum).
   c) Verify that the frequencies have the expected values (not just ratios).
   d) Verify that the components resum to give the input signal.
   e) Experiment on the separate effects of picking different values of the step size h and of enlarging the measurement period $T = Nh$.

2. Sample the odd signal

$$y(t) = \sin(\omega t) + 2\sin(3\omega t) + 3\sin(5\omega t). \tag{12.48}$$

   Decompose this into its components; then check that they are essentially imaginary and in the ratio 1 : 2 : 3 (or 1 : 4 : 9 if a power spectrum is plotted) and that they resum to give the input signal.

3. Sample the mixed-symmetry signal

$$y(t) = 5\sin(\omega t) + 2\cos(3\omega t) + \sin(5\omega t). \tag{12.49}$$

Decompose this into its components; then check that there are three of them in the ratio 5 : 2 : 1 (or 25 : 4 : 1 if a power spectrum is plotted) and that they resum to give the input signal.

4. Sample the signal

$$y(t) = 5 + 10\sin(t + 2)\ .$$

Compare and explain the results obtained by sampling (a) without the 5, (b) as given but without the 2, and (c) without the 5 and without the 2.

5. In our discussion of aliasing, we examined Figure 12.2 showing the functions $\sin(\pi t/2)$ and $\sin(2\pi t)$. Sample the function

$$y(t) = \sin\left(\frac{\pi}{2}t\right) + \sin(2\pi t) \tag{12.50}$$

and explore how aliasing occurs. Explicitly, we know that the true transform contains peaks at $\omega = \pi/2$ and $\omega = 2\pi$. Sample the signal at a rate that leads to aliasing, as well as at a higher sampling rate at which there is no aliasing. Compare the resulting DFTs in each case and check if your conclusions agree with the Nyquist criterion.

Highly nonlinear oscillator Recall the numerical solution for oscillations of a spring with power $p = 12$ (see (12.1)). Decompose the solution into a Fourier series and determine the number of higher harmonics that contribute at least 10%; for example, determine the n for which $|b_n/b_1| < 0.1$. Check that resuming the components reproduces the signal.

Nonlinearly perturbed oscillator Remember the harmonic oscillator with a nonlinear perturbation (8.2):

$$V(x) = \frac{1}{2}kx^2\left(1 - \frac{2}{3}\alpha x\right)\ , \quad F(x) = -kx(1 - \alpha x)\ . \tag{12.51}$$

For very small amplitudes of oscillation ($x \ll 1/\alpha$), the solution $x(t)$ essentially should be only the first term of a Fourier series.

1. We want the signal to contain "approximately 10% nonlinearity." This being the case, fix your value of α so that $\alpha x_{max} \simeq 10\%$, where x_{max} is the maximum amplitude of oscillation. For the rest of the problem, keep the value of α fixed.
2. Decompose your numerical solution into a discrete Fourier spectrum.
3. Plot a graph of the percentage of importance of the first *two*, non-DC Fourier components as a function of the initial displacement for $0 < x_0 < 1/2\alpha$. You should find that higher harmonics are more important as the amplitude increases. Because both even and odd components are present, Y_n should be complex. Because a 10% effect in amplitude becomes a 1% effect in power, make sure that you make a semilog plot of the power spectrum.
4. As always, check that resumations of your transforms reproduce the signal.

(*Warning:* The ω you use in your series must correspond to the *true* frequency of the system, not the ω_0 of small oscillations.)

12.5.4
Nonperiodic Function DFT (Exploration)

Consider a simple model (a wave packet) of a "localized" electron moving through space and time. A good model for an electron initially localized around $x = 5$ is a Gaussian multiplying a plane wave:

$$\psi(x, t = 0) = \exp\left[-\frac{1}{2}\left(\frac{x - 5.0}{\sigma_0}\right)^2\right] e^{ik_0 x} . \tag{12.52}$$

This wave packet is not an eigenstate of the momentum operator[6] $p = i\,d/dx$ and in fact contains a spread of momenta. Your *problem* is to evaluate the Fourier transform

$$\psi(p) = \int_{-\infty}^{+\infty} dx \frac{e^{ipx}}{\sqrt{2\pi}} \psi(x) , \tag{12.53}$$

as a way of determining the momenta components in (12.52).

12.6
Filtering Noisy Signals

You measure a signal $y(t)$ that obviously contains noise. Your *problem* is to determine the frequencies that would be present in the spectrum of the signal if the signal did not contain noise. Of course, once you have a Fourier transform from which the noise has been removed, you can transform it to obtain a signal $s(t)$ with no noise.

In the process of solving this problem, we examine two simple approaches: the use of autocorrelation functions and the use of filters. Both approaches find wide applications in science, with our discussion not doing the subjects justice. We will see filters again in the discussion of wavelets in Chapter 13.

12.7
Noise Reduction via Autocorrelation (Theory)

We assume that the measured signal is the sum of the true signal $s(t)$, which we wish to determine, plus some unwelcome *noise* $n(t)$:

$$y(t) = s(t) + n(t) . \tag{12.54}$$

Our first approach at removing the noise relies on that fact that noise is a random process and thus should not be correlated with the signal. Yet what do we

6) We use natural units in which $\hbar = 1$.

mean when we say that two functions are not *correlated?* Well, if the two tend to oscillate with their nodes and peaks in much the same places, then the two functions are clearly correlated. An analytic measure of the correlation of two arbitrary functions $y(t)$ and $x(t)$ is the *correlation function*

$$c(\tau) = \int_{-\infty}^{+\infty} dt \; y^*(t)x(t+\tau) \equiv \int_{-\infty}^{+\infty} dt\, y^*(t-\tau)x(t) \,, \tag{12.55}$$

where τ, the *lag time*, is a variable. Even if the two signals have different magnitudes, if they have similar time dependences except for one lagging or leading the other, then for certain values of τ the integrand in (12.55) will be positive for all values of t. For those values of τ, the two signals interfere constructively and produce a large value for the correlation function. In contrast, if both functions oscillate independently regardless of the value of τ, then it is just as likely for the integrand to be positive as to be negative, in which case the two signals interfere destructively and produce a small value for the integral.

Before we apply the correlation function to our problem, let us study some of its properties. We use (12.18) to express c, y^*, and x in terms of their Fourier transforms:

$$c(\tau) = \int_{-\infty}^{+\infty} d\omega'' C(\omega'') \frac{e^{i\omega'' t}}{\sqrt{2\pi}} \,, \qquad y^*(t) = \int_{-\infty}^{+\infty} d\omega Y^*(\omega) \frac{e^{-i\omega t}}{\sqrt{2\pi}} \,,$$

$$x(t+\tau) = \int_{-\infty}^{+\infty} d\omega' \; X(\omega') \frac{e^{+i\omega t}}{\sqrt{2\pi}} \,. \tag{12.56}$$

Because ω, ω', and ω'' are dummy variables, other names may be used for these variables without changing any results. When we substitute these representations into definition (12.55) of the correlation function and assume that the resulting integrals converge well enough to be rearranged, we obtain

$$\begin{aligned}
\int_{-\infty}^{+\infty} d\omega C(\omega) e^{i\omega t} &= \int_{-\infty}^{+\infty} \frac{d\omega}{2\pi} \int_{-\infty}^{+\infty} d\omega' Y^*(\omega) X(\omega') e^{i\omega\tau} 2\pi\delta(\omega' - \omega) \\
&= \int_{-\infty}^{+\infty} d\omega Y^*(\omega) X(\omega) e^{i\omega\tau} \,, \\
\Rightarrow \quad C(\omega) &= \sqrt{2\pi}\, Y^*(\omega) X(\omega) \,,
\end{aligned} \tag{12.57}$$

where the last line follows because ω'' and ω are equivalent dummy variables. Equation 12.57 says that the Fourier transform of the correlation function between two signals is proportional to the product of the transform of one signal and the complex conjugate of the transform of the other. (We shall see a related convolution theorem for filters.)

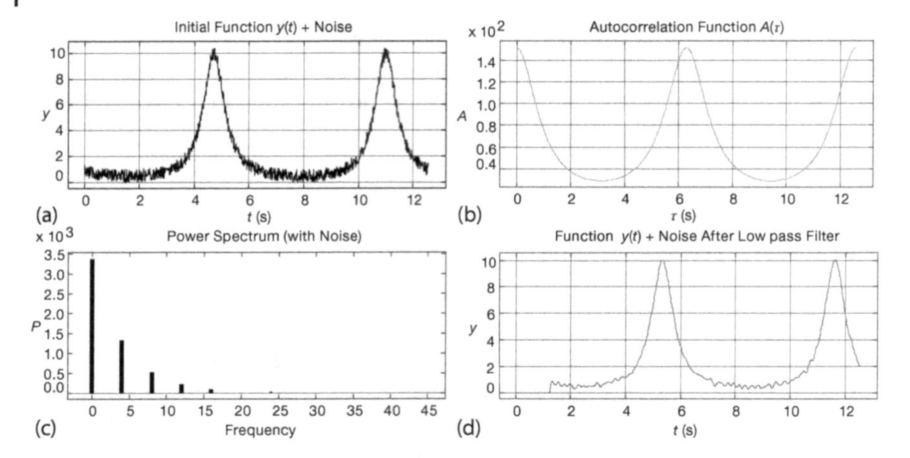

Figure 12.3 (a) A function that is a signal plus noise $s(t) + n(t)$; (b) the autocorrelation function vs. time deduced by processing this signal; (c) the power spectrum obtained from autocorrelation function; (d) the signal plus noise after passage through a low-pass filter.

A special case of the correlation function $c(\tau)$ is the *autocorrelation function* $A(\tau)$. It measures the correlation of a time signal with itself:

$$A(\tau) \stackrel{\text{def}}{=} \int_{-\infty}^{+\infty} \mathrm{d}t\, y^*(t) y(t+\tau) \equiv \int_{-\infty}^{+\infty} \mathrm{d}t\, y(t) y^*(t-\tau) \, . \tag{12.58}$$

This function is computed by taking a signal $y(t)$ that has been measured over some time period and then averaging it over time using $y(t + \tau)$ as a weighting function. This process is also called *folding* a function onto itself (as might be done with dough) or a *convolution*. To see how this folding removes noise from a signal, we go back to the measured signal (12.54), which was the sum of pure signal plus noise $s(t) + n(t)$. As an example, in Figure 12.3a we show a signal that was constructed by adding random noise to a smooth signal. When we compute the autocorrelation function for this signal, we obtain a function (Figure 12.3b) that looks like a broadened, smoothed version of the signal $y(t)$.

We can understand how the noise is removed by taking the Fourier transform of $s(t) + n(t)$ to obtain a simple sum of transforms:

$$Y(\omega) = S(\omega) + N(\omega) \, , \tag{12.59}$$

$$\begin{bmatrix} S(\omega) \\ N(\omega) \end{bmatrix} = \int_{-\infty}^{+\infty} \mathrm{d}t \begin{bmatrix} s(t) \\ n(t) \end{bmatrix} \frac{\mathrm{e}^{-\mathrm{i}\omega t}}{\sqrt{2\pi}} \, . \tag{12.60}$$

Because the autocorrelation function (12.58) for $y(t) = s(t) + n(t)$ involves the second power of y, is not a linear function, that is, $A_y \neq A_s + A_n$, but instead

$$A_y(\tau) = \int_{-\infty}^{+\infty} \mathrm{d}t[s(t)s(t+\tau) + s(t)n(t+\tau) + n(t)n(t+\tau)] \, . \tag{12.61}$$

If we assume that the noise $n(t)$ in the measured signal is truly random, then it should average to zero over long times and be uncorrelated at times t and $t+\tau$. This being the case, both integrals involving the noise vanish, and so

$$A_y(\tau) \simeq \int_{-\infty}^{+\infty} dt\, s(t)\, s(t+\tau) = A_s(\tau)\,. \tag{12.62}$$

Thus, the part of the noise that is random tends to be averaged out of the autocorrelation function, and we are left with an approximation of the autocorrelation function of the pure signal.

So how does this help us? Application of (12.57) with $Y(\omega) = X(\omega) = S(\omega)$ tells us that the Fourier transform $A(\omega)$ of the autocorrelation function is proportional to $|S(\omega)|^2$:

$$A(\omega) = \sqrt{2\pi}\,|S(\omega)|^2\,. \tag{12.63}$$

The function $|S(\omega)|^2$ is the *power spectrum* of the pure signal. Thus, evaluation of the autocorrelation function of the noisy signal gives us the pure signal's power spectrum, which is often all that we need to know. For example, in Figure 12.3a we see a noisy signal, the autocorrelation function (Figure 12.3b), which clearly is smoother than the signal, and finally, the deduced power spectrum (Figure 12.3c). Note that the broadband high-frequency components characteristic of noise are absent from the power spectrum.

You can easily modify the sample program `DFTcomplex.py` in Listing 12.1 to compute the autocorrelation function and then the power spectrum from $A(\tau)$. We present a program `NoiseSincFilter.py` on the instructor's site that does this.

12.7.1 Autocorrelation Function Exercises

1. Imagine that you have sampled the pure signal

$$s(t) = \frac{1}{1 - 0.9\sin t}\,. \tag{12.64}$$

Although there is just a single sine function in the denominator, there is an infinite number of overtones as follows from the expansion

$$s(t) \simeq 1 + 0.9\sin t + (0.9\sin t)^2 + (0.9\sin t)^3 + \cdots \tag{12.65}$$

a) Compute the DFT $S(\omega)$. Make sure to sample just one period but to cover the entire period. Make sure to sample at enough times (fine scale) to obtain good sensitivity to the high-frequency components.
b) Make a semilog plot of the power spectrum $|S(\omega)|^2$.
c) Take your input signal $s(t)$ and compute its autocorrelation function $A(\tau)$ for a full range of τ values (an analytic solution is okay too).

d) Compute the power spectrum indirectly by performing a DFT on the autocorrelation function. Compare your results to the spectrum obtained by computing $|S(\omega)|^2$ directly.

2. Add some random noise to the signal using a random number generator:

$$y(t_i) = s(t_i) + \alpha(2r_i - 1) , \quad 0 \le r_i \le 1 , \tag{12.66}$$

where α is an adjustable parameter. Try several values of α, from small values that just add some fuzz to the signal to large values that nearly hide the signal.

a) Plot your noisy data, their Fourier transform, and their power spectrum obtained directly from the transform with noise.

b) Compute the autocorrelation function $A(\tau)$ and its Fourier transform $A(\omega)$.

c) Compare the DFT of $A(\tau)$ to the true power spectrum and comment on the effectiveness of reducing noise by use of the autocorrelation function.

d) For what value of α do you essentially lose all the information in the input?

12.8
Filtering with Transforms (Theory)

A filter (Figure 12.4) is a device that converts an input signal $f(t)$ to an output signal $g(t)$ with some specific property for $g(t)$. More specifically, an *analog filter* is defined (Hartmann, 1998) as integration over the input function:

$$g(t) = \int_{-\infty}^{+\infty} d\tau\, f(\tau)\, h(t - \tau) \stackrel{\text{def}}{=} f(t) * h(t) . \tag{12.67}$$

The operation indicated in (12.67) occurs often enough that it is given the name *convolution* and is denoted by an asterisk $*$. The function $h(t)$ is called the *response* or *transfer function* of the filter because it is the response of the filter to a unit impulse:

$$h(t) = \int_{-\infty}^{+\infty} d\tau\, \delta(\tau)\, h(t - \tau) . \tag{12.68}$$

Equation 12.67 states that the output $g(t)$ of a filter equals the input $f(t)$ convoluted with the transfer function $h(t - \tau)$. Because the argument of the response function is delayed by a time τ relative to that of the signal in integral (12.67), τ is called the *lag time*. While the integration is over all times, the response of a good detector usually peaks around zero time. In any case, the response must equal zero for $\tau > t$ because events in the future cannot affect the present (causality).

The *convolution theorem* states that the Fourier transform of the convolution $g(t)$ is proportional to the product of the transforms of $f(t)$ and $h(t)$:

$$G(\omega) = \sqrt{2\pi} F(\omega) H(\omega) . \tag{12.69}$$

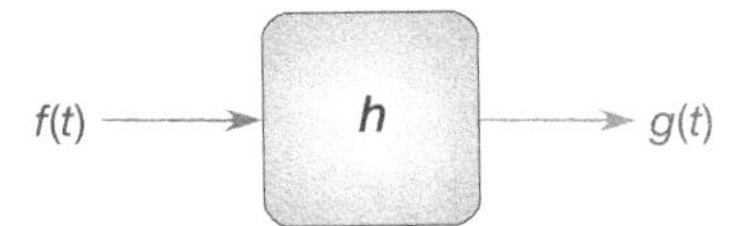

Figure 12.4 A schematic of an input signal $f(t)$ passing through a filter h that outputs the function $g(t)$.

The theorem results from expressing the functions in (12.67) by their transforms and using the resulting Dirac delta function to evaluate an integral (essentially what we did in our discussion of correlation functions).

Regardless of the domain used, filtering as we have defined it is a linear process involving just the first powers of f. This means that the output at one frequency is proportional to the input at that frequency. The constant of proportionality between the two may change with frequency and thus suppress specific frequencies relative to others, but that constant remains fixed in time. Because the law of linear superposition is valid for filters, if the input to a filter is represented as the sum of various functions, then the transform of the output will be the sum of the functions' Fourier transforms.

Filters that remove or decrease high-frequency components more than they do low-frequency ones, are called *low-pass* filters. Those that filter out the low frequencies are called *high-pass filters.* A simple low-pass filter is the RC circuit shown in Figure 12.5a. It produces the transfer function

$$H(\omega) = \frac{1}{1 + \mathrm{i}\omega\tau} = \frac{1 - \mathrm{i}\omega\tau}{1 + \omega^2\tau^2}, \tag{12.70}$$

where $\tau = \mathrm{RC}$ is the time constant. The ω^2 in the denominator leads to a decrease in the response at high frequencies and therefore makes this a low-pass filter (the $\mathrm{i}\omega$ affects only the phase). A simple high-pass filter is the RC circuit shown in Figure 12.5b. It produces the transfer function

$$H(\omega) = \frac{\mathrm{i}\omega\tau}{1 + \mathrm{i}\omega\tau} = \frac{\mathrm{i}\omega\tau + \omega^2\tau^2}{1 + \omega^2\tau^2}. \tag{12.71}$$

$H = 1$ at large ω, yet H vanishes as $\omega \to 0$, as expected for a high-pass filter.

Filters composed of resistors and capacitors are fine for analog signal processing. For digital processing we want a *digital filter* that has a specific response function for each frequency range. A physical model for a digital filter may be constructed from a delay line with taps at various spacing along the line (Figure 12.6)

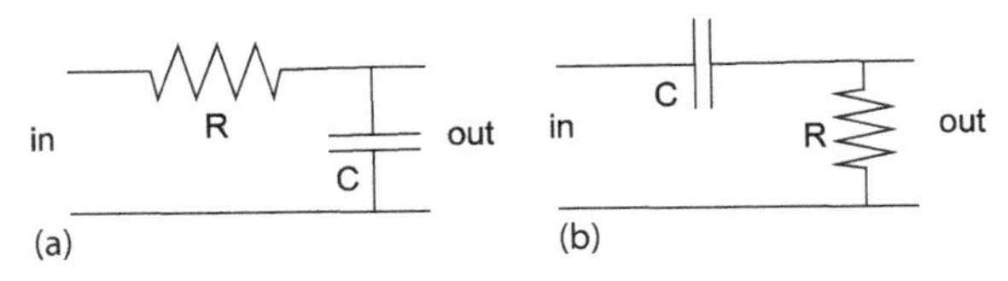

Figure 12.5 (a) An RC circuit arranged as a low-pass filter. (b) An RC circuit arranged as a high-pass filter.

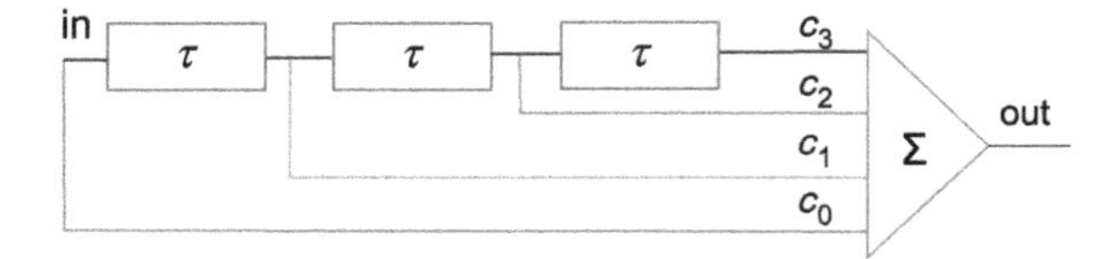

Figure 12.6 A delay-line filter in which the signal at different times is scaled by different amounts c_i.

(Hartmann, 1998). The signal read from tap n is just the input signal delayed by time $n\tau$, where the delay time τ is a characteristic of the particular filter. The output from each tap is described by the transfer function $\delta(t - n\tau)$, possibly with scaling factor c_n. As represented by the triangle in Figure 12.6b, the signals from all taps are ultimately summed together to form the total response function:

$$h(t) = \sum_{n=0}^{N} c_n \delta(t - n\tau) . \qquad (12.72)$$

In the frequency domain, the Fourier transform of a delta function is an exponential, and so (12.72) results in the transfer function

$$H(\omega) = \sum_{n=0}^{N} c_n \mathrm{e}^{-in\omega\tau} , \qquad (12.73)$$

where the exponential indicates the phase shift from each tap.

If a digital filter is given a continuous time signal $f(t)$ as input, its output will be the discrete sum

$$g(t) = \int_{-\infty}^{+\infty} \mathrm{d}t' \, f(t') \sum_{n=0}^{N} c_n \delta(t - t' - n\tau) = \sum_{n=0}^{N} c_n f(t - n\tau) . \qquad (12.74)$$

And of course, if the signal's input is a discrete sum, its output will remain a discrete sum. In either case, we see that knowledge of the filter coefficients c_i provides us with all we need to know about a digital filter. If we look back at our work on the DFT in Section 12.5, we can view a digital filter (12.74) as a Fourier transform in which we use an N-point approximation to the Fourier integral. The c_ns then contain both the integration weights and the values of the response function at the integration points. The transform itself can be viewed as a filter of the signal into specific frequencies.

12.8.1
Digital Filters: Windowed Sinc Filters (Exploration) ⊙

Problem Construct digital versions of high-pass and low-pass filters and determine which filter works better at removing noise from a signal.

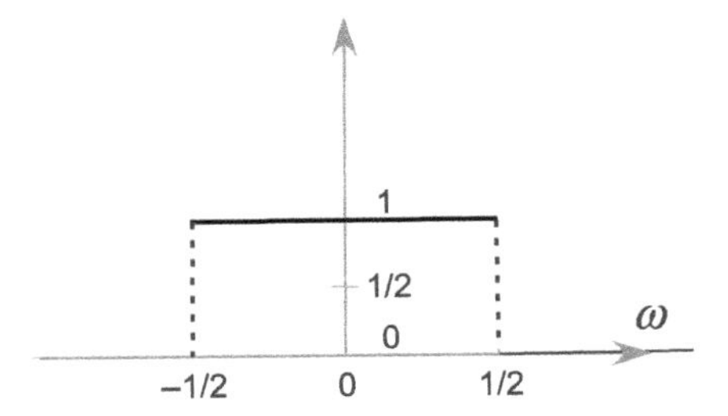

Figure 12.7 The rectangle function rect(ω) that is constant for a finite frequency interval. The Fourier transform of this function is sinc(t).

A popular way to separate the bands of frequencies in a signal is with a *windowed sinc filter* (Smith, 1999). This filter is based on the observation that an ideal *low-pass* filter passes all frequencies below a cutoff frequency ω_c, and blocks all frequencies above this frequency. And because there tends to be more noise at high frequencies than at low frequencies, removing the high frequencies tends to remove more noise than signal, although some signal is inevitably lost. One use for windowed sinc filters is in reducing aliasing in DFTs by removing the high-frequency component of a signal before determining its Fourier components. The graph in Figure 12.1b was obtained by passing our noisy signal through a sinc filter (using the program `NoiseSincFilter.py`).

If both positive and negative frequencies are included, an ideal low-frequency filter will look like the rectangular pulse in frequency space (Figure 12.7):

$$H(\omega, \omega_c) = \text{rect}\left(\frac{\omega}{2\omega_c}\right), \quad \text{rect}(\omega) = \begin{cases} 1, & \text{if } |\omega| \leq \frac{1}{2}, \\ 0, & \text{otherwise}. \end{cases} \tag{12.75}$$

Here rect(ω) is the rectangular function. Although maybe not obvious, a rectangular pulse in the frequency domain has a Fourier transform that is proportional to the *sinc function* in the time domain (Smith, 1991)

$$\int_{-\infty}^{+\infty} d\omega e^{-i\omega t} \text{rect}(\omega) = \text{sinc}\left(\frac{t}{2}\right) \stackrel{\text{def}}{=} \frac{\sin(\pi t/2)}{\pi t/2}, \tag{12.76}$$

where the πs are sometimes omitted. Consequently, we can filter out the high-frequency components of a signal by convoluting it with $\sin(\omega_c t)/(\omega_c t)$, a technique also known as the *Nyquist–Shannon* interpolation formula. In terms of discrete transforms, the time-domain representation of the sinc filter is simply

$$h[i] = \frac{\sin(\omega_c i)}{i\pi}. \tag{12.77}$$

Because all frequencies below the cutoff frequency ω_c are passed with unit amplitude, while all higher frequencies are blocked, we can see the importance of a sinc filter.

In practice, there are a number of problems in using sinc function as the filter. First, as formulated, the filter is *noncausal*; that is, there are coefficients at negative times, which is nonphysical because we do not start measuring the signal until $t = 0$. Second, in order to produce a perfect rectangular response, we would have to sample the signal at an infinite number of times. In practice, we sample at $(M+1)$ points (M even) placed symmetrically around the main lobe of $\sin(\pi t)/\pi t$, and then shift times to purely positive values via

$$h[\mathrm{i}] = \frac{\sin[2\pi\omega_c(\mathrm{i} - M/2)]}{\mathrm{i} - M/2}, \quad 0 \le t \le M . \tag{12.78}$$

As might be expected, a penalty is incurred for making the filter discrete; instead of the ideal rectangular response, we obtain some *Gibbs overshoot*, with rounded corners and oscillations beyond the corner.

There are two ways to reduce the departures from the ideal filter. The first is to increase the length of times for which the filter is sampled, which inevitably leads to longer compute times. The other way is to smooth out the truncation of the sinc function by multiplying it with a smoothly tapered curve like the *Hamming window function*:

$$w[\mathrm{i}] = 0.54 - 0.46\cos(2\pi\mathrm{i}/M) . \tag{12.79}$$

In this way, the filter's kernel becomes

$$h[\mathrm{i}] = \frac{\sin[2\pi\omega_c(\mathrm{i} - M/2)]}{\mathrm{i} - M/2}\left[0.54 - 0.46\cos\left(\frac{2\pi\mathrm{i}}{M}\right)\right] . \tag{12.80}$$

The cutoff frequency ω_c should be a fraction of the sampling rate. The time length M determines the *bandwidth* over which the filter changes from 1 to 0.

Exercise Repeat the exercise that added random noise to a known signal, this time using the sinc filter to reduce the noise. See how small you can make the signal and still be able to separate it from the noise.

12.9 The Fast Fourier Transform Algorithm ⊙

We have seen in (12.37) that a discrete Fourier transform can be written in the compact form as

$$Y_n = \frac{1}{\sqrt{2\pi}} \sum_{k=1}^{N} Z^{nk} y_k \,, \quad Z = e^{-2\pi i/N} \,, \quad n = 0, 1, \ldots, N-1 \,. \tag{12.81}$$

Even if the signal elements y_k to be transformed are real, Z is complex, and therefore we must process both real and imaginary parts when computing transforms. Because both n and k range over N integer values, the $(Z^n)^k y_k$ multiplications in (12.81) require some N^2 multiplications and additions of complex numbers. As N gets large, as happens in realistic applications, this geometric increase in the number of steps leads to long computation times.

In 1965, Cooley and Tukey discovered an algorithm[7] that reduces the number of operations necessary to perform a DFT from N^2 to roughly $N \log_2 N$ (Cooley, 1965; Donnelly and Rust, 2005). Although this may not seem like such a big difference, it represents a 100-fold speedup for 1000 data points, which changes a full day of processing into 15 min of work. Because of its widespread use (including cell phones), the fast Fourier transform algorithm is considered one of the 10 most important algorithms of all time.

The idea behind the FFT is to utilize the periodicity inherent in the definition of the DFT (12.81) to reduce the total number of computational steps. Essentially, the algorithm divides the input data into two equal groups and transforms only one group, which requires $\sim (N/2)^2$ multiplications. It then divides the remaining (nontransformed) group of data in half and transforms them, continuing the process until all the data have been transformed. The total number of multiplications required with this approach is approximately $N \log_2 N$.

Specifically, the FFT's time economy arises from the computationally expensive complex factor $Z^{nk}[= ((Z)^n)^k]$ having values that are repeated as the integers n and k vary sequentially. For instance, for $N = 8$,

$$\begin{aligned}
Y_0 &= Z^0 y_0 + Z^0 y_1 + Z^0 y_2 + Z^0 y_3 + Z^0 y_4 + Z^0 y_5 + Z^0 y_6 + Z^0 y_7 \,, \\
Y_1 &= Z^0 y_0 + Z^1 y_1 + Z^2 y_2 + Z^3 y_3 + Z^4 y_4 + Z^5 y_5 + Z^6 y_6 + Z^7 y_7 \,, \\
Y_2 &= Z^0 y_0 + Z^2 y_1 + Z^4 y_2 + Z^6 y_3 + Z^8 y_4 + Z^{10} y_5 + Z^{12} y_6 + Z^{14} y_7 \,, \\
Y_3 &= Z^0 y_0 + Z^3 y_1 + Z^6 y_2 + Z^9 y_3 + Z^{12} y_4 + Z^{15} y_5 + Z^{18} y_6 + Z^{21} y_7 \,, \\
Y_4 &= Z^0 y_0 + Z^4 y_1 + Z^8 y_2 + Z^{12} y_3 + Z^{16} y_4 + Z^{20} y_5 + Z^{24} y_6 + Z^{28} y_7 \,, \\
Y_5 &= Z^0 y_0 + Z^5 y_1 + Z^{10} y_2 + Z^{15} y_3 + Z^{20} y_4 + Z^{25} y_5 + Z^{30} y_6 + Z^{35} y_7 \,, \\
Y_6 &= Z^0 y_0 + Z^6 y_1 + Z^{12} y_2 + Z^{18} y_3 + Z^{24} y_4 + Z^{30} y_5 + Z^{36} y_6 + Z^{42} y_7 \,, \\
Y_7 &= Z^0 y_0 + Z^7 y_1 + Z^{14} y_2 + Z^{21} y_3 + Z^{28} y_4 + Z^{35} y_5 + Z^{42} y_6 + Z^{49} y_7 \,,
\end{aligned}$$

7) Actually, this algorithm has been discovered a number of times, for instance, in 1942 by Danielson and Lanczos (Danielson and Lanczos , 1942), as well as much earlier by Gauss.

where we include $Z^0(\equiv 1)$ for clarity. When we actually evaluate these powers of Z, we find only four independent values:

$$
\begin{aligned}
Z^0 &= \exp(0) = +1\,, & Z^1 &= \exp(-\frac{2\pi}{8}) = +\frac{\sqrt{2}}{2} - i\frac{\sqrt{2}}{2}\,,\\
Z^2 &= \exp\left(-\frac{2\cdot 2i\pi}{8}\right) = -i\,, & Z^3 &= \exp\left(-\frac{2\pi\cdot 3i}{8}\right) = -\frac{\sqrt{2}}{2} - i\frac{\sqrt{2}}{2}\,,\\
Z^4 &= \exp\left(-\frac{2\pi\cdot 4i}{8}\right) = -Z^0\,, & Z^5 &= \exp\left(-\frac{2\pi\cdot 5i}{8}\right) = -Z^1\,,\\
Z^6 &= \exp\left(-\frac{2\cdot 6i\pi}{8}\right) = -Z^2\,, & Z^7 &= \exp\left(-\frac{2\cdot 7i\pi}{8}\right) = -Z^3\,,\\
Z^8 &= \exp\left(-\frac{2\pi\cdot 8i}{8}\right) = +Z^0\,, & Z^9 &= \exp\left(-\frac{2\pi\cdot 9i}{8}\right) = +Z^1\,,\\
Z^{10} &= \exp\left(-\frac{2\pi\cdot 10i}{8}\right) = +Z^2\,, & Z^{11} &= \exp\left(-\frac{2\pi\cdot 11i}{8}\right) = +Z^3\,,\\
Z^{12} &= \exp\left(-\frac{2\pi\cdot 11i}{8}\right) = -Z^0\,, & &\cdots
\end{aligned}
\tag{12.82}
$$

When substituted into the definitions of the transforms, we obtain

$$Y_0 = Z^0 y_0 + Z^0 y_1 + Z^0 y_2 + Z^0 y_3 + Z^0 y_4 + Z^0 y_5 + Z^0 y_6 + Z^0 y_7\,, \tag{12.83}$$

$$Y_1 = Z^0 y_0 + Z^1 y_1 + Z^2 y_2 + Z^3 y_3 - Z^0 y_4 - Z^1 y_5 - Z^2 y_6 - Z^3 y_7\,, \tag{12.84}$$

$$Y_2 = Z^0 y_0 + Z^2 y_1 - Z^0 y_2 - Z^2 y_3 + Z^0 y_4 + Z^2 y_5 - Z^0 y_6 - Z^2 y_7\,, \tag{12.85}$$

$$Y_3 = Z^0 y_0 + Z^3 y_1 - Z^2 y_2 + Z^1 y_3 - Z^0 y_4 - Z^3 y_5 + Z^2 y_6 - Z^1 y_7\,, \tag{12.86}$$

$$Y_4 = Z^0 y_0 - Z^0 y_1 + Z^0 y_2 - Z^0 y_3 + Z^0 y_4 - Z^0 y_5 + Z^0 y_6 - Z^0 y_7\,, \tag{12.87}$$

$$Y_5 = Z^0 y_0 - Z^1 y_1 + Z^2 y_2 - Z^3 y_3 - Z^0 y_4 + Z^1 y_5 - Z^2 y_6 + Z^3 y_7\,, \tag{12.88}$$

$$Y_6 = Z^0 y_0 - Z^2 y_1 - Z^0 y_2 + Z^2 y_3 + Z^0 y_4 - Z^2 y_5 - Z^0 y_6 + Z^2 y_7\,, \tag{12.89}$$

$$Y_7 = Z^0 y_0 - Z^3 y_1 - Z^2 y_2 - Z^1 y_3 - Z^0 y_4 + Z^3 y_5 + Z^2 y_6 + Z^1 y_7\,, \tag{12.90}$$

$$Y_8 = Y_0\,. \tag{12.91}$$

We see that these transforms now require $8 \times 8 = 64$ multiplications of complex numbers, in addition to some less time-consuming additions. We place these equations in an appropriate form for computing by regrouping the terms into sums and differences of the y's:

$$Y_0 = Z^0(y_0 + y_4) + Z^0(y_1 + y_5) + Z^0(y_2 + y_6) + Z^0(y_3 + y_7)\,, \tag{12.92}$$

$$Y_1 = Z^0(y_0 - y_4) + Z^1(y_1 - y_5) + Z^2(y_2 - y_6) + Z^3(y_3 - y_7)\,, \tag{12.93}$$

$$Y_2 = Z^0(y_0 + y_4) + Z^2(y_1 + y_5) - Z^0(y_2 + y_6) - Z^2(y_3 + y_7)\,, \tag{12.94}$$

$$Y_3 = Z^0(y_0 - y_4) + Z^3(y_1 - y_5) - Z^2(y_2 - y_6) + Z^1(y_3 - y_7) , \quad (12.95)$$

$$Y_4 = Z^0(y_0 + y_4) - Z^0(y_1 + y_5) + Z^0(y_2 + y_6) - Z^0(y_3 + y_7) , \quad (12.96)$$

$$Y_5 = Z^0(y_0 - y_4) - Z^1(y_1 - y_5) + Z^2(y_2 - y_6) - Z^3(y_3 - y_7) , \quad (12.97)$$

$$Y_6 = Z^0(y_0 + y_4) - Z^2(y_1 + y_5) - Z^0(y_2 + y_6) + Z^2(y_3 + y_7) , \quad (12.98)$$

$$Y_7 = Z^0(y_0 - y_4) - Z^3(y_1 - y_5) - Z^2(y_2 - y_6) - Z^1(y_3 - y_7) , \quad (12.99)$$

$$Y_8 = Y_0 . \quad (12.100)$$

Note the repeating factors inside the parentheses, with combinations of the form $y_p \pm y_q$. These symmetries are systematized by introducing the *butterfly operation* (Figure 12.8). This operation takes the y_p and y_q data elements from the left wing and converts them to the $y_p + Zy_q$ elements in the right wings. In Figure 12.9 we show what happens when we apply the butterfly operations to an entire FFT process, specifically to the pairs (y_0, y_4), (y_1, y_5), (y_2, y_6), and (y_3, y_7). Note how the number of multiplications of complex numbers has been reduced: For the first butterfly operation there are 8 multiplications by Z^0; for the second butterfly operation there are 8 multiplications, and so forth, until a total of 24 multiplications are made in four butterflies. In contrast, 64 multiplications are required in the original DFT (12.91).

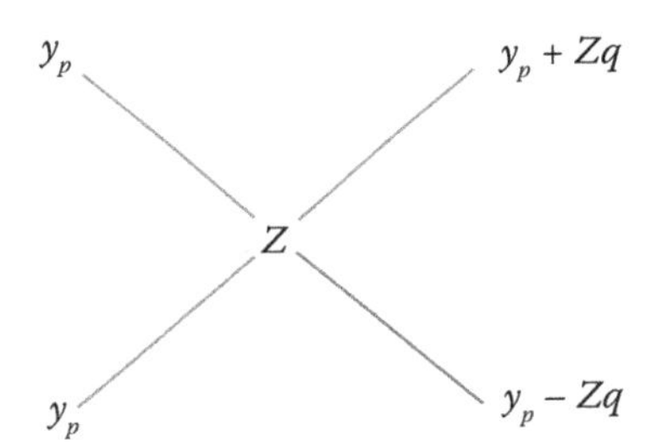

Figure 12.8 The basic butterfly operation in which elements y_p and y_q on the left are transformed into $y_p + Zy_q$ and $y_p - Zy_q$ on the right.

12.9.1 Bit Reversal

The reader may have observed in Figure 12.9 that we started with eight data elements in the order 0–7 and that after three butterfly operators we obtained transforms in the order 0, 4, 2, 6, 1, 5, 3, 7. The astute reader may further have observed that these numbers correspond to the bit-reversed order of 0–7. Let us look into this further. We need 3 bits to give the order of each of the 8 input data elements (the numbers 0–7). Explicitly, on the left in Table 10.1 we give the binary representation for decimal numbers 0–7, their bit reversals, and the corresponding

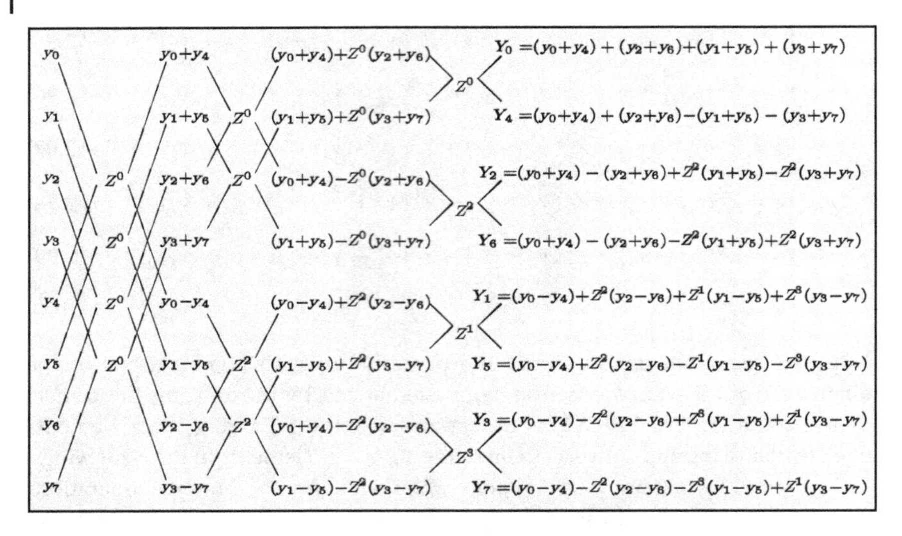

Figure 12.9 The butterfly operations performing a FFT on the eight data on the left leading to eight transforms on the right. The transforms are different linear combinations of the input data.

decimal numbers. On the right we give the ordering for 16 input data elements, where we need 4 bits to enumerate their order. Notice that the order of the first 8 elements differs in the two cases because the number of bits being reversed differs. Also note that after the reordering, the first half of the numbers are all even and the second half are all odd.

The fact that the Fourier transforms are produced in an order corresponding to the bit-reversed order of the numbers 0–7 suggests that if we process the data in the bit-reversed order 0, 4, 2, 6, 1, 5, 3, 7, then the output Fourier transforms will be ordered (see Table 10.1). We demonstrate this conjecture in Figure 12.10, where we see that to obtain the Fourier transform for the eight input data, the butterfly operation had to be applied three times. The number 3 occurs here because it is the power of 2 that gives the number of data; that is, $2^3 = 8$. In general, in order for a FFT algorithm to produce transforms in the proper order, it must reshuffle the input data into bit-reversed order. As a case in point, our sample program starts by reordering the 16 (2^4) data elements given in Table 12.1. Now the four butterfly operations produce sequentially ordered output.

Table 12.1 Reordering for 16 data complex points.

Order	Input data	New order	Order	Input data	New order
0	0.0 + 0.0i	0.0 + 0.0i	8	8.0 + 8.0i	1.0 + 1.0i
1	1.0 + 1.0i	8.0 + 8.0i	9	9.0 + 9.0i	9.0 + 9.0i
2	2.0 + 2.0i	4.0 + 4.0i	10	10.0 + 10.i	5.0 + 5.0i
3	3.0 + 3.0i	12.0 + 12.0i	11	11.0 + 11.0i	13.0 + 13.0i
4	4.0 + 4.0i	2.0 + 2.0i	12	12.0 + 12.0i	3.0 + 3.0i
5	5.0 + 5.0i	10.0 + 10.i	13	13.0 + 13.0i	11.0 + 11.0i
6	6.0 + 6.0i	6.0 + 6.0i	14	14.0 + 14.i	7.0 + 7.0i
7	7.0 + 7.0i	14.0 + 14.0i	15	15.0 + 15.0i	15.0 + 15.0i

		Binary-Reversed 0–7		Binary-Reversed 0–16	
Dec	Bin	Rev	Dec Rev	Rev	Dec Rev
0	000	000	0	0000	0
1	001	100	4	1000	8
2	010	010	2	0100	4
3	011	110	6	1100	12
4	100	001	1	0010	2
5	101	101	5	1010	10
6	110	011	3	0110	6
7	111	111	7	1110	14
8	1000	—	—	0001	1
9	1001	—	—	1001	9
10	1010	—	—	0101	5
11	1011	—	—	1101	13
12	1100	—	—	0011	3
13	1101	—	—	1011	11
14	1101	—	—	0111	7
15	1111	—	—	1111	15

12.10 FFT Implementation

The first FFT program we are aware of was written in Fortran IV by Norman Brenner at MIT's Lincoln Laboratory (Higgins, 1976) and was hard for us to follow. Our (easier-to-follow) Python version is in Listing 12.3. Its input is $N = 2^n$ data to be transformed (FFTs always require that the number of input data are a power of 2). If the number of your input data is not a power of 2, then you can make it so by concatenating some of the initial data to the end of your input until a power of

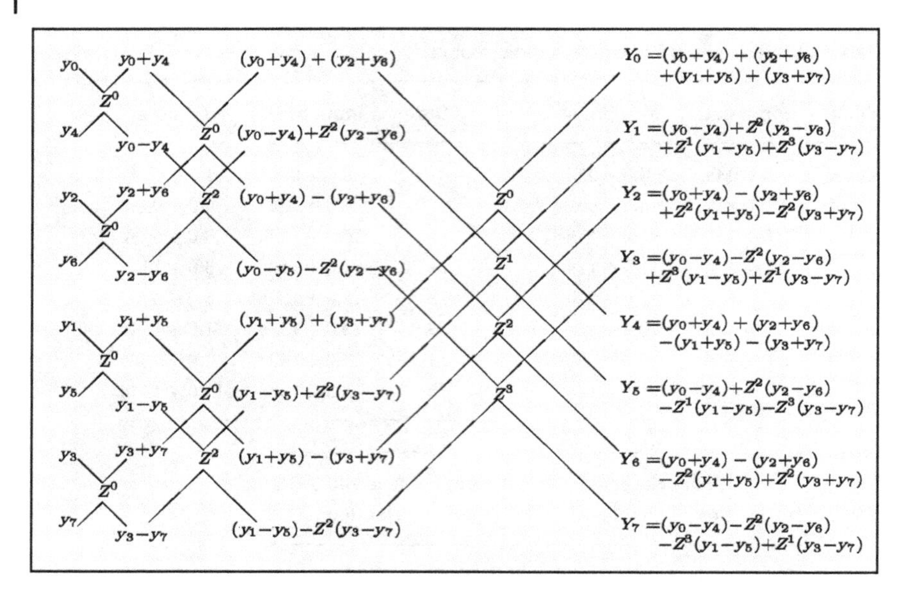

Figure 12.10 A modified FFT in which the eight input data on the left are transformed into eight transforms on the right. The results are the same as in the previous figure, but now the output transforms are in numerical order whereas in the previous figure the input signals were in numerical order.

2 is obtained; because a DFT is always periodic, this just starts the period a little earlier. Our program assigns complex numbers at the 16 data points

$$y_m = m + mi\,, \quad m = 0, \dots, 15\,, \tag{12.101}$$

reorders the data via bit reversal, and then makes four butterfly operations. The data are stored in the array `dtr[max,2]`, with the second subscript denoting real and imaginary parts. We increase the speed further by using the 1D array `data` to make memory access more direct:

$$\mathtt{data}[1] = \mathtt{dtr}[0,1], \quad \mathtt{data}[2] = \mathtt{dtr}[1,1], \quad \mathtt{data}[3] = \mathtt{dtr}[1,0], \dots\,, \tag{12.102}$$

which also provides storage for the output. The FFT transforms `data` using the butterfly operation and stores the results back in `dtr[,]`, where the input data were originally.

12.11 FFT Assessment

1. Compile and execute `FFT.py`. Make sure you understand the output.
2. Take the output from `FFT.py`, inverse-transform it back to signal space, and compare it to your input. (Checking that the double transform is proportional

to itself is adequate, although the normalization factors in (12.37) should make the two equal.)

3. Compare the transforms obtained with a FFT to those obtained with a DFT (you may choose any of the functions studied before). Make sure to compare both precision and execution times.

Listing 12.3 FFT.py computes the FFT or inverse transform depending upon the sign of **isign**.

```
# FFT.py: FFT for complex numbers in dtr[][2], returned in dtr

from numpy import *
from sys import version
max = 2100
points = 1026                                          # Can be increased
data = zeros((max), float)
dtr  = zeros((points,2), float)

def fft(nn,isign):                                     # FFT of dtr[n,2]
    n = 2*nn
    for i in range(0,nn+1):                # Original data in dtr to data
         j = 2*i+1
         data[j] = dtr[i,0]                        # Real dtr, odd data[j]
         data[j+1] = dtr[i,1]                  # Imag dtr, even data[j+1]
    j = 1                                 # Place data in bit reverse order
    for i in range(1,n+2, 2):
        if (i-j) < 0 :                   # Reorder equivalent to bit reverse
            tempr = data[j]
            tempi = data[j+1]
            data[j] = data[i]
            data[j+1] = data[i+1]
            data[i] = tempr
            data[i+1] = tempi
        m = n/2;
        while (m-2 > 0):
            if  (j-m) <= 0 :
                break
            j = j-m
            m = m/2
        j = j+m;

    print(" Bit-reversed data ")

    for i in range(1, n+1, 2):
        print("%2d  data[%2d]  %9.5f "%(i,i,data[i]))         # To see reorder
    mmax = 2
    while (mmax-n) < 0 :                                  # Begin transform
       istep = 2*mmax
       theta = 6.2831853/(1.0* isign *mmax)
       sinth = math.sin(theta/2.0)
       wstpr = -2.0*sinth**2
       wstpi = math.sin(theta)
       wr = 1.0
       wi = 0.0
       for m in range(1,mmax +1,2):
           for i in range(m,n+1,istep):
               j = i+mmax
               tempr = wr*data[j]    -wi *data[j+1]
               tempi = wr*data[j+1] +wi *data[j]
               data[j]   = data[i]   -tempr
               data[j+1] = data[i+1] -tempi
               data[i]   = data[i]   +tempr
               data[i+1] = data[i+1] +tempi
           tempr = wr
```

```
            wr = wr*wstpr - wi*wstpi + wr
            wi = wi*wstpr + tempr*wstpi + wi;
         mmax = istep
    for i in range(0,nn):
        j = 2*i+1
        dtr[i,0] = data[j]
        dtr[i,1] = data[j+1]
nn = 16                                                    # Power of 2
isign = -1                          # -1 transform, +1 inverse transform
print('          INPUT')
print("  i   Re part    Im   part")
for i in range(0,nn ):                                     # Form array
    dtr[i,0] = 1.0*i                                       # Real part
    dtr[i,1] = 1.0*i                                         # Im part
    print(" %2d %9.5f %9.5f" %(i,dtr[i,0],dtr[i,1]))
fft(nn, isign)                         # Call FFT, use global dtr[][]
print('     Fourier transform')
print("  i        Re        Im      ")
for i in range(0,nn):
    print(" %2d  %9.5f  %9.5f "%(i,dtr[i,0],dtr[i,1]))
print("Enter and return any character to quit")
```

13
Wavelet and Principal Components Analyses: Nonstationary Signals and Data Compression

There are a number of techniques that extend Fourier analysis to signals whose forms change in time. This chapter introduces *wavelet analysis*, a field that has seen extensive development and application in areas as diverse as brain waves, stock-market trends, gravitational waves, and compression of photographic images. The first part of the chapter deals with wavelet basics, and covers the essential materials. The second part of the chapter explores the discrete wavelet transform, and is marked as optional as a result of its technical nature. However, it is a beautiful bit of analysis and the basis of much of the digital revolution. The chapter ends with a discussion of principal component analysis, another powerful technique for analyzing signals with space and time correlations.

13.1
Problem: Spectral Analysis of Nonstationary Signals

Problem You have sampled the signal in Figure 13.1 that seems to contain an increasing number of frequencies as time increases. Your *problem* is to undertake a spectral analysis of this signal that tells you, in the most compact way possible, the amount of each frequency present at each instant of time. *Hint:* Although we want the method to be general enough to work with numerical data, for pedagogical purposes it is useful to know that the signal is

$$y(t) = \begin{cases} \sin 2\pi t\,, & \text{for } 0 \le t \le 2\,, \\ 5\sin 2\pi t + 10\sin 4\pi t\,, & \text{for } 2 \le t \le 8\,, \\ 2.5\sin 2\pi t + 6\sin 4\pi t + 10\sin 6\pi t\,, & \text{for } 8 \le t \le 12\,. \end{cases} \tag{13.1}$$

13.2
Wavelet Basics

The Fourier analysis we used in Chapter 12 reveals the amount of the harmonic functions $\sin(\omega t)$ and $\cos(\omega t)$, and their overtones, that are present in a signal. An

Computational Physics, 3rd edition. Rubin H. Landau, Manuel J. Páez, Cristian C. Bordeianu.

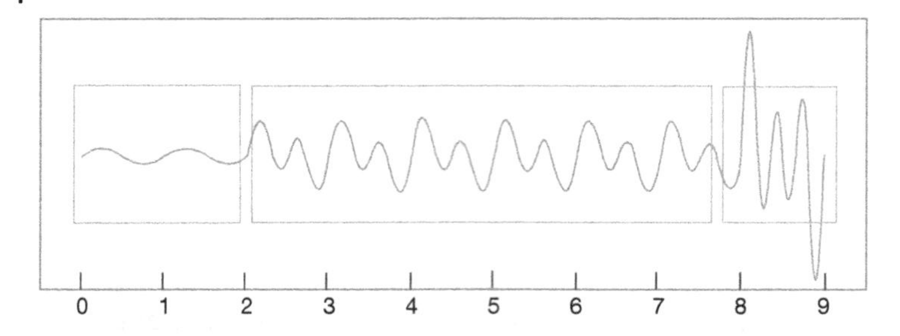

Figure 13.1 The input time signal (13.1) we wish to analyze. The signal is seen to contain additional frequencies as time increases. The boxes are possible placements of windows for short-time Fourier transforms.

expansion in periodic functions is fine for *stationary* signals (those whose forms do not change in time) but has shortcomings for the variable form of our *problem* signal (13.1). One such problem is that the Fourier reconstruction has all constituent frequencies occurring simultaneously and so does not contain *time resolution* information indicating when each frequency occurs. Another shortcoming is that all the Fourier components are correlated, which results in more information being stored than may be needed to reconstruct the measured signal.

There are a number of techniques that extend simple Fourier analysis to nonstationary signals. The idea behind wavelet analysis is to expand a signal in a complete set of functions (wavelets), each of which oscillates for a finite period of time, and each of which is centered at a different time. To give you a preview before we get into the details, we show four sample wavelets in Figure 13.2. Because each wavelet is local in time, it is a wave packet,[1] with its time localization leading to a spectrum with a range of frequencies. These wave packets are called "wavelets" because they exist for only short periods of time (Polikar, 2001).

Although wavelets are required to oscillate in time, they are not restricted to a particular functional form (Addison, 2002; Goswani and Chan, 1999; Graps, 1995). As a case in point, they may be oscillating Gaussians (Morlet: in Figure 13.2a),

$$\Psi(t) = e^{2\pi i t} e^{-t^2/2\sigma^2} = (\cos 2\pi t + i \sin 2\pi t) e^{-t^2/2\sigma^2} \quad \text{(Morlet)} , \tag{13.2}$$

the second derivative of a Gaussian (Mexican hat, Figure 13.2b),

$$\Psi(t) = -\sigma^2 \frac{d^2}{dt^2} e^{-t^2/2\sigma^2} = \left(1 - \frac{t^2}{\sigma^2}\right) e^{-t^2/2\sigma^2} , \tag{13.3}$$

an up-and-down step function (Figure 13.2c), or a fractal shape (Figure 13.2d). All of these wavelets are *localized* in both time and frequency, that is, they are large for just a limited time and contain a limited range of frequencies. As we shall see, translating and scaling a *mother wavelets* generates an entire set of *daughter wavelet* or basis functions, with the daughters covering different frequency ranges at different times.

1) We discuss wave packets further in Section 13.3.

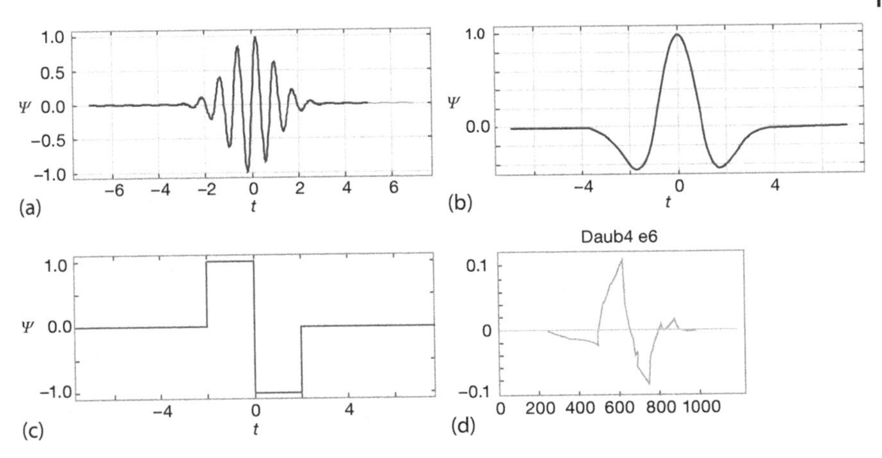

Figure 13.2 Four possible mother wavelets that can be used to generate entire sets of daughter wavelets. (a–d) Morlet (real part), Mexican hat, Daub4 e6 (explained later), and Haar. The daughter wavelets are generated by scaling and translating these mother wavelets.

13.3 Wave Packets and Uncertainty Principle (Theory)

A *wave packet* or *wave train* is a collection of waves of differing frequencies added together in such a way as to produce a pulse of width Δt. As we shall see, the Fourier transform of a wave packet is a pulse in the frequency domain of width $\Delta\omega$. We will first study such wave packets analytically, and then use others numerically. An example of a simple wave packet is just a sine wave that oscillates at frequency ω_0 for N periods (Figure 13.3a) (Arfken and Weber, 2001)

$$y(t) = \begin{cases} \sin \omega_0 t\,, & \text{for} \quad |t| < N\frac{\pi}{\omega_0} \equiv N\frac{T}{2}\,, \\ 0\,, & \text{for} \quad |t| > N\frac{\pi}{\omega_0} \equiv N\frac{T}{2}\,, \end{cases} \tag{13.4}$$

where we relate the frequency to the period via the usual $\omega_0 = 2\pi/T$. In terms of these parameters, the width of the wave packet is

$$\Delta t = NT = N\frac{2\pi}{\omega_0}\,. \tag{13.5}$$

The Fourier transform of the wave packet (13.4) is calculated via a straightforward application of the transform formula (12.19):

$$\begin{aligned} Y(\omega) &= \int_{-\infty}^{+\infty} dt \frac{e^{-i\omega t}}{\sqrt{2\pi}} y(t) = \frac{-i}{\sqrt{2\pi}} \int_0^{N\pi/\omega_0} dt \sin \omega_0 t \sin \omega t \\ &= \frac{(\omega_0 + \omega)\sin\left[(\omega_0 - \omega)\frac{N\pi}{\omega_0}\right] - (\omega_0 - \omega)\sin\left[(\omega_0 + \omega)\frac{N\pi}{\omega_0}\right]}{\sqrt{2\pi}(\omega_0^2 - \omega^2)}\,, \end{aligned} \tag{13.6}$$

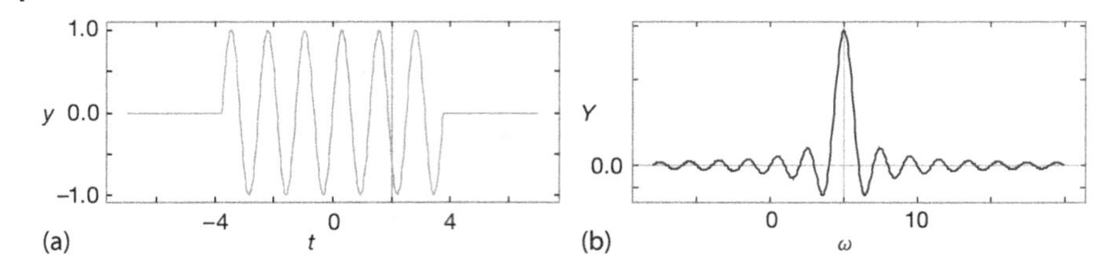

Figure 13.3 (a) A wave packet in time corresponding to the functional form (13.4) with $\omega_0 = 5$ and $N = 6$. (b) The Fourier transform in frequency of this same wave packet.

where we have dropped a factor of $-i$ that affects only the phase. While at first glance (13.6) appears to be singular at $\omega = \omega_0$, it actually just peaks there (Figure 13.3b), reflecting the predominance of frequency ω_0. Note that although the signal $y(t)$ appears to have only one frequency, it does drop off sharply in time (Figure 13.3a), and these corners give $Y(\omega)$ a width $\Delta\omega$.

There is a fundamental relation between the widths Δt and $\Delta\omega$ of a wave packet. Although we use a specific example to determine that relation, it is true in general. While there may not be a precise definition of "width" for all functions, one can usually deduce a good measure of the width (say, within 25%). To illustrate, if we look Figure 13.3b, it makes sense to use the distance between the first zeros of the transform $Y(\omega)$ (13.6) as the frequency width $\Delta\omega$. The zeros occur at

$$\frac{\omega - \omega_0}{\omega_0} = \pm\frac{1}{N} \Rightarrow \quad \Delta\omega \simeq \omega - \omega_0 = \frac{\omega_0}{N}, \tag{13.7}$$

where N is the number of cycles in our original wave packet. Because the wave packet in time makes N oscillations each of period T, a reasonable measure of the time width Δt of the signal $y(t)$ is

$$\Delta t = NT = N\frac{2\pi}{\omega_0}. \tag{13.8}$$

When the products of the frequency width (13.7) and the time width (13.8) are combined, we obtain

$$\Delta t \Delta\omega \geq 2\pi. \tag{13.9}$$

The greater than sign is used to indicate that this is a minimum, that is, that $y(t)$ and $Y(\omega)$ extend beyond Δt and $\Delta\omega$, respectively. Nonetheless, most of the signal and transform should lie within the bound (13.9).

A relation of the form (13.9) also occurs in quantum mechanics, where it is known as the *Heisenberg uncertainty principle*, with Δt and $\Delta\omega$ being called the uncertainties in t and ω. It is true for transforms in general and states that as a signal is made more localized in time (smaller Δt) the transform becomes less localized (larger $\Delta\omega$). Conversely, the sine wave $y(t) = \sin\omega_0 t$ is completely localized in frequency, and consequently has an infinite extent in time, $\Delta t \simeq \infty$.

13.3.1 Wave Packet Assessment

Consider the following wave packets:

$$y_1(t) = e^{-t^2/2}, \quad y_2(t) = \sin(8t)e^{-t^2/2}, \quad y_3(t) = (1 - t^2)e^{-t^2/2}. \tag{13.10}$$

For each wave packet:

1. Estimate the width Δt. A good measure might be the *full width at half-maxima* (FWHM) of $|y(t)|$.
2. Use your DFT program to evaluate and plot the Fourier transform $Y(\omega)$ for each wave packet. Make *both* a linear and a semilog plot (small components are often important, yet not evident in linear plots). Make sure that your transform has a good number of closely spaced frequency values over a range that is large enough to show the periodicity of $Y(\omega)$.
3. What are the units for $Y(\omega)$ and ω in your DFT?
4. For each wave packet, estimate the width $\Delta\omega$. A good measure might be the *full width at half-maxima* of $|Y(\omega)|$.
5. For each wave packet determine approximate value for the constant C of the uncertainty principle

$$\Delta t \Delta\omega \geq 2\pi C. \tag{13.11}$$

13.4 Short-Time Fourier Transforms (Math)

The constant amplitude of the functions $\sin n\omega t$ and $\cos n\omega t$ for all times can limit the usefulness of Fourier analysis for reproducing signals. Because these functions and their overtones extend over all times with a constant amplitude, there is considerable overlap among them, and consequently the information present in various Fourier components are correlated. This is undesirable for data storage and compression, where you want to store a minimum number of data information and also want to adjust the amount stored based on the desired quality of the reconstructed signal.[2] In *lossless compression*, which exactly reproduces the original signal, you save space by storing how many times each data element is repeated, and where each element is located. In *lossy compression*, in addition to removing repeated elements, you also eliminate some transform components consistent with the uncertainty relation (13.9) and with the level of resolution required in the reproduction. This leads to yet greater compression.

2) Wavelets have proven to be a highly effective approach to data compression, with the Joint Photographic Experts Group (JPEG) 2000 standard being based on wavelets.

In Section 12.5, we defined the Fourier transform $Y(\omega)$ of signal $y(t)$ as

$$Y(\omega) = \int_{-\infty}^{+\infty} dt \frac{e^{-i\omega t}}{\sqrt{2\pi}} y(t) \equiv \langle \omega | y \rangle . \tag{13.12}$$

As is true for simple vectors, you can think of (13.12) as giving the overlap or scalar product of the basis function $\exp(i\omega t)/\sqrt{2\pi}$ and the signal $y(t)$ (notice that the complex conjugate of the exponential basis function appears in (13.12)). Another view of (13.12) is as the mapping or projection of the signal y into ω space. In this latter case the overlap projects out the amount of the periodic function $\exp(i\omega t)/\sqrt{2\pi}$ in the signal. In other words, the Fourier component $Y(\omega)$ is also the correlation between the signal $y(t)$ and the basis function $\exp(i\omega t)/\sqrt{2\pi}$, which is the same as what results from filtering the signal $y(t)$ through a frequency-ω filter. If there is no $\exp(i\omega t)$ in the signal, then the integral vanishes and there is no output. If $y(t) = \exp(i\omega t)$, the signal is at only one frequency, and the integral is accordingly singular.

The signal in Figure 13.1 for our problem clearly has different frequencies present at different times and for different lengths of time. In the past, this signal might have been analyzed with a precursor of wavelet analysis known as the *short-time Fourier transform*. With that technique, the signal $y(t)$ is "chopped up" into different segments along the time axis, with successive segments centered about successive times $\tau_1, \tau_2, \dots, \tau_N$. For instance, we show three such segments in the boxes of Figure 13.1. Once we have the dissected signal, a Fourier analysis is made for each segment. We are then left with a sequence of transforms $(Y^{(ST)}_{\tau_1}, Y^{(ST)}_{\tau_2}, \dots, Y^{(ST)}_{\tau_N})$, one for each short-time interval, where the superscript $^{(ST)}$ indicates short time.

Rather than chopping up a signal, we can express short-time Fourier transforming mathematically by imagining translating a *window function* $w(t-\tau)$, which is zero outside of some chosen interval, over the signal in Figure 13.1:

$$Y^{(ST)}(\omega, \tau) = \int_{-\infty}^{+\infty} dt \frac{e^{i\omega t}}{\sqrt{2\pi}} w(t-\tau)\, y(t) . \tag{13.13}$$

Here the values of the translation time τ correspond to different locations of the window w over the signal, and the window function is essentially a transparent box of small size on an opaque background. Any signal within the width of the window is transformed, while the signal lying outside the window is not seen. Note that in (13.13), the extra variable τ in the Fourier transform indicates the location of the time around which the window was placed. Clearly, because the short-time transform is a function of two variables, a surface or 3D plot is needed to view the amplitude as a function of both ω and τ.

13.5 The Wavelet Transform

The wavelet transform of a time signal $y(t)$ is defined as

$$Y(s,\tau) = \int_{-\infty}^{+\infty} dt \psi^*_{s,\tau}(t) y(t) \quad \text{(wavelet transform)} , \tag{13.14}$$

and is similar in concept and notation to a short-time Fourier transform. The difference is rather than using $\exp(i\omega t)$ as the basis functions, here we are using wave packets or wavelets $\psi_{s,\tau}(t)$ localized in time, such as the those shown in Figure 13.2. Because each wavelet is localized in time, each acts as its own window function. Because each wavelet is oscillatory, each contains its own small range of frequencies.

Equation 13.14 says that the wavelet transform $Y(s,\tau)$ is a measure of the amount of basis function $\psi_{s,\tau}(t)$ present in the signal $y(t)$. The τ variable indicates the time portion of the signal being decomposed, while the s variable is equivalent to the frequency present during that time:

$$\omega = \frac{2\pi}{s} , \quad s = \frac{2\pi}{\omega} \quad \text{(scale-frequency relation)} . \tag{13.15}$$

Because it is key to much that follows, it is a good idea to think about (13.15) for a while. If we are interested in the time *details* of a signal, then this is another way of saying that we are interested in what is happening at small values of the *scale s*. Equation 13.15 indicates that small values of s correspond to high-frequency components of the signal. That being the case, the time details of the signal are in the high-frequency, or low-scale, components.

13.5.1 Generating Wavelet Basis Functions

The conceptual discussion of wavelets is over, and it is time to get down to work. We first need a technique for generating wavelet basis functions, and then we need to discretize this technique. As is often the case, the final formulation will turn out to be simple and short, but it will be a while before we get there.

Just as the expansion of an arbitrary function in a complete set of orthogonal functions is not restricted to any particular set, so too is the wavelet transform not restricted to any particular wavelet basis, although some might be better than others for a given signal. The standard way to generate a family of wavelet basis functions starts with $\Psi(t)$, a *mother* or *analyzing* function of the real variable t, and then to use this to generate *daughter* wavelets. As a case in point, we start with the mother wavelet

$$\Psi(t) = \sin(8t)e^{-t^2/2} . \tag{13.16}$$

By scaling, translating, and normalizing this mother wavelet,

$$\psi_{s,\tau}(t) \stackrel{\text{def}}{=} \frac{1}{\sqrt{s}} \Psi\left(\frac{t-\tau}{s}\right) = \frac{1}{\sqrt{s}} \sin\left[\frac{8(t-\tau)}{s}\right] e^{-(t-\tau)^2/2s^2} , \tag{13.17}$$

we generate the four wavelet basis functions (daughters) displayed in Figure 13.4. We see that larger or smaller values of s, respectively, expand or contract the mother wavelet, while different values of τ shift the center of the wavelet. Because the wavelets are inherently oscillatory, the scaling leads to the same number of oscillations occurring in different time spans, which is equivalent to having basis states with differing frequencies. We see that $s < 1$ produces a higher frequency wavelet, while $s > 1$ produces a lower frequency one, both of the same shape. As we shall see, we do not need to store much information to outline the large-time-scale s behavior of a signal (its *smooth envelope*), but we do need more information to specify its short-time-scale s behavior (*details*). And if we want to resolve yet finer features in the signal, then we will need to have more information on yet finer details. Here the division by $\sqrt{s}$ is made to ensure that there is equal "power" (or energy or intensity) in each region of s, although other normalizations can also be found in the literature. After substituting in the definition of daughters, the wavelet transform (13.14) and its inverse (van den Berg, 1999) are

$$\boxed{Y(s,\tau) = \frac{1}{\sqrt{s}} \int_{-\infty}^{+\infty} \mathrm{d}t\, \Psi^*\left(\frac{t-\tau}{s}\right) y(t) \quad \text{(Wavelet Transform)} ,} \tag{13.18}$$

$$\boxed{y(t) = \frac{1}{C} \int_{-\infty}^{+\infty} \mathrm{d}\tau \int_{0}^{+\infty} \mathrm{d}s \frac{\psi^*_{s,\tau}(t)}{s^{3/2}} Y(s,\tau) \quad \text{(Inverse Transform)} ,} \tag{13.19}$$

where the normalization constant C depends on the wavelet used. In summary, wavelet bases are functions of the time variable t, as well as of the two parameters s and τ. The t variable is integrated over to yield a transform that is a function of the time scale s (frequency $2\pi/s$) and window location τ. You can think of scale as being like the scale on a map (also discussed in Section 16.5.2 in relation to fractal analysis) or in terms of *resolution*, as might occur in photographic images. Regardless of the words, as we see in Chapter 16, if we have a fractal, then we have a self-similar object that looks the same at all scales or resolutions. Similarly, each wavelet in a set of basis functions is self-similar to the others, but at a different scale or location. The general requirements for a mother wavelet Ψ are (Addison, 2002; van den Berg, 1999)

1. $\Psi(t)$ is real.
2. $\Psi(t)$ oscillates around zero such that its average is zero:

$$\int_{-\infty}^{+\infty} \Psi(t)\, \mathrm{d}t = 0 . \tag{13.20}$$

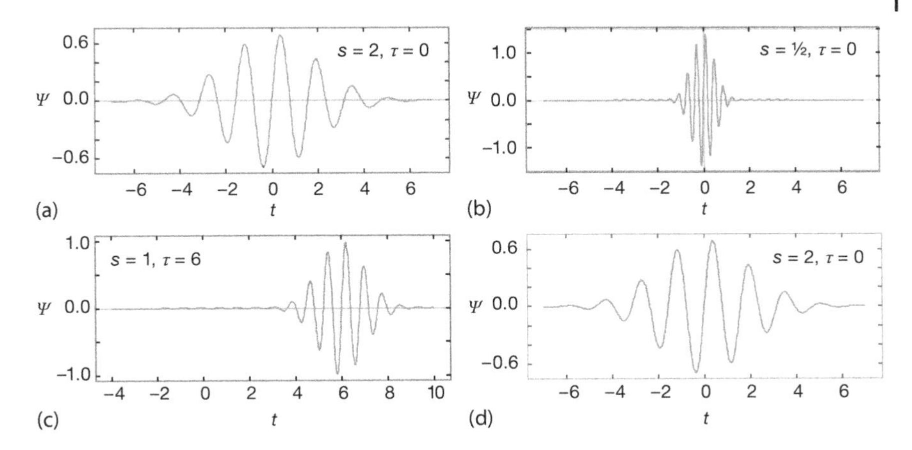

Figure 13.4 Four wavelet basis functions (daughters) generated by scaling (s) and translating (τ) an oscillating Gaussian mother wavelet. (a–d) ($s = 1, \tau = 0$), ($s = 1/2, \tau = 0$), ($s = 1, \tau = 6$), and ($s = 2, \tau = 60$). Note how $s < 1$ is a wavelet with higher frequency, while $s > 1$ has a lower frequency than the $s = 1$ mother. Likewise, the $\tau = 6$ wavelet is just a translated version of the $\tau = 0$ one directly above it.

3. $\Psi(t)$ is local, that is, a wave packet, and is square integrable:

$$\Psi(|t| \to \infty) \to 0 \quad \text{(rapidly)}, \qquad \int_{-\infty}^{+\infty} |\Psi(t)|^2 \, \mathrm{d}t < \infty . \tag{13.21}$$

4. The transforms of low powers of t vanish, that is, the first p moments:

$$\int_{-\infty}^{+\infty} t^0 \, \Psi(t) \, \mathrm{d}t = \int_{-\infty}^{+\infty} t^1 \, \Psi(t) \, \mathrm{d}t = \cdots = \int_{-\infty}^{+\infty} t^{p-1} \, \Psi(t) \, \mathrm{d}t = 0 . \tag{13.22}$$

This makes the transform more sensitive to details than to general shape.

As an example of how we use the s and τ degrees of freedom in a wavelet transform, consider the analysis of a chirp signal $y(t) = \sin(60t^2)$ (Figure 13.5). We see that a slice at the beginning of the signal is compared to our first basis function. (The comparison is carried out via the *convolution* of the wavelet with the signal.) This first comparison is with a narrow version of the wavelet, that is, at low scale, and yields a single coefficient. The comparison at this scale continues with the next signal slice, and eventually ends when the entire signal has been covered (the top row in Figure 13.5). Then in the second row, the wavelet is expanded to larger s values, and comparisons are repeated. Eventually, the data are processed at all scales and at all time intervals. The narrow wavelets correspond to a high-resolution analysis, while the broad wavelets correspond to low resolution. As the scales get larger (lower frequencies, lower resolution), fewer details of the time signal remain visible, but the overall shape or gross features of the signal become clearer.

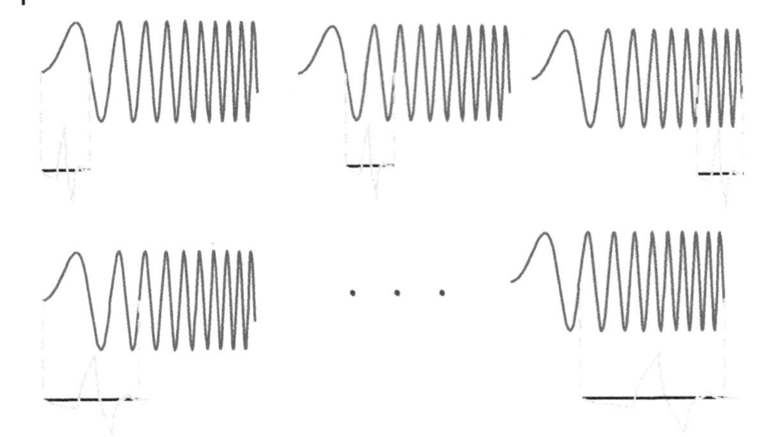

Figure 13.5 A schematic representation of the steps followed in performing a wavelet transformation over all time displacements and scales. The upper signal is first analyzed by evaluating its overlap with a narrow wavelet at the signal's beginning. This produces a coefficient that measures the similarity of the signal to the wavelet. The wavelet is successively shifted over the length of the signal and the overlaps are successively evaluated. After the entire signal is covered, the wavelet is expanded and the entire analysis is repeated.

13.5.2
Continuous Wavelet Transform Implementation

We want to develop some intuition as to what wavelet transforms look like before going on to apply them in unknown situations and to develop a discrete algorithm. Accordingly, modify the program you have been using for the Fourier transform so that it now computes the continuous wavelet transform.

1. You will want to see the effect of using different mother wavelets. Accordingly, write a method that calculates the mother wavelet for
   a) a Morlet wavelet (13.2),
   b) a Mexican hat wavelet (13.3),
   c) a Haar wavelet (the square wave in Figure 13.2).
2. Try out your transform for the following input signals and see if the results make sense:
   a) A pure sine wave $y(t) = \sin 2\pi t$,
   b) A sum of sine waves $y(t) = 2.5 \sin 2\pi t + 6 \sin 4\pi t + 10 \sin 6\pi t$,
   c) The nonstationary signal for our problem (13.1)

$$y(t) = \begin{cases} \sin 2\pi t\,, & \text{for } 0 \le t \le 2\,, \\ 5 \sin 2\pi t + 10 \sin 4\pi t\,, & \text{for } 2 \le t \le 8\,, \\ 2.5 \sin 2\pi t + 6 \sin 4\pi t + 10 \sin 6\pi t\,, & \text{for } 8 \le t \le 12\,. \end{cases} \tag{13.23}$$

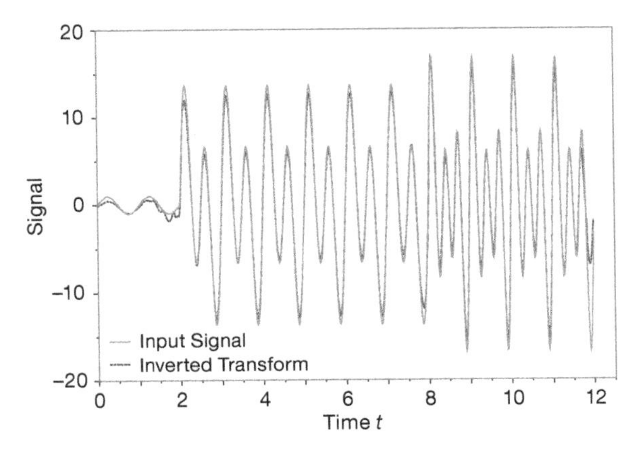

Figure 13.6 Comparison of an input and reconstituted signal (13.23) using Morlet wavelets (the curves overlap nearly perfectly). As expected for Fourier transforms, the reconstruction is least accurate near the endpoints.

d) The half-wave function

$$y(t) = \begin{cases} \sin \omega t\,, & \text{for } 0 < t < \frac{T}{2}\,, \\ 0\,, & \text{for } \frac{T}{2} < t < T\,. \end{cases} \tag{13.24}$$

3. ⊙ Use (13.19) to invert your wavelet transform and compare the reconstructed signal to the input signal (you can normalize the two to each other). In Figure 13.6 we show our reconstruction.

In Listing 11.1, we give our *continuous wavelet transformation* `CWT.py` (Lang and Forinash, 1998). Because wavelets, with their transforms in two variables, are somewhat hard to grasp at first, we suggest that you write your own code and include a portion that does the inverse transform as a check. In the next section we will describe the *discrete wavelet transformation* that makes optimal discrete choices for the scale and time translation parameters s and τ. Figure 13.7 shows a surface plot of the spectrum produced for the input signal (13.1) in Figure 13.1. In realization of our goal, we see predominantly one frequency at short times, two frequencies at intermediate times, and three frequencies at longer times.

13.6 Discrete Wavelet Transforms, Multiresolution Analysis ⊙

As was true for DFTs, if a time signal is measured at only N discrete times,

$$y(t_m) \equiv y_m\,, \quad m = 1, \dots, N\,, \tag{13.25}$$

then we can determine only N independent components of the transform Y. The trick with wavelets is to remain consistent with the uncertainty principle as we

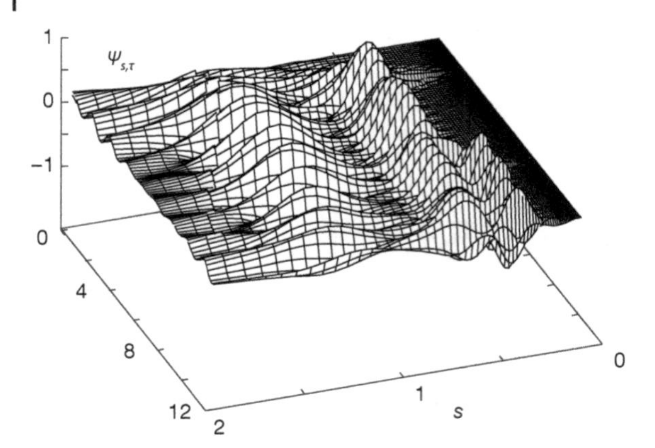

Figure 13.7 The continuous wavelet spectrum obtained by analyzing the input signal with Morelet wavelets. Observe how at small values of time τ there is predominantly one frequency present, how a second, higher frequency (smaller scale) component enters at intermediate times, and how at larger times a still higher frequency components enter (figure courtesy of Z. Dimcovic).

compute only the N independent components required to reproduce the signal. The *discrete wavelet transform* (DWT) evaluates the transforms with discrete values for the scaling parameter s and the time translation parameter τ:

$$\psi_{j,k}(t) = \frac{\Psi\left[(t - k2^j)/2^j\right]}{\sqrt{2^j}} \equiv \frac{\Psi\left(t/2^j - k\right)}{\sqrt{2^j}} \quad \text{(DWT)}, \tag{13.26}$$

$$s = 2^j, \quad \tau = \frac{k}{2^j}, \quad k,\ j = 0, 1, \ldots \tag{13.27}$$

Here j and k are integers whose maximum values are yet to be determined, and we have assumed that the total time interval $T = 1$, so that time is always measured in fractions. This choice of s and τ based on powers of 2 is called a *dyadic grid* arrangement and will be seen to automatically perform the scalings and translations at the different time scales that are at the heart of wavelet analysis.[3] The DWT now becomes

$$Y_{j,k} = \int_{-\infty}^{+\infty} \mathrm{d}t \psi_{j,k}(t) y(t) \simeq \sum_m \psi_{j,k}(t_m) y(t_m) h \quad \text{(DWT)}, \tag{13.28}$$

where the discreteness here refers to the wavelet basis set and *not* the time variable. For an orthonormal wavelet basis, the inverse discrete transform is then

$$y(t) = \sum_{j,k=-\infty}^{+\infty} Y_{j,k} \psi_{j,k}(t) \quad \text{(inverse DWT)}. \tag{13.29}$$

3) Note that some references scale down with increasing j, in contrast to our scaling up.

This inversion will exactly reproduce the input signal at the N input points, but only if we sum over an infinite number of terms (Addison, 2002). Practical calculations will be less exact.

Listing 13.1 CWT.py computes a normalized continuous wavelet transform of the signal data in input (here assigned as a sum of sine functions) using Morlet wavelets (courtesy of Z. Dimcovic). The discrete wavelet transform (DWT) is faster and yields a compressed transform, but is less transparent.

```
# CWT.py Continuous Wavelet TF. Based on program by Zlatko Dimcovic

import matplotlib.pylab as p;
from mpl_toolkits.mplot3d import Axes3D ;
from visual.graph import *;

originalsignal=gdisplay(x=0, y=0, width=600, height=200, \
        title='Input Signal',xmin=0,xmax=12,ymin=-20,ymax=20)
orsigraph=gcurve(color=color.yellow)
invtrgr = gdisplay(x=0, y=200, width=600, height=200,
         title='Inverted Transform',xmin=0,xmax=12,ymin=-20,ymax=20)
invtr   = gcurve(x = list(range(0,240)), display= invtrgr , color=
    color.green)
iT =  0.0;              fT =  12.0;           W = fT - iT;
N =  240;               h = W/N
noPtsSig = N;           noS =  20;            noTau =  90;
iTau =  0.;             iS =  0.1;            tau =  iTau;          s =  iS

# Need *very* small s steps for high frequency;
dTau = W/noTau;         dS =  (W/iS)**(1./noS);
maxY =  0.001;          sig =  zeros((noPtsSig), float)             # Signal

def signal(noPtsSig, y):                                   # Signal function
    t = 0.0;       hs = W/noPtsSig;       t1 = W/6.;      t2 = 4.*W/6.
    for i in range(0, noPtsSig):
        if  t >= iT  and t <=  t1:  y[i] =  sin(2*pi*t)
        elif t >= t1 and t <=  t2: y[i] = 5.*sin(2*pi*t) +
            10.*sin(4*pi*t);
        elif t >= t2 and t <=  fT:
            y[i] = 2.5*sin(2*pi*t) + 6.*sin(4*pi*t) + 10.*sin(6*pi*t)
        else:
            print("In signal(...) : t out of range.")
            sys.exit(1)
        yy=y[i]
        orsigraph.plot(pos=(t,yy))
        t += hs
signal(noPtsSig, sig)                                      # Form signal
Yn =  zeros( (noS+1, noTau+1), float)                      # Transform

def morlet(t, s, tau):                                     # Mother
     T =  (t - tau)/s
     return sin(8*T) * exp( - T*T/2. )

def transform(s, tau, sig):                                # Find wavelet TF
    integral = 0.
    t = iT;
    for i in range(0, len(sig) ):
        t += h
        integral += sig[i]*morlet(t, s, tau)*h
    return integral / sqrt(s)

def invTransform(t, Yn):                                   # Compute inverse
    s = iS                                                      # Transform
    tau = iTau
    recSig_t = 0
```

```
    for i in range (0, noS):
        s *= dS                                                    # Scale graph
        tau = iTau
        for j in range (0, noTau):
            tau += dTau
            recSig_t += dTau*dS *(s**(-1.5))* Yn[i,j] * morlet(t,s,tau)
    return recSig_t

print("working, finding transform, count 20")
for i in range( 0, noS):
    s *= dS                                                        # Scaling
    tau = iT
    print(i)
    for j in range(0, noTau):
        tau += dTau                                                # Translate
        Yn[i, j] = transform(s, tau, sig)
print("transform found")
for i in range( 0, noS):
    for j in range( 0, noTau):
        if Yn[i, j] > maxY or Yn[i, j] < - 1 *maxY :
            maxY = abs( Yn[i, j] )                                 # Find max Y
tau = iT
s = iS
print("normalize")
for i in range( 0, noS):
    s *= dS
    for j in range( 0, noTau):
        tau += dTau                                                # Transform
        Yn[i, j] = Yn[i, j]/maxY
    tau = iT
print("finding inverse transform")                                 # Inverse TF
recSigData = "recSig.dat"
recSig = zeros(len(sig) )
t = 0.0;
print("count to 10")
kco = 0;              j = 0;              Yinv = Yn
for rs in range(0, len(recSig) ):
    recSig[rs] = invTransform(t, Yinv)                             # Find input signal
    xx=rs/20
    yy=4.6*recSig[rs]
    invtr.plot(pos=(xx,yy))
    t += h
    if kco %24 == 0:
        j += 1
        print(j)
    kco += 1
x = list(range(1, noS + 1))
y = list(range(1, noTau + 1))
X,Y = p.meshgrid(x, y)

def functz(Yn):                                                    # Transform function
    z = Yn[X, Y]
    return z

Z = functz(Yn)
fig = p.figure()
ax = Axes3D(fig)
ax.plot_wireframe(X, Y, Z, color = 'r')
ax.set_xlabel('s: scale')
ax.set_ylabel('Tau')
ax.set_zlabel('Transform')
p.show()

print("Done")
```

Note in (13.26) and (13.28) that we have kept the time variable t in the wavelet basis functions continuous, despite the fact that s and τ have been made discrete. This is useful in establishing the orthonormality of the basis functions

$$\int_{-\infty}^{+\infty} \mathrm{d}t \psi^*_{j,k}(t)\psi_{j',k'}(t) = \delta_{jj'}\delta_{kk'} , \tag{13.30}$$

where $\delta_{m,n}$ is the Kronecker delta function. Being normalized to 1 means that each wavelet basis has "unit energy"; being orthogonal means that each basis function is independent of the others. And because wavelets are localized in time, the different transform components have low levels of correlation with each other. Altogether, this leads to efficient and flexible data storage.

The use of a discrete wavelet basis makes it clear that we sample the input signal at the discrete values of time determined by the integers j and k. In general, you want time steps that sample the signal at enough times in each interval to obtain the desired level of precision. A rule of thumb is to start with 100 steps to cover each major feature. Ideally, the needed times correspond to the times at which the signal was sampled, although this may require some forethought.

Consider Figure 13.8. We measure a signal at a number of discrete times within the intervals (k or τ values) corresponding to the vertical columns of fixed width along the time axis. For each time interval we want to sample the signal at a number of scales (frequencies or j values). However, as discussed in Section 13.3, the basic mathematics of Fourier transforms indicates that the width Δt of a wave packet $\psi(t)$ and the width $\Delta\omega$ of its Fourier transform $Y(\omega)$ are related by an uncertainty principle

$$\Delta\omega\,\Delta t \geq 2\pi .$$

This relation places a constraint on the intervals in which can measure times based on the intervals in which we deduce frequencies. So while we may want a high-resolution reproduction of our signal, we do not want to store more data than

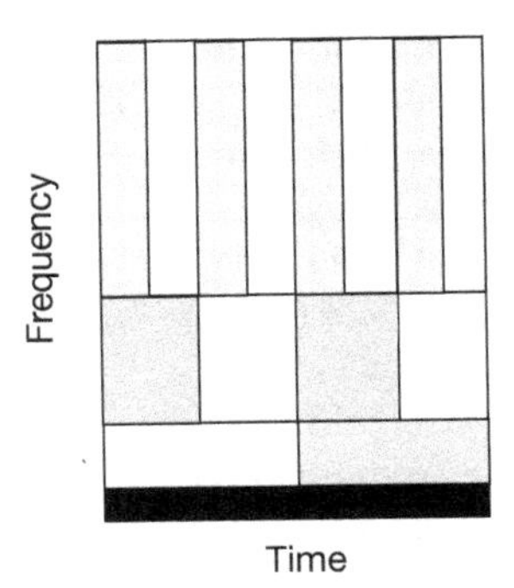

Figure 13.8 A graphical representation of the relation between time and frequency resolutions (the uncertainty relation). Each box represents an equal portion of the time-frequency plane but with different proportions of time and frequency.

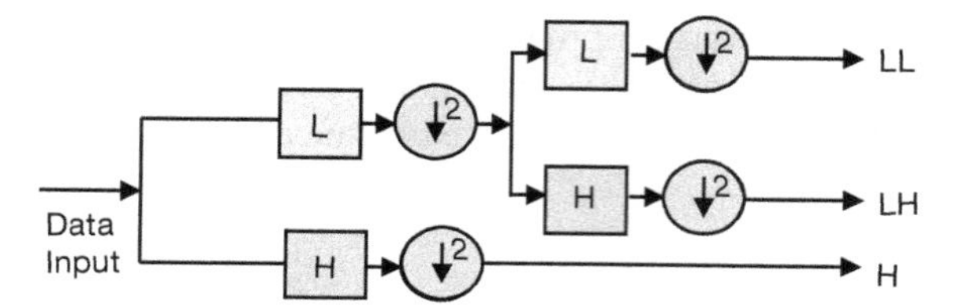

Figure 13.9 A multifrequency dyadic (power-of-2) filter tree used for discrete wavelet transformations. The L boxes represent low-pass filters and the H boxes represent high-pass filters. Each filter performs a convolution (transform). The circles containing "↓ 2" filter out half of the signal that enters them, which is called *subsampling* or *factor-of-2 decimation*. The signal on the left yields a transform with a single low and two high components (less information is needed about the low components for a faithful reproduction).

are needed to obtain that reproduction. If we sample the signal for times centered about some τ in an interval of width $\Delta\tau$ (Figure 13.8) and then compute the transform at a number of scales s or frequencies $\omega = 2\pi/s$ covering a range of height $\Delta\omega$, then the relation between the height and width is restricted by the uncertainty relation, which means that each of the rectangles in Figure 13.8 has the same area $\Delta\omega\ \Delta t = 2\pi$. The increasing heights of the rectangles at higher frequencies means that a larger range of frequencies should be sampled as the frequency increases. The premise here is that the low-frequency components provide the gross or *smooth* outline of the signal which, being smooth, does not require much detail, while the high-frequency components give the details of the signal over a short time interval and so require many components in order to record these details with high resolution.

Industrial-strength wavelet analyses do not compute explicit integrals, but instead apply a technique known as *multiresolution analysis* (MRA) (Mallat, 1989). We give an example of this technique in Figure 13.9 and in the code `DWT.py` in Listing 13.2. It is based on a *pyramid algorithm* that samples the signal at a finite number of times, and then passes it successively through a number of *filters*, with each filter representing a digital version of a wavelet.

Filters were discussed in Section 12.8, where in (12.67) we defined the action of a linear filter as a convolution of the filter response function with the signal. A comparison of the definition of a filter to the definition of a wavelet transform (13.14) shows that the two are essentially the same. Such being the case, the result of the transform operation is a weighted sum over the input signal values, with each weight the product of the integration weight times the value of the wavelet function at the integration point. Therefore, *rather than tabulate explicit wavelet functions, a set of filter coefficients is all that is needed for DWTs.*

Because each filter in Figure 13.9 changes the relative strengths of different frequency components, passing the signal through a series of filters is equivalent, in wavelet language, to analyzing the signal at different scales. This is the origin of the name "multiresolution analysis." Figure 13.9 shows how the pyramid algorithm passes the signal through a series of high-pass filters (H) and then through a series of low-pass filters (L). Each filter changes the scale to that of the level below.

Also note that the circles containing ↓ 2 in Figure 13.9. This operation filters out half of the signal and so is called *subsampling* or *factor-of-2 decimation.* It is the way we keep the areas of each box in Figure 13.8 constant as we vary the scale and translation times. We consider subsampling further when we discuss the pyramid algorithm.

In summary, the DWT process decomposes the signal into *smooth* information stored in the low-frequency components and *detailed* information stored in the high-frequency components. Because *high-resolution* reproductions of signals require more information about details than about gross shape, the pyramid algorithm is an effective way to compress data while still maintaining high resolution. In addition, because components of different resolutions are independent of each other, it is possible to lower the number of data stored by systematically eliminating higher resolution components. And finally, the use of wavelet filters builds in progressive scaling, which is particularly appropriate for fractal-like reproductions.

13.6.1 Pyramid Scheme Implementation ⊙

We now implement the pyramid scheme outlined in Figure 13.9. The H and L filters will be represented by matrices, which is an approximate way to perform the integrations or convolutions. Then there is a decimation of the output by one-half, and finally an interleaving of the output for further filtering. This process simultaneously cuts down on the number of points in the data set and changes the scale and the resolution. The decimation reduces the number of values of the remaining signal by one half, with the low-frequency part discarded because the details are in the high-frequency parts. As indicated in Figure 13.10, the pyramid DWT algorithm follows five steps:

1. Successively applies the (soon-to-be-derived) c matrix (13.41) to the whole N-length vector

$$\begin{pmatrix} Y_0 \\ Y_1 \\ Y_2 \\ Y_3 \end{pmatrix} = \begin{pmatrix} c_0 & c_1 & c_2 & c_3 \\ c_3 & -c_2 & c_1 & -c_0 \\ c_2 & c_3 & c_0 & c_1 \\ c_1 & -c_0 & c_3 & -c_2 \end{pmatrix} \begin{pmatrix} y_0 \\ y_1 \\ y_2 \\ y_3 \end{pmatrix} . \tag{13.31}$$

2. Applies it to the $N/2$-length smooth vector.
3. Repeats the application until only two smooth components remain.
4. After each filtering, the elements are ordered, with the newest two smooth elements on top, the newest detailed elements below, and the older detailed elements below that.
5. The process continues until there are just two smooth elements left.

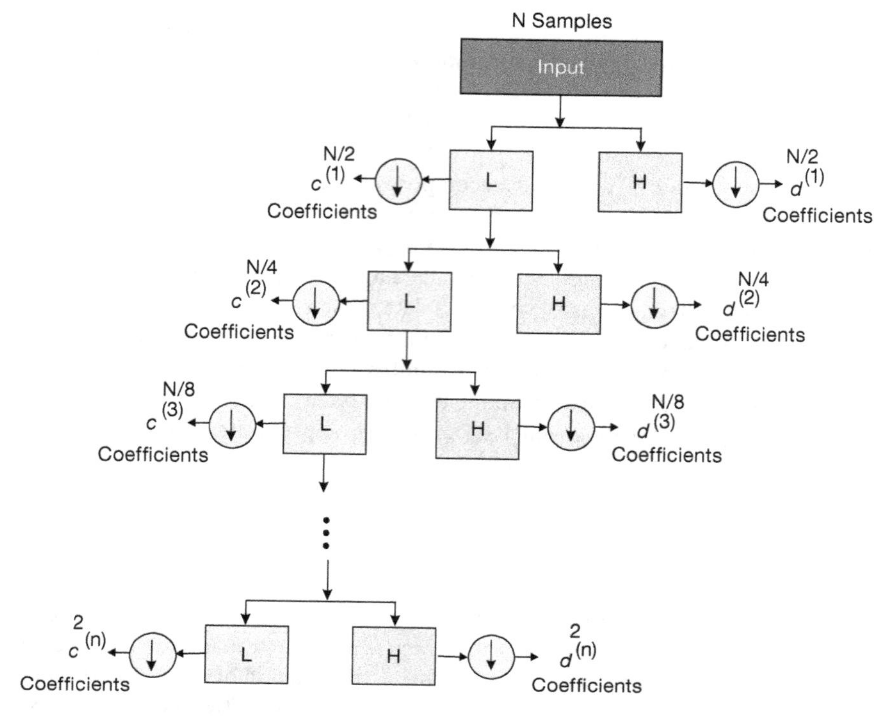

Figure 13.10 An input signal at the top is processed by a tree of high- and low-band filters. The outputs from each filtering are down-sampled with half the data kept. The process continues until there are only two data of high-band filtering and two data of low-band filtering. When complete, the total number of output data equals the total number of signal data. The process of wavelet analysis is thus equivalent to a series of filterings.

To illustrate, here we filter and reorder an initial vector of length $N = 8$:

$$\begin{pmatrix} y_1 \\ y_2 \\ y_3 \\ y_4 \\ y_5 \\ y_6 \\ y_7 \\ y_8 \end{pmatrix} \xrightarrow{\text{filter}} \begin{pmatrix} s_1^{(1)} \\ d_1^{(1)} \\ s_2^{(1)} \\ d_2^{(1)} \\ s_3^{(1)} \\ d_3^{(1)} \\ s_4^{(1)} \\ d_4^{(1)} \end{pmatrix} \xrightarrow{\text{order}} \begin{pmatrix} s_1^{(1)} \\ s_2^{(1)} \\ s_3^{(1)} \\ s_4^{(1)} \\ \hline d_1^{(1)} \\ d_2^{(1)} \\ d_3^{(1)} \\ d_4^{(1)} \end{pmatrix} \xrightarrow{\text{filter}} \begin{pmatrix} s_1^{(2)} \\ d_1^{(2)} \\ s_2^{(2)} \\ d_2^{(2)} \\ \hline d_1^{(1)} \\ d_2^{(1)} \\ d_3^{(1)} \\ d_4^{(1)} \end{pmatrix} \xrightarrow{\text{order}} \begin{pmatrix} s_1^{(2)} \\ s_2^{(2)} \\ d_1^{(2)} \\ d_2^{(2)} \\ \hline d_1^{(1)} \\ d_2^{(1)} \\ d_3^{(1)} \\ d_4^{(1)} \end{pmatrix}. \tag{13.32}$$

The discrete inversion of a transform vector back to a signal vector is made using the transpose (inverse) of the transfer matrix at each stage. For instance,

$$\begin{pmatrix} y_0 \\ y_1 \\ y_2 \\ y_3 \end{pmatrix} = \begin{pmatrix} c_0 & c_3 & c_2 & c_1 \\ c_1 & -c_2 & c_3 & -c_0 \\ c_2 & c_1 & c_0 & c_3 \\ c_3 & -c_0 & c_1 & -c_2 \end{pmatrix} \begin{pmatrix} Y_0 \\ Y_1 \\ Y_2 \\ Y_3 \end{pmatrix} . \tag{13.33}$$

As a more realistic example, imagine that we have sampled the chirp signal $y(t) = \sin(60t^2)$ for 1024 times. The filtering process through which we place this signal is illustrated as a passage from the top to the bottom in Figure 13.10. First the original 1024 samples are passed through a single low band and a single high band (which is mathematically equivalent to performing a series of convolutions). As indicated by the down arrows, the output of the first stage is then downsampled, that is, the number is reduced by a factor of 2. This results in 512 points from the high-band filter as well as 512 points from the low-band filter. This produces the first-level output. The output coefficients from the high-band filters are called $\{d_i^{(1)}\}$ to indicate that they show details, and $\{s_i^{(1)}\}$ to indicate that they show smooth features. The superscript indicates that this is the first level of processing. The detail coefficients $\{d^{(1)}\}$ are stored to become part of the final output.

In the next level down, the 512 smooth data $\{s_i^{(1)}\}$ are passed through new low- and high-band filters using a broader wavelet. The 512 outputs from each are downsampled to form a smooth sequence $\{s_i^{(2)}\}$ of size 256 and a detailed sequence $\{d_i^{(2)}\}$ of size 256. Again the detail coefficients $\{d^{(2)}\}$ are stored to become part of the final output. (Note that this is only half the size of the previously stored details.) The process continues until there are only two numbers left for the detail coefficients and two numbers left for the smooth coefficients. Because this last filtering is carried out with the broadest wavelet, it is of the lowest resolution and therefore requires the least information.

In Figure 13.11, we show the actual effects on the chirp signal of pyramid filtering for various levels in the processing. (The processing is carried out with four-coefficient *Daub4* wavelets, which we will discuss soon.) At the uppermost level, the wavelet is narrow, and so convoluting this wavelet with successive sections of the signal results in smooth components that still contain many large high-frequency parts. The detail components, in contrast, are much smaller in magnitude. In the next stage, the wavelet is dilated to a lower frequency and the analysis is repeated *on just the smooth (low-band) part.* The resulting output is similar, but with coarser features for the smooth coefficients and larger values for the details. Note that in the upper graphs we have connected the points to make the output look continuous, while in the lower graphs, with fewer points, we have plotted the output as histograms to make the points more evident. Eventually the downsampling leads to just two coefficients output from each filter, at which point the filtering ends.

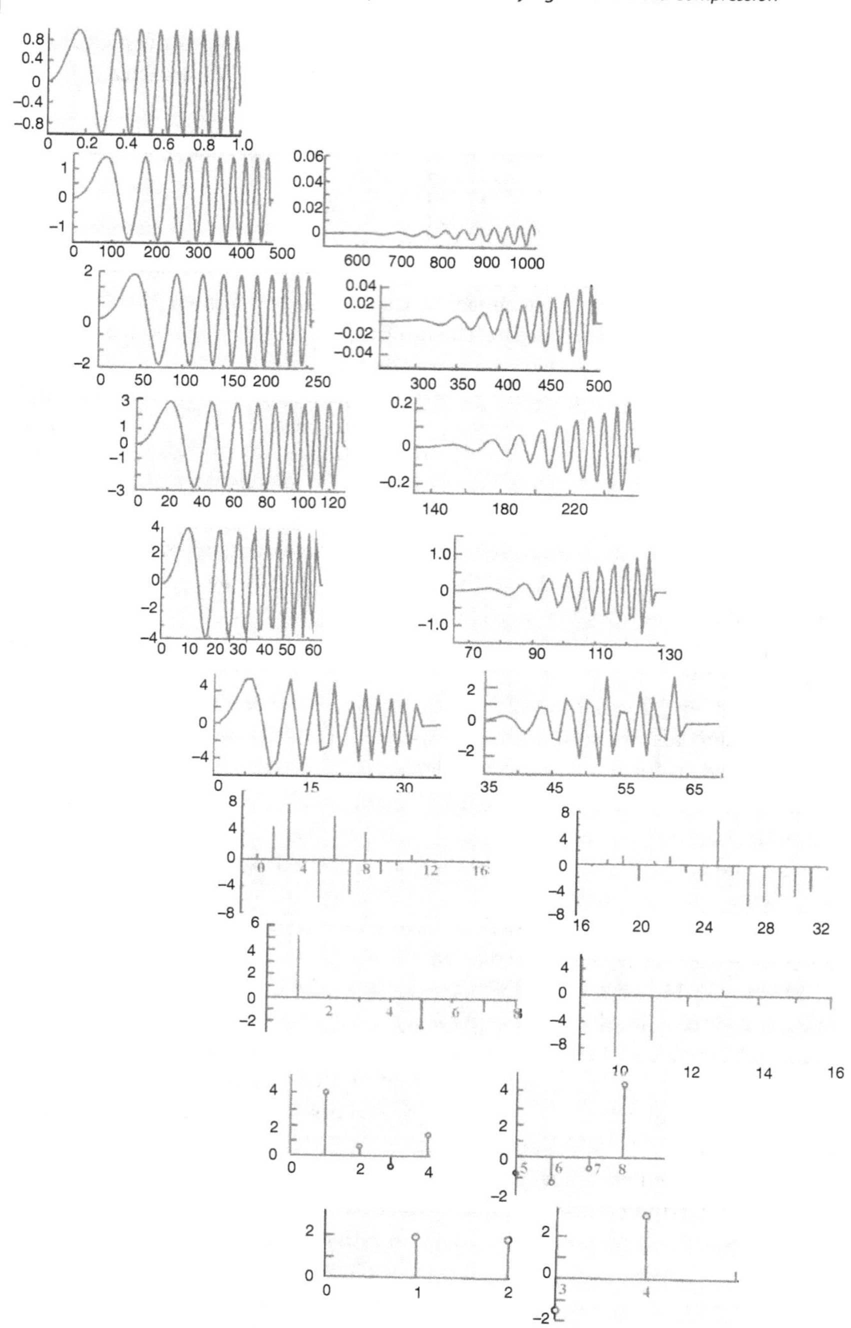

Figure 13.11 In successive passes, the filtering of the original signal at the top goes through the pyramid algorithm and produces the outputs shown. The sampling is reduced by a factor of 2 in each step. Note that in the upper graphs, we have connected the points to emphasize their continuous nature while in the lower graphs we plot the individual output points as histograms.

To reconstruct the original signal (called *synthesis* or *transformation*) a reversed process is followed: Begin with the last sequence of four coefficients, upsample them, pass them through low- and high-band filters to obtain new levels of coefficients, and repeat until all the N values of the original signal are recovered. The inverse scheme is the same as the processing scheme (Figure 13.10), only now the directions of all the arrows are reversed.

13.6.2 Daubechies Wavelets via Filtering

We should now be able to understand that digital wavelet analysis has been standardized to the point where classes of wavelet basis functions are specified not by their analytic forms, but rather by their *wavelet filter coefficients*. In 1988, the Belgian mathematician Ingrid Daubechies discovered an important class of such filter coefficients (Daubechies, 1995; Rowe and Abbott, 1995). We will study just the Daub4 class containing the four coefficients c_0, c_1, c_2, and c_3.

Imagine that our input contains the four elements $\{y_1, y_2, y_3, y_4\}$ corresponding to measurements of a signal at four times. We represent a low-pass filter L and a high-pass filter H in terms of the four filter coefficients as

$$L = \begin{pmatrix} c_0 & +c_1 & c_2 & +c_3 \end{pmatrix}, \tag{13.34}$$

$$H = \begin{pmatrix} c_3 & -c_2 & c_1 & -c_0 \end{pmatrix}. \tag{13.35}$$

To see how this works, we form an input vector by placing the four signal elements in a column and then multiply the input by L and H:

$$L\begin{pmatrix} y_0 \\ y_1 \\ y_2 \\ y_3 \end{pmatrix} = \begin{pmatrix} c_0 & c_1 & c_2 & c_3 \end{pmatrix}\begin{pmatrix} y_0 \\ y_1 \\ y_2 \\ y_3 \end{pmatrix} = c_0 y_0 + c_1 y_1 + c_2 y_2 + c_3 y_3 ,$$

$$H\begin{pmatrix} y_0 \\ y_1 \\ y_2 \\ y_3 \end{pmatrix} = \begin{pmatrix} c_3 & -c_2 & c_1 & -c_0 \end{pmatrix}\begin{pmatrix} y_0 \\ y_1 \\ y_2 \\ y_3 \end{pmatrix} = c_3 y_0 - c_2 y_1 + c_1 y_2 - c_0 y_3 .$$

We see that if we choose the values of the c_i's carefully, the result of L acting on the signal vector is a single number that may be viewed as a weighted average of the four input signal elements. Because an averaging process tends to smooth out data, the low-pass filter may be thought of as a *smoothing filter* that outputs the general shape of the signal.

In turn, we see that if we choose the c_i values carefully, the result of H acting on the signal vector is a single number that may be viewed as the weighted differences of the input signal. Because a differencing process tends to emphasize the variation in the data, the high-pass filter may be thought of as a *detail* filter that produces a large output when the signal varies considerably, and a small output when the signal is smooth.

We have just seen how the individual L and H filters, each represented by a single row of the filter matrix, outputs one number when acting upon an input signal containing four elements in a column. If we want the output of the filtering process Y to contain the same number of elements as the input (four y's in this case), we just stack the L and H filters together:

$$\begin{pmatrix} Y_0 \\ Y_1 \\ Y_2 \\ Y_3 \end{pmatrix} = \begin{pmatrix} L \\ H \\ L \\ H \end{pmatrix} \begin{pmatrix} y_0 \\ y_1 \\ y_2 \\ y_3 \end{pmatrix} = \begin{pmatrix} c_0 & c_1 & c_2 & c_3 \\ c_3 & -c_2 & c_1 & -c_0 \\ c_2 & c_3 & c_0 & c_1 \\ c_1 & -c_0 & c_3 & -c_2 \end{pmatrix} \begin{pmatrix} y_0 \\ y_1 \\ y_2 \\ y_3 \end{pmatrix} . \tag{13.36}$$

Of course, the first and third rows of the $\boldsymbol{Y}$ vector will be identical, as will the second and fourth, but we will take care of that soon.

Now we go about determining the values of the filter coefficients c_i by placing specific demands upon the output of the filter. We start by recalling that in our discussion of discrete Fourier transforms we observed that a transform is equivalent to a rotation from the time domain to the frequency domain. Yet we know from our study of linear algebra that rotations are described by orthogonal matrices, that is, matrices whose inverses are equal to their transposes. In order for the inverse transform to return us to the input signal, the transfer matrix must be orthogonal. For our wavelet transformation to be orthogonal, we must have the 4×4 filter matrix times its transpose equal to the identity matrix:

$$\begin{pmatrix} c_0 & c_1 & c_2 & c_3 \\ c_3 & -c_2 & c_1 & -c_0 \\ c_2 & c_3 & c_0 & c_1 \\ c_1 & -c_0 & c_3 & -c_2 \end{pmatrix} \begin{pmatrix} c_0 & c_3 & c_2 & c_1 \\ c_1 & -c_2 & c_3 & -c_0 \\ c_2 & c_1 & c_0 & c_3 \\ c_3 & -c_0 & c_1 & -c_2 \end{pmatrix} = \begin{pmatrix} 1 & 0 & 0 & 0 \\ 0 & 1 & 0 & 0 \\ 0 & 0 & 1 & 0 \\ 0 & 0 & 0 & 1 \end{pmatrix} ,$$
$$\Rightarrow \quad c_0^2 + c_1^2 + c_2^2 + c_3^2 = 1 , \quad c_2 c_0 + c_3 c_1 = 0 . \tag{13.37}$$

Two equations in four unknowns are not enough for a unique solution, so we now include the further requirement that the detail filter $H = (c_3, -c_0, c_1, -c_2)$ must output a zero if the input is smooth. We define "smooth" to mean that the input is constant or linearly increasing:

$$\begin{pmatrix} y_0 & y_1 & y_2 & y_3 \end{pmatrix} = \begin{pmatrix} 1 & 1 & 1 & 1 \end{pmatrix} \quad \text{or} \quad \begin{pmatrix} 0 & 1 & 2 & 3 \end{pmatrix} . \tag{13.38}$$

This is equivalent to demanding that the moments up to order p are zero, that is, that we have an "approximation of order p." Explicitly,

$$H \begin{pmatrix} y_0 & y_1 & y_2 & y_3 \end{pmatrix} = H \begin{pmatrix} 1 & 1 & 1 & 1 \end{pmatrix} = H \begin{pmatrix} 0 & 1 & 2 & 3 \end{pmatrix} = 0 ,$$
$$\Rightarrow \quad c_3 - c_2 + c_1 - c_0 = 0, \quad 0 \times c_3 - 1 \times c_2 + 2 \times c_1 - 3 \times c_0 = 0 ,$$

$$\Rightarrow \quad c_0 = \frac{1+\sqrt{3}}{4\sqrt{2}} \simeq 0.483\,, \quad c_1 = \frac{3+\sqrt{3}}{4\sqrt{2}} \simeq 0.836\,, \tag{13.39}$$

$$c_2 = \frac{3-\sqrt{3}}{4\sqrt{2}} \simeq 0.224\,, \quad c_3 = \frac{1-\sqrt{3}}{4\sqrt{2}} \simeq -0.129\,. \tag{13.40}$$

These are the basic Daub4 filter coefficients. They are used to create larger filter matrices by placing the row versions of L and H along the diagonal, with successive pairs displaced two columns to the right. For example, for eight elements

$$\begin{pmatrix} Y_0 \\ Y_1 \\ Y_2 \\ Y_3 \\ Y_4 \\ Y_5 \\ Y_6 \\ Y_7 \end{pmatrix} = \begin{pmatrix} c_0 & c_1 & c_2 & c_3 & 0 & 0 & 0 & 0 \\ c_3 & -c_2 & c_1 & -c_0 & 0 & 0 & 0 & 0 \\ 0 & 0 & c_0 & c_1 & c_2 & c_3 & 0 & 0 \\ 0 & 0 & c_3 & -c_2 & c_1 & -c_0 & 0 & 0 \\ 0 & 0 & 0 & 0 & c_0 & c_1 & c_2 & c_3 \\ 0 & 0 & 0 & 0 & c_3 & -c_2 & c_1 & -c_0 \\ c_2 & c_3 & 0 & 0 & 0 & 0 & c_0 & c_1 \\ c_1 & -c_0 & 0 & 0 & 0 & 0 & c_3 & -c_2 \end{pmatrix} \begin{pmatrix} y_0 \\ y_1 \\ y_2 \\ y_3 \\ y_4 \\ y_5 \\ y_6 \\ y_7 \end{pmatrix}. \tag{13.41}$$

Note that in order not to lose any information, the last pair on the bottom two rows is wrapped over to the left. If you perform the actual multiplications indicated in (13.41), you will note that the output has successive *smooth* and *detailed* information. The output is processed with the pyramid scheme.

The time dependences of two Daub4 wavelets is displayed in Figure 13.12. To obtain these from our filter coefficients, first imagine that an elementary wavelet $y_{1,1}(t) \equiv \psi_{1,1}(t)$ is input into the filter. This should result in a transform $Y_{1,1} = 1$. Inversely, we obtain $y_{1,1}(t)$ by applying the inverse transform to a $\boldsymbol{Y}$ vector with a 1 in the first position and zeros in all the other positions. Likewise, the ith member of the Daubechies class is obtained by applying the inverse

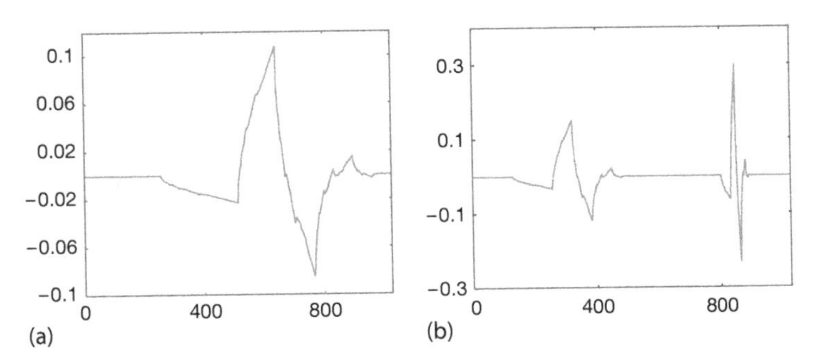

Figure 13.12 (a) The Daub4 e6 wavelet constructed by inverse transformation of the wavelet coefficients. This wavelet has been found to be particularly effective in wavelet analyses. (b) The sum of Daub4 e10 and Daub4 1e58 wavelets of different scale and time displacements.

transform to a Y vector with a 1 in the ith position and zeros in all the other positions.

The wavelet for coefficient 6 (thus the e6 notation) is shown in Figure 13.12a and in Figure 13.12b the sum of two wavelets corresponding to the coefficients 10 and 58. We see that the two wavelets have different levels of scale as well as different time positions. So despite the fact that the time dependence of the wavelets is not evident when wavelet (filter) coefficients are used, it is there.

13.6.3
DWT Implementation and Exercise

Listing 13.2 gives our program for performing a DWT on the chirp signal $y(t) = \sin(60t^2)$. The method `pyram` calls the `daube4` method to perform the DWT or inverse DWT, depending upon the value of `sign`.

1. Modify the program so that you output to a file the values for the input signal that your code has read in. It is always important to check your input.
2. Try to reproduce Figure 13.11 by using various values for the variable `nend` that controls when the filtering ends. A value `nend=1024` should produce just the first step in the downsampling (top row in Figure 13.11). Selecting `nend=512` should produce the next row, while `nend=4` should output just two smooth and detailed coefficients.
3. Reproduce the scale-time diagram shown on the right of Figure 13.11. This diagram shows the output at different scales and serves to interpret the main components of the signal and the time in which they appear. The time line at the bottom of the figure corresponds to a signal of length 1 over which 256 samples were recorded. The low-band (smooth) components are shown on the left, and the high-band components on the right.
   a) The bottommost figure results when `nend` = 256.
   b) The figure in the second row up results from `nend` = 128, and we have the output from two filterings. The output contains 256 coefficients but divides time into four intervals and shows the frequency components of the original signal in more detail.
   c) Continue with the subdivisions for `nend` = 64, 32, 16, 8, and 4.
4. For each of these choices except the topmost, divide the time by 2 and separate the intervals by vertical lines.
5. The topmost spectrum is your final output. Can you see any relation between it and the chirp signal?
6. Change the sign of `sign` and check that the inverse DWT reproduces the original signal.
7. Use the code to visualize the time dependence of the Daubechies mother function at different scales.
   a) Start by performing an inverse transformation on the eight-component signal `[0,0,0,0,1,0,0,0]`. This hould yield a function with a width of about 5 units.

b) Next perform an inverse transformation on a unit vector with $N = 32$ but with all components except the fifth equal to zero. The width should now be about 25 units, a larger scale but still covering the same time interval.
c) Continue this procedure until you obtain wavelets of 800 units.
d) Finally, with $N = 1024$, select a portion of the mother wavelet with data in the horizontal interval [590,800]. This should show some self similarity.

Listing 13.2 DWT.py computes the DWT using the pyramid algorithm for the 2^n signal values stored in f[ ] (here assigned as the chirp signal $\sin 60t^2$). The Daub4 digital wavelets are the basis functions, and **sign** $= \pm 1$ for transform/inverse.

```
# DWT.py: Discrete Wavelet Transform, Daubechies type, global variables

from visual import *
from visual.graph import *

sq3 = sqrt(3);              fsq2 = 4.0*sqrt(2);         N = 1024         # N = 2^n
c0 = (1+sq3)/fsq2;                 c1 = (3+sq3)/fsq2         # Daubechies 4 coeff
c2 = (3-sq3)/fsq2;                 c3 = (1-sq3)/fsq2
transfgr1 = None                                          # Display indicator
transfgr1 = None

def chirp( xi):                                                    # Chirp signal
    y = sin(60.0*xi**2);
    return y;

def daube4(f, n, sign):                   # DWT if sign >= 0, inverse if sign < 0
    global transfgr1, transfgr2
    tr = zeros( (n + 1), float)                                     # Temporary
    if n < 4 : return
    mp = n/2
    mp1 = mp + 1                                                    # midpoint + 1
    if sign >= 0:                                                   # DWT
        j = 1
        i = 1
        maxx   = n/2
        if n > 128:                                                 # Scale
            maxy = 3.0
            miny = - 3.0
            Maxy = 0.2
            Miny = - 0.2
            speed = 50                                               # Fast rate
        else:
            maxy = 10.0
            miny = - 5.0
            Maxy = 7.5
            Miny = - 7.5
            speed = 8                                                # Lower rate
        if transfgr1:
            transfgr1.display.visible = False
            transfgr2.display.visible = False
            del transfgr1
            del transfgr2
        transfgr1 = gdisplay(x=0, y=0, width=600, height=400,\
        title='Wavelet TF, down sample + low pass', xmax=maxx,\
        xmin=0, ymax=maxy, ymin=miny)
        transf   = gvbars(delta=2.*n/N, color=color.cyan, display=transfgr1)
        transfgr2 = gdisplay(x=0, y=400, width=600, height=400,\
       title='Wavelet TF, down sample + high pass',\
          xmax=2*maxx, xmin=0, ymax=Maxy, ymin=Miny)
        transf2 = gvbars(delta=2.*n/N, color=color.cyan, display=transfgr2)
        while j <= n - 3:
```

```
                rate(speed)
                tr[i] = c0*f[j] + c1*f[j+1] + c2*f[j+2] + c3*f[j+3]# low-pass
                transf.plot(pos = (i, tr[i]) )                # c coefficients
                tr[i+mp] = c3*f[j] - c2*f[j+1] + c1*f[j+2] - c0*f[j+3] # high
                transf2.plot(pos = (i + mp, tr[i + mp]) )
                i += 1                                        # d coefficents
                j += 2                                        # downsampling
            tr[i] = c0*f[n-1] + c1*f[n] + c2*f[1] + c3*f[2]       # low-pass
            transf.plot(pos = (i, tr[i]) )                    # c coefficients
            tr[i+mp] = c3*f[n-1] - c2*f[n] + c1*f[1] - c0*f[2]    # high-pass
            transf2.plot(pos = (i+mp, tr[i+mp]) )
        else:                                                 # inverse DWT
            tr[1] = c2*f[mp] + c1*f[n] + c0*f[1] + c3*f[mp1]      # low-pass
            tr[2] = c3*f[mp] - c0*f[n] + c1*f[1] - c2*f[mp1]      # high-pass
            j = 3
            for i in range (1, mp):
                tr[j] = c2*f[i] + c1*f[i+mp] + c0*f[i+1] + c3*f[i+mpl]   # low
                j += 1                                        # upsample
                tr[j] = c3*f[i] - c0*f[i+mp] + c1*f[i+1] - c2*f[i+mpl]  # high
                j += 1;                                       # upsampling
        for i in range(1, n+1):
            f[i] = tr[i]                                      # copy TF to array

    def pyram(f, n, sign):                               # DWT, replaces f by TF
        if (n < 4): return                                    # too few data
        nend = 4                                     # indicates when to stop
        if sign >= 0 :                                        # Transform
            nd = n
            while nd >= nend:                          # Downsample filtering
                daube4(f, nd, sign)
                nd //= 2
        else:                                                 # Inverse TF
            while nd <= n:               # Upsampling, fix thanks to Pavel Snopok
                daube4(f, nd, sign)
                nd *= 2

    f = zeros( (N + 1), float)                                # data vector
    inxi = 1.0/N                                              # for chirp signal
    xi = 0.0
    for i in range(1, N + 1):
        f[i] = chirp(xi)                                      # Function to TF
        xi    += inxi;
    n = N                                                     # must be 2^m
    pyram(f, n, 1)                                            # TF
    # pyram(f, n,   - 1)                                      # Inverse TF
```

13.7
Principal Components Analysis

We have seen that Fourier analysis has a shortcoming of having all of its components correlated. This slows down the calculation of transforms and makes compression and reconstitution of data problematic. Wavelets, on the other hand, appear excellent at data compression, but not appropriate for high-dimensionality data sets or for all physical situations. Principal components analysis (PCA) is excellent for situations in which there are correlation among the variable in the data, and especially for the type of space-time correlations as might be found in brain waves, facial patterns and ocean currents. The same basic PCA approach is used

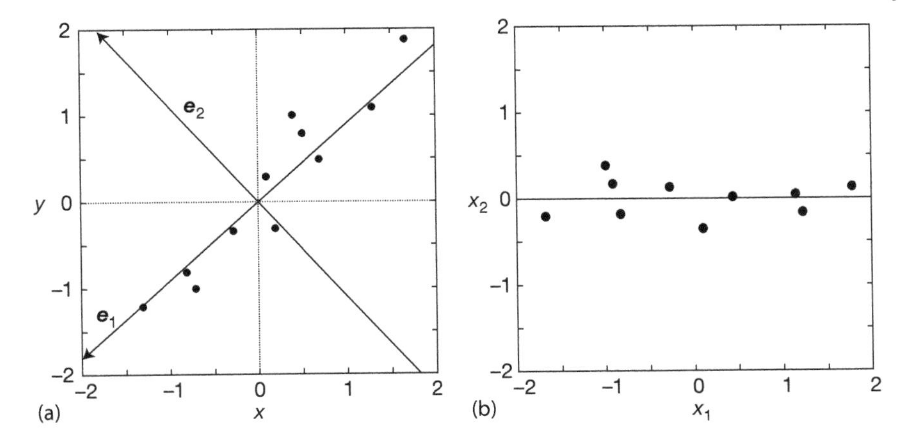

Figure 13.13 (a) The normalized data and eigenvectors of covariance matrix. (b) The normalized data using the PCA eigenvectors as basis.

in many fields with names such as the Karhunen–Loève transform, the Hotelling transform, the proper orthogonal decomposition, singular value decomposition, factor analysis, empirical orthogonal functions, empirical component analysis, empirical modal analysis, and so forth (Wikipedia, 2014).

A key element in PCA is viewing a collection of data as defining a multidimensional *data space*, and then finding a few basic components in those data that contain most of the "power". For example, imagine the output of 32 detectors of magnetic brain waves recorded every tenth of a second for an hour. In this case there will be 33 space-time dimensions to the data and $10 \times 60 \times 60 \times 32 = 1\,152\,000$ data elements. And so we are now motivated to transform from the detector-time interval basis to a new set of basis functions known as the *principal components* which are pretty much guaranteed to concentrate most of the signal strength into a few components (Jackson, 1991; Jolliffe, 1991; Smith, 2002). This is analogous to the principal axes theorem of mechanics, in which the description of the rotation of a solid object is greatly simplified if moments of inertia relative to the principal axes are used.

As follows from the example that we will work out soon, on the left of Figure 13.13, we see a bunch of data point along with the two principal component eigenvectors for these data. This figure shows how the first component accounts for most of the variability of the data (has largest possible variance), while the next component is orthogonal, and thus not correlated with the first component. The second component accounts for much less of the data, namely the most possible variance in the data within the constraint of being orthogonal to the first component.

Table 13.1 PCA data.

Data		Adjusted data		In PCA basis	
x	y	x	y	x_1	x_2
2.5	2.4	0.69	0.49	−0.828	−0.175
0.5	0.7	−1.31	−1.21	1.78	0.143
2.2	2.9	0.39	0.99	−0.992	0.484
1.9	2.2	0.09	0.29	−0.274	0.130
3.1	3.0	1.29	1.09	−1.68	−0.209
2.3	2.7	0.49	0.79	0.913	0.175
2	1.6	0.19	−0.31	0.0991	−0.350
1.0	1.1	−0.81	−0.81	1.14	0.464
1.6	1.6	−0.31	−0.31	0.438	0.0178
1.1	0.9	−0.71	−1.01	1.22	−0.163

13.7.1
Demonstration of Principal Component Analysis

The derivation of PCA and proofs of its theorems can get quite involved. Instead, as our introduction to the subject we take a purely operational approach and work through a simple PCA analysis following the example of (Smith, 2002). We assume that the data have two dimension, and we call these x and y, but they need not be related to spatial positions.

1. Enter Data We start with a data set, in our case the first two columns of Table 13.1.

2. Subtract the Mean PCA analysis assumes that the data in each dimension has zero mean. Accordingly, as shown in columns two and three in Table 13.1, we calculate the mean for each column, $(\bar{x}, \bar{y})$, and then subtract them from the data in each column. The resulting adjusted data are given in the third and fourth columns of the table, and placed here into a data matrix:

$$\mathbf{X} = \begin{pmatrix} 0.69 & 0.49 \\ -1.31 & -1.21 \\ 0.39 & 0.99 \\ 0.09 & 0.29 \\ 1.29 & 1.09 \\ 0.49 & 0.79 \\ 0.19 & -0.31 \\ -0.81 & -0.81 \\ -0.31 & -0.31 \\ -0.71 & -1.01 \end{pmatrix} . \tag{13.42}$$

3. Calculate Covariance Matrix Recall that for a data set with N members, the *variance* is a measure of the deviation of the data from their mean:

$$\text{var}(x) = \frac{1}{N-1}\sum_{i=1}^{N}(x_i - \bar{x})^2 = \frac{1}{N-1}\sum_{i=1}^{N}(x_i - \bar{x})(x_i - \bar{x}) . \tag{13.43}$$

When a data set contains multiple dimensions (variables), there may well be a dependence of the data in one-dimensional variable to that in another. The *covariance* gives a measure of how much the deviation of one variable from the mean varies with the deviation of another variable from the mean:

$$\text{cov}(x, y) = \frac{1}{N-1}\sum_{i=1}^{N}(x_i - \bar{x})(y_i - \bar{y}) , \tag{13.44}$$

so that a positive covariance indicates that the x and y variables tend to change together in the same direction. Also, we see that the variance (13.43) can be viewed as a special case of the covariance, $\text{var}(x) = \text{cov}(x, x)$, and that there is a symmetry here with $\text{cov}(x, y) = \text{cov}(y, x)$. All of these possible covariance values are combined into a symmetric covariance matrix, which for our 2×2 case is just

$$\mathbf{C} = \begin{pmatrix} \text{cov}(x, x) & \text{cov}(x, y) \\ \text{cov}(y, x) & \text{cov}(y, y) \end{pmatrix} . \tag{13.45}$$

The next step in a PSA is to compute the covariance matrix for all of the data, which in our case turns out to be

$$\mathbf{C} = \begin{pmatrix} 0.6166 & 0.6154 \\ 0.6154 & 0.7166 \end{pmatrix} . \tag{13.46}$$

4. Compute Unit Eigenvector and Eigenvalues of C Easy to do with NumPy:

$$\lambda_1 = 1.284 , \quad \lambda_2 = 0.4908 , \tag{13.47}$$

$$\boldsymbol{e}_1 = \begin{pmatrix} -0.6779 \\ -0.7352 \end{pmatrix} , \quad \boldsymbol{e}_2 = \begin{pmatrix} -0.7352 \\ 0.6789 \end{pmatrix} , \tag{13.48}$$

where we have ordered the eigenvalues and eigenvectors so that the largest eigenvalue is first. As we shall see, the eigenvector corresponding to this largest eigenvalue is in fact the principal component in the data, typically with $\sim 80\%$ of the power in it.

In Figure 13.13, we show the normalized data and the two unit eigenvectors of the covariance matrix (scaled to fill frame). Note that the $\boldsymbol{e}_1$ eigenvector looks very much like a straight-line best fit to the data. This is the major trend in the data. The other eigenvector $\boldsymbol{e}_2$ is clearly orthogonal to $\boldsymbol{e}_1$, and contain much less of the signal strength than $\boldsymbol{e}_1$. It is in the direction of the variation from the straight line fit, and cleaerly contains less information about the data. These are the essential ideas behind PSA.

5. Expressing Data in Terms of Principal Components Now that we have the principal components, we need to do something with them, namely, express the data in terms of them. Clearly, one choice is to ignore the eigenvectors corresponding to the smaller eigenvalues (only one in our simple example). This is useful for focusing attention on the key elements in the data, as well as for compressing the data. What one usually does now is to form a *feature vector* $\boldsymbol{F}$ made up of the eigenvectors that we wish to keep, for example,

$$\boldsymbol{F}_2 = \begin{pmatrix} -0.6779 & -0.7352 \\ -0.7352 & 0.6779 \end{pmatrix}, \tag{13.49}$$

$$\boldsymbol{F}_1 = \begin{pmatrix} -0.6779 \\ -0.7352 \end{pmatrix}, \tag{13.50}$$

where $\boldsymbol{F}_1$ keeps just one principal component, while $\boldsymbol{F}_2$ keeps two. The matrix gets its name because by deciding which eigenvectors we wish to keep, we are deciding which features of the data we wish to display.

Next, we form the transpose of feature matrix composed of the eigenvectors we wish to keep, and of the adjusted data matrix:

$$\boldsymbol{F}_2^{\mathrm{T}} = \begin{pmatrix} -0.6779 & -0.7352 \\ -0.7352 & 0.6779 \end{pmatrix}, \tag{13.51}$$

$$\boldsymbol{X}^{\mathrm{T}} = \begin{pmatrix} 0.69 & -1.31 & 0.39 & 0.09 & 1.29 & 0.49 & 0.19 & -0.81 & -0.31 & -0.71 \\ 0.49 & -1.21 & 0.99 & 0.29 & 1.09 & 0.79 & -0.31 & -0.81 & -0.31 & -1.01 \end{pmatrix}. \tag{13.52}$$

To express the data in terms of the principal components, we multiply the transposed feature matrix by the transposed adjusted data matrix:

$$\boldsymbol{X}^{\mathrm{PCA}} = \boldsymbol{F}_2^{\mathrm{T}} \times \boldsymbol{X}^{\mathrm{T}} \tag{13.53}$$

$$= \begin{pmatrix} -0.6779 & -0.7352 \\ -0.7352 & 0.6779 \end{pmatrix} \tag{13.54}$$

$$\times \begin{pmatrix} 0.69 & -1.31 & 0.39 & 0.09 & 1.29 & 0.49 & 0.19 & -0.81 & -0.31 & -0.71 \\ 0.49 & -1.21 & 0.99 & 0.29 & 1.09 & 0.79 & -0.31 & -0.81 & -0.31 & -1.01 \end{pmatrix} \tag{13.55}$$

$$= \begin{pmatrix} 0.828 & 1.78 & -0.992 & -0.274 & -1.68 & -0.913 & 0.0991 & 1.15 & 0.438 & 1.22 \\ -0.175 & 0.143 & 0.384 & 0.130 & -0.209 & 0.175 & -0.350 & 0.464 & 0.178 & -0.162 \end{pmatrix}. \tag{13.56}$$

In Table 13.1, we place the transform data elements back into standard form, along side the original data. On the right of Figure 13.13 we show the normalized data plotted using the eigenvectors $\boldsymbol{e}_1$ and $\boldsymbol{e}_2$ as basis. This plot shows just where each datum point sits relative to the trend in the data. If we use only the principal component, we would have all of the data on a straight line (we leave that as an exercise). Of course, our data are so simple that this example does not show the power of the technique. But if we have millions of data, it would most valuable to be able to categorize them in terms of a few components.

13.7.2 PCA Exercises

1. Use just the principal eigenvector to perform the PCA analysis just completed with two eigenvectors.
2. Store data from ten cycles of the chaotic pendulum studied in Chapter 15, but do not include transients. Perform a PCA of these data and plot the results using principal component axes.

14
Nonlinear Population Dynamics

Nonlinear dynamics is one of the success stories of computational physics. It has been explored by scientists, and engineers with computers as an essential tool, often then followed by mathematicians (Motter and Campbell, 2013). The computations have led to the discovery of new phenomena such as solitons, chaos, and fractals, as you will discover on your own. In addition, because biological systems often have complex interactions and may not be in thermodynamic equilibrium states, models of them are often nonlinear, with properties similar to those of other complex systems. In this chapter, we look at discrete models of population dynamics that are simple yet produce surprising complex behavior. In the next chapter, we explore chaos for a continuous system.

14.1
Bug Population Dynamics

Problem Populations of insects and patterns of weather do not appear to follow any simple laws.[1] At times, the populations patterns appear stable, at other times they vary periodically, and at other times they appear chaotic, with no discernable regularity, only to settle down to something simple again. Your *problem* is to deduce if a simple law can produce such complicated behaviors.

14.2
The Logistic Map (Model)

Imagine a bunch of insects reproducing generation after generation. We start with N_0 bugs, then in the next generation we have to live with N_1 of them, and after i generations there are N_i bugs to bug us. We want to define a model of how N_n varies with the discrete generation number n. Clearly, if the rates of breeding and dying are the same, then a stable population should occur. Yet bugs cannot live

1) Except maybe in Oregon, where storm clouds come to spend their weekends.

Computational Physics, 3rd edition. Rubin H. Landau, Manuel J. Páez, Cristian C. Bordeianu.

on love alone, they must also eat, and bugs not being farmers must compete for the available food supply. This tends to restrict their number to lie below some maximum population. We want to build these observations into our model.

For guidance, we look to the radioactive decay simulation in Chapter 4, where the discrete decay law, $\Delta N/\Delta t = -\lambda N$, led to exponential-like decay. Likewise, if we reverse the sign of λ, we should get exponential-like *growth*, which is a good place to start our modelling. We assume that the bug-breeding rate is proportional to the number of bugs:

$$\frac{\Delta N_i}{\Delta t} = \lambda N_i \, . \tag{14.1}$$

Because we know the empirical fact that exponential growth usually tapers off, we extend the model by incorporating the observation that for a given environment there must be maximum population N_*, which is called the *carrying capacity*. Consequently, we modify the exponential growth model (14.1) by modifiying the growth rate so that it decreases as the population N_i approaches N_*:

$$\lambda \to \lambda'(N_* - N_i) \tag{14.2}$$

$$\Rightarrow \quad \frac{\Delta N_i}{\Delta t} = \lambda'(N_* - N_i)N_i \, . \tag{14.3}$$

We expect that when N_i is small compared to N_*, the population will grow exponentially. We also expect that as N_i approaches N_*, the growth rate will decrease, eventually becoming negative if N_i exceeds the carrying capacity N_*.

Equation 14.3 is a form of the *logistic map*. It is usually written as a relation between the number of bugs in future and present generations:

$$N_{i+1} = N_i + \lambda' \Delta t (N_* - N_i) N_i \tag{14.4}$$

$$= N_i(1 + \lambda' \Delta t N_*) \left[1 - \frac{\lambda' \Delta t}{1 + \lambda' \Delta t N_*} N_i\right] \, . \tag{14.5}$$

This relation looks simpler when expressed in terms of dimensionless variables:

$$\boxed{x_{i+1} = \mu x_i (1 - x_i) \, ,} \tag{14.6}$$

$$\mu \stackrel{\text{def}}{=} 1 + \lambda' \Delta t N_* \, , \tag{14.7}$$

$$x_i \stackrel{\text{def}}{=} \frac{\lambda' \Delta t}{1 + \lambda' \Delta t N_*} N_i \simeq \frac{N_i}{N_*} \, . \tag{14.8}$$

Here μ is a dimensionless growth parameter and x_i is a dimensionless population variable. Observe from (14.7) that the *growth rate* $\mu = 1$ when the breeding rate λ' equals 0, and is otherwise expected to be larger than 1. If the number of bugs born per generation $\lambda' \Delta t$ is large, then $\mu \simeq \lambda' \Delta t N_*$ and $x_i \simeq N_i/N_*$, that is, x_i is essentially the fraction of the maximum population N_*. Consequently, realistic x

values generally lie in the range $0 \leq x_i \leq 1$, with $x = 0$ corresponding to no bugs, and $x = 1$ corresponding to the maximum population.

The map (14.6) is seen to be the sum of a linear and a quadratic dependence on x_i. In general, we employ the term "map" to refer to a function $f(x)$ that converts one number in a sequence to the next

$$x_{i+1} = f(x_i) \,. \tag{14.9}$$

For the logistic map, $f(x) = \mu x(1-x)$, with the quadratic dependence on x making this a nonlinear map, and the dependence on only the one variable x making it a *one-dimensional* map.

Just by looking at (14.6) there is no way of knowing how good a model this will be. Being as simple as it is, the model cannot be expected to be a complete description of the population dynamics of bugs. However, if it exhibits some features similar to those found in nature, then it may well form the foundation for a more complete description.

14.3 Properties of Nonlinear Maps (Theory and Exercise)

Rather than going through some fancy mathematical analysis to learn about the properties of the logistic map (Rasband, 1990), we suggest that you explore it yourself on a computer or a calculator by generating and plotting sequences of values. You should get results similar to those shown in Figures 14.1 and 14.2.

Stable Populations We want to see if the model can produce a stable population, that is, one that remains the same from generation to generation.

1. Calculate and plot x_{i+1} as a function of the generation number i.
2. The initial population x_0 is called the *seed*, and we suggest $x_0 = 0.75$ as a good starting value (the dynamical effects are not sensitive to the seed).

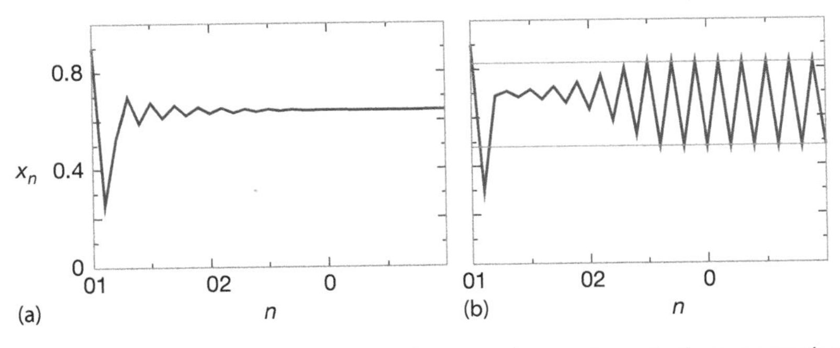

Figure 14.1 The insect population x_n vs. the generation number n for the two growth rates: (a) $\mu = 2.8$, a single attractor; (b) $\mu = 3.3$, a double attractor.

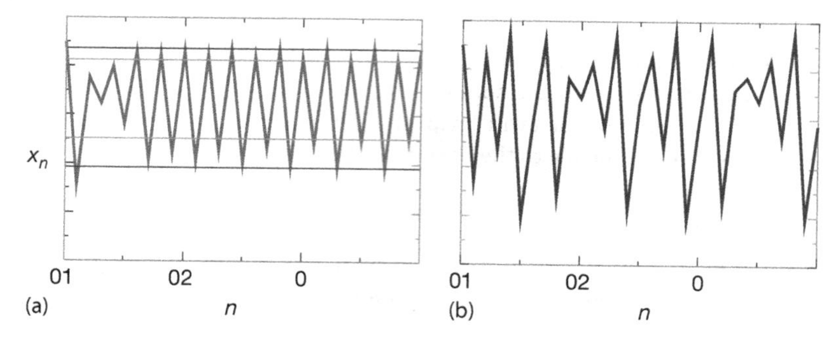

Figure 14.2 The insect population x_n vs. the generation number n for two different growth rates: (a) $\mu = 3.5$, a quadruple attractor; (b) $\mu = 3.8$, a chaotic regime.

3. Your results should be highly sensitive to the value for the growth rate μ. Too large a value may lead to instabilities, while too small a value may lead to extinction. To make sure that the model is behaving reasonably, try some cases for which we can be fairly sure of what the results should be. By trying some negative and zero values for μ (for example, $-1, -0.75, -0.5, -0.25, 0$) we should obtain decaying populations.
4. Now that you have some confidence in the model, see if you can increase the population to some stable values. With the same initial population as before, try $\mu = 0, 0.5, 1.0, 1.5, 2$. Make plots of x_i vs. i for each of these cases.
5. Take note of the *transient* behavior in these plots that occur for early generations before a steady state or more regular behavior sets in. In general, these are not the long-term dynamical behaviors of interest.
6. For a fixed value of μ, try different values for the seed population x_0. Verify that differing values of x_0 do affect the transients, but *not* the values of the stable populations.

You should have found that with positive growth rates μ, this model yields stable populations, with the bugs approaching the maximum population more rapidly as μ gets larger. This is a good validation of the model. Some typical behaviors are shown in Figures 14.1 and 14.2. In Figure 14.1a, we see equilibration into a single population; in Figure 14.1b, we see oscillation between two population levels; in Figure 14.2a we see oscillation among four levels; and in Figure 14.2b we see a chaotic system.

14.3.1 Fixed Points

An important property of the map (14.6) is the possibility of the sequence x_i reaching a *fixed* point x_*, that is, a value of the population at which the system remains. At a *one-cycle* fixed point, there is no change in the population from generation i to generation $i+1$, that is,

$$x_{i+1} = x_i = x_* \,. \tag{14.10}$$

Substituting the logistic map (14.6) into this equation produces an algebraic equation that we can solve

$$\mu x_*(1 - x_*) = x_* \,, \tag{14.11}$$

$$\Rightarrow \quad x_* = 0 \,, \quad \frac{\mu - 1}{\mu} \,. \tag{14.12}$$

The nonzero fixed point $x_* = (\mu - 1)/\mu$ corresponds to a stable population with a balance between birth and death that is reached regardless of the initial population (Figure 14.1a). In contrast, the $x_* = 0$ point is unstable and the population remains static only as long as no bugs exist; if even a few bugs are introduced, exponential growth occurs. Further analysis (Section 14.8) tells us that the stability of a population is determined by the magnitude of the derivative of the mapping function $f(x_i)$ at the fixed point (Rasband, 1990):

$$\left|\frac{\mathrm{d}f}{\mathrm{d}x}\right|_{x_*} < 1 \quad \text{(stable)} \,. \tag{14.13}$$

For the one cycle of the logistic map (14.6), the derivative is

$$\left.\frac{\mathrm{d}f}{\mathrm{d}x}\right|_{x_*} = \mu - 2\mu x_* = \begin{cases} \mu, & \text{stable at } x_* = 0 \quad \text{if} \quad \mu < 1 \,, \\ 2 - \mu, & \text{stable at } x_* = \frac{\mu-1}{\mu} \quad \text{if} \quad \mu < 3 \,. \end{cases} \tag{14.14}$$

14.3.2
Period Doubling, Attractors

Equation 14.14 tells us that while the equation for fixed points (14.12) may be satisfied for all values of μ, the populations will not be stable if $\mu > 3$. In this latter case, the system's long-term population *bifurcates* into two populations, a so called *two-cycle*. The effect is known as *period doubling* and is evident in Figure 14.1b. Because the system now acts as if it were attracted to two populations, these populations are called *attractors* or *cycle points*. We can easily predict the x values for these two-cycle attractors by demanding that generation $i+2$ have the same population as generation i:

$$x_i = x_{i+2} = \mu x_{i+1}(1 - x_{i+1}) \tag{14.15}$$

$$\Rightarrow \quad x_* = \frac{1 + \mu \pm \sqrt{\mu^2 - 2\mu - 3}}{2\mu} \,. \tag{14.16}$$

We see that as long as $\mu > 3$, the square root produces a real number and thus that physical solutions exist (complex or negative x_* values are nonphysical). We leave it to your computer explorations to discover how the system continues to double periods as μ is further increased. In all cases, the pattern repeats: one populations bifurcates into two.

14.4 Mapping Implementation

It is now time to carry out a more careful investigation of the logistic map along the original path followed by (Feigenbaum, 1979):

1. Confirm Feigenbaum's observations of the different patterns shown in Figures 14.1 and 14.2 that occur for $\mu = (0.4, 2.4, 3.2, 3.6, 3.8304)$ and seed $x_0 = 0.75$.
2. Identify the following in your graphs:
   a) *Transients:* Irregular behaviors before reaching a steady state that differ for different seeds.
   b) *Asymptotes:* In some cases, the steady state is reached after only 20 generations, while for larger μ values, hundreds of generations may be needed. These steady-state populations are independent of the seed.
   c) *Extinction:* If the growth rate is too low, $\mu \leq 1$, the population dies off.
   d) *Stable states:* The stable single-population states attained for $\mu < 3$ should agree with the prediction (14.12).
   e) *Multiple cycles:* Examine the map orbits for a growth parameter μ increasing continuously through 3. Observe how the system continues to double periods as μ increases. To illustrate, in Figure 14.2a with $\mu = 3.5$, we notice a steady state in which the population alternates among four attractors (a *four-cycle*).
   f) *Intermittency:* Observe simulations for $3.8264 < \mu < 3.8304$. Here the system appears stable for a finite number of generations and then jumps all around, only to become stable again.
   g) *Chaos: We define chaos as the deterministic behavior of a system displaying no discernible regularity.* This may seem contradictory; if a system is deterministic, it must have step-to-step correlations (which, when added up, mean long-range correlations); but if it is chaotic, the complexity of the behavior may hide the simplicity within. In an operational sense, a chaotic system is one with an extremely high sensitivity to parameters or initial conditions. This sensitivity to even minuscule changes is so high that it is impossible to predict the long-range behavior unless the parameters are known to infinite precision (a physical impossibility).
3. The system's behavior in the chaotic region is critically dependent on the exact values of μ and x_0. Systems starting out with nearly identical values for μ and x_0 may end up with quite different behaviors. In some cases, the complicated behaviors of nonlinear systems will be *chaotic*, but this is not the same as being random.[2)]
   a) Compare the long-term behaviors of starting with the two essentially identical seeds $x_0 = 0.75$ and $x_0' = 0.75(1 + \epsilon)$, where $\epsilon \simeq 2 \times 10^{-14}$.

2) You may recall from Chapter 4 that a truly random sequence of events does not even have next step predictability, while these chaotic systems do.

b) Repeat the simulation with $x_0 = 0.75$ and two essentially identical survival parameters, $\mu = 4.0$ and $\mu' = 4.0(1 - \epsilon)$, where $\epsilon \simeq 2 \times 10^{-14}$. Both simulations should start off the same but eventually diverge.

14.5 Bifurcation Diagram (Assessment)

Watching the population change with generation number gives a good idea of the basic dynamics, at least until it gets too complicated to discern patterns. In particular, as the number of bifurcations keeps increasing and the system becomes chaotic, it may be hard to see a simple underlying structure within the complicated behavior. One way to visualize what is going on is to concentrate on the attractors, that is, those populations that appear to attract the solutions and to which the solutions continuously return (long-term iterates). A plot of these attractors of the logistic map as a function of the growth parameter μ is an elegant way to summarize the results of extensive computer simulations.

A *bifurcation diagram* for the logistic map is shown in Figure 14.3, with one for a Gaussian map given in Figure 14.4. To generate such a diagram you proceed through all values of μ in steps. For each value of μ, you let the system go through hundreds of iterations to establish that the transients have died out, and then write the values (μ, x_*) to a file for hundreds of iterations after that. If the system falls into an n-cycle for this μ value, then there should predominantly be n different values written to the file. Next, the value of the initial populations x_0 is changed slightly, and the entire procedure is repeated to ensure that no fixed points are missed. When finished, your program will have stepped through all the values of μ and x_0. Our sample program `Bugs.py` is shown in Listing 14.1.

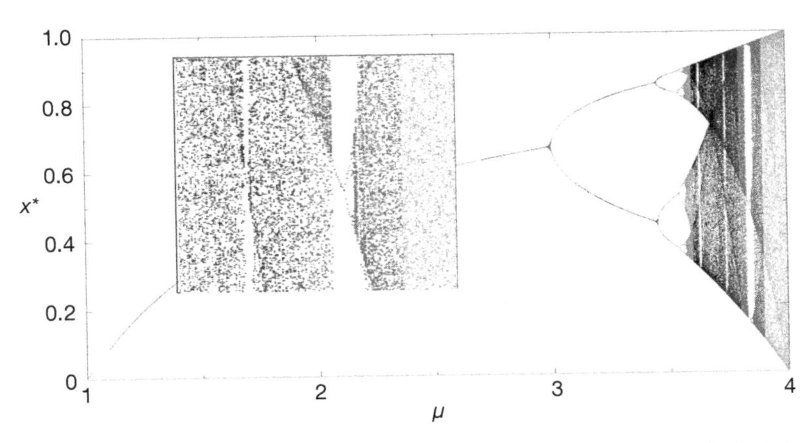

Figure 14.3 Bifurcation plot of attractor population x^* *vs.* growth rate μ for the logistic map. The inset shows some details of a three-cycle window. (The colors or gray scales indicate the regimes over which we distributed the work on different CPUs when run in parallel.)

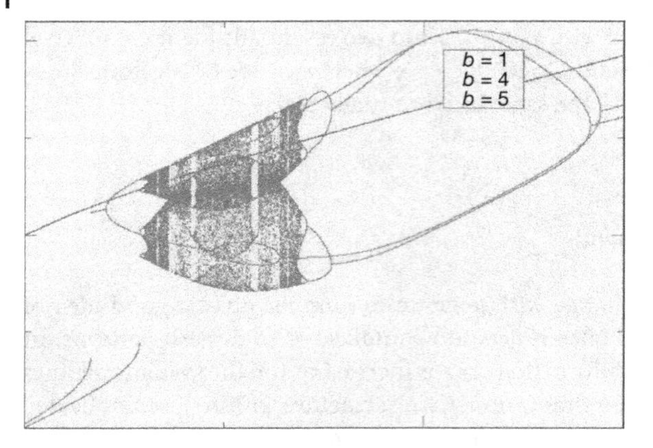

Figure 14.4 Bifurcation plot for the Gaussian map $f(x) = \exp(-bx^2) + \mu$ (courtesy of W. Hager).

Listing 14.1 Bugs.py produces the bifurcation diagram of the logistic map. A proper program requires finer grids, a scan over initial values, and removal of duplicates.

```
# Logistic map
from visual.graph import *

m_min = 1.0;        m_max = 4.0;         step = 0.01

graph1 = gdisplay(width=600, height=400, title='Logistic Map',\
        xtitle='m', ytitle='x', xmax=4.0, xmin=1., ymax=1., ymin=0.)
pts = gdots(shape = 'round', size = 1.5, color = color.green)
lasty = int(1000 * 0.5)                          # Eliminates some points
count = 0                                        # Plot every 2 iterations
for m in arange(m_min, m_max, step):
    y = 0.5
    for i in range(1,201,1):                          # Avoid transients
        y = m*y*(1-y)
    for i in range(201,402,1):
        y = m*y*( 1 - y)
    for i in range(201, 402, 1):                      # Avoid transients
        oldy=int(1000*y)
        y = m*y*(1 - y)
        inty = int(1000 * y)
        if  inty != lasty and count%2 == 0:
            pts.plot(pos=(m,y))                          # Avoid repeats
        lasty = inty
        count  += 1
```

14.5.1 Bifurcation Diagram Implementation

The last part of this problem asks you to reproduce Figure 14.3 at various levels of detail. While the best way to make a visualization of this sort would be with visualization software that permits you to vary the intensity of each individual point on the screen, we simply plot individual points and have the density in each region determined by the number of points plotted there. When thinking about plotting many individual points to draw a figure, it is important to keep in mind that your monitor may have approximately 100 pixels per inch and your laser printer 300 dots per inch. This means that if you plot a point at each pixel, you will be plotting $\sim 3000 \times 3000 \simeq 10$ million elements. *Beware:* This can require some time and may choke a printer. In any case, printing at a finer resolution is a waste of time.

14.5.2
Visualization Algorithm: Binning

1. Break up the range $1 \leq \mu \leq 4$ into 1000 steps. These are the "bins" into which we will place the x_* values.
2. In order not to miss any structures in your bifurcation diagram, loop through a range of initial x_0 values as well.
3. Wait at least 200 generations for the transients to die out, and then print the next several hundred values of (μ, x_*) to a file.
4. Print out your x_* values to no more than three or four decimal places. You will not be able to resolve more places than this on your plot, and this restriction will keep your output files smaller, as will removing duplicate entries. You can use formatted output to control the number of decimal places, or you can do via a simple conversion: multiply the x_i values by 1000 and then throw away the part to the right of the decimal point:

   ```
   Ix[i]= int(1000*x[i]) .
   ```

   You may then divide by 1000 if you want floating-point numbers.
5. Plot your file of x_* vs. μ. Use small symbols for the points and do not connect them.
6. Enlarge (zoom in on) sections of your plot and notice that a similar bifurcation diagram tends to be contained within each magnified portion (this is called *self-similarity*).
7. Look over the series of bifurcations occurring at

   $$\mu_k \simeq 3, 3.449, 3.544, 3.5644, 3.5688, 3.569\,692, 3.569\,89, \ldots \tag{14.17}$$

   The end of this series is a region of chaotic behavior.
8. Inspect the way this and other sequences begin and then end in chaos. The changes sometimes occur quickly, and so you may have to make plots over a very small range of μ values to see the structures.
9. A close examination of Figure 14.3 shows regions where, for a slight increase in μ, a very large number of populations suddenly change to very few populations. Whereas these may appear to be artifacts of the video display, this is a real effect and these regions are called *windows*. Check that at $\mu = 3.828\,427$ chaos moves into a three-cycle population.

14.5.3
Feigenbaum Constants (Exploration)

Feigenbaum discovered that the sequence of μ_k values (14.17) at which bifurcations occur follows a regular pattern (Feigenbaum, 1979). Specifically, the μ values converge geometrically when expressed in terms of the distance between bifurcations δ:

$$\mu_k \to \mu_\infty - \frac{c}{\delta^k} , \tag{14.18}$$

$$\delta = \lim_{k\to\infty} \frac{\mu_k - \mu_{k-1}}{\mu_{k+1} - \mu_k} . \tag{14.19}$$

Use your sequence of μ_k values to determine the constants in (14.18) and compare them to those found by Feigenbaum:

$$\mu_\infty \simeq 3.569\,95 , \quad c \simeq 2.637 , \quad \delta \simeq 4.6692 . \tag{14.20}$$

Amazingly, the value of the *Feigenbaum constant* δ is universal for all second-order maps.

14.6
Logistic Map Random Numbers (Exploration) ⊙

Although we have emphasized that chaos, with its short-term predictability, is different from random, with has no predictability, it is nevertheless possible for the logistic map in the chaotic region ($\mu \geq 4$) to be used to generate pseudo random numbers (Phatak and Rao, 1995). This is carried out in two steps:

$$x_{i+1} \simeq 4x_i(1 - x_i) , \tag{14.21}$$

$$y_i = \frac{1}{\pi} \cos^{-1}(1 - 2x_i) . \tag{14.22}$$

Although successive x_i's are correlated, if the population for every sixth generation or so is examined, the correlation is effectively gone and pseudorandom numbers result. To make the sequence more uniform, the trigonometric transformation is then used.

Exercise Use the random-number tests discussed in Chapter 4, or an actual Monte Carlo simulation, to test this claim.

14.7
Other Maps (Exploration)

Bifurcations and chaos are characteristic properties of nonlinear systems. Yet systems can be nonlinear in a number of ways. The table below lists four maps that generate x_i sequences containing bifurcations.

Name	$f(x)$	Name	$f(x)$
Logistic	$\mu x(1-x)$	Tent	$\mu(1-2\|x-1/2\|)$
Ecology	$xe^{\mu(1-x)}$	Quartic	$\mu[1-(2x-1)^4]$
Gaussian	$e^{-bx^2}+\mu$		

The tent map derives its nonlinear dependence from the absolute value operator, while the logistic map is a subclass of the ecology map. Explore the properties of these other maps and note the similarities and differences.

14.8 Signals of Chaos: Lyapunov Coefficient and Shannon Entropy ⊙

The Lyapunov coefficient or exponent λ_i provides an analytic measure of whether a system is chaotic (Wolf *et al.*, 1985; Ramasubramanian and Sriram, 2000; Williams, 1997). Essentially, the coefficient is a measure of the rate of exponential growth of the solution near an attractor. If the coefficient is positive then the solution moves away from the attractor, which is an indication of chaos, while if the coefficient is negative then the solution moves back toward the attractor, which is an indication of stability. For 1D problems there is only one such coefficient, whereas in general there is a coefficient for each degree of freedom. The essential assumption is that the distance L between neighboring paths x_n near an attractor have an n (generation number or time) dependence $L \propto \exp(\lambda t)$. Consequently, orbits that have $\lambda > 0$ diverge and are chaotic; orbits that have $\lambda = 0$ remain marginally stable, while orbits with $\lambda < 0$ are periodic and stable. Mathematically, the Lyapunov coefficient is defined as

$$\lambda = \lim_{t\to\infty} \frac{1}{t} \log \frac{L(t)}{L(t_0)} . \tag{14.23}$$

As an example, we calculate the Lyapunov exponent for a general 1D map,

$$x_{n+1} = f(x_n) , \tag{14.24}$$

and then apply the result to the logistic map. To determine stability, we examine perturbations about a reference trajectory x_0 by adding a small perturbation and iterating once (Manneville, 1990; Ramasubramanian and Sriram, 2000):

$$\hat{x}_0 = x_0 + \delta x_0 , \quad \hat{x}_1 = x_1 + \delta x_1 . \tag{14.25}$$

We substitute this into (14.24) and expand f in a Taylor series around x_0:

$$x_1 + \delta x_1 = f(x_0 + \delta x_0) \simeq f(x_0) + \left.\frac{\delta f}{\delta x}\right|_{x_0} \delta x_0 = x_1 + \left.\frac{\delta f}{\delta x}\right|_{x_0} \delta x_0 ,$$

$$\Rightarrow \quad \delta x_1 \simeq \left(\frac{\delta f}{\delta x}\right)_{x_0} \delta x_0 . \tag{14.26}$$

This is the proof of our earlier statement that a negative derivative indicates stability. To deduce the general result we examine one iteration:

$$\delta x_2 \simeq \left(\frac{\delta f}{\delta x}\right)_{x_1}, \quad \delta x_1 = \left(\frac{\delta f}{\delta x}\right)_{x_0} \left(\frac{\delta f}{\delta x}\right)_{x_1} \delta x_0 , \tag{14.27}$$

$$\Rightarrow \quad \delta x_n = \prod_{i=0}^{n-1} \left(\frac{\delta f}{\delta x}\right)_{x_i} \delta x_0 . \tag{14.28}$$

This last relation tells us how trajectories differ on the average after n steps:

$$|\delta x_n| = L^n |\delta x_0| , \quad L^n = \prod_{i=0}^{n-1} \left| \left(\frac{\delta f}{\delta x}\right)_{x_i} \right| . \tag{14.29}$$

We now solve for the Lyapunov number L and take its logarithm to obtain the Lyapunov coefficient:

$$\lambda = \ln(L) = \lim_{n\to\infty} \frac{1}{n} \sum_{i=0}^{n-1} \ln \left| \left(\frac{\delta f}{\delta x}\right)_{x_i} \right| . \tag{14.30}$$

For the logistic map, we obtain

$$\lambda = \frac{1}{n} \sum_{i=0}^{n-1} \ln |\mu - 2\mu x_i| , \tag{14.31}$$

where the sum is over iterations.

The code `LyapLog.py` in Listing 14.2 computes the Lyapunov exponents for the bifurcation plot of the logistic map. In Figures 14.5 and 14.6, we show its output, and note the sign changes in λ where the system becomes chaotic, and the

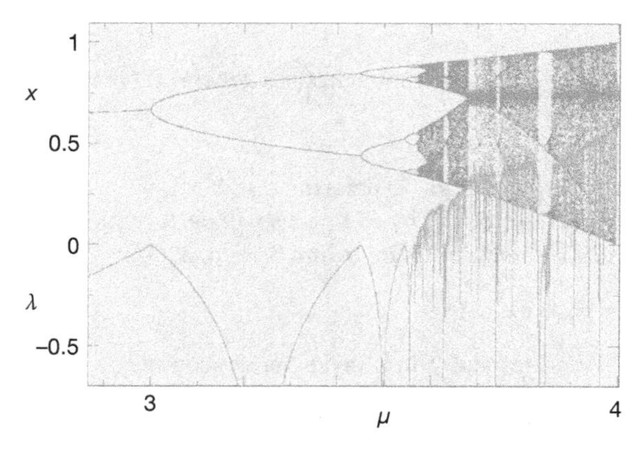

Figure 14.5 Lyapunov coefficient (bottom) and bifurcation values (top) for the logistic map as functions of the growth rate μ. Note how the Lyapunov coefficient, whose value is a measure of chaos, changes abruptly at the bifurcations.

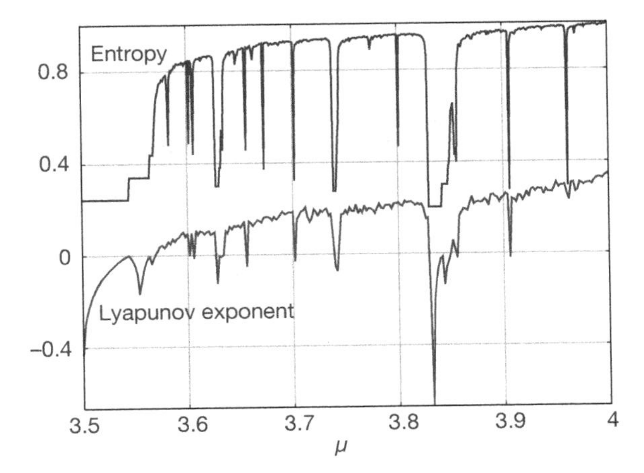

Figure 14.6 Shannon entropy (top) and Lyapunov coefficient (bottom) for the logistic map. Note the close relation between the thermodynamic measure of disorder (entropy) and the nonlinear dynamics measure of chaos (Lyapunov).

abrupt changes in slope at bifurcations. (A similar curve is obtained for the fractal dimension of the logistic map as, indeed, the two are proportional.)

Shannon entropy, like the Lyapunov exponent, is used an indicator of chaos. Entropy is a measure of uncertainty (garbled signal) that has proven useful in communication theory (Shannon, 1948; Ott, 2002; Gould *et al.*, 2006). Imagine that an experiment has N possible outcomes. If the probability of each is $p_1, p_2, \dots, p_N$, with normalization such that $\sum_{i=1}^{N} p_i = 1$, then the Shannon entropy is defined as

$$S_{\text{Shannon}} = -\sum_{i=1}^{N} p_i \ln p_i \,. \tag{14.32}$$

Listing 14.2 LyapLog.py computes Lyapunov coefficient for the bifurcation plot of the logistic map as a function of growth rate. Note the fineness of the μ grid.

```
# LyapLog.py:                         Lyapunov coef for logistic map

from visual import *
from visual.graph import *

m_min = 3.5;              m_max = 4.5;            step = 0.25
graph1 = gdisplay( title = 'Lyapunov coef (blue) for LogisticMap (red)',
                   xtitle = 'm', ytitle = 'x , Lyap',
                   xmax=5.0, xmin=0, ymax = 1.0, ymin = - 0.6)
funct1 = gdots(color = color.red)
funct2 = gcurve(color = color.yellow)

for m in arange(m_min, m_max, step):                                # m loop
    y = 0.5
    suma = 0.0
    for i in range(1, 401, 1):     y = m*y*(1 - y)              # Skip transients
    for i in range(402, 601, 1):
        y = m*y*(1 - y)
        funct1.plot(pos = (m, y) )
```

```
        suma = suma  +  log(abs(m*(1. - 2.*y) ))                        # Lyapunov
    funct2.plot(pos = (m, suma/401) )                                   # Normalize
```

If $p_i \equiv 0$, there is no uncertainty and $S_{\text{Shannon}} = 0$, as you might expect. If all N outcomes have equal probability, $p_i \equiv 1/N$, we obtain the expression familiar from statistical mechanics, $S_{\text{Shannon}} = \ln N$.

The code **Entropy.py** in Listing 14.3 computes the Shannon entropy for the logistic map as a function of the growth parameter μ. The results (Figure 14.6, top) are seen to be quite similar to the Lyapunov exponent, again with discontinuities occurring at the bifurcations.

Listing 14.3 **Entropy.py** computes the Shannon entropy for the logistic map as a function of growth parameter μ.

```
# Entropy.py Shannon Entropy with Logistic map using Tkinter

try:
    from tkinter import *
except:
    from Tkinter import *
import math
from numpy import zeros, arange

global Xwidth, Yheight
root = Tk( )
root.title('Entropy versus mu ')
mumin = 3.5;    mumax = 4.0;    dmu = 0.005;    nbin = 1000;    nmax = 100000
prob = zeros( (1000), float)
minx = mumin;   maxx = mumax;   miny=0; maxy=2.5; Xwidth=500; Yheight=500

c = Canvas(root, width = Xwidth, height = Yheight)
c.pack()

Button(root, text = 'Quit', command = root.quit).pack()     # to begin quit

def world2sc(xl, yt, xr, yb): # x - left, y - top, x - right, y - bottom
    maxx = Xwidth     # canvas width                 ______________________________
    maxy = Yheight    # canvas height                |        |      |tm          |
    lm = 0.10*maxx    # left margin                  |     ___|_____|________ ___|
    rm = 0.90*maxx    # right margin                 |lm  |    |              |   |
    bm = 0.85*maxy    # bottom margin                |___|     |              |   |
    tm = 0.10*maxy    # top margin                   |__ |___|_____________|   |
    mx = (lm - rm)/(xl - xr) #                       |    |  bm          rm|   |
    bx = (xl*rm - xr*lm)/(xl - xr) #                 |    |__|_____________|   |
    my = (tm - bm)/(yt - yb)       #                 |                          |
    by = (yb*tm - yt*bm)/(yb - yt) #                 |__________________________|
    linearTr = [mx, bx, my, by]                                   # (maxx, maxy)
    return linearTr                        # returns 4 element list

def xyaxis(mx, bx, my, by):                  # to be called after call workd2sc
    x1 = (int)(mx*minx + bx)                 # minima and maxima converted to
    x2 = (int)(mx*maxx + bx)                           # canvas coordinades
    y1  = (int)(my*maxy + by)
    y2 = (int)(my*miny + by)
    yc  = (int)(my*0.0 + by)
    c.create_line(x1, yc, x2, yc, fill = "red")                    # plots x axis
    c.create_line(x1, y1, x1, y2, fill = 'red')                  # plots y - axis

    for i in range (7):                                          # to plot x tics
        x = minx + (i - 1)*0.1                               # world coordinates
        x1 = (int)(mx*x + bx)                             # convert to canvas coord
        x2 = (int)(mx*minx + bx)
```

```
        y = miny + i*0.5                                      # real coordinates
        y2 = (int)(my*y + by)                           # convert to canvas coords
        c.create_line(x1, yc - 4, x1, yc + 4, fill = 'red')           # tics  x
        c.create_line(x2 - 4, y2, x2 + 4, y2, fill = 'red')           # tics  y
        c.create_text(x1 + 10, yc + 10, text = '%5.2f'% (x), \
                      fill = 'red', anchor = E)                        # x axis
        c.create_text(x2 + 30, y2, text = '%5.2f'% (y),\
                      fill = 'red', anchor = E)                        # y axis
    c.create_text(70, 30, text = 'Entropy', fill = 'red', anchor = E) # y
    c.create_text(420, yc - 10, text = 'mu', fill = 'red', anchor = E)# x

mx, bx, my, by = world2sc(minx, maxy, maxx, miny)              # return a list
xyaxis(mx, bx, my, by)                                      # give values to axis
mu0 = mumin*mx + bx                                           # for the beginning
entr0 = my*0.0 + by                                  # first coord. mu0, entr0

for mu in arange(mumin, mumax, dmu):                                  # mu loop
    print(mu)
    for j in range(1, nbin):
        prob[j] = 0
    y  = 0.5
    for n in range(1, nmax + 1):
        y = mu*y*(1.0 - y) # Logistic map, Skip transients
        if (n > 30000):
            ibin = int(y*nbin)  + 1
            prob[ibin]  += 1
    entropy = 0.
    for ibin in range(1, nbin):
        if (prob[ibin]>0):
            entropy = entropy -
                (prob[ibin]/nmax)*math.log10(prob[ibin]/nmax)
    entrpc = my*entropy + by
    muc = mx*mu + bx
    c.create_line(mu0, entr0, muc, entrpc, width = 1, fill = 'blue')
    mu0 = muc
    entr0 = entrpc
root.mainloop()
```

14.9
Coupled Predator–Prey Models

At the beginning of this chapter, we saw complicated behavior arising from a model of bug population dynamics in which we imposed a maximum population. We described that system with a discrete logistic map. Now we study models describing predator–prey population dynamics proposed by the American physical chemist Lotka (Lotka, 1925) and the Italian mathematician Volterra (Volterra, 1926; Gurney and Nisbet, 1998). Though simple, versions of these equations are used to model biological systems and neural networks.

Problem Is it possible to use a small number of predators to control a population of pests so that the number of pests remains essentially constant? Include in your considerations the interaction between the populations as well as the competition for food and predation time.

14.10
Lotka–Volterra Model

We extend the logistic map to the Lotka–Volterra Model (LVM) which describes two populations coexisting in the same geographical region. Let

$$p(t) = \text{prey density}\,, \quad P(t) = \text{Predator density}\,. \tag{14.33}$$

In the absence of interactions between the species, we assume that the prey population p breeds at a per-capita rate of a. This would lead to exponential growth:

$$\frac{\Delta p}{\Delta t} = ap \quad \text{(Discrete)}\,, \tag{14.34}$$

$$\frac{\mathrm{d}p}{\mathrm{d}t} = ap \quad \text{(Continuous)}\,, \tag{14.35}$$

$$\Rightarrow \quad p(t) = p(0)\mathrm{e}^{at}\,. \tag{14.36}$$

Here we give both the discrete and continuous versions of the model, with the exponential the solution of the continuous version. Yet exponential growth does not occur because the predators P eat more prey if the number of prey increases. The interaction rate between predator and prey requires both to be present, with the simplest assumption being that it is proportional to their joint probability:

$$\text{Interaction rate} = bpP\,, \tag{14.37}$$

where b is a constant. This leads to a prey growth rate including both predation and breeding:

$$\frac{\Delta p}{\Delta t} = ap - bpP \quad \text{(Discrete LVM-I for prey)}\,, \tag{14.38}$$

$$\boxed{\frac{\mathrm{d}p}{\mathrm{d}t} = ap - bpP \quad \text{(LVM-I for prey)}\,.} \tag{14.39}$$

If left to themselves, predators P will also breed and increase their population. Yet predators need animals to eat, and if there are no other populations to prey upon, we assume that they will eat each other (or their young) at a per-capita mortality rate m:

$$\left.\frac{\mathrm{d}P}{\mathrm{d}t}\right|_{\text{competition}} = -mP \quad \Rightarrow \quad P(t) = P(0)\mathrm{e}^{-mt}\,. \tag{14.40}$$

However, if there are prey also to interact with (read "eat") at the rate bpP, the predator population will grow at the rate

$$\boxed{\frac{\mathrm{d}P}{\mathrm{d}t} = \epsilon bpP - mP \quad \text{(LVM-I for predators)}\,,} \tag{14.41}$$

where ϵ is a constant that measures the efficiency with which predators convert prey interactions into food.

Equations 14.39 and 14.41 are two simultaneous ODEs and are our first model. After placing them in the standard dynamic form, we solve them with the `rk4` algorithm:

$$dy/dt = f(y, t) ,$$
$$y_0 = p , \qquad f_0 = a y_0 - b y_0 y_1 , \tag{14.42}$$
$$y_1 = P , \qquad f_1 = \epsilon b y_0 y_1 - m y_1 .$$

A sample code to solve these equations is `PredatorPrey.py` in Listing 14.4.

Listing 14.4 PredatorPrey.py computes population dynamics for a group of interacting predators and prey.

```
# PredatorPrey.py:          Lotka-Volterra models

from visual import *
from visual.graph import *

Tmin = 0.0
Tmax = 500.0
y = zeros( (2), float)
Ntimes = 1000
y[0] = 2.0
y[1] = 1.3
h = (Tmax - Tmin)/Ntimes
t = Tmin

def f( t, y, F):                      # Modify this function for your problem
        F[0] = 0.2*y[0]*(1 - (y[0]/(20.0) )) - 0.1*y[0]*y[1]
        F[1]  = - 0.1*y[1] + 0.1*y[0]*y[1];

def rk4(t, y, h, Neqs):                         # rk4 method,   DO NOT modify
    F = zeros((Neqs), float)
    ydumb = zeros((Neqs), float)
    k1    = zeros((Neqs), float)
    k2    = zeros((Neqs), float)
    k3    = zeros((Neqs), float)
    k4    = zeros((Neqs), float)
    f(t, y, F)
    for i in range(0, Neqs):
        k1[i] = h*F[i]
        ydumb[i] = y[i] + k1[i]/2.
    f(t + h/2., ydumb, F)
    for i in range(0, Neqs):
        k2[i] = h*F[i]
        ydumb[i] = y[i] + k2[i]/2.
    f(t + h/2., ydumb, F)
    for i in range(0, Neqs):
        k3[i] =  h*F[i]
        ydumb[i] = y[i] + k3[i]
    f(t + h, ydumb, F)
    for i in range(0, Neqs):
        k4[i] = h*F[i]
        y[i] = y[i] + (k1[i] + 2.*(k2[i] + k3[i]) + k4[i])/6.

graph1 = gdisplay(x= 0,y= 0, width = 500, height = 400, \
      title = 'Prey p(green) and predator P(yellow) vs time',xtitle =
          't', \
     ytitle = 'P, p',xmin=0,xmax=500,ymin=0,ymax=3.5)
```

```
funct1 = gcurve(color = color.yellow)
funct2 = gcurve(color = color.green)
graph2 = gdisplay(x= 0,y= 400, width = 500, height = 400,
                  title = 'Predator P vs prey p',
             xtitle = 'P', ytitle = 'p',xmin=0,xmax=2.5,ymin=0,ymax=3.5)
funct3 = gcurve(color = color.red)

for t in arange(Tmin, Tmax + 1, h):
    funct1.plot(pos = (t, y[0]) )
    funct2.plot(pos = (t, y[1]) )
    funct3.plot(pos = (y[0], y[1]) )
    rate(60)
    rk4(t, y, h, 2)
```

14.10.1 Lotka–Volterra Assessment

Results from the code `PredatorPrey.py` are shown in Figure 14.7. In Figure 14.7a, we see that the two populations oscillate out of phase with each other in time; when there are many prey, the predator population eats them and grows; yet then the predators face a decreased food supply and so their population decreases; that in turn permits the prey population to grow, and so forth. In Figure 14.7b, we plot what is called a "phase–space" plot of $P(t)$ vs. $p(t)$. A closed orbit here indicates a limit cycle that repeats indefinitely. Although increasing the initial number of predators does decrease the maximum number of pests, it is not a satisfactory solution to our *problem*, as the large variation in the number of pests cannot be called "control."

1. Explain in words why their is a correlation between extremums in prey and predator populations?
2. Because predators eat prey, one might expect the existence of a large number of predators to lead to the eating of a large number of prey. Explain why the maxima in predator population and the minima in prey population do not occur at the same time.
3. Why do the extreme values of the population just repeat with no change in their values?
4. Explain the meaning of the spirals in the predator–prey phase space diagram.
5. Explain why the phase–space orbits closed?
6. What different initial conditions would lead to different phase–space orbits?
7. Discuss the symmetry and lack of symmetry in the phase–space orbits.

14.11 Predator–Prey Chaos

Mathematical analysis tells us that, in addition to nonlinearity, a system must contain a minimum number of degrees of freedom before chaos will occur. For example, for chaos to occur in a predator–prey model, there must be three or more

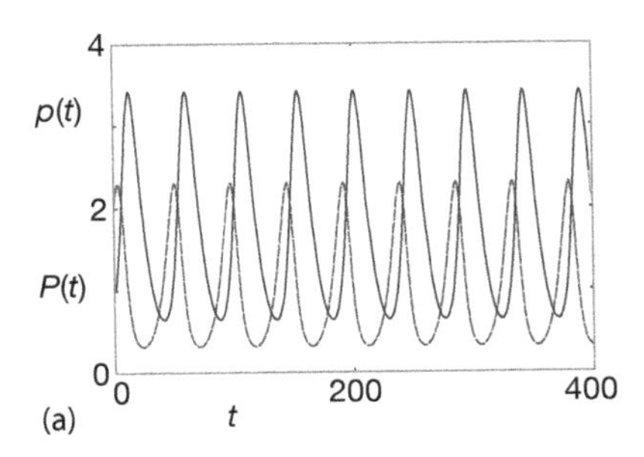

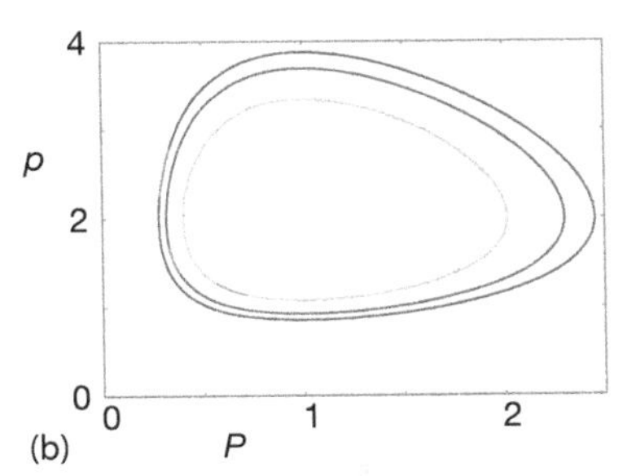

Figure 14.7 (a) The time dependencies of the populations of prey $p(t)$ (solid curve) and of predator $P(t)$ (dashed curve) from the Lotka–Volterra model. (b) A phase–space plot of Prey population p as a function of predator population P. The different orbits correspond to different initial conditions.

species present. And so to produce some real chaos, we must extend our previous treatment. Accordingly, we now extend the predator–prey model to include four species, each with population p_i competing for the same finite set of resources. The differential equation form of the model (Vano *et al.*, 2006; Cencini *et al.*, 2010) extends the Lotka–Volterra model (14.41) to:

$$\frac{\mathrm{d}p_i}{\mathrm{d}t} = a_i p_i \left(1 - \sum_{j=1}^{4} b_{ij} p_j\right), \quad i = 1, 4 . \tag{14.43}$$

Here a_i is a measure of the growth rate of species i, and b_{ij} is a measure of the extent to which species j consumes the resources otherwise used by species i. If we require both $a_i \geq 0$ and $b_{ij} \geq 0$, then all populations should remain in the range $0 \leq p_i \leq 1$.

Because four species covers a very large parameter space, we suggest that you start your exploration using the same parameters that (Vano *et al.*, 2006) found produces chaos:

$$a_i = \begin{pmatrix} 1 \\ 0.72 \\ 1.53 \\ 1.27 \end{pmatrix}, \quad b_{ij} = \begin{pmatrix} 1 & 1.09 & 1.52 & 0 \\ 0 & 1 & 0.44 & 1.36 \\ 2.33 & 0 & 1 & 0.47 \\ 1.21 & 0.51 & 0.35 & 1 \end{pmatrix} . \tag{14.44}$$

But because chaotic systems are hyper sensitive to the exact parameter values, you may need to modify these slightly.

Note that the self-interactions terms $a_{ii} = 1$, which is a consequence of measuring the population of each species in units of its individual carrying capacity. We solve (14.43) with initial conditions corresponding to an equilibrium point at

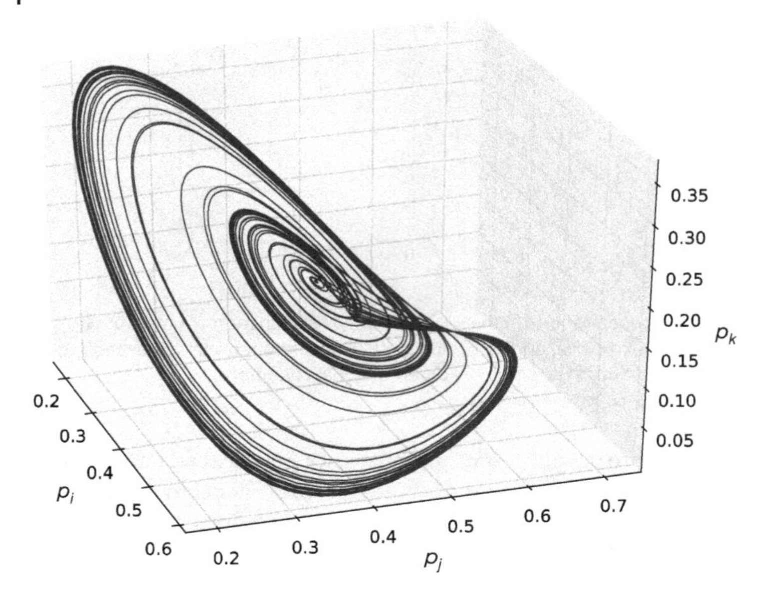

Figure 14.8 A chaotic attractor for the 4D Lotka–Volterra model projected onto three axes.

which all species coexist:

$$p_i(t=0) = \begin{pmatrix} 0.3013 \\ 0.4586 \\ 0.1307 \\ 0.3557 \end{pmatrix}. \tag{14.45}$$

One way to visualize the solution to (14.43) is to make four plots of $p_1(t)$, $p_2(t)$, $p_3(t)$, and $p_4(t)$ vs. time. The results are quite interesting, and we leave that as an exercise for you. A more illuminating visualization would be to create a 4D phase–space plot of $[p_1(t_i), p_2(t_i), p_3(t_i), p_4(t_i)]$ for $i = 1, N$, where N is the number of time steps used in your numerical solution. Even in the presence of chaos, the geometric structure so created may have a smooth and well-defined shape. Unfortunately, we have no way to show such a plot, and so in its stead we must project the 4D structure onto 2D and 3D axes. We show one such 3D plot in Figure 14.8. This is a classic type of chaotic attractor, with the 3D structure folded over into a nearly 2D structure (the Rossler equation is famous for producing such a structure).

14.11.1 Exercises

1. Solve (14.43) for the suggested parameters and initial conditions, and make plots of $p_{1-4}(t)$ as functions of time. Comment on the type of behavior that these plots exhibit.
2. Construct the 4D phase chaotic attractor formed by the solutions of (14.43) by writing to a file the values $[p_1(t_i), p_2(t_i), p_3(t_i), p_4(t_i)]$ where i labels the time step. In order to avoid needlessly long files, you may want to skip a number of time steps before each file output.
   a) Plot all possible 2D phase–space plots, that is, plots of p_i *vs* p_j, $i \neq j = 1-3$.
   b) Plot all possible 3D phase–space plots, that is, plots of p_i *vs* p_j *vs* p_k.

   Note: you have to adjust the parameters or initial conditions slightly to obtain truly chaotic behavior.

14.11.2 LVM with Prey Limit

The initial assumption in the LVM that prey grow without limit in the absence of predators is clearly unrealistic. As with the logistic map, we include a limit on prey numbers that accounts for depletion of the food supply as the prey population grows. Accordingly, we modify the constant growth rate from a to $a(1 - p/K)$ so that growth vanishes when the population reaches a limit K, the *carrying capacity*:

$$\frac{dp}{dt} = ap\left(1 - \frac{p}{K}\right) - bpP\,, \tag{14.46}$$

$$\frac{dP}{dt} = \epsilon bpP - mP \quad \text{(LVM-II)}\,. \tag{14.47}$$

The behavior of this model with prey limitations is shown in Figure 14.9. We see that both populations exhibit damped oscillations as they approach their equilibrium values, and that as hoped for, the equilibrium populations are independent

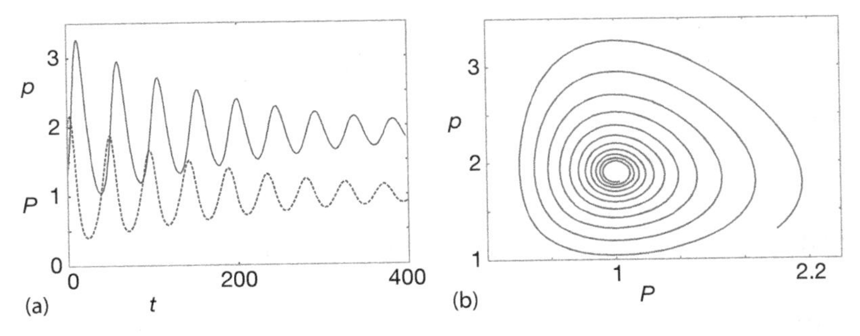

Figure 14.9 (a) The Lotka–Volterra model of prey population $p(t)$ (solid curve) and predator population $P(t)$ (dashed curve) vs. time when a prey population limit is included. (b) Prey population p as a function of predator population P.

of the initial conditions. Note how the phase–space plot spirals inward to a single close limit cycle, on which it remains, with little variation in prey number. This is "control," and we may use it to start a chemical-free pest control business.

14.11.3 LVM with Predation Efficiency

An additional unrealistic assumption in the original LVM is that the predators immediately eat all the prey with which they interact. As anyone who has watched a cat hunt a mouse knows, predators spend some time finding prey and also chasing, killing, eating, and digesting it (all together called *handling*). This extra time decreases the rate bpP at which prey are eliminated. We define the *functional response* p_a as the probability of one predator finding one prey. If a single predator spends time t_{search} searching for prey, then

$$p_a = bt_{\text{search}}p \quad \Rightarrow \quad t_{\text{search}} = \frac{p_a}{bp} . \tag{14.48}$$

If we call t_{h} the time a predator spends handling a single prey, then the effective time a predator spends handling a prey is $p_a t_{\text{h}}$. Such being the case, the total time T that a predator spends finding and handling a single prey is

$$T = t_{\text{search}} + t_{\text{handling}} = \frac{p_a}{bp} + p_a t_{\text{h}} \tag{14.49}$$

$$\Rightarrow \quad \frac{p_a}{T} = \frac{bp}{1 + bpt_{\text{h}}} , \tag{14.50}$$

where p_a/T is the effective *rate* of eating prey. We see that as the number of prey $p \to \infty$, the efficiency in eating them $\to 1$. We include this efficiency in (14.46) by modifying the rate b at which a predator eliminates prey to $b/(1 + bpt_{\text{h}})$:

$$\frac{\mathrm{d}p}{\mathrm{d}t} = ap\left(1 - \frac{p}{K}\right) - \frac{bpP}{1 + bpt_{\text{h}}} , \quad \text{(LVM-III)} . \tag{14.51}$$

To be more realistic about the predator growth, we also place a limit on the predator carrying capacity but make it proportional to the number of prey:

$$\frac{\mathrm{d}P}{\mathrm{d}t} = mP\left(1 - \frac{P}{kp}\right) , \quad \text{(LVM-III)} . \tag{14.52}$$

Solutions for the extended model (14.51) and (14.52) are shown in Figure 14.10. Observe the existence of three dynamic regimes as a function of b:

- small b: no oscillations, no overdamping,
- medium b: damped oscillations that converge to a stable equilibrium,
- large b: limit cycle.

The transition from equilibrium to a limit cycle is called a *phase transition*.

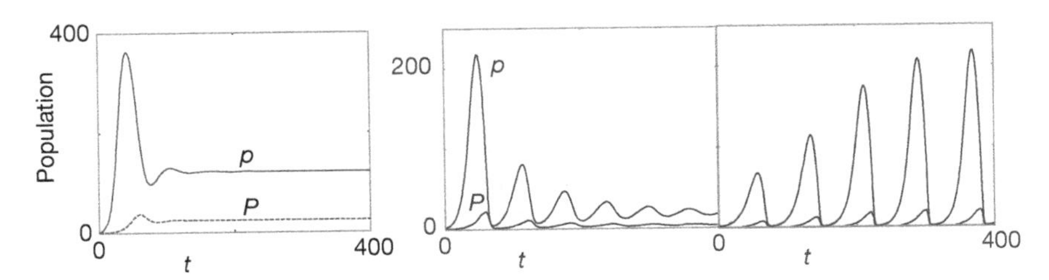

Figure 14.10 Lotka–Volterra model with predation efficiency and prey limitations. From left to right: overdamping, $b = 0.01$; damped oscillations, $b = 0.1$, and limit cycle, $b = 0.3$.

We finally have a satisfactory solution to our *problem*. Although the prey population is not eliminated, it can be kept from getting too large and from fluctuating widely. Nonetheless, changes in the parameters can lead to large fluctuations or to nearly vanishing predators.

14.11.4 LVM Implementation and Assessment

1. Write a program to solve (14.51) and (14.52) using the `rk4` algorithm and the following parameter values.

Model	a	b	ϵ	m	K	k
LVM-I	0.2	0.1	1	0.1	0	—
LVM-II	0.2	0.1	1	0.1	20	—
LVM-III	0.2	0.1	—	0.1	500	0.2

2. For each of the three models, construct
   a) a time series for prey and predator populations,
   b) phase–space plots of predator vs. prey populations.
3. *LVM-I*: Compute the equilibrium values for the prey and predator populations. Do you think that a model in which the cycle amplitude depends on the initial conditions can be realistic? Explain.
4. *LVM-II*: Calculate numerical values for the equilibrium values of the prey and predator populations. Make a series of runs for different values of prey carrying capacity K. Can you deduce how the equilibrium populations vary with prey carrying capacity?
5. Make a series of runs for different initial conditions for predator and prey populations. Do the cycle amplitudes depend on the initial conditions?
6. *LVM-III*: Make a series of runs for different values of b and reproduce the three regimes present in Figure 14.10.
7. Calculate the critical value for b corresponding to a phase transition between the stable equilibrium and the limit cycle.

14.11.5

Two Predators, One Prey (Exploration)

1. Another version of the LVM includes the possibility that two populations of predators P_1 and P_2 may "share" the same prey population p. Investigate the behavior of a system in which the prey population grows logistically in the absence of predators:

$$\frac{\mathrm{d}p}{\mathrm{d}t} = ap\left(1 - \frac{p}{K}\right) - \left(b_1 P_1 + b_2 P_2\right) p \,, \tag{14.53}$$

$$\frac{\mathrm{d}P}{\mathrm{d}t} = \epsilon_1 b_1 p P_1 - m_1 P_1 \,, \quad \frac{\mathrm{d}P_2}{\mathrm{d}t} = \epsilon_2 b_2 p P_2 - m_2 P_2 \,. \tag{14.54}$$

   a) Use the following values for the model parameters and initial conditions: $a = 0.2, K = 1.7, b_1 = 0.1, b_2 = 0.2, m_1 = m_2 = 0.1, \epsilon_1 = 1.0, \epsilon_2 = 2.0, p(0) = P_2(0) = 1.7$, and $P_1(0) = 1.0$.
   b) Determine the time dependence for each population.
   c) Vary the characteristics of the second predator and calculate the equilibrium population for the three components.
   d) What is your answer to the question, "Can two predators that share the same prey coexist?"
2. The nonlinear nature of the Lotka–Volterra model can lead to chaos and fractal behavior. Search for complex behaviors by varying the growth rates.

15
Continuous Nonlinear Dynamics

In Chapter 14, we developed the logistic map and predator–prey models as a means to understand how biological populations achieve dynamic equilibrium. In this chapter, we explore the driven realistic pendulum, a continuous system chaos that can support chaotic behavior. In Chapter 8, we have already seen how useful our tools are at solving nonlinear equations. Our emphasis now is on exploring chaos and on the usefulness of phase space in displaying the simplicity underlying complex behavior.

15.1
Chaotic Pendulum

Problem The plane pendulum is a classic subject for physics. However, it has most often been studied assuming small angle displacements, which is a good approximation only for very large grandfather clocks. However, when the pendulum is driven and the displacements get large, the motion becomes too complicated for analytic solution. Your *problem* is to compute the motion of a driven pendulum with large displacements, to ensure that your calculation is reliable and sensitive, and then to search for the simplicity that may underly complexity.

What we call a *chaotic pendulum* is just a pendulum with friction and a driving torque (Figure 15.1a), but without the assumption of small deflection angle. Newton's laws of rotational motion tell us that the sum of the gravitational torque $-mgl\sin\theta$, the frictional torque $-\beta\dot{\theta}$, and the external torque $\tau_0\cos\omega t$ equals the moment of inertia of the pendulum times its angular acceleration (Rasband, 1990):

$$I\frac{\mathrm{d}^2\theta}{\mathrm{d}t^2} = -mgl\sin\theta - \beta\frac{\mathrm{d}\theta}{\mathrm{d}t} + \tau_0\cos\omega t \tag{15.1}$$

$$\Rightarrow \quad \frac{\mathrm{d}^2\theta}{\mathrm{d}t^2} = -\omega_0^2\sin\theta - \alpha\frac{\mathrm{d}\theta}{\mathrm{d}t} + f\cos\omega t\,, \tag{15.2}$$

$$\omega_0 = \frac{mgl}{I}\,, \quad \alpha = \frac{\beta}{I}\,, \quad f = \frac{\tau_0}{I}\,. \tag{15.3}$$

Computational Physics, 3rd edition. Rubin H. Landau, Manuel J. Páez, Cristian C. Bordeianu.

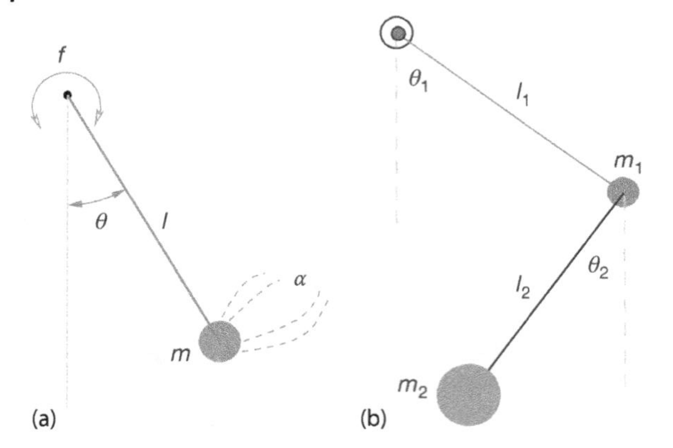

Figure 15.1 (a) A pendulum of length l driven through resistive air (dotted arcs) by an external sinusoidal torque (semicircle). The strength of the external torque is given by f and that of air resistance by α. (b) A double pendulum with neither air resistance nor a driving force. In both cases there is a gravitational torque.

Equation 15.2 is a second-order time-dependent nonlinear differential equation. Its nonlinearity arises from the $\sin\theta$, as opposed to the θ, dependence of the gravitational torque. The parameter ω_0 is the natural frequency of the system arising from the restoring torque, α is a measure of the strength of friction, and f is a measure of the strength of the driving torque. In our standard ODE form, $\mathrm{d}\boldsymbol{y}/\mathrm{d}t = \boldsymbol{f}$ (Chapter 8), we have two simultaneous first-order equations:

$$\frac{\mathrm{d}y^{(0)}}{\mathrm{d}t} = y^{(1)} , \tag{15.4}$$

$$\frac{\mathrm{d}y^{(1)}}{\mathrm{d}t} = -\omega_0^2 \sin y^{(0)} - \alpha y^{(1)} + f \cos\omega t ,$$

$$y^{(0)} = \theta(t) , \quad y^{(1)} = \frac{\mathrm{d}\theta(t)}{\mathrm{d}t} . \tag{15.5}$$

15.1.1 Free Pendulum Oscillations

If we ignore friction and external torques, Newton's law (15.2) takes the simple, yet still nonlinear, form

$$\frac{\mathrm{d}^2\theta}{\mathrm{d}t^2} = -\omega_0^2 \sin\theta . \tag{15.6}$$

If the displacements are small, we can approximate $\sin\theta$ by θ and obtain the linear equation of simple harmonic motion with frequency ω_0:

$$\frac{\mathrm{d}^2\theta}{\mathrm{d}t^2} \simeq -\omega_0^2\theta \quad \Rightarrow \quad \theta(t) = \theta_0 \sin(\omega_0 t + \phi) . \tag{15.7}$$

In Chapter 8, we studied how nonlinearities produce anharmonic oscillations, and indeed (15.6) is another good candidate for such studies. As before, we expect solutions of (15.6) for the free realistic pendulum to be periodic, but with a frequency $\omega \simeq \omega_0$ only for small oscillations. Furthermore, because the restoring torque, $mgl\sin\theta \simeq mgl(\theta - \theta^3/3)$, is less than the $mgl\theta$ assumed in a harmonic oscillator, realistic pendulums swing slower (have longer periods) as their angular displacements are made larger.

15.1.2 Solution as Elliptic Integrals

The analytic solution to the realistic pendulum is a textbook problem (Landau and Lifshitz, 1976; Marion and Thornton, 2003; Scheck, 1994), except that it is hardly a solution and hardly analytic. The "solution" is based on energy being a constant (integral) of the motion. For simplicity, we start the pendulum off at rest at its maximum displacement θ_m. Because the initial energy is all potential, we know that the total energy of the system equals its initial potential energy (Figure 15.1),

$$E = \text{PE}(0) = mgl - mgl\cos\theta_\text{m} = 2mgl\sin^2\left(\frac{\theta_\text{m}}{2}\right) . \tag{15.8}$$

Yet because $E = \text{KE} + \text{PE}$ is a constant, we can write for any value of θ

$$2mgl\sin^2\frac{\theta_\text{m}}{2} = \frac{1}{2}I\left(\frac{d\theta}{dt}\right)^2 + 2mgl\sin^2\frac{\theta}{2},$$

$$\Rightarrow \quad \frac{d\theta}{dt} = 2\omega_0\left[\sin^2\frac{\theta_\text{m}}{2} - \sin^2\frac{\theta}{2}\right]^{1/2}$$

$$\Rightarrow \quad \frac{dt}{d\theta} = \frac{T_0/\pi}{\left[\sin^2(\theta_\text{m}/2) - \sin^2(\theta/2)\right]^{1/2}},$$

$$\Rightarrow \quad \frac{T}{4} = \frac{T_0}{4\pi}\int_0^{\theta_\text{m}} \frac{d\theta}{\left[\sin^2(\theta_\text{m}/2) - \sin^2(\theta/2)\right]^{1/2}}, \tag{15.9}$$

$$\Rightarrow \quad T \simeq T_0\left[1 + \left(\frac{1}{2}\right)^2\sin^2\frac{\theta_\text{m}}{2} + \left(\frac{1\cdot 3}{2\cdot 4}\right)^2\sin^4\frac{\theta_\text{m}}{2} + \cdots\right] . \tag{15.10}$$

Because the motion is still periodic, we have assumed that it takes $T/4$ for the pendulum to travel from $\theta = 0$ to $\theta = \theta_\text{m}$. The integral in (15.9) can be expressed as an *elliptic integral of the first kind* (studied in Section 5.16). If you think of an elliptic integral as a generalization of a trigonometric function, then this is a closed-form solution; otherwise, it is an integral needing computation. The series expansion of the period (15.10) is obtained by expanding the denominator and integrating it term by term. It tells us, for example, that an amplitude of 80° leads to a 10% slowdown of the pendulum relative to the small θ result. In contrast, we will determine the period computationally without the need for any expansions.

15.1.3
Implementation and Test: Free Pendulum

As a preliminary to the solution of the full equation (15.2), modify your `rk4` program to solve (15.6) for the free oscillations of a realistic pendulum.

1. Start your pendulum at $\theta = 0$ with $\dot{\theta}(0) \neq 0$. Gradually increase $\dot{\theta}(0)$ to increase the importance of nonlinear effects.
2. Test your program for the linear case ($\sin\theta \to \theta$) and verify that
   a) your solution is harmonic with frequency $\omega_0 = 2\pi/T_0$, and
   b) the frequency of oscillation is independent of the amplitude.
3. Devise an algorithm to determine the period T of the oscillation by counting the time it takes for three successive passes of the amplitude through $\theta = 0$. (You need *three* passes because a general oscillation may not be symmetric about the origin.) Test your algorithm for simple harmonic motion where you know T_0.
4. For the realistic pendulum, observe the change in period as a function of increasing initial energy or displacement. Plot your observations along with (15.10).
5. Verify that as the initial KE approaches $2mgl$, the motion remains oscillatory but not harmonic (Figure 15.4).
6. At $E = 2mgl$ (the *separatrix*), the motion changes from oscillatory to rotational ("over the top" or "running"). See how close you can get your solution to the separatrix and to its infinite period.
7. ⊙ Use the applet `HearData` (Figure 15.2) to convert your different oscillations to sound and hear the difference between harmonic motion (boring) and anharmonic motion containing overtones (interesting).

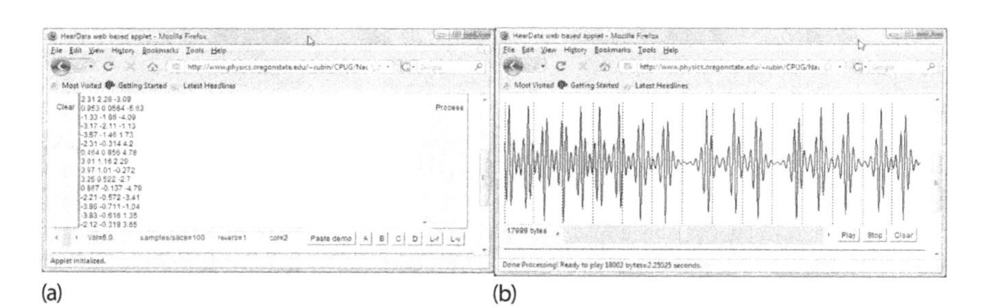

(a) (b)

Figure 15.2 The data screen (a) and the output screen (b) of the applet `HearData` that converts data into sounds. Columns of $[t_i, x(t_i)]$ data are pasted into the data window, processed into the graph in the output window, and then converted to sound data that are played by Java.

15.2
Visualization: Phase-Space Orbits

The conventional solution to an equation of motion is the position $x(t)$ and the velocity $v(t)$ as functions of time. Often behaviors that appear complicated as functions of time appear simpler when viewed in an abstract space called *phase space*, where the ordinate is the velocity $v(t)$ and the abscissa is the position $x(t)$ (Figure 15.3). As we see from the phase-space figures, the solutions form geometric objects that are easy to recognize. (We provide two applets `Pend1` and `Pend2` to help the reader make the connections between phase-space shapes and the corresponding physical motion.)

The position and velocity of a free harmonic oscillator are given by the trigonometric functions

$$x(t) = A\sin(\omega t)\,, \quad v(t) = \frac{\mathrm{d}x}{\mathrm{d}t} = \omega A\cos(\omega t)\,. \tag{15.11}$$

When substituted into the total energy, we obtain two important results:

$$E = \mathrm{KE} + \mathrm{PE} = \left(\frac{1}{2}m\right)v^2 + \left(\frac{1}{2}\omega^2 m^2\right)x^2 \tag{15.12}$$

$$= \frac{m\omega^2 A^2}{2}\cos^2(\omega t) + \frac{m\omega^2 A^2}{2}\sin^2(\omega t) = \frac{1}{2}m\omega^2 A^2\,. \tag{15.13}$$

The first equation, being that of an ellipse, proves that the harmonic oscillator follows closed elliptical orbits in phase space, with the size of the ellipse increasing with the system's energy. The second equation proves that the total energy is a constant of the motion. Different initial conditions having the same energy start at different places on the same ellipse and transverse the same orbits.

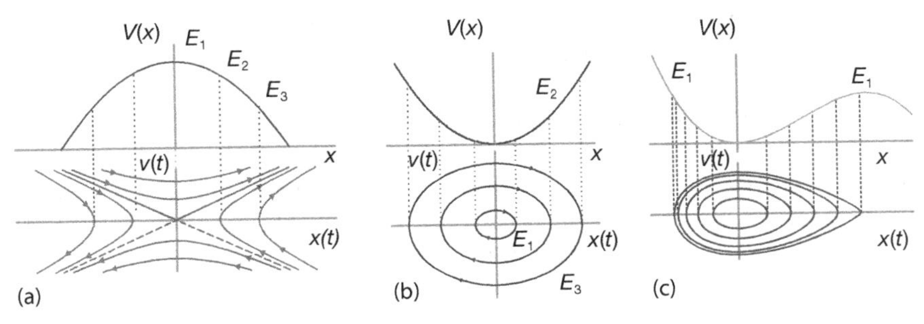

Figure 15.3 Three potentials and their behaviors in phase space. The different orbits below the potentials correspond to different energies, as indicated by the limits of maximum displacements within the potentials (dashed lines). (a) A repulsive potential leads to open orbits characteristic of nonperiodic motion. The trajectories cross at the hyperbolic point in the middle, an unstable equilibrium point. (b) The harmonic oscillator leads to symmetric ellipses; the closed orbits indicate periodic behavior, and the symmetric trajectories indicate a symmetric potential. (c) A nonharmonic oscillator. Notice that the ellipse-like trajectories at the bottom are neither ellipses nor symmetric.

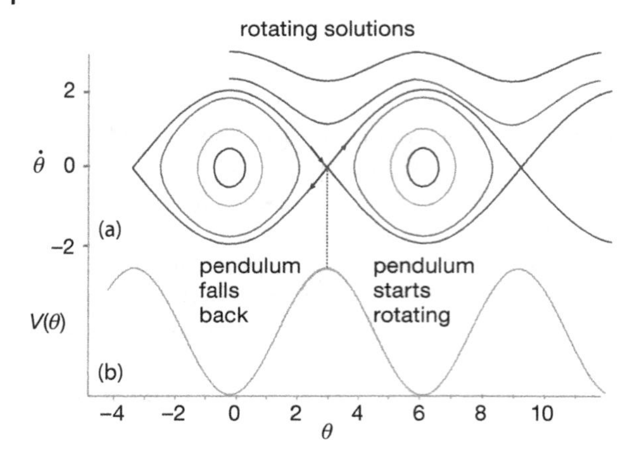

Figure 15.4 (a) Phase-space trajectories for a pendulum including "over the top" motions. (b) The corresponding θ dependence of the potential.

In Figures 15.3–15.7, we show various phase-space structures. *Study these figures and their captions* and note the following:

- The orbits of anharmonic oscillations will still be ellipse like, but with angular corners that become more distinct with increasing nonlinearity.
- Closed trajectories describe periodic oscillations (the same (x, v) occur again and again), with a clockwise motion arising from the restoring torque.
- Open orbits correspond to nonperiodic or "running" motion (a pendulum rotating like a propeller).
- Regions of space where the potential is repulsive lead to open trajectories in phase space (Figure 15.3).
- As seen in Figure 15.4a, the separatrix corresponds to the trajectory in phase space that separates open and closed orbits. Motion on the separatrix itself is indeterminant, as the pendulum may balance, or move either way at the maxima of $V(\theta)$.
- Friction may cause the energy to decrease with time and the phase space orbit to spiral into a *fixed point*. Yet if the system is driven, it would not remain there.
- For certain parameters, a closed *limit cycle* occurs in which the energy pumped in by the external torque exactly balances that lost by friction (Figure 15.5b).
- Because solutions for different initial conditions are unique, different orbits do not cross. Nonetheless, open orbits join at points of unstable equilibrium (*hyperbolic points* in Figure15.3a) where an indeterminacy exists.

15.2.1 Chaos in Phase Space

It is easy to solve the nonlinear ODE (15.4) on the computer using our usual techniques. However, it is not so easy to understand the solutions because they are

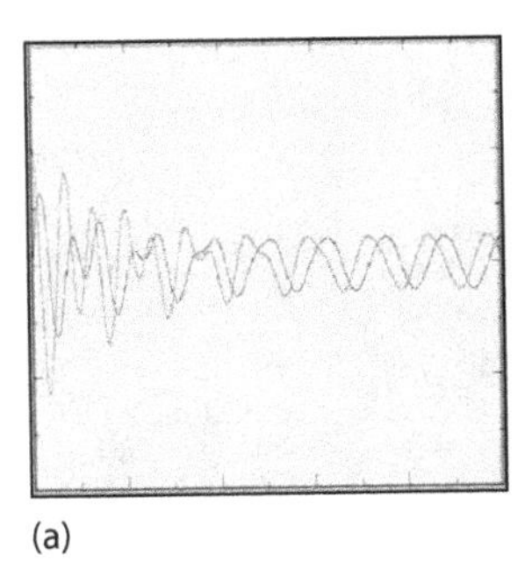

(a)

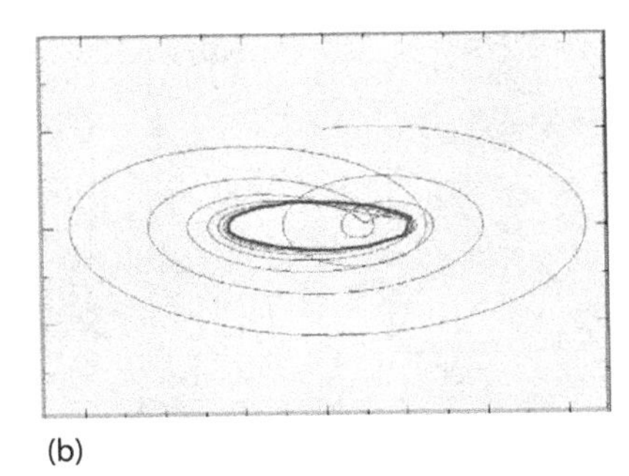

(b)

Figure 15.5 (a) Position vs. time for two initial conditions of a chaotic pendulum that end up with the same limit cycle. (b) A phase space plot of position versus velocity for the limit cycle shown in (a) (courtesy of W. Hager).

so rich in complexity. The solutions are easier to understand in phase space, particularly if you learn to recognize some characteristic structures there. Actually, there are a number of "tools" that can be used to decide if a system is chaotic in contrast to just complex. Geometric structures in phase space is one of them, and determination of the Lyupanov coefficient (discussed in Section 14.8) is another. Both signal the simplicity lying within the complexity.

What may be surprising is that even though the ellipse-like figures in phase space were observed originally for free systems with no friction and no driving torque, similar structures continue to exist for driven systems with friction. The actual trajectories may not remain on a single structure for all times, but they are *attracted* to them and return to them often. In contrast to periodic motion which produces closed figures in phase-space, random motion appears as a diffuse cloud filling the entire energetically accessible region. Complex or chaotic motion falls someplace in between (Figure 15.6b). If viewed for long times and for many initial conditions, chaotic trajectories in phase space, while resembling the familiar geometric figures, may contain dark or diffuse *bands* rather than single lines. The continuity of trajectories within bands means that a continuum of solutions are possible and that the system flows continuously among the different trajectories forming the band. The transitions among different phase-space orbits appear chaotic in normal space. The bands are what makes the solutions hypersensitive to the initial conditions as the slightest change in them causes the system to flow to nearby phase-space trajectories.

Pick out the following phase-space structures in your simulations:

Limit cycles When a chaotic pendulum is driven by a not-too-large driving torque, it is possible to pick the magnitude for this torque such that after the initial transients die off, the average energy put into the system during one period exactly balances the average energy dissipated by friction during that period (Figure 15.5):

$$\langle f \cos \omega t \rangle = \left\langle \alpha \frac{d\theta}{dt} \right\rangle = \left\langle \alpha \frac{d\theta(0)}{dt} \cos \omega t \right\rangle \quad \Rightarrow \quad f = \alpha \frac{d\theta(0)}{dt} . \tag{15.14}$$

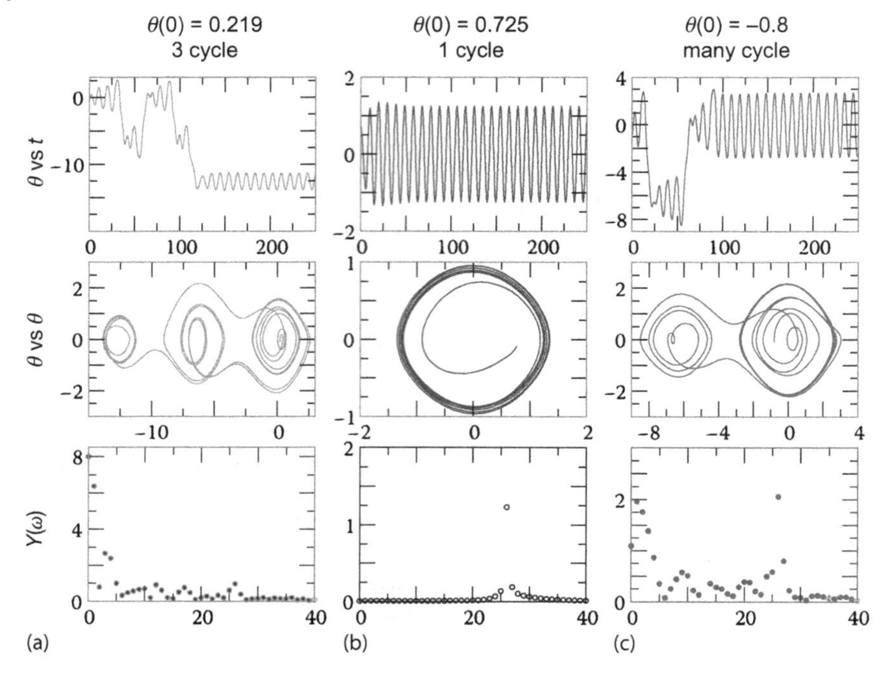

Figure 15.6 Position vs. time, phase-space plot, and Fourier spectrum for a chaotic pendulum with $\omega_0 = 1$, $\alpha = 0.2$, $f = 0.52$, and $\omega = 0.666$ and three different initial conditions. Column (a) displays three dominant cycles, the column (b) only one, while column (c) has multiple cycles. (Examples of chaotic behavior can be seen in Figure 15.7).

This leads to *limit cycles* that appear as closed ellipse-like figures. (Yet unstable solutions may make sporadic jumps between limit cycles.)

Predictable attractors There are orbits, such as fixed points and limit cycles, into which the system settles or returns to often, and that are not particularly sensitive to initial conditions. If your location in phase space is near a predictable attractor, ensuing times will bring you to it.

Strange attractors Well-defined, yet complicated, semiperiodic behaviors that appear to be uncorrelated with the motion at an earlier time. They are distinguished from predictable attractors by being fractal (Chapter 16) chaotic, and highly sensitive to the initial conditions (José and Salatan, 1998). Even after millions of oscillations, the motion remains *attracted* to them.

Chaotic paths Regions of phase space that appear as filled-in bands rather than lines. Continuity within the bands implies complicated behaviors, yet still with simple underlying structure.

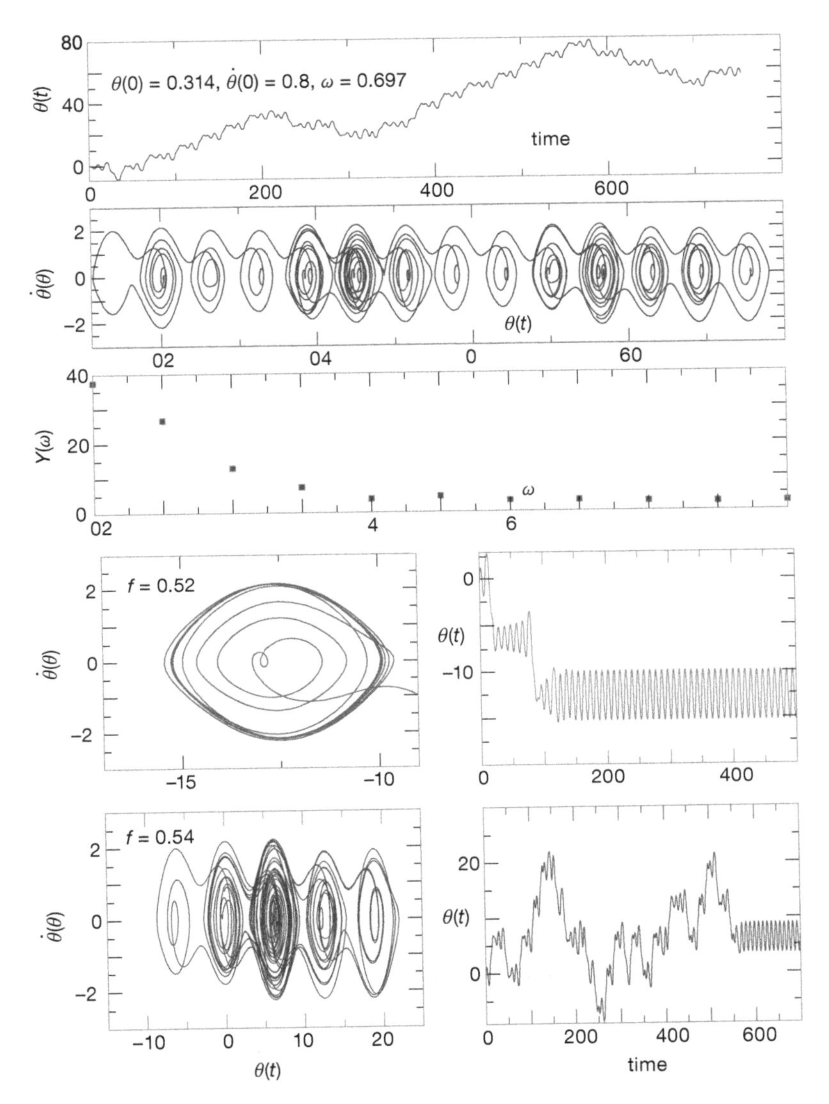

Figure 15.7 Some examples of complicated behaviors of a realistic pendulum. From top to bottom in each column of the figure, the first three rows show $\theta(t)$ behavior, a broadband Fourier spectrum, a phase-space diagram containing regular patterns with dark bands, and a broad Fourier spectrum. These features are characteristic of chaos. In the bottom two rows we see how the behavior changes abruptly after a slight change in the magnitude of the force and that for $f = 0.54$ there occur the characteristic broad bands of chaos.

Mode locking When the magnitude f of the driving torque is larger than that for a limit cycle (15.14), the driving torque can overpower the natural oscillations, resulting in a steady-state motion at the frequency of the driver. This is called *mode locking*. While mode locking can occur for linear or nonlinear systems, for nonlinear systems the driving torque may lock onto the system by exciting an overtone, leading to a rational relation between the driving frequency and the natural frequency:

$$\frac{\omega}{\omega_0} = \frac{n}{m} , \quad n, m = \text{integers} . \tag{15.15}$$

Butterfly effects One of the classic quips about the hypersensitivity of chaotic systems to the initial conditions is that the weather pattern in North America is hard to predict well because it is sensitive to the flapping of butterfly wings in South America. Although this appears to be counterintuitive because we know that systems with essentially identical initial conditions should behave the same, eventually the systems diverge. The applet `Pend2` (Figure15.8c and d) lets you compare two simulations of nearly identical initial conditions. As seen in Figure 15.8b and d, the initial conditions for both pendulums differ by only 1 part in 917, and so the initial paths in phase space are the same. Nonetheless, at just the time shown here, the pendulums balance in the vertical position, and then one falls before the other, leading to differing oscillations and differing phase-space plots from this time onward.

15.2.2 Assessment in Phase Space

The challenge in understanding simulations of the chaotic pendulum (15.4) is that the 4D parameter space $(\omega_0, \alpha, f, \omega)$ is so immense that only sections of it can be studied systematically. We would expect that sweeping through driving frequency ω should show resonances and beating; sweeping through the frictional force α should show underdamping, critical damping, and overdamping; and sweeping through the driving torque f should show mode locking (for the right values of ω). All these behaviors can be found in the solution of your differential equation, yet they are mixed together in complex ways.

In this assessment, you should try to reproduce the behaviors shown in the phase-space diagrams in Figure 15.6 and in the applets in Figure 15.8. *Beware*: Because the system is chaotic, you should expect that your results will be sensitive to the exact values of the initial conditions and to the details of your integration routine. We suggest that you experiment; start with the parameter values we used to produce our plots and then observe the effects of making *very small* changes in parameters until you obtain different modes of behavior.

1. Take your solution to the realistic pendulum and include friction. Run it for a variety of initial conditions, including over-the-top ones. Because no energy is fed to the system, you should see spirals in phase space. Note, if you plot points at uniform time steps without connecting them, then the spacing between points gives an indication of the speed of the pendulum.
2. Next, verify that with no friction, but with a very small driving torque, you obtain a perturbed ellipse in phase space.

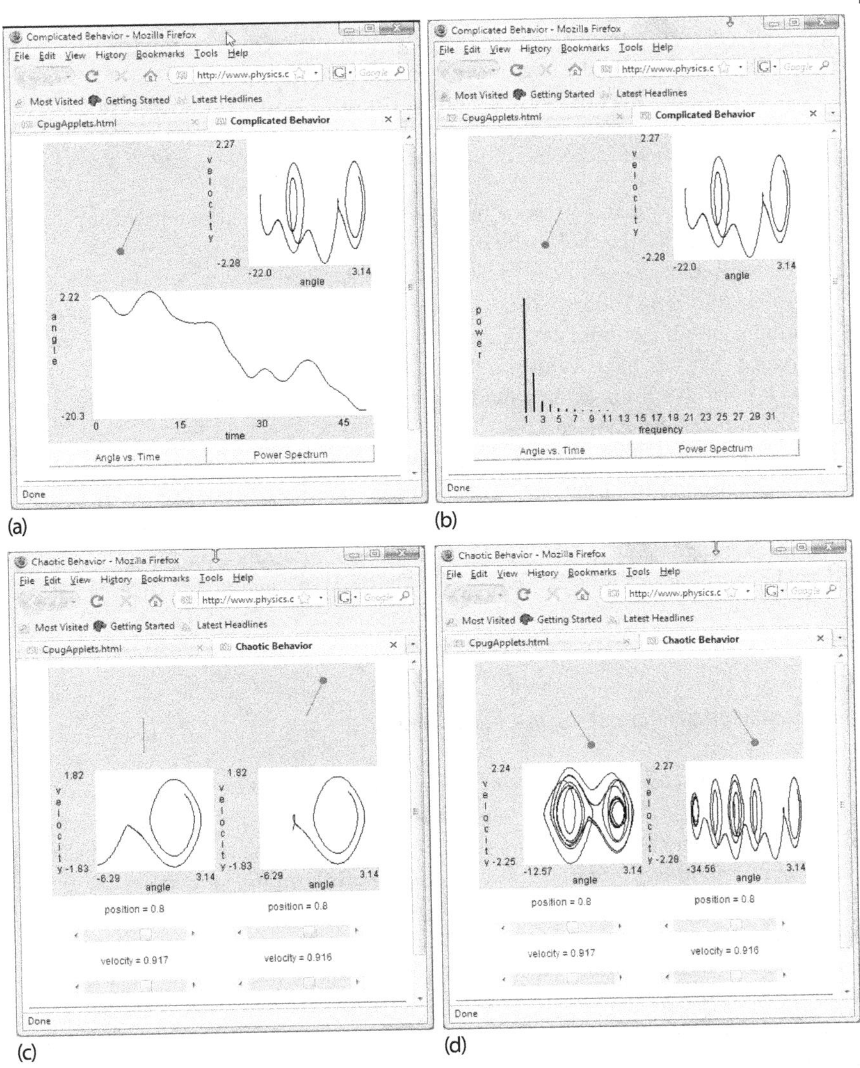

Figure 15.8 (a,b) Output from the applet `Pend1` that produces an animation of a chaotic pendulum, along with the corresponding position vs. time and phase-space plots. (b) The resulting Fourier spectrum produced by `Pend1`. (c,d) outputs from two runs of the applet `Pend2` producing an animation of two chaotic pendula along with the corresponding phase-space plots, and the final output with limit cycles (dark bands).

3. Set the driving torque's frequency close to the natural frequency ω_0 of the pendulum and search for beats (Figure 15.2b). Note that in addition to the frequency, you may need to adjust the magnitude and phase of the driving torque to avoid an "impedance mismatch" between the pendulum and driver.

4. Finally, scan the frequency ω of the driving torque and search for nonlinear resonance (it looks like beating).
5. *Explore chaos*: Start off with the initial conditions we used in Figure 15.6,

$$(x_0, v_0) = (-0.0885, 0.8)\,, \quad (-0.0883, 0.8)\,, \quad (-0.0888, 0.8)\,. \tag{15.16}$$

   To save time and storage, you may want to use a larger time step for plotting than the one used to solve the differential equations.
6. Identify which parts of the phase-space plots correspond to transients. (The applets may help you with this, especially if you watch the phase-space features being built up in time.)
7. Ensure that you have found:
   a) a period-3 limit cycle where the pendulum jumps between three major orbits in phase space,
   b) a running solution where the pendulum keeps going over the top,
   c) chaotic motion in which some paths in the phase space appear as bands.
8. Look for the "butterfly effect" (Figure 15.8c and d). Start two pendulums off with identical positions but with velocities that differ by 1 part in 1000. Notice that the initial motions are essentially identical but eventually diverge.

15.3 Exploration: Bifurcations of Chaotic Pendulums

We have seen that a chaotic system contains a number of dominant frequencies and that the system tends to "jump" from one to another. This means that the dominant frequencies occur sequentially, in contrast to linear systems where they occur simultaneously. We now want to explore this jumping as a computer experiment. If we sample the instantaneous angular velocity $\dot{\theta} = \mathrm{d}\theta/\mathrm{d}t$ of our chaotic simulation at various instances in time, we should obtain a series of values for the frequency, with the major Fourier components occurring more often than others.[1] These are the frequencies to which the system is *attracted*. That being the case, if we make a scatterplot of the sampled $\dot{\theta}$s for many times at one particular value of the driving force and then change the magnitude of the driving force slightly and sample the frequencies again, the resulting plot may show distinctive patterns of frequencies. That a bifurcation diagram similar to the one for bug populations results is one of the mysteries of life.

In the scatterplot in Figure 15.9, we sample $\dot{\theta}$ for the motion of a chaotic pendulum with a vibrating pivot point (in contrast to our usual vibrating external torque). This pendulum is similar to our chaotic one (15.2), but with the driving

1) We refer to this angular velocity as $\dot{\theta}$ because we have already used ω for the frequency of the driver and ω_0 for the natural frequency.

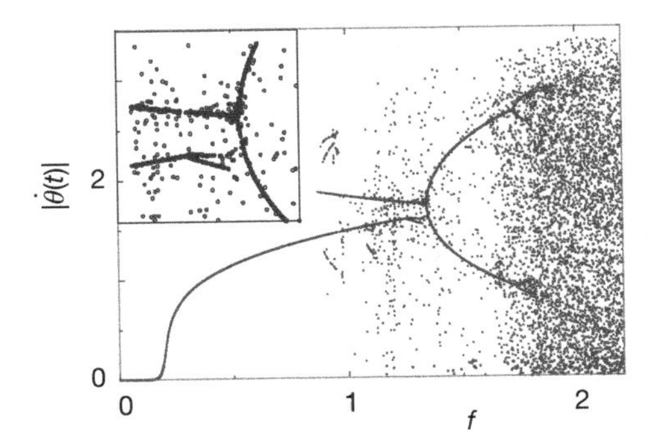

Figure 15.9 A bifurcation diagram for the damped pendulum with a vibrating pivot (see also the similar diagram for a double pendulum, Figure 15.11). The ordinate is $|\,\mathrm{d}\theta/\,\mathrm{d}t|$, the absolute value of the instantaneous angular velocity at the beginning of the period of the driver, and the abscissa is the magnitude of the driving force f. Note that the heavy line results from the overlapping of points, not from connecting the points (see enlargement in the inset).

force depending on $\sin\theta$:

$$\frac{\mathrm{d}^2\theta}{\mathrm{d}t^2} = -\alpha\frac{\mathrm{d}\theta}{\mathrm{d}t} - \left(\omega_0^2 + f\cos\omega t\right)\sin\theta\,. \tag{15.17}$$

Essentially, the acceleration of the pivot is equivalent to a sinusoidal variation of g or ω_0^2. Analytic and numeric studies of this system are in the literature (Landau and Lifshitz, 1976; DeJong, 1992; Gould *et al.*, 2006). To obtain the bifurcation diagram in Figure 15.9:

1. Use the initial conditions $\theta(0) = 1$ and $\dot{\theta}(0) = 1$.
2. Set $\alpha = 0.1$, $\omega_0 = 1$, and $\omega = 2$, and vary $0 \le f \le 2.25$.
3. For each value of f, wait 150 periods of the driver before sampling to permit transients to die off. Sample $\dot{\theta}$ for 150 times at the instant the driving force passes through zero.
4. Plot the 150 values of $|\dot{\theta}|$ vs. f.

15.4 Alternate Problem: The Double Pendulum

For those of you who have already studied a chaotic pendulum, an alternative is to study a double pendulum without any small-angle approximation (Figures 15.1b and 15.10, and animation `DoublePend.mp4`). A double pendulum has a second pendulum connected to the first, and because each pendulum acts as a driving force for the other, we need not include an external driving torque to produce a chaotic system (there are enough degrees of freedom without it).

The equations of motions for the double pendulum are derived most directly from the Lagrangian formulation of mechanics. The Lagrangian is fairly simple

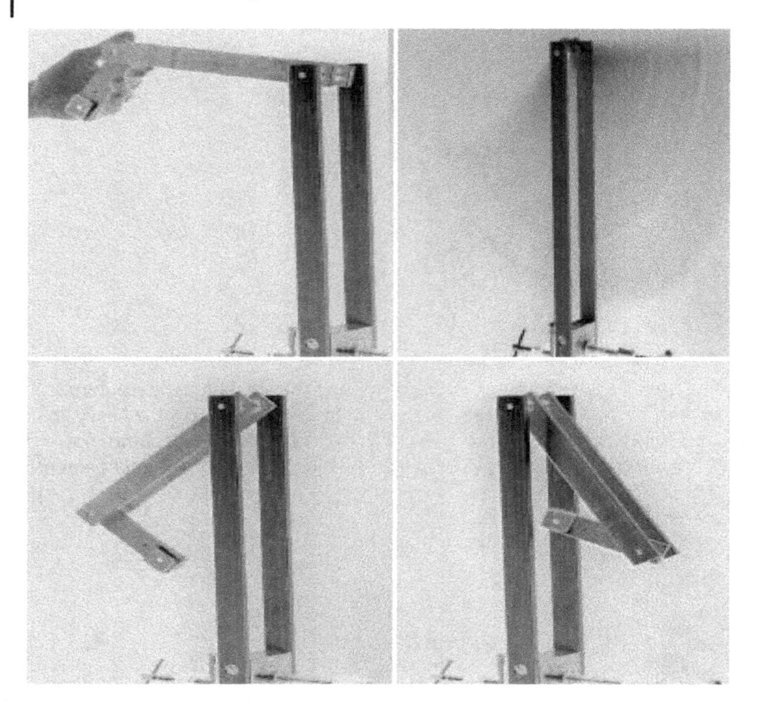

Figure 15.10 Photographs of a double pendulum built by a student in the OSU Physics Department. The upper pendulum consists of two separated shafts so that the lower one can rotate completely around. Both pendula can go over their tops. The first two frames show the pendulum released from rest and then moving quickly. The photography with faster shutter speeds stops the motion in various stages (photograph by R. Landau).

but has the θ_1 and θ_2 motions innately coupled:

$$L = \text{KE} - \text{PE} = \frac{1}{2}(m_1 + m_2)l_1^2\dot{\theta}_1^{\,2} + \frac{1}{2}m_2 l_2^2\dot{\theta}_2^{\,2} + m_2 l_1 l_2 \dot{\theta}_1 \dot{\theta}_2 \cos(\theta_1 - \theta_2) + (m_1 + m_2) g l_1 \cos\theta_1 + m_2 g l_2 \cos\theta_2 \, . \tag{15.18}$$

Usually, textbooks approximate these equations for small oscillations, which diminish the nonlinear effects, and results in "slow" and "fast" modes that look much like regular harmonic motions. What is more interesting is the motion that results without any small-angle restrictions, particularly when the pendula have enough initial energy to go over the top (Figure 15.10). In Figure 15.11a, we see several phase-space plots for the lower pendulum with $m_1 = 10m_2$. When given enough initial kinetic energy to go over the top, the trajectories are seen to flow between two major attractors as energy is transferred back and forth between the pendula.

In Figure 15.11b is a bifurcation diagram for the double pendulum. This was created by sampling and plotting the instantaneous angular velocity $\dot{\theta}_2$ of the lower pendulum at 70 times as the pendulum passed through its equilibrium position.

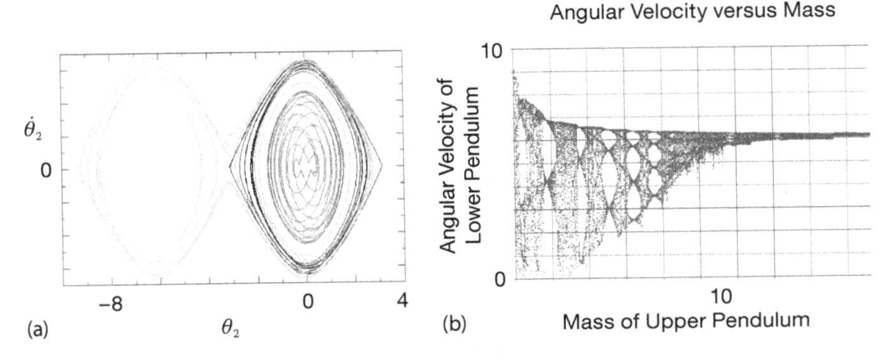

Figure 15.11 (a) Phase-space trajectories for a double pendulum with $m_1 = 10m_2$ and with two dominant attractors. (b) A bifurcation diagram for the double pendulum displaying the instantaneous velocity of the lower pendulum as a function of the mass of the upper pendulum (both plots courtesy of J. Danielson).

The mass of the upper pendulum (a convenient parameter) was then changed, and the process repeated. The resulting structure is fractal and indicates bifurcations in the number of dominant frequencies in the motion. A plot of the Fourier or wavelet spectrum as a function of mass is expected to show similar characteristic frequencies.

15.5 Assessment: Fourier/Wavelet Analysis of Chaos

We have seen that a realistic pendulum experiences a restoring torque, $\tau_g \propto \sin\theta \simeq \theta - \theta^3/3! + \theta^5/5! + \cdots$, that contains nonlinear terms that lead to nonharmonic behavior. In addition, when a realistic pendulum is driven by an external sinusoidal torque, the pendulum may mode-lock with the driver and consequently oscillate at a frequency that is rationally related to the driver's. Consequently, the behavior of the realistic pendulum is expected to be a combination of various periodic behaviors, with discrete jumps between modes.

In this assessment, you should determine the Fourier components present in the pendulum's complicated and chaotic behaviors. You should show that a three-cycle structure, for example, contains three major Fourier components, while a five-cycle structure has five. You should also notice that when the pendulum goes over the top, its spectrum contains a steady-state (DC) component.

1. Dust off your program for analyzing a $y(t)$ into Fourier components.
2. Apply your analyzer to the solution of the chaotic pendulum for the cases where there are one-, three-, and five-cycle structures in phase space. Deduce the major frequencies contained in these structures. Wait for the transients to die out before conducting your analysis.

3. Compare your results with the output of the `Pend1` applet (Figure 15.8a and b).
4. Try to deduce a relation among the Fourier components, the natural frequency ω_0, and the driving frequency ω.
5. A classic signal of chaos is a broadband, although not necessarily flat, Fourier spectrum. Examine your system for parameters that give chaotic behavior and verify this statement by plotting the power spectrum in both linear and semilogarithmic plots. (The power spectrum often varies over several orders of magnitude.)

Wavelet Exploration We saw in Chapter 13 that a wavelet expansion is more appropriate than a Fourier expansion for a signal containing components that occur for finite periods of time. Because chaotic oscillations are just such signals, repeat the Fourier analysis of this section using wavelets instead of sines and cosines. Can you discern the temporal sequence of the various components?

15.6 Exploration: Alternate Phase-Space Plots

Imagine that you have measured the displacement of some system as a function of time. Your measurements appear to indicate characteristic nonlinear behaviors, and you would like to check this by making a phase-space plot but without going to the trouble of measuring the conjugate momenta to plot vs. displacement. Amazingly enough, one may instead plot $x(t+\tau)$ vs. $x(t)$ as a function of time to obtain a phase-space plot (Abarbanel *et al.*, 1993). Here τ is a *lag time* and should be chosen as some fraction of a characteristic time for the system under study. While this may not seem like a valid way to make a phase-space plot, recall the forward difference approximation for the derivative,

$$v(t) = \frac{dx(t)}{dt} \simeq \frac{x(t+\tau) - x(t)}{\tau} . \tag{15.19}$$

We see that plotting $x(t+\tau)$ vs. $x(t)$ is thus similar to plotting $v(t)$ vs. $x(t)$.

Exercise Create a phase-space plot from the output of your chaotic pendulum by plotting $\theta(t+\tau)$ vs. $\theta(t)$ for a large range of t values. Explore how the graphs change for different values of the lag time τ. Compare your results to the conventional phase-space plots you obtained previously.

15.7 Further Explorations

1. The nonlinear behavior in once-common objects such as vacuum tubes and metronomes is described by the *van der Pool Equation,*

$$\frac{d^2x}{dt^2} + \mu\left(x^2 - x_0^2\right)\frac{dx}{dt} + \omega_0^2 x = 0 \; . \tag{15.20}$$

Verify that the behavior predicted for these systems is *self-limiting* because the equation contains a limit cycle that is also a predictable attractor. You can think of (15.20) as describing an oscillator with x-dependent damping (the μ term). If $x > x_0$, friction slows the system down; if $x < x_0$, friction speeds the system up. Orbits internal to the limit cycle spiral out until they reach the limit cycle; orbits external to it spiral in.

2. *Duffing Oscillator:* Another damped, driven nonlinear oscillator is

$$\frac{d^2\theta}{dt^2} - \frac{1}{2}\theta(1-\theta^2) = -\alpha\frac{d\theta}{dt} + f\cos\omega t \; . \tag{15.21}$$

While similar to the chaotic pendulum, it is easier to find multiple attractors for this oscillator. Perform a phase-space analysis for this equation. (Moon and Li, 1985).

3. *Lorenz Attractors:* In 1961, Edward Lorenz was using a simplified atmospheric convection model to predict weather patterns, when, as a shortcut, he entered the decimal 0.506 instead of entering the full 0.506 127 for a parameter in the model (Peitgen *et al.*, 1994; Motter and Campbell, 2013). The results for the two numbers were so different that at first he thought it to be a numerical error, but in time he realized that this was a nonlinear system with chaotic behavior. Now we want you to repeat this discovery.
With simplified variables, the equation used by Lorenz are

$$\dot{x} = \sigma(y - x) \; , \tag{15.22}$$

$$\dot{dy} = \rho x - y - xz \; , \tag{15.23}$$

$$\dot{dz} = -\beta z + xy \; , \tag{15.24}$$

where $x(t)$ is a measure of fluid velocity as a function of time t, $y(t)$ and $z(t)$ are measures of the temperature distributions in two directions, and σ, ρ, and β are parameters. Note that the xz and xy terms make these equations nonlinear.

a) Modify your ODE solver to handle these three, simultaneous Lorenz equations.
b) To start, use parameter values $\sigma = 10$, $\beta = 8/3$, and $\rho = 28$.
c) Make sure to use a small enough step size so that good precision is obtained. You must have confidence that you are seeing chaos and not numerical error.

d) Make plots of x vs. t, y vs. t, and z vs. t.
e) The initial behaviors in these plots are called "transients" and are not considered dynamically interesting. Leave off these transients in the plots to follow.
f) Make a "phase-space" plot of $z(t)$ vs. $x(t)$ (the independent variable t does not appear in such a plot). The distorted, number eight-like figures you obtain (Figure 15.12) are called Lorenz attractors, "attractors" because even chaotic solutions tend to be attracted to them.
g) Make phase-space plots of $y(t)$ vs. $x(t)$, and $z(t)$ vs. $y(t)$.
h) Make a 3D plot of $x(t)$ vs. $y(t)$ vs. $z(t)$.
i) The parameters given to you should lead to chaotic solutions. Check this claim by seeing how small a change you can make in a parameter value and still, eventually, obtain different answers.

4. *A 3D Computer Fly:* Make $x(t)$ vs. $y(t)$, $x(t)$ vs. $z(t)$, and $y(t)$ vs. $z(t)$ phase-space plots of the equations

$$x = \sin a y - z \cos bx\,, \quad y = z \sin cx - \cos dy\,, \quad z = e \sin x\,. \tag{15.25}$$

Here the parameter e controls the degree of apparent randomness.

5. *Hénon–Heiles Potential:* The potential and Hamiltonian

$$V(x,y) = \frac{1}{2}x^2 + \frac{1}{2}y^2 + x^2 y - \frac{1}{3}y^3\,, \quad H = \frac{1}{2}p_x^2 + \frac{1}{2}p_y^2 + V(x,y)\,, \tag{15.26}$$

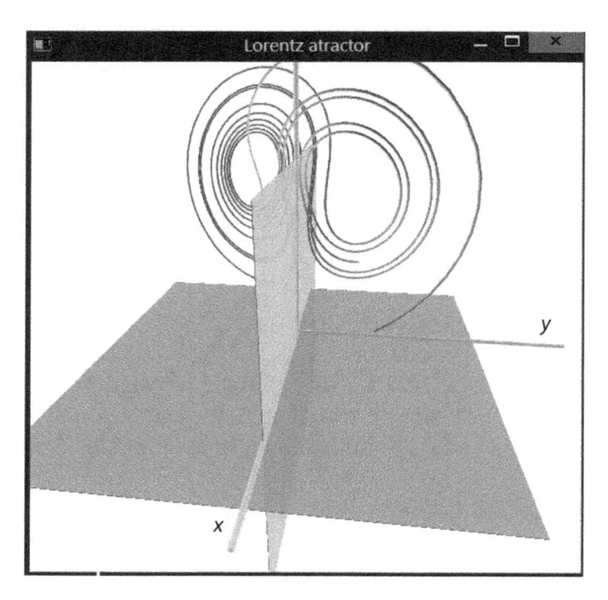

Figure 15.12 A 3D plot of a Lorenz attractor output from the program `LorentzAtract.py` (on the Instructor's disk).

are used to describe three interacting astronomical objects. The potential binds the objects near the origin but releases them if they move far out. The equations of motion follow from the Hamiltonian equations:

$$\frac{\mathrm{d}p_x}{\mathrm{d}t} = -x - 2xy\,, \quad \frac{\mathrm{d}p_y}{\mathrm{d}t} = -y - x^2 + y^2\,, \quad \frac{\mathrm{d}x}{\mathrm{d}t} = p_x\,, \quad \frac{\mathrm{d}y}{\mathrm{d}t} = p_y\,. \tag{15.27}$$

a) Numerically solve for the position $[x(t), y(t)]$ for a particle in the Hénon–Heiles potential.
b) Plot $[x(t), y(t)]$ for a number of initial conditions. Check that the initial condition $E < 1/6$ leads to a bounded orbit.
c) Produce a Poincaré section in the (y, p_y) plane by plotting (y, p_y) each time an orbit passes through $x = 0$.

16
Fractals and Statistical Growth Models

It is common to notice regular and eye-pleasing natural objects, such as plants and sea shells, that do not have well-defined geometric shapes. When analyzed mathematically with a prescription that normally produces integers, some of these objects are found to have a dimension that is a fractional number, and so they are called "fractals." In this chapter, we implement simple, statistical models that grow fractals. To the extent that these models generate structures that look like those in nature, it is reasonable to assume that the natural processes must be following similar rules arising from the basic physics or biology that creates the objects. Detailed applications of fractals can be found in the literature (Mandelbrot, 1982; Armin and Shlomo, 1991; Sander *et al.*, 1994; Peitgen *et al.*, 1994).

16.1
Fractional Dimension (Math)

Benoit Mandelbrot, who first studied fractional-dimension figures with supercomputers at IBM Research, gave them the name *fractals* (Mandelbrot, 1982). Some geometric objects, such as Koch curves, are exact fractals with the same dimension for all their parts. Other objects, such as bifurcation curves, are statistical fractals in which elements of chaos occur and the dimension can be defined only for each part of the object, or on the average.

Consider an abstract object such as the density of charge within an atom. There are an infinite number of ways to measure the "size" of this object. For example, each moment $\langle r^n \rangle$ is a measure of the size, and there is an infinite number of such moments. Likewise, when we deal with complicated objects, there are different definitions of dimension and each may give a somewhat different value.

Our first definition of the dimension d_f, the *Hausdorff–Besicovitch dimension*, is based on our knowledge that a line has dimension 1, a triangle has dimension 2, and a cube has dimension 3. It seems perfectly reasonable to ask if there is some mathematical formula that agrees with our experience with regular objects, yet can also be used for determining fractional dimensions. For simplicity, let us consider objects that have the same length L on each side, as do equilateral triangles

Computational Physics, 3rd edition. Rubin H. Landau, Manuel J. Páez, Cristian C. Bordeianu.

and squares, and that have uniform density. We postulate that the dimension of an object is determined by the dependence of its total mass upon its length:

$$M(L) \propto L^{d_f} , \tag{16.1}$$

where the power d_f is the *fractal dimension*. As you may verify, this rule works with the 1D, 2D, and 3D regular figures in our experience, so it is a reasonable to try it elsewhere. When we apply (16.1) to fractal objects, we end up with fractional values for d_f. Actually, we will find it easier to determine the fractal dimension not from an object's mass, which is *extensive* (depends on size), but rather from its density, which is *intensive*. The density is defined as mass/length for a linear object, as mass/area for a planar object, and as mass/volume for a solid object. That being the case, for a planar object we hypothesize that

$$\rho = \frac{M(L)}{\text{area}} \propto \frac{L^{d_f}}{L^2} \propto L^{d_f - 2} . \tag{16.2}$$

16.2
The Sierpiński Gasket (Problem 1)

To generate our first fractal (Figure 16.1), we play a game of chance in which we place dots at points picked randomly within a triangle (Bunde and Havlin, 1991). Here are the rules (which you should try out in the margins now).

1. Draw an equilateral triangle with vertices and coordinates:

$$\text{vertex 1}: (a_1, b_1) ; \quad \text{vertex 2}: (a_2, b_2) ; \quad \text{vertex 3}: (a_3, b_3) . \tag{16.3}$$

2. Place a dot at an arbitrary point $P = (x_0, y_0)$ within this triangle.
3. Find the next point by selecting randomly the integer 1, 2, or 3:
   a) If 1, place a dot halfway between P and vertex 1.
   b) If 2, place a dot halfway between P and vertex 2.
   c) If 3, place a dot halfway between P and vertex 3.
4. Repeat the process using the last dot as the new P.

Mathematically, the coordinates of successive points are given by the formulas

$$(x_{k+1}, y_{k+1}) = \frac{(x_k, y_k) + (a_n, b_n)}{2} , \quad n = \text{integer}\,(1 + 3r_i) , \tag{16.4}$$

where r_i is a random number between 0 and 1 and where the *integer* function outputs the closest integer smaller than or equal to the argument. After 15 000 points, you should obtain a collection of dots like those in Figure 16.1a.

16.2.1
Sierpiński Implementation

Write a program to produce a Sierpiński gasket. Determine empirically the fractal dimension of your figure. Assume that each dot has mass 1 and that $\rho = CL^{\alpha}$. (You

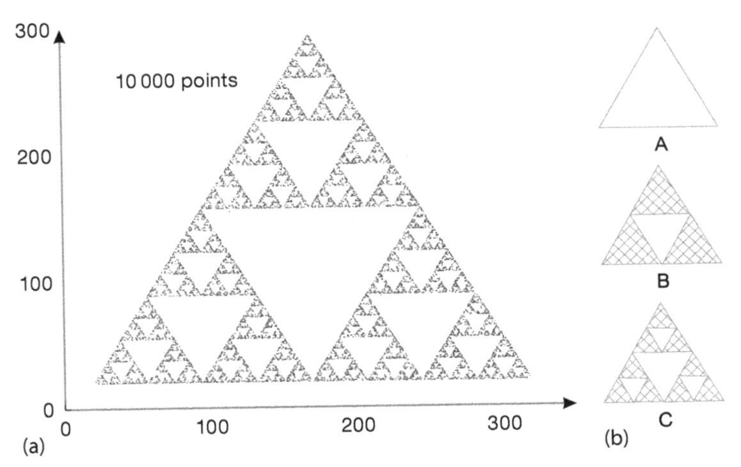

Figure 16.1 (a) A statistical fractal Sierpiński gasket containing 10 000 points. Note the self-similarity at different scales. (b) A geometric Sierpiński gasket constructed by successively connecting the midpoints of the sides of each equilateral triangle. The first three steps in the process are labeled as A, B, C.

can have the computer do the counting by defining an array *box* of all 0 values and then change a 0 to a 1 when a dot is placed there.)

16.2.2
Assessing Fractal Dimension

The topology in Figure 16.1 was first analyzed by the Polish mathematician Sierpiński. Observe that there is the same structure in a small region as there is in the entire figure. In other words, if the figure had infinite resolution, any part of the figure could be scaled up in size and would be similar to the whole. This property is called *self-similarity*.

We construct a nonstatistical form of the Sierpiński gasket by removing an inverted equilateral triangle from the center of all filled equilateral triangles to create the next figure (Figure 16.1b). We then repeat the process ad infinitum, scaling up the triangles so each one has side $r = 1$ after each step. To see what is unusual about this type of object, we look at how its density (mass/area) changes with size, and then apply (16.2) to determine its fractal dimension. Assume that each triangle has mass m and assign unit density to the single triangle:

$$\rho(L = r) \propto \frac{M}{r^2} = \frac{m}{r^2} \stackrel{\text{def}}{=} \rho_0 \quad \text{(Figure 16.1b-A)} . \tag{16.5}$$

Next, for the equilateral triangle with side $L = 2$, the density is

$$\rho(L = 2r) \propto \frac{(M = 3m)}{(2r)^2} = \frac{3}{4} m r^2 = \frac{3}{4}\rho_0 \quad \text{(Figure 16.1b-B)} . \tag{16.6}$$

We see that the extra white space in Figure 16.1b leads to a density that is 3/4 that of the previous stage. For the structure in Figure 16.1c, we obtain

$$\rho(L = 4r) \propto \frac{(M = 9m)}{(4r)^2} = \frac{9}{16}\frac{m}{r^2} = \left(\frac{3}{4}\right)^2 \rho_0 \quad \text{(Figure 16.1b-C)} . \tag{16.7}$$

We see that as we continue the construction process, the density of each new structure is 3/4 that of the previous one. Interesting. Yet in (16.2) we derived that

$$\rho \propto CL^{d_f - 2} . \tag{16.8}$$

Equation 16.8 implies that a plot of the logarithm of the density ρ vs. the logarithm of the length L for successive structures yields a straight line of slope

$$d_f - 2 = \frac{\Delta \log \rho}{\Delta \log L} . \tag{16.9}$$

As applied to our problem,

$$d_f = 2 + \frac{\Delta \log \rho(L)}{\Delta \log L} = 2 + \frac{\log 1 - \log \frac{3}{4}}{log 1 - \log 2} \simeq 1.58496 . \tag{16.10}$$

As is evident in Figure 16.1, as the gasket grows larger (and consequently more massive), it contains more open space. So despite the fact that its mass approaches infinity as $L \to \infty$, its density approaches zero! And because a 2D figure like a solid triangle has a constant density as its length increases, a 2D figure has a slope equal to 0. Because the Sierpiński gasket has a slope $d_f - 2 \simeq -0.41504$, it fills space to a lesser extent than a 2D object but more so than a 1D object; it is a fractal with dimension 1.6.

16.3 Growing Plants (Problem 2)

It seems paradoxical that natural processes subject to chance can produce objects of high regularity and symmetry. For example, it is hard to believe that something as beautiful and graceful as a fern (Figure 16.2a) has random elements in it. Nonetheless, there is a clue here in that much of the fern's beauty arises from the similarity of each part to the whole (self-similarity), with different ferns similar but not identical to each other. These are characteristics of fractals. Your *problem* is to discover if a simple algorithm including some randomness can draw regular ferns. If the algorithm produces objects that resemble ferns, then presumably you have uncovered mathematics similar to that responsible for the shapes of ferns.

16.3.1 Self-Affine Connection (Theory)

In (16.4), which defines mathematically how a Sierpiński gasket is constructed, a *scaling factor* of 1/2 is part of the relation of one point to the next. A more general

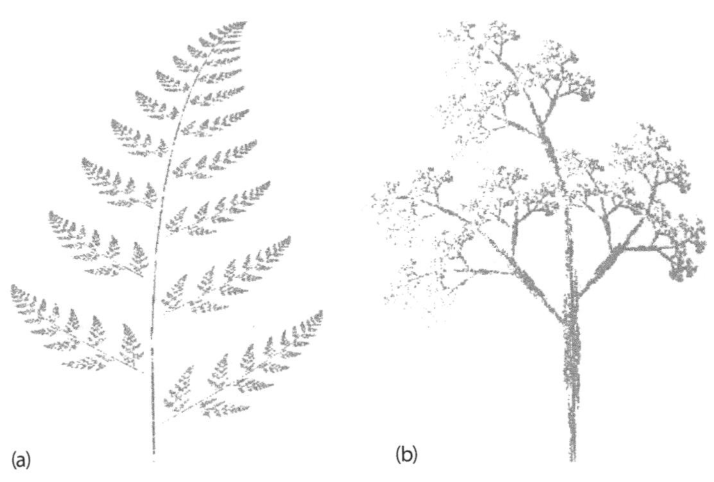

Figure 16.2 (a) A fractal fern generated by 30 000 iterations of the algorithm (16.14). Enlarging this fern shows that each frond with a similar structure. (b) A fractal tree created with the simple algorithm (16.17).

transformation of a point $P = (x, y)$ into another point $P' = (x', y')$ via *scaling* is

$$(x', y') = s(x, y) = (sx, sy) \quad \text{(scaling)} . \tag{16.11}$$

If the scale factor $s > 0$, an amplification occurs, whereas if $s < 0$, a reduction occurs. In our definition (16.4) of the Sierpiński gasket, we also added in a constant a_n. This is a *translation operation*, which has the general form

$$(x', y') = (x, y) + (a_x, a_y) \quad \text{(translation)} . \tag{16.12}$$

Another operation, not used in the Sierpiński gasket, is a *rotation* by angle θ:

$$x' = x\cos\theta - y\sin\theta , \quad y' = x\sin\theta + y\cos\theta \quad \text{(rotation)} . \tag{16.13}$$

The entire set of transformations, scalings, rotations, and translations defines an *affine transformation* (affine denotes a close relation between successive points). The transformation is still considered affine even if it is a more general linear transformation with the coefficients not all related by a single θ (in that case, we can have contractions and reflections). What is important is that the object created with these rules turns out to be self-similar; each step leads to new parts of the object that bear the same relation to the ancestor parts as the ancestors did to theirs. This is what makes the object look similar at all scales.

16.3.2
Barnsley's Fern Implementation

We obtain a Barnsley's fern (Barnsley and Hurd, 1992) by extending the dots game to one in which new points are selected using an affine connection with some elements of chance mixed in:

$$(x, y)_{n+1} = \begin{cases} (0.5, 0.27y_n)\,, & \text{with 2\% probability}\,, \\ (-0.139x_n + 0.263y_n + 0.57 & \\ \quad 0.246x_n + 0.224y_n - 0.036)\,, & \text{with 15\% probability}\,, \\ (0.17x_n - 0.215y_n + 0.408 & \\ \quad 0.222x_n + 0.176y_n + 0.0893)\,, & \text{with 13\% probability}\,, \\ (0.781x_n + 0.034y_n + 0.1075 & \\ \quad -0.032x_n + 0.739y_n + 0.27)\,, & \text{with 70\% probability}\,. \end{cases} \tag{16.14}$$

To select a transformation with probability $\mathcal{P}$, we select a uniform random number $0 \le r \le 1$ and perform the transformation if r is in a range proportional to $\mathcal{P}$:

$$\mathcal{P} = \begin{cases} 2\%\,, & r < 0.02\,, \\ 15\%\,, & 0.02 \le r \le 0.17\,, \\ 13\%\,, & 0.17 < r \le 0.3\,, \\ 70\%\,, & 0.3 < r < 1\,. \end{cases} \tag{16.15}$$

The rules (16.14) and (16.15) can be combined into one:

$$(x, y)_{n+1} = \begin{cases} (0.5, 0.27y_n), & r < 0.02\,, \\ (-0.139x_n + 0.263y_n + 0.57 & \\ \quad 0.246x_n + 0.224y_n - 0.036), & 0.02 \le r \le 0.17\,, \\ (0.17x_n - 0.215y_n + 0.408 & \\ \quad 0.222x_n + 0.176y_n + 0.0893), & 0.17 < r \le 0.3\,, \\ (0.781x_n + 0.034y_n + 0.1075, & \\ \quad -0.032x_n + 0.739y_n + 0.27), & 0.3 < r < 1\,. \end{cases} \tag{16.16}$$

Although (16.14) makes the basic idea clearer, (16.16) is easier to program, which we do in Listing 16.1.

The starting point in Barnsley's fern (Figure 16.2) is $(x_1, y_1) = (0.5, 0.0)$, and the points are generated by repeated iterations. An important property of this fern is that it is not completely self-similar, as you can see by noting how different the stems and the fronds are. Nevertheless, the stem can be viewed as a compressed copy of a frond, and the fractal obtained with (16.14) is still *self-affine*, yet with a dimension that varies from part to part in the fern.

Listing 16.1 Fern3D.pysimulates the growth of ferns in 3D.

```
# Fern3D.py:  Fern in 3D, see Barnsley, "Fractals Everywhere"

from visual import *
from visual.graph import *
import random
```

```
imax = 20000
x = 0.5;      y = 0.0;      z = -0.2;      xn = 0.0;      yn = 0.0

graph1 = display(width=500, height=500, forward=(-3,0,-1),\
          title='3D Fractal Fern (rotate via right mouse button)', range=10)
graph1.show_rendertime = True # Pts/sphs: cycle=27/750 ms, render=6/30

pts = points(color=color.green, size=0.01)
for i in range(1,imax):
    r = random.random();
    if ( r <= 0.1):                                    # 10% probability
        xn = 0.0
        yn = 0.18*y
        zn = 0.0
    elif ( r > 0.1 and r <= 0.7):                      # 60% probability
        xn =  0.85  * x
        yn =  0.85  * y + 0.1  * z + 1.6
        zn = -0.1   * y + 0.85  * z
    elif ( r > 0.7 and r <= 0.85):                     # 15 % probability
        xn =  0.2 * x - 0.2 * y
        yn =  0.2 * x + 0.2 * y + 0.8
        zn=   0.3 * z
    else:
        xn = -0.2 * x +0.2 * y                         # 15% probability
        yn =  0.2 * x +0.2 * y + 0.8
        zn =  0.3 * z
    x = xn
    y = yn
    z = zn
    xc = 4.0*x                                         # linear TF for plot
    yc = 2.0*y-7
    zc = z
    pts.append(pos=(xc,yc,zc))
```

16.3.3 Self-Affinity in Trees Implementation

Now that you know how to grow ferns, look around and notice the regularity in trees (such as in Figure 16.2b). Can it be that this also arises from a self-affine structure? Write a program, similar to the one for the fern, starting at $(x_1, y_1) = (0.5, 0.0)$ and iterating the following self-affine transformation:

$$(x_{n+1}, y_{n+1}) = \begin{cases} (0.05x_n, 0.6y_n)\,, & 10\%\ \text{probability}\,, \\ (0.05x_n, -0.5y_n + 1.0)\,, & 10\%\ \text{probability}\,, \\ (0.46x_n - 0.15y_n, 0.39x_n + 0.38y_n + 0.6)\,, & 20\%\ \text{probability}\,, \\ (0.47x_n - 0.15y_n, 0.17x_n + 0.42y_n + 1.1)\,, & 20\%\ \text{probability}\,, \\ (0.43x_n + 0.28y_n, -0.25x_n + 0.45y_n + 1.0)\,, & 20\%\ \text{probability}\,, \\ (0.42x_n + 0.26y_n, -0.35x_n + 0.31y_n + 0.7)\,, & 20\%\ \text{probability}\,. \end{cases} \tag{16.17}$$

16.4
Ballistic Deposition (Problem 3)

There are a number of physical and manufacturing processes in which particles are deposited on a surface and form a film. Because the particles are evaporated from a hot filament, there is randomness in the emission process yet the produced films turn out to have well-defined, regular structures. Again we suspect fractals. Your *problem* is to develop a model that simulates this growth process and compare your produced structures to those observed.

16.4.1
Random Deposition Algorithm

The idea of simulating random depositions was first reported by Vold (1959), which includes tables of random numbers used to simulate the sedimentation of moist spheres in hydrocarbons. We shall examine a method of simulation that results in the deposition shown in Figure 16.3 (Family and Vicsek, 1985).

Consider particles falling onto and sticking to a horizontal line of length L composed of 200 deposition sites. All particles start from the same height, but to simulate their different velocities, we assume they start at random distances from the left side of the line. The simulation consists of generating uniform random sites between 0 and L, and having a particle stick to the site on which it lands. Because a realistic situation may have columns of aggregates of different heights, the particle may be stopped before it makes it to the line, or it may bounce around until it falls into a hole. We therefore assume that if the column height at which the particle lands is greater than that of both its neighbors, it will add to that height. If the particle lands in a hole, or if there is an adjacent hole, it will fill up the hole. We speedup the simulation by setting the height of the hole equal to the maximum of its neighbors:

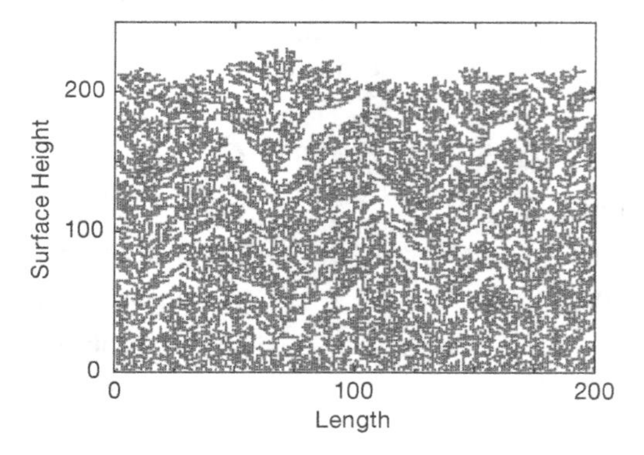

Figure 16.3 A simulation of the ballistic deposition of 20 000 particles onto a substrate of length 200. The vertical height increases in proportion to the length of deposition time, with the top being the final surface.

1. Choose a random site r.
2. Let the array h_r be the height of the column at site r.
3. Make the decision:

$$h_r = \begin{cases} h_r + 1 , & \text{if} \quad h_r \geq h_{r-1} , \quad h_r > h_{r+1} , \\ \max[h_{r-1}, h_{r+1}] , & \text{if} \quad h_r < h_{r-1} , \quad h_r < h_{r+1} . \end{cases} \tag{16.18}$$

Our simulation is **Fractals/Film.py** with the essential loop:

```
spot = int(random)
if (spot == 0)
    if ( coast[spot] < coast[spot+1] )
        coast[spot] = coast[spot+1];
        else coast[spot]++;
    else if (spot == coast.length - 1)
        if (coast[spot] < coast[spot-1]) coast[spot] = coast[spot-1];
        else coast[spot]++;
    else if ( coast[spot]<coast[spot-1] && coast[spot]<coast[spot+1] )
    if ( coast[spot-1] > coast[spot+1] ) coast[spot] = coast[spot-1];
        else coast[spot] = coast[spot+1];
    else coast[spot]++;
```

The results of this type of simulation show several empty regions scattered throughout the line (Figure 16.3), which is an indication of the statistical nature of the process while the film is growing. Simulations by Fereydoon reproduced the experimental observation that the average height increases linearly with time and produced fractal surfaces. (You will be asked to determine the fractal dimension of a similar surface as an exercise.)

Exercise Extend the simulation of random deposition to two dimensions, so rather than making a line of particles we now deposit an entire surface.

16.5 Length of British Coastline (Problem 4)

In 1967, Mandelbrot (Mandelbrot, 1967) asked a classic question, "How long is the coast of Britain?" If Britain had the shape of Colorado or Wyoming, both of which have straight-line boundaries, its perimeter would be a curve of dimension 1 with finite length. However, coastlines are geographic not geometric curves, with each portion of the coast sometimes appearing self-similar to the entire coast. If the perimeter of the coast is in fact a fractal, then its length is either infinite or meaningless. Mandelbrot deduced the dimension of the west coast of Britain to be $d_f = 1.25$, which implies infinite length. In your *problem* we ask you to determine the dimension of the perimeter of one of our fractal simulations.

16.5.1 Coastlines as Fractals (Model)

The length of the coastline of an island is the perimeter of that island. While the concept of perimeter is clear for regular geometric figures, some thought is required to give it meaning for an object that may be infinitely self-similar. Let us assume that a map maker has a ruler of length r. If she walks along the coastline and counts the number of times N that she must place the ruler down in order to *cover* the coastline, she will obtain a value for the length L of the coast as Nr. Imagine now that the map maker keeps repeating her walk with smaller and smaller rulers. If the coast was a geometric figure or a *rectifiable curve*, at some point the length L would become essentially independent of r and would approach a constant. Nonetheless, as discovered empirically by Richardson (1961) for natural coastlines such as those of South Africa and Britain, the perimeter appears to be an unusual function of r:

$$L(r) = Mr^{1-d_f} , \tag{16.19}$$

where M and d_f are empirical constants. For a geometric figure or for Colorado, $d_f = 1$ and the length approaches a constant as $r \to 0$. Yet for a fractal with $d_f > 1$, the perimeter $L \to \infty$ as $r \to 0$. This means that as a consequence of self-similarity, fractals may be of finite size but have infinite perimeters. Physically, at some point there may be no more details to discern as $r \to 0$ (say, at the quantum or Compton size limit), and so the limit may not be meaningful.

16.5.2 Box Counting Algorithm

Consider a line of length L broken up into segments of length r (Figure 16.4a). The number of segments or "boxes" needed to cover the line is related to the size r of the box by

$$N(r) = \frac{L}{r} = \frac{C}{r} , \tag{16.20}$$

where C is a constant. A proposed definition of fractional dimension is the power of r in this expression as $r \to 0$. In our example, it tells us that the line has dimension $d_f = 1$. If we now ask how many little circles of radius r it would take to *cover* or fill a circle of area A (Figure 16.4, middle), we will find that

$$N(r) = \lim_{r \to 0} \frac{A}{\pi r^2} \quad \Rightarrow \quad d_f = 2 , \tag{16.21}$$

as expected. Likewise, counting the number of little spheres or cubes that can be packed within a large sphere tells us that a sphere has dimension $d_f = 3$. In general, if it takes N little spheres or cubes of side $r \to 0$ to cover some object, then the fractal dimension d_f can be deduced as

$$N(r) = C\left(\frac{1}{r}\right)^{d_f} = C' s^{d_f} \quad (\text{as } r \to 0) , \tag{16.22}$$

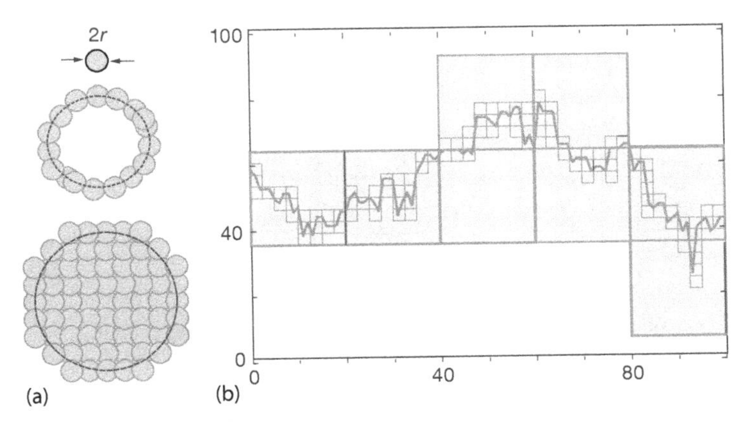

Figure 16.4 Examples of the use of "box" counting to determine fractal dimension. (a) In the top "boxes" are circles and the perimeter is being covered. In the bottom an entire figure is being covered. In (b) a "coastline" is being covered by boxes of two different sizes (scales). The fractal dimension can be deduced by recording the number of box of different scale needed to cover the figures.

$$\log N(r) = \log C - d_f \log(r) \quad (\text{as } r \to 0) , \tag{16.23}$$

$$\Rightarrow \quad d_f = -\lim_{r \to 0} \frac{\Delta \log N(r)}{\Delta \log r} . \tag{16.24}$$

Here $s \propto 1/r$ is called the *scale* in geography, so $r \to 0$ corresponds to an infinite scale. To illustrate, you may be familiar with the low scale on a map being 10 000 m to a centimeter, while the high scale is 100 m to a centimeter. If we want the map to show small details (sizes), we need a map of high scale.

We will use box counting to determine the dimension of a perimeter, not of an entire figure. Once we have a value for d_f, we can determine a value for the length of the perimeter via (16.19). (If you cannot wait to see box counting in action, in the auxiliary online files you will find an applet `Jfracdim` that goes through all the steps of box counting before your eyes and even plots the results.)

16.5.3 Coastline Implementation and Exercise

Rather than ruin our eyes using a geographic map, we use a mathematical one, specifically, the top portion of Figure 16.3 that may look like a natural coastline. Determine d_f by covering this figure, or one you have generated, with a semitransparent piece of graph paper,[1] and counting the number of boxes containing any part of the coastline (Figures 16.4 and 16.5).

1) Yes, we are suggesting a painfully analog technique based on the theory that trauma leaves a lasting impression. If you prefer, you can store your output as a matrix of 1 and 0 values and let the computer do the counting, but this will take more of your time!

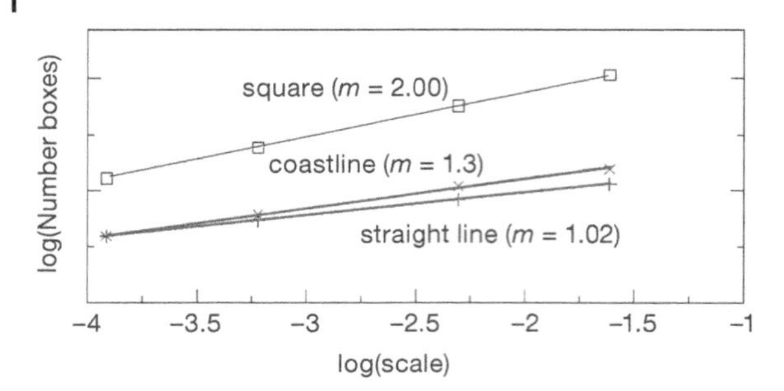

Figure 16.5 Fractal dimensions of a line, box, and coastline determined by box counting. The slope at vanishingly small scale determines the dimension.

1. Print your coastline graph with the same physical scale (*aspect ratio*) for the vertical and horizontal axes. This is required because the graph paper you will use for box counting has square boxes and so you want your graph to also have the same vertical and horizontal scales. Place a piece of graph paper over your printout and look though the graph paper at your coastline. If you do not have a piece of graph paper available, or if you are unable to obtain a printout with the same aspect ratio for the horizontal and vertical axes, add a series of closely spaced horizontal and vertical lines to your coastline printout and use these lines as your graph paper. (Box counting should still be accurate if both your coastline and your graph paper have the same aspect ratios.)
2. The vertical height in our printout was 17 cm, and the largest division on our graph paper was 1 cm. This sets the scale of the graph as 1 : 17, or $s = 17$ for the largest divisions (lowest scale). Measure the vertical height of your fractal, compare it to the size of the biggest boxes on your "piece" of graph paper, and thus determine your lowest scale.
3. With our largest boxes of 1 cm × 1 cm, we found that the coastline passed through $N = 24$ large boxes, that is, that 24 large boxes covered the coastline at $s = 17$. Determine how many of the largest boxes (lowest scale) are needed to cover your coastline.
4. With our next smaller boxes of 0.5 cm × 0.5 cm, we found that 51 boxes covered the coastline at a scale of $s = 34$. Determine how many of the midsize boxes (midrange scale) are needed to cover your coastline.
5. With our smallest boxes of 1 mm × 1 mm, we found that 406 boxes covered the coastline at a scale of $s = 170$. Determine how many of the smallest boxes (highest scale) are needed to cover your coastline.
6. Equation 16.24 tells us that as the box sizes get progressively smaller, we have

$$\log N \simeq \log A + d_f \log s ,$$

$$\Rightarrow \quad d_f \simeq \frac{\Delta \log N}{\Delta \log s} = \frac{\log N_2 - \log N_1}{\log s_2 - \log s_1} = \frac{\log(N_2/N_1)}{\log(s_2/s_1)} .$$

Clearly, only the relative scales matter because the proportionality constants cancel out in the ratio. A plot of $\log N$ vs. $\log s$ should yield a straight line.

In our example we found a slope of $d_f = 1.23$. Determine the slope and thus the fractal dimension for your coastline. Although only two points are needed to determine the slope, use your lowest scale point as an important check. (Because the fractal dimension is defined as a limit for infinitesimal box sizes, the highest scale points are more significant.)

7. As given by (16.19), the perimeter of the coastline

$$L \propto s^{1.23-1} = s^{0.23} . \tag{16.25}$$

If we keep making the boxes smaller and smaller so that we are looking at the coastline at higher and higher scale *and* if the coastline is self-similar at all levels, then the scale s will keep getting larger and larger with no limits (or at least until we get down to some quantum limits). This means

$$L \propto \lim_{s\to\infty} s^{0.23} = \infty . \tag{16.26}$$

Does your fractal implies an infinite coastline? Does it make sense that a small island like Britain, which you can walk around, has an infinite perimeter?

16.6 Correlated Growth, Forests, Films (Problem 5)

It is an empirical fact that in nature there is increased likelihood that a plant will grow if there is another one nearby (Figure 16.6a). This *correlation* is also valid for the simulation of surface films, as in the previous algorithm. Your *problem* is to include correlations in the surface simulation.

16.6.1 Correlated Ballistic Deposition Algorithm

A variation of the ballistic deposition algorithm, known as the *correlated ballistic deposition*, simulates mineral deposition onto substrates on which dendrites form (Tait *et al.*, 1990). In Listing 16.2 we extend the previous algorithm to include the likelihood that a freshly deposited particle will attract another particle. We assume that the probability of sticking $\mathcal{P}$ depends on the distance d that the added particle is from the last one (Figure 16.6b):

$$\mathcal{P} = c d^{\eta} . \tag{16.27}$$

Here η is a parameter and c is a constant that sets the probability scale.[2] For our implementation we choose $\eta = -2$, which means that there is an inverse square attraction between the particles (decreased probability as they get farther apart).

2) The absolute probability, of course, must be less than one, but it is nice to choose c so that the relative probabilities produce a graph with easily seen variations.

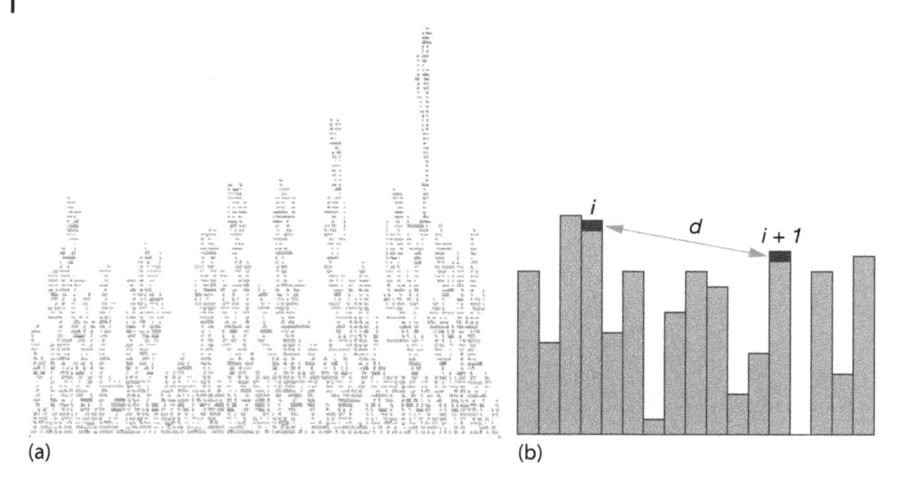

Figure 16.6 (a) A view that might be seen in the undergrowth of a forest or after a correlated ballistic deposition. (b) The probability of particle $i + 1$ sticking in one column depends upon the distance d from the previously deposited particle i.

As in our study of uncorrelated deposition, a uniform random number in the interval $[0, L]$ determines the column in which the particle will be deposited. We use the same rules about the heights as before, but now a second random number is used in conjunction with (16.27) to decide if the particle will stick. For instance, if the computed probability is 0.6 and if $r < 0.6$, the particle will be accepted (sticks), if $r > 0.6$, the particle will be rejected.

16.7
Globular Cluster (Problem 6)

Consider a bunch of grapes on an overhead vine. Your *problem* is to determine how its tantalizing shape arises. In a flash of divine insight, you realize that these shapes, as well as others such as those of dendrites, colloids, and thin-film structure, appear to arise from an aggregation process that is limited by diffusion.

16.7.1
Diffusion-Limited Aggregation Algorithm

A model of diffusion-limited aggregation (DLA) has successfully explained the relation between a cluster's perimeter and mass (Witten and Sander, 1983). We start with a 2D lattice containing a seed particle in the middle, draw a circle around the particle, and place another particle on the circumference of the circle at some random angle. We then release the second particle and have it execute a random walk, much like the one we studied in Chapter 4, but restricted to vertical or horizontal jumps between lattice sites. This is a simulation of a type of *Brownian motion*

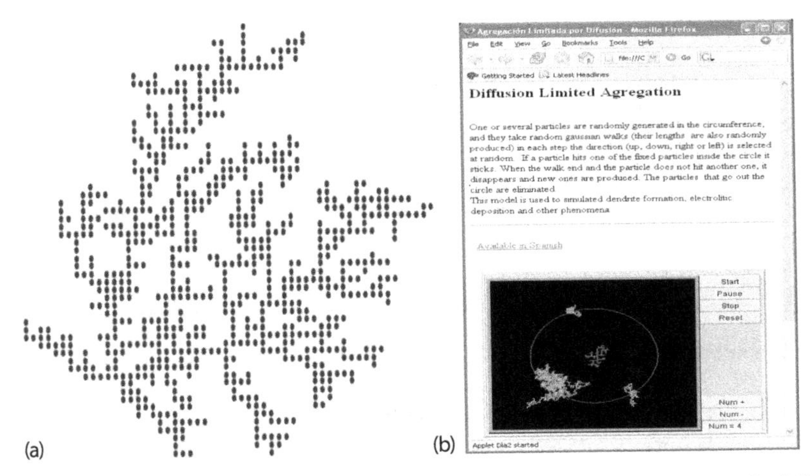

Figure 16.7 (a) A globular cluster of particles of the type that might occur in a colloid. (b) The applet **Dla2en.html** lets you watch these clusters grow. Here the cluster is at the center of the circle, and random walkers are started at random points around the circle.

related to diffusion. To make the model more realistic, we let the length of each step vary according to a random Gaussian distribution. If at some point during its random walk, the particle encounters another particle within one lattice spacing, they stick together and the walk terminates. If the particle passes outside the circle from which it was released, it is lost forever. The process is repeated as often as desired and results in clusters (Figure 16.7 and applet dla).

Listing 16.2 Column.py simulates correlated ballistic deposition of minerals onto substrates on which dendrites form.

```
# Column.py

from visual import *;
import random

maxi = 100000;     npoints = 200                          # Number iterations, spaces
i = 0;    dist = 0;     r = 0;     x = 0;   y = 0
oldx = 0;    oldy = 0;    pp = 0.0;     prob = 0.0
hit = zeros( (200), int)

graph1 = display(width = 500, height = 500, range=250,
                  title = 'Correlated Ballistic Deposition')
pts = points(color=color.green, size=2)

for i in range(0, npoints):   hit[i] = 0                          # Clear array
oldx = 100;             oldy = 0

for i in range(1, maxi + 1):
    r = int(npoints*random.random() )
    x = r - oldx
    y = hit[r] - oldy
    dist = x*x  +  y*y
    if (dist ==   0): prob = 1.0                  # Sticking prob depends on last x
    else: prob = 9.0/dist
    pp = random.random()
```

```
    if (pp < prob):
        if(r>0 and r<(npoints - 1) ):
            if( (hit[r] >=  hit[r - 1]) and (hit[r] >=  hit[r + 1]) ):
                hit[r] = hit[r] + 1
            else:
                if (hit[r - 1] > hit[r + 1]):
                    hit[r] = hit[r - 1]
                else: hit[r] = hit[r + 1]
        oldx = r
        oldy = hit[r]
        olxc = oldx*2 - 200                                    # TF for plot
        olyc = oldy*4 - 200
        pts.append(pos=(olxc,olyc))
```

1. Write a subroutine that generates random numbers with a Gaussian distribution.[3)]
2. Define a 2D lattice of points represented by the array `grid[400,400]` with all elements initially zero.
3. Place the seed at the center of the lattice; that is, set `grid[199,199]=1`.
4. Imagine a circle of radius 180 lattice spacings centered at `grid[199,199]`. This is the circle from which we release particles.
5. Determine the angular position of the new particle on the circle's circumference by generating a uniform random angle between 0 and 2π.
6. Compute the x and y positions of the new particle on the circle.
7. Determine whether the particle moves horizontally or vertically by generating a uniform random number $0 < r_{xy} < 1$ and applying the rule

$$\text{if} \quad \begin{cases} r_{xy} < 0.5\,, & \text{motion is vertical}\,, \\ r_{xy} > 0.5\,, & \text{motion is horizontal}\,. \end{cases} \tag{16.28}$$

8. Generate a Gaussian-weighted random number in the interval $[-\infty, \infty]$. This is the size of the step, with the sign indicating direction.
9. We now know the total distance and direction the particle will move. It jumps one lattice spacing at a time until this total distance is covered.
10. Before a jump, check whether a nearest-neighbor site is occupied:
    a) If occupied, the particle stays at its present position and the walk is over.
    b) If unoccupied, the particle jumps one lattice spacing.
    c) Continue the checking and jumping until the total distance is covered, until the particle sticks, or until it leaves the circle.
11. Once one random walk is over, another particle can be released and the process repeated. This is how the cluster grows.

Because many particles are lost, you may need to generate hundreds of thousands of particles to form a cluster of several hundred particles.

3) We indicated how to do this in Section 5.22.1.

16.7.2
Fractal Analysis of DLA or a Pollock

A cluster generated with the DLA technique is shown in Figure 16.7. We wish to analyze it to see if the structure is a fractal and, if so, to determine its dimension. (As an alternative, you may analyze the fractal nature of the Pollock painting in Figure 16.8, a technique used to determine the authenticity of this sort of art.) As a control, *simultaneously* analyze a geometric figure, such as a square or circle, whose dimension is known. The analysis is a variation of the one used to determine the length of the coastline of Britain.

1. If you have not already done so, use the box counting method to determine the fractal dimension of a simple square.
2. Draw a square of length L, small relative to the size of the cluster, around the seed particle. (Small might be seven lattice spacings to a side.)
3. Count the number of particles within the square.
4. Compute the density ρ by dividing the number of particles by the number of sites available in the box (49 in our example).
5. Repeat the procedure using larger and larger squares.
6. Stop when the cluster is covered.
7. The (box counting) fractal dimension d_f is estimated from a log–log plot of the density ρ vs. L. If the cluster is a fractal, then (16.2) tells us that $\rho \propto L^{d_f-2}$, and the graph should be a straight line of slope $d_f - 2$.

The graph we generated had a slope of -0.36, which corresponds to a fractal dimension of 1.66. Because random numbers are involved, the graph you generate will be different, but the fractal dimension should be similar. (Actually, the structure is multifractal, and so the dimension also varies with position.)

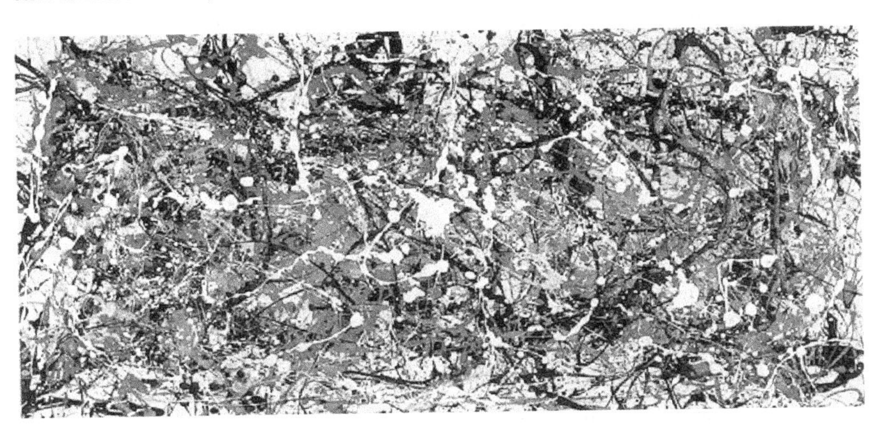

Figure 16.8 Number 8 by the American painter Jackson Pollock. (Used with permission, Neuberger Museum, State University of New York.) Some researchers claim that Pollock's paintings exhibit a characteristic fractal structure, while some other researchers question this (Kennedy, 2006). See if you can determine the fractal dimensions within this painting.

16.8
Fractals in Bifurcation Plot (Problem 7)

Recollect the project involving the logistics map where we plotted the values of the stable population numbers vs. the growth parameter μ. Take one of the bifurcation graphs you produced and determine the fractal dimension of different parts of the graph by using the same technique that was applied to the coastline of Britain.

16.9
Fractals from Cellular Automata

> We have already indicated in places how statistical models may lead to fractals. There is a class of statistical models known as *cellular automata* that produce complex behaviors from very simple systems. Here we study some.

Cellular automata were developed by von Neumann and Ulam in the early 1940s (von Neumann was also working on the theory behind modern computers then). Though very simple, cellular automata have found applications in many branches of science (Peitgen *et al.*, 1994; Sipper, 1996). Their classic definition is (Barnsley and Hurd, 1992):

> *A cellular automaton is a discrete dynamical system in which space, time, and the states of the system are discrete. Each point in a regular spatial lattice, called a cell, can have any one of a finite number of states, and the states of the cells in the lattice are updated according to a local rule. That is, the state of a cell at a given time depends only on its own state one time step previously, and the states of its nearby neighbors at the previous time step. All cells on the lattice are updated synchronously, and so the state of the entice lattice advances in discrete time steps.*

A cellular automaton in two dimensions consists of a number of square cells that grow upon each other. A famous one is *Conway's Game of Life*, which we implement in Listing 16.3. In this, cells with value 1 are alive, while cells with value 0 are dead. Cells grow according to the following rules:

1. If a cell is alive and if two or three of its eight neighbors are alive, then the cell remains alive.
2. If a cell is alive and if more than three of its eight neighbors are alive, then the cell dies because of overcrowding.
3. If a cell is alive and only one of its eight neighbors is alive, then the cell dies of loneliness.
4. If a cell is dead and more than three of its neighbors are alive, then the cell revives.

A variation on the Game of Life is to include a "rule one out of eight" that a cell will be alive if exactly one of its neighbors is alive, otherwise the cell will remain unchanged.

Listing 16.3 Gameoflife.py is an extension of Conway's Game of Life in which cells always revive if one out of eight neighbors is alive.

```
# Gameoflife.py:          Cellular automata in 2 dimensions

'''* Rules: a cell can be either dead (0) or alive (1)
  * If a cell is alive:
  * on next step will remain alive if
  * 2 or 3 of its closer 8 neighbors are alive.
  * If > 3 of 8 neighbors are alive, cell dies of overcrowdedness
  * If less than 2 neighbors are alive the cell dies of loneliness
  * A dead cell will be alive if 3 of its 8 neighbors are alive'''

from visual import *
from visual.graph import * ; import random

scene = display(width= 500,height= 500, title= 'Game of Life')
cell  = zeros((50,50));          cellu = zeros((50,50))
curve(pos=
    [(-49,-49),(-49,49),(49,49),(49,-49),(-49,-49)],color=color.white)
boxes = points(shape='square', size=8, color=color.cyan)

def drawcells(ce):
    boxes.pos = []                                   # Erase previous cells
    for j in range(0,50):
        for i in range(0,50):
            if ce[i,j] == 1:
                xx = 2*i-50
                yy = 2*j-50
                boxes.append(pos=(xx,yy))


def initial():
   for j in range (20,28):
        for  i in range(20, 28):
             r= int(random.random()*2)
             cell[j,i] = r
   return cell

def gameoflife(cell):
    for i in range(1,49):
        for j in range(1,49):
            sum1 = cell[i-1,j-1] + cell[i,j-1] + cell[i+1,j-1] # neighb
            sum2 = cell[i-1,j] + cell[i+1,j] + cell[i-1,j+1] \
               + cell[i,j+1] + cell[i+1,j+1]
            alive = sum1+sum2
            if cell[i,j] == 1:
                 if  alive == 2 or alive == 3:                     # Alive
                     cellu[i,j] = 1                                # Lives
                 if  alive > 3 or alive < 2:       # Overcrowded or solitude
                     cellu[i,j] = 0                                 # dies
            if  cell[i,j] == 0:
                 if  alive == 3:
                     cellu[i,j] = 1                              # Revives
                 else:
                     cellu[i,j] = 0                         # Remains dead
    alive = 0
    return cellu
 temp = initial()
 drawcells(temp)
 while True:
```

```
    rate(6)
    cell = temp
    temp = gameoflife(cell)
    drawcells(cell)
```

Early studies of the statistical mechanics of cellular automata were made by (Wolfram, 1983), who indicated how one can be used to generate a Sierpiński gasket. Because we have already seen that a Sierpiński gasket exhibits fractal geometry (Section 16.2), this represents a microscopic model of how fractals may occur in nature. This model uses eight rules, given graphically at the top of Figure 16.9, to generate new cells from old. We see all possible configurations for three cells in the top row, and the begetted next generation in the row below. At the bottom of Figure 16.9 is a Sierpiński gasket of the type created by the applet `JCellAut`. This plays the game and lets you watch and control the growth of the gasket.

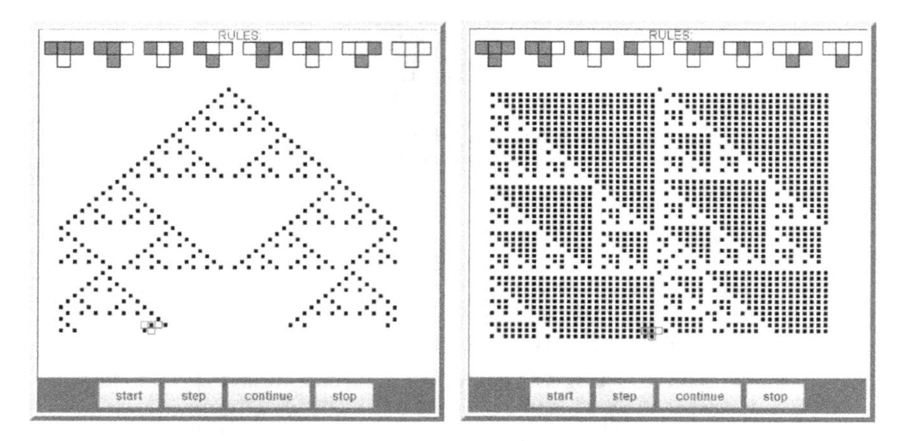

Figure 16.9 The rules for two versions of the Game of Life. The rules, given graphically on the top row, create the gaskets below (output obtained from the applet `JCellAut` in the auxiliary files).

16.10
Perlin Noise Adds Realism ⊙

We have already seen in this chapter how statistical fractals are able to generate objects with a striking resemblance to those in nature. This appearance of realism may be further enhanced by including a type of coherent randomness known as *Perlin noise*. The technique we are about to discuss was developed by Ken Perlin of New York University, who won an Academy Award (an Oscar) in 1997 for it and has continued to improve it (Perlin, 1985). This type of coherent noise has found use in important physics simulations of stochastic media (Tickner, 2004), as well as in video games and motion pictures like *Tron*.

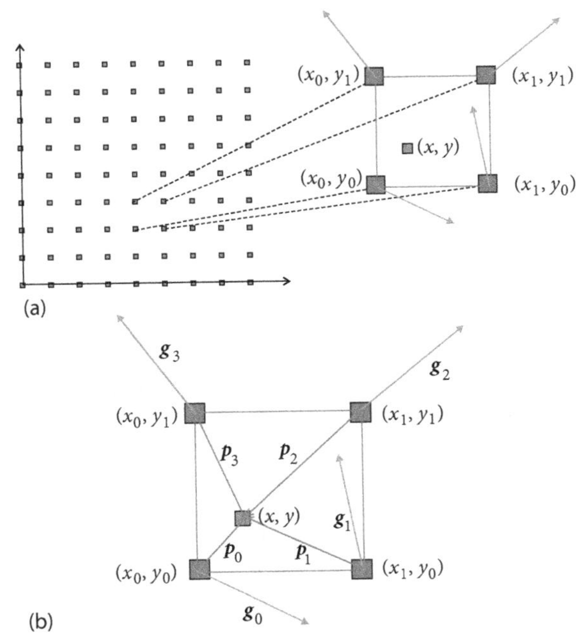

Figure 16.10 The coordinates used in adding Perlin noise. (a) The rectangular grid used to locate a square in space and a corresponding point within the square. As shown with the arrows, unit vectors $\boldsymbol{g}_i$ with random orientation are assigned at each grid point. (b) A point within each square is located by drawing the four $\boldsymbol{p}_i$. The $\boldsymbol{g}_i$ vectors are the same as on the left.

The inclusion of Perlin noise in a simulation adds both randomness and a type of coherence among points in space that tends to make dense regions denser and sparse regions sparser. This is similar to our correlated ballistic deposition simulations (Section 16.6.1) and related to chaos in its long-range randomness and short-range correlations. We start with some known function of x and y and add noise to it. For this purpose, Perlin used the mapping or *ease* function (Figure 16.11b)

$$f(p) = 3p^2 - 2p^3 \,. \tag{16.29}$$

As a consequence of its S shape, this mapping makes regions close to 0 even closer to 0, while making regions close to 1 even closer (in other words, it increases the tendency to clump, which shows up as higher contrast). We then break space up into a uniform rectangular grid of points (Figure 16.10a), and consider a point (x, y) within a square with vertices (x_0, y_0), (x_1, y_0), (x_0, y_1), and (x_1, y_1). We next assign unit gradients vectors $\boldsymbol{g}_0$ to $\boldsymbol{g}_3$ with random orientation at each grid point. A point within each square is located by drawing the four $\boldsymbol{p}_i$ vectors (Figure16.10b):

$$\boldsymbol{p}_0 = (x - x_0)\boldsymbol{i} + (y - y_0)\boldsymbol{j} \,, \quad \boldsymbol{p}_1 = (x - x_1)\boldsymbol{i} + (y - y_0)\boldsymbol{j} \,, \tag{16.30}$$

$$\boldsymbol{p}_2 = (x - x_1)\boldsymbol{i} + (y - y_1)\boldsymbol{j} \,, \quad \boldsymbol{p}_3 = (x - x_0)\boldsymbol{i} + (y - y_1)\boldsymbol{j} \,. \tag{16.31}$$

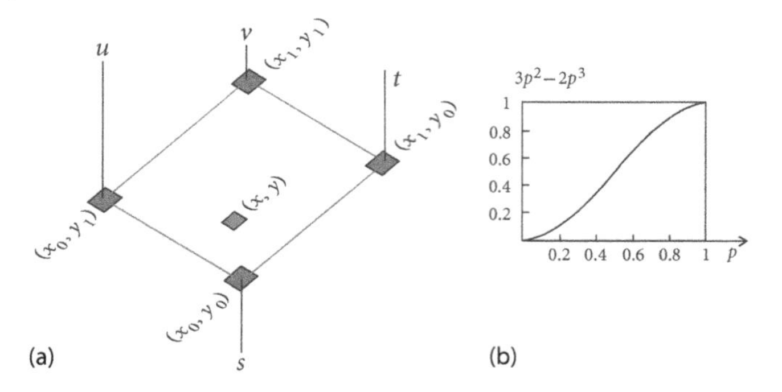

Figure 16.11 The mapping used in adding Perlin noise. (a) The numbers s, t, u, and v are represented by perpendiculars to the four vertices, with lengths proportional to their values. (b) The function $3p^2 - 2p^3$ is used as a map of the noise at a point like (x, y) to others close by.

Next the scalar products of the $\boldsymbol{p}'$s and the $\boldsymbol{g}'$s are formed:

$$s = \boldsymbol{p}_0 \cdot \boldsymbol{g}_0 , \quad t = \boldsymbol{p}_1 \cdot \boldsymbol{g}_1 , \quad v = \boldsymbol{p}_2 \cdot \boldsymbol{g}_2 , \quad u = \boldsymbol{p}_3 \cdot \boldsymbol{g}_3 . \tag{16.32}$$

As shown in Figure 16.11a, the numbers s, t, u, and v are assigned to the four vertices of the square and represented there by lines perpendicular to the square with lengths proportional to the values of s, t, u, and v (which can be positive or negative).

The actual mapping proceeds via a number of steps (Figure 16.12):

1. Transform the point (x, y) to (s_x, s_y),

$$s_x = 3x^2 - 2x^3 , \quad s_y = 3y^2 - 2y^3 . \tag{16.33}$$

2. Assign the lengths s, t, u, and v to the vertices in the mapped square.
3. Obtain the height a (Figure 16.12) via linear interpolation between s and t.
4. Obtain the height b via linear interpolation between u and v.
5. Obtain s_y as a linear interpolation between a and b.
6. The vector $\boldsymbol{c}$ so obtained is now the two components of the noise at (x, y).

16.10.1
Ray Tracing Algorithms

Ray tracing is a technique that renders an image of a scene by simulating the way rays of light travel (Pov-Ray, 2013). To avoid tracing rays that do not contribute to the final image, ray-tracing programs start at the viewer, trace rays backward onto the scene, and then back again onto the light sources. You can vary the location of the viewer and light sources and the properties of the objects being viewed, as well as atmospheric conditions such as fog, haze, and fire.

As an example of what can be carried out, in Figure 16.14a we show the output from the ray-tracing program `Islands.pov` in Listing 16.4, using as input the

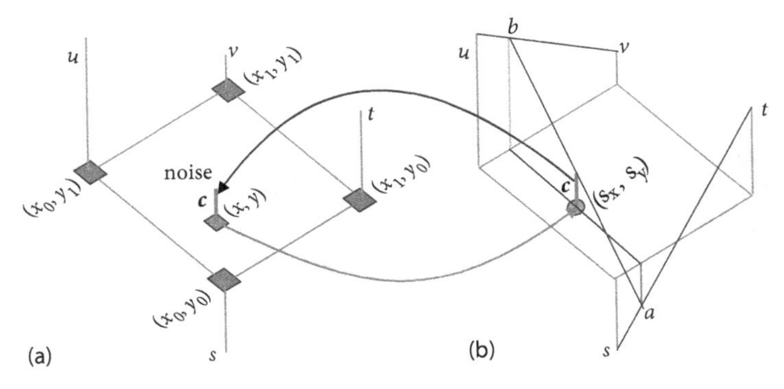

Figure 16.12 Perlin noise mapping. (a) The point (x, y) is mapped to point (s_x, x_y). (b) Using (16.33). Then three linear interpolations are performed to find c, the noise at (x, y).

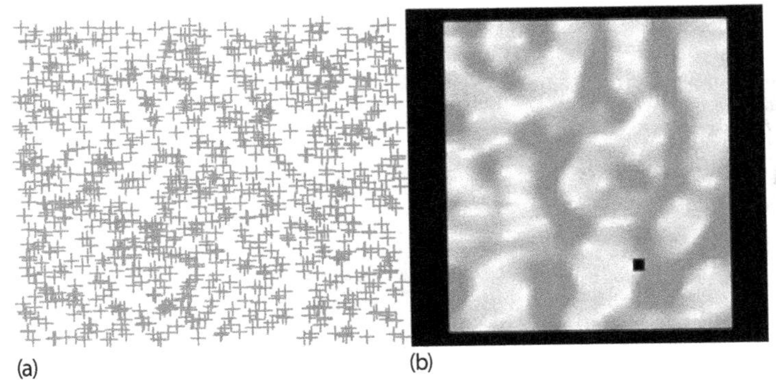

Figure 16.13 After the addition of Perlin noise, the random scatterplot in (a) becomes the clusters on (b).

coherent random noise as shown in Figure 16.13b. The program options we used are given in Listing 16.4 and are seen to include commands to color the islands, to include waves, and to give textures to the sky and the sea. Pov-Ray also allows the possibility of using Perlin noise to give textures to the objects to be created. For example, the stone cup on the right in Figure 16.14 has a marblelike texture produced by Perlin noise.

Listing 16.4 Islands.pov in the Codes/Animations/Fractals directory gives the Pov-Ray ray-tracing commands needed to convert the coherent noise random plot of Figure 16.13 into the mountain like image as shown in Figure 16.14a.

```
// Islands.pov  Pov-Ray program to create Islands, by Manuel J Paez
plane {                                                        // Sky
  <0, 1, 0>, 0
  pigment { color rgb <0, 0, 1> }
  scale 1
  rotate <0, 0, 0>
  translate y*0.2
}
global_settings {
```

(a)

(b)

Figure 16.14 (a) The output from the Pov-Ray ray-tracing program that took as input the 2D coherent random noise plot in Figure 16.13 and added height and fog. (b) An image of a surface of revolution produced by Pov-Ray in which the marble-like texture is created by Perlin noise.

```
  adc_bailout 0.00392157
  assumed_gamma 1.5
  noise_generator 2
}
#declare Island_texture = texture {
  pigment {
    gradient <0, 1, 0>                                    // Vertical direction
    color_map {                                           // Color the islands
      [ 0.15 color rgb <1, 0.968627, 0> ]
      [ 0.2  color rgb <0.886275, 0.733333, 0.180392>   ]
      [ 0.3  color rgb <0.372549, 0.643137, 0.0823529>  ]
      [ 0.4  color rgb <0.101961, 0.588235, 0.184314>   ]
      [ 0.5  color rgb <0.223529, 0.666667, 0.301961>   ]
      [ 0.6  color rgb <0.611765, 0.886275, 0.0196078>  ]
      [ 0.69 color rgb <0.678431, 0.921569, 0.0117647>  ]
      [ 0.74 color rgb <0.886275, 0.886275, 0.317647>   ]
      [ 0.86 color rgb <0.823529, 0.796078, 0.0196078>  ]
      [ 0.93 color rgb <0.905882, 0.545098, 0.00392157> ]
    }
  }
  finish {
    ambient rgbft <0.2, 0.2, 0.2, 0.2, 0.2>
    diffuse 0.8
  }
}
camera {                                  // Camera characteristics and location
  perspective
  location <-15, 6, -20>                                      // Located here
  sky <0, 1, 0>
  direction <0, 0, 1>
  right <1.3333, 0, 0>
  up <0, 1, 0>
  look_at <-0.5, 0, 4>                               //looking at that point
  angle 36
}
light_source {<-10, 20, -25>, rgb <1, 0.733333, 0.00392157>}         // Light

#declare Islands = height_field {             // Takes gif and finds heights
  gif "d:\pov\montania.gif"                       // Windows directory naming
  scale <50, 2, 50>
  translate <-25, 0, -25>
}
object {                                                           // Islands
```

```
  Islands
  texture {
    Island_texture
    scale 2
  }
}
box {                                   // Upper face of the box is the sea
  <-50, 0, -50>, <50, 0.3, 50>          // Location of 2 opposite vertices
  translate <-25, 0, -25>
  texture {                                              // Simulate waves
    normal {
      spotted
      0.4
      scale <0.1, 1, 0.1>
    }
    pigment { color rgb <0.164706, 0.556863, 0.901961> }
  }
}
fog {                                        // A constant fog is defined
  fog_type 1
  distance 30
  rgb <0.984314, 1, 0.964706>
}
```

16.11 Exercises

1. Figure 16.9 gives the rules (at top) and the results (below) for two versions of the Game of Life. These results were produced by the applet `JCellAut`. Write a Python program that runs the same games. Check that you obtain the same results for the same initial conditions.
2. Recall how box counting is used to determine the fractal dimension of an object. Imagine that the result of some experiment or simulation is an interesting geometric figure.
   a) What might be the physical/theoretical importance of determining that this object is a fractal?
   b) What might be the importance of determining its fractal dimension?
   c) Why is it important to use *more* than two sizes of boxes?

17
Thermodynamic Simulations and Feynman Path Integrals

We start this chapter by describing how magnetic materials can be simulated by applying the Metropolis algorithm to the Ising model. This extends the Monte Carlo techniques studied in Chapter 4 to include now thermodynamics. Not only do thermodynamic simulations have important practical applications, but they also give us insight into what is "dynamic" in thermodynamics. Toward the middle of the chapter, we describe a recent Monte Carlo algorithm known as Wang–Landau sampling that has shown itself to be far more efficient than the 60-plus-year-old Metropolis algorithm. Wang–Landau sampling is an active subject in present research, and it is nice to see it fitting well into an elementary textbook. We end the chapter by applying the Metropolis algorithm to Feynman's path integral formulation of quantum mechanics (Feynman and Hibbs, 1965). The theory, while the most advanced to be found in this book, forms the basis for field-theoretic computations of quantum chromodynamics, some of the most fundamental and most time-consuming computations in existence. Basic discussions can be found in (Mannheim, 1983; MacKeown, 1985; MacKeown and Newman, 1987), with a recent review in (Potvin, 1993).

17.1
Magnets via Metropolis Algorithm

Ferromagnets contain finite-size *domains* in which the spins of all the atoms point in the same direction. When an external magnetic field is applied to these materials, the different domains align and the materials become "magnetized." Yet as the temperature is raised, the total magnetism decreases, and at the Curie temperature the system goes through a *phase transition* beyond which all magnetization vanishes. Your *problem* is to explain the thermal behavior of ferromagnets.

Computational Physics, 3rd edition. Rubin H. Landau, Manuel J. Páez, Cristian C. Bordeianu.

17.2
An Ising Chain (Model)

As our model, we consider N magnetic dipoles fixed in place on the links of a linear chain (Figure 17.1). (It is a straightforward generalization to handle 2D and 3D lattices.) Because the particles are fixed, their positions and momenta are not dynamic variables, and we need worry only about their spins. We assume that the particle at site i has spin s_i, which is either up or down:

$$s_i \equiv s_{z,i} = \pm\frac{1}{2} \,. \tag{17.1}$$

Each configuration of the N particles is described by a quantum state vector

$$\left|\alpha_j\right\rangle = \left|s_1, s_2, \ldots, s_N\right\rangle = \left\{\pm\frac{1}{2}, \pm\frac{1}{2}, \ldots\right\} \,, \quad j = 1, \ldots, 2^N \,. \tag{17.2}$$

Because the spin of each particle can assume any one of the *two* values, there are 2^N different possible states for the N particles in the system. Because fixed particles cannot be interchanged, we do not need to concern ourselves with the symmetry of the wave function.

The energy of the system arises from the interaction of the spins with each other and with the external magnetic field B. We know from quantum mechanics that an electron's spin and magnetic moment are proportional to each other, so a magnetic *dipole–dipole* interaction is equivalent to a *spin–spin* interaction. We assume that each dipole interacts with the external magnetic field and with its nearest neighbor through the potential:

$$V_i = -J\boldsymbol{s}_i \cdot \boldsymbol{s}_{i+1} - g\mu_b \boldsymbol{s}_i \cdot \boldsymbol{B} \,. \tag{17.3}$$

Here the constant J is called the *exchange energy* and is a measure of the strength of the spin–spin interaction. The constant g is the gyromagnetic ratio, that is, the proportionality constant between a particle's angular momentum and magnetic moment. The constant $\mu_b = e\hbar/(2m_e c)$ is the Bohr magneton, the basic measure for magnetic moments.

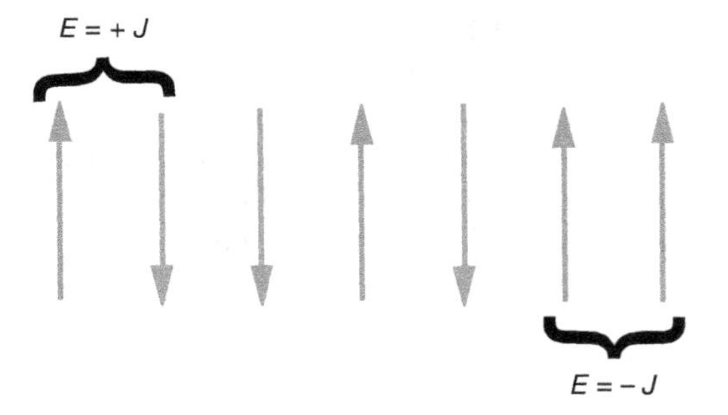

Figure 17.1 The 1D lattice of N spins used in the Ising model of magnetism. The interaction energy between nearest-neighbor pairs $E = \pm J$ is shown for aligned and opposing spins.

Even for small numbers of particles, the 2^N possible spin configurations gets to be very large ($2^{20} > 10^6$), and it is expensive for the computer to examine them all. Realistic samples with $\sim 10^{23}$ particles are beyond imagination. Consequently, statistical approaches are usually assumed, even for moderate values of N. Just how large N must be for this to be accurate is one of the things we want you to explore with your simulations.

The energy of this system in state α_k is the expectation value of the sum of the potential V over the spins of the particles:

$$E_{\alpha_k} = \langle \alpha_k | \sum_i V_i | \alpha_k \rangle = -J \sum_{i=1}^{N-1} s_i s_{i+1} - B\mu_b \sum_{i=1}^{N} s_i \,. \tag{17.4}$$

An apparent paradox in the Ising model occurs when we turn off the external magnetic field and thereby eliminate a preferred direction in space. This means that the average magnetization should vanish despite the fact that the lowest energy state would have all spins aligned. The answer to this paradox is that the system with $B = 0$ is unstable; even if all the spins are aligned, there is nothing to stop their spontaneous reversal. The instabilities are a type of Bloch-wall transitions in which regions of different spin orientations change size. Indeed, natural magnetic materials have multiple domains with all the spins aligned, but with different domains pointing in different directions.

For simplicity, we assume $B = 0$, which means that there are just spin–spin interactions. However, be cognizant of the fact that this means there is no preferred direction in space, and so you may have to be careful how you calculate observables when averaging over domains. For example, you may need to take an absolute value of the total spin when calculating the magnetization, that is, to calculate $\langle | \sum_i s_i | \rangle$ rather than $\langle \sum_i s_i \rangle$.

The equilibrium alignment of the spins depends critically on the sign of the exchange energy J. If $J > 0$, the lowest energy state will tend to have neighboring spins aligned. If the temperature is low enough, the ground state will be a *ferromagnet* with all the spins aligned. Yet if $J < 0$, the lowest energy state will tend to have neighbors with opposite spins. If the temperature is low enough, the ground state will be a *antiferromagnet* with alternating spins.

The simple 1D Ising model has its limitations. Although the model is accurate in describing a system in thermal equilibrium, it is not accurate in describing the *approach* to thermal equilibrium. Also, we have postulated that only one spin is flipped at a time, while real magnetic materials tend to flip many spins at a time. Other limitations are straightforward to improve, for example, the addition of longer range interactions than just nearest neighbors, the motion of the centers, higher multiplicity spin states, and extension to two and three dimensions.

A fascinating aspect of magnetic materials is the existence of a critical temperature, the *Curie temperature*, above which the gross magnetization essentially vanishes. Below the Curie temperature the system is in a quantum state with macroscopic order; above the Curie temperature there is only short-range order extending over atomic dimensions. Although the 1D Ising model predicts realistic temperature dependences for the thermodynamic quantities, the model is too simple to support a phase transition. However, the 2D and 3D Ising models do support the Curie temperature phase transition (Yang, 1952).

17.3
Statistical Mechanics (Theory)

Statistical mechanics starts with the elementary interactions among a system's particles and constructs the macroscopic thermodynamic properties such as specific heats. The essential assumption is that all configurations of the system consistent with the constraints are possible. In some simulations, such as the molecular dynamics ones in Chapter 18, the problem is set up such that the *energy* of the system is fixed. The states of this type of system are described by what is called a *microcanonical ensemble*. In contrast, for the thermodynamic simulations we study in this chapter, the temperature, volume, and number of particles remain fixed, and so we have what is called a *canonical ensemble*.

When we say that an object is *at* temperature T, we mean that the object's atoms are in thermodynamic equilibrium with an average kinetic energy proportional to T. Although this may be an equilibrium state, it is also a dynamic one in which the object's energy fluctuates as it exchanges energy with its environment (it is thermodynamics after all). Indeed, one of the most illuminating aspects of the simulation to follow its visualization of the continual and random interchange of energy that occurs at equilibrium.

The energy E_{α_j} of state α_j in a canonical ensemble is not constant but is distributed with probabilities $P(\alpha_j)$ given by the Boltzmann distribution:

$$\mathcal{P}(E_{\alpha_j}, T) = \frac{e^{-E_{\alpha_j}/k_B T}}{Z(T)}, \quad Z(T) = \sum_{\alpha_j} e^{-E_{\alpha_j}/k_B T}. \tag{17.5}$$

Here k is Boltzmann's constant, T the temperature, and $Z(T)$ the partition function, a weighted sum over the individual *states* or *configurations* of the system. Another formulation, such as the Wang–Landau algorithm discussed in Section 17.5, sums over the *energies* of the states of the system and includes a density-of-states factor $g(E_i)$ to account for degenerate states with the same energy. While the present sum over states is a simpler way to express the problem (one less function), we shall see that the sum over energies is more efficient numerically. In fact, we are even able to ignore the partition function $Z(T)$ because it cancels out in our forming the *ratio* of probabilities.

17.3.1 Analytic Solution

For very large numbers of particles, the thermodynamic properties of the 1D Ising model can be solved analytically and yields (Plischke and Bergersen, 1994)

$$U = \langle E \rangle \,, \tag{17.6}$$

$$\frac{U}{J} = -N \tanh \frac{J}{k_B T} = -N \frac{e^{J/k_B T} - e^{-J/k_B T}}{e^{J/k_B T} + e^{-J/k_B T}} = \begin{cases} N \,, & k_B T \to 0 \,, \\ 0 \,, & k_B T \to \infty \,. \end{cases} \tag{17.7}$$

The analytic results for the specific heat per particle and the magnetization are

$$C(k_B T) = \frac{1}{N}\frac{dU}{dT} = \frac{(J/k_B T)^2}{\cosh^2(J/k_B T)} \,, \tag{17.8}$$

$$M(k_B T) = \frac{N e^{J/k_B T} \sinh(B/k_B T)}{\sqrt{e^{2J/k_B T} \sinh^2(B/k_B T) + e^{-2J/k_B T}}} \,. \tag{17.9}$$

The *2D Ising model* has an analytic solution, but it is not easy to derive it (Yang, 1952; Huang, 1987). Whereas the internal energy and heat capacity are expressed in terms of elliptic integrals, the spontaneous magnetization per particle has the simple form

$$\mathcal{M}(T) = \begin{cases} 0 \,, & T > T_c \,, \\ \frac{(1+z^2)^{1/4}(1-6z^2+z^4)^{1/8}}{\sqrt{1-z^2}} \,, & T < T_c \,, \end{cases} \tag{17.10}$$

$$kT_c \simeq 2.269\,185 J \,, \quad z = e^{-2J/k_B T} \,, \tag{17.11}$$

where the temperature is measured in units of the Curie temperature T_c.

17.4 Metropolis Algorithm

In trying to devise an algorithm that simulates thermal equilibrium, it is important to understand that the Boltzmann distribution (17.5) does not require a system to remain always in the state of lowest energy, but says that it is less likely for the system to be found in a higher energy state than in a lower energy one. Of course, as $T \to 0$ only the lowest energy state will be populated. For finite temperatures we expect the energy to fluctuate by approximately $k_B T$ about the equilibrium value.

In their simulation of neutron transmission through matter, Metropolis, Rosenbluth, Teller, and Teller (Metropolis *et al.*, 1953) invented an algorithm to improve the Monte Carlo calculation of averages. This *Metropolis algorithm* is now a cornerstone of computational physics because the sequence of configurations it produces (a *Markov chain*) accurately simulates the fluctuations that occur during thermal equilibrium. The algorithm randomly changes the individual spins such that, on the average, the probability of a configuration occurring follows a Boltzmann distribution. (We do not find the proof illuminating.)

The Metropolis algorithm is a combination of the variance reduction technique discussed in Section 5.19 and the von Neumann rejection technique discussed in Section 5.21. There we showed how to make Monte Carlo integration more efficient by sampling random points predominantly where the integrand is large and how to generate random points with an arbitrary probability distribution. Now we would like to have spins flip randomly, have a system that can reach any energy in a finite number of steps (*ergodic* sampling), and have a distribution of energies described by a Boltzmann distribution, yet have systems that equilibrate quickly enough to compute in reasonable times.

The Metropolis algorithm is implemented via a number of steps. We start with a fixed temperature and an initial spin configuration, and apply the algorithm until a thermal equilibrium is reached (equilibration). Continued application of the algorithm generates the statistical fluctuations about equilibrium from which we deduce the thermodynamic quantities such as the magnetization $M(T)$. Then the temperature is changed, and the whole process is repeated in order to deduce the T dependence of the thermodynamic quantities. The accuracy of the deduced temperature dependences provides convincing evidence of the validity of the algorithm. Because the 2^N possible configurations of N particles can be a very large number, the amount of computer time needed can be very long. Typically, a small number of iterations $\simeq 10N$ is adequate for equilibration.

The explicit steps of the Metropolis algorithm are:

1. Start with an arbitrary spin configuration $\alpha_k = \{s_1, s_2, \dots, s_N\}$.
2. Generate a trial configuration α_{k+1} by
   a) picking a particle i randomly and
   b) flipping its spin.[1]
3. Calculate the energy $E_{\alpha_{\text{tr}}}$ of the trial configuration.
4. If $E_{\alpha_{\text{tr}}} \le E_{\alpha_k}$, accept the trial by setting $\alpha_{k+1} = \alpha_{\text{tr}}$.
5. If $E_{\alpha_{\text{tr}}} > E_{\alpha_k}$, accept with relative probability $\mathcal{R} = \exp(-\Delta E/k_B T)$:
   a) Choose a uniform random number $0 \le r_i \le 1$.
   b) Set $\alpha_{k+1} = \begin{cases} \alpha_{\text{tr}}\,, & \text{if} \quad \mathcal{R} \ge r_j \quad \text{(accept)}\,, \\ \alpha_k\,, & \text{if} \quad \mathcal{R} < r_j \quad \text{(reject)}\,. \end{cases}$

1) Large-scale, practical computations make a full sweep in which every spin is updated once, and then use this as the new trial configuration. This is found to be efficient and useful in removing some autocorrelations.

The heart of this algorithm is its generation of a random spin configuration α_j (17.2) with probability

$$\mathcal{P}(E_{\alpha_j}, T) \propto e^{-E_{\alpha_j}/k_B T} . \tag{17.12}$$

The technique is a variation of von Neumann rejection (stone throwing) in which a random *trial* configuration is either accepted or rejected depending upon the value of the Boltzmann factor. Explicitly, the ratio of probabilities for a trial configuration of energy E_t to that of an initial configuration of energy E_i is

$$\mathcal{R} = \frac{\mathcal{P}_{tr}}{\mathcal{P}_i} = e^{-\Delta E/k_B T} , \quad \Delta E = E_{\alpha_{tr}} - E_{\alpha_i} . \tag{17.13}$$

If the trial configuration has a lower energy ($\Delta E \leq 0$), the relative probability will be greater than 1 and we will accept the trial configuration as the new initial configuration without further ado. However, if the trial configuration has a higher energy ($\Delta E > 0$), we will not reject it out of hand, but instead accept it with relative probability $\mathcal{R} = \exp(-\Delta E/k_B T) < 1$. To accept a configuration with a probability, we pick a uniform random number between 0 and 1, and if the probability is greater than this number, we accept the trial configuration; if the probability is smaller than the chosen random number, we reject it. (You can remember which way this goes by letting $E_{\alpha_{tr}} \to \infty$, in which case $\mathcal{P} \to 0$ and nothing is accepted.) When the trial configuration is rejected, the next configuration is identical to the preceding one.

How do you start? One possibility is to start with random values of the spins (a "hot" start). Another possibility (Figure 17.2) is a "cold" start in which you start with all spins parallel ($J > 0$) or antiparallel ($J < 0$). In general, one tries to remove the importance of the starting configuration by letting the calculation run a

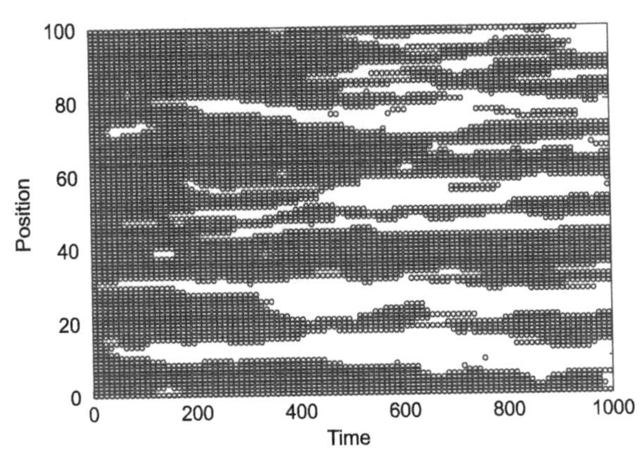

Figure 17.2 An Ising model simulation on a 1D lattice of 100 initially aligned spins (on the left). Up spins are indicated by circles, and down spins by blank spaces. Although the system starts with all up spins (a "cold" start), the system is seen to form domains of up and down spins as time progresses.

while ($\simeq 10N$ rearrangements) before calculating the equilibrium thermodynamic quantities. You should get similar results for hot, cold, or arbitrary starts, and by taking their average you remove some of the statistical fluctuations.

17.4.1
Metropolis Algorithm Implementation

1. Write a program that implements the Metropolis algorithm, that is, that produces a new configuration α_{k+1} from the present configuration α_k. (Alternatively, use the program **IsingViz.py** shown in Listing 17.1.)
2. Make the key data structure in your program an array **s[N]** containing the values of the spins s_i. For debugging, print out + and − to give the spin at each lattice point and examine the pattern for different trial numbers.
3. The value for the exchange energy J fixes the energy scale. Keep it fixed at $J = 1$. (You may also wish to study antiferromagnets with $J = -1$, but first examine ferromagnets whose domains are easier to understand.)
4. The thermal energy $k_B T$ is in units of Joule and is the independent variable. Use $k_B T = 1$ for debugging.
5. Use periodic boundary conditions on your chain to minimize end effects. This means that the chain is a circle with the first and last spins adjacent to each other.
6. Try $N \simeq 20$ for debugging, and larger values for production runs.
7. Use the printout to check that the system equilibrates for
   a) a totally ordered initial configuration (cold start); your simulation should resemble Figure 17.2.
   b) a random initial configuration (hot start).

Listing 17.1 IsingViz.py implements the Metropolis algorithm for a 1D Ising chain.

```
# IsingViz.py: Ising model

from visual import *
import random
from visual.graph import *

# Display for the arrows
scene = display(x=0,y=0,width=700,height=200, range=40,title='Spins')
engraph = gdisplay(y=200,width=700,height=300, title='E of Spin System',\
          xtitle='iteration', ytitle='E',xmax=500, xmin=0, ymax=5, ymin=-5)
enplot = gcurve(color=color.yellow)
N     = 30
B     = 1.
mu    = .33                                          # g mu
J     = .20
k     = 1.                                           # Boltzmann
T     = 100.
state = zeros((N))                          # spins up(1), down (0)
S     = zeros((N) ,float)
test  = state
random.seed()                                        # Seed generator

def energy ( S) :
```

```
    FirstTerm = 0.
    SecondTerm = 0.
    for  i in range(0,N-2):   FirstTerm += S[i]*S[i + 1]
    FirstTerm *= -J
    for i in range(0,N-1):   SecondTerm += S[i]
    SecondTerm *= -B*mu;
    return (FirstTerm + SecondTerm);

ES = energy(state)

def spstate(state):                                          # Plots spins
    for obj in scene.objects: obj.visible=0                # Erase old arrows
    j=0
    for i in range(-N,N,2):
        if state[j]==-1:  ypos = 5                            # Spin down
        else:             ypos = 0
        if  5*state[j]<0: arrowcol = (1,1,1)          # White arrow if down
        else:             arrowcol =(0.7,0.8,0)
        arrow(pos=(i,ypos,0),axis=(0,5*state[j],0),color=arrowcol)
        j +=1

for  i in range(0 ,N):  state[i] = -1               # Initial spins all down

for obj in scene.objects:    obj.visible=0
spstate(state)
ES = energy(state)

for  j in range (1,500):
     rate(3)
     test = state
     r = int(N*random.random());     # Flip spin randomly
     test[r] *= -1
     ET = energy(test)
     p = math.exp((ES-ET)/(k*T))     #  Boltzmann test
     enplot.plot(pos=(j,ES))         # Adds segment to curve
     if p >= random.random():
          state = test
          spstate(state)
          ES = ET
```

17.4.2
Equilibration, Thermodynamic Properties (Assessment)

1. Watch a chain of N atoms attain thermal equilibrium when in contact with a heat bath. At high temperatures, or for small numbers of atoms, you should see large fluctuations, while at lower temperatures you should see smaller fluctuations.
2. Look for evidence of instabilities in which there is a spontaneous flipping of a large number of spins. This becomes more likely for larger $k_B T$ values.
3. Note how at thermal equilibrium the system is still quite dynamic, with spins flipping all the time. It is this energy exchange that determines the thermodynamic properties.
4. You may well find that simulations at small $k_B T$ (say, $k_B T \simeq 0.1$ for $N = 200$) are slow to equilibrate. Higher $k_B T$ values equilibrate faster yet have larger fluctuations.

5. Observe the formation of domains and the effect they have on the total energy. Regardless of the direction of spin within a domain, the atom–atom interactions are attractive and so contribute negative amounts to the energy of the system when aligned. However, the ↑↓ or ↓↑ interactions between domains contribute positive energy. Therefore, you should expect a more negative energy at lower temperatures where there are larger and fewer domains.
6. Make a graph of average domain size vs. temperature.

Thermodynamic Properties For a given spin configuration α_j, the energy and magnetization are given by

$$E_{\alpha_j} = -J \sum_{i=1}^{N-1} s_i s_{i+1} , \quad \mathcal{M}_j = \sum_{i=1}^{N} s_i . \tag{17.14}$$

The internal energy $U(T)$ is just the average value of the energy,

$$U(T) = \langle E \rangle , \tag{17.15}$$

where the average is taken over a system in equilibrium. At high temperatures, we expect a random assortment of spins and so a vanishing magnetization. At low temperatures when all the spins are aligned, we expect $\mathcal{M}$ to approach $N/2$. Although the specific heat can be computed from the elementary definition,

$$C = \frac{1}{N} \frac{\mathrm{d}U}{\mathrm{d}T} , \tag{17.16}$$

the numerical differentiation may be inaccurate because U has statistical fluctuations. A better way to calculate the specific heat is to first calculate the fluctuations in energy occurring during M trials and then determine the specific heat from the fluctuations:

$$U_2 = \frac{1}{M} \sum_{t=1}^{M} (E_t)^2 , \tag{17.17}$$

$$C = \frac{1}{N^2} \frac{U_2 - (U)^2}{k_B T^2} = \frac{1}{N^2} \frac{\langle E^2 \rangle - \langle E \rangle^2}{k_B T^2} . \tag{17.18}$$

1. Extend your program to calculate the internal energy U and the magnetization $\mathcal{M}$ for the chain. Do not recalculate entire sums when only one spin changes.
2. Make sure to wait for your system to equilibrate before you calculate thermodynamic quantities. (You can check that U is fluctuating about its average.) Your results should resemble Figure 17.3.
3. Reduce statistical fluctuations by running the simulation a number of times with different seeds and taking the average of the results.

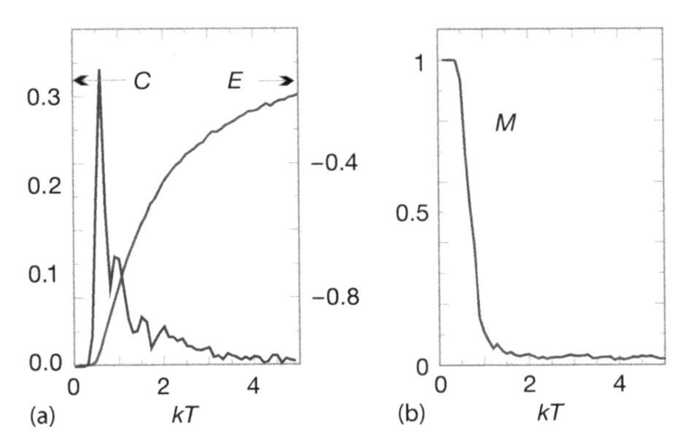

Figure 17.3 Simulation results from a 1D Ising model of 100 spins. (a) Energy and specific heat as functions of temperature; (b) magnetization as a function of temperature.

4. The simulations you run for small N may be realistic but may not agree with statistical mechanics, which assumes $N \simeq \infty$ (you may assume that $N \simeq 2000$ is close to infinity). Check that agreement with the analytic results for the thermodynamic limit is better for large N than small N.
5. Check that the simulated thermodynamic quantities are independent of initial conditions (within statistical uncertainties). In practice, your cold and hot start results should agree.
6. Make a plot of the internal energy U as a function of $k_B T$ and compare it to the analytic result (17.7).
7. Make a plot of the magnetization $\mathcal{M}$ as a function of $k_B T$ and compare it to the analytic result. Does this agree with how you expect a heated magnet to behave?
8. Compute the energy fluctuations U_2 (17.17) and the specific heat C (17.18). Compare the simulated specific heat to the analytic result (17.8).

17.4.3
Beyond Nearest Neighbors, 1D (Exploration)

- Extend the model so that the spin–spin interaction (17.3) extends to next-nearest neighbors as well as nearest neighbors. For the ferromagnetic case this should lead to more binding and less fluctuation because we have increased the couplings among spins, and thus increased the thermal inertia.
- Extend the model so that the ferromagnetic spin–spin interaction (17.3) extends to nearest neighbors in two dimensions, and for the truly ambitious, three dimensions. Continue using periodic boundary conditions and keep the number of particles small, at least to start with (Gould *et al.*, 2006).

1. Form a square lattice and place $\sqrt{N}$ spins on each side.
2. Examine the mean energy and magnetization as the system equilibrates.

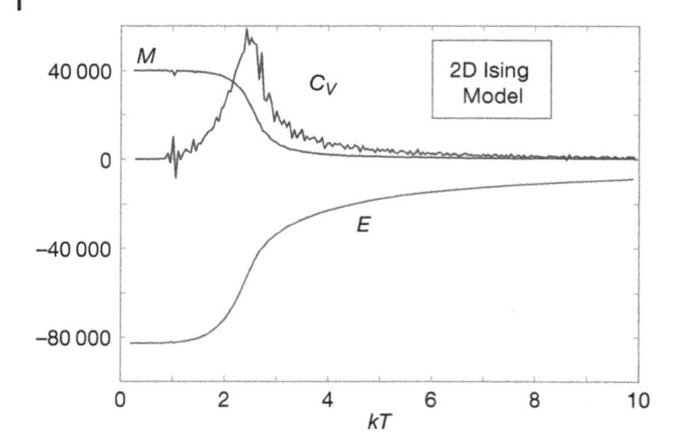

Figure 17.4 The energy, specific heat, and magnetization as a function of temperature from a 2D Ising model simulation with 40 000 spins. Evidence of a phase transition at the Curie temperature $kT = \simeq 2.5$ is seen in all three functions. The values of C and E have been scaled to fit on the same plot as M (courtesy of J. Wetzel).

3. Is the temperature dependence of the average energy qualitatively different from that of the 1D model?
4. Make a print out of the spin configuration for small N, and identify domains.
5. Once your system appears to be behaving properly, calculate the heat capacity and magnetization of the 2D Ising model with the same technique used for the 1D model. Use a total number of particles of $100 \leq N \leq 2000$.
6. Look for a phase transition from ordered to unordered configurations by examining the heat capacity and magnetization as functions of temperature. The former should diverge, while the latter should vanish at the phase transition (Figure 17.4).

Exercise Three fixed spin-1/2 particles interact with each other at temperature $T = 1/k_b$ such that the energy of the system is

$$E = -(s_1 s_2 + s_2 s_3) . \tag{17.19}$$

The system starts in the configuration $\uparrow\downarrow\uparrow$. Do a simulation by hand that uses the Metropolis algorithm and the series of random numbers 0.5, 0.1, 0.9, 0.3 to determine the results of just two thermal fluctuations of these three spins.

17.5 Magnets via Wang–Landau Sampling ⊙

We have used a Boltzmann distribution to simulate the thermal properties of an Ising model. We described the probabilities for explicit spin *states* α with energy E_α for a system at temperature T, and summed over various configurations. An equivalent formulation describes the probability that the system will have the explicit *energy* E at temperature T:

$$\mathcal{P}(E_i, T) = g(E_i)\frac{e^{-E_i/k_B T}}{Z(T)}, \quad Z(T) = \sum_{E_i} g(E_i)e^{-E_i/k_B T}. \tag{17.20}$$

Here k_B is Boltzmann's constant, T is the temperature, $g(E_i)$ is the number of states of energy E_i ($i = 1, \ldots, M$), $Z(T)$ is the partition function, and the sum is still over all M states of the system, but now with states of the same energy entering just once owing to $g(E_i)$ accounting for their degeneracy. Because we again apply the theory to the Ising model with its discrete spin states, the energy assumes only discrete values. If the physical system had an energy that varied continuously, then the number of states in the interval $E \to E + dE$ would be given by $g(E)\,dE$ and $g(E)$ would be called the *density of states*. As a matter of convenience, we call $g(E_i)$ the density of states even when dealing with discrete systems, although the term "degeneracy factor" may be more precise.

Even as the Metropolis algorithm has been providing excellent service for more than 60 years, recent literature shows increasing use of Wang–Landau sampling (WLS)[2] (Landau and Wang, 2001; STP, 2011). Because WLS determines the density of states and the associated partition function, it is not a direct substitute for the Metropolis algorithm and its simulation of thermal fluctuations. However, we will see that WLS provides an equivalent simulation for the Ising model.

The advantages of WLS is that it requires much shorter simulation times than the Metropolis algorithm and provides a direct determination of $g(E_i)$. For these reasons, it has shown itself to be particularly useful for first-order phase transitions where systems spend long times trapped in metastable states, as well as in areas as diverse as spin systems, fluids, liquid crystals, polymers, and proteins. The time required for a simulation becomes crucial when large systems are modeled. Even a spin lattice as small as 8×8 has $2^{64} \simeq 1.84 \times 10^{19}$ configurations, which makes visiting them all too expensive.

In our discussion of the Ising model, we ignored the partition function when applying the Metropolis algorithm. Now we focus on the partition function $Z(T)$ and the density-of-states function $g(E)$. Because $g(E)$ is a function of energy but not temperature, once it has been deduced, $Z(T)$ and all thermodynamic quantities can be calculated from it without having to repeat the simulation for each temperature. For example, the internal energy and the entropy are calculated directly as

$$U(T) \stackrel{\text{def}}{=} \langle E \rangle = \frac{\sum_{E_i} E_i g(E_i)e^{-E_i/k_B T}}{\sum_{E_i} g(E_i)e^{-E_i/k_B T}}, \tag{17.21}$$

$$S = k_B \ln g(E_i). \tag{17.22}$$

The density of states $g(E_i)$ will be determined by taking the equivalent of a random walk in *energy* space. We flip a randomly chosen spin, record the energy of

2) We thank Oscar A. Restrepo of the Universidad de Antioquia for letting us use some of his material here.

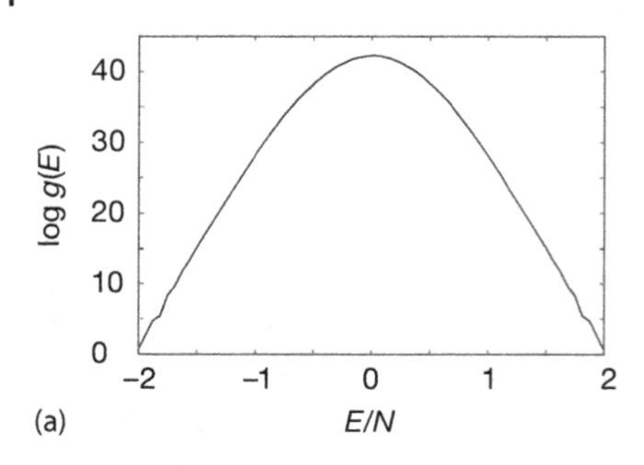

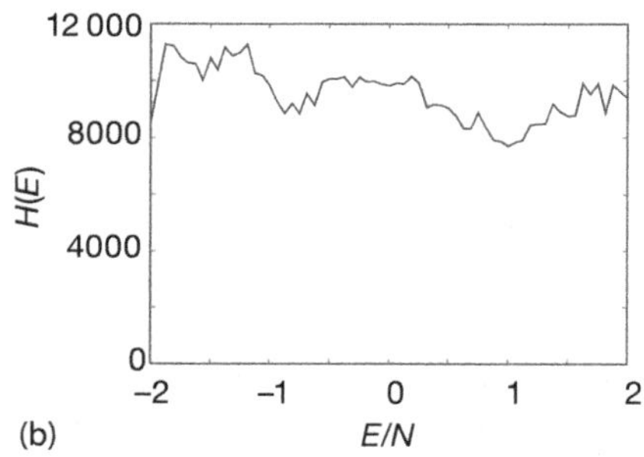

Figure 17.5 Wang–Landau sampling used in the Ising model 2D Ising model on an 8×8 lattice. (a) Logarithm of the density of states $\log g(E) \propto S$ vs. the energy per particle. (b) The histogram $H(E)$ showing the number of states visited as a function of the energy per particle. The aim of WLS is to make this function flat.

the new configuration, and then keep flipping more spins to change the energy. The table $H(E_i)$ of the number of times each energy E_i is attained is called the energy *histogram* (Figure 17.5b). If the walk were continued for a very long time, the histogram $H(E_i)$ would converge to the density of states $g(E_i)$. Yet with 10^{19}–10^{30} steps required even for small systems, this direct approach is unrealistically inefficient because the walk would rarely ever get away from the most probable energies.

Clever idea number 1 behind the Wang–Landau algorithm is to explore more of the energy space by increasing the likelihood of walking into less probable configurations. This is carried out by increasing the acceptance of less likely configurations while simultaneously decreasing the acceptance of more likely ones. In other words, we want to accept more states for which the density $g(E_i)$ is small and reject more states for which $g(E_i)$ is large (fret not these words, the equations are simple). To accomplish this trick, we accept a new energy E_i with a probability inversely proportional to the (initially unknown) density of states,

$$\mathcal{P}(E_i) = \frac{1}{g(E_i)} \,, \tag{17.23}$$

and then build up a histogram of visited states via a random walk.

The problem with clever idea number 1 is that $g(E_i)$ is unknown. WLS's clever idea 2 is to determine the unknown $g(E_i)$ simultaneously with the construction of the random walk. This is accomplished by improving the value of $g(E_i)$ via the multiplication $g(E_i) \rightarrow f g(E_i)$, where $f > 1$ is an empirical factor. When this works, the resulting histogram $H(E_i)$ becomes "flatter" because making the small $g(E_i)$ values larger makes it more likely to reach states with small $g(E_i)$ values. As the histogram gets flatter, we keep decreasing the multiplicative factor f until it is satisfactory close to 1. At that point, we have a flat histogram and a determination of $g(E_i)$.

At this point you may be asking yourself, "Why does a flat histogram mean that we have determined $g(E_i)$?" Flat means that all energies are visited equally, in contrast to the peaked histogram that would be obtained normally without the

$1/g(E_i)$ weighting factor. Thus, if by including this weighting factor we produce a flat histogram, then we have perfectly counteracted the actual peaking in $g(E_i)$, which means that we have arrived at the correct $g(E_i)$.

17.6 Wang–Landau Algorithm

The steps in WLS are similar to those in the Metropolis algorithm, but now with use of the density-of-states function $g(E_i)$ rather than a Boltzmann factor:

1. Start with an arbitrary spin configuration $\alpha_k = \{s_1, s_2, \dots, s_N\}$ and with arbitrary values for the density of states $g(E_i) = 1,\ i = 1, \dots, M$, where $M = 2^N$ is the number of states of the system.
2. Generate a trial configuration α_{k+1} by
   a) picking a particle i randomly and
   b) flipping i's spin.
3. Calculate the energy $E_{\alpha_{tr}}$ of the trial configuration.
4. If $g(E_{\alpha_{tr}}) \le g(E_{\alpha_k})$, accept the trial, that is, set $\alpha_{k+1} = \alpha_{tr}$.
5. If $g(E_{\alpha_{tr}}) > g(E_{\alpha_k})$, accept the trial with probability $\mathcal{P} = g(E_{\alpha_k})/(g(E_{\alpha_{tr}})$:
   a) choose a uniform random number $0 \le r_i \le 1$.
   b) set $\alpha_{k+1} = \begin{cases} \alpha_{tr}\,, & \text{if} \quad \mathcal{P} \ge r_j \quad \text{(accept)}\,, \\ \alpha_k\,, & \text{if} \quad \mathcal{P} < r_j \quad \text{(reject)}\,. \end{cases}$

   This acceptance rule can be expressed succinctly as

$$\mathcal{P}(E_{\alpha_k} \to E_{\alpha_{tr}}) = \min\left[1, \frac{g(E_{\alpha_k})}{g(E_{\alpha_{tr}})}\right] , \tag{17.24}$$

   which manifestly always accepts low-density (improbable) states.
6. One we have a new state, we modify the current density of states $g(E_i)$ via the multiplicative factor f:

$$g(E_{\alpha_{k+1}}) \to f g(E_{\alpha_{k+1}}) , \tag{17.25}$$

   and add 1 to the bin in the histogram corresponding to the new energy:

$$H(E_{\alpha_{k+1}}) \to H(E_{\alpha_{k+1}}) + 1 . \tag{17.26}$$

7. The value of the multiplier f is empirical. We start with Euler's number $f = e = 2.718\,28$, which appears to strike a good balance between very large numbers of small steps (small f) and too rapid a set of jumps through energy space (large f). Because the entropy $S = k_B \ln g(E_i) \to k_B[\ln g(E_i) + \ln f]$, (17.25) corresponds to a uniform increase by k_B in entropy.
8. Even with reasonable values for f, the repeated multiplications in (17.25) lead to exponential growth in the magnitude of g. This may cause floating-point overflows and a concordant loss of information (in the end, the magnitude

of $g(E_i)$ does not matter because the function is normalized). These overflows are avoided by working with logarithms of the function values, in which case the update of the density of states (17.25) becomes

$$\ln g(E_i) \to \ln g(E_i) + \ln f \ . \tag{17.27}$$

9. The difficulty with storing $\ln g(E_i)$ is that we need the ratio of $g(E_i)$ values to calculate the probability in (17.24). This is circumvented by employing the identity $x \equiv \exp(\ln x)$ to express the ratio as

$$\frac{g(E_{\alpha_k})}{g(E_{\alpha_{tr}})} = \exp\left[\ln \frac{g(E_{\alpha_k})}{g(E_{\alpha_{tr}})}\right] = \exp\left[\ln g(E_{\alpha_k})\right] - \exp\left[\ln g(E_{\alpha_{tr}})\right] \ . \tag{17.28}$$

In turn, $g(E_k) = f \times g(E_k)$ is modified to $\ln g(E_k) \to \ln g(E_k) + \ln f$.

10. The random walk in E_i continues until a flat histogram of visited energy values is obtained. The flatness of the histogram is tested regularly (every 10 000 iterations), and the walk is terminated once the histogram is sufficiently flat. The value of f is then reduced so that the next walk provides a better approximation to $g(E_i)$. Flatness is measured by comparing the variance in $H(E_i)$ to its average. Although 90–95% flatness can be achieved for small problems like ours, we demand only 80% (Figure 17.5):

$$\text{If} \quad \frac{H_{\max} - H_{\min}}{H_{\max} + H_{\min}} < 0.2 \text{ , stop, let } f \to \sqrt{f}(\ln f \to \ln f/2) \ . \tag{17.29}$$

11. Keep the generated $g(E_i)$ and reset the histogram values $h(E_i)$ to zero.
12. The walks are terminated and new ones initiated until no significant correction to the density of states is obtained. This is measured by requiring the multiplicative factor $f \simeq 1$ within some level of tolerance, for example, $f \leq 1 + 10^{-8}$. If the algorithm is successful, the histogram should be flat (Figure 17.5) within the bounds set by (17.29).
13. The final step in the simulation is normalization of the deduced density of states $g(E_i)$. For the Ising model with N up or down spins, a normalization condition follows from knowledge of the total number of states (STP, 2011):

$$\sum_{E_i} g(E_i) = 2^N \quad \Rightarrow \quad g^{(\text{norm})}(E_i) = \frac{2^N}{\sum_{E_i} g(E_i)} g(E_i) \ . \tag{17.30}$$

Because the sum in (17.30) is most affected by those values of energy where $g(E_i)$ is large, it may not be precise for the low-E_i densities that contribute little to the sum. Accordingly, a more precise normalization, at least if your simulation has performed a good job in occupying all energy states, is to require that there are just two grounds states with energies $E = -2N$ (one with all spins up and one with all spins down):

$$\sum_{E_i=-2N} g(E_i) = 2 \ . \tag{17.31}$$

In either case, it is good practice to normalize $g(E_i)$ with one condition and then use the other as a check.

17.6.1
WLS Ising Model Implementation

We assume an Ising model with spin–spin interactions between nearest neighbors located in an $L \times L$ lattice (Figure 17.6). To keep the notation simple, we set $J = 1$ so that

$$E = -\sum_{i \leftrightarrow j}^{N} \sigma_i \sigma_j , \tag{17.32}$$

where $\leftrightarrow$ indicates nearest neighbors. Rather than recalculate the energy each time a spin is flipped, only the difference in energy is computed. For example, for eight spins in a 1D array,

$$-E_k = \sigma_0\sigma_1 + \sigma_1\sigma_2 + \sigma_2\sigma_3 + \sigma_3\sigma_4 + \sigma_4\sigma_5 + \sigma_5\sigma_6 + \sigma_6\sigma_7 + \sigma_7\sigma_0 , \tag{17.33}$$

where the 0–7 interaction arises because we assume periodic boundary conditions. If spin 5 is flipped, the new energy is

$$-E_{k+1} = \sigma_0\sigma_1 + \sigma_1\sigma_2 + \sigma_2\sigma_3 + \sigma_3\sigma_4 - \sigma_4\sigma_5 - \sigma_5\sigma_6 + \sigma_6\sigma_7 + \sigma_7\sigma_0 , \tag{17.34}$$

and the difference in energy is

$$\Delta E = E_{k+1} - E_k = 2(\sigma_4 + \sigma_6)\sigma_5 . \tag{17.35}$$

This is cheaper to compute than calculating and then subtracting two energies.

When we advance to two dimensions with the 8×8 lattice in Figure 17.6, the change in energy when a spin $\sigma_{i,j}$ at site (i, j) flips is

$$\Delta E = 2\sigma_{i,j}(\sigma_{i+1,j} + \sigma_{i-1,j} + \sigma_{i,j+1} + \sigma_{i,j-1}) , \tag{17.36}$$

which can assume the values $-8, -4, 0, 4$, and 8. There are two states of minimum energy $E = -2N$ for a 2D system with N spins, and they are ones with all spins pointing in the same direction (either up or down). The maximum energy is $2N$, and it corresponds to alternating spin directions on neighboring sites. Each spin flip on the lattice changes the energy by four units between these limits, and so the values of the energies are

$$E_i = -2N , \quad -2N+4 , \quad -2N+8, \dots , 2N-8 , \quad 2N-4 , \quad 2N . \tag{17.37}$$

These energies can be stored in a uniform 1D array via the simple mapping

$$E' = \frac{E + 2N}{4} \quad \Rightarrow \quad E' = 0, 1, 2, \dots , N . \tag{17.38}$$

Listing 17.2 displays our implementation of Wang–Landau sampling to calculate the density of states and internal energy $U(T)$ (17.21). We used it to obtain the

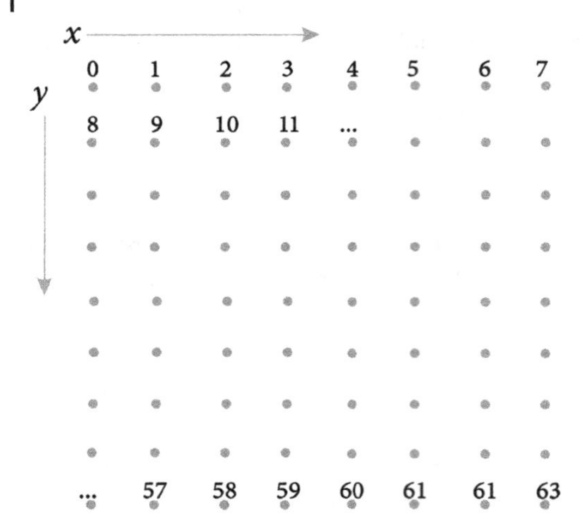

Figure 17.6 The numbering scheme used in our WLS implementation of the 2D Ising model with an 8 × 8 lattice of spins.

entropy $S(T)$ and the energy histogram $H(E_i)$ illustrated in Figure 17.5. Other thermodynamic functions can be obtained by replacing the E in (17.21) with the appropriate variable. The results look like those in Figure 17.4.

A problem that may be encountered when calculating these variables is that the sums in (17.21) can become large enough to cause overflows, although the ratio would not. You work around that by factoring out a common large factor, for example,

$$\sum_{E_i} X(E_i) g(E_i) e^{-E_i/k_B T} = e^{\lambda} \sum_{E_i} X(E_i) e^{\ln g(E_i) - E_i/k_B T - \lambda} , \tag{17.39}$$

where λ is the largest value of $\ln g(E_i) - E_i/k_B T$ at each temperature. The factor e^{λ} does not actually need to be included in the calculation of the variable because it is present in both the numerator and denominator and so eventually cancels out.

Listing 17.2 WangLandau.py simulates the 2D Ising model using Wang–Landau sampling to compute the density of states and from that the internal energy.

```
# WangLandau.py: Wang Landau algorithm for 2D spin system

""" Author in Java: Oscar A. Restrepo,
 Universidad de Antioquia, Medellin, Colombia
 Each time fac changes, a new histogrm is generated.
 Only the first Histogram plotted to reduce computational time"""
from visual import *
import random;
from visual.graph import *

L = 8;  N  = (L*L)

# Set up graphics
entgr = gdisplay(x=0,y=0,width=500,height=250,title='Density of States',\
      xtitle= 'E/N', ytitle='log g(E)', xmax=2., xmin=-2.,ymax=45,ymin=0)
```

```
entrp     = gcurve(color = color.yellow, display = entgr)
energygr = gdisplay(x=0, y=250, width=500, height=250, title='E vs T',\
        xtitle = 'T', ytitle='U(T)/N', xmax=8.,xmin=0, ymax =0.,ymin=-2.)
energ     = gcurve(color = color.cyan, display = energygr)
histogr   = display(x = 0, y = 500, width = 500, height = 300,\
       title = '1st histogram: H(E) vs. E/N, corresponds to log(f) = 1')
histo = curve(x = list(range(0, N+1)), color=color.red, display=histogr)
xaxis     = curve(pos = [( - N, - 10), (N, - 10)])
minE      = label(text = ' - 2', pos = ( - N + 3, - 15), box = 0)
maxE      = label(text = '2', pos = (N - 3, - 15), box = 0)
zeroE     = label(text = '0', pos = (0, - 15), box = 0)
ticm      = curve(pos = [( - N, - 10), ( - N, - 13)])
tic0      = curve(pos = [(0, - 10), (0, - 13)])
ticM      = curve(pos = [(N, - 10), (N, - 13)])
enr       = label(text = 'E/N', pos = (N/2, - 15), box = 0)

sp      = zeros( (L, L) )                                   # Grid size, spins
hist    = zeros( (N + 1) )
prhist = zeros( (N + 1) )                                         # Histograms
S       = zeros( (N + 1), float)                          # Entropy = log g(E)

def iE(e):  return int((e + 2*N)/4)

def IntEnergy():
    exponent = 0.0
    for T in arange (0.2, 8.2, 0.2 ):                      # Select lambda max
        Ener = - 2*N
        maxL = 0.0                                                # Initialize
        for i in range(0, N + 1):
            if S[i]!= 0 and (S[i] - Ener/T)>maxL:
                maxL = S[i] - Ener/T
                Ener = Ener + 4
        sumdeno = 0
        sumnume = 0
        Ener     = -2*N
        for i in range(0, N):
            if S[i] != 0:
                exponent = S[i] - Ener/T - maxL
            sumnume += Ener*exp(exponent)
            sumdeno += exp(exponent)
            Ener      = Ener + 4.0
        U = sumnume/sumdeno/N                             # internal energy U(T)/N
        energ.plot(pos = (T, U) )

def WL():                                               # Wang - Landau sampling
    Hinf    = 1.e10                                 # initial values for Histogram
    Hsup    = 0.
    tol     = 1.e-3                               # tolerance, stops the algorithm
    ip      = zeros(L)
    im      = zeros(L)                                  # BC R or down, L or up
    height = abs(Hsup - Hinf)/2.                           # Initialize histogram
    ave = (Hsup + Hinf)/2.                            # about average of histogram
    percent = height / ave
    for i in range(0, L):
        for j in range(0, L): sp[i, j] = 1                         # Initial spins
    for i in range(0, L):
        ip[i] = i + 1
        im[i] = i - 1                                         # Case plus, minus
    ip[L - 1] = 0
    im[0]     = L - 1                                                 # Borders
    Eold = - 2*N                                             # Initialize energy
    for j in range(0, N + 1): S[j] = 0                     # Entropy initialized
    iter = 0
    fac  = 1
    while fac > tol :

        i = int(N*random.random() )                           # Select random spin
```

```
            xg = i%L
            # Must be i//L, not i/L for Python 3:
            yg     = i//L                                  # Localize x, y, grid point
            Enew   = Eold + 2*(sp[ip[xg],yg] + sp[im[xg],yg] + sp[xg,ip[yg]]
                + sp[xg, im[yg]] ) * sp[xg, yg]                  # Change energy
            deltaS = S[iE(Enew)]  -  S[iE(Eold)]
            if  deltaS <= 0 or random.random() < exp( - deltaS):
                Eold = Enew;
                sp[xg, yg] *=  - 1                                  # Flip spin
            S[iE(Eold)]   += fac;                              # Change entropy
            if iter%10000 == 0:              # Check flatness every 10000 sweeps
                for j in range( 0, N + 1):
                    if  j == 0 :
                        Hsup = 0
                        Hinf = 1e10                       # Initialize new histogram
                    if  hist[j] == 0 :  continue         # Energies never visited
                    if  hist[j] > Hsup: Hsup = hist[j]
                    if  hist[j] < Hinf: Hinf = hist[j]
                height = Hsup - Hinf
                ave = Hsup + Hinf
                percent = 1.0* height/ave               # 1.0 to make it float number
                if percent < 0.3 :                               # Histogram flat?
                    print(" iter ", iter, "   log(f) ", fac)
                    for j in range(0, N + 1):
                        prhist[j] = hist[j]                              # to plot
                        hist[j]   = 0                                  # Save hist
                    fac *= 0.5                        # Equivalent to log(sqrt(f))
            iter += 1
            hist[iE(Eold)]   += 1                # Change histogram, add 1, update
            if fac >= 0.5:                         # just show the first histogram
              # Speed up by using array calculations:
              histo.x = 2.0*arange(0,N+1) - N
              histo.y = 0.025*hist-10
deltaS = 0.0
print("wait because iter > 13 000 000")                       # not always the same
WL()                                                    # Call Wang Landau algorithm
deltaS = 0.0
for j in range(0, N + 1):
    rate(150)
    order    = j*4  -  2*N
    deltaS   = S[j]  -  S[0] + log(2)
    if S[j] != 0 : entrp.plot(pos = (1.*order/N, deltaS))        # plot entropy
IntEnergy();
print("Done")
```

17.6.2
WLS Ising Model Assessment

Repeat the assessment conducted in Section 17.4.2 for the thermodynamic properties of the Ising model using WLS in place of the Metropolis algorithm.

17.7 Feynman Path Integral Quantum Mechanics ⊙

Problem As is well known, a classical particle attached to a linear spring undergoes simple harmonic motion with a position as a function of time given by $x(t) = A\sin(\omega_0 t + \phi)$. Your *problem* is to take this classical space–time trajectory $x(t)$ and use it to generate the quantum wave function $\psi(x,t)$ for a particle bound in a harmonic oscillator potential.

17.8 Feynman's Space–Time Propagation (Theory)

Feynman was looking for a formulation of quantum mechanics that gave a more direct connection to classical mechanics than does Schrödinger theory, and that made the statistical nature of quantum mechanics evident from the start. He followed a suggestion by Dirac that Hamilton's principle of least action, which can be used to derive classical dynamics, may also be the $\hbar \to 0$ limit of a quantum least-action principle. Seeing that Hamilton's principle deals with the paths of particles through space–time, Feynman postulated that the quantum-mechanical wave function describing the propagation of a free particle from the space–time point $a = (x_a, t_a)$ to the point $b = (x_b, t_b)$ can expressed as (Feynman and Hibbs, 1965; Mannheim, 1983)

$$\psi(x_b, t_b) = \int dx_a\, G(x_b, t_b; x_a, t_a)\psi(x_a, t_a)\,, \tag{17.40}$$

where G is *Green's function* or *propagator*

$$G(x_b, t_b; x_a, t_a) \equiv G(b, a) = \sqrt{\frac{m}{2\pi i(t_b - t_a)}}\exp\left[i\frac{m(x_b - x_a)^2}{2(t_b - t_a)}\right]. \tag{17.41}$$

Equation 17.40 is a form of Huygens's wavelet principle in which each point on the wavefront $\psi(x_a, t_a)$ emits a spherical wavelet $G(b;a)$ that propagates forward in space and time. It states that the new wavefront $\psi(x_b, t_b)$ is created by summation over and interference with all the other wavelets.

Feynman imagined that another way of interpreting (17.40) is as a form of Hamilton's principle in which the probability amplitude, that is the wave function ψ, for a particle to be at B equals the sum over all *paths* through space–time originating at time A and ending at B (Figure 17.7). This view incorporates the statistical nature of quantum mechanics by having different probabilities for travel along the different paths. All paths are possible, but some are more likely than others. (When you realize that Schrödinger theory solves for wave functions and considers paths a classical concept, you can appreciate how different it is from Feynman's view.) The values for the probabilities of the paths derive from *Hamilton's classical principle of least action*:

> *The most general motion of a physical particle moving along the classical trajectory $\bar{x}(t)$ from time t_a to t_b is along a path such that the action $S[\bar{x}(t)]$ is an*

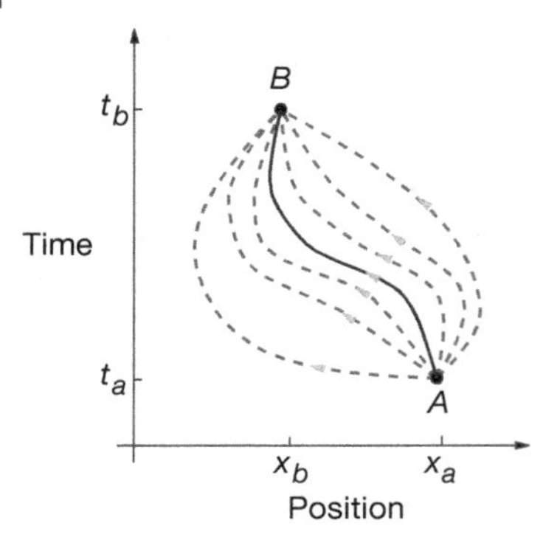

Figure 17.7 In the Feynman path-integral formulation of quantum mechanics a collection of paths connect the initial space–time point A to the final point B. The solid line is the trajectory followed by a classical particle, while the dashed lines are additional paths sampled by a quantum particle. A classical particle somehow "knows" ahead of time that travel along the classical trajectory minimizes the action S.

extremum:

$$\delta S[\bar{x}(t)] = S[\bar{x}(t) + \delta x(t)] - S[\bar{x}(t)] = 0 , \tag{17.42}$$

with the paths constrained to pass through the endpoints:

$$\delta(x_a) = \delta(x_b) = 0 .$$

This formulation of classical mechanics, which is based on the calculus of variations, is equivalent to Newton's differential equations if the action S is taken as the line integral of the Lagrangian along the path:

$$S[\bar{x}(t)] = \int_{t_a}^{t_b} \mathrm{d}t L[x(t), \dot{x}(t)] , \quad L = T[x, \dot{x}] - V[x] . \tag{17.43}$$

Here T is the kinetic energy, V is the potential energy, $\dot{x} = \mathrm{d}x/\mathrm{d}t$, and square brackets indicate a *functional*[3] of the function $x(t)$ and $\dot{x}(t)$.

Feynman observed that the classical action for a free ($V = 0$) particle,

$$S[b, a] = \frac{m}{2}(\dot{x})^2(t_b - t_a) = \frac{m}{2}\frac{(x_b - x_a)^2}{t_b - t_a} , \tag{17.44}$$

is related to the free-particle propagator (17.41) by

$$G(b, a) = \sqrt{\frac{m}{2\pi \mathrm{i}(t_b - t_a)}} \mathrm{e}^{\mathrm{i}S[b,a]/\hbar} . \tag{17.45}$$

3) A *functional* is a number whose value depends on the complete behavior of some function and not just on its behavior at one point. For example, the derivative $f'(x)$ depends on the value of f at x, yet the integral $I[f] = \int_a^b \mathrm{d}x f(x)$ depends on the entire function, and is therefore a functional of f.

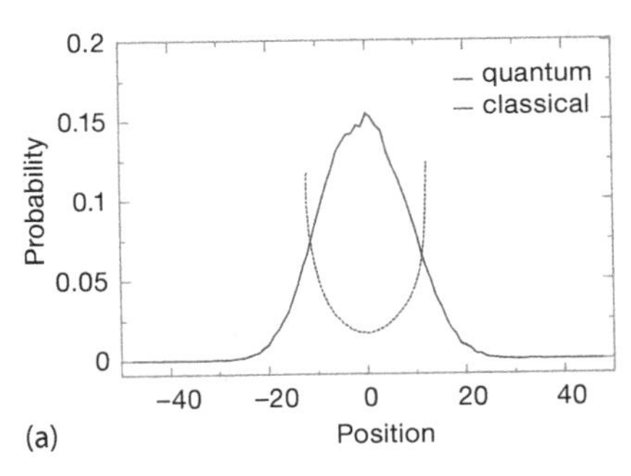

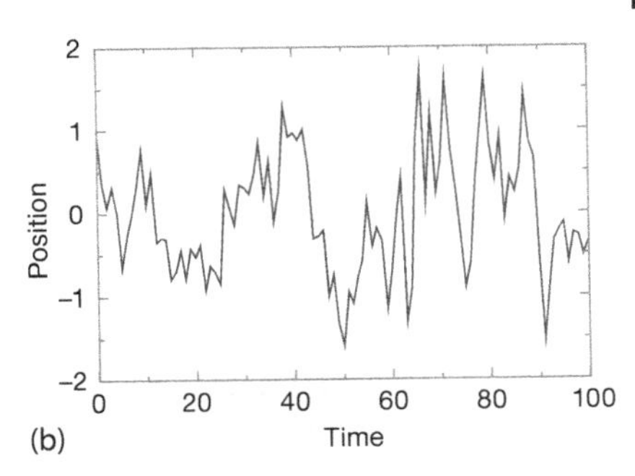

Figure 17.8 (a) The probability distribution for the harmonic oscillator ground state as determined with a path-integral calculation (the classical result has maxima at the two turning points). (b) A space–time quantum path resulting from applying the Metropolis algorithm.

This is the much sought-after connection between quantum mechanics and Hamilton's principle. Feynman went on to postulate a reformulation of quantum mechanics that incorporated its statistical aspects by expressing $G(b, a)$ as the weighted sum over all *paths* connecting a to b,

$$G(b, a) = \sum_{\text{paths}} e^{iS[b,a]/\hbar} \quad \text{(path integral)} . \tag{17.46}$$

Here the classical action S (17.43) is evaluated along different paths (Figure 17.7), and the exponential of the action is summed over paths. The sum (17.46) is called a *path integral* because it sums over actions $S[b, a]$, each of which is an integral (on the computer an integral and a sum are the same anyway).

The essential connection between classical and quantum mechanics is the realization that in units of $\hbar \simeq 10^{-34}$ J s, the action is a very large number, $S/\hbar \geq 10^{20}$, and so even though all paths enter into the sum (17.46), the main contributions come from those paths adjacent to the classical trajectory $\bar{x}$. In fact, because S is an extremum for the classical trajectory, it is a constant to first order in the variation of paths, and so nearby paths have phases that vary smoothly and relatively slowly. In contrast, paths far from the classical trajectory are weighted by a rapidly oscillating $\exp(iS/\hbar)$, and when many are included they tend to cancel each other out. In the classical limit, $\hbar \to 0$, only the single classical trajectory contributes and (17.46) becomes Hamilton's principle of least action. In Figure 17.8, we show an example of an actual trajectory used in path-integral calculations.

17.8.1
Bound-State Wave Function (Theory)

Although you may be thinking that you have already seen enough expressions for Green's function, there is yet another one we need for our computation. Let us assume that the Hamiltonian operator $\tilde{H}$ supports a spectrum of eigenfunctions,

$$\tilde{H}\psi_n = E_n \psi_n , \tag{17.47}$$

each labeled by the index n. Because $\tilde{H}$ is Hermitian, its wave functions form a complete orthonormal set in which we may expand a general solution:

$$\psi(x,t) = \sum_{n=0}^{\infty} c_n e^{-iE_n t} \psi_n(x) , \quad c_n = \int_{-\infty}^{+\infty} dx \psi_n^*(x) \psi(x, t=0) , \tag{17.48}$$

where the value for the expansion coefficients c_n follows from the orthonormality of the ψ_n's. If we substitute this c_n back into the wave function expansion (17.48), we obtain the identity

$$\psi(x,t) = \int_{-\infty}^{+\infty} dx_0 \sum_n \psi_n^*(x_0) \psi_n(x) e^{-iE_n t} \psi(x_0, t=0) . \tag{17.49}$$

Comparison with (17.40) yields the eigenfunction expansion for G:

$$G(x,t;x_0,t_0=0) = \sum_n \psi_n^*(x_0)\psi_n(x) e^{-iE_n t} . \tag{17.50}$$

We relate this to the bound-state wave function (recall that our *problem* is to calculate that) by (1) requiring all paths to start and end at the space position $x_0 = x$, (2) by taking $t_0 = 0$, and (3) by making an analytic continuation of (17.50) to negative imaginary time (permissable for analytic functions):

$$G(x, -i\tau; x, 0) = \sum_n |\psi_n(x)|^2 e^{-E_n \tau} = |\psi_0|^2 e^{-E_0\tau} + |\psi_1|^2 e^{-E_1 \tau} + \cdots , \tag{17.51}$$

$$\Rightarrow \quad |\psi_0(x)|^2 = \lim_{\tau\to\infty} e^{E_0 \tau} G(x, -i\tau; x, 0) . \tag{17.52}$$

The limit here corresponds to long imaginary times τ, after which the parts of ψ with higher energies decay more quickly, leaving only the ground state ψ_0.

Equation 17.52 provides a closed-form solution for the ground-state wave function directly in terms of the propagator G. Although we will soon describe how to compute this equation, look now at Figure 17.8 showing some results of a computation. Although we start with a probability distribution that peaks near the classical turning points at the edges of the well, after a large number of iterations we end up with a distribution that resembles the expected Gaussian, which indicates that the formulation appears to be working. On the right, we see a trajectory that has been generated via statistical variations about the classical trajectory $x(t) = A \sin(\omega_0 t + \phi)$.

17.8.2
Lattice Path Integration (Algorithm)

Because both time and space need to be integrated over when evaluating a path integral, our simulation starts with a lattice of discrete space–time points. We

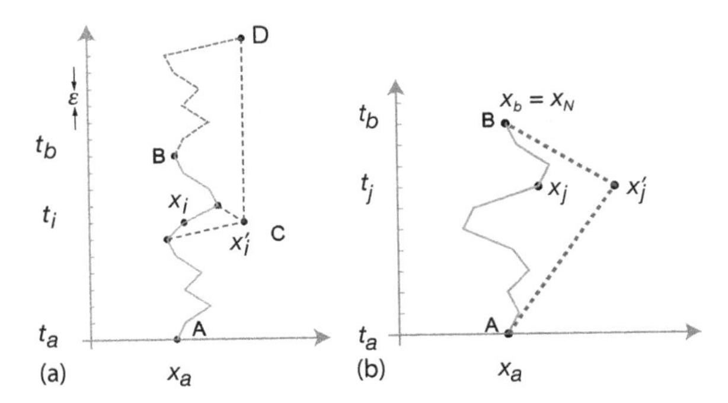

Figure 17.9 (a) A path through a space–time lattice that starts and ends at $x = x_a = x_b$. The action is an integral over this path, while the *path integral* is a sum of integrals over all paths. The dotted path BD is a transposed replica of path AC. (b) The dashed path joins the initial and final times in two equal time steps; the solid curve uses N steps each of size ε. The position of the curve at time t_j defines the position x_j.

visualize a particle's trajectory as a series of straight lines connecting one time to the next (Figure 17.9). We divide the time between points A and B into N equal steps of size ε, and label them with the index j:

$$\varepsilon \stackrel{\text{def}}{=} \frac{t_b - t_a}{N} \quad \Rightarrow \quad t_j = t_a + j\varepsilon \,, \quad (j = 0, N) \,. \tag{17.53}$$

Although it is more precise to use the actual positions $x(t_j)$ of the trajectory at the times t_j to determine the x_j's (as in Figure 17.9), in practice we discretize space uniformly and have the links end at the nearest regular points. Once we have a lattice, it is easy to evaluate derivatives or integrals on a link[4]:

$$\frac{\mathrm{d}x_j}{\mathrm{d}t} \simeq \frac{x_j - x_{j-1}}{t_j - t_{j-1}} = \frac{x_j - x_{j-1}}{\varepsilon} \,, \tag{17.54}$$

$$S_j \simeq L_j \Delta t \simeq \frac{1}{2} m \frac{(x_j - x_{j-1})^2}{\varepsilon} - V(x_j)\varepsilon \,, \tag{17.55}$$

where we have assumed that the Lagrangian is constant over each link.

Lattice path integration is based on the *composition theorem* for propagators:

$$G(b,a) = \int \mathrm{d}x_j G(x_b, t_b; x_j, t_j) G(x_j, t_j; x_a, t_a) \,, \quad (t_a < t_j, \; t_j < t_b) \,. \tag{17.56}$$

4) Although Euler's rule has a large error, it is often use in lattice calculations because of its simplicity. However, if the Lagrangian contains second derivatives, you should use the more precise central-difference method to avoid singularities.

For a free particle, this yields

$$G(b,a) = \sqrt{\frac{m}{2\pi i(t_b - t_j)}}\sqrt{\frac{m}{2\pi i(t_j - t_a)}}\int dx_j e^{i(S[b,j]+S[j,a])}$$
$$= \sqrt{\frac{m}{2\pi i(t_b - t_a)}}\int dx_j e^{iS[b,a]} , \tag{17.57}$$

where we have added the actions because line integrals combine as $S[b,j] + S[j,a] = S[b,a]$. For the N-linked path in Figure 17.9, (17.56) becomes

$$G(b,a) = \int dx_1 \cdots dx_{N-1} e^{iS[b,a]} , \quad S[b,a] = \sum_{j=1}^{N} S_j , \tag{17.58}$$

where S_j is the value of the action for link j. At this point the integral over the *single* path shown in Figure 17.9 has become an N-term sum that becomes an infinite sum as the time step ε approaches zero.

To summarize, Feynman's path-integral postulate (17.46) means that we sum over all paths connecting A to B to obtain the Green's function $G(b,a)$. This means that we must sum not only over the links in one path but *also* over all the different paths in order to produce the variation in paths required by Hamilton's principle. The sum is constrained such that paths must pass through A and B and cannot double back on themselves (causality requires that particles move only forward in time). This is the essence of *path integration*. Because we are integrating over functions as well as along paths, the technique is also known as *functional integration*.

The propagator (17.46) is the sum over all paths connecting A to B, with each path weighted by the exponential of the action along that path, explicitly:

$$G(x,t;x_0,t_0) = \sum\!\!\!\!\!\!\int dx_1\, dx_2 \cdots dx_{N-1} e^{iS[x,x_0]} , \tag{17.59}$$

$$S[x,x_0] = \sum_{j=1}^{N-1} S[x_{j+1},x_j] \simeq \sum_{j=1}^{N-1} L\left(x_j,\dot{x}_j\right)\varepsilon , \tag{17.60}$$

where $L(x_j,\dot{x}_j)$ is the average value of the Lagrangian on link j at time $t = j\varepsilon$. The computation is made simpler by assuming that the potential $V(x)$ is independent of velocity and does not depend on other x values (local potential). Next we observe that G is evaluated with a negative imaginary time in expression (17.52) for the ground-state wave function. Accordingly, we evaluate the Lagrangian with $t = -i\tau$:

$$L(x,\dot{x}) = T - V(x) = +\frac{1}{2}m\left(\frac{dx}{dt}\right)^2 - V(x) \tag{17.61}$$

$$\Rightarrow \quad L\left(x,\frac{dx}{-i\,d\tau}\right) = -\frac{1}{2}m\left(\frac{dx}{d\tau}\right)^2 - V(x) . \tag{17.62}$$

We see that the reversal of the sign of the kinetic energy in L means that L now equals the negative of the Hamiltonian evaluated at a real positive time $t = \tau$:

$$H\left(x, \frac{dx}{d\tau}\right) = \frac{1}{2}m\left(\frac{dx}{d\tau}\right)^2 + V(x) = E \tag{17.63}$$

$$\Rightarrow \quad L\left(x, \frac{dx}{-id\tau}\right) = -H\left(x, \frac{dx}{d\tau}\right) . \tag{17.64}$$

In this way, we rewrite the t-path integral of L as a τ-path integral of H, and so express the action and Green's function in terms of the Hamiltonian:

$$S[j+1, j] = \int_{t_j}^{t_{j+1}} L(x, t)\, dt = -i \int_{\tau_j}^{\tau_{j+1}} H(x, \tau)\, d\tau \tag{17.65}$$

$$\Rightarrow \quad G(x, -i\tau; x_0, 0) = \int dx_1 \ldots dx_{N-1} e^{-\int_0^\tau H(\tau')d\tau'} , \tag{17.66}$$

where the line integral of H is over an entire trajectory. Next we express the path integral in terms of the average energy of the particle on each link, $E_j = T_j + V_j$, and then sum over links to obtain the summed energy $\mathcal{E}$[5]:

$$\int H(\tau)\, d\tau \simeq \sum_j \varepsilon E_j = \varepsilon \mathcal{E}(\{x_j\}) , \tag{17.67}$$

$$\mathcal{E}(\{x_j\}) \stackrel{\text{def}}{=} \sum_{j=1}^{N} \left[\frac{m}{2}\left(\frac{x_j - x_{j-1}}{\varepsilon}\right)^2 + V\left(\frac{x_j + x_{j-1}}{2}\right)\right] . \tag{17.68}$$

In (17.68), we have approximated each path link as a *straight line*, used Euler's derivative rule to obtain the velocity, and evaluated the potential at the midpoint of each link. We now substitute this G into our solution (17.52) for the ground-state wave function in which the initial and final points in space are the same:

$$\lim_{\tau\to\infty} \frac{G(x, -i\tau, x_0 = x, 0)}{\int dx G(x, -i\tau, x_0 = x, 0)} = \frac{\int dx_1 \cdots dx_{N-1} \exp\left[-\int_0^\tau H\, d\tau'\right]}{\int dx\, dx_1 \cdots dx_{N-1} \exp\left[-\int_0^\tau H\, d\tau'\right]}$$

$$\Rightarrow \quad |\psi_0(x)|^2 = \frac{1}{Z} \lim_{\tau\to\infty} \int dx_1 \cdots dx_{N-1} e^{-\varepsilon\mathcal{E}} , \tag{17.69}$$

$$Z = \lim_{\tau\to\infty} \int dx\, dx_1 \cdots dx_{N-1} e^{-\varepsilon\mathcal{E}} . \tag{17.70}$$

The similarity of these expressions to thermodynamics, even with a partition function Z, is no accident; by making the time parameter of quantum mechanics

5) In some cases, such as for an infinite square well, this can cause problems if the trial link causes the energy to be infinite. In that case, one can modify the algorithm to use the potential at the beginning of a link.

imaginary, we have converted the time-dependent Schrödinger equation to the heat diffusion equation

$$\mathrm{i}\frac{\partial \psi}{\partial(-\mathrm{i}\tau)} = \frac{-\nabla^2}{2m}\psi \quad \Rightarrow \quad \frac{\partial \psi}{\partial \tau} = \frac{\nabla^2}{2m}\psi\,. \tag{17.71}$$

It is not surprising then that the sum over paths in Green's function has each path weighted by the Boltzmann factor $\mathcal{P} = \mathrm{e}^{-\varepsilon\mathcal{E}}$ usually associated with thermodynamics. We make the connection complete by identifying the temperature with the inverse time step:

$$\mathcal{P} = \mathrm{e}^{-\varepsilon\mathcal{E}} = \mathrm{e}^{-\mathcal{E}/k_B T} \quad \Rightarrow \quad k_B T = \frac{1}{\varepsilon} \equiv \frac{\hbar}{\varepsilon}\,. \tag{17.72}$$

Consequently, the $\varepsilon \to 0$ limit, which makes time continuous, is a "high-temperature" limit. The $\tau \to \infty$ limit, which is required to project the ground-state wave function, means that we must integrate over a path that is long in imaginary time, that is, long compared to a typical time $\hbar/\Delta E$. Just as our simulation of the Ising model required us to wait a long time for the system to equilibrate, so the present simulation requires us to wait a long time so that all but the ground-state wave function has decayed. Alas, this is the solution to our *problem* of finding the ground-state wave function.

To summarize, we have expressed Green's function as a path integral that requires integration of the Hamiltonian along paths and a summation over all the paths (17.69). We evaluate this path integral as the sum over all the trajectories in a space–time lattice. Each trial path occurs with a probability based on its action, and we use the Metropolis algorithm to include statistical fluctuation in the links, as if they are in thermal equilibrium. This is similar to our work with the Ising model, however now, rather than reject or accept a *flip in spin* based on the change in energy, we reject or accept a *change in a link* based on the change in energy. The more iterations we let the algorithm run for, the more time the deduced wave function has to equilibrate to the ground state.

In general, Monte Carlo Green's function techniques work best if we start with a good guess at the correct answer, and then have the algorithm calculate variations on our guess. For the present problem this means that if we start with a path in space–time close to the classical trajectory, the algorithm may be expected to do a good job at simulating the quantum fluctuations about the classical trajectory. However, it does not appear to be good at finding the classical trajectory from arbitrary locations in space–time. We suspect that the latter arises from $\delta S/\hbar$ being so large that the weighting factor $\exp(\delta S/\hbar)$ fluctuates wildly (essentially averaging out to zero) and so loses its sensitivity.

17.8.2.1 A Time-Saving Trick

As we have formulated the computation, we pick a value of x and perform an expensive computation of line integrals over all space and time to obtain $|\psi_0(x)|^2$ at one x. To obtain the wave function at another x, the entire simulation must be repeated from scratch. Rather than go through all that trouble again and again, we will compute the entire x dependence of the wave function in one fell swoop. The trick is to insert a delta function into the probability integral (17.69), thereby fixing the initial position to be x_0, and then to integrate over all values for x_0:

$$|\psi_0(x)|^2 = \int dx_1 \cdots dx_N e^{-\varepsilon\mathcal{E}(x,x_1,\ldots)} \tag{17.73}$$

$$= \int dx_0 \cdots dx_N \delta(x - x_0) e^{-\varepsilon\mathcal{E}(x,x_1,\ldots)} . \tag{17.74}$$

This equation expresses the wave function as an average of a delta function over all paths, a procedure that might appear totally inappropriate for numerical computation because there is tremendous error in representing a singular function on a finite-word-length computer. Yet when we simulate the sum over all paths with (17.74), there will always be some x value for which the integral is nonzero, and we need to accumulate only the solution for various (discrete) x values to determine $|\psi_0(x)|^2$ for all x.

To understand how this works in practice, consider path AB in Figure 17.9 for which we have just calculated the summed energy. We form a new path by having one point on the chain jump to point C (which changes two links). If we replicate section AC and use it as the extension AD to form the top path, we see that the path CBD has the same summed energy (action) as path ACB, and in this way can be used to determine $|\psi(x'_j)|^2$. That being the case, once the system is equilibrated, we determine new values of the wave function at new locations x'_j by flipping links to new values and calculating new actions. The more frequently some x_j is accepted, the greater the wave function at that point.

17.8.3 Lattice Implementation

The program `QMC.py` in Listing 17.3 evaluates the integral (17.46) by finding the average of the integrand $\delta(x_0 - x)$ with paths distributed according to the weighting function $\exp[-\varepsilon\mathcal{E}(x_0, x_1, \ldots, x_N)]$. The physics enters via (17.76), the calculation of the summed energy $\mathcal{E}(x_0, x_1, \ldots, x_N)$. We evaluate the action integral for the harmonic oscillator potential

$$V(x) = \frac{1}{2}x^2 , \tag{17.75}$$

and for a particle of mass $m = 1$. Using a convenient set of natural units, we measure lengths in $\sqrt{1/m\omega} \equiv \sqrt{\hbar/m\omega} = 1$ and times in $1/\omega = 1$. Correspondingly, the oscillator has a period $T = 2\pi$. Figure 17.8 shows results from an application of the Metropolis algorithm. In this computation, we started with an initial path close to the classical trajectory and then examined a half million variations about this path. All paths are constrained to begin and end at $x = 1$ (which turns out to be somewhat less than the maximum amplitude of the classical oscillation). When the time difference $t_b - t_a$ equals a short time like $2T$, the system has not had enough time to equilibrate to its ground state and the wave function looks like the probability distribution of an excited state (nearly classical with the probability highest for the particle to be near its turning points where its velocity vanishes). However, when $t_b - t_a$ equals the longer time $20T$, the system has had enough time to decay to its ground state and the wave function looks like the expected Gaussian distribution. In either case (Figure 17.8a), the trajectory through space–time fluctuates about the classical trajectory. This fluctuation is a consequence of the Metropolis algorithm occasionally going uphill in its search; if you modify the program so that searches go only downhill, the space–time trajectory will be a very smooth trigonometric function (the classical trajectory), but the wave function, which is a measure of the fluctuations about the classical trajectory, will vanish! The explicit steps of the calculation are (MacKeown, 1985; MacKeown and Newman, 1987):

1. Construct a grid of N time steps of length ε (Figure 17.9). Start at $t = 0$ and extend to time $\tau = N\varepsilon$ [this means N time intervals and $(N + 1)$ lattice points in time]. Note that time always increases monotonically along a path.
2. Construct a grid of M space points separated by steps of size δ. Use a range of x values several time larger than the characteristic size or range of the potential being used and start with $M \simeq N$.
3. When calculating the wave function, any x or t value falling between lattice points should be assigned to the closest lattice point.
4. Associate a position x_j with each time τ_j, subject to the boundary conditions that the initial and final positions always remain the same, $x_N = x_0 = x$.
5. Choose a path of straight-line links connecting the lattice points corresponding to the classical trajectory. Observe that the x values for the links of the path may have values that increase, decrease, or remain unchanged (in contrast to time, which always increases).
6. Evaluate the energy $\mathcal{E}$ by summing the kinetic and potential energies for each link of the path starting at $j = 0$:

$$\mathcal{E}(x_0, x_1, \ldots, x_N) \simeq \sum_{j=1}^{N} \left[\frac{m}{2} \left(\frac{x_j - x_{j-1}}{\varepsilon} \right)^2 + V\left(\frac{x_j + x_{j-1}}{2} \right) \right] . \tag{17.76}$$

7. Begin a sequence of repetitive steps in which a random position x_j associated with time t_j is changed to the position x'_j (point C in Figure 17.9). This changes *two* links in the path.
8. For the coordinate that is changed, use the Metropolis algorithm to weigh the change with the Boltzmann factor.
9. For each lattice point, establish a running sum that represents the value of the wave function squared at that point.
10. After each single-link change (or decision not to change), increase the running sum for the new x value by 1. After a sufficiently long running time, the sum

divided by the number of steps is the simulated value for $|\psi(x_j)|^2$ at each lattice point x_j.

11. Repeat the entire link-changing simulation starting with a different seed. The average wave function from a number of intermediate-length runs is better than that from one very long run.

Listing 17.3 QMC.py determines the ground-state probability via a Feynman path integration using the Metropolis algorithm to simulate variations about the classical trajectory.

```
# QMC.py:        Quantum MonteCarlo (Feynman path integration)

from visual import *
import random
from visual.graph import *

N = 100;              M = 101;          xscale = 10.
path = zeros([M], float);               prob = zeros([M], float)    # Initialize

trajec = display(width = 300,height=500, title='Spacetime Trajectories')
trplot = curve(y = range(0, 100), color=color.magenta, display = trajec)

def trjaxs():                                                        # Axis
   trax = curve(pos = [(-97,-100),(100,-100)], color = color.cyan,
        display = trajec)
   label(pos = (0,-110),  text = '0', box = 0, display = trajec)
   label(pos = (60,-110), text = 'x', box = 0, display = trajec)

wvgraph = display(x=340,y=150,width=500,height=300, title='Ground State')
wvplot = curve(x = range(0, 100), display = wvgraph)
wvfax = curve(color = color.cyan)

def wvfaxs():                                        # Axis for probability
   wvfax = curve(pos =[(-600,-155),(800,-155)],
        display=wvgraph,color=color.cyan)
   curve(pos = [(0,-150), (0,400)], display=wvgraph, color=color.cyan)
   label(pos = (-80,450), text='Probability', box = 0, display = wvgraph)
   label(pos = (600,-220), text='x', box=0, display=wvgraph)
   label(pos = (0,-220), text='0', box=0, display=wvgraph)

trjaxs();                wvfaxs()                                # Plot axes

def energy(path):                                                # HO energy
    sums = 0.
    for i in range(0,N-2):sums += (path[i+1]-path[i])*(path[i+1]-path[i])
    sums += path[i+1]*path[i+1];
    return sums

def plotpath(path):                                        # Plot trajectory
   for j in range (0, N):
        trplot.x[j] = 20*path[j]
        trplot.y[j] = 2*j - 100

def plotwvf(prob):                                              # Plot prob
    for i in range (0, 100):
        wvplot.color = color.yellow
        wvplot.x[i] = 8*i - 400                            # For centered fig

oldE = energy(path)

while True:                                           # Pick random element
    rate(10)                                              # Slow paintings
    element = int(N*random.random() )                # Metropolis algorithm
    change = 2.0*(random.random() - 0.5)
```

```
path[element] += change                                    # Change path
newE = energy(path);                                       # Find new E
if  newE > oldE and math.exp( - newE + oldE)<= random.random():
    path[element] -= change                                # Reject
    plotpath(path)                                         # Plot trajectory
elem = int(path[element]*16 + 50)                          # if path = 0, elem = 50

# elem = m *path[element] + b is the linear transformation
# if path=-3, elem=2 if path=3., elem=98 => b=50, m=16 linear TF.
# this way x = 0 correspond to prob[50]

if elem < 0: elem = 0,
if elem > 100:  elem = 100                                 # If exceed max
prob[elem] += 1                                            # increase probability
plotwvf(prob)                                              # Plot prob
oldE = newE
```

17.8.4 Assessment and Exploration

1. Plot some of the actual space–time paths used in the simulation along with the classical trajectory.
2. For a more continuous picture of the wave function, make the x lattice spacing smaller; for a more precise value of the wave function at any particular lattice site, sample more points (run longer) and use a smaller time step ε.
3. Because there are no sign changes in a ground-state wave function, you can ignore the phase, assume $\psi(x) = \sqrt{\psi^2(x)}$, and then estimate the energy via

$$E = \frac{\langle\psi|H|\psi\rangle}{\langle\psi|\psi\rangle} = \frac{\omega}{2\langle\psi|\psi\rangle}\int_{-\infty}^{+\infty}\psi^*(x)\left(-\frac{d^2}{dx^2}+x^2\right)\psi(x)\,dx\,, \qquad (17.77)$$

   where the space derivative is evaluated numerically.
4. Explore the effect of making $\hbar$ larger and thus permitting greater fluctuations around the classical trajectory. Do this by decreasing the value of the exponent in the Boltzmann factor. Determine if this makes the calculation more or less robust in its ability to find the classical trajectory.
5. Test your ψ for the gravitational potential (see quantum bouncer below):

$$V(x) = mg|x|\,, \quad x(t) = x_0 + v_0 t + \frac{1}{2}gt^2\,. \qquad (17.78)$$

17.9 Exploration: Quantum Bouncer's Paths ⊙

Another problem for which the classical trajectory is well known is that of a *quantum bouncer.*[6] Here we have a particle dropped in a uniform gravitational field, hitting a hard floor, and then bouncing. When treated quantum mechanically,

6) Oscar A. Restrepo assisted in the preparation of this section.

quantized levels for the particle result (Gibbs, 1975; Goodings and Szeredi, 1992; Whineray, 1992; Banacloche, 1999; Vallée, 2000). In 2002, an experiment to discern this gravitational effect at the quantum level was performed by Nesvizhevsky *et al.* (2002) and described by Shaw (1992). It consisted of dropping ultracold neutrons from a height of 14 μm unto a neutron mirror and watching them bounce. It found a neutron ground state at 1.4 peV.

We start by determining the analytic solution to this problem for stationary states and then generalize it to include time dependence. The time-independent Schrödinger equation for a particle in a uniform gravitation potential is

$$-\frac{\hbar^2}{2m}\frac{\mathrm{d}^2\psi(x)}{\mathrm{d}x^2} + mxg\psi(x) = E\psi(x) , \tag{17.79}$$

$$\psi(x \le 0) = 0 \quad \text{(boundary condition)} . \tag{17.80}$$

The boundary condition (17.80) is a consequence of the hard floor at $x = 0$. A change of variables converts (17.79) to a dimensionless form

$$\frac{\mathrm{d}^2\psi}{\mathrm{d}z^2} - (z - z_\mathrm{E})\psi = 0 , \tag{17.81}$$

$$z = x\left(\frac{2gm^2}{\hbar^2}\right)^{1/3} , \quad z_\mathrm{E} = E\left(\frac{2}{\hbar^2 mg^2}\right)^{1/3} . \tag{17.82}$$

This equation has an analytic solution in terms of Airy functions $\mathrm{Ai}(z)$:

$$\psi(z) = N_\mathrm{n}\,\mathrm{Ai}(z - z_\mathrm{E}) , \tag{17.83}$$

where N_n is a normalization constant and z_E is the scaled value of the energy. The boundary condition $\psi(0) = 0$ implies that

$$\psi(0) = N_\mathrm{E}\,\mathrm{Ai}(-z_\mathrm{E}) = 0 , \tag{17.84}$$

which means that the allowed energies of the system are discrete and correspond to the zeros z_n of the Airy functions (Press *et al.*, 1994) at negative argument. To simplify the calculation, we take $\hbar = 1$, $g = 2$, and $m = 1/2$, which leads to $z = x$ and $z_\mathrm{E} = E$.

The time-dependent solution for the quantum bouncer is constructed by forming the infinite sum over all the discrete eigenstates, each with a time dependence appropriate to its energy:

$$\psi(z, t) = \sum_{n=1}^{\infty} C_n N_\mathrm{n}\mathrm{Ai}(z - z_n)\mathrm{e}^{-\mathrm{i}E_n t/\hbar} , \tag{17.85}$$

where the C_n's are constants.

Figure 17.10 shows the results of solving for the quantum bouncer's ground-state probability $|\psi_0(z)|^2$ using Feynman's path integration.The time increment $\mathrm{d}t$ and the total time t were selected by trial and error in such a way as to make

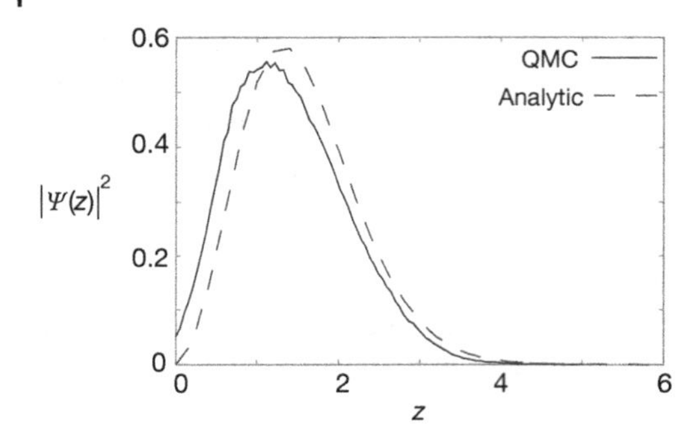

Figure 17.10 The analytic and quantum Monte Carlo solution for the quantum bouncer. The continuous line is the Airy function squared and the dashed line $|\psi_0(z)|^2$ after a million trajectories.

$|\psi(0)|^2 \simeq 0$ (the boundary condition). To account for the fact that the potential is infinite for negative x values, we selected trajectories that have positive x values over all their links. This incorporates the fact that the particle can never penetrate the floor. Our program is given in Listing 17.4, and it yields the results in Figure 17.10 after using 10^6 trajectories and a time step $\varepsilon = d\tau = 0.05$. Both results were normalized via a trapezoid integration. As can be seen, the agreement between the analytic result and the path integration is satisfactory.

Listing 17.4 QMCbouncer.py uses Feynman path integration to compute the path of a quantum particle in a gravitational field.

```
# QMCbouncer.py:          g.s. wavefunction via path integration

from visual import *
import random
from visual.graph import *

# Parameters
N = 100;    dt = 0.05;          g = 2.0;        h = 0.00;       maxel = 0
path = zeros([101], float);   arr = path; prob = zeros([201],float)

trajec = display(width = 300, height=500,title = 'Spacetime Trajectory')
trplot = curve(y = range(0, 100), color=color.magenta, display = trajec)

def trjaxs():                                    # plot axis for trajectories
    trax=curve(pos=[(-97,-100),(100,-100)],color=color.cyan,display=trajec)
    curve(pos = [(-65, -100),(-65, 100)], color=color.cyan,display=trajec)
    label(pos = (-65,110), text = 't', box = 0, display = trajec)
    label(pos = (-85, -110), text = '0', box = 0, display = trajec)
    label(pos = (60, -110), text = 'x', box = 0, display = trajec)
wvgraph = display(x=350, y=80, width=500, height=300, title = 'GS Prob')
wvplot  = curve(x = range(0, 50), display = wvgraph)  # wave function plot
wvfax   = curve(color = color.cyan)

def wvfaxs():                                    # plot axis for wavefunction
    wvfax = curve(pos
        =[(-200,-155),(800,-155)],display=wvgraph,color=color.cyan)
    curve(pos = [(-200,-150),(-200,400)],display=wvgraph,color=color.cyan)
    label(pos = (-70, 420),text = 'Probability', box = 0, display=wvgraph)
    label(pos = (600, -220),text = 'x', box = 0, display = wvgraph)
    label(pos = (-200, -220),text = '0', box = 0, display = wvgraph)
```

```
trjaxs();    wvfaxs()                                          # plot axes

def energy (arr):                                              # Energy of path
    esum = 0.
    for i in range(0,N):
      esum += 0.5*((arr[i+1]-arr[i])/dt)**2+g*(arr[i]+arr[i+1])/2
    return esum

def plotpath(path):                                            # Plot xy trajectory
   for j in range (0, N):
        trplot.x[j] = 20*path[j] - 65
        trplot.y[j] = 2*j - 100

def plotwvf(prob):                                             # Plot wave function
    for i in range (0, 50):
        wvplot.color = color.yellow
        wvplot.x[i] = 20*i - 200
        wvplot.y[i] = 0.5*prob[i] - 150

oldE = energy(path)
counter = 1
norm = 0.                                                      # plot psi every 100
maxx = 0.0

while 1:                                                       # "Infinite" loop
    rate(100)
    element = int(N*random.random() )
    if element != 0 and element!= N:                           # Ends not allowed
        change = ( (random.random() - 0.5)*20.)/10.
        if  path[element] + change > 0.:                       # No negative paths
          path[element]    += change
          newE = energy(path)                                  # New trajectory E
          if newE > oldE and exp( - newE + oldE) <= random.random() :
              path[element]    -= change                       # Link rejected
              plotpath(path)
          ele = int(path[element]*1250./100.)                  # Scale changed
          if  ele >= maxel:  maxel = ele                       # Scale change 0 to N
          if  element != 0:  prob[ele]   += 1
          oldE = newE;
    if counter%100 == 0:                                       # plot psi every 100
        for  i in range(0, N):                                 # max x of path
            if  path[i] >= maxx:  maxx = path[i]
        h = maxx/maxel                                         # space step
        firstlast = h*0.5*(prob[0] +  prob[maxel])   # for trap. extremes
        for  i in range(0, maxel + 1):  norm = norm + prob[i]         # norm
        norm = norm*h + firstlast                              # Trap rule
        plotwvf(prob)                                          # plot probability
    counter   += 1
```

18
Molecular Dynamics Simulations

You may have seen in introductory chemistry or physics classes that the ideal gas law can be derived from first principles when the interactions of the molecules with each other are ignored, and only reflections off the walls of the surrounding box are considered. We extend that model so that we can solve for the motion of every molecule in a box interacting with every other molecule in the box (but not with the walls). While our example is a simple one, molecular dynamics is of key importance in many fields, including material science and biology.

Your *problem* is to determine whether a collection of argon molecules placed in a box will form an ordered structure at low temperature.

18.1
Molecular Dynamics (Theory)

Molecular dynamics (MD) is a powerful simulation technique for studying the physical and chemical properties of solids, liquids, amorphous materials, and biological molecules. Although we know that quantum mechanics is the proper theory for molecular interactions, MD uses Newton's laws as the basis of the technique and focuses on bulk properties, which do not depend much on small-r behaviors. In 1985, Car and Parrinello showed how MD can be extended to include quantum mechanics by applying density functional theory to calculate the force (Car and Parrinello, 1985). This technique, known as *quantum MD*, is an active area of research but is beyond the realm of this chapter.[1] For those with more interest in the subject, there are full texts (Allan and Tildesley, 1987; Rapaport, 1995; Hockney, 1988) on MD and good discussions (Gould *et al.*, 2006; Thijssen, 1999; Fosdick *et al.*, 1996), as well as primers (Ercolessi, 1997) and codes, (Nelson *et al.*, 1996; Refson, 2000; Anderson *et al.*, 2008) available online.

MD's solution of Newton's laws is conceptually simple, yet when applied to a very large number of particles becomes the "high school physics problem from hell." Some approximations must be made in order not to have to solve

1) We thank Satoru S. Kano for pointing this out to us.

the 10^{23}–10^{25} equations of motion describing a realistic sample, but instead to limit the problem to $\sim 10^6$ particles for protein simulations, and $\sim 10^8$ particles for materials simulations. If we have some success, then it is a good bet that the model will improve if we incorporate more particles or more quantum mechanics, something that becomes easier as computing power continues to increase.

In a number of ways, MD simulations are similar to the thermal Monte Carlo simulations we studied in Chapter 17. Both typically involve a large number N of interacting particles that start out in some set configuration and then equilibrate into some dynamic state on the computer. However, in MD we have what statistical mechanics calls a *microcanonical ensemble* in which the energy E and volume V of the N particles are fixed. We then use Newton's laws to generate the dynamics of the system. In contrast, Monte Carlo simulations do not start with first principles, but instead, incorporate an element of chance and have the system in contact with a heat bath at a fixed temperature rather than keeping the energy E fixed. This is called a *canonical ensemble.*

Because a system of molecules in an MD simulation is dynamic, the velocities and positions of the molecules change continuously. After the simulation has run long enough to stabilize, we will compute time averages of the dynamic quantities in order to deduce the thermodynamic properties. We apply Newton's laws with the assumption that the net force on each molecule is the sum of the two-body forces with all other $(N-1)$ molecules:

$$m\frac{\mathrm{d}^2\boldsymbol{r}_i}{\mathrm{d}t^2} = \boldsymbol{F}_i(\boldsymbol{r}_0, \dots, \boldsymbol{r}_{N-1}), \tag{18.1}$$

$$m\frac{\mathrm{d}^2\boldsymbol{r}_i}{\mathrm{d}t^2} = \sum_{i<j=0}^{N-1} \boldsymbol{f}_{ij}, \quad i = 0, \dots, (N-1). \tag{18.2}$$

In writing these equations, we have ignored the fact that the force between argon atoms really arises from the particle–particle interactions of the 18 electrons and the nucleus that constitute each atom (Figure 18.1). Although it may be possible to ignore this internal structure when deducing the long-range properties of inert elements, it matters for systems such as polyatomic molecules that display rotational, vibrational, and electronic degrees of freedom as the temperature is raised.[2)]

The force on molecule i derives from the sum of molecule–molecule potentials:

$$\boldsymbol{F}_i(\boldsymbol{r}_0, \boldsymbol{r}_1, \dots, \boldsymbol{r}_{N-1}) = -\boldsymbol{\nabla}_{\boldsymbol{r}_i} U(\boldsymbol{r}_0, \boldsymbol{r}_1, \dots, \boldsymbol{r}_{N-1}), \tag{18.3}$$

$$U(\boldsymbol{r}_0, \boldsymbol{r}_1, \dots, \boldsymbol{r}_{N-1}) = \sum_{i<j} u(r_{ij}) = \sum_{i=0}^{N-2} \sum_{j=i+1}^{N-1} u(r_{ij}) \tag{18.4}$$

$$\Rightarrow \quad \boldsymbol{f}_{ij} = -\frac{\mathrm{d}u(r_{ij})}{\mathrm{d}r_{ij}} \left(\frac{x_i - x_j}{r_{ij}} \hat{\boldsymbol{e}}_x + \frac{y_i - y_j}{r_{ij}} \hat{\boldsymbol{e}}_y + \frac{z_i - z_j}{r_{ij}} \hat{\boldsymbol{e}}_z \right). \tag{18.5}$$

2) We thank Saturo Kano for clarifying this point.

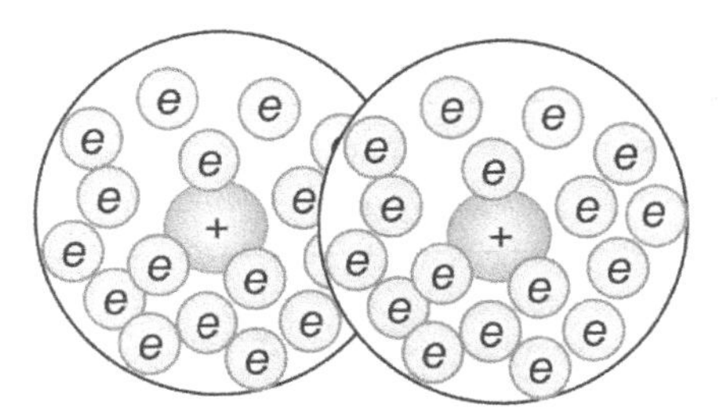

Figure 18.1 The molecule–molecule effective interaction arises from the many-body interaction of the electrons and nucleus in one molecule (circle) with the electrons and nucleus in another molecule (other circle). Note, the size of the nucleus at the center of each molecule is highly exaggerated, and real electrons have no size.

Here $r_{ij} = |\boldsymbol{r}_i - \boldsymbol{r}_j| = r_{ji}$ is the distance between the centers of molecules i and j, and the limits on the sums are such that no interaction is counted twice. Because we have assumed a *conservative* potential, the total energy of the system, that is, the potential plus kinetic energies summed over all particles, should be conserved over time. Nonetheless, in a practical computation we "cut the potential off" (assume $u(r_{ij}) = 0$) when the molecules are far apart. Because the derivative of the potential produces an infinite force at this cutoff point, energy will no longer be precisely conserved. Yet because the cutoff radius is large, the cutoff occurs only when the forces are minuscule, and so the violation of energy conservation should be small relative to approximation and round-off errors.

In a first-principles calculation, the potential between any two argon atoms arises from the sum over approximately 1000 electron–electron and electron–nucleus Coulomb interactions. A more practical calculation would derive an effective potential based on a form of many-body theory, such as Hartree–Fock or density functional theory. Our approach is simpler yet. We use the Lennard–Jones potential,

$$u(r) = 4\epsilon\left[\left(\frac{\sigma}{r}\right)^{12} - \left(\frac{\sigma}{r}\right)^{6}\right], \tag{18.6}$$

$$\boldsymbol{f}(r) = -\frac{\mathrm{d}u}{\mathrm{d}r}\frac{\boldsymbol{r}}{r} = \frac{48\epsilon}{r^2}\left[\left(\frac{\sigma}{r}\right)^{12} - \frac{1}{2}\left(\frac{\sigma}{r}\right)^{6}\right]\boldsymbol{r}. \tag{18.7}$$

Here the parameter ϵ governs the strength of the interaction, and σ determines the length scale. Both are deduced by fits to data, which is why this potential is called a "phenomenological" potential.

Some typical values for the parameters and scales for the variables are given in Table 18.1. In order to make the program simpler and to avoid under- and overflows, it is helpful to measure all variables in the natural units formed by these constants. The interparticle potential and force then take the forms

$$u(r) = 4\left[\frac{1}{r^{12}} - \frac{1}{r^6}\right], \quad f(r) = \frac{48}{r}\left[\frac{1}{r^{12}} - \frac{1}{2r^6}\right]. \tag{18.8}$$

Table 18.1 Parameter and scales for the Lennard–Jones potential.

Quantity	Mass	Length	Energy	Time	Temperature
Unit	m	σ	ϵ	$\sqrt{m\sigma^2/\epsilon}$	ϵ/k_B
Value	6.7×10^{-26} kg	3.4×10^{-10} m	1.65×10^{-21} J	4.5×10^{-12} s	119 K

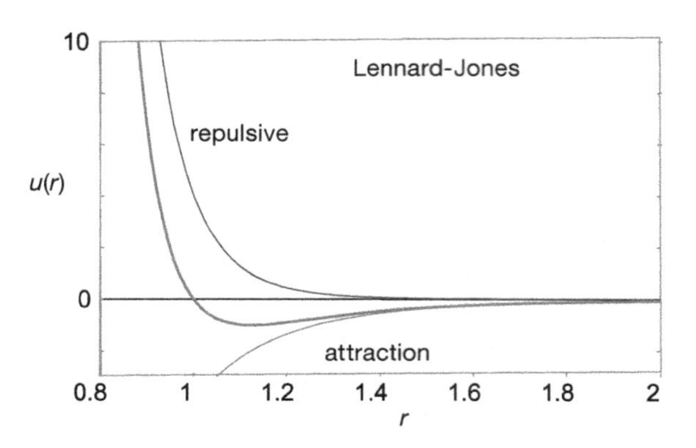

Figure 18.2 The Lennard–Jones effective potential used in many MD simulations. Note the sign change at $r = 1$ and the minimum at $r \simeq 1.1225$ (natural units). Also note that because the r-axis does not extend to $r = 0$, the infinitely high central repulsion is not shown.

The Lennard–Jones potential is seen in Figure 18.2 to be the sum of a long-range attractive interaction $\propto 1/r^6$ and a short-range repulsive one $\propto 1/r^{12}$. The change from repulsion to attraction occurs at $r = \sigma$. The minimum of the potential occurs at $r = 2^{1/6}\sigma = 1.1225\sigma$, which would be the atom–atom spacing in a solid bound by this potential. The repulsive $1/r^{12}$ term in the Lennard–Jones potential (18.6) arises when the electron clouds from two atoms overlap, in which case the Coulomb interaction and the Pauli exclusion principle force the electrons apart. The $1/r^{12}$ term dominates at short distances and makes atoms behave like hard spheres. The precise value of 12 is not of theoretical significance (although it being large is) and was probably chosen because it is 2×6.

The $1/r^6$ term that dominates at large distances models the weak *van der Waals* induced dipole–dipole attraction between two molecules. The attraction arises from fluctuations in which at some instant in time a molecule on the right tends to be more positive on the left side, like a dipole $\Leftarrow$. This in turn attracts the negative charge in a molecule on its left, thereby inducing a dipole $\Leftarrow$ in it. As long as the molecules stay close to each other, the polarities continue to fluctuate in synchronization $\Leftarrow\Leftarrow$ so that the attraction is maintained. The resultant dipole–dipole attraction behaves like $1/r^6$, and although much weaker than a Coulomb force, it is responsible for the binding of neutral, inert elements, such as argon for which the Coulomb force vanishes.

18.1.1 Connection to Thermodynamic Variables

We assume that the number of particles is large enough to use statistical mechanics to relate the results of our simulation to the thermodynamic quantities. The simulation is valid for any number of particles, but the use of statistics requires large numbers. The equipartition theorem tells us that, on the average, for molecules in thermal equilibrium at temperature T, each degree of freedom has an energy $k_B T/2$ associated with it, where $k_B = 1.38 \times 10^{-23}$ J/K is Boltzmann's constant. A simulation provides the kinetic energy of translation[3]:

$$\mathrm{KE} = \frac{1}{2}\left\langle \sum_{i=0}^{N-1} v_i^2 \right\rangle . \tag{18.9}$$

The time average of KE (three degrees of freedom) is related to temperature by

$$\langle \mathrm{KE} \rangle = N\frac{3}{2}k_B T \quad \Rightarrow \quad T = \frac{2\langle \mathrm{KE} \rangle}{3k_B N} . \tag{18.10}$$

The system's pressure P is determined by a version of the *Virial theorem*,

$$PV = Nk_B T + \frac{w}{3} , \quad w = \left\langle \sum_{i<j}^{N-1} \boldsymbol{r}_{ij} \cdot \boldsymbol{f}_{ij} \right\rangle , \tag{18.11}$$

where the Virial w is a weighted average of the forces. Note that because ideal gases have no intermolecular forces, their Virial vanishes and we have the ideal gas law. The pressure is thus

$$P = \frac{\rho}{3N}(2\langle \mathrm{KE} \rangle + w) , \tag{18.12}$$

where $\rho = N/V$ is the density of the particles.

18.1.2 Setting Initial Velocities

Although we start the system off with a velocity distribution characteristic of a definite temperature, this is not a true temperature of the system because the system is not in equilibrium initially, and there will a redistribution of energy between KE and PE (Thijssen, 1999). Note that this initial randomization is the only place where chance enters into our MD simulation, and it is there to speed the simulation along. Indeed, in Figure 18.3 we show results of simulations in which the molecules are initially at rest and equally spaced. Once started, the time evolution is determined by Newton's laws, in contrast to Monte Carlo simulations which are inherently stochastic. We produce a Gaussian (Maxwellian) velocity distribution with the methods discussed in Chapter 4. In our sample code, we take the average $1/12 \sum_{i=1}^{12} r_i$ of uniform random numbers $0 \le r_i \le 1$ to produce a Gaussian distribution with mean $\langle r \rangle = 0.5$. We then subtract this mean value to obtain a distribution about 0.

3) Unless the temperature is very high, argon atoms, being inert spheres, have no rotational energy.

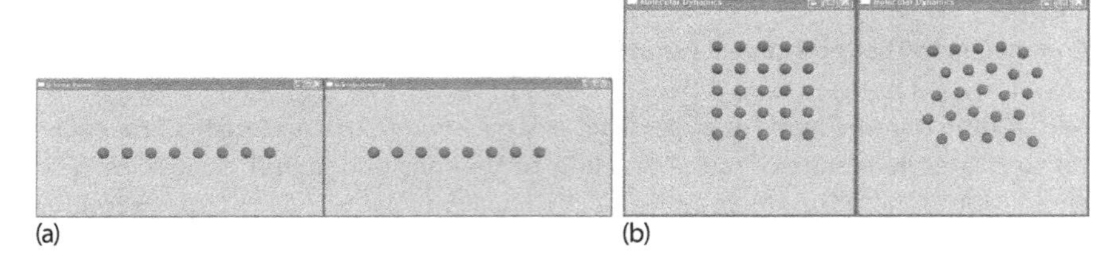

(a) (b)

Figure 18.3 (a) Two frames from an animation showing the results of a 1D MD simulation that starts with uniformly spaced atoms. Note the unequal spacing resulting from an image atom moving in from the left after an atom left from the right. (b) Two frames from the animation of a 2D MD simulation showing the initial and an equilibrated state. Note how the atoms start off in a simple cubic arrangement but then equilibrate to a face-centered-cubic lattice. In both the cases, the atoms remain confined as a result of the interatomic forces.

18.1.3 Periodic Boundary Conditions and Potential Cutoff

It is easy to believe that a simulation of 10^{23} molecules should predict bulk properties well, but with typical MD simulations employing only 10^3–10^6 particles, one must be clever to make less seem like more. Furthermore, because computers are finite, the molecules in the simulation are constrained to lie within a finite box, which inevitably introduces artificial *surface effects* arising from the walls. Surface effects are particularly significant when the number of particles is small because then a large fraction of the molecules reside near the walls. For example, if 1000 particles are arranged in a $10 \times 10 \times 10 \times 10$ cube, there are $10^3 - 8^3 = 488$ particles one unit from the surface, that is, 49% of the molecules, while for 10^6 particles this fraction falls to 6%.

The imposition of *periodic boundary conditions* (PBCs) strives to minimize the shortcomings of both the small numbers of particles and of artificial boundaries. Although we limit our simulation to an $L_x \times L_y \times L_z$ box, we imagine this box being replicated to infinity in all directions (Figure 18.4). Accordingly, after each time-integration step we examine the position of each particle and check if it has left the simulation region. If it has, then we bring an *image* of the particle back through the opposite boundary (Figure 18.4):

$$x \quad \Rightarrow \quad \begin{cases} x + L_x \, , & \text{if} \quad x \le 0 \, , \\ x - L_x \, , & \text{if} \quad x > L_x \, . \end{cases} \tag{18.13}$$

Consequently, each box looks the same and has continuous properties at the edges. As shown by the one-headed arrows in Figure 18.4, if a particle exits the simulation volume, its image enters from the other side, and so balance is maintained.

In principle, a molecule interacts with all others molecules and their images, so despite the fact that there is a finite number of atoms in the interaction volume, there is an effective infinite number of interactions (Ercolessi, 1997). Nonetheless, because the Lennard–Jones potential falls off so rapidly for large r, $V(r = 3\sigma) \simeq$

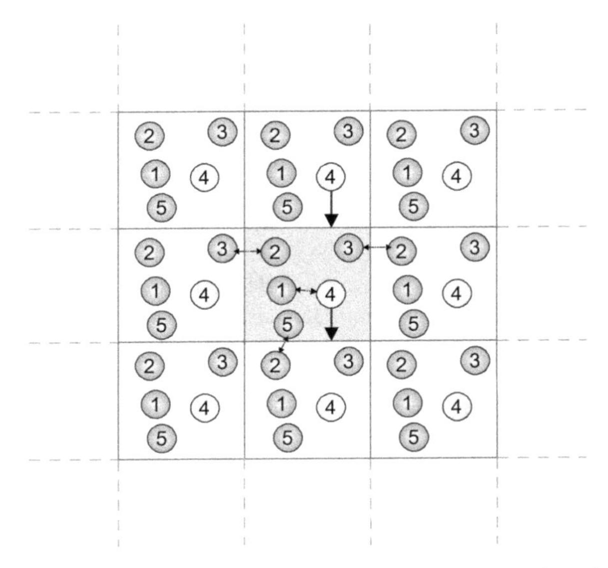

Figure 18.4 The infinite space generated by imposing periodic boundary conditions on the particles within the simulation volume (shaded box). The two-headed arrows indicate how a particle interacts with the nearest version of another particle, be that within the simulation volume or an image. The vertical arrows indicate how the image of particle 4 enters when the actual particle 4 exits.

$V(1.13\sigma)/200$, far-off molecules do not contribute significantly to the motion of a molecule, and we pick a value $r_{\text{cut}} \simeq 2.5\sigma$ beyond which we ignore the effect of the potential:

$$u(r) = \begin{cases} 4(r^{-12} - r^{-6}) , & \text{for} \quad r < r_{\text{cut}} , \\ 0 , & \text{for} \quad r > r_{\text{cut}} . \end{cases} \tag{18.14}$$

Accordingly, if the simulation region is large enough for $u(r > L_i/2) \simeq 0$, an atom interacts with only the *nearest image* of another atom.

As already indicated, a shortcoming with the cutoff potential (18.14) is that because the derivative $\mathrm{d}u/\mathrm{d}r$ is singular at $r = r_{\text{cut}}$, the potential is no longer conservative and thus energy conservation is no longer ensured. However, because the forces are already very small at r_{cut}, the violation will also be very small.

18.2
Verlet and Velocity–Verlet Algorithms

A realistic MD simulation may require integration of the 3D equations of motion for 10^{10} time steps for each of 10^3–10^6 particles. Although we could use our standard `rk4` ODE solver for this, time is saved by using a simple rule embedded in the program. The Verlet algorithm uses the central-difference approximation (Chapter 5) for the second derivative to advance the solutions by a single time step h for all N particles simultaneously:

$$F_i[r(t), t] = \frac{d^2 r_i}{dt^2} \simeq \frac{r_i(t+h) + r_i(t-h) - 2r_i(t)}{h^2} \tag{18.15}$$

$$\Rightarrow \quad r_i(t+h) \simeq 2r_i(t) - r_i(t-h) + h^2 F_i(t) + O(h^4) , \tag{18.16}$$

where we have set $m = 1$. (Improved algorithms may vary the time step depending upon the speed of the particle.) Note that although the atom–atom force does not have an explicit time dependence, we include a t dependence in it as a way of indicating its dependence upon the atoms' positions at a particular time. Because this is really an implicit time dependence, energy remains conserved.

Part of the efficiency of the Verlet algorithm (18.16) is that it solves for the position of each particle without requiring a separate solution for the particle's velocity. However, once we have deduced the position for various times, we can use the central-difference approximation for the first derivative of r_i to obtain the velocity:

$$v_i(t) = \frac{dr_i}{dt} \simeq \frac{r_i(t+h) - r_i(t-h)}{2h} + O(h^2) . \tag{18.17}$$

Finally, note that because the Verlet algorithm needs r from two previous steps, it is not self-starting and so we start it with the forward difference

$$r(t = -h) \simeq r(0) - hv(0) + \frac{h^2}{2} F(0) . \tag{18.18}$$

Velocity–Verlet Algorithm Another version of the Verlet algorithm, which we recommend because of its increased stability, uses a forward-difference approximation for the derivative to advance *both* the position and velocity simultaneously:

$$r_i(t+h) \simeq r_i(t) + hv_i(t) + \frac{h^2}{2} F_i(t) + O(h^3) , \tag{18.19}$$

$$v_i(t+h) \simeq v_i(t) + h\overline{a(t)} + O(h^2) \tag{18.20}$$

$$\simeq v_i(t) + h \left[\frac{F_i(t+h) + F_i(t)}{2} \right] + O(h^2) . \tag{18.21}$$

Although this algorithm appears to be of lower order than (18.16), the use of updated positions when calculating velocities, and the subsequent use of these velocities, make both algorithms of similar precision.

Of interest is that (18.21) approximates the average force during a time step as $[F_i(t+h) + F_i(t)]/2$. Updating the velocity is a little tricky because we need the force at time $t+h$, which depends on the particle positions at $t+h$. Consequently, we must update all the particle positions and forces to $t+h$ before we update any velocities, while saving the forces at the earlier time for use in (18.21). As soon as the positions are updated, we impose periodic boundary conditions to establish that we have not lost any particles, and then we calculate the forces.

18.3
1D Implementation and Exercise

In the supplementary materials to this book, you will find a number of 2D animations (movies) of solutions to the MD equations. Some frames from these animations are shown in Figure 18.3. We recommend that you look at the movies in order to better visualize what the particles do during an MD simulation. In particular, these simulations use a potential and temperature that should lead to a solid or liquid system, and so you should see the particles binding together.

Listing 18.1 MD.py performs a 2D MD simulation with a small number of rather large time steps for just a few particles. To be realistic the user should change the parameters and the number of random numbers added to form the Gaussian distribution.

```
# MD.py:                    Molecular dynamics in 2D

from visual import *
from visual.graph import *
import random

scene = display(x=0,y=0,width=350,height=350, title='Molecular Dynamics',
                range=10)
sceneK = gdisplay (x=0,y=350,width=600,height=150,title='Average KE',
              ymin=0.0,ymax=5.0,xmin=0,xmax=500,xtitle='time',ytitle='KE avg')
Kavegraph=gcurve(color= color.red)
sceneT = gdisplay (x=0,y=500,width=600,height=150,title='Average PE',
              ymin=-60,ymax=0.,xmin=0,xmax=500,xtitle='time',ytitle='PE avg')
Tcurve = gcurve(color=color.cyan)
Natom = 25
Nmax =  25
Tinit = 2.0

dens = 1.0                                          # Density (1.20 for fcc)
t1 = 0
x  = zeros( (Nmax),      float)
y  = zeros( (Nmax),      float)
vx = zeros( (Nmax),      float)
vy = zeros( (Nmax),      float)
fx = zeros( (Nmax, 2),   float)
fy = zeros( (Nmax, 2),   float)
L = int(1.*Natom**0.5)                                   # Side of lattice
atoms=[]

def twelveran():                           # Average 12 rands for Gaussian
    s=0.0
    for i in range (1,13):
        s += random.random()
    return s/12.0-0.5

def initialposvel():                                            # Initialize
    i = -1
    for ix in range(0, L):                      # x->   0  1  2  3  4
        for iy in range(0, L):                  # y=0   0  5  10 15 20
            i = i + 1                           # y=1   1  6  11 16 21
            x[i]  = ix                          # y=2   2  7  12 17 22
            y[i]  = iy                          # y=3   3  8  13 18 23
            vx[i] = twelveran()                 # y=4   4  9  14 19 24
            vy[i] = twelveran()                 # numbering of 25 atoms
            vx[i] = vx[i]*sqrt(Tinit)
            vy[i] = vy[i]*sqrt(Tinit)
    for j in range(0,Natom):
```

```
        xc = 2*x[j] - 4
        yc = 2*y[j] - 4
        atoms.append(sphere(pos=(xc,yc), radius=0.5,color=color.red))

def sign(a, b):
    if (b >= 0.0):
        return abs(a)
    else:
        return - abs(a)

def Forces(t, w, PE, PEorW):                                        # Forces
    # invr2 = 0.
    r2cut = 9.                                  # Switch: PEorW = 1 for PE
    PE = 0.
    for i in range(0, Natom):
        fx[i][t] = fy[i][t] = 0.0
    for i in range( 0, Natom-1 ):
        for j in range(i + 1, Natom):
            dx = x[i] - x[j]
            dy = y[i] - y[j]
            if (abs(dx) > 0.50*L):
                dx = dx - sign(L, dx)     # Interact with closer image
            if (abs(dy) > 0.50*L):
                dy = dy - sign(L, dy)
            r2 = dx*dx + dy*dy
            if (r2 < r2cut):
                if (r2 == 0.):                     # To avoid 0 denominator
                    r2 = 0.0001
                invr2 = 1./r2
                wij = 48.*(invr2**3 - 0.5) *invr2**3
                fijx = wij*invr2*dx
                fijy = wij*invr2*dy
                fx[i][t] = fx[i][t] + fijx
                fy[i][t] = fy[i][t] + fijy
                fx[j][t] = fx[j][t] - fijx
                fy[j][t] = fy[j][t] - fijy
                PE = PE + 4.*(invr2**3)*((invr2**3) - 1.)
                w = w + wij
    if (PEorW == 1):
        return PE
    else:
        return w

def timevolution():
    avT = 0.0
    avP = 0.0
    Pavg = 0.0
    avKE = 0.0
    avPE = 0.0
    t1 = 0
    PE = 0.0
    h = 0.031                                                          # step
    hover2 = h/2.0
    # initial KE & PE via Forces
    KE = 0.0
    w = 0.0
    initialposvel()
    for i in range(0, Natom):
        KE = KE+(vx[i]*vx[i]+vy[i]*vy[i])/2.0
  # System.out.println(""+t+" PE= "+PE+" KE = "+KE+" PE+KE = "+(PE+KE));
    PE = Forces(t1,w,PE,1)
    time =1
    while 1:
        rate(100)
        for i in range(0, Natom):
            PE = Forces(t1,w,PE,1)
            x[i] = x[i] + h*(vx[i] + hover2*fx[i][t1])
```

```
        y[i] = y[i] + h*(vy[i] + hover2*fy[i][t1]);
        if x[i] <= 0.:
            x[i] = x[i] + L                    # Periodic boundary conditions
        if x[i] >= L :
            x[i] = x[i] - L
        if y[i] <= 0.:
            y[i] = y[i] + L
        if y[i] >= L:
            y[i] = y[i] - L
        xc = 2*x[i] - 4
        yc = 2*y[i] - 4
        atoms[i].pos=(xc,yc)
    PE = 0.
    t2=1
    PE = Forces(t2, w, PE, 1)
    KE = 0.
    w = 0.
    for  i in range(0 , Natom):
        vx[i] = vx[i] + hover2*(fx[i][t1] + fx[i][t2])
        vy[i] = vy[i] + hover2*(fy[i][t1] + fy[i][t2])
        KE = KE + (vx[i]*vx[i] + vy[i]*vy[i])/2
    w = Forces(t2, w, PE, 2)
    P=dens*(KE+w)
    T=KE/(Natom)
    # increment averages
    avT = avT + T
    avP = avP + P
    avKE = avKE + KE
    avPE = avPE + PE
    time += 1
    t=time
    if (t==0):
        t=1
    Pavg  = avP /t
    eKavg = avKE /t
    ePavg = avPE /t
    Tavg  = avT /t
    pre = (int)(Pavg*1000)
    Pavg = pre/1000.0
    kener = (int)(eKavg*1000)
    eKavg = kener/1000.0
    Kavegraph.plot(pos=(t,eKavg))
    pener = (int)(ePavg*1000)
    ePavg = pener/1000.0
    tempe = (int)(Tavg*1000000)
    Tavg = tempe/1000000.0
    Tcurve.plot(pos=(t,ePavg),display=sceneT)

timevolution()
```

The program MD.py in Listings 18.1 implements an MD simulation in 1D using the velocity-Verlet algorithm. Use it as a model and do the following:

1. Establish that you can run and visualize the 1D simulation.
2. Place the particles initially at the sites of a simple cubic lattice. The equilibrium configuration for a Lennard–Jones system at low temperature is a face-centered-cubic, and if your simulation is running properly, then the particles should migrate from SC to FCC. An FCC lattice has four quarters of a particle per unit cell, so an L^3 box with a lattice constant L/N contains (parts of) $4N^3 = 32, 108, 256, \dots$ particles.

3. To save computing time, assign initial particle velocities corresponding to a fixed-temperature Maxwellian distribution.
4. Print the code and indicate on it which integration algorithm is used, where the periodic boundary conditions are imposed, where the nearest image interaction is evaluated, and where the potential is cut off.
5. A typical time step is $\Delta t = 10^{-14}$ s, which in our natural units equals 0.004. You probably will need to make 10^4–10^5 such steps to equilibrate, which corresponds to a total time of only 10^{-9} s (a lot can happen to a speedy molecule in 10^{-9} s). Choose the *largest* time step that provides stability and gives results similar to Figure 18.5.
6. The PE and KE change with time as the system equilibrates. Even after that, there will be fluctuations because this is a dynamic system. Evaluate the time-averaged energies for an equilibrated system.
7. Compare the final temperature of your system to the initial temperature. Change the initial temperature and look for a simple relation between it and the final temperature (Figure 18.6).

18.4 Analysis

1. Modify your program so that it outputs the coordinates and velocities of a few particles throughout the simulation. Note that you do not need as many time steps to follow a trajectory as you do to compute it, and so you may want to use the *mod* operator `%100` for output.
   a) Start your assessment with a 1D simulation at zero temperature. The particles should remain in place without vibration. Increase the temperature and note how the particles begin to move about and interact.
   b) Try starting off all your particles at the minima in the Lennard–Jones potential. The particles should remain bound within the potential until you raise the temperature.
   c) Repeat the simulations for a 2D system. The trajectories should resemble billiard ball-like collisions.
   d) Create an animation of the time-dependent locations of several particles.
   e) Calculate and plot the root-mean-square displacement of molecules as a function of temperature:

$$R_{\rm rms} = \sqrt{\langle |\boldsymbol{r}(t + \Delta t) - \boldsymbol{r}(t)|^2 \rangle}\,, \tag{18.22}$$

   where the average is over all the particles in the box. Determine the approximate time dependence of $R_{\rm rms}$.
   f) Test your system for time-reversal invariance. Stop it at a fixed time, reverse all velocities, and see if the system retraces its trajectories back to the initial configuration after this same fixed time.

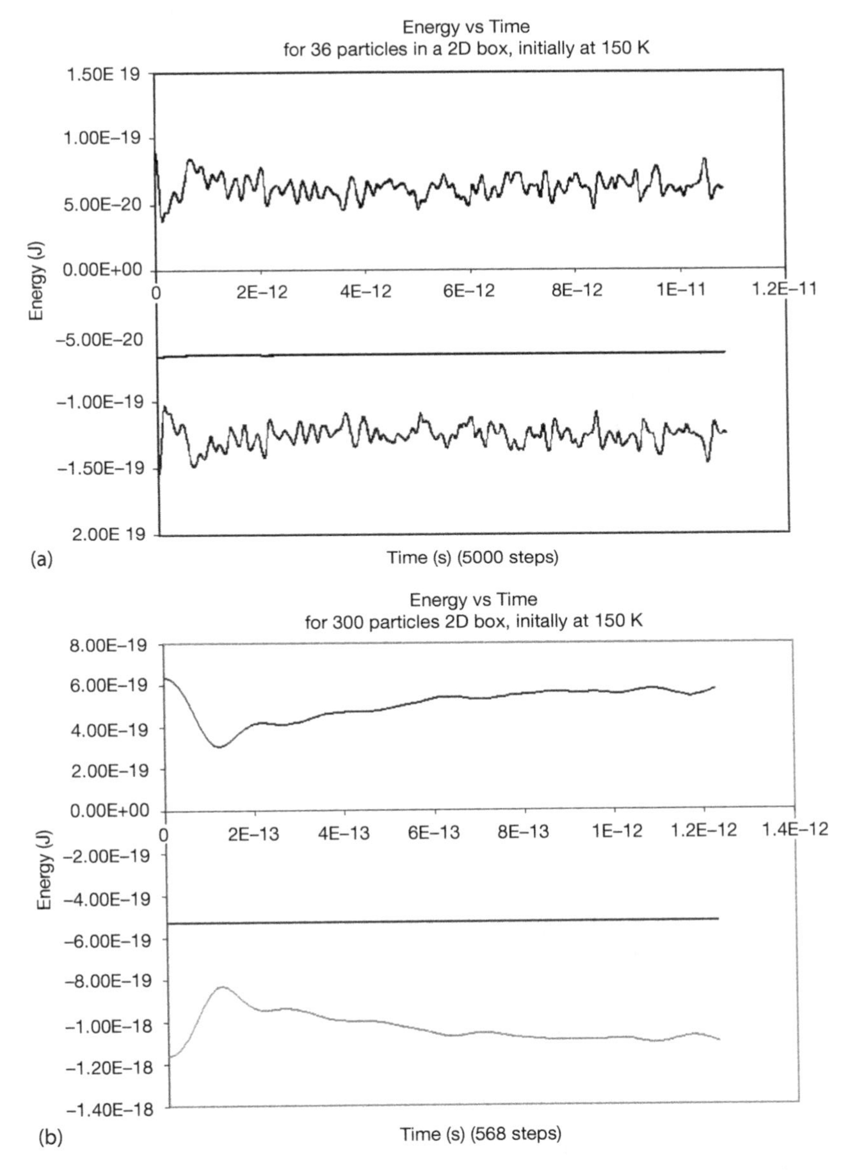

Figure 18.5 The kinetic, potential, and total energy as a function of time or number of steps for a 2D MD simulation with 36 particles (a), and 300 particles (b), both with an initial temperature of 150 K. The potential energy is negative, the kinetic energy is positive, and the total energy is seen to be conserved (flat).

2. *Hand Computation* We wish to make an MD simulation *by hand* of the positions of particles 1 and 2 that are in a 1D box of side 8. For an origin located at the center of the box, the particles are initially at rest and at loca-

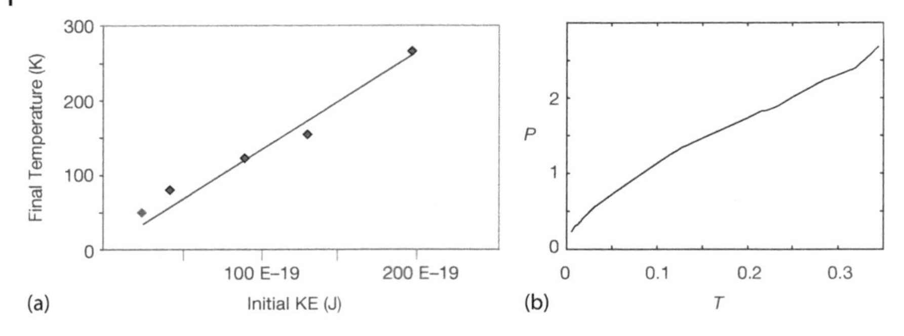

Figure 18.6 (a) The temperature after equilibration as a function of initial kinetic energy for a 2D MD simulation with 36 particles. The two are nearly proportional. (b) The pressure vs. temperature for a simulation with several hundred particles. An ideal gas (noninteracting particles) would yield a straight line (courtesy of J. Wetzel).

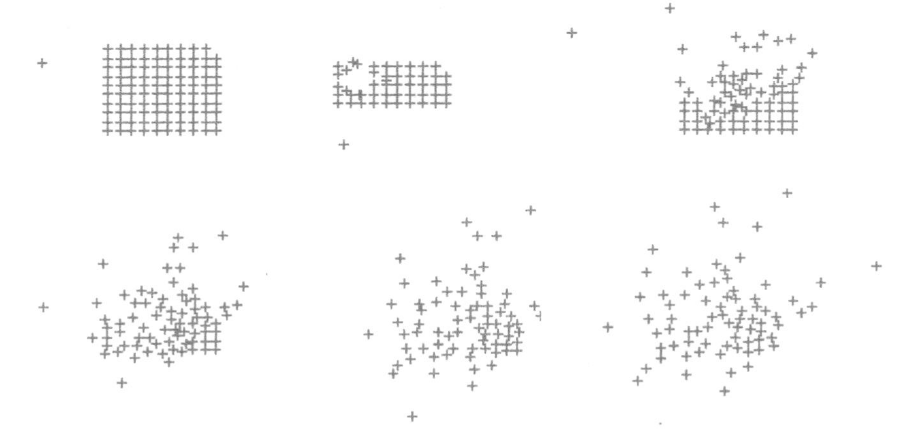

Figure 18.7 A simulation of a projectile shot into a group of particles. The energy introduced by the projectile is seen to lead to evaporation of the particles (courtesy of J. Wetzel).

tions $x_i(0) = -x_2(0) = 1$. The particles are subject to the force

$$F(x) = \begin{cases} 10\,, & \text{for} \quad |x_1 - x_2| \leq 1\,, \\ -1\,, & \text{for} \quad 1 \leq |x_1 - x_2| \leq 3\,, \\ 0\,, & \text{otherwise}\,. \end{cases} \tag{18.23}$$

Use a simple algorithm to determine the positions of the particles up until the time they leave the box. Make sure to apply periodic boundary conditions. *Hint:* Because the configuration is symmetric, you know the location of particle 2 by symmetry and do not need to solve for it. We suggest the Verlet algorithm (no velocities) with a forward-difference algorithm to initialize it. To speed things along, use a time step of $h = 1/\sqrt{2}$.

3. *Diffusion* It is well known that light molecules diffuse more quickly than heavier ones. See if you can simulate diffusion with your MD simulation using a Lennard–Jones potential and periodic boundary conditions (Satoh, 2011).
   a) Generalize the velocity-Verlet algorithm so that it can be used for molecules of different masses.
   b) Modify the simulation code so that it can be used for five heavy molecules of mass $M = 10$ and five light molecules of mass $m = 1$.
   c) Start with the molecules placed randomly near the center of the square simulation region.
   d) Assign random initial velocities to the molecules.
   e) Run the simulation several times and verify visually that the lighter molecules tend to diffuse more quickly than the heavier ones.
   f) For each ensemble of molecules, calculate the RMS velocity at regular instances of time, and then plot the RMS velocities as functions of time. Do the lighter particles have a greater RMS velocity?
4. As shown in Figure 18.7, simulate the impact of a projectile with a block of material.

19
PDE Review and Electrostatics via Finite Differences and Electrostatics via Finite Differences

This chapter is the first of several dealing with partial differential equations (PDEs); several because PDEs are more complex than ODEs, and because each type of PDE requires its own algorithm. We start the chapter with a discussion of PDEs in general, and the requirements for a unique solution of each type to exist. Then we get down to business and examine the simple, but powerful, *finite-differences* method for solving Poisson's and Laplace's equations on a lattice in space. Chapter 23 covers the more complicated, but ultimately more efficient, *finite elements* method for solving the same equations.

19.1
PDE Generalities

Physical quantities such as temperature and pressure vary continuously in both space and time. Such being our world, the function or *field* $U(x, y, z, t)$ used to describe these quantities must contain independent space and time variations. As time evolves, the changes in $U(x, y, z, t)$ at any one position affect the field at neighboring points. This means that the dynamic equations describing the dependence of U on four independent space–time variables must be written in terms of partial derivatives, and therefore the equations must be *partial differential equations* (PDEs), in contrast to ordinary differential equations (ODEs).

The most general form for a two-independent variable PDE is

$$A\frac{\partial^2 U}{\partial x^2} + 2B\frac{\partial^2 U}{\partial x \partial y} + C\frac{\partial^2 U}{\partial y^2} + D\frac{\partial U}{\partial x} + E\frac{\partial U}{\partial y} = F \,, \tag{19.1}$$

where A, B, C, and F are arbitrary functions of the variables x and y. In Table 19.1, we define the classes of PDEs by the value of the discriminant $d = AC - B^2$ in row two (Arfken and Weber, 2001), and give examples in rows three and four.

We usually think of an elliptic equation as containing the second-order derivatives of all the variables, with all having the same sign when placed on the same side of the equal sign; a parabolic equation as containing a first-order derivative in one variable and a second-order derivative in the other; and a hyperbolic equa-

Computational Physics, 3rd edition. Rubin H. Landau, Manuel J. Páez, Cristian C. Bordeianu.

Table 19.1 Three categories of PDE based on the value of their discriminant d.

Elliptic	**Parabolic**	**Hyperbolic**
$d = AC - B^2 > 0$	$d = AC - B^2 = 0$	$d = AC - B^2 < 0$
$\nabla^2 U(x) = -4\pi\rho(x)$	$\nabla^2 U(\boldsymbol{x}, t) = a\partial U/\partial t$	$\nabla^2 U(\boldsymbol{x}, t) = c^{-2}\partial^2 U/\partial t^2$
Poisson's	Heat	Wave

tion as containing second-order derivatives of all the variables, with opposite signs when placed on the same side of the equal sign.

After solving enough problems, one often develops some physical intuition as to whether one has sufficient *boundary conditions* for there to exist a unique solution for a given physical situation (this, of course, is in addition to requisite *initial conditions*). Table 19.2 gives the requisite boundary conditions for a unique solution to exist for each type of PDE. For instance, a string tied at both ends and a heated bar placed in an infinite heat bath are physical situations for which the boundary conditions are adequate. If the boundary condition is the value of the solution on a surrounding closed surface, we have a *Dirichlet boundary condition*. If the boundary condition is the value of the normal derivative on the surrounding surface, we have a *Neumann boundary condition*. If the value of both the solution and its derivative are specified on a closed boundary, we have a *Cauchy boundary condition*. Although having an adequate boundary condition is necessary for a unique solution, having too many boundary conditions, for instance, both Neumann and Dirichlet, may be an overspecification for which no solution exists.[1)]

Solving PDEs numerically differs from solving ODEs in a number of ways. First, because we are able to write all ODEs in a standard form

$$\frac{\mathrm{d}\boldsymbol{y}(t)}{\mathrm{d}t} = \boldsymbol{f}(\boldsymbol{y}, t) , \tag{19.2}$$

with t the single independent variable, we are able to use a standard algorithm such as `rk4` to solve all such equations. Yet because PDEs have several independent variables, for example, $\rho(x, y, z, t)$, we would have to apply (19.2) simultaneously and independently to each variable, which would be very complicated. Second, because there are more equations to solve with PDEs than with ODEs, we need more information than just the two *initial conditions* $[x(0), \dot{x}(0)]$. In addition, because each PDE often has its own particular set of boundary conditions, we have to develop a special algorithm for each particular problem.

1) Although conclusions drawn for exact PDEs may differ from those drawn for the finite-difference equations, we use for our algorithms, they are usually the same. In fact, Morse and Feshbach (Morse and Feshbach, 1953) use the finite-difference form to derive the relations between boundary conditions and uniqueness for each type of equation shown in Table 19.2 (Jackson, 1988).

19.2 Electrostatic Potentials

Your *problem* is to find the electric potential for all points *inside* the charge-free square shown in Figure 19.1. The bottom and sides of the region are made up of wires that are "grounded" (kept at 0 V). The top wire is connected to a voltage source that keeps it at a constant 100 V.

19.2.1 Laplace's Elliptic PDE (Theory)

We consider the entire square in Figure 19.1 as our boundary with the voltages prescribed upon it. If we imagine infinitesimal insulators placed at the top corners of the box, then we will have a closed boundary. Because the values of the potential are given on all sides, we have Neumann conditions on the boundary and, according to Table 19.2, a unique and stable solution.

Table 19.2 The relation between boundary conditions and uniqueness for PDEs.

Boundary Condition	Elliptic (Poisson equation)	Hyperbolic (Wave equation)	Parabolic (Heat equation)
Dirichlet open surface	Underspecified	Underspecified	*Unique and stable (1D)*
Dirichlet closed surface	*Unique and stable*	Overspecified	Overspecified
Neumann open surface	Underspecified	Underspecified	*Unique and stable (1D)*
Neumann closed surface	*Unique and stable*	Overspecified	Overspecified
Cauchy open surface	Nonphysical	*Unique and stable*	Overspecified
Cauchy closed surface	Overspecified	Overspecified	Overspecified

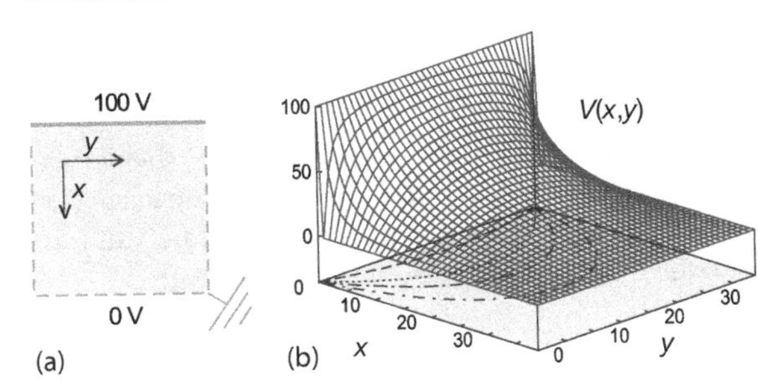

Figure 19.1 (a) The shaded region of space within a square in which we determine the electric potential by solving Laplace's equation. There is a wire at the top kept at a constant 100 V and a grounded wire (dashed) at the sides and bottom. (b) The computed electric potential as a function of x and y. The projections onto the shaded xy plane are equipotential (contour) lines.

It is known from classical electrodynamics that the electric potential $U(\boldsymbol{x})$ arising from static charges satisfies Poisson's PDE (Jackson, 1988):

$$\nabla^2 U(\boldsymbol{x}) = -4\pi\rho(\boldsymbol{x}) , \tag{19.3}$$

where $\rho(\boldsymbol{x})$ is the charge density. In charge-free regions of space, that is, regions where $\rho(\boldsymbol{x}) = 0$, the potential satisfies *Laplace's equation*:

$$\nabla^2 U(\boldsymbol{x}) = 0 . \tag{19.4}$$

Both these equations are elliptic PDEs of a form that occurs in various applications. We solve them in 2D rectangular coordinates:

$$\frac{\partial^2 U(x,y)}{\partial x^2} + \frac{\partial^2 U(x,y)}{\partial y^2} = 0 , \quad \text{Laplace's equation} , \tag{19.5}$$

$$\frac{\partial^2 U(x,y)}{\partial x^2} + \frac{\partial^2 U(x,y)}{\partial y^2} = -4\pi\rho(\boldsymbol{x}) , \quad \text{Poisson's equation} . \tag{19.6}$$

In both cases, we see that the potential depends simultaneously on x and y. For Laplace's equation, the charges, which are the source of the field, enter indirectly by specifying the potential values in some region of space; for Poisson's equation they enter directly.

19.3 Fourier Series Solution of a PDE

For the simple geometry of Figure 19.1, an analytic solution of Laplace's equation (19.5) exists in the form of an infinite series. If we assume that the solution is the product of independent functions of x and y and substitute the product into (19.5), we obtain

$$U(x,y) = X(x)Y(y) \quad \Rightarrow \quad \frac{\mathrm{d}^2 X(x)/\,\mathrm{d}x^2}{X(x)} + \frac{\mathrm{d}^2 Y(y)/\,\mathrm{d}y^2}{Y(y)} = 0 . \tag{19.7}$$

Because $X(x)$ is a function of only x, and $Y(y)$ is a function of only y, the derivatives in (19.7) are *ordinary* as opposed to *partial* derivatives. Because $X(x)$ and $Y(y)$ are assumed to be independent, the only way (19.7) can be valid for *all* values of x and y is for each term in (19.7) to be equal to a constant:

$$\frac{\mathrm{d}^2 Y(y)/\,\mathrm{d}y^2}{Y(y)} = -\frac{\mathrm{d}^2 X(x)/\,\mathrm{d}x^2}{X(x)} = k^2 \tag{19.8}$$

$$\Rightarrow \quad \frac{\mathrm{d}^2 X(x)}{\mathrm{d}x^2} + k^2 X(x) = 0 , \quad \frac{\mathrm{d}^2 Y(y)}{\mathrm{d}y^2} - k^2 Y(y) = 0 . \tag{19.9}$$

We shall see that this choice of sign for the constant matches the boundary conditions and gives us periodic behavior in x. The other choice of sign would give periodic behavior in y, and that would not work with these boundary conditions.

The solutions for $X(x)$ are periodic, and those for $Y(y)$ are exponential:

$$X(x) = A \sin kx + B \cos kx \,, \qquad Y(y) = Ce^{ky} + De^{-ky} \,. \tag{19.10}$$

The $x = 0$ boundary condition $U(x = 0, y) = 0$ can be met only if $B = 0$. The $x = L$ boundary condition $U(x = L, y) = 0$ can be met only for

$$kL = n\pi \,, \qquad n = 1, 2, \ldots \tag{19.11}$$

Such being the case, for each value of n there is the solution

$$X_n(x) = A_n \sin\left(\frac{n\pi}{L}x\right) \,. \tag{19.12}$$

For each value of k_n, $Y(y)$ must satisfy the y boundary condition $U(x, 0) = 0$, which requires $D = -C$:

$$Y_n(y) = C(e^{k_n y} - e^{-k_n y}) \equiv 2C \sinh\left(\frac{n\pi}{L}y\right) \,. \tag{19.13}$$

Because we are solving linear equations, the principle of linear superposition holds, which means that the most general solution is the sum of the products:

$$U(x, y) = \sum_{n=1}^{\infty} E_n \sin\left(\frac{n\pi}{L}x\right) \sinh\left(\frac{n\pi}{L}y\right) \,. \tag{19.14}$$

The E_n values are arbitrary constants and are fixed by requiring the solution to satisfy the remaining boundary condition at $y = L$, $U(x, y = L) = 100\,\text{V}$:

$$\sum_{n=1}^{\infty} E_n \sin\frac{n\pi}{L}x \sinh n\pi = 100\,\text{V} \,. \tag{19.15}$$

We determine the constants E_n by projection: multiply both sides of the equation by $\sin m\pi/Lx$, with m an integer, and integrate from 0 to L:

$$\sum_{n}^{\infty} E_n \sinh n\pi \int_0^L dx \sin\frac{n\pi}{L}x \sin\frac{m\pi}{L}x = \int_0^L dx 100 \sin\frac{m\pi}{L}x \,. \tag{19.16}$$

The integral on the LHS is nonzero only for $n = m$, which yields

$$E_n = \begin{cases} 0 \,, & \text{for} \quad n \text{ even} \,, \\ \frac{4(100)}{n\pi \sinh n\pi} \,, & \text{for} \quad n \text{ odd} \,. \end{cases} \tag{19.17}$$

Finally, we obtain an infinite series (analytic solution) for the potential at any point (x, y):

$$U(x, y) = \sum_{n=1,3,5,\ldots}^{\infty} \frac{400}{n\pi} \sin\left(\frac{n\pi x}{L}\right) \frac{\sinh(n\pi y/L)}{\sinh(n\pi)} \,. \tag{19.18}$$

19.3.1
Polynomial Expansion as an Algorithm

If we try to use (19.18) as an algorithm, we must terminate the sum at some point. Yet in practice the convergence of the series is so painfully slow that many terms are needed for good accuracy, and so the round-off error may become a problem. In addition, the sinh functions in (19.18) overflow for large n, which can be avoided somewhat by expressing the quotient of the two sinh functions in terms of exponentials and then taking a large n limit:

$$\frac{\sinh(n\pi y/L)}{\sinh(n\pi)} = \frac{e^{n\pi(y/L-1)} - e^{-n\pi(y/L+1)}}{1 - e^{-2n\pi}} \xrightarrow[n\to\infty]{} e^{n\pi(y/L-1)} . \tag{19.19}$$

A third problem with the "analytic" solution is that a Fourier series converges only in the *mean square* (Figure 19.2). This means that it converges to the *average* of the left- and right-hand limits in the regions where the solution is discontinuous (Kreyszig, 1998), such as in the corners of the box. Explicitly, what you see in Figure 19.2 is a phenomenon known as the *Gibbs overshoot* that occurs when a Fourier series with a finite number of terms is used to represent a discontinuous function. Rather than fall off abruptly, the series develops large oscillations that tend to overshoot the function at the corner. To obtain a smooth solution, we had to sum 40 000 terms, where, in contrast, the numerical solution required only several hundred steps.

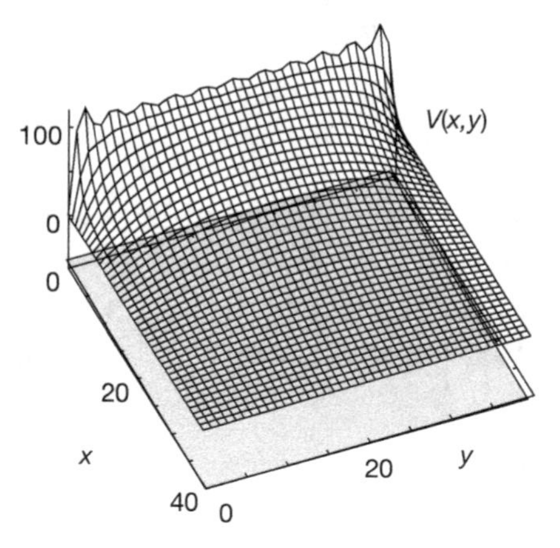

Figure 19.2 The analytic (Fourier series) solution of Laplace's equation summing 21 terms. Gibbs overshoot leads to the oscillations near $x = 0$, and persist even if a large number of terms is summed.

19.4 Finite-Difference Algorithm

To solve our 2D PDE numerically, we divide space up into a lattice (Figure 19.3) and solve for U at each site on the lattice. Because we will express derivatives in terms of the finite differences in the values of U at the lattice sites, this is called a *finite-difference* method. A numerically more efficient method, but with more complicated set up, is the *finite-element* method (FEM) that solves the PDE for small geometric elements and then matches the solution over the elements. We discuss FEM in Chapter 23.

To derive the finite-difference algorithm for the numeric solution of (19.5), we follow the same path taken in Section 5.1 to derive the forward-difference algorithm for differentiation. We start by adding the two Taylor expansions of the potential to the right and left of (x, y) and the two for above and below (x, y):

$$U(x + \Delta x, y) = U(x, y) + \frac{\partial U}{\partial x}\Delta x + \frac{1}{2}\frac{\partial^2 U}{\partial x^2}(\Delta x)^2 + \cdots , \tag{19.20}$$

$$U(x - \Delta x, y) = U(x, y) - \frac{\partial U}{\partial x}\Delta x + \frac{1}{2}\frac{\partial^2 U}{\partial x^2}(\Delta x)^2 - \cdots , \tag{19.21}$$

$$U(x, y + \Delta y) = U(x, y) + \frac{\partial U}{\partial y}\Delta y + \frac{1}{2}\frac{\partial^2 U}{\partial y^2}(\Delta y)^2 + \cdots , \tag{19.22}$$

$$U(x, y - \Delta y) = U(x, y) - \frac{\partial U}{\partial y}\Delta y + \frac{1}{2}\frac{\partial^2 U}{\partial y^2}(\Delta y)^2 - \cdots . \tag{19.23}$$

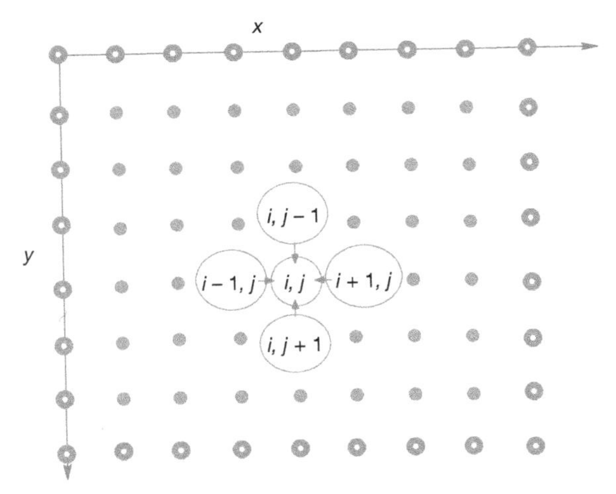

Figure 19.3 The algorithm for Laplace's equation in which the potential at the point $(x, y) = (i, j)\Delta$ equals the average of the potential values at the four nearest-neighbor points. The nodes with white centers correspond to fixed values of the potential along the boundaries.

All odd terms cancel when we add these equations in pairs, and we obtain a central-difference approximation for the second partial derivative good to order Δ^4:

$$\frac{\partial^2 U(x,y)}{\partial x^2} \simeq \frac{U(x+\Delta x, y) + U(x-\Delta x, y) - 2U(x,y)}{(\Delta x)^2} , \tag{19.24}$$

$$\frac{\partial^2 U(x,y)}{\partial y^2} \simeq \frac{U(x, y+\Delta y) + U(x, y-\Delta y) - 2U(x,y)}{(\Delta y)^2} . \tag{19.25}$$

Substitution of both these approximations in Poisson's equation (19.6) leads us to a finite-difference form of the PDE:

$$\frac{U(x+\Delta x, y) + U(x-\Delta x, y) - 2U(x,y)}{(\Delta x)^2} \tag{19.26}$$

$$+\frac{U(x, y+\Delta y) + U(x, y-\Delta y) - 2U(x,y)}{(\Delta y)^2} = -4\pi\rho . \tag{19.27}$$

We take the x and y grids to be of equal spacings $\Delta x = \Delta y = \Delta$, and thus obtain a simple form for the equation

$$U(x+\Delta, y) + U(x-\Delta, y) + U(x, y+\Delta) + U(x, y-\Delta) - 4U(x,y) = -4\pi\rho . \tag{19.28}$$

The reader will notice that this equation shows a relation among the solutions at five points in space. When $U(x, y)$ is evaluated for the N_x x values on the lattice and for the N_y y values, we obtain a set of $N_x \times N_y$ simultaneous linear algebraic equations for `U[i,j]` to solve. One approach is to solve these equations explicitly as a (big) matrix problem. This is attractive, as it is a direct solution, but it requires a great deal of memory and accounting. The approach we use follows from the algebraic solution of (19.28) for $U(x, y)$:

$$\begin{aligned} 4U(x,y) \simeq{} & U(x+\Delta, y) + U(x-\Delta, y) + U(x, y+\Delta) + U(x, y-\Delta) \\ & + 4\pi\rho(x,y)\Delta^2 , \end{aligned} \tag{19.29}$$

where we would omit the $\rho(x)$ term for Laplace's equation. In terms of discrete locations on our lattice, the x and y variables are

$$x = x_0 + i\Delta, \quad y = y_0 + j\Delta , \quad i, j = 0, \dots, N_{\max -1} , \tag{19.30}$$

where we have placed our lattice in a square of side L. The finite-difference algorithm (19.29) becomes

$$\boxed{U_{i,j} = \frac{1}{4}\left[U_{i+1,j} + U_{i-1,j} + U_{i,j+1} + U_{i,j-1}\right] + \pi\rho(i\Delta, j\Delta)\Delta^2 .} \tag{19.31}$$

This equation says that when we have a proper solution, it will be the average of the potential at the four nearest neighbors (Figure 19.3) plus a contribution from the local charge density. As an algorithm, (19.31) does not provide a direct solution to Poisson's equation, but rather must be repeated many times to converge upon the solution. We start with an initial guess for the potential, improve it by sweeping through all space taking the average over nearest neighbors at each node, and keep repeating the process until the solution no longer changes to some level of precision or until failure is evident. When converged, the initial guess is said to have *relaxed* into the solution.

A reasonable question with this simple an approach is, "Does it always converge, and if so, does it converge fast enough to be useful?" In some sense, the answer to the first question is not an issue; if the method does not converge, then we will know it; otherwise we have ended up with a solution and the path we followed to get there is no body's business! The answer to the question of speed is that relaxation methods may converge slowly (although still faster than a Fourier series), yet we will show you two clever tricks to accelerate the convergence.

At this point, it is important to remember that our algorithm arose from expressing the Laplacian ∇^2 in rectangular coordinates. While this does not restrict us from solving problems with circular symmetry, there may be geometries where it is better to develop an algorithm based on expressing the Laplacian in cylindrical or spherical coordinates in order to have grids that fit the geometry better.

19.4.1 Relaxation and Over-relaxation

There are a number of ways in which algorithm (19.31) can be iterated so as to convert the boundary conditions to a solution. Its most basic form is the *Jacobi method* and is one in which the potential values are not changed until an entire sweep of applying (19.31) at each point is completed. This maintains the symmetry of the initial guess and boundary conditions. A rather obvious improvement on the Jacobi method utilizes the updated guesses for the potential in (19.31) as soon as they are available. As a case in point, if the sweep starts in the upper left-hand corner of Figure 19.3, then the leftmost `U([ -1, j]` and topmost `U[i,j-1]` values of the potential used will be from the present generation of guesses, while the other two values of the potential will be from the previous generation: (Gauss–Seidel method)

$$U_{i,j}^{(\text{new})} = \frac{1}{4}\left[U_{i+1,j}^{(\text{old})} + U_{i-1,j}^{(\text{new})} + U_{i,j+1}^{(\text{old})} + U_{i,j-1}^{(\text{new})}\right] . \tag{19.32}$$

This technique, known as the *Gauss–Seidel method*, usually leads to accelerated convergence, which in turn leads to less round-off error. It also uses less memory as there is no need to store two generations of guesses. However, it does distort the symmetry of the boundary conditions, which one hopes is insignificant when convergence is reached.

A less obvious improvement in the relaxation technique, known as *successive over-relaxation* (SOR), starts by writing algorithm (19.31) in a form that determines the new values of the potential $U^{(\text{new})}$ as the old values $U^{(\text{old})}$ plus a correction or residual r:

$$U_{i,j}^{(\text{new})} = U_{i,j}^{(\text{old})} + r_{i,j}. \tag{19.33}$$

While the Gauss–Seidel technique may still be used to incorporate the updated values in $U^{(\text{old})}$ to determine r, we rewrite the algorithm here in the general form

$$\begin{aligned} r_{i,j} &\equiv U_{i,j}^{(\text{new})} - U_{i,j}^{(\text{old})} \\ &= \frac{1}{4}\left[U_{i+1,j}^{(\text{old})} + U_{i-1,j}^{(\text{new})} + U_{i,j+1}^{(\text{old})} + U_{i,j-1}^{(\text{new})}\right] - U_{i,j}^{(\text{old})} . \end{aligned} \tag{19.34}$$

The successive over-relaxation technique (Press *et al.*, 1994; Garcia, 2000) proposes that if convergence is obtained by adding r to U, then more rapid convergence might be obtained by adding more or less of r:

$$\boxed{U_{i,j}^{(\text{new})} = U_{i,j}^{(\text{old})} + \omega r_{i,j}} \quad (\text{SOR}) , \tag{19.35}$$

where ω is a parameter that amplifies or reduces the residual. The nonaccelerated relaxation algorithm (19.32) is obtained with $\omega = 1$, accelerated convergence (over-relaxation) is obtained with $\omega \geq 1$, and underrelaxation is obtained with $\omega < 1$. Values of $1 \leq \omega \leq 2$ often work well, yet $\omega > 2$ may lead to numerical instabilities. Although a detailed analysis of the algorithm is needed to predict the optimal value for ω, we suggest that you explore different values for ω to see which one works best for your particular problem.

19.4.2 Lattice PDE Implementation

In Listing 19.1, we present the code **LaplaceLine.py** that solves the square-wire problem (Figure 19.1). Here we have kept the code simple by setting the length of the box $L = N_{\text{max}}\Delta = 100$ and by taking $\Delta = 1$:

$$\begin{aligned} &U(i, N_{\text{max}}) = 99 \quad (\text{top}) , \quad U(1, j) = 0 \quad (\text{left}) , \\ &U(N_{\text{max}}, j) = 0 \quad (\text{right}) , \quad U(i, 1) = 0 \quad (\text{bottom}) . \end{aligned} \tag{19.36}$$

We run algorithm (10.19) for a fixed 1000 iterations. A better code would vary Δ and the dimensions and would quit iterating once the solution converges to some tolerance. Study, compile, and execute the basic code.

Listing 19.1 LaplaceLine.py solves Laplace's equation via relaxation. Various parameters should be adjusted for an accurate solution.

```
# LaplaceLine.py:  Solve Laplace's eqtn, 3D matplot, close shell to quit


import matplotlib.pylab as p;
from mpl_toolkits.mplot3d import Axes3D
from numpy import *;
import numpy;
```

```
print("Initializing")
Nmax = 100; Niter = 70; V = zeros((Nmax, Nmax), float)

print ("Working hard, wait for the figure while I count to 60")
for k in range(0, Nmax-1):  V[k,0] = 100.0                    # Line at 100V

for iter in range(Niter):
    if iter%10 == 0: print(iter)
    for i in range(1, Nmax-2):
        for j in range(1,Nmax-2):
            V[i,j] = 0.25*(V[i+1,j]+V[i-1,j]+V[i,j+1]+V[i,j-1])
x = range(0, Nmax-1, 2);  y = range(0, 50, 2)
X, Y = p.meshgrid(x,y)

def functz(V):                                                # V(x, y)
    z = V[X,Y]
    return z

Z = functz(V)
fig = p.figure()                                              # Create figure
ax = Axes3D(fig)                                              # Plot axes
ax.plot_wireframe(X, Y, Z, color = 'r')                       # Red wireframe
ax.set_xlabel('X')
ax.set_ylabel('Y')
ax.set_zlabel('Potential')
p.show()                                                      # Show fig
```

19.5
Assessment via Surface Plot

After executing `LaplaceLine.py`, you should see a surface plot like Figure 19.1. Study this file in order to understand how to make surface plots with Matplotlib in Python. It is important to visualize your output as a surface plot to establish the reasonableness of the solution.

19.6
Alternate Capacitor Problems

We give you (or your instructor) a choice now. You can carry out the assessment using our wire-plus-grounded-box problem, or you can replace that problem with a more interesting one involving a realistic capacitor or nonplanar capacitors. We now describe the capacitor problem and then move on to the assessment and exploration.

Elementary textbooks solve the capacitor problem for the uniform field confined between two infinite plates. The field in a finite capacitor varies near the edges (edge effects) and extends beyond the edges of the capacitor (fringe fields). We model the realistic capacitor in a grounded box (Figure 19.4) as two plates (wires) of finite length. Write your simulation such that it is convenient to vary the grid spacing Δ and the geometry of the box and plate. We pose three versions of this

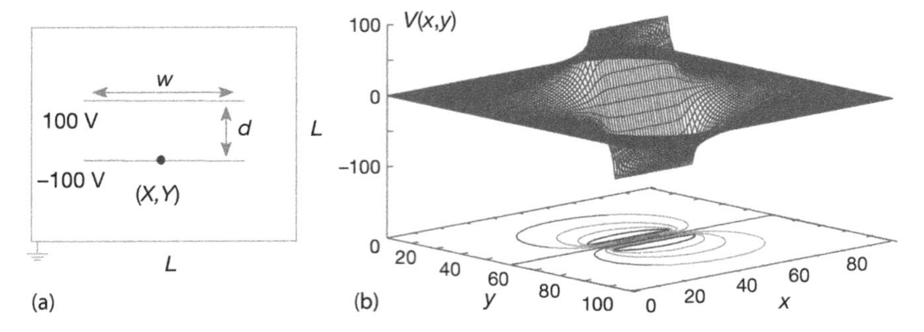

Figure 19.4 (a) A simple model of a parallel-plate capacitor within a box. A realistic model would have the plates close together, in order to condense the field, and the enclosing grounded box so large that it has no effect on the field near the capacitor. (b) A numerical solution for the electric potential for this geometry. The projection on the xy plane gives the equipotential lines.

problem, each displaying somewhat different physics. In each case, the boundary condition $U = 0$ on the surrounding box must be imposed for all iterations in order to obtain a unique solution.

1. For the simplest version, assume that the plates are very thin sheets of conductors, with the top plate maintained at 100 V and the bottom at −100 V. Because the plates are conductors, they must be equipotential surfaces, and a battery can maintain them at constant voltages. Write or modify the given program to solve Laplace's equation such that the plates have fixed voltages.
2. For the next version of this problem, assume that the plates are composed of a line of dielectric material with uniform charge densities ρ on the top and $-\rho$ on the bottom. Solve Poisson's equation (19.3) in the region including the plates, and Laplace's equation elsewhere. Experiment until you find a numerical value for ρ that gives a potential similar to that shown in Figure 19.6 for plates with fixed voltages.
3. For the final version of this problem investigate how the charges on a capacitor with finite-thickness conducting plates (Figure 19.5) distribute themselves. Because the plates are conductors, they are still equipotential surfaces at 100 and −100 V, only now you should make them have a thickness of at least 2Δ (so we can see the difference between the potential near the top and the bottom surfaces of the plates). Such being the case, we solve Laplace's equation (19.4) much as before to determine $U(x, y)$. Once we have $U(x, y)$, we substitute it into Poisson's equation (19.3) and determine how the charge density distributes itself along the top and bottom surfaces of the plates. *Hint:* Because the electric field is no longer uniform, we know that the charge distribution also will no longer be uniform. In addition, because the electric field now extends beyond the ends of the capacitor and because field lines begin and end on charge, some charge may end up on the edges and outer surfaces of the plates (Figure 19.4).

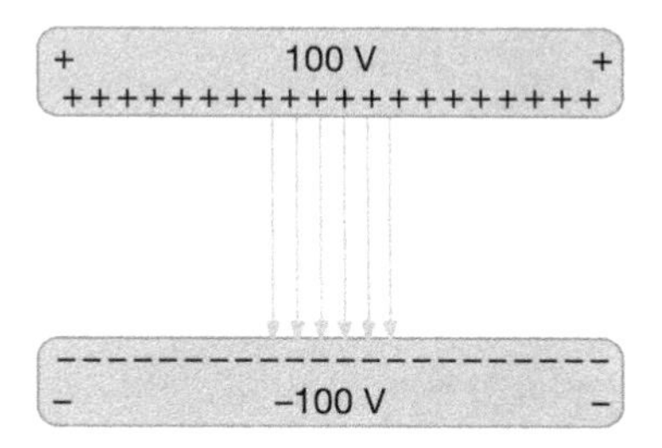

Figure 19.5 A guess as to how charge may rearrange itself on finite conducting plates.

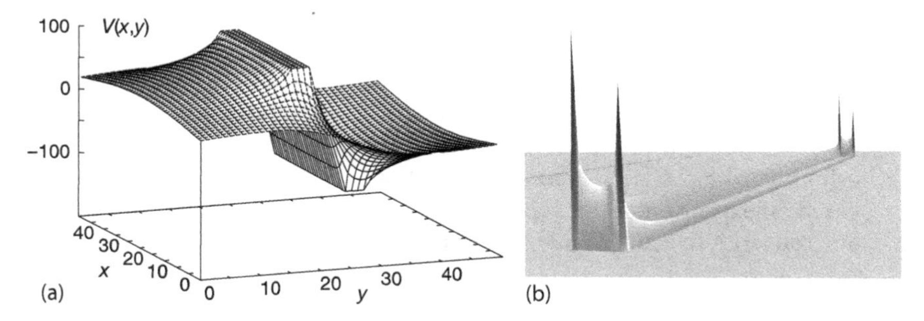

Figure 19.6 (a) A visualization of the computed electric potential for a capacitor with finite width plates. (b) A visualization of the charge distribution along one plate determined by evaluating $\nabla^2 U(x, y)$ (courtesy of J. Wetzel). Note the "lightening rod" effect of charge accumulating at corners and points.

4. The numerical solution to our PDE can be applied to arbitrary boundary conditions. Two boundary conditions to explore are triangular and sinusoidal:

$$U(x) = \begin{cases} 200x/w\,, & x \leq w/2\,, \\ 100(1 - x/w)\,, & x \geq w/2\,, \end{cases} \quad \text{or} \quad U(x) = 100 \sin\left(\frac{2\pi x}{w}\right)\,. \tag{19.37}$$

5. *Square conductors:* You have designed a piece of equipment consisting of a small metal box at 100 V within a larger grounded one (Figure 19.7). You find that sparking occurs between the boxes, which means that the electric field is too large. You need to determine where the field is greatest so that you can change the geometry and eliminate the sparking. Modify the program to satisfy these boundary conditions and to determine the field between the boxes. Gauss's law tells us that the field vanishes within the inner box because it contains no charge. Plot the potential and equipotential surfaces and sketch in the electric field lines. Deduce where the electric field is most intense and try redesigning the equipment to reduce the field.
6. *Cracked cylindrical capacitor:* You have designed the cylindrical capacitor containing a long outer cylinder surrounding a thin inner cylinder (Figure 19.7b). The cylinders have a small crack in them in order to connect

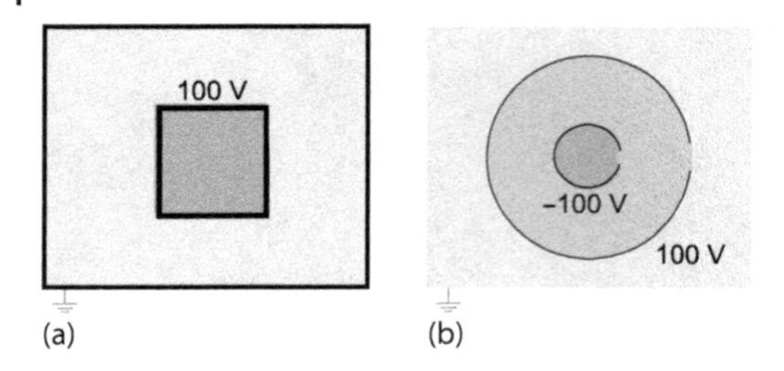

Figure 19.7 (a) The geometry of a capacitor formed by placing two long, square cylinders within each other. (b) The geometry of a capacitor formed by placing two long, circular cylinders within each other. The cylinders are cracked on the side so that wires can enter the region.

them to the battery that maintains the inner cylinder at −100 V and outer cylinder at 100 V. Determine how this small crack affects the field configuration. In order for a unique solution to exist for this problem, place both cylinders within a large grounded box. Note that because our algorithm is based on expansion of the Laplacian in rectangular coordinates, you cannot just convert it to a radial and angle grid.

19.7 Implementation and Assessment

1. Write or modify the program to find the electric potential for a capacitor within a grounded box. Use the labeling scheme as shown in Figure 19.4a.
2. To start, have your program undertake 1000 iterations and then quit. During debugging, examine how the potential changes in some key locations as you iterate toward a solution.
3. Repeat the process for different step sizes Δ and draw conclusions regarding the stability and accuracy of the solution.
4. Once your program produces reasonable solutions, modify it so that it stops iterating after convergence is reached, or if the number of iterations becomes too large. Rather than trying to discern small changes in highly compressed surface plots, use a numerical measure of precision, for example,

$$\texttt{trace} = \sum_i |\texttt{U[i,i]}| \,, \tag{19.38}$$

   which samples the solution along the diagonal. Remember, this is a simple algorithm and so may require many iterations for high precision. You should be able to obtain changes in the trace that are less than 1 part in 10^4. (The `break` command or a `while` loop is useful for this type of test.)
5. Equation 19.35 expresses the *successive over-relaxation* technique in which convergence is accelerated by using a judicious choice of ω. Determine by trial and error a best value of ω. This should let you double the speed of the algorithm.

6. Now that your code is accurate, modify it to simulate a more realistic capacitor in which the plate separation is approximately 1/10 of the plate length. You should find the field more condensed and more uniform between the plates.
7. If you are working with the wire-in-the-box problem, compare your numerical solution to the analytic one (19.18). Do not be surprised if you need to sum thousands of terms before the analytic solution converges!

19.8 Electric Field Visualization (Exploration)

Plot the equipotential surfaces on a separate 2D plot. Start with a crude, hand-drawn sketch of the electric field by drawing curves orthogonal to the equipotential lines, beginning and ending on the boundaries (where the charges lie). The regions of high density are regions of high electric field. Physics tells us that the electric field $\boldsymbol{E}$ is the negative gradient of the potential:

$$\boldsymbol{E} = -\nabla U(x, y) = -\frac{\partial U(x, y)}{\partial x}\hat{\epsilon}_x - \frac{\partial U(x, y)}{\partial y}\hat{\epsilon}_y , \tag{19.39}$$

where $\hat{\epsilon}_i$ is a unit vector in the i direction. While at first it may seem that some work is involved in determining these derivatives, once you have a solution for $U(x, y)$ on a grid, it is simple to use the central-difference approximation for the derivative to determine the field, for example:

$$E_x \simeq \frac{U(x+\Delta, y) - U(x-\Delta, y)}{2\Delta} = \frac{U_{i+1,j} - U_{i-1,j}}{2\Delta} . \tag{19.40}$$

Once you have a data file representing such a vector field, it can be visualized by plotting arrows of varying lengths and directions, or with just lines (Figure 19.8). In Section 1.5.6, we have shown how to do this with Mayavi.

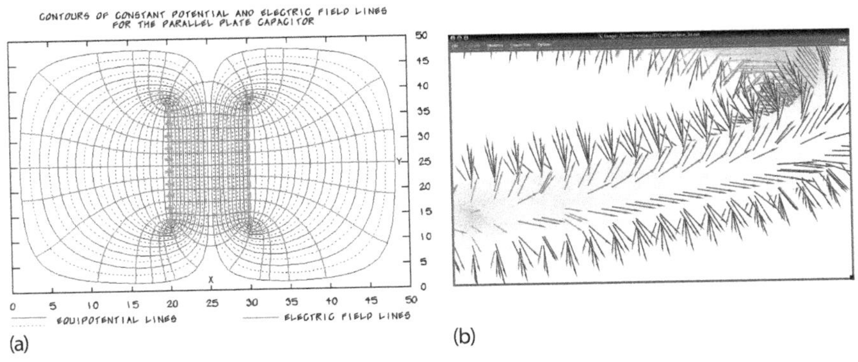

Figure 19.8 (a) Computed equipotential surfaces and electric field lines for a realistic capacitor. (b) Equipotential surfaces and electric field lines mapped onto the surface for a 3D capacitor constructed from two tori.

19.9 Review Exercise

You are given a simple Laplace-type equation

$$\frac{\partial u}{\partial x} + \frac{\partial u}{\partial y} = -\rho(x, y) , \tag{19.41}$$

where x and y are the Cartesian spatial coordinates and $\rho(x, y)$ is the charge density in space.

1. Develop a simple algorithm that will permit you to solve for the potential u between two square conductors kept at fixed u, with a charge density ρ between them.
2. Make a simple sketch that shows with arrows how your algorithm works.
3. Make sure to specify how you start and terminate the algorithm.
4. *Thinking outside the box*☺: Find the electric potential for all points *outside* the charge-free square shown in Figure 19.1. Is your solution unique?

20
Heat Flow via Time Stepping

As the present now
Will later be past
The order is
Rapidly fadin'

And the first one now
Will later be last
For the times they are a-changin'.

Bob Dylan

This chapter examines the heat equation and develops the leapfrog method for solving it on a space–time lattice. We also develop an improved Crank–Nicolson method that determines the solution over all of space in a single step. Time stepping is simple, yet important, and we will see it again when we attack various wave equations.

20.1
Heat Flow via Time-Stepping (Leapfrog)

Problem You are given an aluminum bar of length $L = 1$ m and width w aligned along the x-axis (Figure 20.1). It is insulated along its length but not at its ends. Initially the bar is at a uniform temperature of 100 °C, and then both ends are placed in contact with ice water at 0 °C. Heat flows out of the noninsulated ends only. Your *problem* is to determine how the temperature will vary as we move along the length of the bar at later times.

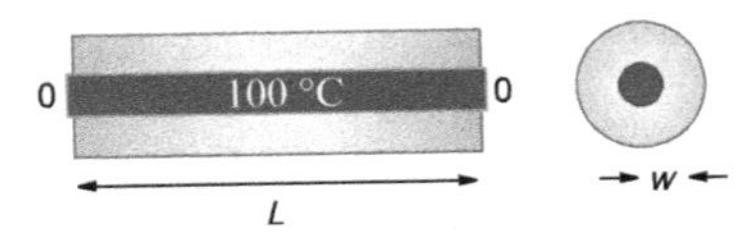

Figure 20.1 A metallic bar insulated along its length with its ends in contact with ice. The bar is displayed in dark gray and the insulation is of lighter gray.

20.2
The Parabolic Heat Equation (Theory)

A basic fact of nature is that heat flows from hot to cold, that is, from regions of high temperature to regions of low temperature. We give these words mathematical expression by stating that the rate of heat flow $\mathbf{H}$ through a material is proportional to the gradient of the temperature T across the material:

$$\mathbf{H} = -K\nabla T(\boldsymbol{x}, t) , \tag{20.1}$$

where K is the thermal conductivity of the material. The total amount of heat $Q(t)$ in the material at any one time is proportional to the integral of the temperature over the material's volume:

$$Q(t) = \int \mathrm{d}\boldsymbol{x} C\rho(\boldsymbol{x}) T(\boldsymbol{x}, t) , \tag{20.2}$$

where C is the specific heat of the material and ρ is its density. Because energy is conserved, the rate of decrease in Q with time must equal the amount of heat flowing out of the material. After this energy balance is struck and the divergence theorem applied, there results the *heat equation*

$$\frac{\partial T(\boldsymbol{x}, t)}{\partial t} = \frac{K}{C\rho} \nabla^2 T(\boldsymbol{x}, t) . \tag{20.3}$$

The heat equation (20.3) is a parabolic PDE with space and time as independent variables. The specification of this problem implies that there is no temperature variation in directions perpendicular to the bar (y and z), and so we have only one spatial coordinate in the Laplacian:

$$\frac{\partial T(x, t)}{\partial t} = \frac{K}{C\rho} \frac{\partial^2 T(x, t)}{\partial x^2} . \tag{20.4}$$

As given, the initial temperature of the bar and the boundary conditions are

$$T(x, t = 0) = 100\,^\circ\mathrm{C} , \quad T(x = 0, t) = T(x = L, t) = 0\,^\circ\mathrm{C} . \tag{20.5}$$

20.2.1
Solution: Analytic Expansion

Analogous to Laplace's equation, the analytic solution starts with the assumption that the solution separates into the product of functions of space and time:

$$T(x, t) = X(x)\mathcal{T}(t) . \tag{20.6}$$

When (20.6) is substituted into the heat equation (20.4) and the resulting equation is divided by $X(x)\mathcal{T}(t)$, two noncoupled ODEs result:

$$\frac{\mathrm{d}^2 X(x)}{\mathrm{d}x^2} + k^2 X(x) = 0 , \quad \frac{\mathrm{d}\mathcal{T}(t)}{\mathrm{d}t} + k^2 \frac{C}{C\rho} \mathcal{T}(t) = 0 , \tag{20.7}$$

where k is a constant still to be determined. The boundary condition that the temperature equals zero at $x = 0$ requires a sine function for X:

$$X(x) = A \sin kx \,. \tag{20.8}$$

The boundary condition that the temperature equals zero at $x = L$ requires the sine function to vanish there:

$$\sin kL = 0 \quad \Rightarrow \quad k = k_n = \frac{n\pi}{L} \,, \quad n = 1, 2, \ldots \tag{20.9}$$

To avoid blow up, the time function must be a decaying exponential with k in the exponent:

$$\mathcal{T}(t) = \mathrm{e}^{-k_n^2 t/C\rho} \quad \Rightarrow \quad T(x,t) = A_n \sin k_n x \mathrm{e}^{-k_n^2 t/C\rho} \,, \tag{20.10}$$

where n can be any integer and A_n is an arbitrary constant. Because (20.4) is a linear equation, the most general solution is a linear superposition of $X_n(x)T_n(t)$ products for all values of n:

$$T(x,t) = \sum_{n=1}^{\infty} A_n \sin k_n x \mathrm{e}^{-k_n^2 t/C\rho} \,. \tag{20.11}$$

The coefficients A_n are determined by the initial condition that at time $t = 0$ the entire bar has temperature $T = 100\,\mathrm{K}$:

$$T(x, t = 0) = 100 \quad \Rightarrow \quad \sum_{n=1}^{\infty} A_n \sin k_n x = 100 \,. \tag{20.12}$$

Projecting the sine functions determines $A_n = 4T_0/n\pi$ for n odd, and so

$$T(x,t) = \sum_{n=1,3,\ldots}^{\infty} \frac{4T_0}{n\pi} \sin k_n x \mathrm{e}^{-k_n^2 K t/(C\rho)} \,. \tag{20.13}$$

20.2.2
Solution: Time Stepping

As we did with Laplace's equation, the numerical solution is based on converting the differential equation to a finite-difference ("difference") equation. We discretize space and time on a lattice (Figure 20.2) and solve for solutions on the lattice sites. The sites along the top with white centers correspond to the known values of the temperature for the initial time, while the sites with white centers along the sides correspond to the fixed temperature along the boundaries. If we *also* knew the temperature for times along the bottom row, then we could use a relaxation algorithm as we did for Laplace's equation. However, with only the top and side rows known, we shall end up with an algorithm that steps forward in time one row at a time, as in the children's game *leapfrog*.

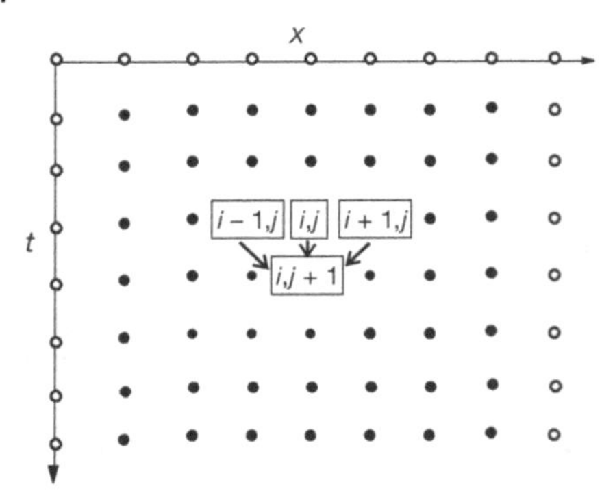

Figure 20.2 The algorithm for the heat equation in which the temperature at the location $x = i\Delta x$ and time $t = (j+1)\Delta t$ is computed from the temperature values at three points of an earlier time. The nodes with white centers correspond to known initial and boundary conditions (the boundaries are placed artificially close for illustrative purposes).

As is often the case with PDEs, the algorithm is customized for the equation being solved and for the constraints imposed by the particular set of initial and boundary conditions. With only one row of times to start with, we use a forward-difference approximation for the time derivative of the temperature:

$$\frac{\partial T(x,t)}{\partial t} \simeq \frac{T(x,t+\Delta t) - T(x,t)}{\Delta t} . \tag{20.14}$$

Because we know the spatial variation of the temperature along the entire top row and the left and right sides, we are less constrained with the space derivative as with the time derivative. Consequently, as we did with the Laplace equation, we use the more accurate central-difference approximation for the space derivative:

$$\frac{\partial^2 T(x,t)}{\partial x^2} \simeq \frac{T(x+\Delta x,t) + T(x-\Delta x,t) - 2T(x,t)}{(\Delta x)^2} . \tag{20.15}$$

Substitution of these approximations into (20.4) yields the heat difference equation

$$\frac{T(x,t+\Delta t) - T(x,t)}{\Delta t} = \frac{K}{C\rho}\frac{T(x+\Delta x,t) + T(x-\Delta x,t) - 2T(x,t)}{\Delta x^2} . \tag{20.16}$$

We reorder (20.16) into a form in which T can be stepped forward in t:

$$T_{i,j+1} = T_{i,j} + \eta\left[T_{i+1,j} + T_{i-1,j} - 2T_{i,j}\right] , \quad \eta = \frac{K\Delta t}{C\rho\Delta x^2} , \tag{20.17}$$

where $x = i\Delta x$ and $t = j\Delta t$. This algorithm is *explicit* because it provides a solution in terms of known values of the temperature. If we tried to solve for the temperature at all lattice sites in Figure 20.2 simultaneously, then we would have an *implicit* algorithm that requires us to solve equations involving unknown values of the temperature. We see that the temperature at space–time point $(i, j+1)$

is computed from the three temperature values at an earlier time j and at adjacent space values $i \pm 1, i$. We start the solution at the top row, moving it forward in time for as long as we want and keeping the temperature along the ends fixed at 0 K (Figure 20.2).

20.2.3 von Neumann Stability Assessment

When we solve a PDE by converting it to a difference equation, we hope that the solution of the latter is a good approximation to the solution of the former. If the difference-equation solution diverges, then we know we have a bad approximation, but if it converges, then we may feel confident that we have a good approximation to the PDE. The *von Neumann stability analysis* is based on the assumption that eigenmodes of the difference equation can be expressed as

$$T_{m,j} = \xi(k)^j e^{ikm\Delta x} , \tag{20.18}$$

where $x = m\Delta x$ and $t = j\Delta t$, and $i = \sqrt{-1}$ is the imaginary number. The constant k in (20.18) is an unknown wave vector $(2\pi/\lambda)$, and $\xi(k)$ is an unknown complex function. View (20.18) as a basis function that oscillates in space (the exponential) with an amplitude or *amplification factor* $\xi(k)^j$ that increases by a power of ξ for each time step. If the general solution to the difference equation can be expanded in terms of these eigenmodes, then the general solution will be stable if the eigenmodes are stable. Clearly, for an eigenmode to be stable, the amplitude ξ cannot grow in time j, which means $|\xi(k)| < 1$ for all values of the parameter k (Press *et al.*, 1994; Ancona, 2002).

Application of a stability analysis is more straightforward than it might appear. We just substitute expression (20.18) into the difference equation (20.17):

$$\begin{aligned} \xi^{j+1} e^{ikm\Delta x} &= \xi^{j+} e^{ikm\Delta x} \\ &+ \eta \left[\xi^j e^{ik(m+1)\Delta x} + \xi^{j+} e^{ik(m-1)\Delta x} - 2\xi^{j+} e^{ikm\Delta x} \right] . \end{aligned} \tag{20.19}$$

After canceling some common factors, it is easy to solve for ξ:

$$\xi(k) = 1 + 2\eta[\cos(k\Delta x) - 1] . \tag{20.20}$$

In order for $|\xi(k)| < 1$ for all possible k values, we must have

$$\eta = \frac{K\Delta t}{C\rho\Delta x^2} < \frac{1}{2} . \tag{20.21}$$

This equation tells us that if we make the time step Δt smaller, we will always improve the stability, as we would expect. But if we decrease the space step Δx without a simultaneous quadratic *increase* in the time step, we will worsen the stability. The lack of space–time symmetry arises from our use of stepping in time, but not in space.

In general, you should perform a stability analysis for every PDE you have to solve, although it can get complicated (Press *et al.*, 1994). Yet even if you do not, the lesson here is that you may have to try different *combinations* of Δx and Δt variations until a stable, reasonable solution is obtained. You may expect, nonetheless, that there are choices for Δx and Δt for which the numerical solution fails and that simply decreasing an individual Δx or Δt, in the hope that this will increase precision, may not improve the solution.

Listing 20.1 EqHeat.py solves the 1D space heat equation on a lattice by leapfrogging (time-stepping) the initial conditions forward in time. You will need to adjust the parameters to obtain a solution like those in the figures.

```
# EqHeat.py: solves heat equation via finite differences, 3D plot

from numpy import *
import matplotlib.pylab as p
from mpl_toolkits.mplot3d import Axes3D

Nx = 101;          Nt = 3000;       Dx = 0.03;        Dt = 0.9
KAPPA = 210.; SPH = 900.; RHO = 2700. # Conductivity, specf heat, density
T = zeros( (Nx, 2), float);  Tpl = zeros( (Nx, 31), float)

print("Working, wait for figure after count to 10")

for ix in range (1, Nx - 1):  T[ix, 0] = 100.0;                  # Initial T
T[0,0] = 0.0 ;    T[0,1] = 0.                                   # 1st & last T = 0
T[Nx-1,0] = 0. ; T[Nx-1,1] = 0.0
cons = KAPPA/(SPH*RHO)*Dt/(Dx*Dx);                                   # constant
m = 1                                                                 # counter

for t in range (1, Nt):
   for ix in range (1, Nx - 1):
      T[ix, 1] = T[ix, 0] +  cons*(T[ix+1, 0] + T[ix-1, 0] - 2.*T[ix,0])
   if t%300 == 0 or t == 1:                                     # Every 300 steps
        for ix in range (1, Nx - 1, 2): Tpl[ix, m] = T[ix, 1]
        print(m)
        m = m + 1
   for ix in range (1, Nx - 1):  T[ix, 0] = T[ix, 1]
x = list(range(1, Nx - 1, 2))                                # Plot alternate pts
y = list(range(1, 30))
X, Y = p.meshgrid(x, y)

def functz(Tpl):
    z = Tpl[X, Y]
    return z

Z = functz(Tpl)
fig = p.figure()                                                # Create figure
ax = Axes3D(fig)
ax.plot_wireframe(X, Y, Z, color = 'r')
ax.set_xlabel('Position')
ax.set_ylabel('time')
ax.set_zlabel('Temperature')
p.show()
print("finished")
```

20.2.4
Heat Equation Implementation

Recall that we want to solve for the temperature distribution within an aluminum bar of length $L = 1$ m subject to the boundary and initial conditions

$$T(x = 0, t) = T(x = L, t) = 0\,\text{K}\,, \quad T(x, t = 0) = 100\,\text{K}\,. \tag{20.22}$$

The thermal conductivity, specific heat, and density for Al are

$$K = 237\,\text{W/(mK)}\,, \quad C = 900\,\text{J/(kg K)}\,, \quad \rho = 2700\,\text{kg/m}^3\,. \tag{20.23}$$

1. Write or modify `EqHeat.py` in Listing 20.1 to solve the heat equation.
2. Define a 2D array `T[101,2]` for the temperature as a function of space and time. The first index is for the 100 space divisions of the bar, and the second index is for present and past times (because you may have to make thousands of time steps, you save memory by saving only two times).
3. For time $t = 0$ ($j = 1$), initialize `T` so that all points on the bar except the ends are at 100 K. Set the temperatures of the ends to 0 K.
4. Apply (20.14) to obtain the temperature at the next time step.
5. Assign the present-time values of the temperature to the past values: `T[i,1] = T[i,2], i = 1,...,101`.
6. Start with 50 time steps. Once you are confident the program is running properly, use thousands of steps to see the bar cool smoothly with time. For approximately every 500 time steps, print the time and temperature along the bar.

20.3
Assessment and Visualization

1. Check that your program gives a temperature distribution that varies smoothly along the bar and agrees with the boundary conditions, as in Figure 20.3.

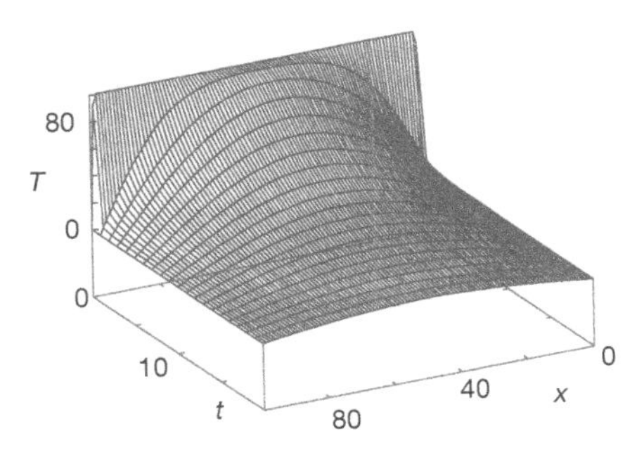

Figure 20.3 A numerical calculation of the temperature vs. position and vs. time, with isotherm contours projected onto the horizontal plane.

2. Check that your program gives a temperature distribution that varies smoothly with time and reaches equilibrium. You may have to vary the time and space steps to obtain stable solutions.
3. Compare the analytic and numeric solutions (and the wall times needed to compute them). If the solutions differ, suspect the one that does not appear smooth and continuous.
4. Make a surface plot of temperature vs. position vs. time.
5. Plot the *isotherms* (contours of constant temperature).
6. *Stability test:* Verify the stability condition (20.21) by observing how the temperature distribution diverges if $\eta > 1/4$.
7. *Material dependence:* Repeat the calculation for iron. Note that the stability condition requires you to change the size of the time step.
8. *Initial sinusoidal distribution* $\sin(\pi x/L)$*:* Compare to the analytic solution,

$$T(x,t) = \sin\left(\frac{\pi x}{L}\right) e^{-\pi^2 K t/(L^2 C\rho)} . \tag{20.24}$$

9. *Two bars in contact:* Two identical bars 0.25 m long are placed in contact along one of their ends with their other ends kept at 0 K. One is kept in a heat bath at 100 K, and the other at 50 K. Determine how the temperature varies with time and location (Figure 20.4).
10. *Radiating bar (Newton's cooling):* Imagine now that instead of being insulated along its length, a bar is in contact with an environment at a temperature T_e. Newton's law of cooling (radiation) says that the rate of temperature change as a result of radiation is

$$\frac{\partial T}{\partial t} = -h(T - T_e) , \tag{20.25}$$

where h is a positive constant. This leads to the modified heat equation

$$\frac{\partial T(x,t)}{\partial t} = \frac{K}{C\rho}\frac{\partial^2 T}{\partial^2 x} - hT(x,t) . \tag{20.26}$$

Modify the algorithm to include Newton's cooling and compare the cooling of a radiating bar with that of the insulated bar.

20.4 Improved Heat Flow: Crank–Nicolson Method

The Crank–Nicolson method (Crank and Nicolson, 1946) provides a higher degree of precision for the heat equation (20.3) than the simple leapfrog method we have just discussed. This method calculates the time derivative with a central-difference approximation, in contrast to the forward-difference approximation used previously. In order to avoid introducing error for the initial time step where

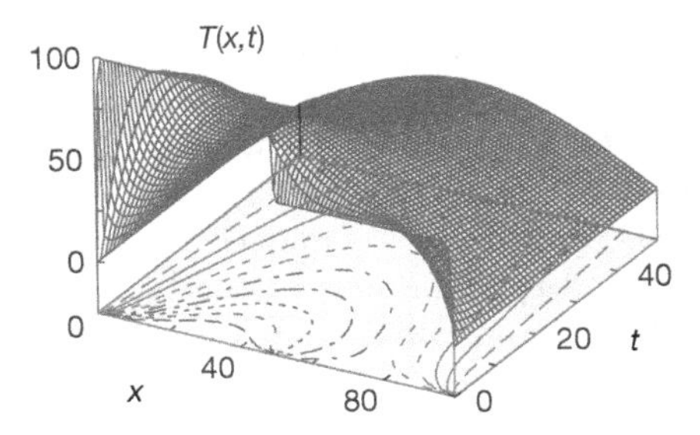

Figure 20.4 Temperature vs. position and time when two bars at differing temperatures are placed in contact at $t = 0$. The projected contours show the isotherms.

only a single time value is known, the method uses a *split time step,*[1] so that time is advanced from time t to $t + \Delta t/2$:

$$\frac{\partial T}{\partial t}\left(x, t + \frac{\Delta t}{2}\right) \simeq \frac{T(x, t+\Delta t) - T(x,t)}{\Delta t} + O(\Delta t^2)\,. \tag{20.27}$$

Yes, we know that this looks just like the forward-difference approximation for the derivative at time $t + \Delta t$, for which it would be a bad approximation; regardless, it is a better approximation for the derivative at time $t + \Delta t/2$, although it makes the computation more complicated. Likewise, in (20.14) we gave the central-difference approximation for the second space derivative for time t. For $t = t + \Delta t/2$, that becomes

$$\begin{aligned} 2(\Delta x)^2 \frac{\partial^2 T}{\partial x^2}\left(x, t + \frac{\Delta t}{2}\right) &\simeq + [T(x - \Delta x, t) - 2T(x,t) + T(x + \Delta x, t)] \\ &+ [T(x - \Delta x, t + \Delta t) - 2T(x, t+\Delta t) + T(x + \Delta x, t + \Delta t)] + O(\Delta x^2)\,. \end{aligned} \tag{20.28}$$

In terms of these expressions, the heat difference equation is

$$\begin{aligned} T_{i,j+1} - T_{i,j} &= \frac{\eta}{2}\left[T_{i-1,j+1} - 2T_{i,j+1} + T_{i+1,j+1} + T_{i-1,j} - 2T_{i,j} + T_{i+1,j}\right] , \\ x &= i\Delta x\,, \quad t = j\Delta t\,, \quad \eta = \frac{K\Delta t}{C\rho\Delta x^2}\,. \end{aligned} \tag{20.29}$$

We group together terms involving the same temperature to obtain an equation with future times on the LHS and present times on the RHS:

$$-T_{i-1,j+1} + \left(\frac{2}{\eta} + 2\right) T_{i,j+1} - T_{i+1,j+1} = T_{i-1,j} + \left(\frac{2}{\eta} - 2\right) T_{i,j} + T_{i+1,j}\,. \tag{20.30}$$

1) In Section 22.2.1 we develop a different split-time algorithm for the solution of the Schrödinger equation. There the real and imaginary parts of the wave function are computed at times that differ by $\Delta t/2$.

This equation represents an *implicit* scheme for the temperature $T_{i,j}$, where the word "implicit" means that we must solve simultaneous equations to obtain the full solution for all space. In contrast, an *explicit* scheme requires iteration to arrive at the solution. It is possible to solve (20.30) simultaneously for all unknown temperatures ($1 \le i \le N$) at times j and $j+1$. We start with the initial temperature distribution throughout all of space, the boundary conditions at the ends of the bar for all times, and the approximate values from the first derivative:

$$\begin{aligned} &T_{i,0}\,, \quad \text{known}\,, && T_{0,j}\,, \quad \text{known}\,, && T_{N,j}\,, \quad \text{known}\,, \\ &T_{0,j+1} = T_{0,j} = 0\,, && T_{N,j+1} = 0\,, && T_{N,j} = 0\,. \end{aligned} \tag{20.31}$$

We rearrange (20.30) so that we can use these known values of T to step the $j = 0$ solution forward in time by expressing (20.30) as a set of simultaneous linear equations (in the matrix form):

$$\begin{pmatrix} \left(\frac{2}{\eta}+2\right) & -1 & & & & \\ -1 & \left(\frac{2}{\eta}+2\right) & -1 & & & \\ & -1 & \left(\frac{2}{\eta}+2\right) & -1 & & \\ & & \ddots & \ddots & \ddots & \\ & & & -1 & \left(\frac{2}{\eta}+2\right) & -1 \\ & & & & -1 & \left(\frac{2}{\eta}+2\right) \end{pmatrix} \begin{pmatrix} T_{1,j+1} \\ T_{2,j+1} \\ T_{3,j+1)} \\ \vdots \\ T_{n-2,j+1} \\ T_{n-1,j+1} \end{pmatrix}$$

$$= \begin{pmatrix} T_{0,j+1} + T_{0,j} + \left(\frac{2}{\eta}-2\right) T_{1,j} + T_{2,j} \\ T_{1,j} + \left(\frac{2}{\eta}-2\right) T_{2,j} + T_{3,j} \\ T_{2,j} + \left(\frac{2}{\eta}-2\right) T_{3,j} + T_{4,j} \\ \vdots \\ T_{n-3,j} + \left(\frac{2}{\eta}-2\right) T_{n-2,j} + T_{n-1,j} \\ T_{n-2,j} + \left(\frac{2}{\eta}-2\right) T_{n-1,j} + T_{n,j} + T_{n,j+1} \end{pmatrix}. \tag{20.32}$$

Observe that the Ts on the RHS are all at the present time j for various positions, and at future time $j+1$ for the two ends (whose Ts are known for all times via the boundary conditions). We start the algorithm with the $T_{i,j=0}$ values of the initial conditions, then solve a matrix equation to obtain $T_{i,j=1}$. With that we know all the terms on the RHS of the equations ($j = 1$ throughout the bar and $j = 2$ at the ends) and so can repeat the solution of the matrix equations to obtain the temperature throughout the bar for $j = 2$. So again, we time-step forward, only now we solve matrix equations at each step. That gives us the spatial solution at all locations directly.

Not only is the Crank–Nicolson method more precise than the low-order time-stepping method, but it also is stable for all values of Δt and Δx. To prove that, we apply the von Neumann stability analysis discussed in Section 20.2.3 to the

Crank–Nicolson algorithm by substituting (20.17) into (20.30). This determines an amplification factor

$$\xi(k) = \frac{1 - 2\eta \sin^2(k\Delta x/2)}{1 + 2\eta \sin^2(k\Delta x/2)} . \tag{20.33}$$

Because $\sin^2()$ is positive definite, this proves that $|\xi| \leq 1$ for all Δt, Δx, and k.

20.4.1 Solution of Tridiagonal Matrix Equations ⊙

The Crank–Nicolson equations (20.32) are in the standard $[\mathbf{A}]\boldsymbol{x} = \boldsymbol{b}$ form for linear equations, and so we can use our previous methods to solve them. Nonetheless, because the coefficient matrix $[\mathbf{A}]$ is tridiagonal (zero elements except for the main diagonal and two diagonals on either side of it),

$$\begin{pmatrix} d_1 & c_1 & 0 & 0 & \cdots & \cdots & \cdots & 0 \\ a_2 & d_2 & c_2 & 0 & \cdots & \cdots & \cdots & 0 \\ 0 & a_3 & d_3 & c_3 & \cdots & \cdots & \cdots & 0 \\ \cdots & \cdots & \cdots & \cdots & \cdots & \cdots & \cdots & \cdots \\ 0 & 0 & 0 & 0 & \cdots & a_{N-1} & d_{N-1} & c_{N-1} \\ 0 & 0 & 0 & 0 & \cdots & 0 & a_N & d_N \end{pmatrix} \begin{pmatrix} x_1 \\ x_2 \\ x_3 \\ \ddots \\ x_{N-1} \\ x_N \end{pmatrix} = \begin{pmatrix} b_1 \\ b_2 \\ b_3 \\ \ddots \\ b_{N-1} \\ b_N \end{pmatrix}, \tag{20.34}$$

a more robust and faster solution exists that makes this implicit method as fast as an explicit one. Because tridiagonal systems occur frequently, we now outline the specialized technique for solving them (Press *et al.*, 1994). If we store the matrix elements $a_{i,j}$ using both subscripts, then we will need N^2 locations for elements and N^2 operations to access them. However, for a tridiagonal matrix, we need to store only the vectors $\{d_i\}_{i=1,N}$, $\{c_i\}_{i=1,N}$, and $\{a_i\}_{i=1,N}$, along, above, and below the diagonals. The single subscripts on a_i, d_i, and c_i reduce the processing from N^2 to $(3N - 2)$ elements.

We solve the matrix equation by manipulating the individual equations until the coefficient matrix is *upper triangular* with all the elements of the main diagonal equal to 1. We start by dividing the first equation by d_1, then subtract a_2 times the first equation,

$$\begin{pmatrix} 1 & \frac{c_1}{d_1} & 0 & 0 & \cdots & \cdots & \cdots & 0 \\ 0 & d_2 - \frac{a_2 c_1}{d_1} & c_2 & 0 & \cdots & \cdots & \cdots & 0 \\ 0 & a_3 & d_3 & c_3 & \cdots & \cdots & \cdots & 0 \\ \cdots & \cdots & \cdots & \cdots & \cdots & \cdots & \cdots & \cdots \\ 0 & 0 & 0 & 0 & \cdots & a_{N-1} & d_{N-1} & c_{N-1} \\ 0 & 0 & 0 & 0 & \cdots & 0 & a_N & d_N \end{pmatrix} \begin{pmatrix} x_1 \\ x_2 \\ x_3 \\ \ddots \\ \cdots \\ x_N \end{pmatrix} = \begin{pmatrix} \frac{b_1}{d_1} \\ b_2 - \frac{a_2 b_1}{d_1} \\ b_3 \\ \ddots \\ \cdots \\ b_N \end{pmatrix}, \tag{20.35}$$

and then dividing the second equation by the second diagonal element,

$$\begin{pmatrix} 1 & \frac{c_1}{d_1} & 0 & 0 & \cdots & \cdots & \cdots & 0 \\ 0 & 1 & \frac{c_2}{d_2 - a_2 \frac{c_1}{a_1}} & 0 & \cdots & \cdots & \cdots & 0 \\ 0 & a_3 & d_3 & c_3 & \cdots & \cdots & \cdots & 0 \\ \cdots & \cdots & \cdots & \cdots & \cdots & \cdots & \cdots & \cdots \\ 0 & 0 & 0 & 0 & \cdots & a_{N-1} & d_{N-1} & c_{N-1} \\ 0 & 0 & 0 & 0 & \cdots & 0 & a_N & d_N \end{pmatrix} \begin{pmatrix} x_1 \\ x_2 \\ x_3 \\ \ddots \\ \cdots \\ x_N \end{pmatrix} = \begin{pmatrix} \frac{b_1}{d_1} \\ \frac{b_2 - a_2 \frac{b_1}{d_1}}{d_2 - a_2 \frac{c_1}{d_1}} \\ b_3 \\ \ddots \\ \cdots \\ b_N \end{pmatrix} . \tag{20.36}$$

Assuming that we can repeat these steps without ever dividing by zero, the system of equations will be reduced to upper triangular form,

$$\begin{pmatrix} 1 & h_1 & 0 & 0 & \cdots & 0 \\ 0 & 1 & h_2 & 0 & \cdots & 0 \\ 0 & 0 & 1 & h_3 & \cdots & 0 \\ 0 & \cdots & \cdots & \ddots & \ddots & \cdots \\ 0 & 0 & 0 & 0 & \cdots & \cdots \\ 0 & 0 & 0 & \cdots & 0 & 1 \end{pmatrix} \begin{pmatrix} x_1 \\ x_2 \\ x_3 \\ \ddots \\ \cdots \\ x_N \end{pmatrix} = \begin{pmatrix} p_1 \\ p_2 \\ p_3 \\ \ddots \\ \cdots \\ p_N \end{pmatrix} , \tag{20.37}$$

where $h_1 = c_1/d_1$ and $p_1 = b_1/d_1$. We then recur for the others elements:

$$h_i = \frac{c_i}{d_i - a_i h_{i-1}} , \quad p_i = \frac{b_i - a_i p_{i-1}}{d_i - a_i h_{i-1}} . \tag{20.38}$$

Finally, back substitution leads to the explicit solution for the unknowns:

$$x_i = p_i - h_i x_{i-1} ; \quad i = n - 1, n - 2, \ldots, 1 , \quad x_N = p_N . \tag{20.39}$$

In Listing 20.2, we give the program HeatCNTridiag.py that solves the heat equation using the Crank–Nicolson algorithm via a triadiagonal reduction.

Listing 20.2 HeatCNTridiag.py is the complete program for solution of the heat equation in one space dimension and time via the Crank–Nicolson method. The resulting matrix equations are solved via a technique specialized to tridiagonal matrices.

```
# HeatCNTridiag.py: solution of heat eqtn via CN method

import matplotlib.pylab as p;
from mpl_toolkits.mplot3d import Axes3D ;
from numpy import *;
import numpy;

Max = 51; n   = 50;    m = 50
Ta  = zeros((Max),float); Tb =zeros((Max),float); Tc = zeros((Max),float)
Td  = zeros((Max),float); a = zeros((Max),float); b = zeros((Max),float)
c   = zeros((Max),float); d = zeros((Max),float); x = zeros((Max),float)
```

```
t   = zeros( (Max, Max),float)

def Tridiag(a, d, c, b, Ta, Td, Tc, Tb, x, n):
    Max = 51
    h = zeros( (Max), float )
    p = zeros( (Max), float )
    for i in range(1,n+1):
        a[i] = Ta[i]
        b[i] = Tb[i]
        c[i] = Tc[i]
        d[i] = Td[i]
    h[1] = c[1]/d[1]
    p[1] = b[1]/d[1]
    for i in range(2,n+1):
        h[i] = c[i] / (d[i]-a[i]*h[i-1])
        p[i] = (b[i] - a[i]*p[i-1]) / (d[i]-a[i]*h[i-1])
    x[n] = p[n]
    for i in range( n - 1, 1,-1 ): x[i] = p[i] - h[i]*x[i+1]

width = 1.0; height = 0.1; ct = 1.0
for i in range(0, n):   t[i,0]  = 0.0
for i in range( 1, m):  t[0][i] = 0.0
h  = width  / ( n - 1 )
k  = height / ( m - 1 )
r  = ct * ct * k / ( h * h )

for j in range(1,m+1):
    t[1,j] = 0.0
    t[n,j] = 0.0                                               # BCs
for i in range( 2, n):   t[i][1] = sin( pi * h *i)            # ICs
for i in range(1, n+1):  Td[i] = 2. + 2./r
Td[1] = 1.; Td[n] = 1.
for i in range(1,n ): Ta[i] = -1.0;       Tc[i] = -1.0;       # Off diagonal
Ta[n-1] = 0.0;    Tc[1] = 0.0; Tb[1] = 0.0; Tb[n] = 0.0
print("I'm working hard, wait for fig while I count to 50")

for j in range(2,m+1):
    print(j)
    for i in range(2,n): Tb[i] = t[i-1][j-1] + t[i+1][j-1] \
            + (2/r-2) * t[i][j-1]
    Tridiag(a, d, c, b, Ta, Td, Tc, Tb, x, n)                  # Solve system
    for i in range(1, n+1):  t[i][j] = x[i]
print("Finished")
x = list(range(1, m+1))                                        # Plot every other x
y = list(range(1, n+1))                                        # every other y
X, Y = p.meshgrid(x,y)

def functz(t):                                                 # Potential
    z = t[X, Y]
    return z

Z = functz(t)
fig = p.figure()
ax = Axes3D(fig)
ax.plot_wireframe(X, Y, Z, color= 'r')
ax.set_xlabel('t')
ax.set_ylabel('x')
ax.set_zlabel('T')
p.show()                                                       # Display figure
```

20.4.2 Crank–Nicolson Implementation, Assessment

1. Write a program using the Crank–Nicolson method to solve for the heat flow in the metal bar of Section 20.1 for at least 100 time steps.
2. Solve the linear system of equations (20.32) using either Matplotlib or the special tridiagonal algorithm.
3. Check the stability of your solution by choosing different values for the time and space steps.
4. Construct a contoured surface plot of temperature vs. position and vs. time.
5. Compare the implicit and explicit algorithms used in this chapter for relative precision and speed. You may assume that a stable answer that uses very small time steps is accurate.

21
Wave Equations I: Strings and Membranes

In this chapter and in the next, we explore the numerical solution of several PDEs that yield waves as solutions. If you have skipped the discussion of the heat equation in Chapter 20, then this chapter will be the first example of how initial conditions are propagated forward in time with a time-stepping or leapfrog algorithm. An important aspect of this chapter is its demonstration that once you have a working algorithm for solving a wave equation, you can include considerably more physics than is possible with analytic treatments. First, we deal with a number of aspects of 1D waves on a string, and then with 2D waves on a membrane. In the next chapter, we look at quantum wave packets and E&M waves. In Chapter 24, we look at shock and solitary waves.

21.1
A Vibrating String

Problem Recall the demonstration from elementary physics in which a string tied down at both ends is plucked "gently" at one location and a pulse is observed to travel along the string. Likewise, if the string has one end free and you shake it just right, a standing-wave pattern is set up in which the nodes remain in place and the antinodes move just up and down. Your *problem* is to develop an accurate model for wave propagation on a string, and to see if you can set up traveling- and standing-wave patterns.[1]

21.2
The Hyperbolic Wave Equation (Theory)

Consider a string of length L tied down at both ends (Figure 21.1a). The string has a constant density ρ per unit length, a constant tension T, no frictional forces acting on it, and a tension that is so high that we may ignore sagging as a result of

1) Some similar but independent studies can also be found in Rawitscher *et al.* (1996).

Computational Physics, 3rd edition. Rubin H. Landau, Manuel J. Páez, Cristian C. Bordeianu.

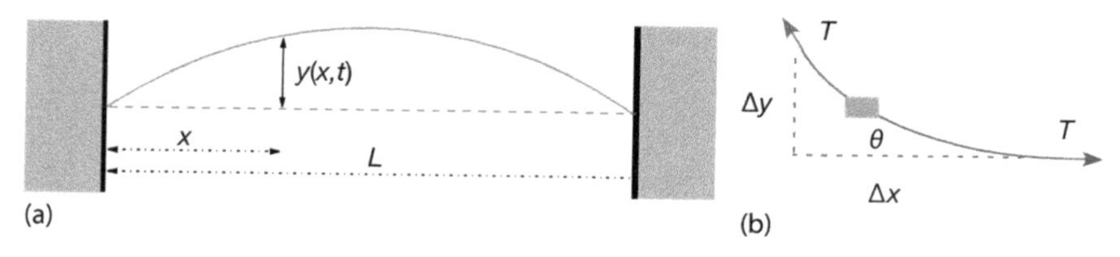

Figure 21.1 (a) A stretched string of length L tied down at both ends and under high enough tension so that we can ignore gravity. The vertical disturbance of the string from its equilibrium position is $y(x, t)$. (b) A differential element of the string showing how the string's displacement leads to the restoring force.

gravity. We assume that the displacement of the string from its rest position $y(x, t)$ is in the vertical direction only and that it is a function of the horizontal location along the string x and the time t.

To obtain a simple linear equation of motion (nonlinear wave equations are discussed in Chapters 24 and 25, we assume that the string's relative displacement $y(x, t)/L$ and slope $\partial y/\partial x$ are small. We isolate an infinitesimal section Δx of the string (Figure 21.1b) and see that the difference in the vertical components of the tension at either end of the string produces the restoring force that accelerates this section of the string in the vertical direction. By applying Newton's laws to this section, we obtain the familiar wave equation:

$$\sum F_y = \rho \Delta x \frac{\partial^2 y}{\partial t^2} , \tag{21.1}$$

$$= T \sin\theta(x + \Delta x) - T \sin\theta(x) \tag{21.2}$$

$$= T \left.\frac{\partial y}{\partial x}\right|_{x+\Delta x} - T \left.\frac{\partial y}{\partial x}\right|_{x} \simeq T \frac{\partial^2 y}{\partial x^2} \tag{21.3}$$

$$\Rightarrow \boxed{\frac{\partial^2 y(x,t)}{\partial x^2} = \frac{1}{c^2}\frac{\partial^2 y(x,t)}{\partial t^2} , \quad c = \sqrt{\frac{T}{\rho}} ,} \tag{21.4}$$

where we have assumed that θ is small enough for $\sin\theta \simeq \tan\theta = \partial y/\partial x$. The existence of two independent variables x and t makes this a PDE. The constant c is the velocity with which a disturbance travels along the wave, and is seen to decrease for a denser string and increase for a tighter one. Note that this signal velocity c is *not* the same as the velocity of a string element $\partial y/\partial t$.

The initial condition for our problem is that the string is plucked gently and released. We assume that the "pluck" places the string in a triangular shape with the center of triangle 8/10 of the way down the string and with a height of 1:

$$y(x, t = 0) = \begin{cases} 1.25x/L , & x \le 0.8L , \\ (5 - 5x/L) , & x > 0.8L , \end{cases} \quad \text{(initial condition 1)} . \tag{21.5}$$

Because (21.4) is second order in time, a second initial condition (beyond initial displacement) is needed to determine the solution. We interpret the "gentleness" of the pluck to mean that the string is released from rest:

$$\frac{\partial y}{\partial t}(x, t = 0) = 0 \quad \text{(initial condition 2)} . \tag{21.6}$$

The boundary conditions have both ends of the string tied down for all times:

$$y(0, t) \equiv 0 , \quad y(L, t) \equiv 0 \quad \text{(boundary conditions)}. \tag{21.7}$$

21.2.1 Solution via Normal-Mode Expansion

The analytic solution to (21.4) is obtained via the familiar separation-of-variables technique. We assume that the solution is the product of a function of space and a function of time:

$$y(x, t) = X(x)T(t) . \tag{21.8}$$

We substitute (21.8) into (21.4), divide by $y(x, t)$, and are left with an equation that has a solution only if there are solutions to the two ODEs:

$$\frac{d^2T(t)}{dt^2} + \omega^2 T(t) = 0 , \quad \frac{d^2X(x)}{dx^2} + k^2 X(x) = 0 , \quad k \stackrel{\text{def}}{=} \frac{\omega}{c} . \tag{21.9}$$

The angular frequency ω and the wave vector k are determined by demanding that the solutions satisfy the boundary conditions. Specifically, the string being attached at both ends demands

$$X(x = 0, t) = X(x = l, t) = 0 \tag{21.10}$$

$$\Rightarrow \quad X_n(x) = A_n \sin k_n x , \quad k_n = \frac{\pi(n+1)}{L} , \quad n = 0, 1, \ldots \tag{21.11}$$

The time solution is

$$T_n(t) = C_n \sin \omega_n t + D_n \cos \omega_n t , \quad \omega_n = nck_0 = n\frac{2\pi c}{L} , \tag{21.12}$$

where the frequency of this nth *normal mode* is also fixed. In fact, it is the single frequency of oscillation that defines a normal mode. The *initial condition* (21.5) of zero velocity, $\partial y/\partial t(t = 0) = 0$, requires the C_n values in (21.12) to be zero. Putting the pieces together, the normal-mode solutions are

$$y_n(x, t) = \sin k_n x \cos \omega_n t , \quad n = 0, 1, \ldots \tag{21.13}$$

Because the wave equation (21.4) is linear in y, the principle of linear superposition holds and the most general solution for waves on a string with fixed ends can be written as the sum of normal modes:

$$y(x, t) = \sum_{n=0}^{\infty} B_n \sin k_n x \cos \omega_n t . \tag{21.14}$$

(Yet we will lose linear superposition once we include nonlinear terms in the wave equation.) The Fourier coefficient B_n is determined by the second initial condition (21.5), which describes how the wave is plucked:

$$y(x, t = 0) = \sum_n^{\infty} B_n \sin nk_0 x . \tag{21.15}$$

We multiply both sides by $\sin mk_0 x$, substitute the value of $y(x, 0)$ from (21.5), and integrate from 0 to l to obtain

$$B_m = 6.25 \frac{\sin(0.8m\pi)}{m^2\pi^2} . \tag{21.16}$$

You will be asked to compare the Fourier series (21.14) to our numerical solution. While it is in the nature of the approximation that the precision of the numerical solution depends on the choice of step sizes, it is also revealing to realize that the precision of the so-called analytic solution depends on summing an infinite number of terms, which can be carried out only approximately.

21.2.2
Algorithm: Time Stepping

As with Laplace's equation and the heat equation, we look for a solution $y(x, t)$ only for discrete values of the independent variables x and t on a grid (Figure 21.2):

$$x = i\Delta x , \quad i = 1, \dots, N_x , \quad t = j\Delta t , \quad j = 1, \dots, N_t , \tag{21.17}$$

$$y(x, t) = y(i\Delta x, i\Delta t) \stackrel{\text{def}}{=} y_{i,j} . \tag{21.18}$$

In contrast to Laplace's equation where the grid was in two space dimensions, the grid in Figure 21.2 is in both space and time. That being the case, moving across a row corresponds to increasing x values along the string for a fixed time, while moving down a column corresponds to increasing time steps for a fixed position. Although the grid in Figure 21.2 may be square, we cannot use a relaxation technique like we did for the solution of Laplace's equation because we do not know the solution on all four sides. The boundary conditions determine the solution along the right- and left-sides, while the initial time condition determines the solution along the top.

As with the Laplace equation, we use the central-difference approximation to *discretize* the wave equation into a difference equation. First, we express the second derivatives in terms of finite differences:

$$\frac{\partial^2 y}{\partial t^2} \simeq \frac{y_{i,j+1} + y_{i,j-1} - 2y_{i,j}}{(\Delta t)^2} , \quad \frac{\partial^2 y}{\partial x^2} \simeq \frac{y_{i+1,j} + y_{i-1,j} - 2y_{i,j}}{(\Delta x)^2} . \tag{21.19}$$

Substituting (21.19) in the wave equation (21.4) yields the difference equation

$$\frac{y_{i,j+1} + y_{i,j-1} - 2y_{i,j}}{c^2(\Delta t)^2} = \frac{y_{i+1,j} + y_{i-1,j} - 2y_{i,j}}{(\Delta x)^2} . \tag{21.20}$$

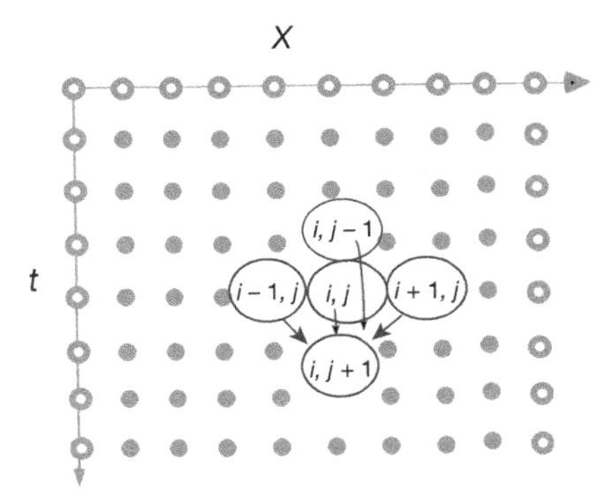

Figure 21.2 The solutions of the wave equation for four earlier space–time points are used to obtain the solution at the present time. The boundary and initial conditions are indicated by the white-centered dots.

Note that this equation contains three time values: $j + 1$ = the future, j = the present, and $j - 1$ = the past. Consequently, we rearrange it into a form that permits us to predict the future solution from the present and past solutions:

$$y_{i,j+1} = 2y_{i,j} - y_{i,j-1} + \frac{c^2}{c'^2}\left[y_{i+1,j} + y_{i-1,j} - 2y_{i,j}\right] , \quad c' \stackrel{\text{def}}{=} \frac{\Delta x}{\Delta t} . \tag{21.21}$$

Here c' is a combination of numerical parameters with the dimension of velocity whose size relative to c determines the stability of the algorithm. The algorithm (21.21) propagates the wave from the two earlier times, j and $j - 1$, and from three nearby positions, $i - 1$, i, and $i + 1$, to a later time $j + 1$ and a single space position i (Figure 21.2).

As you have seen in our discussion of the heat equation, a leapfrog method is quite different from a relaxation technique. We start with the solution along the topmost row and then move down one step at a time. If we write the solution for present times to a file, then we need to store only three time values on the computer, which saves memory. In fact, because the time steps must be quite small to obtain high precision, you may want to store the solution only for every fifth or tenth time.

Initializing the recurrence relation is a bit tricky because it requires displacements from two earlier times, whereas the initial conditions are for only one time. Nonetheless, the rest condition (21.5) when combined with the *central-difference* approximation lets us extrapolate to negative time:

$$\frac{\partial y}{\partial t}(x, 0) \simeq \frac{y(x, \Delta t) - y(x, -\Delta t)}{2\Delta t} = 0 \quad \Rightarrow \quad y_{i,0} = y_{i,2} . \tag{21.22}$$

Here we take the initial time as $j = 1$, and so $j = 0$ corresponds to $t = -\Delta t$. Substituting this relation into (21.21) yields for the initial step

$$y_{i,2} = y_{i,1} + \frac{c^2}{2c'^2}[y_{i+1,1} + y_{i-1,1} - 2y_{i,1}] \quad (\text{for } j = 2 \text{ only}) . \tag{21.23}$$

Equation 21.23 uses the solution throughout all space at the initial time $t = 0$ to propagate (leapfrog) it forward to a time Δt. Subsequent time steps use (21.21) and are continued for as long as you like.

As is also true with the heat equation, the success of the numerical method depends on the relative sizes of the time and space steps. If we apply a von Neumann stability analysis to this problem by substituting $y_{m,j} = \xi^j \exp(ikm\ \Delta x)$, as we did in Section 20.2.3, a complicated equation results. Nonetheless, Press *et al.* (1994) shows that the difference-equation solution will be stable for the general class of transport equations if (Courant *et al.*, 1928)

$$c \leq c' = \frac{\Delta x}{\Delta t} \quad \text{(Courant condition)}\ . \tag{21.24}$$

Equation 21.24 means that the solution gets better with smaller *time* steps but gets worse for smaller space steps (unless you simultaneously make the time step smaller). Having different sensitivities to the time and space steps may appear surprising because the wave equation (21.4) is symmetric in x and t; yet the symmetry is broken by the nonsymmetric initial and boundary conditions.

Exercise Figure out a procedure for solving for the wave equation for all times in just one step. Estimate how much memory would be required.

Exercise Try to figure out a procedure for solving for the wave motion with a relaxation technique. What would you take as your initial guess, and how would you know when the procedure has converged?

21.2.3 Wave Equation Implementation

The program **EqStringAnimate.py** in Listing 21.1 solves the wave equation for a string of length $L = 1$ m with its ends fixed and with the gently plucked initial conditions. Note that our use of $L = 1$ violates our assumption that $y/L \ll 1$ but makes it easy to display the results; you should try $L = 1000$ to be realistic. The values of density and tension are entered as constants, $\rho = 0.01$ kg/m and $T = 40$ N, with the space grid set at 101 points, corresponding to $\Delta = 0.01$ cm.

Listing 21.1 EqStringAnimate.py solves the wave equation via time stepping for a string of length $L = 1$ m with its ends fixed and with the gently plucked initial conditions. You will need to modify this code to include new physics.

```
# EqStringAnimate.py:     Animated leapfrog solution of wave equation

from visual import *

# Set up curve
g = display(width = 600, height = 300, title = 'Vibrating string')
vibst = curve(x = list(range(0, 100)), color = color.yellow)
ball1 = sphere(pos = (100, 0), color = color.red, radius = 2)
ball2 = sphere(pos = ( - 100, 0), color = color.red, radius = 2)
ball1.pos
```

```
ball2.pos
vibst.radius = 1.0

# Parameters
rho    = 0.01
ten    = 40.
c      = sqrt(ten/rho)
c1     = c                                              # CFL criterium
ratio =  c*c/(c1*c1)

# Initialization
xi = zeros((101,3), float)
for i in range(0, 81):      xi[i, 0] = 0.00125*i;
for i in range (81, 101):   xi[i, 0] = 0.1 - 0.005*(i - 80)
for i in range(0, 100):                                 # 1st t step
    vibst.x[i] = 2.0*i - 100.0
    vibst.y[i] = 300.*xi[i, 0]
vibst.pos                                               # Draw string

# Later time steps
for i in range(1, 100): xi[i,1] = xi[i,0] +
    0.5*ratio*(xi[i+1,0]+xi[i-1,0]-2*xi[i,0])
while 1:
    rate(50)                                            # Plotting delay
    for i in range(1, 100):
        xi[i,2] = 2.*xi[i,1] - xi[i,0] + ratio *
              (xi[i+1,1]+xi[i-1,1]-2*xi[i, 1])
    for i in range(1, 100):
         vibst.x[i] = 2.*i - 100.0                      # Scale for plot
         vibst.y[i] = 300.*xi[i, 2]
    vibst.pos
    for i in range(0, 101):
        xi[i, 0] = xi[i, 1]
        xi[i, 1] = xi[i, 2]

print("Done!")
```

21.2.4 Assessment, Exploration

1. Solve the wave equation and make a surface plot of displacement vs. time and position.
2. Explore a number of space and time step combinations. In particular, try steps that satisfy and that do not satisfy the Courant condition (21.24). Does your exploration confirm the stability condition?
3. Compare the analytic and numeric solutions, summing at least 200 terms in the analytic solution.
4. Use the plotted time dependence to estimate the peak's propagation velocity c. Compare the deduced c to (21.4).
5. Our solution of the wave equation for a plucked string leads to the formation of a wave packet that corresponds to the sum of multiple normal modes of the string. On the right in Figure 21.3 we show the motion resulting from the string initially placed in a single normal mode (standing wave),

$$y(x, t = 0) = 0.001 \sin 2\pi x \,, \quad \frac{\partial y}{\partial t}(x, t = 0) = 0 \,, \tag{21.25}$$

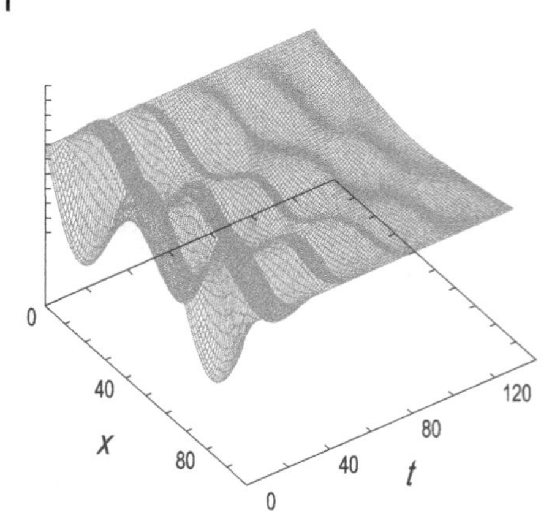

Figure 21.3 The vertical displacement as a function of position x and time t for a string initially placed in a standing mode when friction is included. Notice how the wave moves up and down with time (courtesy of J. Wiren).

but with friction (to be discussed soon) included.

Modify your program to incorporate this initial condition and see if a normal mode results.

6. Observe the motion of the wave for initial conditions corresponding to the sum of two adjacent normal modes. Does beating occur?
7. When a string is plucked near its end, a pulse reflects off the ends and bounces back and forth. Change the initial conditions of the model program to one corresponding to a string plucked exactly in its middle and see if a traveling or a standing wave results.
8. Figure 21.4 shows the wave packets that result as a function of time for initial conditions corresponding to the double pluck. Verify that initial conditions of the form

$$\frac{y(x,t=0)}{0.005} = \begin{cases} 0\,, & 0.0 \le x \le 0.1\,, \\ 10x-1\,, & 0.1 \le x \le 0.2\,, \\ -10x+3\,, & 0.2 \le x \le 0.3\,, \\ 0\,, & 0.3 \le x \le 0.7\,, \\ 10x-7\,, & 0.7 \le x \le 0.8\,, \\ -10x+9\,, & 0.8 \le x \le 0.9\,, \\ 0\,, & 0.9 \le x \le 1.0 \end{cases} \tag{21.26}$$

lead to this type of a repeating pattern. In particular, observe whether the pulses move or just oscillate up and down.

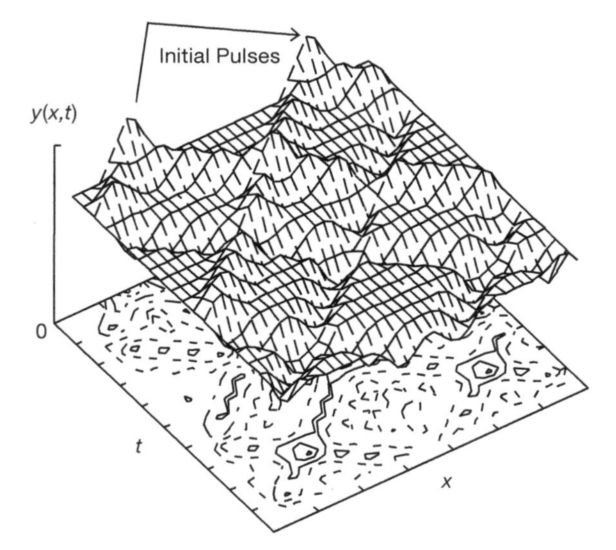

Figure 21.4 The vertical displacement as a function of position and time of a string initially plucked simultaneously at two points, as shown by arrows. Note that each initial peak breaks up into waves traveling to the right and to the left. The traveling waves invert on reflection from the fixed ends. As a consequence of these inversions, the $t \simeq 12$ wave is an inverted $t = 0$ wave.

21.3 Strings with Friction (Extension)

The string problem we have investigated so far can be handled by either a numerical or an analytic technique. We now wish to extend the theory to include some more realistic physics. These extensions have only numerical solutions.

Plucked strings do not vibrate forever because there is friction in the real world. Consider again the element of a string between x and $x + \mathrm{d}x$ (Figure 21.1b) but now imagine that this element is moving in a viscous fluid such as air. An approximate model for the frictional force has it pointing in a direction opposite the vertical velocity of the string and proportional to that velocity, as well as proportional to the length of the string element:

$$F_{\mathrm{f}} \simeq -2\kappa \Delta x \frac{\partial y}{\partial t} , \tag{21.27}$$

where κ is a constant that is proportional to the viscosity of the medium in which the string is vibrating. Including this force in the equation of motion changes the wave equation to

$$\frac{\partial^2 y}{\partial t^2} = c^2 \frac{\partial^2 y}{\partial x^2} - \frac{2\kappa}{\rho} \frac{\partial y}{\partial t} . \tag{21.28}$$

In Figure 21.3, we show the resulting motion of a string plucked in the middle when friction is included. Observe how the initial pluck breaks up into waves traveling to the right and to the left that are reflected and inverted by the fixed ends. Because those parts of the wave with the higher velocity experience greater friction, the peak tends to be smoothed out the most as time progresses.

Exercise Generalize the algorithm used to solve the wave equation now to include friction and check if the wave's behavior seems physical (damps in time). Start with $T = 40\,\text{N}$ and $\rho = 10\,\text{kg/m}$, and pick a value of κ large enough to cause a noticeable effect but not so large as to stop the oscillations. As a check, reverse the sign of κ and see if the wave grows in time (which would eventually violate our assumption of small oscillations).

21.4
Strings with Variable Tension and Density

We have derived the propagation velocity for waves on a string as $c = \sqrt{T/\rho}$. This says that waves move slower in regions of high density and faster in regions of high tension. If the density of the string varies, for instance, by having the ends thicker in order to support the weight of the middle, then c will no longer be a constant and our wave equation will need to be extended. In addition, if the density increases, so will the tension because it takes greater tension to accelerate a greater mass. If gravity acts, then we will also expect the tension at the ends of the string to be higher than in the middle because the ends must support the entire weight of the string.

To derive the equation for wave motion with variable density and tension, consider again the element of a string (Figure 21.1b) used in our derivation of the wave equation. If we do not assume the tension T is constant, then Newton's second law gives

$$F = ma \tag{21.29}$$

$$\Rightarrow \quad \frac{\partial}{\partial x}\left[T(x)\frac{\partial y(x,t)}{\partial x}\right]\Delta x = \rho(x)\Delta x\frac{\partial^2 u(x,t)}{\partial t^2} \tag{21.30}$$

$$\Rightarrow \quad \frac{\partial T(x)}{\partial x}\frac{\partial y(x,t)}{\partial x} + T(x)\frac{\partial^2 y(x,t)}{\partial x^2} = \rho(x)\frac{\partial^2 y(x,t)}{\partial t^2}\,. \tag{21.31}$$

If $\rho(x)$ and $T(x)$ are known functions, then these equations can be solved with just a small modification of our algorithm.

In Section 21.4.1, we will solve for the tension in a string as a result of gravity. Readers interested in an *alternate easier problem* that still shows the new physics may assume that the density and tension are proportional:

$$\rho(x) = \rho_0 e^{\alpha x}\,, \quad T(x) = T_0 e^{\alpha x}\,. \tag{21.32}$$

While we would expect the tension to be greater in regions of higher density (more mass to move and support), being proportional is clearly just an approximation. Substitution of these relations into (21.31) yields the new wave equation:

$$\frac{\partial^2 y(x,t)}{\partial x^2} + \alpha \frac{\partial y(x,t)}{\partial x} = \frac{1}{c^2}\frac{\partial^2 y(x,t)}{\partial t^2}, \quad c^2 = \frac{T_0}{\rho_0}. \tag{21.33}$$

Here c is a constant that would be the wave velocity if $\alpha = 0$. This equation is similar to the wave equation with friction; only now the first derivative is with respect to x and not t. The corresponding difference equation follows from using central-difference approximations for the derivatives:

$$\begin{aligned} y_{i,j+1} &= 2y_{i,j} - y_{i,j-1} + \frac{\alpha c^2 (\Delta t)^2}{2\Delta x}[y_{i+1,j} - y_{i,j}] \\ &\quad + \frac{c^2}{c'^2}[y_{i+1,j} + y_{i-1,j} - 2y_{i,j}], \\ y_{i,2} &= y_{i,1} + \frac{c^2}{c'^2}[y_{i+1,1} + y_{i-1,1} - 2y_{i,1}] + \frac{\alpha c^2(\Delta t)^2}{2\Delta x}[y_{i+1,1} - y_{i,1}]. \end{aligned} \tag{21.34}$$

21.4.1 Waves on Catenary

Up until this point we have been ignoring the effect of gravity upon our string's shape and tension. This is a good approximation if there is very little sag in the string, as might happen if the tension is very high and the string is light. Even if there is some sag, our solution for $y(x,t)$ could still be used as the disturbance about the equilibrium shape. However, if the string is massive, say, like a chain or heavy cable, then the sag in the middle caused by gravity could be quite large (Figure 21.5), and the resulting variation in shape and tension needs to be incorporated into the wave equation. Because the tension is no longer uniform, waves travel faster near the ends of the string, which are under greater tension because they must support the entire weight of the string.

21.4.2 Derivation of Catenary Shape

Consider a string of uniform density ρ acted upon by gravity. To avoid confusion with our use of $y(x)$ to describe a disturbance on a string, we call $u(x)$ the equilibrium shape of the string (Figure 21.5). The statics problem we need to solve is to determine the shape $u(x)$ and the tension $T(x)$. The inset in Figure 21.5 is a free-body diagram of the midpoint of the string and shows that the weight W of this section of arc length s is balanced by the vertical component of the tension T. The horizonal tension T_0 is balanced by the horizontal component of T:

$$T(x)\sin\theta = W = \rho g s, \quad T(x)\cos\theta = T_0, \tag{21.35}$$

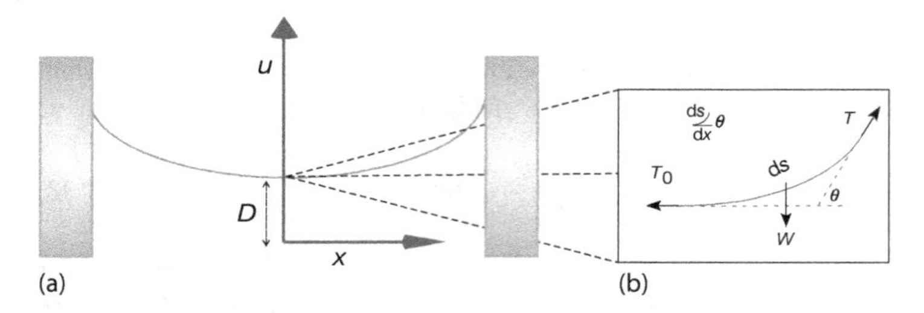

Figure 21.5 (a) A uniform string suspended from its ends in a gravitational field assumes a catenary shape. (b) A force diagram of a section of the catenary at its lowest point. Because the ends of the string must support the entire weight of the string, the tension now varies along the string.

$$\Rightarrow \quad \tan\theta = \frac{\rho g s}{T_0} \,. \tag{21.36}$$

The trick is to convert (21.36) to a differential equation that we can solve. We do that by replacing the slope $\tan\theta$ by the derivative $\mathrm{d}u/\mathrm{d}x$, and taking the derivative with respect to x:

$$\frac{\mathrm{d}u}{\mathrm{d}x} = \frac{\rho g}{T_0} s \,, \quad \Rightarrow \quad \frac{\mathrm{d}^2 u}{\mathrm{d}x^2} = \frac{\rho g}{T_0}\frac{\mathrm{d}s}{\mathrm{d}x} \,. \tag{21.37}$$

Yet because $\mathrm{d}s = \sqrt{\mathrm{d}x^2 + \mathrm{d}u^2}$, we have our differential equation

$$\frac{\mathrm{d}^2 u}{\mathrm{d}x^2} = \frac{1}{D}\frac{\sqrt{\mathrm{d}x^2 + \mathrm{d}u^2}}{\mathrm{d}x} = \frac{1}{D}\sqrt{1 + \left(\frac{\mathrm{d}u}{\mathrm{d}x}\right)^2} \,, \tag{21.38}$$

$$D = \frac{T_0}{\rho g} \,, \tag{21.39}$$

where D is a combination of constants with the dimension of length. Equation 21.38 is the equation for the *catenary* and has the solution (Becker, 1954)

$$u(x) = D\cosh\frac{x}{D} \,. \tag{21.40}$$

Here we have chosen the x-axis to lie a distance D below the bottom of the catenary (Figure 21.5) so that $x = 0$ is at the center of the string where $y = D$ and $T = T_0$. Equation 21.37 tells us the arc length $s = D\,\mathrm{d}u/\mathrm{d}x$, so we can solve for $s(x)$ and for the tension $T(x)$ via (21.35):

$$s(x) = D\sinh\frac{x}{D} \quad \Rightarrow \quad T(x) = T_0\frac{\mathrm{d}s}{\mathrm{d}x} = \rho g u(x) = T_0\cosh\frac{x}{D} \,. \tag{21.41}$$

It is this variation in tension that causes the wave velocity to change for different positions on the string.

21.4.3 Catenary and Frictional Wave Exercises

We have given you the program `EqStringAnimate.py` (Listing 21.1) that solves the wave equation. Modify it to produce waves on a catenary including friction, or for the assumed density and tension given by (21.32) with $\alpha = 0.5$, $T_0 = 40\,\text{N}$, and $\rho_0 = 0.01\,\text{kg/m}$. (The instructor's site contains the programs `CatFriction.py` and `CatString.py` that do this.)

1. Look for some interesting cases and create surface plots of the results.
2. Describe in words how the waves dampen and how a wave's velocity appears to change.
3. *Normal modes:* Search for normal-mode solutions of the variable-tension wave equation, that is, solutions that vary as

   $$u(x,t) = A\cos(\omega t)\sin(\gamma x)\,. \tag{21.42}$$

   Try using this form to start your program and see if you can find standing waves. Use large values for ω.
4. When conducting physics demonstrations, we set up standing-wave patterns by driving one end of the string periodically. Try doing the same with your program; that is, build into your code the condition that for all times

   $$y(x=0,t) = A\sin\omega t\,. \tag{21.43}$$

   Try to vary A and ω until a normal mode (standing wave) is obtained.
5. (For the exponential density case.) If you were able to find standing waves, then verify that this string acts like a high-frequency filter, that is, there is a frequency below which no waves occur.
6. For the catenary problem, plot your results showing *both* the disturbance $u(x,t)$ about the catenary and the actual height $y(x,t)$ above the horizontal for a plucked string initial condition.
7. Try the first two normal modes for a uniform string as the initial conditions for the catenary. These should be close to, but not exactly, normal modes.
8. We derived the normal modes for a uniform string after assuming that $k(x) = \omega/c(x)$ is a constant. For a catenary without too much x variation in the tension, we should be able to make the approximation

   $$c(x)^2 \simeq \frac{T(x)}{\rho} = \frac{T_0\cosh(x/d)}{\rho}\,. \tag{21.44}$$

   See if you get a better representation of the first two normal modes if you include some x dependence in k.

21.5
Vibrating Membrane (2D Waves)

Problem An elastic membrane is stretched across the top of a square box of sides π and attached securely. The tension per unit length in the membrane is T. Initially, the membrane is placed in the asymmetrical shape

$$u(x, y, t = 0) = \sin 2x \sin y\,, \quad 0 \leq x \leq \pi\,, \quad 0 \leq y \leq \pi\,, \tag{21.45}$$

where u is the vertical displacement from equilibrium. Your *problem* is to describe the motion of the membrane when it is released from rest (Kreyszig, 1998).

The description of wave motion on a membrane is basically the same as that of 1D waves on a string discussed in Section 21.2, only now we have wave propagation in two directions. Consider Figure 21.6 showing a square section of the membrane under tension T. The membrane moves only vertically in the z direction, yet because the restoring force arising from the tension in the membrane varies in both the x and y directions, there is wave motion along the surface of the membrane.

Although the tension is constant over the small area in Figure 21.6, there will be a net vertical force on the segment if the angle of incline of the membrane varies as we move through space. Accordingly, the net force on the membrane in the z direction as a result of the change in y is

$$\sum F_z(x) = T\Delta x \sin\theta - T\Delta x \sin\phi\,, \tag{21.46}$$

where θ is the angle of incline at $y + \Delta y$ and ϕ is the angle at y. Yet if we assume that the displacements and the angles are small, then we can make the approximations:

$$\sin\theta \approx \tan\theta = \left.\frac{\partial u}{\partial y}\right|_{y+\Delta y}\,, \quad \sin\phi \approx \tan\phi = \left.\frac{\partial u}{\partial y}\right|_{y}\,, \tag{21.47}$$

$$\Rightarrow \quad \sum F_z(x_{\text{fixed}}) = T\Delta x \left(\left.\frac{\partial u}{\partial y}\right|_{y+\Delta y} - \left.\frac{\partial u}{\partial y}\right|_{y} \right) \approx T\Delta x \frac{\partial^2 u}{\partial y^2}\Delta y\,. \tag{21.48}$$

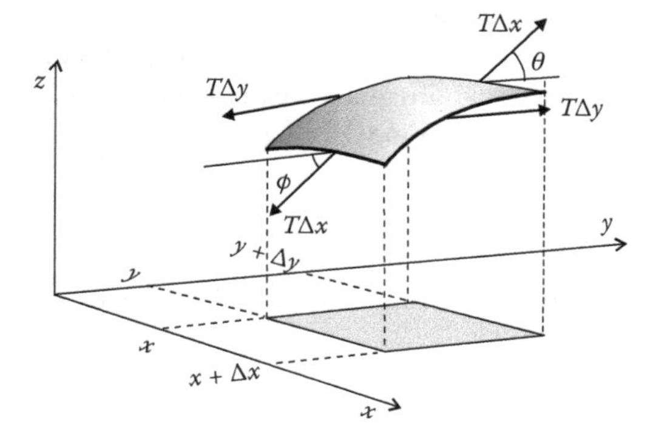

Figure 21.6 A small part of an oscillating membrane and the forces that act on it.

Similarly, the net force in the z direction as a result of the variation in y is

$$\sum F_z(y_{\text{fixed}}) = T\Delta y\left(\left.\frac{\partial u}{\partial x}\right|_{x+\Delta x} - \left.\frac{\partial u}{\partial x}\right|_{x}\right) \approx T\Delta y\frac{\partial^2 u}{\partial x^2}\Delta x \,. \tag{21.49}$$

The membrane section has mass $\rho\Delta x\Delta y$, where ρ is the membrane's mass per unit area. We now apply Newton's second law to determine the acceleration of the membrane section in the z direction as a result of the sum of the net forces arising from both the x and y directions:

$$\rho\Delta x\Delta y\frac{\partial^2 u}{\partial t^2} = T\Delta x\frac{\partial^2 u}{\partial y^2}\Delta y + T\Delta y\frac{\partial^2 u}{\partial x^2}\Delta x \,, \tag{21.50}$$

$$\Rightarrow \quad \boxed{\frac{1}{c^2}\frac{\partial^2 u}{\partial t^2} = \frac{\partial^2 u}{\partial x^2} + \frac{\partial^2 u}{\partial y^2}\,,} \quad c = \sqrt{T/\rho}\,. \tag{21.51}$$

This is the 2D version of the wave equation (21.4) that we studied previously in one dimension. Here c, the propagation velocity, is still the square root of tension over density; only now it is tension per unit length and mass per unit area.

21.6 Analytical Solution

The analytic or numerical solution of the partial differential equation (21.51) requires us to know both the boundary and the initial conditions. The boundary conditions hold for all times and were given when we were told that the membrane is attached securely to a square box of side π:

$$u(x=0, y, t) = u(x=\pi, y, t) = 0\,, \tag{21.52}$$

$$u(x, y=0, t) = u(x, y=\pi, t) = 0\,. \tag{21.53}$$

As required for a second-order equation, the initial conditions has two parts, the shape of the membrane at time $t = 0$, and the velocity of each point of the membrane. The initial configuration is

$$u(x, y, t=0) = \sin 2x \sin y\,, \quad 0 \le x \le \pi\,, \quad 0 \le y \le \pi\,. \tag{21.54}$$

Second, we are told that the membrane is released from rest, which means

$$\left.\frac{\partial u}{\partial t}\right|_{t=0} = 0\,, \tag{21.55}$$

where we write partial derivative because there are also spatial variations.

The analytic solution is based on the guess that because the wave equation (21.51) has separate derivatives with respect to each coordinate and time, the full solution $u(x, y, t)$ is the product of separate functions of x, y, and t:

$$u(x, y, t) = X(x)Y(y)T(t)\,. \tag{21.56}$$

After substituting this into (21.51) and dividing by $X(x)Y(y)T(t)$, we obtain

$$\frac{1}{c^2}\frac{1}{T(t)}\frac{d^2T(t)}{dt^2} = \frac{1}{X(x)}\frac{d^2X(x)}{dx^2} + \frac{1}{Y(y)}\frac{d^2Y(y)}{dy^2} . \tag{21.57}$$

The only way that the LHS of (21.57) can be true for all time while the RHS is also true for all coordinates, is if both sides are constant:

$$\frac{1}{c^2}\frac{1}{T(t)}\frac{d^2T(t)}{dt^2} = -\xi^2 = \frac{1}{X(x)}\frac{d^2X(x)}{dx^2} + \frac{1}{Y(y)}\frac{d^2Y(y)}{dy^2} \tag{21.58}$$

$$\Rightarrow \quad \frac{1}{X(x)}\frac{d^2X(x)}{dx^2} = -k^2 , \tag{21.59}$$

$$\frac{1}{Y(y)}\frac{d^2Y(y)}{dy^2} = -q^2 , \quad (q^2 = \xi^2 - k^2) . \tag{21.60}$$

In (21.59) and (21.60), we have included further deduction that because each term on the RHS of (21.58) depends on either x or y, then the only way for their sum to be constant is if each term is a constant, in this case $-k^2$. The solutions of these equations are standing waves in the x and y directions, which of course are all sinusoidal function,

$$X(x) = A\sin kx + B\cos kx , \tag{21.61}$$

$$Y(y) = C\sin qy + D\cos qy , \tag{21.62}$$

$$T(t) = E\sin c\xi t + F\cos c\xi t . \tag{21.63}$$

We now apply the boundary conditions:

$$\begin{aligned} &u(x=0,y,t) = u(x=\pi,y,z) = 0 \quad \Rightarrow \quad B=0 , \quad k = 1,2,\dots , \\ &u(x,y=0,t) = u(x,y=\pi,t) = 0 \quad \Rightarrow \quad D=0 , \quad q = 1,2,\dots , \\ &\Rightarrow \quad X(x) = A\sin kx , \quad Y(y) = C\sin qy . \end{aligned} \tag{21.64}$$

The fixed values for the eigenvalues m and n describing the modes for the x and y standing waves are equivalent to fixed values for the constants q^2 and k^2. Yet because $q^2 + k^2 = \xi^2$, we must also have a fixed value for ξ^2:

$$\xi^2 = q^2 + k^2 \quad \Rightarrow \quad \xi_{kq} = \pi\sqrt{k^2+q^2} . \tag{21.65}$$

The full space–time solution now takes the form

$$u_{kq} = \left[G_{kq}\cos c\xi t + H_{kq}\sin c\xi t\right]\sin kx \sin qy , \tag{21.66}$$

where k and q are integers. Because the wave equation is linear in u, its most general solution is a linear combination of the eigenmodes (21.66):

$$u(x,y,t) = \sum_{k=1}^{\infty}\sum_{q=1}^{\infty}\left[G_{kq}\cos c\xi t + H_{kq}\sin c\xi t\right]\sin kx \sin qy . \tag{21.67}$$

While an infinite series is not a good algorithm, the initial and boundary conditions means that only the $k = 2, q = 1$ term contributes, and we have a closed form solution:

$$u(x, y, t) = \cos c\sqrt{5} \sin 2x \sin y \;, \tag{21.68}$$

where c is the wave velocity. You should verify that initial and boundary conditions are indeed satisfied.

Listing 21.2 Waves2D.py solves the wave equation numerically for a vibrating membrane.

```
# Waves2D.py:  Solve Helmholtz equation for rectangular vibrating membrane

import matplotlib.pylab as p; from numpy import *
from mpl_toolkits.mplot3d import Axes3D

#
tim = 15;    N = 71
c = sqrt(180./390)                          # Speed = sqrt(ten[]/den[kg/m2;])
u = zeros((N,N,N),float);        v = zeros((N,N),float)
incrx = pi/N;                  incry = pi/N
cprime = c;
covercp = c/cprime;        ratio = 0.5*covercp*covercp   # c/c' 0.5 for stable

def vibration(tim):
    y = 0.0
    for j in range(0,N):                                    # Initial position
        x = 0.0
        for i in range(0,N):
            u[i][j][0] = 3*sin(2.0*x)*sin(y)                  # Initial shape
            x += incrx
        y += incry

    for j in range(1,N-1):                                  # First time step
        for i in range(1,N-1):
            u[i][j][1] = u[i][j][0] + 0.5*ratio*(u[i+1][j][0]+u[i-1][j][0]
                 + u[i][j+1][0]+u[i][j-1][0]-4.*u[i][j][0])

    for k in range(1,tim):                                  # Later time steps
        for j in range(1,N-1):
            for i in range(1,N-1):
                u[i][j][2] = 2.*u[i][j][1] - u[i][j][0] + ratio*(u[i+1][j][1]
                  + u[i-1][j][1] +u[i][j+1][1]+u[i][j-1][1] - 4.*u[i][j][1])
        u[:][:][0] = u[:][:][1]                                   # Reset past
        u[:][:][1] = u[:][:][2]                                # Reset present
        for j in range(0,N):
            for i in range(0,N):
                v[i][j] = u[i][j][2]             # Convert to 2D for matplotlib
    return v

v = vibration(tim)
x1 = range(0, N)
y1 = range(0, N)
X, Y = p.meshgrid(x1,y1)

def functz(v):
    z = v[X,Y]; return z

Z = functz(v)
fig = p.figure()
ax = Axes3D(fig)
ax.plot_wireframe(X, Y, Z, color = 'r')
ax.set_xlabel('x')
```

```
ax.set_ylabel('y')
ax.set_zlabel('u(x,y)')
p.show()
```

21.7
Numerical Solution for 2D Waves

The development of an algorithm for the solution of the 2D wave equation (21.51) follows that of the 1D equation in Section 21.2.2. We start by expressing the second derivatives in terms of central differences:

$$\frac{\partial^2 u(x,y,t)}{\partial t^2} = \frac{u(x,y,t+\Delta t) + u(x,y,t-\Delta t) - 2u(x,y,t)}{(\Delta t)^2}, \quad (21.69)$$

$$\frac{\partial^2 u(x,y,t)}{\partial x^2} = \frac{u(x+\Delta x,y,t) + u(x-\Delta x,y,t) - 2u(x,y,t)}{(\Delta x)^2}, \quad (21.70)$$

$$\frac{\partial^2 u(x,y,t)}{\partial y^2} = \frac{u(x,y+\Delta y,t) + u(x,y-\Delta y,t) - 2u(x,y,t)}{(\Delta y)^2}. \quad (21.71)$$

After discretizing the variables, $u(x = i\Delta, y = i\Delta y, t = k\Delta t) \equiv u^k_{i,j}$, we obtain our time-stepping algorithm by solving for the future solution in terms of the present and past ones:

$$\boxed{u^{k+1}_{i,j} = 2u^k_{i,j} - u^{k-1}_{i,j}\frac{c^2}{c'^2}\left[u^k_{i+1,j} + u^k_{i-1,j} - 4u^k_{i,j} + u^k_{i,j+1} + u^k_{i,j-1}\right],} \quad (21.72)$$

where as before $c' \stackrel{\text{def}}{=} \Delta x/\Delta t$. Whereas the present ($k$) and past ($k-1$) solutions are known after the first step, to initiate the algorithm we need to know the solution at $t = -\Delta t$, that is, before the initial time. To find that, we use the fact that the membrane is released from rest:

$$0 = \frac{\partial u(t=0)}{\partial t} \approx \frac{u^1_{i,j} - u^{-1}_{i,j}}{2\Delta t} \quad \Rightarrow \quad u^{-1}_{i,j} = u^1_{i,j}. \quad (21.73)$$

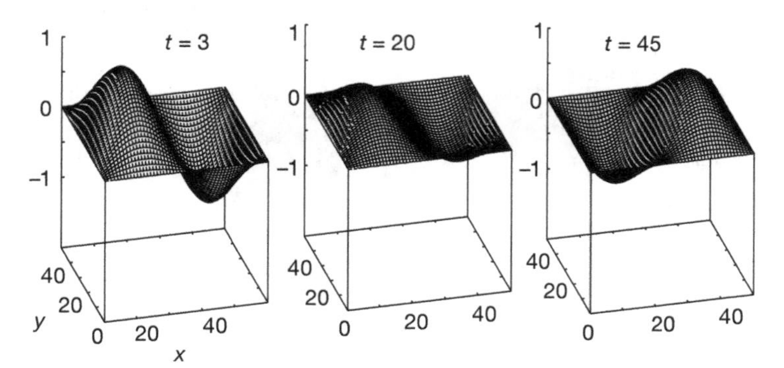

Figure 21.7 The standing wave pattern on a square box top at three different times.

After substituting (21.73) into (21.72) and solving for u^1, we obtain the algorithm for the first step:

$$u^1_{i,j} = u^0_{i,j} + \frac{c^2}{2c'^2}\left[u^0_{i+1,j} + u^0_{i-1,j} - 4u^0_{i,j} + u^0_{i,j+1} + u^0_{i,j-1}\right] . \tag{21.74}$$

Because the displacement $u^0_{i,j}$ is known at time $t = 0$ ($k = 0$), we compute the solution for the first time step with (21.74) and for subsequent steps with (21.72).

The program `Wave2D.py` in Listing 21.2 solves the 2D wave equation using the time-stepping (leapfrog) algorithm. The program `Waves2Danal.py` computes the analytic solution. The shape of the membrane at three different times are shown in Figure 21.7.

22
Wave Equations II: Quantum Packets and Electromagnetic

This chapter continues the discussion of the numerical solution of wave equations begun in Chapter 21, now to equations that require algorithms with a bit more sophistication. First, we explore quantum wave packets, which have their real and imaginary parts solved for at slightly differing (split) times. Then, we explore electromagnetic waves, which have the extra complication of being vector waves with interconnected E and H fields, which also get solved for at split times.

22.1
Quantum Wave Packets

Problem An experiment places an electron with a definite momentum and position in a 1D region of space the size of an atom. It is confined to that region by some kind of attractive potential. Your *problem* is to determine the resultant electron behavior in time and space.

22.2
Time-Dependent Schrödinger Equation (Theory)

Because the region of confinement is the size of an atom, we must solve this problem quantum mechanically. Because the particle has both a definite momentum and position, it is best described as a wave packet, which implies that we must now solve the time-dependent Schrödinger equation for both the spatial and time dependence of the wave packet. Accordingly, this is a different problem from the bound state one of a particle confined to a box, considered in Chapters 7 and 9, where we solved the eigenvalue problem for stationary states of the time-independent Schrödinger equation.

We model an electron initially localized in space at $x = 5$ with momentum k_0 ($\hbar = 1$ in our units) by a wave function that is a wave packet consisting of a Gaus-

Computational Physics, 3rd edition. Rubin H. Landau, Manuel J. Páez, Cristian C. Bordeianu.

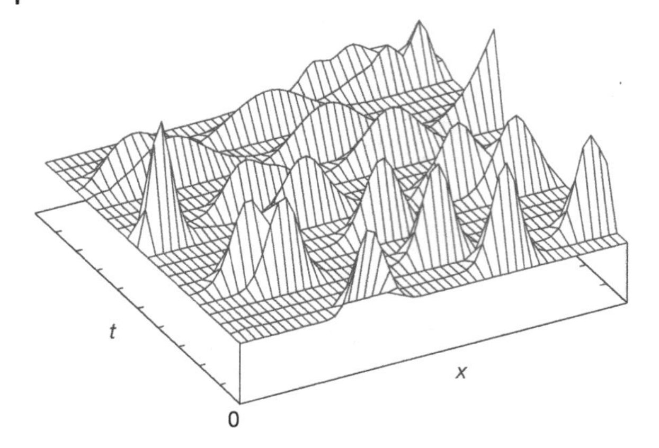

Figure 22.1 The position as a function of time of a localized electron confined to a square well (computed with the code `SqWell.py` available in the instructor's site). The electron is initially on the left with a Gaussian wave packet. In time, the wave packet spreads out and collides with the walls.

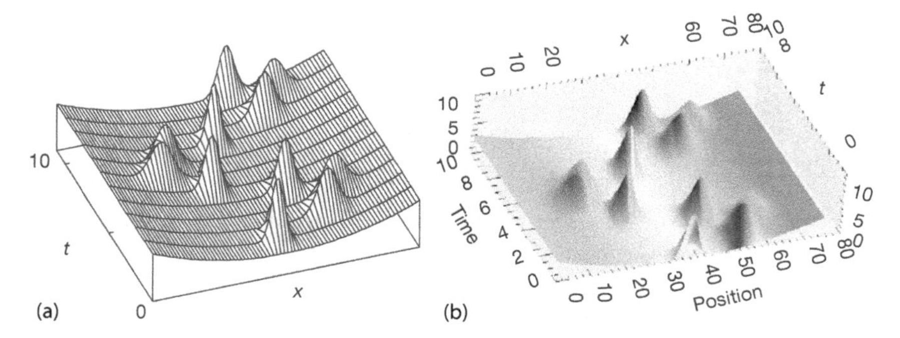

Figure 22.2 The probability density as a function of time for an electron confined to a 1D harmonic oscillator potential well. (a) A conventional surface plot, (b) a color visualization.

sian multiplying a plane wave:

$$\psi(x, t = 0) = \exp\left[-\frac{1}{2}\left(\frac{x-5}{\sigma_0}\right)^2\right] e^{ik_0 x} . \tag{22.1}$$

To solve the *problem*, we must determine the wave function for all later times. If (22.1) was an eigenstate of the Hamiltonian, its $\exp(-i\omega t)$ time dependence can be factored out of the Schrödinger equation (as is usually carried out in textbooks). However, $\tilde{H}\psi \neq E\psi$ for this ψ, and so we must solve the full time-dependent Schrödinger equation. To show you where we are going, the resulting wave packet behavior is shown in Figures 22.1 and 22.2.

The time and space evolution of a quantum particle is described by the 1D time-dependent Schrödinger equation,

$$i\frac{\partial\psi(x,t)}{\partial t} = \tilde{H}\psi(x,t) , \tag{22.2}$$

$$i\frac{\partial\psi(x,t)}{\partial t} = -\frac{1}{2m}\frac{\partial^2\psi(x,t)}{\partial x^2} + V(x)\psi(x,t) , \tag{22.3}$$

where we have set $2m = 1$ to keep the equations simple. Because the initial wave function is complex (in order to have a definite momentum associated with it), the wave function will be complex for all times. Accordingly, we decompose the wave function into its real and imaginary parts:

$$\psi(x,t) = R(x,t) + iI(x,t) , \tag{22.4}$$

$$\psi(x,t) = R(x,t) + iI(x,t) , \tag{22.5}$$

$$\Rightarrow \quad \frac{\partial R(x,t)}{\partial t} = -\frac{1}{2m}\frac{\partial^2 I(x,t)}{\partial x^2} + V(x)I(x,t) , \tag{22.6}$$

$$\frac{\partial I(x,t)}{\partial t} = +\frac{1}{2m}\frac{\partial^2 R(x,t)}{\partial x^2} - V(x)R(x,t) , \tag{22.7}$$

where $V(x)$ is the potential acting on the particle.

22.2.1 Finite-Difference Algorithm

The time-dependent Schrödinger equation can be solved with both implicit (large-matrix) and explicit (leapfrog) methods. The extra challenge with the Schrödinger equation is to establish that the integral of the probability density $\int_{-\infty}^{+\infty} dx\rho(x,t)$ remains constant (conserved) to a high level of precision for all time. For our project, we use an *explicit* method that improves the numerical conservation of probability by solving for the real and imaginary parts of the wave function at slightly different or "staggered" times (Askar and Cakmak, 1977; Visscher, 1991; Maestri *et al.*, 2000). Explicitly, the real part R is determined at times 0, $\Delta t, \ldots$, and the imaginary part I at $1/2\Delta t$, $3/2\Delta t, \ldots$. The algorithm is based on (what else?) the Taylor expansions of R and I:

$$\begin{aligned} R\left(x, t + \frac{1}{2}\Delta t\right) = R\left(x, t - \frac{1}{2}\Delta t\right) &+ [4\alpha + V(x)\Delta t]I(x,t) \\ &- 2\alpha[I(x+\Delta x, t) + I(x-\Delta x, t)] , \end{aligned} \tag{22.8}$$

where $\alpha = \Delta t/2(\Delta x)^2$. In the discrete form with $R^{t=n\Delta t}_{x=i\Delta x}$, we have

$$R_i^{n+1} = R_i^n - 2\left\{\alpha\left[I_{i+1}^n + I_{i-1}^n\right] - 2\left[\alpha + V_i\Delta t\right] I_i^n\right\} , \tag{22.9}$$

$$I_i^{n+1} = I_i^n + 2\left\{\alpha\left[R_{i+1}^n + R_{i-1}^n\right] - 2\left[\alpha + V_i\Delta t\right] R_i^n\right\} , \tag{22.10}$$

where the superscript n indicates the time and the subscript i the position.

The probability density ρ is defined in terms of the wave function evaluated at three different times:

$$\rho(t) = \begin{cases} R^2(t) + I\left(t + \frac{\Delta t}{2}\right) I\left(t - \frac{\Delta t}{2}\right) , & \text{for integer } t , \\ I^2(t) + R\left(t + \frac{\Delta t}{2}\right) R\left(t - \frac{\Delta t}{2}\right) , & \text{for half-integer } t . \end{cases} \tag{22.11}$$

Although probability is not conserved exactly with this algorithm, the error is two orders higher than that in the wave function, and this is usually quite satisfactory. If it is not, then we need to use smaller steps. While this definition of ρ may seem strange, it reduces to the usual one for $\Delta t \to 0$ and so can be viewed as part of the art of numerical analysis. We will ask you to investigate just how well probability is conserved. We refer the reader to Koonin (1986) and Visscher (1991) for details on the stability of the algorithm.

22.2.2
Wave Packet Implementation, Animation

In Listing 22.1, you will find the program `HarmosAnimate.py` that solves for the motion of the wave packet (22.1) inside a harmonic oscillator potential. The program `Slit.py` on the instructor's site solves for the motion of a Gaussian wave packet as it passes through a slit (Figure 22.3). You should solve for a wave packet confined to the square well:

$$V(x) = \begin{cases} \infty , & x < 0 , \quad \text{or} \quad x > 15 , \\ 0 , & 0 \le x \le 15 . \end{cases}$$

1. Define arrays `psr[751,2]` and `psi[751,2]` for the real and imaginary parts of ψ, and `Rho[751]` for the probability. The first subscript refers to the x position on the grid, and the second to the present and future times.
2. Use the values $\sigma_0 = 0.5$, $\Delta x = 0.02$, $k_0 = 17\pi$, and $\Delta t = 1/2\Delta x^2$.

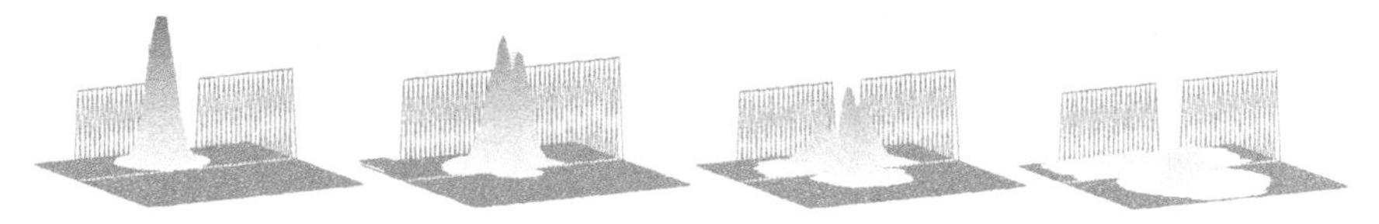

Figure 22.3 The probability density as a function of position and time for an electron incident upon and passing through a slit. Significant reflection is seen to occur.

3. Use (22.1) for the initial wave packet to define psr[j,1] for all j at $t = 0$ and to define psi[j,1] at $t = 1/2\Delta t$.
4. Set Rho[1] = Rho[751] = 0.0 because the wave function must vanish at the infinitely high well walls.
5. Increment time by $1/2\Delta t$. Use (22.9) to compute psr[j,2] in terms of psr[j,1], and (22.10) to compute psi[j,2] in terms of psi[j,1].
6. Repeat the steps through all of space, that is, for $i = 2–750$.
7. Throughout all of space, replace the present wave packet (second index equal to 1) by the future wave packet (second index 2).
8. After you are sure that the program is running properly, repeat the time-stepping for $\sim$ 5000 steps.

Listing 22.1 HarmosAnimate.py solves the time-dependent Schrödinger equation for a particle described by a Gaussian wave packet moving within a harmonic oscillator potential.

```
# HarmonsAnimate: Soltn of t-dependent Sch Eqt fro HO with animation

from visual import *

dx = 0.04;      dx2 = dx*dx;   k0 = 5.5*pi;   dt = dx2/20.0;   xmax = 6.0
xs = arange(-xmax,xmax+dx/2,dx)

g = display(width=500, height=250, title='Wave packet in HO Well')
PlotObj = curve(x=xs, color=color.yellow, radius=0.1)
g.center = (0,2,0)                                    # Scene center
psr = exp(-0.5*(xs/0.5)**2) * cos(k0*xs)             # Initial RePsi
psi = exp(-0.5*(xs/0.5)**2) * sin(k0*xs)             # Initial ImPsi
v   = 15.0*xs**2

while True:
    rate(500)
    psr[1:-1] = psr[1:-1] - (dt/dx2)*(psi[2:] + psi[:-2]\
                - 2*psi[1:-1]) + dt*v[1:-1]*psi[1:-1]
    psi[1:-1] = psi[1:-1] + (dt/dx2)*(psr[2:] + psr[:-2]\
                - 2*psr[1:-1]) - dt*v[1:-1]*psr[1:-1]
    PlotObj.y = 4*(psr**2 + psi**2)
```

1. *Animation:* Output the probability density after every 200 steps for use in animation.
2. Make a surface plot of probability vs. position vs. time. This should look like Figure 22.1 or 22.2.
3. Make an animation showing the wave function as a function of time.
4. Check how well the probability is conserved for early and late times by determining the integral of the probability over all of space, $\int_{-\infty}^{+\infty} dx \rho(x)$, and seeing by how much it changes in time (its specific value doesn't matter because that is just normalization).
5. What might be a good explanation of why collisions with the walls cause the wave packet to broaden and break up? (*Hint:* The collisions do not appear so disruptive when a Gaussian wave packet is confined within a harmonic oscillator potential well.)

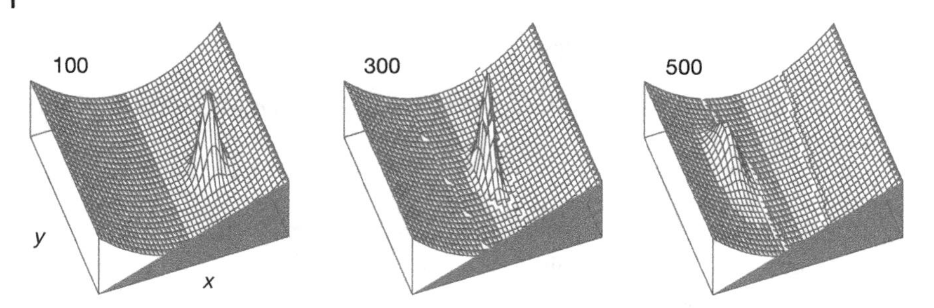

Figure 22.4 The probability density as a function of x and y at three different times for an electron confined to a 2D parabolic tube (infinite in the y direction). The electron is initially placed in a Gaussian wave packet in both the x and y directions, and it is to be noted how there is spreading of the wave packet in the y direction, but not in the x direction.

22.2.3
Wave Packets in Other Wells (Exploration)

1D Well Now confine the electron to a harmonic oscillator potential:

$$V(x) = \frac{1}{2}x^2 \quad (-\infty \leq x \leq \infty) . \tag{22.12}$$

Take the momentum $k_0 = 3\pi$, the space step $\Delta x = 0.02$, and the time step $\Delta t = \frac{1}{4}\Delta x^2$. Note that the wave packet broadens yet returns to its initial shape!

2D Well Now confine the electron to a 2D parabolic tube (Figure 22.4):

$$V(x, y) = 0.9x^2 , \quad -9.0 \leq x \leq 9.0 , \quad 0 \leq y \leq 18.0 . \tag{22.13}$$

The extra degree of freedom means that we must solve the 2D PDE:

$$\mathrm{i}\frac{\partial \psi(x, y, t)}{\partial t} = -\left(\frac{\partial^2 \psi}{\partial x^2} + \frac{\partial^2 \psi}{\partial y^2}\right) + V(x, y)\psi . \tag{22.14}$$

Assume that the electron's initial localization is described by the 2D Gaussian wave packet:

$$\psi(x, y, t = 0) = \mathrm{e}^{\mathrm{i}k_{0x}x}\mathrm{e}^{\mathrm{i}k_{0y}y} \exp\left[-\frac{(x - x_0)^2}{2\sigma_0^2}\right] \exp\left[-\frac{(y - y_0)^2}{2\sigma_0^2}\right] . \tag{22.15}$$

Note that you can solve the 2D equation by extending the method we just used in 1D or you can look at the next section where we develop a special algorithm.

22.3 Algorithm for the 2D Schrödinger Equation

One way to develop an algorithm for solving the time-dependent Schrödinger equation in 2D is to extend the 1D algorithm to another dimension. Rather than that, we apply quantum theory directly to obtain a more powerful algorithm (Maestri *et al.*, 2000). First, we note that Equation 22.14 can be integrated in a formal sense (Landau and Lifshitz, 1976) to obtain the operator solution:

$$\psi(x, y, t) = U(t)\psi(x, y, t = 0) , \tag{22.16}$$

$$U(t) = \mathrm{e}^{-\mathrm{i}\tilde{H}t} , \tag{22.17}$$

$$\tilde{H} = -\left(\frac{\partial^2}{\partial x^2} + \frac{\partial^2}{\partial y^2}\right) + V(x, y) ,$$

where $U(t)$ is an operator that translates a wave function by an amount of time t and $\tilde{H}$ is the Hamiltonian operator. From this formal solution, we deduce that a wave packet can be translated ahead by time Δt via

$$\psi^{n+1}_{i,j} = U(\Delta t)\psi^{n}_{i,j} , \tag{22.18}$$

where the superscripts denote time $t = n\Delta t$ and the subscripts denote the two spatial variables $x = i\Delta x$ and $y = j\Delta y$. Likewise, the inverse of the time evolution operator moves the solution back one time step:

$$\psi^{n-1} = U^{-1}(\Delta t)\psi^{n} = \mathrm{e}^{+\mathrm{i}\tilde{H}\Delta t}\psi^{n} . \tag{22.19}$$

While it would be nice to have an algorithm based on a direct application of (22.19), the references show that the resulting algorithm would not be stable. That being so, we base our algorithm on an indirect application (Askar and Cakmak, 1977), namely, the relation between the difference in ψ^{n+1} and ψ^{n-1}:

$$\psi^{n+1} = \psi^{n-1} + [\mathrm{e}^{-\mathrm{i}\tilde{H}\Delta t} - \mathrm{e}^{+\mathrm{i}\tilde{H}\Delta t}]\psi^{n} , \tag{22.20}$$

where the difference in sign of the exponents is to be noted. The algorithm derives from combining the $O(\Delta x^2)$ expression for the second derivative obtained from the Taylor expansion,

$$\frac{\partial^2\psi}{\partial x^2} \simeq -\frac{1}{2}\left[\psi^{n}_{i+1,j} + \psi^{n}_{i-1,j} - 2\psi^{n}_{i,j}\right] , \tag{22.21}$$

with the corresponding-order expansion of the evolution equation (22.20). Substituting the resulting expression for the second derivative into the 2D time-dependent Schrödinger equation results in[1)]

$$\psi_{i,j}^{n+1} = \psi_{i,j}^{n-1} - 2\mathrm{i}\left[\left(4\alpha + \frac{1}{2}\Delta t V_{i,j}\right)\psi_{i,j}^{n} \right.$$
$$\left. -\alpha\left(\psi_{i+1,j}^{n} + \psi_{i-1,j}^{n} + \psi_{i,j+1}^{n} + \psi_{i,j-1}^{n}\right)\right], \tag{22.22}$$

where $\alpha = \Delta t/2(\Delta x)^2$. We convert this complex equations to coupled real equations by substituting them into the wave function $\psi = R + \mathrm{i}I$,

$$R_{i,j}^{n+1} = R_{i,j}^{n-1} + 2\left[\left(4\alpha + \frac{1}{2}\Delta t V_{i,j}\right)I_{i,j}^{n} - \alpha\left(I_{i+1,j}^{n} + I_{i-1,j}^{n} + I_{i,j+1}^{n} + I_{i,j-1}^{n}\right)\right], \tag{22.23}$$

$$I_{i,j}^{n+1} = I_{i,j}^{n-1} - 2\left[\left(4\alpha + \frac{1}{2}\Delta t V_{i,j}\right)R_{i,j}^{n} + \alpha\left(R_{i+1,j}^{n} + R_{i-1,j}^{n} + R_{i,j+1}^{n} + R_{i,j-1}^{n}\right)\right]. \tag{22.24}$$

This is the algorithm we use to integrate the 2D Schrödinger equation. To determine the probability, we use the same expression (22.11) used in 1D.

22.3.1 Exploration: Bound and Diffracted 2D Packet

1. Determine the motion of a 2D Gaussian wave packet within a 2D harmonic oscillator potential:

$$V(x, y) = 0.3(x^2 + y^2), \quad -9.0 \le x \le 9.0, \quad -9.0 \le y \le 9.0. \tag{22.25}$$

2. Center the initial wave packet at $(x, y) = (3.0, -3)$ and give it momentum $(k_{0x}, k_{0y}) = (3.0, 1.5)$.
3. Young's single-slit experiment has a wave passing through a small slit with the transmitted wave showing interference effects. In quantum mechanics, where we represent a particle by a wave packet, this means that an interference pattern should be formed when a particle passes through a small slit. Pass a Gaussian wave packet of width 3 through a slit of width 5 (Figure 22.3), and look for the resultant quantum interference.

22.4 Wave Packet–Wave Packet Scattering[2)]

We have just seen how to represent a quantum particle as a wave packet and how to compute the interaction of that wave packet/particle with an external potential.

1) For reference sake, note that the constants in the equation change as the dimension of the equation changes; that is, there will be different constants for the 3D equation, and therefore our constants are different from the references!

2) This section is based on the Master of Science thesis of Jon Maestri. It is included in his memory.

Although external potentials do exist in nature, realistic scattering often involves the interaction of one particle with another, which in turn would be represented as the interaction of a wave packet with a *different* wave packet. We have already done the hard work needed to compute wave packet–wave packet scattering in our implementation of wave packet–potential scattering even in 2D. We now need only to generalize it a bit.

Two interacting particles are described by the time-dependent Schrödinger equation in the coordinates of the two particles

$$\mathrm{i}\frac{\partial\psi(x_1,x_2,t)}{\partial t} = -\frac{1}{2m_1}\frac{\partial^2\psi(x_1,x_2,t)}{\partial x_1^2} - \frac{1}{2m_2}\frac{\partial^2\psi(x_1,x_2,t)}{\partial x_2^2} + V(x_1,x_2)\psi(x_1,x_2,t)\,. \tag{22.26}$$

where, for simplicity, we assume a one-dimensional space and again set $\hbar = 1$. Here m_i and x_i are the mass and position of particle $i = 1, 2$. Knowledge of the two-particle wavefunction $\psi(x_1, x_2, t)$ at time t permits the calculation of the probability density of particle 1 being at x_1 and particle 2 being at x_2:

$$\rho(x_1,x_2,t) = |\psi(x_1,x_2,t)|^2\,. \tag{22.27}$$

The fact that particles 1 and 2 must be located somewhere in space leads to the normalization constraint on the wavefunction

$$\int_{-\infty}^{+\infty}\int_{-\infty}^{+\infty} \mathrm{d}x_1\,\mathrm{d}x_2 \rho(x_1,x_2,t) = 1\,. \tag{22.28}$$

The description of a particle within a multiparticle system by a single-particle wavefunction is an approximation unless the system is uncorrelated, in which case the total wavefunction can be written in product form. However, it is possible to deduce meaningful one-particle probabilities (also denoted by ρ) from the two-particle density by integrating over the other particle:

$$\rho(x_i,t) = \int_{-\infty}^{+\infty} \mathrm{d}x_j \rho(x_1,x_2,t)\,, \quad (i \neq j = 1,2)\,. \tag{22.29}$$

Of course, the true solution of the Schrödinger equation is $\psi(x_1, x_2, t)$, but we find it hard to unravel the physics in a three-variable complex function, and so will usually view $\rho(x_1, t)$ and $\rho(x_2, t)$ as two separate wave packets colliding.

If particles 1 and 2 are identical, then their total wavefunction must be symmetric (s) or antisymmetric (a) under interchange of the particles. We impose this condition on our numerical solution $\psi(x_1, x_2)$, by forming the combinations

$$\psi^{s,a}(x_1,x_2) = \frac{1}{\sqrt{2}}[\psi(x_1,x_2) \pm \psi(x_2,x_1)] \quad \Rightarrow \tag{22.30}$$

$$2\rho(x_1,x_2) = |\psi(x_1,x_2)|^2 + |\psi(x_2,x_1)|^2 \pm 2\,\mathrm{Re}\left[\psi^*(x_1,x_2)\psi(x_2,x_1)\right]\,. \tag{22.31}$$

The cross term in (22.31) places an additional correlation into the wave packets.

22.4.1 Algorithm

The algorithm for the solution of the two-particle Schrödinger equation (22.26) in one dimension x, is similar to the one we outlined for the solution for a single particle in the two dimensions x and y, only now with $x \to x_1$ and $y \to x_2$. As usual, we assume discrete space and time

$$x_1 = l\Delta x_1\,, \quad x_2 = m\Delta x_2\,, \quad t = n\Delta t\,. \tag{22.32}$$

We employ the improved algorithm for the time derivative introduced by Askar and Cakmak (1977), which uses a central difference algorithm applied to the formal solution (22.16)

$$\psi^{n+1}_{l,m} - \psi^{n-1}_{l,m} = (\mathrm{e}^{-\mathrm{i}\Delta t H} - \mathrm{e}^{\mathrm{i}\Delta t H})\psi^n_{l,m} \simeq -2\mathrm{i}\Delta t H \psi^n_{l,m}\,, \tag{22.33}$$

$$\begin{aligned}\Rightarrow \quad \psi^{n+1}_{l,m} \simeq \psi^{n-1}_{l,m} - 2\mathrm{i}\Bigg[&\left\{\left(\frac{1}{m_1}+\frac{1}{m_2}\right)4\lambda + \Delta x V_{l,m}\right\}\psi^n_{l,m} \\ &-\lambda\left\{\frac{1}{m_1}\left(\psi^n_{l+l,m}+\psi^n_{l-1,m}\right)+\frac{1}{m_2}(\psi^n_{l,m+1}+\psi^n_{l,m-1})\right\}\Bigg]\,,\end{aligned} \tag{22.34}$$

where we have assumed $\Delta x_1 = \Delta x_2 = \Delta x$ and formed the ratio $\lambda = \Delta t/\Delta x^2$. Again we will take advantage of the extra degree of freedom provided by the complexity of the wavefunction

$$\psi^{n+1}_{l,m} = R^{n+1}_{l,m} + \mathrm{i}I^{n+1}_{l,m}\,, \tag{22.35}$$

and staggered times to preserve probability better (Visscher, 1991). The algorithm (22.34) then separates into the pair of coupled equations

$$\begin{aligned}R^{n+1}_{l,m} = R^{n-1}_{l,m} + 2\Bigg[&\left\{\left(\frac{1}{m_1}+\frac{1}{m_2}\right)4\lambda + \Delta t V_{l,m}\right\}I^n_{l,m} \\ &-\lambda\left\{\frac{1}{m_1}(I^n_{l+1,m}+I^n_{l-1,m})+\frac{1}{m_2}(I^n_{l,m+1}+I^n_{l,m-1})\right\}\Bigg]\,,\end{aligned} \tag{22.36}$$

$$\begin{aligned}I^{n+1}_{l,m} = I^{n-1}_{l,m} - 2\Bigg[&\left\{\left(\frac{1}{m_1}+\frac{1}{m_2}\right)4\lambda + \Delta t V_{l,m}\right\}R^n_{l,m} \\ &-\lambda\left\{\frac{1}{m_1}(R^n_{l+1,m}+R^n_{l-1,m})\frac{1}{m_2}(R^n_{l,m+1}+R^n_{l,m-1})\right\}\Bigg]\,.\end{aligned} \tag{22.37}$$

22.4.2 Implementation

We assume that the particle–particle potential is central and depends only on the relative distance between particles 1 and 2 (the method can handle any x_1 and

x_2 functional dependences if needed). For example, we have used both a "soft" potential with a Gaussian dependence and a "hard" potential with a square-well dependence:

$$V(x_1, x_2) = \begin{cases} V_0 \exp\left[-\frac{|x_1 - x_2|^2}{2\alpha^2}\right] & \text{(Gaussian)} \\ V_0 \theta(\alpha - |x_1 - x_2|) & \text{(Square)} \end{cases} , \tag{22.38}$$

where α is the range parameter and V_0 is the depth parameter.

Because we are solving a PDE, we must specify both initial and boundary conditions. For scattering, we assume that particle 1 is initially at x_1^0 with momentum k_1, and that particle 2 is initially far away at $x_2^0 \simeq \infty$ with momentum k_2. Since the particles are initially too far apart to be interacting, we assume that the initial wave packet is a product of independent wave packets for particles 1 and 2

$$\psi(x_1, x_2, t = 0) = \mathrm{e}^{\mathrm{i}k_1 x_1} \mathrm{e}^{-\frac{(x_1 - x_1^0)^2}{4\sigma^2}} \times \mathrm{e}^{\mathrm{i}k_2 x_2} \mathrm{e}^{-\frac{\left(x_2 - x_2^0\right)^2}{4\sigma^2}} . \tag{22.39}$$

Because of the Gaussian factors here, ψ is not an eigenstate of the momentum operator $-\mathrm{i}\partial/\partial x_i$ for either particle 1 or 2, but instead contains a spread of momenta about the mean, initial momenta k_1 and k_2. If the wave packet is made very broad ($\sigma \to \infty$), we would obtain momentum eigenstates, but then would have effectively eliminated the wave packets. Note that while the Schrödinger equation may separate into one equation in the relative coordinate $x = x_1 - x_2$ and another in the center-of-mass coordinate $X = (m_1 x_1 + m_2 x_2)/(m_1 + m_2)$, the initial condition (22.39), or more general ones, cannot be written as a product of separate functions of x and X, and accordingly, a solution of the partial differential equation in two variables is required (Landau, 1996).

We start the staggered-time algorithm with the real part the wavefunction (22.39) at $t = 0$ and the imaginary part at $t = \Delta t/2$. The initial imaginary part follows by assuming that $\Delta t/2$ is small enough and σ is large enough for the initial time dependence of the wave packet to be that of the plane wave parts:

$$\begin{aligned} I(x_1, x_2, t = \frac{\Delta t}{2}) &\simeq \sin\left[k_1 x_1 + k_2 x_2 - \left(\frac{k_1^2}{2m_1} + \frac{k_2^2}{2m_2}\right)\frac{\Delta t}{2}\right] \\ &\quad \times \exp - \left[\frac{\left(x_1 - x_1^0\right)^2 + \left(x_2 - x_2^0\right)^2}{2\sigma^2}\right] . \end{aligned} \tag{22.40}$$

In an actual scattering experiment, a projectile starts at infinity and the scattered particles are observed also at infinity. We model that by solving our partial differential equation within a box of side L that in ideal world would be much larger than both the range of the potential and the width of the initial wave packet. This leads to the boundary conditions

$$\psi(0, x_2, t) = \psi(x_1, 0, t) = \psi(L, x_2, t) = \psi(x_1, L, t) = 0 . \tag{22.41}$$

The largeness of the box minimizes the effects of the boundary conditions during the collision of the wave packets, although at large times there will be artificial collisions with the box that do not correspond to actual experimental conditions.

Some typical parameters we used are $(\Delta x, \Delta t) = (0.001, 2.5 \times 10^{-7})$, $(k_1, k_2) = (157, -157)$, $\sigma = 0.05$, $(x_1^0, x_2^0) = (467, 934)$, $N_1 = N_2 = 1399$, $(L, T) = (1.401, 0.005)$, and $(V_0, \alpha) = (-100\,000, 0.062)$. The original C code is available on the Web (Maestri *et al.*, 2000). Note that our space step size is 1/1400th of the size of the box L, and 1/70th of the size of the wave packet. Our time step is 1/20 000th of the total time T, and 1/2000th of a typical time for the wave packet. In all cases, the potential and wave packet parameters are chosen to be similar to those used in the one-particle studies by Goldberg *et al.* (1967). The time and space step sizes were determined by trial and error until values were found that provided stability and precision. Importantly, ripples during interactions found in earlier studies essentially disappear when (more accurate) small values of Δx are employed. In general, stability is obtained by making Δt small enough, with simultaneous changes in Δt and Δx made to keep $\lambda = \Delta t/\Delta x^2$ constant. Total probability, as determined by a double Simpson's rule integration of (22.28), is typically conserved to 13 decimal places, impressively close to machine precision. In contrast, the mean energy, for which we do not use an optimized algorithm, is conserved only to 3 places.

22.4.3
Results and Visualization

We solve our problem in the center-of-momentum system by taking $k_2 = -k_1$ (particle 1 moving to larger x values and particle 2 to smaller x). Because the results are time dependent, we make movies of them, and since the movies show the physics much better than the still images, we recommend that the reader look at them (Maestri *et al.*, 2000). We first tested the procedure by emulating the one-particle collisions with barriers and wells studied by Goldberg *et al.* (1967) and presented by Schiff. We made particle 2 ten times heavier than particle 1, which means that particle 2's initial wave packet moves at 1/10th the speed of particle 1's, and so is similar to a barrier. On the left of Fig. 22.5, we see six frames from an animation of the two-particle density $\rho(x_1, x_2, t)$ as a simultaneous function of the particle positions x_1 and x_2. On the right of Fig. 22.5 we show, for this same collision, the *single-particle* densities $\rho_1(x_1, t)$ and $\rho_2(x_2, t)$ extracted from $\rho(x_1, x_2, t)$ by integrating out the dependence on the other particle via (22.29). Because the mean kinetic energy equals twice the maximum height of the potential barrier, we expect complete penetration of the packets, and indeed, at time 18 we see on the right that the wave packets have large overlap, with the repulsive interaction "squeezing" particle 2 (it gets narrower and taller). During time 22–40, we see part of wave packet 1 reflecting off wave packet 2 and then moving back to smaller x (the left). From times 26–55, we also see that a major part of wave packet 1 gets "trapped" inside of wave packet 2 and then leaks out rather slowly.

When looking at the two-particle density $\rho(x_1, x_2, t)$ on the left of Fig. 22.5, we see that for times 1-26, the x_2 position of the peak of changes very little with time, which is to be expected since particle 2 is heavy. In contrast, the x_1 dependence in $\rho(x_1, x_2, t)$ gets broader with time, develops into two peaks at time 26, separates into two distinct parts by time 36, and then, at time 86 after reflecting off the walls,

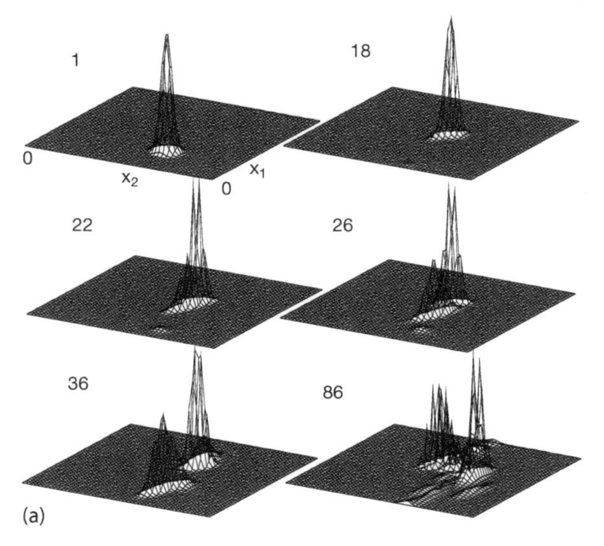

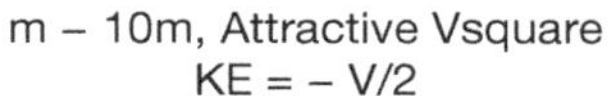

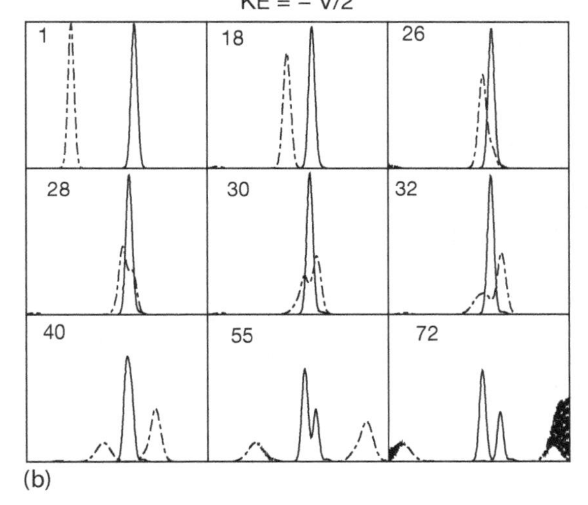

Figure 22.5 (a) Six frames from an animation of the two-particle density $\rho(x_1, x_2, t)$ as a function of the position of particle 1 with mass m and of the position of particle 2 with mass $10m$. (b) This same collision as seen with the single-particle densities $\rho(x_1, t)$ and $\rho(x_2, t)$. The numbers in the left-hand corners are the times in units of $100\Delta t$. Note that each plot ends at the walls of the containing box, and that particle 1 "bounces off" a wall between times 36 and 86.

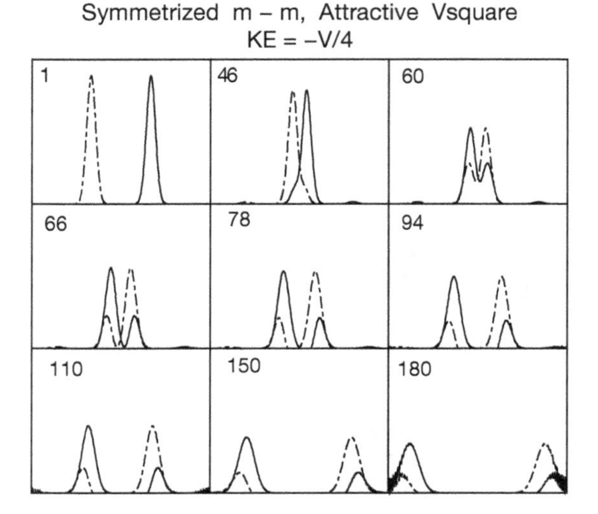

Figure 22.6 A time sequence of two single-particle wave packets scattering from each other. The particles have equal mass, a mean kinetic energy equal to a quarter of the well's depth, and the wavefunction has been symmetrized.

returns to particle 2's position. We also notice in both these figures that at time 40 and thereafter, particle 2 (our "barrier") fissions into reflected and transmitted waves. As this comparison of the visualizations on the right and left of Figures 22.5 demonstrates, it seems easier to understand the physics by superimposing two single-particle densities (thereby discarding information on correlations) than by examining the two-particle density.

In Fig. 22.6, we see nine frames from the movie of an attractive m–m collision in which the mean energy equals one-quarter of the well depth. The initial packets speed up as they approach each other, and at time 60, the centers have already passed through each other. After that, a transmitted and reflected wave for each packet is seen to develop (times 66–78). Although this may be just an artifact of having two particles of equal mass, from times 110–180, we see that each packet appears to capture or "pick up" a part of the other packet and move off with it.

Note in Fig. 22.6 that at time 180, the wave packets are seen to be interacting with the wall (the edges of the frames), as indicated by the interference ripples between incident and reflected waves. Also note that at time 46 and thereafter two additional small wave packets are seen to be traveling in opposite directions to the larger initial wave packets. These small packets are numerical artifacts and arise because outgoing waves satisfy the same differential equations as incoming waves.

22.5 E&M Waves via Finite-Difference Time Domain

Simulations of electromagnetic (EM) waves are of tremendous practical importance. Indeed, the fields of nanotechnology and spintronics rely heavily on such simulations. The basic techniques used to solve for EM waves are essentially the same as those we used for string and quantum waves: set up a grid in space and time and then step the initial solution forward in time one step at a time. When used for *E&M* simulations, this technique is known as the *finite difference time domain* (FDTD) method. What is new for E&M waves is that they are vector fields, with the variations of one vector field generating the other. Our treatment of FDTD does not do justice to the wealth of physics that can occur, and we recommend Sullivan (2000) for a more complete treatment and Ward *et al.* (2005) (and their Web site) for modern applications.

Problem You are given a region in space in which the E and H fields are known to have a sinusoidal spatial variation

$$E_x(z, t = 0) = 0.1 \sin \frac{2\pi z}{100}, \tag{22.42}$$

$$H_y(z, t = 0) = 0.1 \sin \frac{2\pi z}{100}, \quad 0 \le z \le 200, \tag{22.43}$$

with all other components vanishing. Determine the fields for all z values at all subsequent times.

22.6 Maxwell's Equations

The description of EM waves via Maxwell's equations is given in many textbooks. For propagation in just one dimension (z) and for free space with no sinks or sources, four coupled PDEs result:

$$\nabla \cdot \mathbf{E} = 0 \quad \Rightarrow \quad \frac{\partial E_x(z,t)}{\partial x} = 0, \tag{22.44}$$

$$\nabla \cdot \mathbf{H} = 0 \quad \Rightarrow \quad \frac{\partial H_y(z,t)}{\partial y} = 0, \tag{22.45}$$

$$\frac{\partial \mathbf{E}}{\partial t} = +\frac{1}{\epsilon_0} \nabla \times \mathbf{H} \quad \Rightarrow \quad \frac{\partial E_x}{\partial t} = -\frac{1}{\epsilon_0} \frac{\partial H_y(z,t)}{\partial z}, \tag{22.46}$$

$$\frac{\partial \mathbf{H}}{\partial t} = -\frac{1}{\mu_0} \nabla \times \mathbf{E} \quad \Rightarrow \quad \frac{\partial H_y}{\partial t} = -\frac{1}{\mu_0} \frac{\partial E_x(z,t)}{\partial z}. \tag{22.47}$$

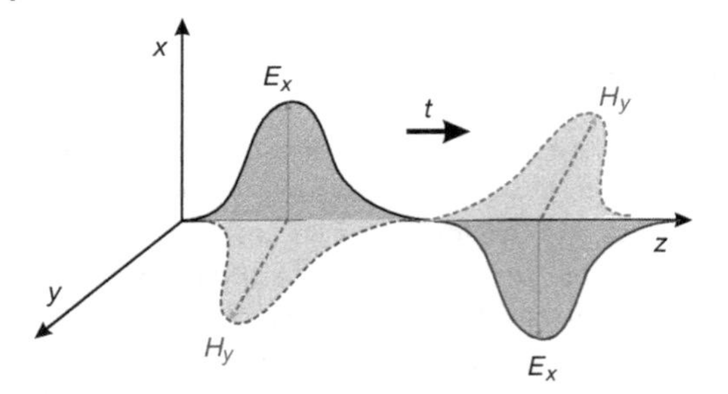

Figure 22.7 A single EM pulse traveling along the z-axis. The coupled E and H pulses are indicated by solid and dashed curves, respectively, and the pulses at different z values correspond to different times.

As indicated in Figure 22.7, we have chosen the electric field $\mathbf{E}(z,t)$ to oscillate (be polarized) in the x direction and the magnetic field $\mathbf{H}(z,t)$ to be polarized in the y direction. As indicated by the bold arrow in Figure 22.7, the direction of power flow for the assumed transverse electromagnetic (TEM) wave is given by the right-hand rule applied to $\mathbf{E} \times \mathbf{H}$. Note that although we have set the initial conditions such that the EM wave is traveling in only one dimension (z), its electric field oscillates in a perpendicular direction (x), and its magnetic field oscillates in yet a third direction (y); so while some may call this a 1D wave, the vector nature of the fields means that the wave occupies all three dimensions.

22.7 FDTD Algorithm

We need to solve the two coupled PDEs (22.46) and (22.47) appropriate for our problem. As is usual for PDEs, we approximate the derivatives via the central-difference approximation, here in both time and space. For example,

$$\frac{\partial E(z,t)}{\partial t} \simeq \frac{E(z,t+\frac{\Delta t}{2}) - E(z,t-\frac{\Delta t}{2})}{\Delta t}, \tag{22.48}$$

$$\frac{\partial E(z,t)}{\partial z} \simeq \frac{E(z+\frac{\Delta z}{2},t) - E(z-\frac{\Delta z}{2},t)}{\Delta z}. \tag{22.49}$$

We next substitute the approximations into Maxwell's equations and rearrange the equations into the form of an algorithm that advances the solution through time. Because only first derivatives occur in Maxwell's equations, the equations are simple, although the electric and magnetic fields are intermixed.

As introduced by Yee (Yee, 1966), we set up a space–time lattice (Figure 22.8) in which there are half-integer time steps as well as half-integer space steps. The magnetic field will be determined at integer time sites and half-integer space sites (open circles), and the electric field will be determined at half-integer time sites

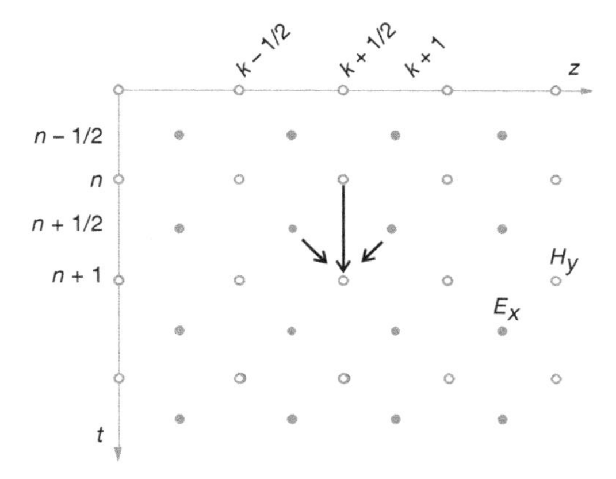

Figure 22.8 The algorithm for using the known values of E_x and H_y at three earlier times and three different space positions to obtain the solution at the present time. Note that the values of E_x are determined on the lattice of filled circles, corresponding to integer space indices and half-integer time indices. In contrast, the values of H_y are determined on the lattice of open circles, corresponding to half-integer space indices and integer time indices.

and integer space sites (filled circles). While this is an extra level of complication, the transposed lattices do lead to an accurate and robust algorithm. Because the fields already have subscripts indicating their vector nature, we indicate the lattice position as superscripts, for example,

$$E_x(z,t) \to E_x(k\Delta z, n\Delta t) \to E_x^{k,n} . \tag{22.50}$$

Maxwell's equations (22.46) and (22.47) now become the discrete equations

$$\frac{E_x^{k,n+1/2} - E_x^{k,n-1/2}}{\Delta t} = -\frac{H_y^{k+1/2,n} - H_y^{k-1/2,n}}{\epsilon_0 \Delta z} ,$$

$$\frac{H_y^{k+1/2,n+1} - H_y^{k+1/2,n}}{\Delta t} = -\frac{E_x^{k+1,n+1/2} - E_x^{k,n+1/2}}{\mu_0 \Delta z} .$$

To repeat, this formulation solves for the electric field at integer space steps (k) but half-integer time steps (n), while the magnetic field is solved for at half-integer space steps but integer time steps.

We convert these equations into two simultaneous algorithms by solving for E_x at time $n + 1/2$, and H_y at time n:

$$E_x^{k,n+1/2} = E_x^{k,n-1/2} - \frac{\Delta t}{\epsilon_0 \Delta z}\left(H_y^{k+1/2,n} - H_y^{k-1/2,n}\right) , \tag{22.51}$$

$$H_y^{k+1/2,n+1} = H_y^{k+1/2,n} - \frac{\Delta t}{\mu_0 \Delta z}\left(E_x^{k+1,n+1/2} - E_x^{k,n+1/2}\right) . \tag{22.52}$$

The algorithms must be applied simultaneously because the space variation of H_y determines the time derivative of E_x, while the space variation of E_x determines the time derivative of H_y (Figure 22.8). These algorithms are more involved than our usual time-stepping ones in that the electric fields (filled circles in Figure 22.8) at future times $t = n + 1/2$ are determined from the electric fields at one time step past $t = n - 1/2$, and the magnetic fields at half a time step past $t = n$. Likewise, the magnetic fields (open circles in Figure 22.8) at future times $t = n + 1$ are determined from the magnetic fields at one time step past $t = n$, and the electric field at half a time step past $t = n + 1/2$. In other words, it is as if we have two interleaved lattices, with the electric fields determined for half-integer times on lattice 1 and the magnetic fields at integer times on lattice 2.

Although these half-integer times appear to be the norm for FDTD methods (Taflove and Hagness , 1989; Sullivan, 2000), it may be easier for some readers to understand the algorithm by doubling the index values and referring to even and odd times:

$$E_x^{k,n} = E_x^{k,n-2} - \frac{\Delta t}{\epsilon_0 \Delta z}\left(H_y^{k+1,n-1} - H_y^{k-1,n-1}\right) , \quad k \text{ even, odd} , \tag{22.53}$$

$$H_y^{k,n} = H_y^{k,n-2} - \frac{\Delta t}{\mu_0 \Delta z}\left(E_x^{k+1,n-1} - E_x^{k-1,n-1}\right) , \quad k \text{ odd, even} . \tag{22.54}$$

This makes it clear that E is determined for even space indices and odd times, while H is determined for odd space indices and even times.

We simplify the algorithm and make its stability analysis simpler by renormalizing the electric fields to have the same dimensions as the magnetic fields,

$$\tilde{E} = \sqrt{\frac{\epsilon_0}{\mu_0}} E . \tag{22.55}$$

The algorithms (22.51) and (22.52) now become

$$\tilde{E}_x^{k,n+1/2} = \tilde{E}_x^{k,n-1/2} + \beta\left(H_y^{k-1/2,n} - H_y^{k+1/2,n}\right) , \tag{22.56}$$

$$H_y^{k+1/2,n+1} = H_y^{k+1/2,n} + \beta\left(\tilde{E}_x^{k,n+1/2} - \tilde{E}_x^{k+1,n+1/2}\right) , \tag{22.57}$$

$$\beta = \frac{c}{\Delta z/\Delta t} , \quad c = \frac{1}{\sqrt{\epsilon_0 \mu_0}} . \tag{22.58}$$

Here, c is the speed of light in vacuum and β is the ratio of the speed of light to grid velocity $\Delta z/\Delta t$.

The space step Δz and the time step Δt must be chosen so that the algorithms are stable. The scales of the space and time dimensions are set by the wavelength and frequency, respectively, of the propagating wave. As a minimum, we want at least 10 grid points to fall within a wavelength:

$$\Delta z \leq \frac{\lambda}{10} . \tag{22.59}$$

The time step is then determined by the Courant stability condition (Taflove and Hagness , 1989; Sullivan, 2000) to be

$$\beta = \frac{c}{\Delta z/\Delta t} \leq \frac{1}{2} . \tag{22.60}$$

As we have seen before, (22.60) implies that making the time step smaller improves precision and maintains stability, but making the space step smaller must be accompanied by a simultaneous decrease in the time step in order to maintain stability (you should check this).

Listing 22.2 FDTD.py solves Maxwell's equations via FDTD time stepping (finite-difference time domain) for linearly polarized wave propagation in the z direction in free space.

```
# FDTD.py  FDTD solution of Maxwell's equations in 1D

from visual import *
xmax=201
ymax=100
zmax=100
scene = display(x=0,y=0,width= 800, height= 500, \
        title= 'E: cyan, H: red. Periodic BC',forward=(-0.6,-0.5,-1))
Efield =
    curve(x=list(range(0,xmax)),color=color.cyan,radius=1.5,display=scene)
Hfield = curve(x=list(range(0,xmax)),color=color.red,
    radius=1.5,display=scene)
vplane= curve(pos=[(-xmax,ymax),(xmax,ymax),(xmax,-ymax),(-xmax,-ymax),
                   (-xmax,ymax)],color=color.cyan)
zaxis=curve(pos=[(-xmax,0),(xmax,0)],color=color.magenta)
hplane=curve(pos=[(-xmax,0,zmax),(xmax,0,zmax),(xmax,0,-zmax),
                   (-xmax,0,-zmax),(-xmax,0,zmax)],color=color.magenta)
ball1 = sphere(pos = (xmax+30, 0,0), color = color.black, radius = 2)
ts = 2                                  # time switch
beta = 0.01
Ex = zeros((xmax,ts),float)             # init E
Hy = zeros((xmax,ts),float)             # init H
Exlabel1 = label( text = 'Ex', pos = (-xmax-10, 50), box = 0 )
Exlabel2 = label( text = 'Ex', pos = (xmax+10, 50), box = 0 )
Hylabel  = label( text = 'Hy', pos = (-xmax-10, 0,50), box = 0 )
zlabel   = label( text = 'Z',  pos = (xmax+10, 0), box = 0 )
ti=0

def inifields():
    k = arange(xmax)
    Ex[:xmax,0] = 0.1*sin(2*pi*k/100.0)
    Hy[:xmax,0] = 0.1*sin(2*pi*k/100.0)

def plotfields(ti):                                # screen coordinates
    k = arange(xmax)
    Efield.x = 2*k-xmax                            # world to screen coords
    Efield.y = 800*Ex[k,ti]
    Hfield.x = 2*k-xmax
    Hfield.z = 800*Hy[k,ti]

inifields()                                        # initial field
plotfields(ti)
while True:
    rate(600)
    Ex[1:xmax-1,1] = Ex[1:xmax-1,0] + beta*(Hy[0:xmax-2,0]-Hy[2:xmax,0])
    Hy[1:xmax-1,1] = Hy[1:xmax-1,0] + beta*(Ex[0:xmax-2,0]-Ex[2:xmax,0])
    Ex[0,1]        = Ex[0,0]        + beta*(Hy[xmax-2,0]  -Hy[1,0]) # BC
    Ex[xmax-1,1]   = Ex[xmax-1,0]   + beta*(Hy[xmax-2,0]  -Hy[1,0])
    Hy[0,1]        = Hy[0,0]        + beta*(Ex[xmax-2,0]  -Ex[1,0]) # BC
    Hy[xmax-1,1]   = Hy[xmax-1,0]   + beta*(Ex[xmax-2,0] - Ex[1,0])
    plotfields(ti)
    Ex[:xmax,0] = Ex[:xmax,1]                                  # New->old
    Hy[:xmax,0] = Hy[:xmax,1]
```

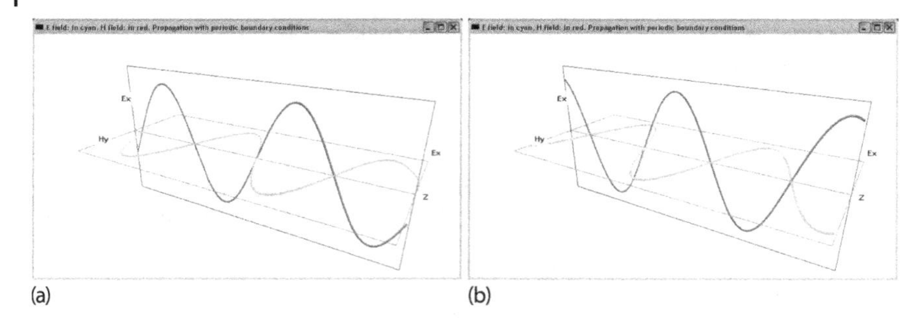

(a) (b)

Figure 22.9 The E field (light) and the H field (dark) at the initial time (a) and at a later time (b). Periodic boundary conditions are used at the ends of the spatial region, which means that the large z wave continues into the $z = 0$ wave.

22.7.1 Implementation

In Listing 22.2, we provide a simple implementation of the FDTD algorithm for a z lattice of 200 sites and in Figure 22.9 we show some results. The initial conditions correspond to a sinusoidal variation of the E and H fields for all z values in for $0 \le z \le 200$:

$$E_x(z, t = 0) = 0.1 \sin \frac{2\pi z}{100}, \quad H_y(z, t = 0) = 0.1 \sin \frac{2\pi z}{100}. \tag{22.61}$$

The algorithm then steps out in time for as long as the user desires. The discrete form of Maxwell equations used are:

```
Ex[k, 1] =  Ex[k, 0]  + beta * (Hy[k-1, 0] - Hy[k+1, 0])
Hy[k, 1] =  Hy[k, 0] + beta * (Ex[k-1, 0] - Ex[k+1, 0])
```

where $1 \le k \le 200$, and beta is a constant. The second index takes the values 0 and 1, with 0 being the old time and 1 the new. At the end of each iteration, the new field throughout all of space becomes the old one, and a new one is computed. With this algorithm, the spatial endpoints `k=0` and `k=xmax-1` remain undefined. We define them by assuming periodic boundary conditions:

```
Ex[0, 1] =    Ex[0, 0]   + beta* (Hy[xmax-2, 0] - Hy[1,0])
Ex[xmax-1, 1] =   Ex[xmax-1, 0]  + beta* (Hy[xmax-2, 0] - Hy[1,0])
Hy[0, 1]   =   Hy[0, 0]  + beta* (Ex[xmax-2, 0] -  Ex[1,0])
Hy[xmax-1, 1] =   Hy[xmax-1, 0] + beta* (Ex[xmax-2, 0] -  Ex[1,0])
```

22.7.2 Assessment

1. Impose boundary conditions such that all fields vanish on the boundaries. Compare the solutions so obtained to those without explicit conditions for times less than and greater than those at which the pulses hit the walls.

2. Examine the stability of the solution for different values of Δz and Δt and thereby test the Courant condition (22.60).
3. Extend the algorithm to include the effect of entering, propagating through, and exiting a dielectric material placed within the z integration region.
4. Ensure that you see both transmission and reflection at the boundaries.
5. Investigate the effect of varying the dielectric's index of refraction.
6. The direction of propagation of the pulse is in the direction of $\mathbf{E} \times \mathbf{H}$, which depends on the relative phase between the **E** and **H** fields. (With no initial **H** field, we obtain pulses both to the right and the left.)
7. Modify the program so that there is an initial **H** pulse as well as an initial **E** pulse, both with a Gaussian times a sinusoidal shape.
8. Verify that the direction of propagation changes if the **E** and **H** fields have relative phases of 0 or π.
9. Investigate the resonator modes of a wave guide by picking the initial conditions corresponding to plane waves with nodes at the boundaries.
10. Investigate standing waves with wavelengths longer than the size of the integration region.
11. Simulate unbounded propagation by building in periodic boundary conditions into the algorithm.
12. Place a medium with periodic permittivity in the integration volume. This should act as a frequency-dependent filter, which does not propagate certain frequencies at all.

22.7.3 Extension: Circularly Polarized Waves

We now extend our treatment to EM waves in which the **E** and **H** fields, while still transverse and propagating in the z direction, are not restricted to linear polarizations along just one axis. Accordingly, we add to (22.46) and (22.47):

$$\frac{\partial H_x}{\partial t} = \frac{1}{\mu_0}\frac{\partial E_y}{\partial z} , \tag{22.62}$$

$$\frac{\partial E_y}{\partial t} = \frac{1}{\epsilon_0}\frac{\partial H_x}{\partial z} . \tag{22.63}$$

When discretized in the same way as (22.51) and (22.52), we obtain

$$H_x^{k+1/2,n+1} = H_x^{k+1/2,n} + \frac{\Delta t}{\mu_0 \Delta z}\left(E_y^{k+1,n+1/2} - E_y^{k,n+1/2}\right) , \tag{22.64}$$

$$E_y^{k,n+1/2} = E_y^{k,n-1/2} + \frac{\Delta t}{\epsilon_0 \Delta z}\left(H_y^{k+1/2,n} - H_y^{k-1/2,n}\right) . \tag{22.65}$$

To produce a circularly polarized traveling wave, we set the initial conditions

$$E_x = \cos\left(t - \frac{z}{c} + \phi_y\right) , \quad H_x = \sqrt{\frac{\epsilon_0}{\mu_0}}\cos\left(t - \frac{z}{c} + \phi_y\right) , \tag{22.66}$$

$$E_y = \cos\left(t - \frac{z}{c} + \phi_x\right), \quad H_y = \sqrt{\frac{\epsilon_0}{\mu_0}} \cos\left(t - \frac{z}{c} + \phi_x + \pi\right). \tag{22.67}$$

We take the phases to be $\phi_x = \pi/2$ and $\phi_y = 0$, so that their difference $\phi_x - \phi_y = \pi/2$, which leads to circular polarization. We include the initial conditions in the same manner as we did the Gaussian pulse, only now with these cosine functions.

Listing 22.3 gives our implementation EMcirc.py for waves with transverse two-component **E** and **H** fields. Some results of the simulation are shown in Figure 22.10, where you will note the difference in phase between **E** and **H**.

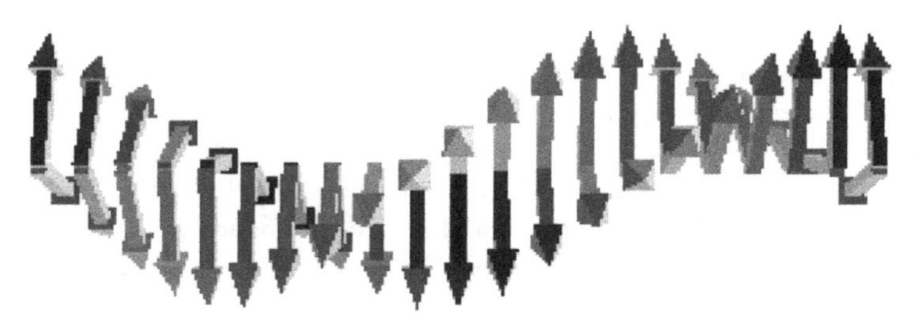

Figure 22.10 E and H fields at $t = 100$ for a circularly polarized wave in free space.

Listing 22.3 CircPolartzn.py solves Maxwell's equations via FDTD time-stepping for circularly polarized wave propagation in the z direction in free space.

```
# CircPolarztn.py: solves Maxwell eqs. using FDTD

from visual import *

scene = display(x=0, y=0, width=600, height=400,range=200,
    title='Circular polarization, E field in white, H field in yellow')
global phy, pyx
max = 201
c = 0.01                                  # Courant stable if c < 0.1

Ex = zeros((max+2,2),float)
Hy = zeros((max+2,2),float)
Ey = zeros((max+2,2),float)
Hx = zeros((max+2,2),float)

arrowcol= color.white
Earrows = []
Harrows = []
for i in range(0,max,10):
    Earrows.append(arrow(pos=(0,i-100,0), axis=(0,0,0), color=arrowcol))
    Harrows.append(arrow(pos=(0,i-100,0), axis=(0,0,0),
         color=color.yellow))

def plotfields(Ex,Ey,Hx,Hy):
    for n, arr in enumerate(Earrows):
        arr.axis = (35*Ey[10*n,1],0,35*Ex[10*n,1])
    for n, arr in enumerate(Harrows):
        arr.axis = (35*Hy[10*n,1],0,35*Hx[10*n,1])

def inifields():                                          # Initial E & H
    phx = 0.5*pi
    phy = 0.0
    k = arange(0,max)
```

```
    Ex[:-2,0] = cos(-2*pi*k/200 + phx)
    Ey[:-2,0] = cos(-2*pi*k/200 + phy)
    Hx[:-2,0] = cos(-2*pi*k/200 + phy + pi)
    Hy[:-2,0] = cos(-2*pi*k/200 + phx)

def newfields():
    while True:                                        # Time stepping
        rate(1000)
        Ex[1:max-1,1] =  Ex[1:max-1,0]+c*(Hy[:max-2,0]-Hy[2:max,0])
        Ey[1:max-1,1] =  Ey[1:max-1,0] + c*(Hx[2:max,0]-Hx[:max-2,0])
        Hx[1:max-1,1] =  Hx[1:max-1,0] + c*(Ey[2:max,0]-Ey[:max-2,0])
        Hy[1:max-1,1] =  Hy[1:max-1,0] + c*(Ex[:max-2,0]-Ex[2:max,0])
        Ex[0,1]   = Ex[0,0]   + c*(Hy[200-1,0]-Hy[1,0])    # Periodic BC
        Ex[200,1] = Ex[200,0] + c*(Hy[200-1,0]-Hy[1,0])
        Ey[0,1]   = Ey[0,0]   + c*(Hx[1,0]- Hx[200-1,0])
        Ey[200,1] = Ey[200,0] + c*(Hx[1,0]- Hx[200-1,0])
        Hx[0,1]   = Hx[0,0]   + c*(Ey[1,0]- Ey[200-1,0])
        Hx[200,1] = Hx[200,0] + c*(Ey[1,0]- Ey[200-1,0])
        Hy[0,1]   = Hy[0,0]   + c*(Ex[200-1,0]-Ex[1,0])
        Hy[200,1] = Hy[200,0] + c*(Ex[200-1,0]-Ex[1,0])
        plotfields(Ex,Ey,Hx,Hy)

        Ex[:max,0] = Ex[:max,1]                        # Update fields
        Ey[:max,0] = Ey[:max,1]
        Hx[:max,0] = Hx[:max,1]
        Hy[:max,0] = Hy[:max,1]

inifields()                                            # Initial field
newfields()                                            # Subsequent field
```

22.8 Application: Wave Plates

Problem Develop a numerical model for a wave plate that convert a linearly polarized EM wave into a circularly polarized one.

As can be seen in Figure 22.11a wave plate is an optical device that alters the polarization of light traveling through it by shifting the relative phase of the components of the polarization vector. A quarter-wave plate introduces a relative phase of $\lambda/4$, where λ is the wavelength of the light. Physically, a wave plate is often a birefringent crystal in which the different propagation velocities of waves in two

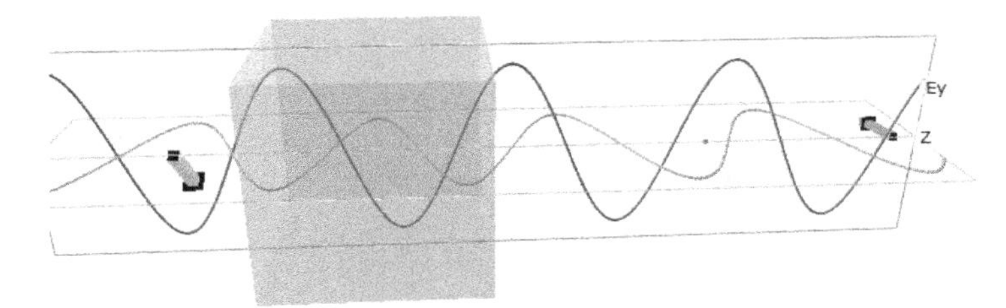

Figure 22.11 One frame from the program `quarterwave.py` (on Instructor's site) showing a linearly polarized EM wave entering a quarter-wave plate from the left and leaving as a circularly polarized wave on the right (the arrow on the left oscillates back and forth at 45° while the one on the right rotates).

orthogonal directions leads to the phase change. The amount of phase change is adjusted by varying the thickness of the plate.

To attack this problem, we apply our FDTD method of solving Maxwell equations. We start with a linear polarized wave with both E_x and E_y components propagating along the z direction. The wave enters the plate and emerges from it still traveling in the z direction, but now with the relative phase of these fields shifted. Of course, because this is an EM wave, there will also be coupled magnetic field components present, in this case H_x and H_y, and they too will need be computed.

Theory Maxwell equations for a wave propagating along the z-axis are:

$$\frac{\partial H_x}{\partial t} = +\frac{1}{\mu_0}\frac{\partial E_y}{\partial z}, \quad \frac{\partial H_y}{\partial t} = -\frac{1}{\mu_0}\frac{\partial E_x}{\partial z}, \tag{22.68}$$

$$\frac{\partial E_x}{\partial t} = -\frac{1}{\epsilon_0}\frac{\partial H_y}{\partial z}, \quad \frac{\partial E_y}{\partial t} = +\frac{1}{\epsilon_0}\frac{\partial H_x}{\partial z}. \tag{22.69}$$

We take as initial conditions a wave incident from the left along the z-axis, linearly polarized (electric field direction of 45°), with corresponding, and perpendicular, **H** components:

$$E_x(t=0) = 0.1\cos\frac{2\pi x}{\lambda}, \quad E_y(t=0) = 0.1\cos\frac{2\pi y}{\lambda}, \tag{22.70}$$

$$H_x(t=0) = 0.1\cos\frac{2\pi x}{\lambda}, \quad H_y(t=0) = 0.1\cos\frac{2\pi y}{\lambda}. \tag{22.71}$$

Because only the relative phases matter, we simplify the calculation by assuming that the E_y and H_x components do not have their phases changed, but that the E_x and H_y components do (in this case by $\lambda/4$ when they leave the plate). Of course, after leaving the plate and traveling in free space, there are no further changes in the relative phase.

22.9 Algorithm

As in Section 22.7 and Figure 22.8, we follow the FDTD approach of using known values of E_x and H_y at three earlier times and three different space positions to obtain the solution at the present time. With the renormalized electric fields as in (22.55), this leads to the beautifully symmetric equations:

$$E_x^{k,n+1} = E_x^{k,n} + \beta\left(H_y^{k+1,n} - H_y^{k,n}\right), \tag{22.72}$$

$$E_y^{k,n+1} = E_y^{k,n} + \beta\left(H_x^{k+1,n} - H_x^{k,n}\right), \tag{22.73}$$

$$H_x^{k,n+1} = H_x^{k,n} + \beta\left(E_y^{k+1,n} - E_y^{k,n}\right), \tag{22.74}$$

$$H_y^{k,n+1} = H_y^{k,n} + \beta\left(E_x^{k+1,n} - E_x^{k,n}\right). \tag{22.75}$$

22.10 FDTD Exercise and Assessment

1. Modify the FDTD program of Listing 22.2 so that it solves the algorithm (22.72)–(22.75). Use $\beta = 0.01$.
2. After each time step, impose a gradual increment of the phase so that the total phase change will be one-quarter of a wavelength. Our program for this, `quarterplat.py`, is on the instructor's page.
3. Verify that the plate converts an initially linearly polarized wave into a circularly polarized one.
4. Verify that the plate converts an initially circularly polarized wave into a linearly polarized one.
5. What happens if you put two plates together? Three? Four? (Verify!)

23
Electrostatics via Finite Elements

In Chapter 19, we discussed the simple, but powerful, *finite-differences* method for solving Poisson's and Laplace's equations on a lattice in space. In this chapter, we provide a basic outline of the *finite-element* method (FEM) for solving PDEs. Our usual approach to solving PDEs uses the *finite-difference* method to approximate various derivatives in terms of the finite differences of a function evaluated upon a fixed grid. The FEM, in contrast, breaks space up into multiple geometric objects (elements), determines an approximate form for the solution appropriate to each element, and then matches the solutions up at the elements' edges. The FEM is ultimately more efficient and powerful than the finite-differences method; however, much more work is required to derive the algorithm. In practice, it is rare to solve a PDE from scratch by deriving the FEM for a particular problem. Rather, and for good reasons, many FEM applications use highly developed FEM packages that get customized for an individual problem. (Python's finite element library is *FiPy*.) Our aim is to give the reader some basic understanding of the FEM, not to develop a practitioner. Accordingly, we examine a 1D problem in some detail, and then outline the similar steps followed for the same equation extended to 2D.

23.1
Finite-Element Method ⊙

The theory and practice of FEM as a numerical method for solving partial differential equations have been developed over the last 30 years and is still an active field of research. One of the theoretical strengths of FEM is that its mathematical foundations allow for elegant proofs of the convergence of its solutions. One of the practical strengths of FEM is that it offers great flexibility for problems on irregular domains, or for problems with highly varying conditions or even singularities. Although finite-difference methods are simpler to implement than FEM, they are less robust mathematically and for big problems less efficient in terms of computer time. Finite elements, in turn, are more complicated to implement, but more appropriate and precise for complicated equations and complicated geometries. In addition, the same basic finite-element technique can be applied to

Computational Physics, 3rd edition. Rubin H. Landau, Manuel J. Páez, Cristian C. Bordeianu.

many problems with only minor modifications, and yields solutions that may be evaluated throughout all space, not just on a grid. In fact, the FEM with various preprogrammed multigrid packages has very much become the standard for large-scale engineering applications. Our discussion is based upon Shaw (1992); Li (2014); Otto (2011).

23.2
Electric Field from Charge Density (Problem)

As shown in Figure 23.1, you are given two conducting plates a distance $b - a$ apart, with the lower one kept at potential U_a, the upper plate at potential U_b, and a uniform charge density $\rho(x)$ placed between them. Your *problem* is to compute the electric potential between the plates.

23.3
Analytic Solution

The relation between charge density $\rho(\boldsymbol{x})$ and potential $U(\boldsymbol{x})$ is given by Poisson's equation (19.6). For our problem, the potential U changes only in the x direction, and so the PDE becomes the ODE:

$$\frac{d^2U(x)}{dx^2} = -4\pi\rho(x) = -1\,, \quad 0 < x < 1\,, \tag{23.1}$$

where we have set $\rho(x) = 1/4\pi$ to simplify the programming. The solution we want is subject to the Dirichlet boundary conditions:

$$U(x = a = 0) = 0\,, \quad U(x = b = 1) = 1\,, \tag{23.2}$$

$$\Rightarrow \quad U(x) = -\frac{x}{2}(x-3)\,. \tag{23.3}$$

Although, we know the analytic solution (23.3), we shall develop the FEM for solving the ODE as if it was a PDE (it would be in 2D), and as if we did not know the solution. Although we will not demonstrate it, this method works equally well for any charge density $\rho(x)$.

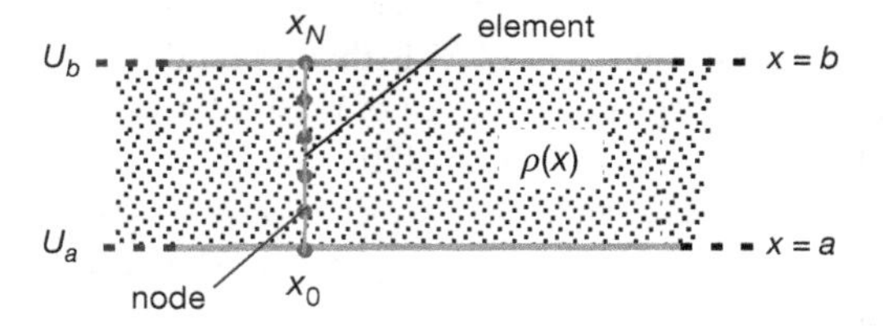

Figure 23.1 A finite element solution to Laplace's equation for two metal plates with a charge density between them. The dots are the nodes x_i, and the lines connecting the nodes are the finite elements.

23.4 Finite-Element (Not Difference) Methods, 1D

In an FEM, the domain in which the PDE is solved is split into finite subdomains, called *elements*, and a *trial solution* to the PDE in each subdomain is hypothesized. Then the parameters of the trial solution are adjusted to obtain a *best fit* (in the sense of Chapter 7) to the exact solution. Essentially, this approach converts a given PDE into an integral equation known as the *weak* or *variational* form ("weak" because there is no longer the requirement that the second derivative of the solution exists). A trial solution on each element is then postulated, and this leads to the numerically intensive work of finding the best values for the parameters in the trial solution, and in matching up the various trial solutions from different subdomains.

In general, an FEM solution follows six steps (Li, 2014):

1. Derivation of a *weak form* of the PDE. This is equivalent to a least-squares minimization of the integral of the difference between the approximate and exact solutions.
2. Discretization of the computational domains.
3. Generation of interpolating or trial functions.
4. Conversion of the "weak form" integral equation into a set of linear equations.
5. Implementation of the boundary conditions.
6. Solution of the resulting linear system of equations.

23.4.1 Weak Form of PDE

Finite-difference methods yield an approximate solution of an approximate PDE. Finite-element methods yield the best possible global agreement between an approximate solution and the exact solution. We start the FEM with the differential equation we want to solve,

$$-\frac{\mathrm{d}^2 U(x)}{\mathrm{d}x^2} = 4\pi\rho(x) . \tag{23.4}$$

We form an integral of the product of the exact solution $U(x)$ and the approximate solution or *trial solution* $\Phi(x)$ over the solution domain. This will be used as a measure of overall agreement between the two solutions. We assume that the trial vanishes at the endpoints, $\Phi(a) = \Phi(b) = 0$ (we satisfy general boundary conditions later). We next multiply both sides of the differential equation (23.1) by Φ and integrate by parts from a to b:

$$-\frac{\mathrm{d}^2 U(x)}{\mathrm{d}x^2}\Phi(x) = 4\pi\rho(x)\Phi(x) , \tag{23.5}$$

$$-\int_a^b \mathrm{d}x \frac{\mathrm{d}^2 U(x)}{\mathrm{d}x^2}\Phi(x) = \int_a^b \mathrm{d}x 4\pi\rho(x)\Phi(x) \tag{23.6}$$

$$-\frac{\mathrm{d}U(x)}{\mathrm{d}x}\Phi(x)|_a^b + \int_a^b \mathrm{d}x\frac{\mathrm{d}U(x)}{\mathrm{d}x}\Phi'(x) = \int_a^b \mathrm{d}x 4\pi\rho(x)\Phi(x) \tag{23.7}$$

$$\Rightarrow \quad \int_a^b \mathrm{d}x\frac{\mathrm{d}U(x)}{\mathrm{d}x}\Phi'(x) = \int_a^b \mathrm{d}x 4\pi\rho(x)\Phi(x) \ . \tag{23.8}$$

Equation 23.8 is the *weak* form of the PDE, "weak" in the sense that it does not require the existence of the second derivative of U, or the continuity of ρ. Because the approximate and exact solutions are related by the integral of their difference over the entire domain, the algorithm provides the global best agreement between the two.

23.4.2
Galerkin Spectral Decomposition

The approximate solution to the weak PDE proceeds via three steps. First, we split the full domain of the PDE into subdomains called *elements*, then we find approximate solutions within each element, and finally we match the elemental solutions onto each other. For our 1D problem, the subdomain elements are straight lines of equal length, while for the 2D problem to be considered soon, the elements are triangles (Figure 23.4).

The critical step in the FEM is the expansion of the solution U in terms of a set of basis functions φ_i:

$$U(x) \simeq \sum_{j=0}^{N-1} \alpha_j \varphi_j(x) \ . \tag{23.9}$$

Even when the basis functions are not sines or cosines, this expansion is still called a *spectral* decomposition. We will choose φ_i's that are convenient for computation, and so the solution reduces to determining the unknown expansion coefficients α_j. Later, in order to satisfy the boundary conditions, we will add an additional term to this expansion.

Considerable study has gone into determining the effectiveness of different basis functions φ_i that are used to represent the solution on the finite elements. If the sizes of the finite elements are made sufficiently small, then good accuracy is obtained with simple piecewise-continuous basis functions φ_i. For our 1D problem, we use finite *elements* that are line segments between x_i and x_{i+1}, and we use *basis functions*, representing the solution on each line segment, that have the form of triangle or "hats" between x_{i-1} and x_{i+1} (Figure 23.2). We also require that each basis function equals 1 at the x_i vertex, $\varphi_i(x_i) = 1$:

$$\varphi_i(x) = \begin{cases} 0 \ , & \text{for} \quad x < x_{i-1} \ , \quad \text{or} \quad x > x_{i+1} \ , \\ \frac{x - x_{i-1}}{h_{i-1}} \ , & \text{for} \quad x_{i-1} \le x \le x_i \ , \\ \frac{x_{i+1} - x}{h_i} \ , & \text{for} \quad x_i \le x \le x_{i+1} \ . \end{cases} \qquad (h_i = x_{i+1} - x_i) \ , \tag{23.10}$$

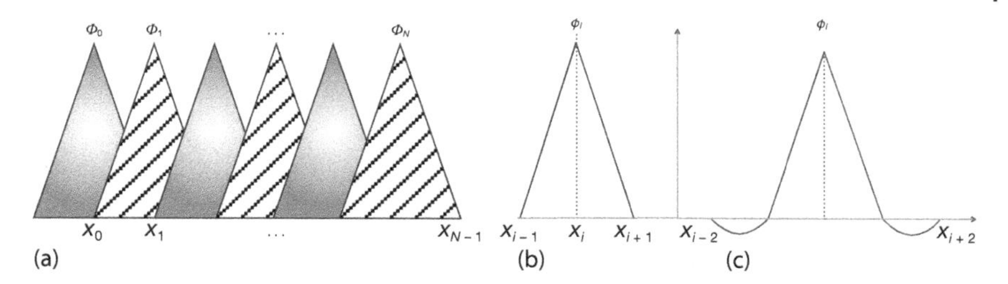

Figure 23.2 Basis functions used in finite-elements solution of the 1D Laplace equation. (a) A set of overlapping basis functions ϕ_i. Each function is a triangle from x_{i-1} to x_{i+1}. (b) A Piecewise-linear function. (c) A piecewise-quadratic function.

Because this choice means that each basis function equals 0 or 1 at the nodes,

$$\phi_i(x_j) = \delta_{ij} \, , \tag{23.11}$$

the values of the expansion coefficients α_i must equal the values of the (still unknown) solution at the nodes:

$$U(x_i) \simeq \sum_{i=0}^{N-1} \alpha_i \phi_i(x_i) = \alpha_i \phi_i(x_i) = \alpha_i \, , \tag{23.12}$$

$$\Rightarrow \quad U(x) \simeq \sum_{j=0}^{N-1} U(x_j)\phi_j(x) \, . \tag{23.13}$$

Equation 23.13 makes it clear that the expansion in terms of basis functions is essentially an interpolation between the actual solution at the nodes.

23.4.2.1 Solution via Linear Equations

Because the basis functions ϕ_i in (23.9) are known, solving for $U(x)$ involves determining the coefficients α_j, which, as we just said, are the unknown values of the true solution $U(x)$ on the nodes. We determine those values by substituting the expansions for $U(x)$ and $\Phi(x)$ into the weak form of the PDE (23.8). This converts the integral equation into a set of simultaneous linear equations. As discussed in Chapter 6, there is a standard matrix form for a set of linear equations,

$$A\boldsymbol{y} = \boldsymbol{b} \, . \tag{23.14}$$

Our equations fit that, with $\boldsymbol{y}$ a vector of unknowns, and where we still need to specify the known *stiffness matrix* $\mathbf{A}$ and the known load matrix $\boldsymbol{b}$. To that end, we substitute the expansion $U(x) \simeq \sum_{j=0}^{N-1} \alpha_j \phi_j(x)$ into the weak form (23.8) to obtain:

$$\int_a^b dx \frac{d}{dx}\left(\sum_{j=0}^{N-1} \alpha_j \phi_j(x)\right) \frac{d\Phi}{dx} = \int_a^b dx 4\pi\rho(x)\Phi(x) \, .$$

By successively selecting $\Phi(x) = \phi_0, \phi_1, \dots, \phi_{N-1}$, we obtain N simultaneous linear equations for the unknown α_j's:

$$\int_a^b dx \frac{d}{dx}\left(\sum_{j=0}^{N-1} \alpha_j \phi_j(x)\right) \frac{d\phi_i}{dx} = \int_a^b dx 4\pi\rho(x)\phi_i(x) , \quad i = 0, N-1 . \tag{23.15}$$

We factor out the unknown α_j's and write the equations out explicitly:

$$\alpha_0 \int_a^b \phi_0'\phi_0' \, dx + \alpha_1 \int_a^b \phi_0'\phi_1' \, dx + \cdots + \alpha_{N-1} \int_a^b \phi_0'\phi_{N-1}' \, dx = \int_a^b 4\pi\rho\phi_0 \, dx ,$$

$$\alpha_0 \int_a^b \phi_1'\phi_0' \, dx + \alpha_1 \int_a^b \phi_1'\phi_1' \, dx + \cdots + \alpha_{N-1} \int_a^b \phi_1'\phi_{N-1}' \, dx = \int_a^b 4\pi\rho\phi_1 \, dx ,$$

$$\ddots$$

$$\alpha_0 \int_a^b \phi_{N-1}'\phi_0' \, dx + \alpha_1 \int \cdots + \alpha_{N-1} \int_a^b \phi_{N-1}'\phi_{N-1}' \, dx = \int_a^b 4\pi\rho\phi_{N-1} \, dx .$$

Because we have chosen the ϕ_i's to be the simple hat functions, the derivatives are easy to evaluate analytically (for other bases they can be carried out numerically):

$$\frac{d\phi_{i,i+1}}{dx} = \begin{cases} 0 , & x < x_{i-1} , \quad \text{or} \quad x_{i+1} < x , \\ \frac{1}{h_{i-1}} , & x_{i-1} \le x \le x_i , \\ \frac{-1}{h_i} , & x_i \le x \le x_{i+1} , \\ 0 , & x < x_i , \quad \text{or} \quad x_{i+2} < x , \\ \frac{1}{h_i} , & x_i \le x \le x_{i+1} , \\ \frac{-1}{h_{i+1}} , & x_{i+1} \le x \le x_{i+2} . \end{cases} \tag{23.16}$$

The integrations are now fairly simple:

$$\int_{x_{i-1}}^{x_{i+1}} dx (\phi_i')^2 = \int_{x_{i-1}}^{x_i} dx \frac{1}{(h_{i-1})^2} + \int_{x_i}^{x_{i+1}} dx \frac{1}{h_i^2} = \frac{1}{h_{i-1}} + \frac{1}{h_i}, \tag{23.17}$$

$$\int_{x_{i-1}}^{x_{i+1}} dx \phi_i' \phi_{i+1}' = \int_{x_{i-1}}^{x_{i+1}} dx \phi_{i+1}' \phi_i' = \int_{x_i}^{x_{i+1}} dx \frac{-1}{h_i^2} = -\frac{1}{h_i} , \tag{23.18}$$

$$\int_{x_{i-1}}^{x_{i+1}} dx (\phi_{i+1}')^2 = \int_{x_i}^{x_{i+1}} dx (\phi_{i+1}')^2 = \int_{x_i}^{x_{i+1}} dx \frac{+1}{h_i^2} = +\frac{1}{h_i} . \tag{23.19}$$

We rewrite these equations in the standard matrix form (23.14) with $\boldsymbol{y}$ constructed from the unknown α_j's, and the tridiagonal stiffness matrix $\mathbf{A}$ constructed from the integrals over the derivatives:

$$\boldsymbol{y} = \begin{bmatrix} \alpha_0 \\ \alpha_1 \\ \ddots \\ \alpha_{N-1} \end{bmatrix}, \quad \boldsymbol{b} = \begin{bmatrix} \int_{x_0}^{x_1} dx 4\pi\rho(x)\phi_0(x) \\ \int_{x_1}^{x_2} dx 4\pi\rho(x)\phi_1(x) \\ \ddots \\ \int_{x_{N-1}}^{x_N} dx 4\pi\rho(x)\phi_{N-1}(x) \end{bmatrix}, \tag{23.20}$$

$$\mathbf{A} = \begin{bmatrix} \frac{1}{h_0}+\frac{1}{h_1} & -\frac{1}{h_1} & -\frac{1}{h_0} & 0 & \dots \\ -\frac{1}{h_1} & \frac{1}{h_1}+\frac{1}{h_2} & -\frac{1}{h_2} & 0 & \dots \\ 0 & -\frac{1}{h_2} & \frac{1}{h_2}+\frac{1}{h_3} & -\frac{1}{h_3} & \dots \\ \ddots & \ddots & -\frac{1}{h_{N-1}} & -\frac{1}{h_{N-2}} & \frac{1}{h_{N-2}}+\frac{1}{h_{N-1}} \end{bmatrix}. \tag{23.21}$$

The elements in $\mathbf{A}$ are just combinations of inverse step sizes and so do not change for different charge densities $\rho(x)$. This is part of what makes FEM so efficient once set up. The elements in $\boldsymbol{b}$ do change for different ρ's, but the required integrals can be performed analytically or with Gaussian quadrature (Chapter 5). Once $\mathbf{A}$ and $\boldsymbol{b}$ are computed, highly efficient methods from a linear algebra package are used to solve the matrix equations for $\boldsymbol{y}$, and thus the expansion coefficients α_j.

23.4.2.2 Dirichlet Boundary Conditions

Because the basis functions vanish at the endpoints, a solution expanded in them must also vanishes there. This will not do in general, and so we must add to our general solution $U(x)$, a particular solution $U_a\phi_0(x)$ that satisfies the boundary conditions (Li, 2014):

$$U(x) = \sum_{j=0}^{N-1} \alpha_j\phi_j(x) + U_a\phi_N(x) \quad \text{(satisfies boundary conditions)}, \tag{23.22}$$

where $U_a = U(x_a)$. We substitute $U(x) - U_a\phi_0(x)$ into the weak form of the PDE to obtain $(N+1)$ simultaneous equations, still of the form $\boldsymbol{Ay} = \boldsymbol{b}$, but now with

$$\mathbf{A} = \begin{bmatrix} A_{0,0} & \cdots & A_{0,N-1} & 0 \\ & \ddots & & \\ A_{N-1,0} & \cdots & A_{N-1,N-1} & 0 \\ 0 & 0 & \cdots & 1 \end{bmatrix}, \quad \boldsymbol{b}' = \begin{bmatrix} b_0 - A_{0,0}U_a \\ \ddots \\ b_{N-1} - A_{N-1,0}U_a \\ U_a \end{bmatrix}. \tag{23.23}$$

This is equivalent to adding a unit element to $\mathbf{A}$ and adding a new load vector element:

$$b'_i = b_i - A_{i,0}U_a, \quad i = 1, \dots, N-1, \quad b'_N = U_a. \tag{23.24}$$

To impose the boundary condition at $x = b$, we again add a particular solution $U_b \phi_{N-1}(x)$ and substitute it into the weak form to obtain

$$b'_i = b_i - A_{i,N-1} U_b \,, \quad i = 1, \dots, N-1 \,, \quad b'_N = U_b \,. \tag{23.25}$$

So now we need to solve the linear equations $\mathbf{A}\boldsymbol{y} = \boldsymbol{b}'$. For 1D problems, 100–1000 equations are common, while for 3D problems there may be millions. Because the number of calculations varies approximately as N^2, it is important to utilize an efficient and accurate algorithm, because otherwise the round-off error can easily dominate the solution.

23.5
1D FEM Implementation and Exercises

In Listing 23.1, we give our program `LaplaceFEM_1D.py` that determines the 1D FEM solution, and in Figure 23.3 we show that solution. We see on the left of the figure that three elements do not provide even visual agreement with the analytic result, whereas $N = 11$ elements do.

1. Examine the FEM solution for the choice of parameters

$$a = 0 \,, \quad b = 1 \,, \quad U_a = 0 \,, \quad U_b = 1 \,. \tag{23.26}$$

2. Generate your own triangulation by assigning explicit x values at the nodes over the interval [0, 1].
3. Start with $N = 3$ and solve the equations for N values up to 1000.
4. Examine the stiffness matrix $\mathbf{A}$ and ensure that it is triangular.
5. Verify that the integrations used to compute the load vector $\boldsymbol{b}$ are accurate.
6. Verify that the solution of the linear equation $\mathbf{A}\boldsymbol{y} = \boldsymbol{b}$ is correct.
7. Plot the numerical solution for $U(x)$ for $N = 10$, 100, and 1000, and compare with the analytic solution.

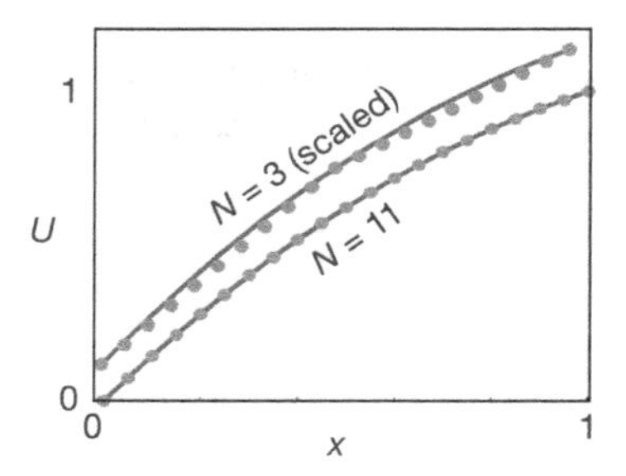

Figure 23.3 Exact (line) vs. FEM solution (points) for the two-plate problem for $N = 3$ and $N = 11$ finite elements ($N = 3$ displaced upwards for clarity). On this scale, the $N = 11$ solution IS identical to the exact one.

8. The log of the relative global error (number of significant figures) is

$$\mathcal{E} = \log_{10} \left| \frac{1}{b-a} \int_a^b dx \frac{U_{\text{FEM}}(x) - U_{\text{exact}}(x)}{U_{\text{exact}}(x)} \right| . \tag{23.27}$$

Plot the global error vs. x for $N = 10, 100$, and 1000.

Listing 23.1 LaplaceFEM_1D.py provides an FEM solution of the 1D Laplace equation via a Galerkin spectral decomposition. The resulting matrix equations are solved with Matplotlib.

```
# LaplaceFEM_1D.py:  Solutn 1D Laplace Eq via finite elements; utf8 coding

from visual import *
from visual.graph import *
from numpy import *
from numpy.linalg import solve

N = 11
h = 1. / (N - 1)
u = zeros(N, float)
A = zeros((N, N), float)
b = zeros((N, N), float)
x2 = zeros(21, float)
u_fem = zeros(21, float)
u_exact = zeros(21, float)
error = zeros(21, float)
x = zeros(N, float)

graph1 = gdisplay(width=500,height=500,title='Analytic (Blue) vs FEM',\
            xtitle='x',ytitle='U',xmax=1, ymax=1, xmin=0, ymin=0)
funct1 = gcurve(color=color.blue)
funct2 = gdots(color=color.red)
funct3 = gcurve(color=color.cyan)

for i in range(0, N):
    x[i] = i * h
for i in range(0, N):                                      # Initialize
    b[i, 0] = 0.
    for j in range(0, N):
        A[i][j] = 0.

def lin1(x, x1, x2):                                       # Hat func
    return (x-x1)/(x2-x1)

def lin2(x, x1, x2):
    return (x2-x)/(x2-x1)

def f(x):
    return 1.

def int1(min, max):                                        # Simpson
    no = 1000
    sum = 0.
    interval = (max - min) / (no - 1)
    for n in range(2, no, 2):                              # Loop odd points
        x = interval * (n - 1)
        sum += 4 * f(x) * lin1(x, min, max)
    for n in range(3, no, 2):                              # Loop even points
        x = interval * (n - 1)
        sum += 2 * f(x) * lin1(x, min, max)
    sum += f(min)*lin1(min, min, max) + f(max)*lin1(max, min, max)
    sum *= interval/6.
```

```
    return sum

def int2(min, max):                                              # Simpson
    no = 1000
    sum = 0.
    interval = (max - min) / (no - 1)
    for n in range(2, no, 2):                                    # Loop odd points
        x = interval * (n - 1)
        sum += 4 * f(x) * lin2(x, min, max)
    for n in range(3, no, 2):                                    # Loop even points
        x = interval * (n - 1)
        sum += 2 * f(x) * lin2(x, min, max)
    sum += f(min) * lin2(min, min, max) + f(max) * lin2(max, min, max)
    sum *= interval / 6.
    return sum

def numerical(x, u, xp):
    N = 11                                                       # Interpolate solution
    y = 0.
    for i in range(0, N - 1):
        if xp >= x[i] and xp <= x[i + 1]:
            y = lin2(xp,x[i],x[i+1])*u[i] + lin1(xp,x[i],x[i+1])*u[i+1]
    return y

def exact(x):                                                    # Analytic solution
    u = -x * (x - 3.) / 2.
    return u

for i in range(1, N):
    A[i - 1, i - 1] = A[i - 1, i - 1] + 1. / h
    A[i - 1, i] = A[i - 1, i] - 1. / h
    A[i, i - 1] = A[i - 1, i]
    A[i, i] = A[i, i] + 1. / h
    b[i - 1, 0] = b[i - 1, 0] + int2(x[i - 1], x[i])
    b[i, 0] = b[i, 0] + int1(x[i - 1], x[i])

for i in range(1, N):                                            # Dirichlet BC left end
    b[i, 0] = b[i, 0] - 0. * A[i, 0]
    A[i, 0] = 0.
    A[0, i] = 0.
A[0, 0] = 1.
b[0, 0] = 0.

for i in range(1, N):                                            # Dirichlet BC right end
    b[i, 0] = b[i, 0] - 1. * A[i, N - 1]
    A[i, N - 1] = 0.
    A[N - 1, i] = 0.
A[N - 1, N - 1] = 1.
b[N - 1, 0] = 1.
sol = solve(A, b)

for i in range(0, N):
    u[i] = sol[i, 0]

for i in range(0, 21):
    x2[i] = 0.05 * i

for i in range(0, 21):
    u_fem[i] = numerical(x, u, x2[i])
    u_exact[i] = exact(x2[i])
    funct1.plot(pos=(0.05 * i, u_exact[i]))
    funct2.plot(pos=(0.05 * i, u_fem[i]))
    error[i] = u_fem[i] - u_exact[i]                             # Global error
```

23.5.1 1D Exploration

1. Modify your program to use piecewise-quadratic functions for interpolation, and compare the results obtained to those obtained with the linear functions.
2. Explore the resulting electric potential and check that the charge distribution between the plates has the explicit x dependence

$$\rho(x) = \frac{1}{4\pi} \begin{cases} \frac{1}{2} - x\,, \\ \sin x\,, \\ 1 \quad \text{at} \quad x = 0\,, \quad -1 \quad \text{at} \quad x = 1 \quad \text{(a capacitor)}\,. \end{cases} \tag{23.28}$$

23.6 Extension to 2D Finite Elements

The steps followed to derive the 2D finite elements method are similar to those for the 1D method, with the big difference being that the finite elements are now 2D triangles as opposed to 1D lines. Figure 23.4 shows how an arbitrarily shaped domain might be decomposed into triangles. Although life is simpler if all the finite elements are of the same size and shape, this is not necessary, and, indeed, as we have shown in the figure, higher precision and faster run times maybe obtained by picking smaller domains in regions where the solution is known to vary rapidly, and picking larger domains in regions of slow variation. As you can imagine, 2D

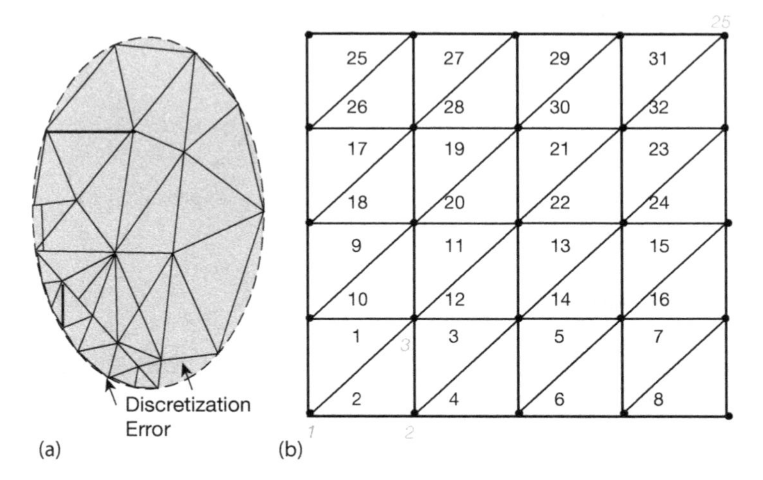

Figure 23.4 (a) Decomposition of a domain into triangular elements. Smaller triangles are used in regions of rapid variation and larger triangles are used in regions of slow variation. Discretization errors occur at boundaries. (b) A decomposition of a rectangular domain into 32 right triangles on a mesh with 25 nodes.

and 3D FEMs can get rather complicated, but not to worry, we will just outline the 2D method and refer the interested reader to Polycarpou (2006) and Reddy (1993) for fuller discussions.

We extend our previous 1D method to solve the 2D version of the Laplace equation (19.4),

$$\frac{\partial^2 U}{\partial x^2} + \frac{\partial^2 U}{\partial y^2} = 0 . \tag{23.29}$$

There are now 2D Dirichlet boundary conditions:

$$U(x,0) = 0 , \quad U(x,h) = U_0 , \quad U(0,y) = 0 , \quad U(w,y) = 0 . \tag{23.30}$$

Here, h is the height and w is the width of the rectangular domain in which we desire a solution. Because our problem domain is now rectangular, it is easy to divide it into right triangles, as we have shown in Figure 23.4b.

23.6.1 Weak Form of PDE

For the 2D problem, the weak form of the PDE again follows from multiplying both sides of the PDE by the trial solution Φ, and then integrating (Polycarpou, 2006):

$$\iint_\Omega \left(\frac{\partial \Phi}{\partial x}\frac{\partial U}{\partial x} + \frac{\partial \Phi}{\partial y}\frac{\partial U}{\partial y} \right) \mathrm{d}x\,\mathrm{d}y = \oint_\Gamma \left(\frac{\partial U}{\partial x} n_x + \frac{\partial U}{\partial y} n_y \right) \mathrm{d}l . \tag{23.31}$$

Here, Ω is a surface boundary of the domain in which we seek a solution, Γ is a perimeter around the surface, U is the solution of the PDE, and n_x and n_y are outward-facing unit normal to Γ. For Dirichlet boundary conditions the contribution of the line integral on the RHS vanishes.

23.6.2 Galerkin's Spectral Decomposition

As in the 1D method, the approximate solution $U(x, y)$ is expanded in a set $\phi_i(x, y)$ of basis functions, in this case 2D functions:

$$U(x,y) = \sum_{j=0}^{N-1} \alpha_j \phi_j(x,y). \tag{23.32}$$

After setting $U = \phi_j$ for $j = 1, 2, \ldots, N-1$, the weak form of the PDE becomes a set of linear equations:

$$\iint_\Omega \left[\left(\frac{\partial \phi_i}{\partial x}\right) \left(\sum_{j=0}^{N-1} \alpha_j \frac{\partial \phi_j}{\partial x} \right) + \left(\frac{\partial \phi_i}{\partial y}\right) \left(\sum_{j=0}^{N-1} \alpha_j \frac{\partial \phi_j}{\partial y} \right) \right] \mathrm{d}x\,\mathrm{d}y$$
$$= \oint_\Gamma \left(\frac{\partial U}{\partial x} n_x + \frac{\partial U}{\partial y} n_y \right) \mathrm{d}l . \tag{23.33}$$

We rewrite these equations in the standard matrix form (23.14) for linear equations:

$$\begin{bmatrix} A_{11} & A_{12} & \cdots & A_{1N} \\ A_{21} & A_{22} & \cdots & A_{2N} \\ \vdots & \vdots & \ddots & \vdots \\ A_{N1} & A_{N2} & \cdots & A_{NN} \end{bmatrix} \begin{bmatrix} U_1 \\ U_2 \\ \vdots \\ U_n \end{bmatrix} = \begin{bmatrix} b_1 \\ b_2 \\ \vdots \\ b_n \end{bmatrix} . \tag{23.34}$$

Here the U_i's are the vectors of unknowns, the *stiffness matrix* elements are

$$A_{ij} = -\iint\limits_{\Omega} \left(\frac{\partial \phi_i}{\partial x} \frac{\partial \phi_j}{\partial x} + \frac{\partial \phi_i}{\partial y} \frac{\partial \phi_j}{\partial y} \right) \mathrm{d}x\, \mathrm{d}y , \tag{23.35}$$

and the load vector $\boldsymbol{b}$ is given by (23.20).

23.6.3 Triangular Elements

Triangular elements are often used in 2D FEM because they can be fit into many arbitrary geometries with little overlap and with little discretization error at the boundary edges (see Figure 23.4). As we see in Figure 23.5a, we take these elements to be triangles of arbitrary shape, with the CCW arrow indicating the direction in which the nodes are numbered. While it is easier to fit arbitrary shaped triangles into a general region, it is easier to do mathematics with right triangles, such as the master triangle shown in Figure 23.5b. The latter have their orthogonal sides lying along the ξ and η axes, with the (x, y) and (ξ, η) coordinates related by a linear coordinate transformation.

These master triangles are the linear interpolation functions in the ξ and η variables that are used in 2D FEM. For example, the linear function for node 1 has the form

$$\phi_1(\xi, \eta) = a + b\xi + c\eta . \tag{23.36}$$

The constants are determined by evaluating the functions at each node, for example,

$$\phi_1(0,0) = a = 1 , \quad \phi_1(1,0) = 1 + b = 0 \Rightarrow b = -1 , \tag{23.37}$$

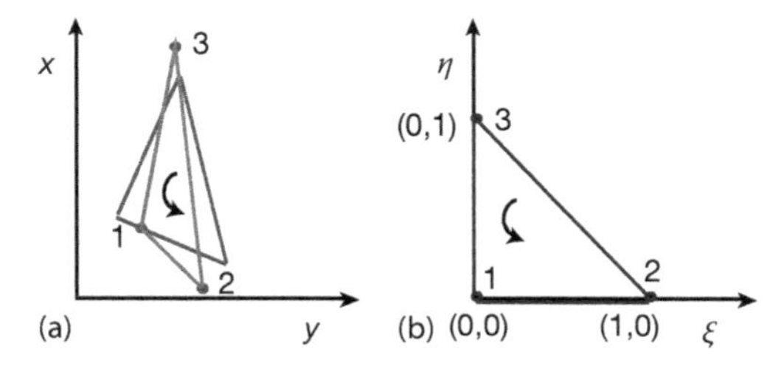

Figure 23.5 (a) Linear triangular elements in the x–y plane. (b) Linear triangular element (master element) in the $\xi\eta$ plane.

$$\phi_1(0,1) = 1 + c = 0 \Rightarrow c = -1 , \tag{23.38}$$

$$\Rightarrow \quad \phi_1(\xi, \eta) = 1 - \xi - \eta . \tag{23.39}$$

Similar evaluations at the other nodes yield (Polycarpou, 2006):

$$\phi_2 = \xi , \quad \phi_3 = \eta . \tag{23.40}$$

With these interpolation functions in hand, it is possible to express the x and y coordinates of any point inside an element in terms of the master coordinates:

$$x = x_1 + \bar{x}_{21}\xi + \bar{x}_{31}\eta , \tag{23.41}$$

$$y = y_1 + \bar{y}_{21}\xi + \bar{y}_{31}\eta , \tag{23.42}$$

$$\bar{x}_{ij} \stackrel{\text{def}}{=} x_i - x_j , \quad \bar{y}_{ij} \stackrel{\text{def}}{=} y_i - y_j . \tag{23.43}$$

Next, we take these discrete forms for the interpolation functions, return to the Galerkin spectral decomposition, and use (23.35) to evaluate the **A** matrix. The required derivatives are evaluated using the chain rule:

$$\frac{\partial \phi}{\partial \xi} = \frac{\partial \phi}{\partial x}\frac{\partial x}{\partial \xi} + \frac{\partial \phi}{\partial y}\frac{\partial y}{\partial \xi} , \tag{23.44}$$

$$\frac{\partial \phi}{\partial \eta} = \frac{\partial \phi}{\partial x}\frac{\partial x}{\partial \eta} + \frac{\partial \phi}{\partial y}\frac{\partial y}{\partial \eta} . \tag{23.45}$$

We write these equations in the matrix form as

$$\begin{bmatrix} \partial\phi/\partial\xi \\ \partial\phi/\partial\eta \end{bmatrix} = \begin{bmatrix} \partial x/\partial\xi & \partial y/\partial\xi \\ \partial x/\partial\eta & \partial y/\partial\eta \end{bmatrix} \begin{bmatrix} \partial\phi/\partial x \\ \partial\phi/\partial y \end{bmatrix} , \tag{23.46}$$

where the 2×2 matrix that defines the coordinate transformation between the (x, y) and the (ξ, η) derivatives is called the *Jacobian* matrix J. After substitution of the explicit forms for the ϕ's, the Jacobian takes the simple form:

$$J = \begin{bmatrix} \bar{x}_{21} & \bar{y}_{21} \\ \bar{x}_{31} & \bar{y}_{31} \end{bmatrix} . \tag{23.47}$$

Likewise, the derivatives in the **A** matrix can be expressed in terms of the x and y derivatives by using the inverse of the Jacobain matrix:

$$\begin{bmatrix} \frac{\partial \phi}{\partial x} \\ \frac{\partial \Phi}{\partial y} \end{bmatrix} = J^{-1} \begin{bmatrix} \frac{\partial \phi}{\partial \xi} \\ \frac{\partial \Phi}{\partial \eta} \end{bmatrix} , \tag{23.48}$$

$$J^{-1} = \frac{1}{|J|} \begin{bmatrix} \bar{y}_{31} & -\bar{y}_{21} \\ -\bar{x}_{31} & \bar{x}_{21} \end{bmatrix} , \quad |J| \equiv \det(J) = \bar{x}_2 \bar{y}_{31} - \bar{x}_{31} \bar{y}_{21} . \tag{23.49}$$

Continued evaluation of the Galerkin matrix elements yields

$$\begin{bmatrix} \frac{\partial \phi_1}{\partial x} \\ \frac{\partial \phi_1}{\partial y} \end{bmatrix} = \frac{1}{|J|} \begin{bmatrix} \bar{y}_{31} & -\bar{y}_{21} \\ -\bar{x}_{31} & \bar{x}_{21} \end{bmatrix} \begin{bmatrix} \frac{\partial \phi_1}{\partial \xi} \\ \frac{\partial \phi_1}{\partial \eta} \end{bmatrix} = \frac{1}{|J|} \begin{bmatrix} \bar{y}_{31} & -\bar{y}_{21} \\ -\bar{x}_{31} & \bar{x}_{21} \end{bmatrix} \begin{bmatrix} -1 \\ -1 \end{bmatrix} \tag{23.50}$$

$$= \frac{1}{|J|} \begin{bmatrix} \bar{y}_{21} & -\bar{y}_{31} \\ \bar{x}_{31} & -\bar{x}_{21} \end{bmatrix} = \frac{1}{|J|} \begin{bmatrix} \bar{y}_{23} \\ \bar{x}_{32} \end{bmatrix} . \tag{23.51}$$

After similar evaluations for ϕ_2 and ϕ_3, we obtain the six needed derivatives:

$$\frac{\partial \phi_1}{\partial x} = \frac{\bar{y}_{23}}{|J|} , \quad \frac{\partial \phi_1}{\partial y} = \frac{\bar{x}_{32}}{|J|} , \tag{23.52}$$

$$\frac{\partial \phi_2}{\partial x} = \frac{\bar{y}_{31}}{|J|} , \quad \frac{\partial \phi_2}{\partial y} = \frac{\bar{x}_{13}}{|J|} , \tag{23.53}$$

$$\frac{\partial \phi_3}{\partial x} = \frac{\bar{y}_{12}}{|J|} , \quad \frac{\partial \phi_3}{\partial y} = \frac{\bar{x}_{21}}{|J|} . \tag{23.54}$$

23.6.4
Solution as Linear Equations

The final evaluations of the stiffness matrix elements are made using the ξ and η coordinates, for example,

$$A_{11} = -\int_0^1 \int_0^{1-\eta} \left[\frac{\bar{y}_{23}^2 + \bar{x}_{32}^2}{|J|^2} \right] |J| \, \mathrm{d}\xi \, \mathrm{d}\eta . \tag{23.55}$$

These elements are found to form a symmetric matrix with values:

$$A_{12} = A_{21} = -\frac{\bar{y}_{23}\bar{y}_{31} + \bar{x}_{32}\bar{x}_{13}}{2|J|} , \quad A_{11} = -\frac{\bar{y}_{23}^2 + \bar{x}_{32}^2}{2|J|} , \tag{23.56}$$

$$A_{13} = A_{31} = -\frac{\bar{y}_{23}\bar{y}_{12} + \bar{x}_{32}\bar{x}_{21}}{2|J|} , \quad A_{22} = -\frac{\bar{y}_{31}^2 + \bar{x}_{13}^2}{2J} , \tag{23.57}$$

$$A_{23} = A_{32} = -\frac{\bar{y}_{31}\bar{y}_{12} + \bar{x}_{13}\bar{x}_{21}}{2|J|} , \quad A_{33} = -\frac{\bar{y}_{12}^2 + \bar{x}_{21}^2}{2J} . \tag{23.58}$$

Next we evaluates the coordinate transformations:

$$\begin{bmatrix} x - x_1 \\ y - y_1 \end{bmatrix} = \begin{bmatrix} \bar{x}_{21} & \bar{x}_{31} \\ \bar{y}_{21} & \bar{y}_{31} \end{bmatrix} \begin{bmatrix} \xi \\ \eta \end{bmatrix} , \tag{23.59}$$

$$\begin{bmatrix} \xi \\ \eta \end{bmatrix} = \frac{1}{\bar{x}_{21}\bar{y}_{31} - \bar{x}_{31}\bar{y}_{21}} \begin{bmatrix} \bar{y}_{31} & -\bar{x}_{31} \\ -\bar{y}_{21} & \bar{x}_{21} \end{bmatrix} \begin{bmatrix} x - x_1 \\ y - y_1 \end{bmatrix} . \tag{23.60}$$

After substituting for ξ and η, we are left with the desired interpolation functions expressed in terms of just x and y.

23.6.5
Imposing Boundary Conditions

The procedure to impose Dirichlet's boundary conditions for the 2D case is essentially the same as that for the 1D case (Section 23.4.2.2), with it now applied to all nodes that lie on the boundary Γ.

Listing 23.2 The code **LaplaceFEM_2D.py** solves the 2D Laplace equation using a finite elements method.

```
# LaplaceFEM_2D.py solve 2D Laplace Eq via Finite elements method;
    utf-8coding

from numpy import *
from numpy.linalg import solve
import pylab as p
from mpl_toolkits.mplot3d import Axes3D

# Num squares, nodes, triangles, mesh coords, Initialization

Width = 1.;      Height = 1.;   Nx = 20;     Ny = 20; U0 = 100
Xurc = Width;   Yurc = Height;     Yllc = 0;     Xllc = 0
Ns = Nx * Ny;   Nn = (Nx + 1)*(Ny + 1)
Dx = (Xurc-Xllc)/Nx;      Dy = (Yurc-Yllc)/Ny;     Ne = 2 * Ns
ge = zeros(Ne, float)
x = zeros(Ne, float);         y = zeros(Ne, float)
Ebcnod = zeros(Ne, int);     Ebcval = zeros(Ne, int)
node = zeros((Ne + 1, Ne + 1), int)

for i in range(1, Nn + 1):
    x[i] = (i - 1) % (Nx + 1) * Dx
    y[i] = floor((i - 1) / (Nx + 1)) * Dy

# Connectivity Information
for i in range(1, Ns + 1):
    node[2 * i - 1, 1] = i + floor((i - 1) / Nx)
    node[2 * i - 1, 2] = node[2 * i - 1, 1] + 1 + Nx + 1
    node[2 * i - 1, 3] = node[2 * i - 1, 1] + 1 + Nx + 1 - 1
    node[2 * i, 1] = i + floor((i - 1) / Nx)
    node[2 * i, 2] = node[2 * i, 1] + 1
    node[2 * i, 3] = node[2 * i, 1] + 1 + Nx + 1

# Dirichlet Boundary Conditions
Tnebc = 0
for i in range(0, Nn):
    if x[i] == Xllc or x[i] == Xurc or y[i] == Yllc:
        Tnebc = Tnebc + 1
        Ebcnod[Tnebc] = i
        Ebcval[Tnebc] = 0
    elif y[i] == Yurc:
        Tnebc = Tnebc + 1
        Ebcnod[Tnebc] = i
        Ebcval[Tnebc] = U0

# Initialize A matrix, b vector, form matrix
A = zeros((Nn + 1, Nn + 1), float)
b = zeros((Nn + 1, 1), float)
for e in range(1, Ne):
    x21 = x[node[e, 2]] - x[node[e, 1]]
    x31 = x[node[e, 3]] - x[node[e, 1]]
    x32 = x[node[e, 3]] - x[node[e, 2]]
    x13 = x[node[e, 1]] - x[node[e, 3]]
    y12 = y[node[e, 1]] - y[node[e, 2]]
```

```
        y21 = y[node[e, 2]] - y[node[e, 1]]
        y31 = y[node[e, 3]] - y[node[e, 1]]
        y23 = y[node[e, 2]] - y[node[e, 3]]
        J = x21 * y31 - x31 * y21

# Evaluate A matrix, element vector ge
        A[1, 1] = -(y23 * y23 + x32 * x32) / (2 * J)
        A[1, 2] = -(y23 * y31 + x32 * x13) / (2 * J)
        A[2, 1] = A[1, 2]
        A[1, 3] = -(y23 * y12 + x32 * x21) / (2 * J)
        A[3, 1] = A[1, 3]
        A[2, 2] = -(y31 * y31 + x13 * x13) / (2 * J)
        A[2, 3] = -(y31 * y12 + x13 * x21) / (2 * J)
        A[3, 2] = A[2, 3]
        A[3, 3] = -(y12 * y12 + x21 * x21) / (2 * J)
        ge[1] = 0
        ge[2] = 0
        ge[3] = 0

# Evaluate element pe & update A matrix
        for i in range(1, 4):
            for j in range(1, 4):
                A[node[e, i], node[e, j]] = A[node[e, i], node[e, j]] \
                    + A[i, j]
            b[node[e, i]] = b[node[e, i]] + ge[i]

# Imposition of Dirichlet boundary conditions
for i in range(1, Tnebc):
    for j in range(1, Nn + 1):
        if j != Ebcnod[i]:
            b[j] = b[j] - A[j, Ebcnod[i]] * Ebcval[i]
    A[Ebcnod[i], :] = 0
    A[:, Ebcnod[i]] = 0
    A[Ebcnod[i], Ebcnod[i]] = 1
    b[Ebcnod[i]] = Ebcval[i]

# Solution, place on grid, plot
V = linalg.solve(A, b)
(X, Y) = p.meshgrid(arange(Xllc, Xurc + 0.1, 0.1 * (Xurc - Xllc)),
                    arange(Yllc, Yurc + 0.1, 0.1 * (Yurc - Yllc)))
Vgrid = zeros((11, 11), float)
for i in arange(1, 11):
    for j in arange(1, 11):
        for e in range(0, Ne):
            x2p = x[node[e, 2]] - X[i, j]
            x3p = x[node[e, 3]] - X[i, j]
            y2p = y[node[e, 2]] - Y[i, j]
            y3p = y[node[e, 3]] - Y[i, j]
            A1 = 0.5 * abs(x2p * y3p - x3p * y2p)
            x2p = x[node[e, 2]] - X[i, j]
            x1p = x[node[e, 1]] - X[i, j]
            y2p = y[node[e, 2]] - Y[i, j]
            y1p = y[node[e, 1]] - Y[i, j]
            A2 = 0.5 * abs(x2p * y1p - x1p * y2p)
            x1p = x[node[e, 1]] - X[i, j]
            y21 = y[node[e, 2]] - y[node[e, 1]]
            y1p = y[node[e, 1]] - Y[i, j]
            x21 = x[node[e, 2]] - x[node[e, 1]]
            A3 = 0.5 * abs(x1p * y3p - x3p * y1p)
            y3p = y[node[e, 3]] - Y[i, j]
            x31 = x[node[e, 3]] - x[node[e, 1]]
            x3p = x[node[e, 3]] - X[i, j]
            y31 = y[node[e, 3]] - y[node[e, 1]]
            J = x21 * y31 - x31 * y21
            if abs(J / 2 - (A1 + A2 + A3)) < 0.00001 * J / 2:
                ksi = (y31 * (X[i, j] - x[node[e, 1]]) - x31 * (Y[i, j]
                      - y[node[e, 1]])) / J
```

```
            ita = (-y21 * (X[i, j] - x[node[e, 1]]) + x21 * (Y[i,
                j] - y[node[e, 1]])) / J
            N1 = 1 - ksi - ita
            N2 = ksi
            N3 = ita
            Vgrid[i, j] = N1 * V[node[e, 1]] + N2 * V[node[e, 2]] \
                + N3 * V[node[e, 3]]

# Plot the finite element solution of V using a contour plot
fig = p.figure()
ax = Axes3D(fig)
ax.plot_wireframe(X, Y, Vgrid, color='r')
ax.set_xlabel('X')
ax.set_ylabel('Y')
ax.set_zlabel('Potential')
p.show()
```

23.6.6 FEM 2D Implementation and Exercise

As shown in Figure 23.4b, our application of 2D FEM has the solution domain covered by a collection of triangular elements. Each triangle in the mesh is numbered, in this case from 1 to 32. In addition, the three vertices of each triangle are numbered in a counter-clockwise direction from 1 to 3. Next, each node in the mesh (the dark circles in Figure 23.4 where lines intersect) is numbered, in this case from 1 to 25. Accordingly, the stiffness matrix **A** in (23.34) has dimension 25×25, while the load vector $\boldsymbol{b}$ has dimension 25×1.

Listing 23.2 presents our implementation of the 2D FEM solution to the 2D Laplace equation, based on the Matlab code of Polycarpou (2006). It utilizes 800 elements and 441 nodes. The output of this code is essentially the same as our solution to the same problem using the finite-differences method.

23.6.7 FEM 2D Exercises

1. Examine the effect of varying the domain height and width, as well as the number of elements.
2. Compare this numerical solution to the analytic one (the Fourier series in Section 19.3) and determine how the precision changes as the number of elements is varied.
3. Modify the program so that it solves the parallel plate capacitor problem and compare to the finite-difference solution.

24
Shocks Waves and Solitons

In the first part of this chapter, we extend the discussion of waves in Chapters 21 and 22 by progressively including nonlinearities, dispersion, and hydrodynamic effects. We end up with the Korteweg–de Vries equation and shallow-water solitons. In the second part of this chapter, we explore the inclusion of related nonlinear physics for the pendulum chain, and end up with the Sine–Gordon equation and solitons in solids.

24.1
Shocks and Solitons in Shallow Water

In 1834, J. Scott Russell (Russell, 1944) observed on the Edinburgh–Glasgow canal (repeated recently in Figure 24.1a):

> *I was observing the motion of a boat which was rapidly drawn along a narrow channel by a pair of horses, when the boat suddenly stopped – not so the mass of water in the channel which it had put in motion; it accumulated round the prow of the vessel in a state of violent agitation, then suddenly leaving it behind, rolled forward with great velocity, assuming the form of a large solitary elevation, a rounded, smooth and well-defined heap of water, which continued its course along the channel apparently without change of form or diminution of speed. I followed it on horseback, and overtook it still rolling on at a rate of some eight or nine miles an hour, preserving its original figure some thirty feet long and a foot to a foot and a half in height. Its height gradually diminished, and after a chase of one or two miles I lost it in the windings of the channel. Such, in the month of August 1834, was my first chance interview with that singular and beautiful phenomenon...*

Russell also noticed that an initial, arbitrary waveform set in motion in the channel evolves into two or more waves that move at different velocities and progressively move apart until they form individual solitary waves. In Figure 24.2b, we see a single step-like wave breaking up into approximately eight of these solitary waves (now called *solitons*). These eight solitons occur so frequently that some consider them the equivalent of normal modes for nonlinear systems. Russell went on to

Computational Physics, 3rd edition. Rubin H. Landau, Manuel J. Páez, Cristian C. Bordeianu.

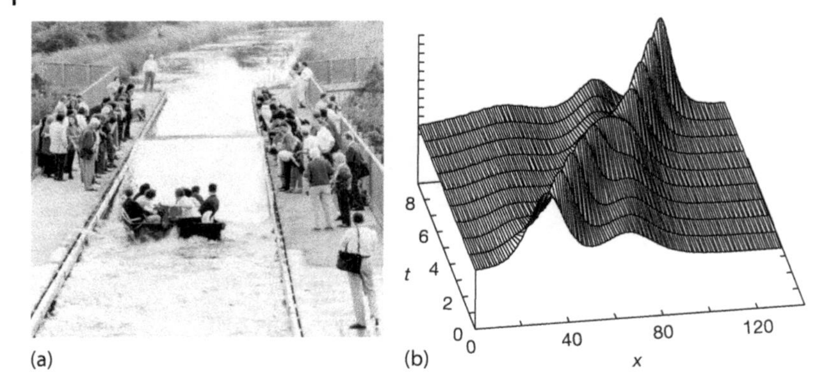

Figure 24.1 (a) A re-creation on the Union Canal near Edinburgh of Russel's soliton (www.ma.hw.ac.uk/solitons/press.html, Nature v. 376, 3 August 1995, p. 373). (b) Two shallow-water solitary waves crossing each other computed with the code Soliton.py. The taller soliton on the left catches up with and overtakes the shorter one at $t \simeq 5$. The waves resume their original shapes after the collision.

produce these solitary waves in a laboratory and to empirically deduced that their speed c is related to the depth h of the water in the canal and to the amplitude A of the wave by

$$c^2 = g(h + A) , \tag{24.1}$$

where g is the acceleration as a result of gravity. Equation 24.1 implies an effect not found in linear systems, namely, that waves with greater amplitudes A travel faster than those with smaller amplitudes.Observe that this is similar to the formation of shock waves, but different from *dispersion* in which waves of different wavelengths have different velocities. The dependence of c on the amplitude A is illustrated in Figure 24.2, where we see a taller soliton catching up with and passing through a shorter one.

Problem Explain Russell's observations and see if they relate to the formation of *tsunamis*. The latter are ocean waves that form from sudden changes in the level of the ocean floor, and then travel over long distances without dispersion or attenuation, possibly reeking havoc on distant shores.

24.2
Theory: Continuity and Advection Equations

The motion of a fluid is described by the continuity equation and the Navier–Stokes equation (Landau and Lifshitz, 1976). We will discuss the former here and the latter in Section 25.2. The continuity equation describes the conservation of mass:

$$\frac{\partial \rho(\boldsymbol{x}, t)}{\partial t} + \boldsymbol{\nabla} \cdot \boldsymbol{j} = 0 , \quad \boldsymbol{j} \stackrel{\text{def}}{=} \rho \boldsymbol{v}(\boldsymbol{x}, t) . \tag{24.2}$$

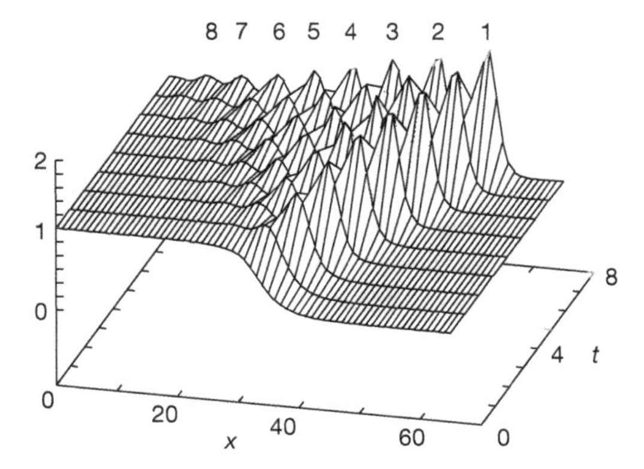

Figure 24.2 The formation of a tsunami. A single two-level waveform at time zero progressively breaks up into eight solitons (labeled) as time increases. The tallest soliton (1) is narrower and faster in its motion to the right. You can generate an animation of this with the program **SolitonAnimate.py**.

Here, $\rho(\boldsymbol{x}, t)$ is the mass density, $\boldsymbol{v}(\boldsymbol{x}, t)$ is the velocity of the fluid, and the product $\boldsymbol{j} = \rho\boldsymbol{v}$ is the mass current. As its name implies, the divergence $\boldsymbol{\nabla} \cdot \boldsymbol{j}$ describes the spreading of the current in a region of space, as might occur if there were a current source there. Physically, the continuity equation (24.2) states that changes in the density of the fluid within some region of space arise from the flow of current in and out of that region.

For 1D flow in the x direction, and for a fluid that is moving with a constant velocity $v = c$, the continuity equation (24.2) takes the simple form

$$\frac{\partial \rho}{\partial t} + c\frac{\partial \rho}{\partial x} = 0 . \tag{24.3}$$

This equation is known as the *advection equation*, where the term "advection" is used to describe the horizontal transport of a quantity from one region of space to another as a result of a flow velocity field. For instance, advection describes dissolved salt transported in water.

The advection equation looks like a first-derivative form of the wave equation, and indeed, the two are related. A simple substitution proves that any function with the form of a traveling wave,

$$u(x, t) = f(x - ct) , \tag{24.4}$$

will be a solution of the advection equation. If we consider a surfer riding along the crest of a traveling wave, that is, remaining at the same position relative to the wave's shape as time changes, then the surfer does not see the shape of the wave change in time, which implies that

$$x - ct = \text{constant} \quad \Rightarrow \quad x = ct + \text{constant} . \tag{24.5}$$

The speed of the surfer is, therefore $\mathrm{d}x/\mathrm{d}t = c$, which is a constant. Any function $f(x - ct)$ is clearly a traveling wave solution in which an arbitrary pulse is carried along by the fluid at velocity c without changing shape.

24.2.1 Advection Implementation

Although the advection equation is simple, trying to solve it by a simple differencing scheme (the leapfrog method) may lead to unstable numerical solutions. As we shall see when we look at the nonlinear version of this equation, there are better ways to solve it. Listing 24.1 presents our code `AdvecLax.py` for solving the advection equation using the Lax–Wendroff method (a better method).

Listing 24.1 AdvecLax.py solves the advection equation via the Lax–Wendroff scheme.

```
# AdvecLax.py:          Solve advection eqnt via Lax--Wendroff scheme
# du/dt+ c*d(u**2/2)/dx=0;    u(x,t=0)=exp(-300(x-0.12)**2)

from visual.graph import *
m = 100                                                 # No steps in x
c = 1.;        dx = 1./m;      beta = 0.8               # beta = c*dt/dx
u = [0]*(m+1);                                          # Initial Numeric
u0 = [0]*(m+1);
uf = [0]*(m+1)
dt = beta*dx/c;
T_final = 0.5;
n = int(T_final/dt)                                      # N time steps

graph1 = gdisplay(width=600, height=500, xtitle = 'x', xmin=0, xmax=1,
         ymin=0, ymax=1, ytitle = 'u(x), Cyan=exact, Yellow=Numerical',
         title='Advection Eqn: Initial (red), Exact (cyan),\
         Numerical Lax--Wendroff (yellow)')
initfn = gcurve(color = color.red);
exactfn = gcurve(color = color.cyan)
numfn = gcurve(color = color.yellow)                     # Numerical solution

def plotIniExac():                              # Plot initial & exact solution
    for i in range(0, m):
        x = i*dx
        u0[i] = exp(-300.* (x - 0.12)**2)                    # Gaussian initial
        initfn.plot(pos = (0.01*i, u0[i]) )                  # Initial function
        uf[i] = exp(-300.*(x - 0.12 - c*T_final)**2)         # Exact in cyan
        exactfn.plot(pos = (0.01*i, uf[i]) )
        rate(50)
plotIniExac()

def numerical():                                # Finds Lax--Wendroff solution
   for j in range(0, n+1):                                     # Time loop
      for i in range(0, m - 1):                                #    x loop
          u[i + 1] = (1.-beta*beta)*u0[i+1]-(0.5*beta)*(1.-beta)*u0[i+2] \
                      +(0.5*beta)*(1. + beta)*u0[i]             # Algorithm
          u[0] = 0.;      u[m-1] = 0.;      u0[i] = u[i]
numerical()
for j in range(0, m-1 ):
    rate(50)
    numfn.plot(pos = (0.01*j, u[j]) )             # Plot numerical Solution
```

24.3
Theory: Shock Waves via Burgers' Equation

In a later section, we will examine the Korteweg–de Vries equation's description of solitary waves. In order to understand the physics contained in that equation, we study, one at a time, some of the terms in it. To start, consider Burgers' equation (Burgers, 1974):

$$\frac{\partial u}{\partial t} + \epsilon u \frac{\partial u}{\partial x} = 0 , \tag{24.6}$$

$$\frac{\partial u}{\partial t} + \epsilon \frac{\partial (u^2/2)}{\partial x} = 0 , \tag{24.7}$$

where the second equation is the *conservative form*. This equation can be viewed as a variation on the advection equation (24.3) in which the wave speed $c = \epsilon u$ is proportional to the amplitude of the wave, as Russell found for his waves. The second, nonlinear, term in Burgers' equation leads to some unusual behaviors. Indeed, von Neumann studied this equation as a simple model for turbulence (Falkovich and Sreenivasan, 2006).

In the advection equation (24.3), all points on the wave move at the same speed c, and so the shape of the wave remains unchanged in time. In Burgers' equation (24.6), the points on the wave move ("advect") themselves such that the local speed depends on the local wave's amplitude, with the high parts of the wave moving progressively faster than the low parts. This changes the shape of the wave in time; if we start with a wave packet that has a smooth variation in height, the high parts will speed up and push their way to the front of the packet, thereby forming a sharp leading edge known as a *shock wave* (Tabor, 1989). A shock wave solution to Burgers' equation with $\epsilon = 1$ is shown in Figure 24.3.

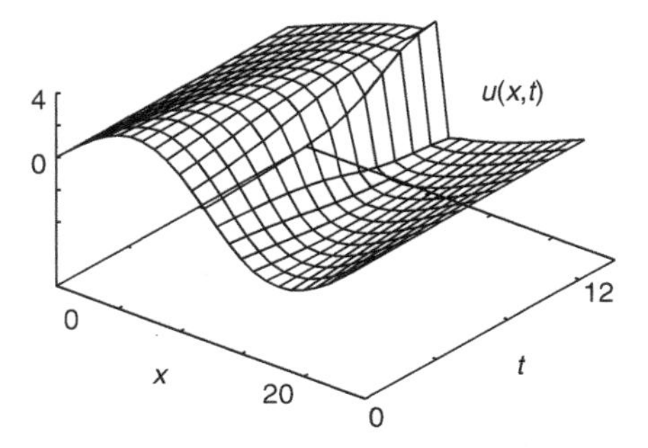

Figure 24.3 A visualization showing the wave height vs. position for increasing times showing the formation of a shock wave (sharp edge) from an initial sine wave.

24.3.1
Lax–Wendroff Algorithm for Burgers' Equation

We first solve Burgers' equation (24.3) via the usual approach in which we express the derivatives as central differences. This leads to a leapfrog scheme for the future solution in terms of present and past ones:

$$u(x, t + \Delta t) = u(x, t - \Delta t) - \beta \left[\frac{u^2(x + \Delta x, t) - u^2(x - \Delta x, t)}{2} \right] ,$$

$$u_{i,j+1} = u_{i,j-1} - \beta \left[\frac{u^2_{i+1,j} - u^2_{i-1,j}}{2} \right] , \quad \beta = \frac{\epsilon}{\Delta x / \Delta t} . \tag{24.8}$$

Here, u^2 is the square of u and is not its second derivative, and β is a ratio of constants known as the *Courant–Friedrichs–Lewy* (CFL) *number*. As you should prove for yourself, $\beta < 1$ is required for stability.

While we have used a leapfrog method successfully in the past, its low-order approximation for the derivative becomes inaccurate when the gradients can get large, as happens with shock waves, and so the leapfrog algorithm may become unstable (Press *et al.*, 1994). The *Lax–Wendroff method* attains better stability and accuracy by retaining second-order differences for the time derivative:

$$u(x, t + \Delta t) \simeq u(x, t) + \frac{\partial u}{\partial t} \Delta t + \frac{1}{2} \frac{\partial^2 u}{\partial t^2} \Delta t^2 . \tag{24.9}$$

To covert (24.9) to an algorithm, we use Burgers' equation $\partial u / \partial t = -\epsilon \partial (u^2/2)/\partial x$ for the first-order time derivative. Likewise, we use Burger's equation to express the second-order time derivative in terms of space derivatives:

$$\frac{\partial^2 u}{\partial t^2} = \frac{\partial}{\partial t} \left[-\epsilon \frac{\partial}{\partial x} \left(\frac{u^2}{2} \right) \right] = -\epsilon \frac{\partial}{\partial x} \frac{\partial}{\partial t} \left(\frac{u^2}{2} \right) \tag{24.10}$$

$$= -\epsilon \frac{\partial}{\partial x} \left(u \frac{\partial u}{\partial t} \right) = \epsilon^2 \frac{\partial}{\partial x} \left[u \frac{\partial}{\partial x} \left(\frac{u^2}{2} \right) \right] . \tag{24.11}$$

We next substitute these derivatives into the Taylor expansion (24.9) to obtain

$$u(x, t + \Delta t) = u(x, t) - \Delta t \epsilon \frac{\partial}{\partial x} \left(\frac{u^2}{2} \right) + \frac{(\Delta t)^2}{2} \epsilon^2 \frac{\partial}{\partial x} \left[u \frac{\partial}{\partial x} \left(\frac{u^2}{2} \right) \right] . \tag{24.12}$$

We now replace the outer x derivatives by central differences of spacing $\Delta x/2$:

$$\begin{aligned} u(x, t + \Delta t) = u(x, t) &- \frac{\Delta t \epsilon}{2} \frac{u^2(x + \Delta x, t) - u^2(x - \Delta x, t)}{2\Delta x} + \frac{(\Delta t)^2 \epsilon^2}{2} \\ &\times \frac{1}{2\Delta x} \left[u \left(x + \frac{\Delta x}{2}, t \right) \frac{\partial}{\partial x} u^2 \left(x + \frac{\Delta x}{2}, t \right) - u \left(x - \frac{\Delta x}{2}, t \right) \right. \\ &\left. \times \frac{\partial}{\partial x} u^2 \left(x - \frac{\Delta x}{2}, t \right) \right] . \end{aligned} \tag{24.13}$$

Next we approximate $u(x \pm \Delta x/2, t)$ by the average of adjacent grid points,

$$u(x \pm \frac{\Delta x}{2}, t) \simeq \frac{u(x,t) + u(x \pm \Delta x, t)}{2}, \tag{24.14}$$

and apply a central-difference approximation to the second derivatives:

$$\frac{\partial u^2(x \pm \Delta x/2, t)}{\partial x} = \frac{u^2(x \pm \Delta x, t) - u^2(x,t)}{\pm \Delta x}. \tag{24.15}$$

Finally, putting all these derivatives together yields the discrete form

$$\begin{aligned} u_{i,j+1} = u_{i,j} &- \frac{\beta}{4}\left(u^2_{i+1,j} - u^2_{i-1,j}\right) + \frac{\beta^2}{8} \\ &\times \left[(u_{i+1,j} + u_{i,j})\left(u^2_{i+1,j} - u^2_{i,j}\right) - (u_{i,j} + u_{i-1,j})\left(u^2_{i,j} - u^2_{i-1,j}\right)\right], \end{aligned} \tag{24.16}$$

where we have substituted the CFL number β. This Lax–Wendroff scheme is explicit, centered upon the grid points, and stable for $\beta < 1$ (small nonlinearities).

24.3.2 Implementation and Assessment of Burgers' Shock Equation

1. Write a program to solve Burgers' equation via the leapfrog method.
2. Define arrays `u0[100]` and `u[100]` for the initial data and the solution.
3. Take the initial wave to be sinusoidal, `u0[i]`$= 3\sin(3.2x)$, with speed $c = 1$.
4. Incorporate the boundary conditions `u[0]=0` and `u[100]=0`.
5. Keep the CFL number $\beta < 1$ for stability.
6. Now modify your program to solve Burgers' shock equation (24.7) using the Lax–Wendroff method (24.16).
7. Save the initial data and the solutions for a number of times in separate files for plotting.
8. Plot the initial wave and the solution for several time values on the same graph in order to see the formation of a shock wave (see Figure 24.3).
9. Run the code for several increasingly large CFL numbers. Is the stability condition $\beta < 1$ correct for this nonlinear problem?
10. Compare the leapfrog and Lax–Wendroff methods. With the leapfrog method you should see shock waves forming but breaking up into ripples as the square edge develops. The ripples are numerical artifacts. The Lax–Wendroff method should give a better shock wave (square edge), although some ripples may still occur.

Listing 24.1 presents our implementation of the Lax–Wendroff method.

24.4 Including Dispersion

We have just seen that Burgers' equation can turn an initially smooth wave into a square-edged shock wave. An inverse wave phenomenon is *dispersion*, in which a waveform disperses or broadens as it travels through a medium. Dispersion does not cause waves to lose energy and attenuate, but rather to lose information with time. Physically, dispersion may arise when the propagating medium has structures with a spatial regularity equal to some fraction of a wavelength. Mathematically, dispersion may arise from terms in the wave equation that contain higher order space derivatives. For example, consider the waveform

$$u(x,t) = \mathrm{e}^{\pm \mathrm{i}(kx-\omega t)} \tag{24.17}$$

corresponding to a plane wave traveling to the right ("traveling" because the phase $kx - \omega t$ remains unchanged if you increase x with time). When this $u(x,t)$ is substituted into the advection equation (24.3), we obtain

$$\omega = ck \ . \tag{24.18}$$

This equation is an example of a *dispersion relation*, that is, a relation between frequency ω and wave vector k. Because the *group velocity* of a wave

$$v_g = \frac{\partial \omega}{\partial k}, \tag{24.19}$$

the linear dispersion relation (24.18) leads to all frequencies having the same group velocity c and thus *dispersionless* propagation.

Let us now imagine that a wave is propagating with a small amount of *dispersion*, that is, with a frequency that has somewhat less than a linear increase with the wave number k:

$$\omega \simeq ck - \beta k^3 \ . \tag{24.20}$$

Note that we skip the even powers in (24.20), so that the group velocity,

$$v_g = \frac{\mathrm{d}\omega}{\mathrm{d}k} \simeq c - 3\beta k^2 \ , \tag{24.21}$$

is the same for waves traveling to the left the or the right. Now we work backward. If plane-wave solutions like (24.17) were to arise from a wave equation, then (as verified by substitution) the ω term of the dispersion relation (24.20) would arise from a first-order time derivative, the ck term from a first-order space derivative, and the k^3 term from a third-order space derivative:

$$\frac{\partial u(x,t)}{\partial t} + c\frac{\partial u(x,t)}{\partial x} + \beta\frac{\partial^3 u(x,t)}{\partial x^3} = 0 \ . \tag{24.22}$$

We leave it as an exercise to show that solutions to this equation do indeed have waveforms that disperse in time.

24.5
Shallow-Water Solitons: The KdeV Equation

In this section, we put all of the pieces together that are needed to generate shallow-water solitary waves. This is a subject for which the computer has been absolutely essential for discovery and understanding. In addition, we recommend that you look at some of the soliton animations we provide online.

We want to understand the unusual water waves that occur in shallow, narrow channels such as canals (Abarbanel *et al.*, 1993; Tabor, 1989). The analytic description of this “heap of water” was given by Korteweg and deVries (1895) (KdeV) with the partial differential equation

$$\boxed{\frac{\partial u(x,t)}{\partial t} + \varepsilon u(x,t)\frac{\partial u(x,t)}{\partial x} + \mu\frac{\partial^3 u(x,t)}{\partial x^3} = 0\ .} \tag{24.23}$$

As we discussed in Section 24.1 in our study of Burgers' equation, the nonlinear term $\varepsilon u \partial u/\partial t$ leads to a sharpening of the wave and ultimately a *shock* wave. In contrast, as we discussed in our study of dispersion, the $\partial^3 u/\partial x^3$ term produces broadening, while the $\partial u/\partial t$ term produces traveling waves. For the proper parameters and initial conditions, the dispersive broadening exactly balances the nonlinear narrowing, and a stable traveling wave is formed.

Korteweg and de Vries solved (24.23) analytically and proved that the speed (24.1) given by Russell is in fact correct. Seventy years after its discovery, the KdeV equation was rediscovered by Zabusky and Kruskal (1965), who solved it numerically and found that a $\cos(x/L)$ initial condition broke up into eight solitary waves (Figure 24.2). They also found that the parts of the wave with larger amplitudes moved faster than those with smaller amplitudes, which is why the higher peaks tend to be on the right in Figure 24.2. As if wonders never cease, Zabusky and Kruskal, who coined the name *soliton* for these solitary waves, also observed that a faster peak passed through a slower one unscathed (Figure 24.1).

24.5.1
Analytic Soliton Solution

The trick in analytic approaches to these types of nonlinear equations is to substitute a guessed solution that has the form of a traveling wave

$$u(x,t) = u(\xi = x - ct)\ . \tag{24.24}$$

This form means that if we move with a constant speed c, we will see a constant wave form (but now the speed will depend on the magnitude of u). There is no guarantee that this form of a solution exists, but it is a lucky guess because substituting it into the KdeV equation produces a solvable ODE:

$$-c\frac{\partial u}{\partial \xi} + \epsilon u\frac{\partial u}{\partial \xi} + \mu\frac{d^3 u}{d\xi^3} = 0\ , \tag{24.25}$$

$$\Rightarrow \quad u(x,t) = \frac{-c}{2}\operatorname{sech}^2\left[\frac{1}{2}\sqrt{c}(x - ct - \xi_0)\right] , \tag{24.26}$$

where ξ_0 is the initial phase. We see in (24.26) an amplitude that is proportional to the wave speed c, and a sech^2 function that gives a single lump-like wave. This is an analytic form for a soliton.

24.5.2 Algorithm for KdeV Solitons

The KdeV equation is solved numerically using a finite-difference scheme with the time and space derivatives given by central-difference approximations:

$$\frac{\partial u}{\partial t} \simeq \frac{u_{i,j+1} - u_{i,j-1}}{2\Delta t} , \quad \frac{\partial u}{\partial x} \simeq \frac{u_{i+1,j} - u_{i-1,j}}{2\Delta x} . \tag{24.27}$$

To approximate $\partial^3 u(x,t)/\partial x^3$, we expand $u(x,t)$ to $\mathcal{O}(\Delta t)^3$ about the four points $u(x \pm 2\Delta x, t)$ and $u(x \pm \Delta x, t)$,

$$u(x \pm \Delta x, t) \simeq u(x,t) \pm (\Delta x)\frac{\partial u}{\partial x} + \frac{(\Delta x)^2}{2!}\frac{\partial^2 u}{\partial^2 x} \pm \frac{(\Delta x)^3}{3!}\frac{\partial^3 u}{\partial x^3} , \tag{24.28}$$

which we solve for $\partial^3 u(x,t)/\partial x^3$. Finally, the factor $u(x,t)$ in the second term of (24.23) is taken as the average of three x values all with the same t:

$$u(x,t) \simeq \frac{u_{i+1,j} + u_{i,j} + u_{i-1,j}}{3} . \tag{24.29}$$

We substitute these approximations to obtain the algorithm for the KdeV equation:

$$\begin{aligned} u_{i,j+1} \simeq{}& u_{i,j-1} - \frac{\epsilon}{3}\frac{\Delta t}{\Delta x}\left[u_{i+1,j} + u_{i,j} + u_{i-1,j}\right]\left[u_{i+1,j} - u_{i-1,j}\right] \\ &- \mu\frac{\Delta t}{(\Delta x)^3}\left[u_{i+2,j} + 2u_{i-1,j} - 2u_{i+1,j} - u_{i-2,j}\right] . \end{aligned} \tag{24.30}$$

To apply this algorithm to predict future times, we need to know $u(x,t)$ at present and past times. The initial time solution $u_{i,1}$ is known for all positions i via the initial condition. To find $u_{i,2}$, we use a forward-difference scheme in which we expand $u(x,t)$, keeping only two terms for the time derivative:

$$\begin{aligned} u_{i,2} \simeq{}& u_{i,1} - \frac{\epsilon\Delta t}{6\Delta x}\left[u_{i+1,1} + u_{i,1} + u_{i-1,1}\right]\left[u_{i+1,1} - u_{i-1,1}\right] \\ &- \frac{\mu}{2}\frac{\Delta t}{(\Delta x)^3}\left[u_{i+2,1} + 2u_{i-1,1} - 2u_{i+1,1} - u_{i-2,1}\right] . \end{aligned} \tag{24.31}$$

The keen observer will note that there are still some undefined columns of points, namely, $u_{1,j}$, $u_{2,j}$, $u_{N_{\max}-1,j}$, and $u_{N_{\max},j}$, where $N_{\max}$ is the total number of grid points. A simple technique for determining their values is to assume that $u_{1,2} = 1$ and $u_{N_{\max},2} = 0$. To obtain $u_{2,2}$ and $u_{N_{\max}-1,2}$, assume that

$u_{i+2,2} = u_{i+1,2}$ and $u_{i-2,2} = u_{i-1,2}$ (avoid $u_{i+2,2}$ for $i = N_{\max} - 1$, and $u_{i-2,2}$ for $i = 2$). To carry out these steps, approximate (24.31) so that

$$u_{i+2,2} + 2u_{i-1,2} - 2u_{i+1,2} - u_{i-2,2} \to u_{i-1,2} - u_{i+1,2} \,. \tag{24.32}$$

The truncation error and stability condition for our algorithm are related:

$$\mathcal{E}(u) = \mathcal{O}[(\Delta t)^3] + \mathcal{O}[\Delta t(\Delta x)^2] \,, \tag{24.33}$$

$$\frac{1}{(\Delta x/\Delta t)}\left[\epsilon|u| + 4\frac{\mu}{(\Delta x)^2}\right] \le 1 \,. \tag{24.34}$$

The first equation shows that smaller time and space steps lead to a smaller approximation error, yet because the round-off error increases with the number of steps, the total error does not necessarily decrease (Chapter 3). Yet, we are also limited in how small the steps can be made by the stability condition (24.34), which indicates that making Δx too small always leads to instability. Care and experimentation are required.

24.5.3
Implementation: KdeV Solitons

Modify or run the program `Soliton.py` in Listing 24.2 that solves the KdeV equation (24.23) for the initial condition

$$u(x, t = 0) = \frac{1}{2}\left[1 - \tanh\left(\frac{x - 25}{5}\right)\right] \,, \tag{24.35}$$

with parameters $\epsilon = 0.2$ and $\mu = 0.1$. Start with $\Delta x = 0.4$ and $\Delta t = 0.1$. These constants are chosen to satisfy (24.33) with $|u| = 1$.

1. Define a 2D array `u[131,3]` with the first index corresponding to the position x and the second to the time t. With our choice of parameters, the maximum value for x is $130 \times 0.4 = 52$.
2. Initialize the time to $t = 0$ and assign values to `u[i,1]`.
3. Assign values to `u[i,2]`, $i = 3, 4, \ldots, 129$, corresponding to the next time interval. Use (24.31) to advance the time but note that you cannot start at $i = 1$ or end at $i = 131$ because (24.31) would include `u[132,2]` and `u[-1,1]`, which are beyond the limits of the array.
4. Increment the time and assume that `u[1,2]` = 1 and `u[131,2]` = 0. To obtain `u[2,2]` and `u[130,2]`, assume that `u[i+2,2]` = `u[i+1,2]` and `u[i-2,2]` = `u[i-1,2]`. Avoid `u[i+2,2]` for `i = 130`, and `u[i-2,2]` for `i = 2`. To do this, approximate (24.31) so that (24.33) is satisfied.
5. Increment time and compute `u[i, j]` for `j = 3` and for `i = 3, 4,..., 129`, using (24.30). Again follow the same procedures to obtain the missing array elements `u[2, j]` and `u[130, j]` (set `u[1, j] = 1.` and `u[131, j] = 0`). As you print out the numbers during the iterations, you will be convinced that it was a good choice.
6. Set `u[i,1]` = `u[i,2]` and `u[i,2]` = `u[i,3]` for all `i`. In this way, you are ready to find the next `u[i,j]` in terms of the previous two rows.

7. Repeat the previous two steps about 2000 times. Write your solution to a file after approximately every 250 iterations.
8. Use your favorite graphics tool to plot your results as a 3D graph of disturbance u vs. position *and vs.* time.
9. Observe the wave profile as a function of time and try to confirm Russell's observation that a taller soliton travels faster than a smaller one.

Listing 24.2 Soliton.py solves the KdeV equation for 1D solitons corresponding to a "bore" initial conditions.

```
# Soliton.py:          Korteweg de Vries equation for a soliton

from visual import *
import matplotlib.pylab as p;
from mpl_toolkits.mplot3d import Axes3D ;
import numpy

ds = 0.4;      dt = 0.1;      max = 2000;  mu = 0.1;  eps = 0.2;     mx = 131
u   = zeros( (mx, 3), float); spl = zeros( (mx, 21), float); m = 1

for  i in range(0, 131):                                        # Initial wave
    u[i, 0] = 0.5*(1
       -((math.exp(2*(0.2*ds*i-5.))-1)/(math.exp(2*(0.2*ds*i-5.))+1)))
u[0,1] = 1. ; u[0,2] = 1.; u[130,1] = 0. ; u[130,2] = 0.      # End points

for i in range (0, 131, 2): spl[i, 0] = u[i, 0]
fac = mu*dt/(ds**3)
print("Working. Please hold breath and wait while I count to 20")
for  i in range (1, mx-1):                                   # First time step
    a1 = eps*dt*(u[i + 1, 0] + u[i, 0] + u[i - 1, 0])/(ds*6.)
    if i > 1 and  i < 129: a2 = u[i+2,0]+2.*u[i-1,0]-2.*u[i+1,0]-u[i-2,0]
    else:  a2 = u[i-1, 0] - u[i+1, 0]
    a3 = u[i+1, 0] - u[i-1, 0]
    u[i, 1] = u[i, 0] - a1*a3 - fac*a2/3.
for j in range (1, max+1):                                   # Next time steps
    for i in range(1, mx-2):
        a1 = eps*dt*(u[i + 1, 1]  +  u[i, 1]  +  u[i - 1, 1])/(3.*ds)
        if i > 1 and i < mx-2:
            a2 = u[i+2,1] + 2.*u[i-1,1] - 2.*u[i+1,1] - u[i-2,1]
        else:  a2 = u[i-1, 1] - u[i+1, 1]
        a3        = u[i+1, 1] - u[i-1, 1]
        u[i, 2] = u[i,0] - a1*a3 - 2.*fac*a2/3.
    if j%100 ==   0:                          # Plot every 100 time steps
        for i in range (1, mx - 2): spl[i, m] = u[i, 2]
        print(m)
        m = m + 1
    for k in range(0, mx):                   # Recycle array saves memory
        u[k, 0] = u[k, 1]
        u[k, 1] = u[k, 2]

x = list(range(0, mx, 2))                            # Plot every other point
y = list(range(0, 21))                      # Plot 21 lines every 100 t steps
X, Y = p.meshgrid(x, y)

def functz(spl):
    z = spl[X, Y]
    return z

fig   = p.figure()                                           # create figure
ax = Axes3D(fig)                                                 # plot axes
ax.plot_wireframe(X, Y, spl[X, Y], color = 'r')              # red wireframe
ax.set_xlabel('Positon')                                        # label axes
```

```
ax.set_ylabel('Time')
ax.set_zlabel('Disturbance')
p.show()                              # Show figure, close Python shell
print("That's all folks!")
```

The code **SolitonAnimate.py** produces an animation.

24.5.4 Exploration: Solitons in Phase Space, Crossing

1. Explore what happens when a tall soliton collides with a short one.
   a) Start by placing a tall soliton of height 0.8 at $x = 12$ and a smaller soliton in front of it at $x = 26$:

$$u(x, t=0) = 0.8\left[1 - \tanh^2\left(\frac{3x}{12} - 3\right)\right] + 0.3\left[1 - \tanh^2\left(\frac{4.5x}{26} - 4.5\right)\right]. \tag{24.36}$$

   b) Do they reflect from each other? Do they go through each other? Do they interfere? Does the tall soliton still move faster than the short one after the collision (Figure 24.1)?
2. Construct phase-space plots ($\dot{u}(t)$ vs. $u(t)$) of the KdeV equation for various parameter values. Note that only very specific sets of parameters produce solitons. In particular, by correlating the behavior of the solutions with your phase-space plots, show that the soliton solutions correspond to the *separatrix* solutions to the KdeV equation. In other words, the stability in time for solitons is analogous to the infinite period for a pendulum balanced straight upward.

24.6 Solitons on Pendulum Chain

In 1955, Fermi, Ulam, and Pastu were investigating how a 1D chain of coupled oscillators disperses waves. Because waves of differing frequencies traveled through the chain with differing speeds, a pulse, which inherently includes a range of frequencies, broadens as time progresses. Surprisingly, when the oscillators were made more realistic by introducing a nonlinear term into Hooke's law

$$F(x) \simeq -k(x + \alpha x^2), \tag{24.37}$$

Fermi, Ulam, and Pastu found that even in the presence of dispersion, a sharp pulse in the chain would survive indefinitely. Your *problem* is to explain how this combination of dispersion and nonlinearity can combine to produce a stable pulse.

In Chapter 14, we studied nonlinear effects in a single pendulum arising from large oscillations. Now we go further and couple a number of those pendulums

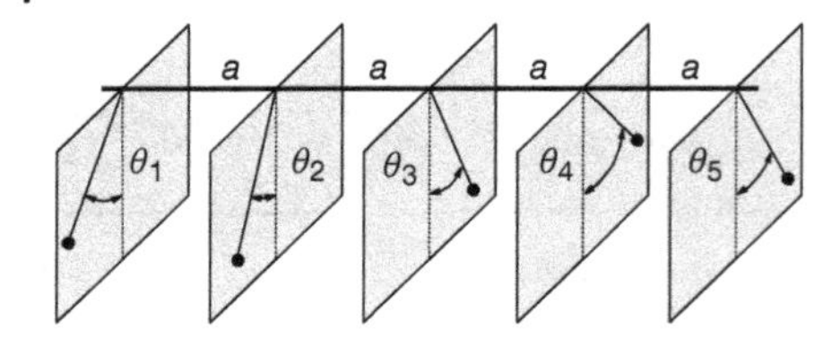

Figure 24.4 A 1D chain of pendulums coupled with a torsion bar on top. The pendulums swing in planes perpendicular to the length of the bar.

together. As shown in Figure 24.4, we take as our model a 1D chain of identical, equally spaced pendulums connected by a torsion bar that twists as the pendulums swing. The angle θ_i measures the displacement of pendulum i from its equilibrium position, and a the distance between pivot points. If all the pendulums are set off swinging together, $\theta_i \equiv \theta_j$, the coupling torques would vanish and we would have our old friend, the equation for a realistic pendulum. We assume that three torques act on each pendulum, a gravitational torque trying to return the pendulum to its equilibrium position, and the two torques from the twisting of the bar to the right and to the left of the pendulum. The equation of motion for pendulum j follows from Newton's law for rotational motion:

$$\sum_{j\neq i} \tau_{ji} = I\frac{d^2\theta_j(t)}{dt^2}, \tag{24.38}$$

$$-\kappa(\theta_j - \theta_{j-1}) - \kappa(\theta_j - \theta_{j+1}) - mgL\sin\theta_j = I\frac{d^2\theta_j(t)}{dt^2}, \tag{24.39}$$

$$\Rightarrow \boxed{\kappa(\theta_{j+1} - 2\theta_j + \theta_{j-1}) - mgL\sin\theta_j = I\frac{d^2\theta_j(t)}{dt^2},} \tag{24.40}$$

where I is the moment of inertia of each pendulum, L is the length of the pendulum, and κ is the torque constant of the bar. The nonlinearity in (24.40) arises from the $\sin\theta \simeq \theta - \theta^3/6 + \dots$ dependence of the gravitational torque. As it stands, (24.40) is a set of coupled nonlinear equations, with the number of equations equal to the number of oscillators, which would be large for a realistic solid.

24.6.1 Including Dispersion

Consider a surfer remaining on the crest of a wave. Since he/she does not see the wave form change with time, his/her position is given by a function of the form $f(kx - \omega t)$. Consequently, to his/her the wave has a constant phase

$$kx - \omega t = \text{constant} \quad \Rightarrow \quad x = \omega t/k = \text{constant}. \tag{24.41}$$

The surfer's (phase) velocity is the rate of change of x with respect to time,

$$v_p = \frac{dx}{dt} = \frac{\omega}{k}, \tag{24.42}$$

which is constant. In general, the frequency ω may be a nonlinear function of k, in which case the phase velocity varies with frequency and we have *dispersion*. If the wave contained just one frequency, then you would not observe any dispersion, but if the wave was a pulse composed of a range of Fourier components, then it would broaden and change shape in time as each frequency moved with a differing phase velocity. So although dispersion does not lead to an energy loss, it may well lead to a loss of information content as pulses broaden and overlap.

The functional relation between frequency ω and the wave vector k is called a *dispersion relation*. If the Fourier components in a wave packet are centered around a mean frequency ω_0, then the pulse's information travels, not with the phase velocity, but with the *group velocity*

$$v_g = \left.\frac{\partial\omega}{\partial k}\right|_{\omega_0} . \tag{24.43}$$

A comparison of (24.42) and (24.43) makes it clear that when there is dispersion the group and phase velocities may well differ.

To isolate the dispersive aspect of (24.40), we examine its linear version

$$\frac{d^2\theta_j(t)}{dt^2} + \omega_0^2\theta_j(t) = \frac{\kappa}{I}(\theta_{j+1} - 2\theta_j + \theta_{j-1}) , \tag{24.44}$$

where $\omega_0 = \sqrt{mgL/I}$ is the natural frequency for any one pendulum. Because we want to determine if a wave with a single frequency propagates on this chain, we start with a traveling wave with frequency ω and wavelength λ,

$$\theta_j(t) = A e^{i(\omega t - kx_j)} , \quad k = \frac{2\pi}{\lambda} , \tag{24.45}$$

and then test if it is a solution. Substitution of (24.45) into the wave equation (24.44) produces the *dispersion relation* (Figure 24.5):

$$\omega^2 = \omega_0^2 - \frac{2\kappa}{I}(1 - \cos ka) \quad \text{(dispersion relation)} . \tag{24.46}$$

To have dispersionless propagation (all frequencies propagate with the same velocity), we need a linear relation between ω and k:

$$\lambda = c\frac{2\pi}{\omega} \quad \Rightarrow \quad \omega = ck \quad \text{(dispersionless propagation)} . \tag{24.47}$$

This will be true for the chain only if ka is small, because then $\cos ka \simeq 1$ and $\omega \simeq \omega_0$.

Not only does the dispersion relation (24.46) change the speed of waves, but it also limits which frequencies can propagate on the chain. In order to have real k solutions, ω must lie in the range

$$\omega_0 \leq \omega \leq \omega^* \quad \text{(waves propagation)} . \tag{24.48}$$

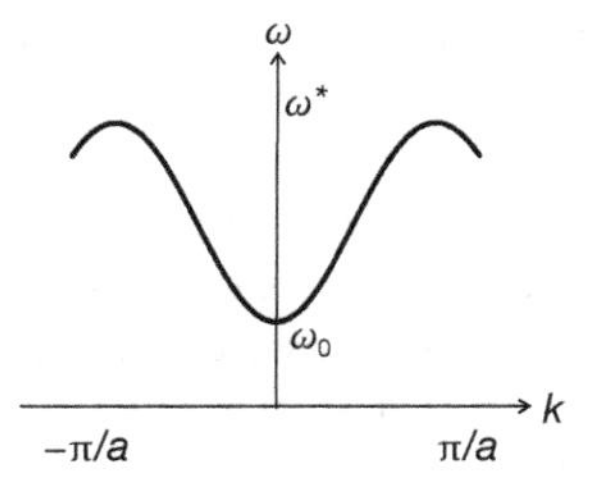

Figure 24.5 The dispersion relation for a linearized chain of pendulums.

The minimum frequency ω_0 and the maximum frequency ω^* are related through the limits of $\cos ka$ in (24.46),

$$(\omega^*)^2 = \omega_0^2 + \frac{4\kappa}{I} . \tag{24.49}$$

Waves with $\omega < \omega_0$ do not propagate, while those with $\omega > \omega^*$ are nonphysical because they correspond to wavelengths $\lambda < 2a$, that is, oscillations where there are no particles. These high and low ω cutoffs change the shape of a propagating pulse, that is, cause dispersion.

24.6.2
Continuum Limit, the Sine-Gordon Equation

If the wavelengths in a pulse are much longer than the pendulum–pendulum repeat distance a, that is, if $ka \ll 1$, the chain can be approximated as a continuous medium. In this limit, a becomes the continuous variable x, and the system of coupled ordinary differential equations becomes a single, partial differential equation:

$$\theta_{j+1} \simeq \theta_j + \frac{\partial \theta}{\partial x}\Delta x \tag{24.50}$$

$$\Rightarrow \quad (\theta_{j+1} - 2\theta_j + \theta_{j-1}) \simeq \frac{\partial^2 \theta}{\partial x^2}\Delta x^2 \to \frac{\partial^2 \theta}{\partial x^2}a^2 \tag{24.51}$$

$$\Rightarrow \quad \frac{\partial^2 \theta}{\partial t^2} - \frac{\kappa a^2}{I}\frac{\partial^2 \theta}{\partial x^2} = \frac{mgL}{I}\sin\theta . \tag{24.52}$$

If we measure time in units of $\sqrt{I/mgL}$ and distances in units of $\sqrt{\kappa a/(mgLb)}$, we obtain the standard form of the sine-Gordon equation (SGE)[1]:

$$\boxed{\frac{1}{c^2}\frac{\partial^2 \theta}{\partial t^2} - \frac{\partial^2 \theta}{\partial x^2} = \sin\theta} \quad \text{(SGE)} , \tag{24.53}$$

where the $\sin\theta$ on the RHS introduces the nonlinear effects.

1) The name "sine-Gordon" is either a reminder that the SGE is like the Klein–Gordon equation of relativistic quantum mechanics with a $\sin u$ added to the RHS, or a reminder of how clever one can be in thinking up names.

24.6.3
Analytic SGE Solution

The nonlinearity of the SGE (24.53) makes it hard to solve analytically. There is, however, a trick. Guess a functional form of a traveling wave, substitute it into (24.53) and thereby convert the PDE into a solvable ODE:

$$\theta(x,t) \stackrel{?}{=} \theta(\xi = t \pm x/v) \quad \Rightarrow \quad \frac{d^2\theta}{d\xi^2} = \frac{v^2}{v^2-1}\sin\theta\,. \tag{24.54}$$

You should recognize (24.54), the equation of motion for the realistic pendulum with no driving force and no friction, as our old friend. The constant v is a velocity in natural units, and separates different regimes of the motion:

$$\begin{aligned} v<1: &\quad \text{pendulums initially down} \quad \downarrow\downarrow\downarrow\downarrow\downarrow \quad \text{(stable)},\\ v>1: &\quad \text{pendulums initially up} \quad \uparrow\uparrow\uparrow\uparrow\uparrow \quad \text{(unstable)}. \end{aligned} \tag{24.55}$$

Although the equation is familiar, it does not mean that an analytic solution exists. However, for an energy $E = \pm 1$, we have motion along the separatrix, and have a solution with the characteristic *soliton* form,

$$\theta(x-vt) = \begin{cases} 4\tan^{-1}\left(\exp\left[+\frac{x-vt}{\sqrt{1-v^2}}\right]\right)\,, & \text{for} \quad E=1\,,\\ 4\tan^{-1}\left(\exp\left[-\frac{x-vt}{\sqrt{1-v^2}}\right]\right)+\pi\,, & \text{for} \quad E=-1\,. \end{cases} \tag{24.56}$$

This soliton corresponds to a solitary *kink* traveling with velocity $v = -1$ that flips the pendulums around by 2π as it moves down the chain. There is also an *antikink* in which the initial $\theta = \pi$ values are flipped to final $\theta = -\pi$.

24.6.4
Numeric Solution: 2D SGE Solitons

We have already solved for 1D solitons arising from the KdeV equation. The elastic-wave solitons that arise from the SGE can be easily generalized to two dimensions, as we do here with the 2D generalization of the SGE equation (24.53):

$$\frac{1}{c^2}\frac{\partial^2 u}{\partial t^2} - \frac{\partial^2 u}{\partial x^2} - \frac{\partial^2 u}{\partial y^2} = \sin u \quad \text{(2D SGE)}\,. \tag{24.57}$$

Whereas the 1D SGE describes wave propagation along a chain of connected pendulums, the 2D form might describe wave propagation in nonlinear elastic media. Interestingly enough, the same 2D SGE also occurs in quantum field theory, where the soliton solutions have been suggested as models for elementary particles (Christiansen and Lomdahl, 1981; Christiansen and Olsen, 1978; Argyris, 1991). The idea is that, like elementary particles, the solutions are confined to a region of space for a long period of time by nonlinear forces, and do not radiate away their energy. We solve (24.57) in a finite region of 2D space and for positive times:

$$-x_0 < x < x_0\,, \quad -y_0 < y < y_0\,, \quad t \geq 0\,. \tag{24.58}$$

We take $x_0 = y_0 = 7$ and impose the *boundary conditions* that the derivative of the displacement vanishes at the ends of the region:

$$\frac{\partial u}{\partial x}(-x_0, y, t) = \frac{\partial u}{\partial x}(x_0, y, t) = \frac{\partial u}{\partial y}(x, -y_0, t) = \frac{\partial u}{\partial y}(x, y_0, t) = 0\,. \tag{24.59}$$

We also impose the *initial condition* that at time $t = 0$ the waveform is that of a pulse (Figure 24.6) with its surface at rest:

$$u(x, y, t = 0) = 4\tan^{-1}(\mathrm{e}^{3-\sqrt{x^2+y^2}})\,, \quad \frac{\partial u}{\partial t}(x, y, t = 0) = 0\,. \tag{24.60}$$

We discretize the equation first by looking for solutions on a space–time lattice:

$$x = m\Delta x\,, \quad y = l\Delta x\,, \quad t = n\Delta t\,, \tag{24.61}$$

$$u^n_{m,l} \stackrel{\text{def}}{=} u(m\Delta x, l\Delta x, n\Delta t)\,. \tag{24.62}$$

Next we replace the derivatives in (24.57) by their finite-difference approximations:

$$\begin{aligned} u^{n+1}_{m,l} \simeq & -u^{n-1}_{m,l} + 2\left[1 - 2\left(\frac{\Delta t}{\Delta x}\right)^2\right] u^n_{m,l} \\ & + \left(\frac{\Delta t}{\Delta x}\right)^2 \left(u^n_{m+1,l} + u^n_{m-1,l} + u^n_{m,l+1} + u^n_{m,l-1}\right) \\ & - \Delta t^2 \sin\left[\frac{1}{4}\left(u^n_{m+1,l} + u^n_{m-1,l} + u^n_{m,l+1} + u^n_{m,l-1}\right)\right]\,. \end{aligned} \tag{24.63}$$

To make the algorithm simpler and establish stability, if we make the time and space steps proportional, $\Delta t = \Delta x/\sqrt{2}$. This leads to all of the $u^n_{m,l}$ terms dropping out:

$$\begin{aligned} u^2_{m,l} \simeq & \frac{1}{2}\left(u^1_{m+1,l} + u^1_{m-1,l} + u^1_{m,l+1} + u^1_{m,l-1}\right) \\ & - \frac{\Delta t^2}{2} \sin\left[\frac{1}{4}\left(u^1_{m+1,l} + u^1_{m-1,l} + u^1_{m,l+1} + u^1_{m,l-1}\right)\right]\,. \end{aligned} \tag{24.64}$$

Likewise, the discrete form of vanishing initial velocity (24.60) becomes

$$\frac{\partial u(x, y, 0)}{\partial t} = 0 \quad \Rightarrow \quad u^2_{m,l} = u^0_{m,l}\,. \tag{24.65}$$

This will be useful in setting the initial conditions. The lattice points on the edges and corners cannot be obtained from these relations, but must be obtained by

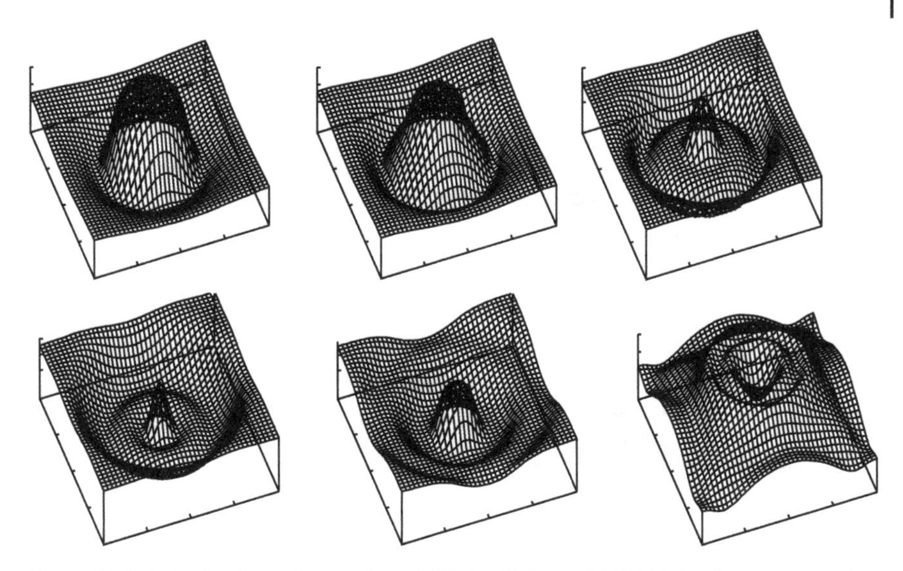

Figure 24.6 A circular ring soliton at times 8, 20, 40, 60, 80, and 120. This has been proposed as a model for an elementary particle.

applying the boundary conditions (24.59):

$$\frac{\partial u}{\partial z}(x_0, y, t) = \frac{u(x + \Delta x, y, t) - u(x, y, t)}{\Delta x} = 0 \tag{24.66}$$

$$\Rightarrow \quad u^n_{1,l} = u^n_{2,l} \, . \tag{24.67}$$

Similarly, the other derivatives in (24.59) give

$$u^n_{N_{\max},l} = u^n_{N_{\max}-1,l} \, , \quad u^n_{m,2} = u^n_{m,1} \, , \quad u^n_{m,N_{\max}} = u^n_{m,N_{\max}-1} \, , \tag{24.68}$$

where $N_{\max}$ is the number of grid points used for one space dimension.

24.6.5 2D Soliton Implementation

1. Define an array $u[N_{\max}, N_{\max}, 3]$ with $N_{\max} = 201$ for the space slots and the 3 for the time slots.
2. The solution (24.60) for the initial time $t = 0$ is placed in $u[m, l, 1]$.
3. The solution for the second time Δt is placed in $u(m, l, 2)$, and the solution for the next time, $2\Delta t$, is placed in $u[m, l, 3]$.
4. Assign the constants, $\Delta x = \Delta y = 7/100$, $\Delta t = \Delta x/\sqrt{2}$, $y_0 = x_0 = 7$.
5. Start off at $t = 0$ with the initial conditions and impose the boundary conditions to this initial solution. This is the solution for the first time step, defined over the entire 201×201 grid.
6. For the second time step, increase time by Δt and use (24.64) for all points in the plane. Do not include the edge points.

7. At the edges, for $i = 1, 2, \dots, 200$, set

$$u[i,1,2] = u[i,2,2]\,, \quad u[i,N_{\max},2] = u[i,N_{\max-1},2]$$
$$u[1,i,2] = u[2,i,2]\,, \quad u[N_{\max},i,2] = u[N_{\max-1},i,2]\,.$$

8. To find values for the four points in the corners for the second time step, again use initial condition (24.64):

$$u[1,1,2] = u[2,1,2]\,, \quad u[N_{\max},N_{\max},2] = u[N_{\max-1},N_{\max-1},2]$$
$$u[1,1,N_{\max}] = u[2,N_{\max},2]\,, \quad u[N_{\max},1,2] = u[N_{\max-1},1,2]\,.$$

9. For the third time step (the future), use (24.64).
10. Continue the propagation forward in time, reassigning the future to the present, and so forth. In this way, we need to store the solutions for only three time steps.

24.6.6
SGE Soliton Visualization

We see in Figure 24.6 the time evolution of a circular ring soliton that results for the stated initial conditions. We note that the ring at first shrinks in size, then expands, and then shrinks back into another (but not identical) ring soliton. A small amount of the particle does radiate away, and in the last frame we can notice some interference between the radiation and the boundary conditions. An animation of this sequence can be found online.

25
Fluid Dynamics

We have already introduced some fluid dynamics in the previous chapter's discussion of shallow-water solitons. This chapter confronts the more general equations of computational fluid dynamics (CFD) and their solutions.[1] The mathematical description of the motion of fluids, although not a new subject, remains a challenging one. Equations are complicated and nonlinear, there are many degrees of freedom, the nonlinearities may lead to instabilities, analytic solutions are rare, and the boundary conditions for realistic geometries (like airplanes) are not intuitive. These difficulties may explain why fluid dynamics is often absent from undergraduate and even graduate physics curricula. Nonetheless, as an essential element of the physical world that also has tremendous practical importance, we encourage its study. We recommend Fetter and Walecka (1980); Landau and Lifshitz (1987) for those interested in the derivations, and Shaw (1992) for more details about the computations.

25.1
River Hydrodynamics

Problem In order to give migrating salmon a place to rest during their arduous upstream journey, the Oregon Department of Environment is placing objects in several deep, wide, fast-flowing streams. One such object is a long beam of rectangular cross section (Figure 25.1a), and another is a set of plates (Figure 19.4b). The objects are to be placed far enough below the water's surface so as not to disturb the surface flow, and far enough from the bottom of the stream so as not to disturb the flow there either. Your *problem* is to determine the spatial dependence of the stream's velocity when the objects are in place, and, specifically, whether the wake of the object will be large enough to provide a resting place for a meter-long salmon.

1) We acknowledge some helpful reading of by Satoru S. Kano.

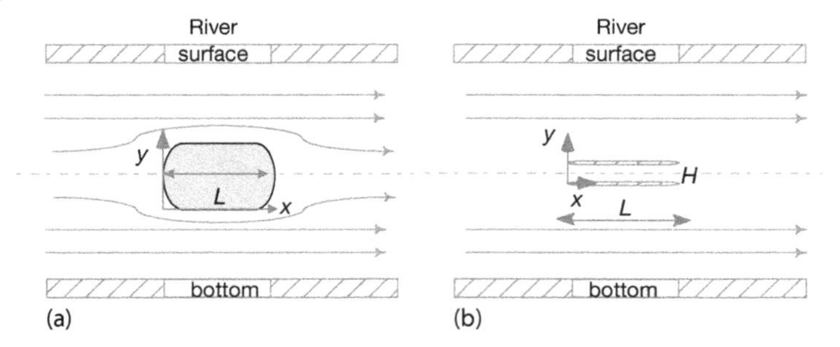

Figure 25.1 Side view of the flow of a stream around a submerged beam (a) and around two parallel plates (b). Both beam and plates have length L along the direction of flow. The flow is seen to be symmetric about the centerline and to be unaffected at the bottom and at the surface by the submerged object.

25.2
Navier–Stokes Equation (Theory)

As with our study of shallow-water waves, we assume that water is *incompressible* and thus that its density ρ is constant. We also simplify the theory by looking only at steady-state situations, that is, ones in which the velocity is not a function of time. However, to understand how water flows around objects, like our beam, it is essential to include the complication of frictional forces (*viscosity*).

For the sake of completeness, we repeat here the first equation of hydrodynamics, the continuity equation (24.2):

$$\frac{\partial \rho(\boldsymbol{x}, t)}{\partial t} + \boldsymbol{\nabla} \cdot \boldsymbol{j} = 0 , \quad \boldsymbol{j} \stackrel{\text{def}}{=} \rho \boldsymbol{v}(\boldsymbol{x}, t) . \tag{25.1}$$

Before proceeding to the second equation, we introduce a special time derivative, the *hydrodynamic* derivative $D\boldsymbol{v}/Dt$, which is appropriate for a quantity contained in a moving fluid (Fetter and Walecka, 1980):

$$\frac{D\boldsymbol{v}}{Dt} \stackrel{\text{def}}{=} (\boldsymbol{v} \cdot \boldsymbol{\nabla})\boldsymbol{v} + \frac{\partial \boldsymbol{v}}{\partial t} . \tag{25.2}$$

This derivative gives the rate of change, as viewed from a stationary frame, of the velocity of material in *an element of flowing fluid* and so incorporates changes as a result of the motion of the fluid (first term) as well as any explicit time dependence of the velocity. Of particular interest is that $D\boldsymbol{v}/Dt$ is of second order in the velocity, and so its occurrence introduces nonlinearities into the theory. You may think of these nonlinearities as related to the fictitious (inertial) forces that occur when we describe the motion in the fluid's rest frame (an accelerating frame to us).

The material derivative is the leading term in the *Navier–Stokes equation*,

$$\frac{D\boldsymbol{v}}{Dt} = \nu\nabla^2\boldsymbol{v} - \frac{1}{\rho}\boldsymbol{\nabla}P(\rho, T, x) , \tag{25.3}$$

$$\begin{aligned}
\frac{\partial v_x}{\partial t} + \sum_{j=x}^{z} v_j \frac{\partial v_x}{\partial x_j} &= \nu \sum_{j=x}^{z} \frac{\partial^2 v_x}{\partial x_j^2} - \frac{1}{\rho}\frac{\partial P}{\partial x} , \\
\frac{\partial v_y}{\partial t} + \sum_{j=x}^{z} v_j \frac{\partial v_y}{\partial x_j} &= \nu \sum_{j=x}^{z} \frac{\partial^2 v_y}{\partial x_j^2} - \frac{1}{\rho}\frac{\partial P}{\partial y} , \\
\frac{\partial v_z}{\partial t} + \sum_{j=x}^{z} v_j \frac{\partial v_z}{\partial x_j} &= \nu \sum_{j=x}^{z} \frac{\partial^2 v_z}{\partial x_j^2} - \frac{1}{\rho}\frac{\partial P}{\partial z} .
\end{aligned} \tag{25.4}$$

Here, ν is the kinematic viscosity, P is the pressure, and (25.4) gives the derivatives in (25.3) in Cartesian coordinates. This equation describes transfer of the momentum of the fluid within some region of space as a result of forces and flow (think $\mathrm{d}\boldsymbol{p}/\mathrm{d}t = \boldsymbol{F}$). There is a simultaneous equation for each of the three velocity components. The $\boldsymbol{v} \cdot \nabla \boldsymbol{v}$ term in $D\boldsymbol{v}/Dt$ describes the transport of momentum in some region of space resulting from the fluid's flow and is often called the *convection* or *advection* term.[2] The $\boldsymbol{\nabla}P$ term describes the velocity change as a result of pressure changes, and the $\nu\nabla^2\boldsymbol{v}$ term describes the velocity change resulting from viscous forces (which tend to dampen the flow).

The explicit functional dependence of the pressure on the fluid's density and temperature $P(\rho, T, x)$ is known as the *equation of state of the fluid*, and would have to be known before trying to solve the Navier–Stokes equation. To keep our problem simple, we assume that the pressure is independent of density and temperature. This leaves the four simultaneous partial differential equations to solve, (25.1) and (25.3). Because we are interested in *steady-state* flow around an object, we assume that all time derivatives of the velocity vanish. Because we assume that the fluid is incompressible, the time derivative of the density also vanishes, and (25.1) and (25.3) become

$$\boldsymbol{\nabla} \cdot \boldsymbol{v} \equiv \sum_i \frac{\partial v_i}{\partial x_i} = 0 , \tag{25.5}$$

$$(\boldsymbol{v} \cdot \boldsymbol{\nabla})\boldsymbol{v} = \nu\nabla^2\boldsymbol{v} - \frac{1}{\rho}\boldsymbol{\nabla}P . \tag{25.6}$$

The first equation expresses the equality of inflow and outflow and is known as the *condition of incompressibility*. In as much as the stream in our problem is much wider than the width (z direction) of the beam, and because we are staying away

2) We discuss pure advection in Section 24.1. In oceanology or meteorology, convection implies the transfer of mass in the vertical direction where it overcomes gravity, whereas advection refers to transfer in the horizontal direction.

from the banks, we will ignore the z dependence of the velocity. The explicit PDEs we need to solve then reduce to

$$\frac{\partial v_x}{\partial x} + \frac{\partial v_y}{\partial y} = 0 \,, \tag{25.7}$$

$$\nu \left(\frac{\partial^2 v_x}{\partial x^2} + \frac{\partial^2 v_x}{\partial y^2} \right) = v_x \frac{\partial v_x}{\partial x} + v_y \frac{\partial v_x}{\partial y} + \frac{1}{\rho} \frac{\partial P}{\partial x} \,, \tag{25.8}$$

$$\nu \left(\frac{\partial^2 v_y}{\partial x^2} + \frac{\partial^2 v_y}{\partial y^2} \right) = v_x \frac{\partial v_y}{\partial x} + v_y \frac{\partial v_y}{\partial y} + \frac{1}{\rho} \frac{\partial P}{\partial y} \,. \tag{25.9}$$

25.2.1 Boundary Conditions for Parallel Plates

The plate problem is relatively easy to solve analytically, and so we will do it! This will give us some experience with the equations as well as a check for our numerical solution. To find a unique solution to the PDEs (25.7)–(25.9), we need to specify boundary conditions. As far as we can tell, picking boundary conditions is somewhat of an acquired skill.

We assume that the submerged parallel plates are placed in a stream that is flowing with a constant velocity V_0 in the horizontal direction (Figure 25.1b). If the velocity V_0 is not too high or the kinematic viscosity ν is sufficiently large, then the flow should be smooth and without turbulence. We call such flow *laminar*. Typically, a fluid undergoing laminar flow moves in smooth paths that do not close on themselves, like the smooth flow of water from a faucet. If we imagine attaching a vector to each element of the fluid, then the path swept out by that vector is called a *streamline* or *line of motion* of the fluid. These streamlines can be visualized experimentally by adding colored dye to the stream. We assume that the plates are so thin that the flow remains laminar as it passes around and through them.

If the plates are thin, then the flow upstream of them is not affected, and we can limit our solution space to the rectangular region in Figure 25.2. We assume that the length L and separation H of the plates are small compared to the size of the stream, so the flow returns to uniform as we get far downstream from the plates. As shown in Figure 25.2, there are boundary conditions at the *inlet* where the fluid enters the solution space, at the *outlet* where it leaves, and at the stationary plates. In addition, because the plates are far from the stream's bottom and surface, we assume that the dotted-dashed centerline is a plane of symmetry, with identical flow above and below the plane. We thus have four different types of boundary conditions to impose on our solution:

Solid plates Because there is friction (viscosity) between the fluid and the plate surface, the only way to have laminar flow is to have the fluid's velocity equal to the plate's velocity, which means both are zero:

$$v_x = v_y = 0 \,. \tag{25.10}$$

Such being the case, we have smooth flow in which the negligibly thin plates lie along streamlines of the fluid (like a "streamlined" vehicle).

Inlet The fluid enters the integration domain at the inlet with a horizontal velocity V_0. Because the inlet is far upstream from the plates, we assume that the fluid velocity at the inlet is unchanged by the presence of the plates:

$$v_x = V_0 , \quad v_y = 0 . \tag{25.11}$$

Outlet Fluid leaves the integration domain at the outlet. While it is totally reasonable to assume that the fluid returns to its unperturbed state there, we are not told what that is. So, instead, we assume that there is a physical outlet at the end with the water just shooting out of it. Consequently, we assume that the water pressure equals zero at the outlet (as at the end of a garden hose) and that the velocity does not change in a direction normal to the outlet:

$$P = 0 , \quad \frac{\partial v_x}{\partial x} = \frac{\partial v_y}{\partial x} = 0 . \tag{25.12}$$

Symmetry plane If the flow is symmetric about the $y = 0$ plane, then there cannot be flow through the plane, and the spatial derivatives of the velocity components normal to the plane must vanish:

$$v_y = 0 , \quad \frac{\partial v_y}{\partial y} = 0 . \tag{25.13}$$

This condition follows from the assumption that the plates are along streamlines and that they are negligibly thin. It means that all the streamlines are parallel to the plates as well as to the water surface, and so it must be that $v_y = 0$ everywhere. The fluid enters in the horizontal direction, the plates do not change the vertical y component of the velocity, and the flow remains symmetric about the centerline. There is a retardation of the flow around the plates because of the viscous nature of the flow and that of the $\boldsymbol{v} = 0$ boundary layers formed on the plates, but there are no actual v_y components.

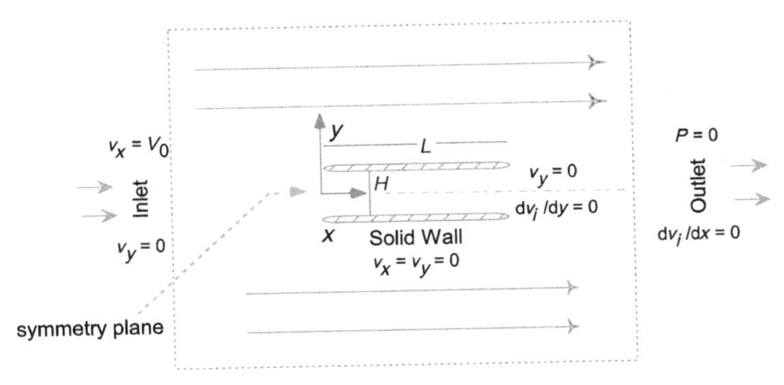

Figure 25.2 The boundary conditions for two thin submerged plates. The surrounding box is the integration volume within which we solve the PDEs and upon whose surface we impose the boundary conditions. In practice, the box is much larger than L and H.

25.2.2 Finite-Difference Algorithm and Overrelaxation

Now we develop an algorithm for solution of the Navier–Stokes and continuity PDEs using successive over-relaxation. This is a variation of the method used in Chapter 19, to solve Poisson's equation. We divide space into a rectangular grid with the spacing h in both the x and y directions:

$$x = ih\,, \quad i = 0, \ldots, N_x\,; \quad y = jh\,, \quad j = 0, \ldots, N_y\,.$$

We next express the derivatives in (25.7)–(25.9) as finite differences of the values of the velocities at the grid points using central-difference approximations. For $\nu = 1\,\mathrm{m}^2/\mathrm{s}$ and $\rho = 1\,\mathrm{kg/m}^3$, this yields

$$v^x_{i+1,j} - v^x_{i-1,j} + v^y_{i,j+1} - v^y_{i,j-1} = 0\,, \tag{25.14}$$

$$\begin{aligned} v^x_{i+1,j} + v^x_{i-1,j} + v^x_{i,j+1} + v^x_{i,j-1} - 4v^x_{i,j} &= \frac{h}{2} v^x_{i,j}\left[v^x_{i+1,j} - v^x_{i-1,j}\right] \\ &+ \frac{h}{2} v^y_{i,j}\left[v^x_{i,j+1} - v^x_{i,j-1}\right] + \frac{h}{2}\left[P_{i+1,j} - P_{i-1,j}\right]\,, \end{aligned} \tag{25.15}$$

$$\begin{aligned} v^y_{i+1,j} + v^y_{i-1,j} + v^y_{i,j+1} + v^y_{i,j-1} - 4v^y_{i,j} &= \frac{h}{2} v^x_{i,j}\left[v^y_{i+1,j} - v^y_{i-1,j}\right] \\ &+ \frac{h}{2} v^y_{i,j}\left[v^y_{i,j+1} - v^y_{i,j-1}\right] + \frac{h}{2}\left[P_{i,j+1} - P_{i,j-1}\right]\,. \end{aligned} \tag{25.16}$$

Because $v^y \equiv 0$ for this problem, we rearrange terms to obtain for v^x:

$$\begin{aligned} 4v^x_{i,j} &= v^x_{i+1,j} + v^x_{i-1,j} + v^x_{i,j+1} + v^x_{i,j-1} - \frac{h}{2} v^x_{i,j}\left[v^x_{i+1,j} - v^x_{i-1,j}\right] \\ &- \frac{h}{2} v^y_{i,j}\left[v^x_{i,j+1} - v^x_{i,j-1}\right] - \frac{h}{2}\left[P_{i+1,j} - P_{i-1,j}\right]\,. \end{aligned} \tag{25.17}$$

We recognize in (25.17) an algorithm similar to the one we used in solving Laplace's equation by relaxation. Indeed, as we did there, we can accelerate the convergence by writing the algorithm with the new value of v^x given as the old value plus a correction (residual):

$$v^x_{i,j} = v^x_{i,j} + r_{i,j}\,, \quad r \overset{\text{def}}{=} v^{x(\text{new})}_{i,j} - v^{x(\text{old})}_{i,j} \tag{25.18}$$

$$\begin{aligned} \Rightarrow \quad r = \frac{1}{4}\Big\{ & v^x_{i+1,j} + v^x_{i-1,j} + v^x_{i,j+1} + v^x_{i,j-1} - \frac{h}{2} v^x_{i,j}\left[v^x_{i+1,j} - v^x_{i-1,j}\right] \\ &- \frac{h}{2} v^y_{i,j}\left[v^x_{i,j+1} - v^x_{i,j-1}\right] - \frac{h}{2}\left[P_{i+1,j} - P_{i-1,j}\right]\Big\} - v^x_{i,j}\,. \end{aligned} \tag{25.19}$$

As performed with the Poisson equation algorithm, successive iterations sweep the interior of the grid, continuously adding in the residual (25.18) until the change becomes smaller than some set level of tolerance, $|r_{i,j}| < \varepsilon$.

A variation of this method, *successive over-relaxation*, increases the speed at which the residuals approach zero via an amplifying factor ω:

$$v^x_{i,j} = v^x_{i,j} + \omega r_{i,j} \quad \text{(SOR)} . \tag{25.20}$$

The standard relaxation algorithm (25.18) is obtained with $\omega = 1$, an accelerated convergence (*over-relaxation*) is obtained with $\omega \geq 1$, and *underrelaxation* occurs for $\omega < 1$. Values $\omega > 2$ are found to lead to numerical instabilities. Although a detailed analysis of the algorithm is necessary to predict the optimal value for ω, we suggest that you test different values for ω to see which one provides the fastest convergence for your problem.

25.2.3 Successive Overrelaxation Implementation

1. Modify the program `Beam.py`, or write your own, to solve the Navier–Stokes equation for the velocity of a fluid in a 2D flow. Represent the x and y components of the velocity by the arrays `vx[Nx,Ny]` and `vy[Nx,Ny]`.
2. Specialize your solution to the rectangular domain and boundary conditions indicated in Figure 25.2.
3. Use of the following parameter values,

$$\begin{aligned} \nu &= 1\,\text{m}^2/\text{s} , \quad \rho = 10^3\,\text{kg/m}^3 , \quad &&\text{(flow parameters)} , \\ N_x &= 400 , \quad N_y = 40 , \quad h = 1 , \quad &&\text{(grid parameters)} , \end{aligned}$$

   leads to the analytic solution

$$\frac{\partial P}{\partial x} = -12, \quad \frac{\partial P}{\partial y} = 0 , \quad v^x = \frac{3j}{20}\left(1 - \frac{j}{40}\right) , \quad v^y = 0 . \tag{25.21}$$

4. For the relaxation method, output the iteration number and the computed v^x and then compare the analytic and numeric results.
5. Repeat the calculation and see if SOR speeds up the convergence.

25.3 2D Flow over a Beam

Now that the comparison with an analytic solution has shown that our CFD simulation works, we return to determining if the beam in Figure 25.1 might produce a good resting place for salmon. While we have no analytic solution with which to compare, our canoeing and fishing adventures have taught us that *standing waves* with fish in them are often formed behind rocks in streams, and so we expect there will be a standing wave formed behind the beam.

25.4
Theory: Vorticity Form of Navier–Stokes Equation

We have seen how to solve numerically the hydrodynamics equations

$$\nabla \cdot \boldsymbol{v} = 0 \,, \tag{25.22}$$

$$(\boldsymbol{v} \cdot \nabla)\boldsymbol{v} = -\frac{1}{\rho}\nabla P + \nu \nabla^2 \boldsymbol{v} \,. \tag{25.23}$$

These equations determine the components of a fluid's velocity, pressure, and density as functions of position. In analogy to electrostatics, where one usually solves for the simpler scalar potential and then takes its gradient to determine the more complicated vector field, we now recast the hydrodynamic equations into forms that permit us to solve two simpler equations for simpler functions, from which the velocity is obtained via a gradient operation.[3)]

We introduce the *stream function* $\boldsymbol{u}(\boldsymbol{x})$ from which the velocity is determined by the curl operator:

$$\boldsymbol{v} \stackrel{\text{def}}{=} \nabla \times \boldsymbol{u}(\boldsymbol{x}) = \hat{\epsilon}_x \left(\frac{\partial u_z}{\partial y} - \frac{\partial u_y}{\partial z} \right) + \hat{\epsilon}_y \left(\frac{\partial u_x}{\partial z} - \frac{\partial u_z}{\partial x} \right) \,. \tag{25.24}$$

Note the absence of the z component of velocity $\boldsymbol{v}$ for our problem. Because $\nabla \cdot (\nabla \times \boldsymbol{u}) \equiv 0$, we see that any $\boldsymbol{v}$ that can be written as the curl of $\boldsymbol{u}$ automatically satisfies the continuity equation $\nabla \cdot \boldsymbol{v} = 0$. Further, because $\boldsymbol{v}$ for our problem has only x and y components, $\boldsymbol{u}(\boldsymbol{x})$ needs have only a z component:

$$u_z \equiv u \quad \Rightarrow \quad v_x = \frac{\partial u}{\partial y} \,, \quad v_y = -\frac{\partial u}{\partial x} \,. \tag{25.25}$$

It is worth noting that in 2D flows, the contour lines u = constant are the *streamlines*.

The second simplifying function is the *vorticity* field $\boldsymbol{w}(\boldsymbol{x})$, which is related physically and alphabetically to the angular velocity $\boldsymbol{\omega}$ of the fluid. Vorticity is defined as the curl of the velocity (sometimes with a $-$ sign):

$$\boldsymbol{w} \stackrel{\text{def}}{=} \nabla \times \boldsymbol{v}(\boldsymbol{x}) \,. \tag{25.26}$$

Because the velocity in our problem does not change in the z direction, we have

$$w_z = \left(\frac{\partial v_y}{\partial x} - \frac{\partial v_x}{\partial y} \right) \,. \tag{25.27}$$

Physically, we see that the vorticity is a measure of how much the fluid's velocity curls or rotates, with the direction of the vorticity determined by the right-hand

3) If we had to solve only the simpler problem of *irrotational flow* (no turbulence), then we would be able to use a scalar velocity potential, in close analogy to electrostatics (Lamb, 1993). For the more general *rotational flow*, two vector potentials are required.

rule for rotations. In fact, if we could pluck a small element of the fluid into space (so it would not feel the internal strain of the fluid), we would find that it is rotating like a solid with angular velocity $\boldsymbol{\omega} \propto \boldsymbol{w}$ (Lamb, 1993). That being the case, it is useful to think of the vorticity as giving the local value of the fluid's angular velocity vector. If $\boldsymbol{w} = 0$, we have *irrotational* flow.

The field lines of w are continuous and move as if they are attached to the particles of the fluid. A uniformly flowing fluid has vanishing curl, while a nonzero vorticity indicates that the current curls back on itself or rotates. From the definition of the stream function (25.24), we see that the vorticity $\boldsymbol{w}$ is related to it by

$$\boldsymbol{w} = \boldsymbol{\nabla} \times \boldsymbol{v} = \boldsymbol{\nabla} \times (\boldsymbol{\nabla} \times \boldsymbol{u}) = \boldsymbol{\nabla}(\boldsymbol{\nabla} \cdot \boldsymbol{u}) - \nabla^2 \boldsymbol{u} \,, \tag{25.28}$$

where we have used a vector identity for $\boldsymbol{\nabla} \times (\boldsymbol{\nabla} \times \boldsymbol{u})$. Yet the divergence $\boldsymbol{\nabla} \cdot \boldsymbol{u} = 0$ because $\boldsymbol{u}$ has a component only in the z-direction and that component is independent of z (or because there is no source for $\boldsymbol{u}$). We have now obtained the basic relation between the stream function $\boldsymbol{u}$ and the vorticity $\boldsymbol{w}$:

$$\boldsymbol{\nabla}^2 \boldsymbol{u} = -\boldsymbol{w} \,. \tag{25.29}$$

Equation 25.29 is analogous to Poisson's equation of electrostatics, $\nabla^2 \phi = -4\pi\rho$, only now each component of vorticity $\boldsymbol{w}$ is a source for the corresponding component of the stream function $\boldsymbol{u}$. If the flow is irrotational, that is, if $\boldsymbol{w} = 0$, then we need only to solve Laplace's equation for each component of u. Rotational flow, with its coupled nonlinearities equations, leads to more interesting behavior.

As is to be expected from the definition of $\boldsymbol{w}$, the vorticity form of the Navier–Stokes equation is obtained by taking the curl of the velocity form, that is, by operating on both sides with $\boldsymbol{\nabla}\times$. After significant manipulations one obtains

$$\nu \nabla^2 \boldsymbol{w} = [(\boldsymbol{\nabla} \times \boldsymbol{u}) \cdot \boldsymbol{\nabla}]\boldsymbol{w} \,. \tag{25.30}$$

This and (25.29) are the two simultaneous PDEs that we need to solve. In 2D, with $\boldsymbol{u}$ and $\boldsymbol{w}$ having only z components, they are

$$\frac{\partial^2 u}{\partial x^2} + \frac{\partial^2 u}{\partial y^2} = -w \,, \tag{25.31}$$

$$\nu \left(\frac{\partial^2 w}{\partial x^2} + \frac{\partial^2 w}{\partial y^2} \right) = \frac{\partial u}{\partial y}\frac{\partial w}{\partial x} - \frac{\partial u}{\partial x}\frac{\partial w}{\partial y} \,. \tag{25.32}$$

So after all that work, we end up with two simultaneous, nonlinear, elliptic PDEs that look like a mixture of Poisson's equation with the wave equation. The equation for u is Poisson's equation with source w, and must be solved simultaneously with the second. It is this second equation that contains mixed products of the derivatives of u and w and thus introduces the nonlinearity.

25.4.1
Finite Differences and the SOR Algorithm

We solve (25.31) and (25.32) on an $N_x \times N_y$ grid of uniform spacing h with

$$x = i\Delta x = ih\,, \quad i = 0, \ldots, N_x\,, \quad y = j\Delta y = jh\,, \quad j = 0, \ldots, N_y\,. \quad (25.33)$$

Because the beam is symmetric about its centerline (Figure 25.1a), we need the solution only in the upper half-plane. We apply the now familiar central-difference approximation to the Laplacians of u and w to obtain the difference equation

$$\frac{\partial^2 u}{\partial x^2} + \frac{\partial^2 u}{\partial y^2} \simeq \frac{u_{i+1,j} + u_{i-1,j} + u_{i,j+1} + u_{i,j-1} - 4u_{i,j}}{h^2}\,. \quad (25.34)$$

Likewise, for the product of the first derivatives,

$$\frac{\partial u}{\partial y}\frac{\partial w}{\partial x} \simeq \frac{u_{i,j+1} - u_{i,j-1}}{2h}\frac{w_{i+1,j} - w_{i-1,j}}{2h}\,. \quad (25.35)$$

The difference form of the vorticity of the Navier–Stokes equation (25.31) becomes

$$u_{i,j} = \frac{1}{4}\left(u_{i+1,j} + u_{i-1,j} + u_{i,j+1} + u_{i,j-1} + h^2 w_{i,j}\right)\,, \quad (25.36)$$

$$w_{i,j} = \frac{1}{4}(w_{i+1,j} + w_{i-1,j} + w_{i,j+1} + w_{i,j-1}) - \frac{R}{16}\left\{\left[u_{i,j+1} - u_{i,j-1}\right] \right.$$
$$\left. \times \left[w_{i+1,j} - w_{i-1,j}\right] - \left[u_{i+1,j} - u_{i-1,j}\right]\left[w_{i,j+1} - w_{i,j-1}\right]\right\}\,, \quad (25.37)$$

$$R = \frac{1}{\nu} = \frac{V_0 h}{\nu}\text{(in normal units)}\,. \quad (25.38)$$

Note that we have placed $u_{i,j}$ and $w_{i,j}$ on the LHS of the equations in order to obtain an algorithm appropriate to a solution by relaxation.

The parameter R in (25.38) is related to the *Reynolds number*. When we solve the problem in natural units, we measure distances in units of grid spacing h, velocities in units of initial velocity V_0, stream functions in units of $V_0 h$, and vorticity in units of V_0/h. The second form is in regular units and is dimensionless. This R is known as the *grid Reynolds number* and differs from the physical R, which has a pipe diameter in place of the grid spacing h.

The grid Reynolds number is a measure of the strength of the coupling of the nonlinear terms in the equation. When the physical R is small, the viscosity acts as a frictional force that damps out fluctuations and keeps the flow smooth. When R is large ($R \simeq 2000$), physical fluids undergo phase transitions from laminar to turbulent flow in which turbulence occurs at a cascading set of smaller and smaller space scales (Reynolds, 1883). However, simulations that produce the onset of turbulence have been a research problem because Reynolds first experiments in 1883 (Falkovich and Sreenivasan, 2006), possibly because the laminar flow is stable against small perturbations and some large-scale "kick" may be needed to change laminar to turbulent flow. Recent research along these lines have been able to find unstable, traveling-wave solutions to the Navier–Stokes equations, and the hope is that these may lead to a turbulent transition (Fitzgerald, 2004).

As discussed in Section 25.2.2, the finite-difference algorithm can have its convergence accelerated by the use of successive over-relaxation (25.36):

$$u_{i,j} = u_{i,j} + \omega r^{(1)}_{i,j}, \quad w_{i,j} = w_{i,j} + \omega r^{(2)}_{i,j} \quad \text{(SOR)}. \tag{25.39}$$

Here, ω is the over-relaxation parameter and should lie in the range $0 < \omega < 2$ for stability. The residuals are just the changes in a single step, $r^{(1)} = u^{\text{new}} - u^{\text{old}}$ and $r^{(2)} = w^{\text{new}} - w^{\text{old}}$:

$$\begin{aligned} r^{(1)}_{i,j} &= \frac{1}{4}\left(u_{i+1,j} + u_{i-1,j} + u_{i,j+1} + u_{i,j-1} + w_{i,j}\right) - u_{i,j}, \\ r^{(2)}_{i,j} &= \frac{1}{4}\Big(w_{i+1,j} + w_{i-1,j} + w_{i,j+1} + w_{i,j-1} - \frac{R}{4}\{[u_{i,j+1} - u_{i,j-1}] \\ &\quad \times [w_{i+1,j} - w_{i-1,j}] - [u_{i+1,j} - u_{i-1,j}][w_{i,j+1} - w_{i,j-1}]\}\Big) - w_{i,j}. \end{aligned} \tag{25.40}$$

25.4.2 Boundary Conditions for a Beam

A well-defined solution of these elliptic PDEs requires a combination of (less than obvious) boundary conditions on u and w. Consider Figure 25.3, based on the analysis of Koonin (1986). We assume that the inlet, outlet, and surface are far from the beam, which may not be evident from the not-to-scale figure.

Freeflow If there were no beam in the stream, then we would have free flow with the entire fluid possessing the inlet velocity:

$$v_x \equiv V_0, \quad v_y = 0, \quad \Rightarrow \quad u = V_0 y, \quad w = 0. \tag{25.41}$$

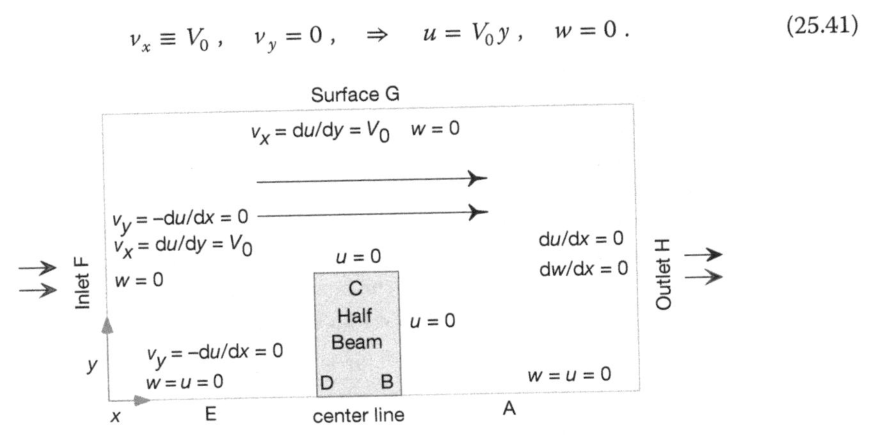

Figure 25.3 Boundary conditions for flow around the beam in Figure 25.1. The flow is symmetric about the centerline, and the beam has length L in the x direction (along flow).

(Recall that we can think of $w = 0$ as indicating no fluid rotation.) The centerline divides the system along a symmetry plane with identical flow above and below it. If the velocity is symmetric about the centerline, then its y component must vanish there:

$$v_y = 0 , \quad \Rightarrow \quad \frac{\partial u}{\partial x} = 0 \quad \text{(centerline AE)} . \tag{25.42}$$

Centerline The centerline is a streamline with $u =$ constant because there is no velocity component perpendicular to it. We set $u = 0$ according to (25.41). Because there cannot be any fluid flowing into or out of the beam, the normal component of velocity must vanish along the beam surfaces. Consequently, the streamline $u = 0$ is the entire lower part of Figure 25.3, that is, the centerline and the beam surfaces. Likewise, the symmetry of the problem permits us to set the vorticity $w = 0$ along the centerline.

Inlet At the inlet, the fluid flow is horizontal with uniform x component V_0 at all heights and with no rotation:

$$v_y = -\frac{\partial u}{\partial x} = 0 , \quad w = 0 \quad \text{(inlet F)} , \quad v_x = \frac{\partial u}{\partial y} = V_0 . \tag{25.43}$$

Surface We are told that the beam is sufficiently submerged so as not to disturb the flow on the surface of the stream. Accordingly, we have free-flow conditions on the surface:

$$v_x = \frac{\partial u}{\partial y} = V_0 , \quad w = 0 \quad \text{(surface G)} . \tag{25.44}$$

Outlet Unless something truly drastic happens, the conditions on the far downstream outlet have little effect on the far upstream flow. A convenient choice is to require the stream function and vorticity to be constant:

$$\frac{\partial u}{\partial x} = \frac{\partial w}{\partial x} = 0 \quad \text{(outlet H)} . \tag{25.45}$$

Beamsides We have already noted that the normal component of velocity v_x and stream function u vanish along the beam surfaces. In addition, because the flow is viscous, it is also true that the fluid "sticks" to the beam somewhat and so the tangential velocity also vanishes along the beam's surfaces. While these may all be true conclusions regarding the flow, specifying them as boundary conditions would over-restrict the solution (see Table 19.2 for elliptic equations) to the point where no solution may exist. Accordingly, we simply impose the *no-slip* boundary condition on the vorticity w. Consider a grid point (x, y) on the upper surface of the beam. The stream function u at a point $(x, y + h)$ above it can be related via a Taylor series in y:

$$u(x, y + h) = u(x, y) + \frac{\partial u}{\partial y}(x, y)h + \frac{\partial^2 u}{\partial y^2}(x, y)\frac{h^2}{2} + \cdots \tag{25.46}$$

Because $\boldsymbol{w}$ has only a z component, it has a simple relation to $\boldsymbol{\nabla} \times \boldsymbol{v}$:

$$w \equiv w_z = \frac{\partial v_y}{\partial x} - \frac{\partial v_x}{\partial y} . \tag{25.47}$$

Because of the fluid's viscosity, the velocity is stationary along the beam top:

$$v_x = \frac{\partial u}{\partial y} = 0 \quad \text{(beam top)} . \tag{25.48}$$

Because the current flows smoothly along the top of the beam, v_y must also vanish. In addition, because there is no x variation, we have

$$\frac{\partial v_y}{\partial x} = 0 \quad \Rightarrow \quad w = -\frac{\partial v_x}{\partial y} = -\frac{\partial^2 u}{\partial y^2} . \tag{25.49}$$

After substituting these relations into the Taylor series (25.46), we can solve for w and obtain the finite-difference version of the top boundary condition:

$$w \simeq -2\frac{u(x, y+h) - u(x, y)}{h^2} \quad \Rightarrow \quad w_{i,j} = -2\frac{u_{i,j+1} - u_{i,j}}{h^2} \quad \text{(top)} . \tag{25.50}$$

Similar treatments applied to other surfaces yield the boundary conditions.

$$\begin{array}{lll}
u = 0 & w = 0 & \text{Centerline EA} \\
u = 0 & w_{i,j} = -2(u_{i+1,j} - u_{i,j})/h^2 & \text{Beam back B} \\
u = 0 & w_{i,j} = -2(u_{i,j+1} - u_{i,j})/h^2 & \text{Beam top C} \\
u = 0 & w_{i,j} = -2(u_{i-1,j} - u_{i,j})/h^2 & \text{Beam front D} \\
\partial u/\partial x = 0 & w = 0 & \text{Inlet F} \\
\partial u/\partial y = V_0 & w = 0 & \text{Surface G} \\
\partial u/\partial x = 0 & \partial w/\partial x = 0 & \text{Outlet H}
\end{array} \tag{25.51}$$

25.4.3 SOR on a Grid

Beam.py in Listing 25.1 is our program for the solution of the vorticity form of the Navier–Stokes equation. You will notice that while the relaxation algorithm is rather simple, some care is needed in implementing many boundary conditions. Relaxation of the stream function and of the vorticity are performed by separate functions.

Listing 25.1 Beam.py solves the Navier–Stokes equation for the flow over a plate.

```
# Beam.py: solves Navier--Stokes equation for flow around beam

import matplotlib.pylab as p;
from mpl_toolkits.mplot3d import Axes3D;
from numpy import *;

print("Working, wait for the figure after 100 iterations")

Nxmax = 70;    Nymax = 20;    IL = 10;    H = 8;    T = 8;    h = 1.
u = zeros((Nxmax+1, Nymax+1), float)                          # Stream
w = zeros((Nxmax+1, Nymax+1), float)                          # Vorticity
V0 = 1.0;     omega = 0.1;     nu = 1.;     iter = 0;   R = V0 * h/nu
```

```
def borders():
    for i in range(0, Nxmax+1):                          # Init stream
        for j in range(0, Nymax+1):                      # Init vorticity
              w[i, j] = 0.
              u[i, j] = j * V0
    for i in range(0, Nxmax+1 ):                         # Fluid surface
      u[i, Nymax] = u[i, Nymax-1] + V0*h
      w[i, Nymax-1] = 0.
    for j in range(0, Nymax+1):
        u[1, j] = u[0, j]
        w[0, j] = 0.                                     # Inlet
    for  i in range(0, Nxmax+1):                         # Centerline
        if i <=  IL and i>= IL+T:
            u[i, 0] = 0.
            w[i, 0] = 0.
    for j in range(1, Nymax ):                           # Outlet
        w[Nxmax, j] = w[Nxmax-1, j]
        u[Nxmax, j] = u[Nxmax-1, j]

def beam():                                              # BC for beam
   for  j in range  (0, H+1):                            # Sides
      w[IL, j] =  - 2 * u[IL-1, j]/(h*h)                 # Front
      w[IL+T, j] =  - 2 * u[IL  +  T  +  1, j]/(h*h)     # Back
   for  i in range(IL, IL+T + 1): w[i, H - 1] =  - 2 * u[i, H]/(h*h);
   for i in range(IL, IL+T+1):
     for j in range(0, H+1):
         u[IL, j] = 0.                                   # Front
         u[IL+T, j] = 0.                                 # Back
         u[i, H] = 0;                                    # Top

def relax():                                             # Relax stream
    beam()                                               # Reset conditions
    for i in range(1, Nxmax):                            # Relax stream
      for  j in range (1, Nymax):
        r1 = omega*((u[i+1,j]+u[i-1,j]+u[i,j+1]+u[i,j-1] +
               h*h*w[i,j])/4-u[i,j])
        u[i, j] +=  r1
    for  i in range(1, Nxmax):                           # Relax vorticity
       for j in range(1, Nymax):
           a1 = w[i+1, j]  +  w[i-1,j]  +  w[i,j+1] + w[i,j-1]
           a2 = (u[i,j+1] - u[i,j-1])*(w[i+1,j] - w[i - 1, j])
           a3 = (u[i+1,j] - u[i-1,j])*(w[i,j+1] - w[i, j - 1])
           r2 = omega *( (a1 - (R/4.)*(a2 - a3) )/4. - w[i,j])
           w[i, j] +=  r2

borders()
while (iter <=  100):
    iter +=  1
    if iter%10 ==  0: print (iter)
    relax()
for i in range (0, Nxmax+1):
    for  j in range(0, Nymax+ 1):  u[i,j] = u[i,j]/V0/h     # V0h units
x = range(0, Nxmax-1);          y = range(0, Nymax-1)
X, Y = p.meshgrid(x, y)

def functz(u):                                           # Stream flow
    z = u[X, Y]
    return z

Z = functz(u)
fig = p.figure()
ax = Axes3D(fig)
ax.plot_wireframe(X, Y, Z, color = 'r')
ax.set_xlabel('X')
ax.set_ylabel('Y')
ax.set_zlabel('Stream Function')
p.show()
```

25.4.4
Flow Assessment

1. Use `Beam.py` as a basis for your solution for the stream function u and the vorticity w using the finite-differences algorithm (25.36).
2. A good place to start your simulation is with a beam of size $L = 8h$, $H = h$, Reynolds number $R = 0.1$, and intake velocity $V_0 = 1$. Keep your grid small during debugging, say, $N_x = 24$ and $N_y = 70$.
3. Explore the *convergence* of the algorithm.
   a) Print out the iteration number and u values upstream from, above, and downstream from the beam.
   b) Determine the number of iterations necessary to obtain three-place convergence for successive relaxation ($\omega = 0$).
   c) Determine the number of iterations necessary to obtain three-place convergence for successive over-relaxation ($\omega \simeq 0.3$). Use this number for future calculations.
4. Change the beam's horizontal placement so that you can see the undisturbed current entering on the left and then developing into a standing wave. Note that you may need to increase the size of your simulation volume to see the effect of all the boundary conditions.
5. Make surface plots including contours of the stream function u and the vorticity w. Explain the behavior seen.
6. Is there a region where a big fish can rest behind the beam?
7. The results of the simulation (Figure 25.4) are for the one-component stream function u. Make several visualizations showing the fluid velocity throughout

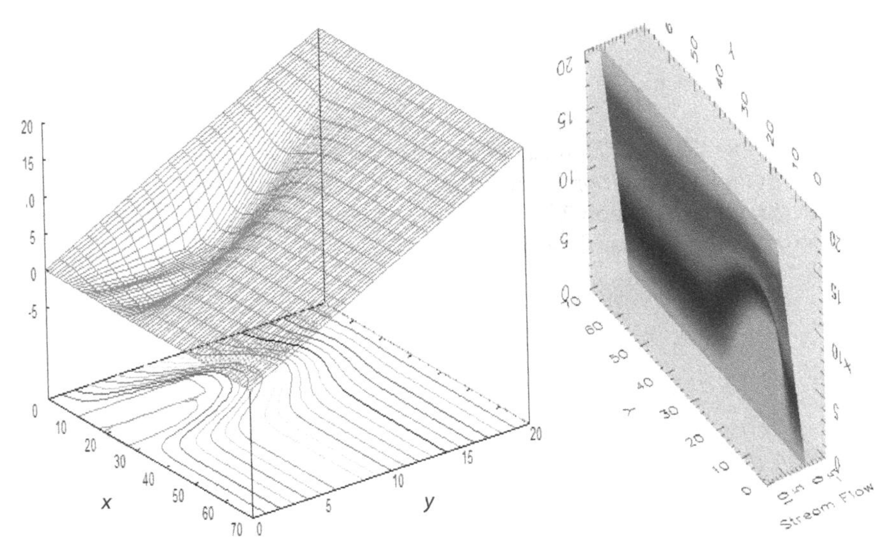

Figure 25.4 Two visualizations of the stream function u for Reynold's number $R = 5$. The one on the left uses contours, while the one on the right uses color.

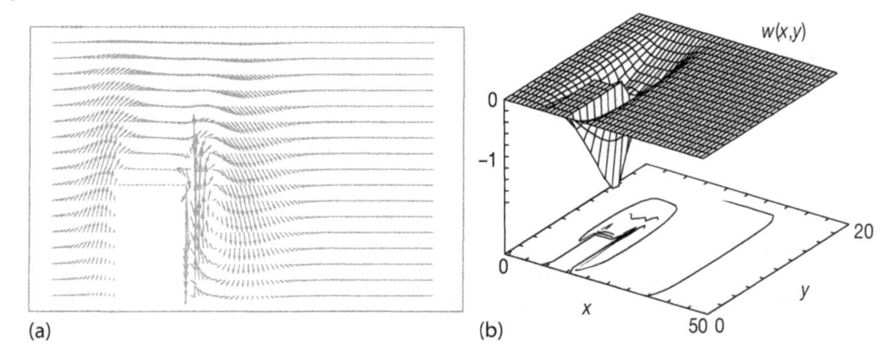

Figure 25.5 (a) The velocity field around the beam as represented by vectors. (b) The vorticity as a function of x and y for the same velocity field. Rotation is seen to be largest behind the beam.

the simulation region. Note that the velocity is a vector with two components (or a magnitude and direction), and both degrees of freedom are interesting to visualize. A plot of vectors would work well here.

8. Explore how increasing the Reynolds number R changes the flow pattern. Start at $R = 0$ and gradually increase R while watching for numeric instabilities. To overcome the instabilities, reduce the size of the relaxation parameter ω and continue to larger R values.
9. Verify that the flow around the beam is smooth for small R values, but that it separates from the back edge for large R, at which point a small vortex develops, as can be seen in Figure 25.5.

25.4.5
Exploration

1. Determine the flow behind a circular rock in the stream.
2. The boundary condition at an outlet far downstream should not have much effect on the simulation. Explore the use of other boundary conditions there.
3. Determine the pressure variation around the beam.

26
Integral Equations of Quantum Mechanics

> There are people – amongst whom I would include myself – who detest happy endings.
>
> *Vlasimir Nabokov, Pnin*

We have put this chapter off till last because it is the one in which we have the closest personal connections. The power and accessibility of high-speed computers have changed the view as to what kind of equations are soluble. We have seen how even nonlinear differential equations can be solved easily and can give new insight into the physical world. In this chapter, we examine how the integral equations of quantum mechanics can be solved for both bound and scattering states. We start by extending our treatment of the eigenvalue problem, earlier solved as a coordinate-space differential equation, to the equivalent integral-equation problem in momentum space. Then, we treat the singular integral equations for scattering, a problem whose multiple challenges have been met well by computational physics.

26.1
Bound States of Nonlocal Potentials

Problem A particle undergoes a many-body interaction with a medium (Figure 26.1). Although, we can write down the many-body Schrödinger equation that describes this interaction as the sum of many potentials, the number of coordinates involved is too large for a practical solution. Instead theorists have derived an effective single-particle potential that accounts for the many particles present by having the potential that is *nonlocal*, that is, the effective potential at $\boldsymbol{r}$ that depends on the wave function at the $\boldsymbol{r}'$ values of the other particles (Landau, 1996):

$$V(r)\psi(r) \to \int \mathrm{d}r'\, V(r,r')\psi(r') \,. \tag{26.1}$$

Computational Physics, 3rd edition. Rubin H. Landau, Manuel J. Páez, Cristian C. Bordeianu.

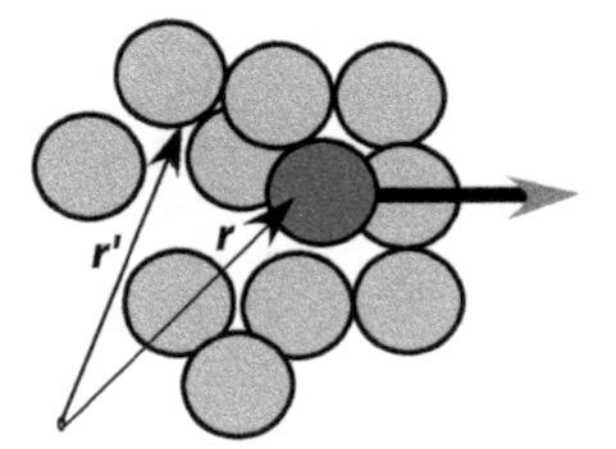

Figure 26.1 A dark particle moving in a dense medium in which it interacts with all particles present. The nonlocality of the potential felt by the dark particle at $\boldsymbol{r}$ arises from the interactions at all $\boldsymbol{r}'$.

This type of interaction leads to a Schrödinger equation which is a combined integral and differential ("integrodifferential") equation:

$$-\frac{1}{2\mu}\frac{\mathrm{d}^2\psi(r)}{\mathrm{d}r^2} + \int \mathrm{d}r'\, V(r, r')\psi(r') = E\psi(r) \,. \tag{26.2}$$

Your *problem* is to figure out how to find the bound-state energies E_n and wave functions ψ_n for the equation in (26.2).[1]

26.2 Momentum–Space Schrödinger Equation (Theory)

One way of dealing with (26.2) is by going to momentum space where it becomes the integral equation (Landau, 1996)

$$\boxed{\frac{k^2}{2\mu}\psi_n(k) + \frac{2}{\pi}\int_0^\infty \mathrm{d}p\, p^2 V(k, p)\psi_n(p) = E_n\psi_n(k) \,.} \tag{26.3}$$

We restrict our solution to angular momentum $l = 0$ partial waves, but for simplicity of notation we do not include an index to indicate that. In (26.3), $V(k, p)$ is the momentum–space representation (double Fourier transform) of the coordinate-space potential,

$$V(k, p) = \frac{1}{kp}\int_0^\infty \mathrm{d}r \sin(kr)V(r)\sin(pr) \,. \tag{26.4}$$

$\psi_n(k)$ is the momentum–space wave function (the probability amplitude for finding the particle with momentum k), and is the Fourier transform of $\psi_n(r)$:

$$\psi_n(k) = \int_0^\infty \mathrm{d}r kr\psi_n(r)\sin(kr) \,. \tag{26.5}$$

1) We use natural units in which $\hbar \equiv 1$.

Equation 26.3 is an integral equation for $\psi_n(k)$. It differs from an integral representation of $\psi_n(k)$, because the integral in it cannot be evaluated until $\psi_n(p)$ is known. Although this may seem like an insurmountable barrier, we will transform this equation into a matrix equation that can be solved with the matrix techniques discussed in Chapter 6.

26.2.1 Integral to Matrix Equations

We approximate the integral over the potential as a weighted sum over N integration points (usually Gauss quadrature points[2]) for $p = k_j,\ j = 1, N$:

$$\int_0^\infty \mathrm{d}p\, p^2 V(k,p)\psi_n(p) \simeq \sum_{j=1}^{N} w_j k_j^2 V(k,k_j)\psi_n(k_j) \,. \tag{26.6}$$

This converts the integral equation (26.3) to the algebraic equation

$$\frac{k^2}{2\mu}\psi_n(k) + \frac{2}{\pi}\sum_{j=1}^{N} w_j k_j^2 V(k,k_j)\psi_n(k_j) = E_n \,. \tag{26.7}$$

Equation 26.7 contains the N unknown function values $\psi_n(k_j)$, the unknown E_n, as well as the unknown functional dependence of $\psi_n(k)$. We eliminate the functional dependence of $\psi_n(k)$ by restricting the solution to the same values of $k = k_i$ as used in the approximation of the integral, which means that we are solving for wave function for k values on the grid of Figures 26.2. This leads to a set of N coupled linear equations in $(N+1)$ unknowns:

$$\boxed{\frac{k_i^2}{2\mu}\psi_n(k_i) + \frac{2}{\pi}\textstyle\sum_{j=1}^{N} w_j k_j^2\, V(k_i,k_j)\psi_n(k_j) = E_n\psi_n(k_i)\,, \quad i = 1, N\,.} \tag{26.8}$$

As a concrete example, for $N = 2$ we have two simultaneous linear equations

$$\frac{k_1^2}{2\mu}\psi_n(k_1) + \frac{2}{\pi}w_1 k_1^2 V(k_1,k_1)\psi_n(k_1) + w_2 k_2^2 V(k_1,k_2)\psi_n(k_2) = E_n\psi_n(k_1)\,,$$

$$\frac{k_2^2}{2\mu}\psi_n(k_2) + \frac{2}{\pi}w_1 k_1^2 V(k_2,k_1)\psi_n(k_1) + w_2 k_2^2 V(k_2,k_2)\psi_n(k_2) = E_n\psi_n(k_2)\,.$$

$k_1 \quad k_2 \quad k_3 \quad \cdots \quad k_N$

Figure 26.2 The grid of momentum values on which the integral equation is solved.

2) See Chapter 5, for a discussion of numerical integration.

Of course, realistic examples require more than two integration points for precision.

We write our coupled equations (26.8) in the matrix form as

$$[\mathbf{H}][\psi_n] = E_n[\psi_n] \tag{26.9}$$

or as explicit matrices

$$\begin{pmatrix} \frac{k_1^2}{2\mu} + \frac{2}{\pi}V(k_1,k_1)k_1^2 w_1 & \frac{2}{\pi}V(k_1,k_2)k_2^2 w_2 & \cdots & \frac{2}{\pi}V(k_1,k_N)k_N^2 w_N \\ \frac{2}{\pi}V(k_2,k_1)k_1^2 w_1 & \frac{2}{\pi}V(k_2,k_2)k_2^2 w_2 + \frac{k_2^2}{2\mu} & \cdots & \cdots \\ \ddots & \ddots & \ddots & \ddots \\ \cdots & \cdots & \cdots & \frac{k_N^2}{2\mu} + \frac{2}{\pi}V(k_N,k_N)k_N^2 w_N \end{pmatrix}$$
$$\times \begin{pmatrix} \psi_n(k_1) \\ \psi_n(k_2) \\ \ddots \\ \psi_n(k_N) \end{pmatrix} = E_n \begin{pmatrix} \psi_n(k_1) \\ \psi_n(k_2) \\ \ddots \\ \psi_n(k_N) \end{pmatrix} . \tag{26.10}$$

Equation 26.9 is the matrix representation of the Schrödinger equation (26.3). The wave function $\psi_n(k)$ on the grid is the $N \times 1$ vector

$$[\psi_n(k_i)] = \begin{pmatrix} \psi_n(k_1) \\ \psi_n(k_2) \\ \ddots \\ \psi_n(k_N) \end{pmatrix} . \tag{26.11}$$

The astute reader may be questioning the possibility of solving N equations for $(N+1)$ unknowns, $\psi_n(k_i)$, and E_n. Only sometimes, and only for certain values of E_n (eigenvalues), will the computer be able to find solutions. To see how this arises, we try to apply the matrix inversion technique (which we will use successfully for scattering in Section 26.3). We rewrite (26.9) as

$$[\mathbf{H} - E_n I][\psi_n] = [0] \tag{26.12}$$

and multiply both sides by the inverse of $[\mathbf{H} - E_n I]$ to obtain the formal solution

$$[\psi_n] = [\mathbf{H} - E_n I]^{-1}[0] . \tag{26.13}$$

This equation tells us that (1) if the inverse exists, then we have the *trivial* solution $\psi_n \equiv 0$, which is not revealing, and (2) for a nontrivial solution to exist, our assumption that the inverse exists must be incorrect. Yet we know from the theory of linear equations that the inverse fails to exist when the determinant vanishes:

$$\boxed{\det[\mathbf{H} - E_n I] = 0} \quad \text{(bound-state condition)} . \tag{26.14}$$

Equation 26.14 is the $N + 1$th equation needed to find unique solutions to the eigenvalue problem. Although, there is no guarantee that solutions of (26.14) can always be found, if they are found they are the desired *eigenvalues* of (26.9).

26.2.2 Delta-Shell Potential (Model)

To keep things simple and to have an analytic answer to compare with, we consider the local delta-shell potential:

$$V(r) = \frac{\lambda}{2\mu}\delta(r-b)\,. \tag{26.15}$$

This might be a good model for an interaction that occurs when two particles are predominantly a fixed distance b apart. We use (26.4) to determine its momentum–space representation:

$$V(k',k) = \int_0^\infty \frac{\sin(k'r')}{k'k}\frac{\lambda}{2\mu}\delta(r-b)\sin(kr)\,\mathrm{d}r = \frac{\lambda}{2\mu}\frac{\sin(k'b)\sin(kb)}{k'k}\,. \tag{26.16}$$

Beware: We have chosen this potential because it is easy to evaluate the momentum–space matrix element of the potential. However, its singular nature in r space leads to (26.16) having a very slow falloff in k space, and this causes the integrals to converge so slowly that numerics are not as precise as we would like.

If the energy is parameterized in terms of a wave vector κ by $E_n = -\kappa^2/2\mu$, then for this potential there is, at most, one bound state and it satisfies the transcendental equation (Gottfried, 1966)

$$\mathrm{e}^{-2\kappa b} - 1 = \frac{2\kappa}{\lambda}\,. \tag{26.17}$$

Note that bound states occur only for attractive potentials. For the present case, this requires $\lambda < 0$.

Exercise Pick some values of b and λ and verify that (26.17) can be solved for κ.

26.2.3 Binding Energies Solution

An actual computation may follow two paths. The first calls subroutines to evaluate the determinant of the $[\mathbf{H} - E_n I]$ matrix in (26.14), and then to *search* for those values of energy for which the computed determinant vanishes. This provides E_n, but not wave functions. The other approach calls an eigenproblem solver that may give some or all eigenvalues and eigenfunctions. In both the cases, the solution is obtained iteratively, and you may be required to guess starting values for both the eigenvalues and eigenvectors. In Listing 26.1, we present our solution of the integral equation for bound states of the delta-shell potential using the NumPy matrix library and the `gauss` method for Gaussian quadrature points and weights.

Listing 26.1 Bound.py solves the Lippmann–Schwinger integral equation for bound states within a delta-shell potential. The integral equations are converted to matrix equations using Gaussian grid points, and they are solved with linalg.

```
# Bound.py: Bound state solutn of Lippmann--Schwinger equation in p space
from visual import *
from numpy import*
from numpy.linalg import*

min1 =0.;      max1 =200.;  u =0.5;     b =10.

def gauss(npts,a,b,x,w):
    pp = 0.;   m = (npts + 1)//2;   eps = 3.E-10              # Accuracy: ADJUST!

    for i in range(1,m+1):
        t = cos(math.pi*(float(i)-0.25)/(float(npts) + 0.5))
        t1 = 1
        while((abs(t-t1)) >= eps):
            p1 = 1. ;   p2 = 0.;
            for j in range(1,npts+1):
                p3 = p2
                p2 = p1
                p1=((2*j-1)*t*p2-(j-1)*p3)/j
            pp = npts*(t*p1-p2)/(t*t-1.)
            t1 = t;  t = t1 - p1/pp
        x[i-1] = -t
        x[npts-i] = t
        w[i-1] = 2./((1.-t*t)*pp*pp)
        w[npts-i] = w[i-1]
    for i in range(0,npts):
        x[i] = x[i]*(b-a)/2. + (b + a)/2.
        w[i] = w[i]*(b-a)/2.

for M in range(16, 32, 8):
    z=[-1024, -512, -256, -128, -64, -32, -16, -8, -4, -2]
    for lmbda in z:
        A = zeros((M,M), float)                                        # Hamiltonian
        WR = zeros((M), float)                            # Eigenvalues, potential
        k = zeros((M), float); w = zeros((M),float);                   # Pts & wts
        gauss(M, min1, max1, k, w)                               # Call gauss points
        for i in range(0,M):                                       # Set Hamiltonian
            for j in range(0,M):
                VR = lmbda/2/u*sin(k[i]*b)/k[i]*sin(k[j]*b)/k[j]
                A[i,j] = 2./math.pi*VR*k[j]*k[j]*w[j]
                if (i == j):
                    A[i,j] += k[i]*k[i]/2/u
        Es, evectors = eig(A)
        realev = Es.real                                          # Real eigenvalues
        for j in range(0,M):
            if (realev[j]<0):
                print(" M (size), lmbda, ReE = ",M," ",lmbda,"
                     ",realev[j])
                break
```

1. Write a program, or modify ours, to solve the integral equation (26.9) for the delta-shell potential (26.16). Either find the E_n's for which the determinant vanishes *or*, find the eigenvalues and eigenvectors for this **H**.
2. Set the scale by setting $2\mu = 1$ and $b = 10$.
3. Set up the potential and Hamiltonian matrices $\mathbf{V}(i, j)$ and $\mathbf{H}(i, j)$ for Gaussian quadrature integration with at least $N = 16$ grid points.

4. Adjust the value and sign of λ for bound states. A good approach is to start with a large negative value for λ and then make it less negative. You should find that the eigenvalue moves up in energy.
5. *Note:* Your eigenenergy solver may return several eigenenergies. The true bound state will be at negative energy and change little as the number of grid points changes. The others are numerical artifacts.
6. Try increasing the number of grid points in steps of 8, for example, 16, 24, 32, 64, …, and see how the energy changes.
7. Extract the best value for the bound-state energy and estimate its precision by seeing how it changes with the number of grid points.
8. If you are solving the eigenvalue problem, check your solution by comparing the RHS and LHS in the matrix multiplication $[\mathbf{H}][\psi_n] = E_n[\psi_n]$.
9. Verify that, regardless of the potential's strength, there is only a single bound state and that it gets deeper as the magnitude of λ increases. Compare with (26.17).

26.2.4 Wave Function (Exploration)

1. Determine the momentum–space wave function $\psi_n(k)$ using an eigen problem solver. Does $\psi_n(k)$ fall off at $k \to \infty$? Does it oscillate? Is it well behaved at the origin?
2. Using the same points and weights as used to evaluate the integral in the integral equation, determine the coordinate-space wave function via the Bessel transform

$$\psi_n(r) = \int_0^\infty \mathrm{d}k \psi_n(k) \frac{\sin(kr)}{kr} k^2 \,. \tag{26.18}$$

   Does $\psi_n(r)$ fall off as you would expect for a bound state? Does it oscillate? Is it well behaved at the origin?
3. Compare the r dependence of this $\psi_n(r)$ to the analytic wave function:

$$\psi_n(r) \propto \begin{cases} \mathrm{e}^{-\kappa r} - \mathrm{e}^{\kappa r} \,, & \text{for} \quad r < b \,, \\ \mathrm{e}^{-\kappa r} \,, & \text{for} \quad r > b \,. \end{cases} \tag{26.19}$$

26.3 Scattering States of Nonlocal Potentials ⊙

Problem Again we have a particle interacting with the nonlocal potential discussed for bound states (Figure 26.3a), only now the particle has sufficiently high energy that it scatters from rather than binds with the medium. Your *problem* is to determine the scattering cross section for scattering from a nonlocal potential.

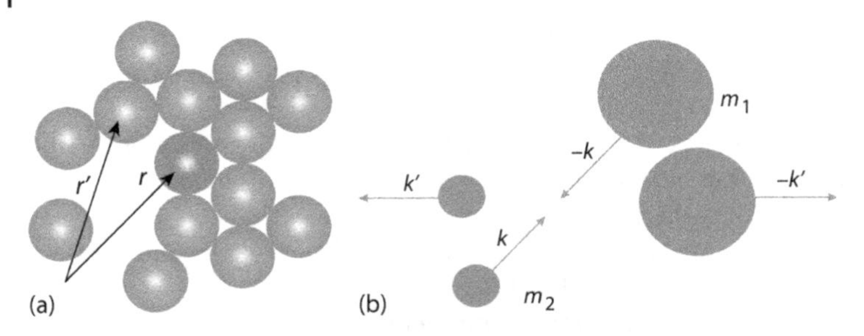

Figure 26.3 (a) A projectile (dark particle at r) scattering from a dense medium. (b) The same process viewed in the CM system where the projectile and target always have equal and opposite momenta.

26.4 Lippmann–Schwinger Equation (Theory)

Because experiments measure scattering amplitudes and not wave functions, it is more direct to have a theory dealing with amplitudes rather than wave functions.[3] An integral form of the Schrödinger equation dealing with the scattering amplitude R is the *Lippmann–Schwinger equation*:

$$R(k', k) = V(k', k) + \frac{2}{\pi}\mathcal{P}\int_0^\infty \mathrm{d}p \frac{p^2 V(k', p)R(p, k)}{(k_0^2 - p^2)/2\mu} , \tag{26.20}$$

where the symbol $\mathcal{P}$ in (26.20) indicates the Cauchy principal-value prescription for avoiding the singularity arising from the zero of the denominator (we discuss how to do that next). As for the bound-state problem, this equation is for partial wave $l = 0$ and $\hbar = 1$. In (26.20), the momentum k_0 is related to the energy E and the reduced mass μ by

$$E = \frac{k_0^2}{2\mu} , \quad \mu = \frac{m_1 m_2}{m_1 + m_2} . \tag{26.21}$$

The initial and final COM momenta k and k' are the momentum–space variables. The experimental observable that results from a solution of (26.20) is the diagonal matrix element $R(k_0, k_0)$, which is related to the scattering phase shift δ_0 and thus the cross section:

$$R(k_0, k_0) = -\frac{\tan \delta_l}{\rho} , \quad \rho = 2\mu k_0 . \tag{26.22}$$

Note that (26.20) is not just the evaluation of an integral, it is an integral equation in which $R(p, k)$ is integrated over all p. Yet because $R(p, k)$ is unknown, the integral cannot be evaluated until after the equation is solved!

3) To make the presentation simpler, but still perfectly valid, we solve for the *reaction matrix R*, but call it the scattering amplitude.

26.4.1 Singular Integrals (Math)

A *singular* integral

$$\mathcal{G} = \int_a^b g(k)\,\mathrm{d}k\,, \tag{26.23}$$

is one in which the integrand $g(k)$ is singular at a point k_0 within the integration interval $[a, b]$, yet with the integral $\mathcal{G}$ remaining finite. (If the integral itself were infinite, we could not compute it.) Unfortunately, computers are notoriously bad at dealing with infinite numbers, and if an integration point gets too near the singularity, overwhelming subtractive cancellation or overflow may occur. Consequently, we apply some results from complex analysis before evaluating singular integrals numerically.[4)]

In Figure 26.4, we show three ways to avoid the singularity of an integrand. The paths in Figure 26.4a and b move the singularity slightly off the real k-axis by giving the singularity a small imaginary part $\pm i\epsilon$. The Cauchy principal-value prescription $\mathcal{P}$ in Figure 26.4c is seen to follow a path that "pinches" both sides of the singularity at k_0, but does not to pass through it:

$$\mathcal{P}\int_{-\infty}^{+\infty} f(k)\,\mathrm{d}k = \lim_{\epsilon\to 0}\left[\int_{-\infty}^{k_0-\epsilon} f(k)\,\mathrm{d}k + \int_{k_0+\epsilon}^{+\infty} f(k)\,\mathrm{d}k\right]. \tag{26.24}$$

The preceding three prescriptions are related by the identity

$$\int_{-\infty}^{+\infty} \frac{f(k)\,\mathrm{d}k}{k-k_0 \pm i\epsilon} = \mathcal{P}\int_{-\infty}^{+\infty} \frac{f(k)\,\mathrm{d}k'}{k-k_0} \mp i\pi f(k_0)\,, \tag{26.25}$$

which follows from Cauchy's residue theorem.

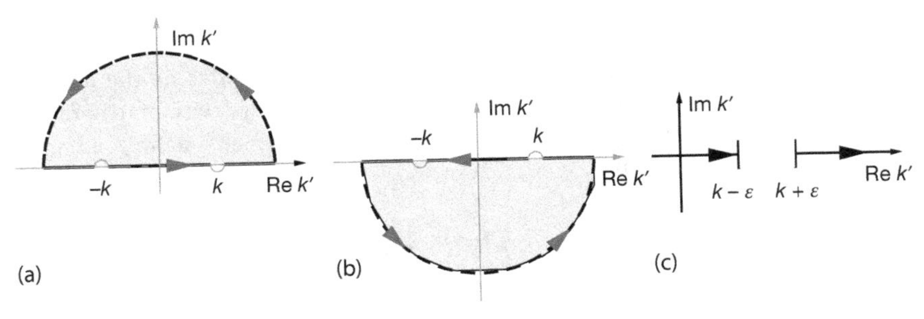

Figure 26.4 Three different paths in the complex k' plane used to evaluate line integrals when there are singularities. Here the singularities are at k and $-k$, and the integration variable is k'.

4) Singh and Thompson (1993) describe a different approach using Maple and Mathematica.

26.4.2
Numerical Principal Values

A direct numerical evaluation of the principal value limit (26.24) is troublesome because of the large cancellations that occur near the singularity. A better algorithm follows from the mathematical theorem

$$\mathcal{P}\int_{-\infty}^{+\infty}\frac{\mathrm{d}k}{k-k_0}=0\,. \tag{26.26}$$

This equation says that the curve of $1/(k-k_0)$ as a function of k has equal and opposite areas on both sides of the singular point k_0. If we break the integral up into one over positive k and one over negative k, a change of variable $k\to -k$ permits us to rewrite (26.26) as

$$\mathcal{P}\int_{0}^{+\infty}\frac{\mathrm{d}k}{k^2-k_0^2}=0\,. \tag{26.27}$$

We observe that the principal-value exclusion of the singular point's contribution to the integral is equivalent to a simple subtraction of the zero integral (26.27):

$$\mathcal{P}\int_{0}^{+\infty}\frac{f(k)\,\mathrm{d}k}{k^2-k_0^2}=\int_{0}^{+\infty}\frac{[f(k)-f(k_0)]\,\mathrm{d}k}{k^2-k_0^2}\,. \tag{26.28}$$

Note that there is no $\mathcal{P}$ on the RHS of (26.28) because the integrand is no longer singular at $k=k_0$ (it is proportional to the $\mathrm{d}f/\mathrm{d}k$) and can therefore be evaluated numerically using the usual rules. The integral (26.28) is called the *Hilbert transform* of f and also arises in subjects like inverse problems.

26.4.3
Reducing Integral Equations to Matrix Equations (Method)

Now that we can handle singular integrals, we go back to reducing the integral equation (26.20) to a set of linear equations that can be solved with matrix methods. We start by rewriting the principal-value prescription as a definite integral (Haftel and Tabakin, 1970):

$$R(k',k)=V(k',k)+\frac{2}{\pi}\int_{0}^{\infty}\mathrm{d}p\frac{p^2V(k',p)R(p,k)-k_0^2V(k',k_0)R(k_0,k)}{(k_0^2-p^2)/2\mu}\,. \tag{26.29}$$

We convert this integral equation to linear equations by approximating the integral as a sum over N integration points (usually Gaussian) k_j with weights w_j:

$$R(k,k_0) \simeq V(k,k_0) + \frac{2}{\pi}\sum_{j=1}^{N}\frac{k_j^2 V(k,k_j)R(k_j,k_0)w_j}{(k_0^2-k_j^2)/2\mu}$$
$$-\frac{2}{\pi}k_0^2 V(k,k_0)R(k_0,k_0)\sum_{m=1}^{N}\frac{w_m}{(k_0^2-k_m^2)/2\mu}\,. \tag{26.30}$$

We note that the last term in (26.30) implements the principal-value prescription and cancels the singular behavior of the previous term. Equation 26.30 contains the $(N+1)$ unknowns $R(k_j,k_0)$ for $j=0,N$. We turn it into $(N+1)$ simultaneous equations by evaluating it for $(N+1)$ k values on a grid (Figure 26.2) consisting of the observable momentum k_0 and the integration points:

$$k = k_i = \begin{cases} k_j\,, & j=1,N \quad \text{(quadrature points)}\,,\\ k_0\,, & i=0 \quad \text{(observable point)}\,. \end{cases} \tag{26.31}$$

There are now $(N+1)$ linear equations for $(N+1)$ unknowns $R_i \equiv R(k_i,k_0)$:

$$R_i = V_i + \frac{2}{\pi}\sum_{j=1}^{N}\frac{k_j^2 V_{ij}R_j w_j}{(k_0^2-k_j^2)/2\mu} - \frac{2}{\pi}k_0^2 V_{i0}R_0\sum_{m=1}^{N}\frac{w_m}{(k_0^2-k_m^2)/2\mu}\,. \tag{26.32}$$

We express these equations in the matrix form by combining the denominators and weights into a single denominator vector D:

$$D_i = \begin{cases} +\frac{2}{\pi}\frac{w_i k_i^2}{(k_0^2-k_i^2)/2\mu}\,, & \text{for} \quad i=1,N\,,\\ -\frac{2}{\pi}\sum_{j=1}^{N}\frac{w_j k_0^2}{(k_0^2-k_j^2)/2\mu}\,, & \text{for} \quad i=0\,. \end{cases} \tag{26.33}$$

The linear equations (26.32) now assume the matrix form

$$R - DVR = [1-DV]R = V\,, \tag{26.34}$$

where R and V are *column vectors* of length $N+1$:

$$[R] = \begin{pmatrix} R_{0,0}\\ R_{1,0}\\ \ddots\\ R_{N,0}\end{pmatrix}, \quad [V] = \begin{pmatrix} V_{0,0}\\ V_{1,0}\\ \ddots\\ V_{N,0}\end{pmatrix}. \tag{26.35}$$

We call the matrix $[1-DV]$ in (26.34) the wave matrix F and write the integral equation as the matrix equation

$$[F][R] = [V]\,, \quad F_{ij} = \delta_{ij} - D_j V_{ij}\,. \tag{26.36}$$

With R the unknown vector, (26.36) is in the standard form $AX = B$, which can be solved by the mathematical subroutine libraries discussed in Chapter 6.

26.4.4 Solution via Inversion, Elimination

An elegant (but alas not most efficient) solution to (26.36) is by matrix inversion:

$$[R] = [F]^{-1}[V] . \tag{26.37}$$

Because the inversion of even complex matrices is a standard routine in mathematical libraries, (26.37) is a *direct solution* for the R amplitude. Unless you need the inverse for other purposes (like calculating wave functions), a more efficient approach is to use Gaussian *elimination* to find an $[R]$ that solves $[F][R] = [V]$ without computing the inverse.

Listing 26.2 Scatt.py solves the Lippmann–Schwinger integral equation for scattering from a delta-shell potential. The singular integral equations are regularized by a subtraction, converted to matrix equations using Gaussian grid points, and then solved with matrix library routines.

```
# Scatt.py:     Soln p space Lippmann Schwinger for scattering

from visual import *
from visual.graph import *
import numpy.linalg as lina                          # Numpy's LinearAlgebra

def gauss(npts, job, a, b, x, w):
    m = i = j = t = t1 = pp = p1 = p2 = p3 = 0.
    eps = 3.E-14                      # Accuracy: ******ADJUST THIS*******!
    m = (npts + 1)/2
    for i in arange(1, m + 1):
        t = cos(math.pi*(float(i) - 0.25)/(float(npts) + 0.5) )
        t1 = 1
        while( (abs(t - t1) ) >= eps):
            p1 = 1. ;  p2 = 0.
            for j in range(1, npts + 1):
                p3 = p2;    p2 = p1
                p1 = ((2.*float(j)-1)*t*p2 - (float(j)-1.)*p3)/(float(j))
            pp = npts*(t*p1 - p2)/(t*t - 1.)
            t1 = t; t = t1  -  p1/pp
        x[i - 1] = - t;    x[npts - i] = t
        w[i - 1] = 2./( (1. - t*t)*pp*pp)
        w[npts - i] = w[i - 1]
    if (job == 0):
        for i in range(0, npts):
            x[i] = x[i]*(b - a)/2. + (b + a)/2.
            w[i] = w[i]*(b - a)/2.
    if (job == 1):
        for i in range(0, npts):
            xi   = x[i]
            x[i] = a*b*(1. + xi) / (b + a - (b - a)*xi)
            w[i] = w[i]*2.*a*b*b/( (b + a - (b-a)*xi)*(b + a - (b-a)*xi))
    if (job == 2):
        for i in range(0, npts):
            xi = x[i]
            x[i] = (b*xi +  b + a + a) / (1. - xi)
            w[i] = w[i]*2.*(a + b)/( (1. - xi)*(1. - xi) )

graphscatt = gdisplay(x=0, y=0, xmin=0, xmax=6,ymin=0, ymax=1, width=600,
    height=400,
```

```
title='S Wave Cross Section vs E', xtitle='kb', ytitle='[sin(delta)]**2')
sin2plot = gcurve(color=color.yellow)
M = 27;                  b = 10.0;           n = 26
k = zeros((M),float);          x = zeros((M),float);          w =
    zeros((M),float)
Finv = zeros((M,M),float);     F = zeros((M,M), float);     D =
    zeros((M),float)
V = zeros((M), float);         Vvec = zeros((n+1,1),float)
scale = n/2;                   lambd = 1.5

gauss(n, 2, 0., scale, k, w)                        # Set up points & wts
ko = 0.02
for m in range(1,901):
    k[n] = ko
    for i in range (0, n):  D[i]=2/pi*w[i]*k[i]*k[i]/(k[i]*k[i]-ko*ko) #D
    D[n] = 0.
    for  j in range(0,n):   D[n]=D[n]+w[j]*ko*ko/(k[j]*k[j]-ko*ko)
    D[n] = D[n]*(-2./pi)
    for i in range(0,n+1):                                 # Set up F & V
        for j in range(0,n+1):
            pot = -b*b * lambd * sin(b*k[i])*sin(b*k[j])/(k[i]*b*k[j]*b)
            F[i][j] = pot*D[j]
            if i==j: F[i][j] = F[i][j] + 1.
        V[i] = pot
    for  i in range(0,n+1):  Vvec[i][0]= V[i]
    Finv = lina.inv(F)                         # LinearAlgebra for inverse
    R = dot(Finv, Vvec)                                # Matrix multiply
    RN1 = R[n][0]
    shift = atan(-RN1*ko)
    sin2 = (sin(shift))**2
    sin2plot.plot(pos = (ko*b,sin2))                   # Plot sin**2(delta)
    ko = ko + 0.2*pi/1000.
print("Done")
```

26.4.5 Scattering Implementation

For the scattering problem, we use the same delta-shell potential (26.16) discussed in Section 26.2.2 for bound states:

$$V(k', k) = \frac{-|\lambda|}{2\mu k' k} \sin(k' b) \sin(kb) . \tag{26.38}$$

This is one of the few potentials for which the Lippmann–Schwinger equation (26.20) has an analytic solution (Gottfried, 1966) with which to check:

$$x \tan \delta_0 = \frac{\lambda b \sin^2(kb)}{kb - \lambda b \sin(kb) \cos(kb)} . \tag{26.39}$$

Our results were obtained with $2\mu = 1$, $\lambda b = 15$, and $b = 10$, the same as in (Gottfried, 1966). In Figure 26.5, we give a plot of $\sin^2 \delta_0$ vs. kb, which is proportional to the scattering cross section arising from the angular momentum $l = 0$ phase shift. It is seen to reach its maximum values at energies corresponding to resonances. In Listing 26.2, we present our program for solving the scattering integral equation using the NumPy Linear Algebra matrix library and the gauss method for quadrature points. For your implementation:

1. Set up the matrices `V[]`, `D[]`, and `F[,]`. Use at least $N = 16$ Gaussian quadrature points for your grid.
2. Calculate the matrix F^{-1} using a library subroutine.
3. Calculate the vector $\boldsymbol{R}$ by matrix multiplication $\boldsymbol{R} = F^{-1}V$.
4. Deduce the phase shift δ from the $i = 0$ element of R:

$$R(k_0, k_0) = R_{0,0} = -\frac{\tan\delta}{\rho}, \quad \rho = 2\mu k_0 . \tag{26.40}$$

5. Estimate the precision of your solution by increasing the number of grid point in steps of two (we found the best answer for $N = 26$). If your phase shift changes in the second or third decimal place, you probably have that much precision.
6. Plot $\sin^2\delta$ vs. energy $E = k_0^2/2\mu$ starting at zero energy and ending at energies where the phase shift is again small. Your results should be similar to those in Figure 26.5. Note that a *resonance* occurs when δ_l increases rapidly through $\pi/2$, that is, when $\sin^2\delta_0 = 1$.
7. Check your answer against the analytic results (26.39).

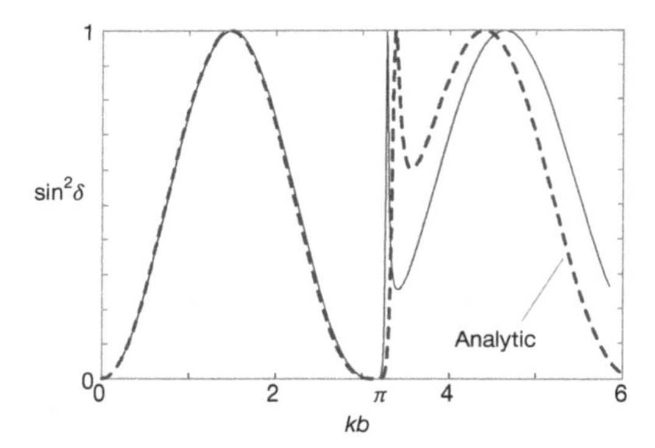

Figure 26.5 The energy dependence of the cross section for $l = 0$ scattering from an attractive delta-shell potential with $\lambda b = 15$. The dashed curve is the analytic solution (26.39), and the solid curve results from numerically solving the integral Schrödinger equation, either via direct matrix inversion or via LU decomposition.

26.4.6
Scattering Wave Function (Exploration)

The F^{-1} matrix that occurred in our solution to the integral equation

$$R = F^{-1}V = (1 - VG)^{-1}V \tag{26.41}$$

is actually quite useful. In scattering theory, it is known as the *wave matrix* because it is used in expansion of the wave function:

$$u(r) = N_0 \sum_{i=1}^{N} \frac{\sin(k_i r)}{k_i r} F(k_i, k_0)^{-1} . \tag{26.42}$$

Here N_0 is a normalization constant and the R amplitude is appropriate for standing-wave boundary conditions. So once we know F^1, we also know the wave function.

1. Plot $u(r)$ and compare it to a free wave.

A
Codes, Applets, and Animations

Table A.1 Python codes, Chapters 1–26.

Name	Listing No.	Description	Name	Listing No.	Description
EasyVisual	1.1	Visual easy plot	3GraphVisual	1.2	Visual multiplots
3Dshapes	1.3	Visual 3D shapes	EasyMatPlot	1.4	Matplot 2D
GradesMatPlot	1.5	Matplot multiplots	MatPlot2figs	1.6	Matplot multiplots
PondMatPlot	1.7	Matplot scatter plot	Simple3Dplot	1.8	Matplot surface
Scatter3dPlot	1.9	Matplot 3D scatter	EqHeatAnimate	1.10	Matplot animation
MayaLines	1.12	Mayavi flow lines	MayaYlm	1.10	Mayavi surface
Area	2.1	Simple screen I/O	AreaFormatted	2.2	Formatted I/O
Directives	2.3	I/O directives, escape	Limits	2.4	Machine precision
Bessel	3.1	Downward recursion	Walk	4.1	Random walk
Walk3D	4.1	3D random walk	DecaySound	4.2	Spontaneous decay
TrapMethods	5.1	Trapezoid rule	IntegGauss	5.2	Gaussian quadrature
vonNeuman	5.3	von Neumann rejection	Eigen	6.34	Matrix eigenvalues
Matrix	6.5	Matrix array mult	NewtonNDanimate	6.1	N-D Newton–Raphson
Bisection	7.1	Bisection algorithm	NewtonCD	7.2	Newton–Raphson search
Lagrange	7.5	Lagrange interpolation	Spline	7.3	Spline fitting
SplineInteract	7.3	Interactive splines	Fit	7.4	Least-squares fitting
rk4	8.1	rk4 ODE solver	rk45	8.2	Adaptive step rk4
ABM	8.3	ABM ODE solver	QuantumNumerov	9.1	Schrödinger equation
QuantumEigen	9.2	Quantum eigen rk4	ProjectileAir	9.3	Projectile with drag
TuneNumPy	11.5	NumPy vectors	Tune.f90	11.6	Fortran tuning
Tune	11.7	Python tuning	Tune4	11.8	Python loop unroll
tune4.f95	11.9	Fortran loop unrolling	SumArraysCuda	11.16	CUDA GPU program
SumArraysCuda2	11.17	GPU blocks	DFTcomplex	12.1	Complex DFT
DFTreal	12.2	Real DFT	FourierMatplot	12.2	Interactive DFT
NoiseSincFilter	12.4	Fourier filtering	FFT.py	12.3	Fast Fourier transform
FFTappl	12.3	FFT + graphs	CWT	13.1	Continuous wavelets
DWT	13.2	Discrete wavelets	Bugs	14.1	Logistic bifurcations

Computational Physics, 3rd edition. Rubin H. Landau, Manuel J. Páez, Cristian C. Bordeianu.

Table A.1 (continued).

Name	Listing No.	Description	Name	Listing No.	Description
LyapLog	14.2	Lyapunov coefficient	Entropy	14.3	Shannon entropy
PredatorPrey	14.4	Population dynamics	Fern3D	16.1	3D fern
Column	16.2	Column growth	Gameoflife	16.3	Game of life
Islands.pov	16.4	Ray tracing	Film	online	Film deposition
Coastline	online	Box counting	DLA	16.7	Aggregation
Fern	16.1	1D fern	Sierpin	16.2	Sierpinsky gasket
IsingViz	17.1	Ising model	WangLandau	17.2	Wang–Landau MC
QMC	17.3	Quantum MC	QMCbouncer	17.4	QMC bouncer
MD	18.1	1D MD	MD2D	18.1	2D MD
LaplaceLine	19.1	Laplace equation	EqHeat	20.1	Heat equation solution
EqHeat	20.1	Heat equation	HeatCNTridiag	20.2	Better heat
EqStringAnimate	21.1	Wave equation	EqStringMatPlot.py	online	Waves with Matplot
Waves2D	21.2	2D wave equation	Waves2Danal	online	Analytic membrane
HarmosAnimate	22.1	Quantum packet	FDTD	22.2	Finite difference time domain
CircPolarztn	22.3	FDTD circular			
LaplaceFEM_1D	23.1	Finite element 1D	LaplaceFEM_2D	23.2	Finite element 2D
AdvecLax	24.1	Advection equation	Soliton	24.2	KdeV solitons
SolitonAnimate	online	Soliton movie	Beam	25.1	Navier–Stokes equation
BeamContour	25.4	Flow contours	Bound	26.1	Integral equation eigen
Scatt	26.2	Integral equation scatter			

Table A.2 Animations (A player such as VLC or QuickTime required for mpeg and avi, and a Web browser for gifs).

Directory	Chapter	Directory	Chapter
DoublePendulum (see also applets)	15	Fractals (see also applets)	16
MapleWaveMovie (requires Maple)	21	Laplace (DX movie)	19
MD	18	TwoSlits	22
2D solitons	21,25	Utilities (scripts, colormaps)	
Waves (animated gifs need browser)	21		

Bibliography

Abarbanel, H.D.I., Rabinovich, M.I., and Sushchik, M.M. (1993) *Introduction to Nonlinear Dynamics for Physicists*, World Scientific, Singapore.

Abramowitz, M. and Stegun, I.A. (1972) *Handbook of Mathematical Functions*, 10th edn, US Govt. Printing Office, Washington.

Addison, P.S. (2002) *The Illustrated Wavelet Transform Handbook*, Institute of Physics Publishing, Bristol and Philadelphia.

Allan, M.P. and Tildesley, J.P. (1987) *Computer Simulations of Liquids*, Oxford Science Publications, Oxford.

Amdahl, G. (1967) *Validity of the Single-Processor Approach to Achieving Large-Scale Computing Capabilities*, Proc. AFIPS, p. 483.

Ancona, M.G. (2002) *Computational Methods for Applied Science and Engineering*, Rinton Press, Princeton.

Anderson, E., Bai, Z., Bischof, C., Demmel, J., Dongarra, J., Du Croz, J., Greenbaum, A., Hammarling, S., McKenney, A., Ostrouchov, S., and Sorensen, D. (2013) *LAPACK Users' Guide*, 3rd edn, SIAM, Philadelphia, www.netlib.org (accessed 22 March 2015).

Anderson, J.A., Lorenz, C.D., and Travesset, A. (2008) HOOMD-blue, general purpose molecular dynamics simulations. *J. Comput. Phys.*, **227** (10), 5342, codeblue.umich.edu/hoomd-blue (accessed 22 March 2015).

Arfken, G.B. and Weber, H.J. (2001) *Mathematical Methods for Physicists*, Harcourt/Academic Press, San Diego.

Argyris, J., Haase, M., and Heinrich, J.C. (1991) *Comput. Methods Appl. Mech. Eng.*, **86**, 1.

Armin, B. and Shlomo, H. (eds) (1991) *Fractals and Disordered Systems*, Springer, Berlin.

Askar, A. and Cakmak, A.S. (1977) *J. Chem. Phys.*, **68**, 2794.

Banacloche, J.G. (1999) A quantum bouncing ball. *Am. J. Phys.*, **67**, 776.

Barnsley, M.F. and Hurd, L.P. (1992) *Fractal Image Compression*, A.K. Peters, Wellesley.

Beazley, D.M. (2009) *Python Essential Reference*, 4th edn, Addison-Wesley, Reading, MA, USA.

Becker, R.A. (1954) *Introduction to Theoretical Mechanics*, McGraw-Hill, New York.

Bevington, P.R. and Robinson, D.K. (2002) *Data Reduction and Error Analysis for the Physical Sciences*, 3rd edn, McGraw-Hill, New York.

Bleher, S., Grebogi, C., and Ott, E. (1990) Bifurcations in chaotic scattering. *Physica D*, **46**, 87.

Briggs, W.L. and Henson, V.E. (1995) *The DFT, An Owner's Manual*, SIAM, Philadelphia.

Bunde, A. and Havlin, S. (eds) (1991) *Fractals and Disordered Systems*, Springer, Berlin.

Burgers, J.M. (1974) *The Non-Linear Diffusion Equation; Asymptotic Solutions and Stattistical Problems*, Reidel, Boston.

Car, R. and Parrinello, M. (1985) *Phys. Rev. Lett.*, **55**, 2471.

Cencini, M., Ceconni, F. and Vulpiani, A. (2010) *Chaos From Simple Models To Complex Systems*, World Scientific, Singapore.

Christiansen, P.L. and Lomdahl, P.S. (1981) *Physica D*, **2**, 482.

Christiansen, P.L. and Olsen, O.H. (1978) *Phys. Lett. A*, **68**, 185; Christiansen, P.L. and Olsen, O.H. (1979) *Phys. Scr.*, **20**, 531.

Clark University (2011) *Statistical and Thermal Physics Curriculum Development Project*, stp.clarku.edu/ (accessed 22 March 2015); *Density of States of the 2D Ising Model.*

CPUG, Computational Physics degree program for Undergraduates (2009), physics.oregonstate.edu/CPUG (accessed 22 March 2015).

Crank, J. and Nicolson, P. (1946) *Proc. Cambridge Philos. Soc.*, **43**, 50.

Cooley, J.W. and Tukey, J.W. (1965) *Math. Comput.*, **19**, 297.

Courant, R., Friedrichs, K., and Lewy, H. (1928) *Math. Ann.*, **100**, 32.

Critchley, S. (2014) *The Dangers of Certainty: A Lesson from Auschwitz*, New York Times, New York.

Danielson, G.C. and Lanczos, C. (1942) *J. Franklin Inst.*, **233**, 365.

Daubechies, I. (1995) *Wavelets and other phase domain localization methods*, Proc. Int. Congr. Math., **1, 2**, Basel, 56, Birkhäuser, Basel.

DeJong, M.L. (1992) Chaos and the simple pendulum. *Phys. Teach.*, **30**, 115.

Dongarra, J. (2011) *On the Future of High Performance Computing: How to Think for Peta and Exascale Computing*, Conference on Computational Physics 2011, Gatlinburg; *Emerging Technologies for High Performance Computing*, GPU Club presentation, University of Manchester, www.netlib.org/utk/people/JackDongarra/SLIDES/gpu-0711.pdf (accessed 22 March 2015).

Dongarra, J., Sterling, T., Simon, H., and Strohmaier, E. (2005) High-performance computing. *Comput. Sci. Eng.*, **7**, 51.

Dongarra, J., Hittinger, J., Bell, J., Chacson, L., Falgout, R., Heroux, M., Hovland, P., Ng, E., Webster, C., and Wild, S. (2014) *Applied Mathematics Research for Exascale Computing*, US Department of Energy Report, http://www.osti.gov/bridge (accessed 22 March 2015).

Donnelly, D. and Rust, B. (2005) The fast Fourier transform for experimentalists. *Comput. Sci. Eng.*, **7**, 71.

Eclipse an open development platform (2014) www.eclipse.org (accessed 22 March 2015).

Ercolessi, F. (1997) A molecular dynamics primer, www.ud.infn.it/~ercolessi/md/ (accessed 22 March 2015).

Faber, R. (2010) CUDA, Supercomputing for the Masses: Part 15, www.drdobbs.com/architecture-and-design/cuda-supercomputing-for-the-masses-part/222600097 (accessed 22 March 2015).

Falkovich, G. and Sreenivasan, K.R. (2006) Lesson from hydrodynamic turbulence. *Phys. Today*, **59**, 43.

Family, F. and Vicsek, T. (1985) *J. Phys. A*, **18**, L75.

Feigenbaum, M.J. (1979) *J. Stat. Phys.*, **21**, 669.

Fetter, A.L. and Walecka, J.D. (1980) *Theoretical Mechanics of Particles and Continua*, McGraw-Hill, New York.

Feynman, R.P. and Hibbs, A.R. (1965) *Quantum Mechanics and Path Integrals*, McGraw-Hill, New York.

Fitzgerald, R. (2004) New experiments set the scale for the onset of turbulence in pipe flow. *Phys. Today*, **57**, 21.

Fosdick L.D., Jessup, E.R. Schauble, C.J.C., and Domik, G. (1996) *An Introduction to High Performance Scientific Computing*, MIT Press, Cambridge.

Fox, G. (1994) *Parallel Computing Works*! Morgan Kaufmann, San Diego.

Gara, A., Blumrich, M.A., Chen, D., Chiu, G.L.-T., Coteus, P., Giampapa, M.E., Haring, R.A., Heidelberger, P., Hoenicke, D., Kopcsay, G.V., Liebsch, T.A., Ohmacht, M., Steinmacher-Burow, B.D., Takken, T., and Vranas, P. (2005) Overview of the Blue Gene/L system architecure. *IBM J. Res Dev.*, **49**, 195; Feldman, M., IBM Specs Out Blue Gene/Q Chip, (2011) *HPC Wire*, August 22 2011.

Garcia, A.L. (2000) *Numerical Methods for Physics*, 2nd edn, Prentice-Hall, Upper Saddle River, NJ, USA.

Gibbs, R.L. (1975) The quantum bouncer. *Am. J. Phys.*, **43**, 25.

Gnuplot (2014) gnuplot homepage www.gnuplot.info (accessed 22 March 2015).

Goldberg, A., Schey, H.M., and Schwartz, J.L. (1967) Computer-generated motion pictures of one-dimensional quantum-mechanical transmission and reflection phenomena. *Am. J. Phys.*, **35**, 177–186.

Goodings, D.A. and Szeredi, T. (1992) The quantum bouncer by the path integral method. *Am. J. Phys.*, **59**, 924.

Goswani, J.C. and Chan, A.K. (1999) *Fundamentals of Wavelets*, John Wiley & Sons, New York.

Gottfried, K. (1966) *Quantum Mechanics*, Benjamin, New York.

Gould, H., Tobochnik, J., and Christian, W. (2006) *An Introduction to Computer Simulations Methods*, 3rd edn, Addison-Wesley, Reading, USA.

Graps, A. (1995) An introduction to wavelets. *Comput. Sci. Eng.*, **2**, 50.

Gurney, W.S.C. and Nisbet, R.M. (1998) *Ecological Dynamics*, Oxford University Press, Oxford.

Haftel, M.I. and Tabakin, F. (1970) *Nucl. Phys.*, **158**, 1.

Hardwich, J. (1996) Rules for Optimization, www.cs.cmu.edu/~jch/java (accessed 22 March 2015).

Hartmann, W.M. (1998) *Signals, Sound, and Sensation*, AIP Press, Springer, New York.

Higgins, R.J. (1976) Fast Fourier transform: An introduction with some minicomputer experiments. *Am. J. Phys.*, **44**, 766.

Hildebrand, F.B. (1956) *Introduction to Numerical Analysis*, McGraw-Hill, New York.

Hinsen, K. (2013) Software development for reproducible research. *Comput. Sci. Eng*, **4** (15), 60–63, www.computer.org/portal/web/cise/home (accessed 22 March 2015).

History of Python (2009) The History of Python python-history.blogspot.com/2009/01/brief-timeline-of-python.html (accessed 22 March 2015).

Hockney, R.W. and J.W. Eastwood (1988) *Computer Simulation Using Particles*, Adam Hilger, Bristol.

Hubble, E. (1929) A relation between distance and radial velocity among extra-galactic nebulae. *Proc. Natl. Acad. Sci. USA*, **15** (3), 168.

Hunag, K. (1987) *Statistical Mechanics*, John Wiley & Sons, New York.

Jackson, J.D. (1988) *Classical Electrodynamics*, 3rd edn, John Wiley & Sons, New York.

Jackson, J.E. (1988) *A User's Guide to Principal Components*, John Wiley & Sons, New York.

Jolliffe, I.Y. (2001) *Principal Component Analysis*, 2nd edn, Springer, New York.

José, J.V. and Salatan, E.J. (1988) *Classical Dynamics*, Cambridge University Press, Cambridge.

Kennedy, R. (2006) *The case of Pollock's Fractals Focuses on Physics*, New York Times, 2, 5 December 2006.

Kirk, D. and Wen-Mei, W.H. (2013) *Programming Massively Parallel Processors*, 2nd edn, Morgan Kauffman, Waltham.

Kittel, C. (2005) *Introduction to Solid State Physics*, 8th edn, John Wiley & Sons, Inc., Hoboken.

Klöckner, A. (2014) PyCUDA, mathema.tician.de/software/pycuda (accessed 22 March 2015).

Koonin, S.E. (1986) *Computational Physics*, Benjamin, Menlo Park, CA.

Korteweg, D.J. and deVries, G. (1895) *Philos. Mag.*, **39**, 4.

Kreyszig, E. (1998) *Advanced Engineering Mathematics*, 8th edn, John Wiley & Sons, New York.

Lamb, H. (1993) *Hydrodynamics*, 6th edn, Cambridge University Press, Cambridge.

Landau, D.P. and Wang, F. (2001) Determining the density of states for classical statistical models: A random walk algorithm to produce a flat histogram. *Phys. Rev. E*, **64**, 056101; Landau, D.P., Tsai, S.-H., and Exler, M. (2004) A new approach to Monte Carlo simulations in statistical physics: Wang–Landau sampling. *Am. J. Phys.*, **72**, 1294.

Landau, L.D. and Lifshitz, E.M. (1987) *Fluid Mechanics*, 2nd edn, Butterworth-Heinemann, Oxford.

Landau, L.D. and Lifshitz, E.M. (1976) *Quantum Mechanics*, Pergamon, Oxford.

Landau, L.D. and Lifshitz, E.M. (1976) *Mechanics*, 3rd edn, Butterworth-Heinemann, Oxford.

Landau, R.H. (2008) Resource letter CP-2: Computational physics. *Am. J. Phys.*, **76**, 296.

Landau, R.H. (2005) *A First Course in Scientific Computing*, Princeton University Press, Princeton.

Landau, R.H. (1996) *Quantum Mechanics II, A Second Course in Quantum Theory*, 2nd edn, John Wiley & Sons, New York.

Lang, W.C. and Forinash, K. (1998) Time-frequency analysis with the continuous wavelet transform. *Am. J. Phys.*, **66**, 794.

Langtangen, H.P. (2008) *Python Scripting for Computational Science*, Springer, Heidelberg.

Langtangen, H.P. (2009) *A Primer on Scientific Programming with Python*, Springer, Heidelberg.

Li, Z. (2014) Numerical Methods for Partial Differential Equations – Finite Element Method, www4.ncsu.edu/~zhilin/ (accessed 22 March 2015).

Lorenz, E.N. (1963) Deterministic nonperiodic flow. *J. Atmos. Sci.*, **20**,130.

Lotka, A.J. (1925) *Elements of Physical Biology*, Williams and Wilkins, Baltimore.

MacKeown, P.K. (1985) *Am. J. Phys.*, **53**, 880.

MacKeown, P.K. and Newman, D.J. (1987) *Computational Techniques in Physics*, Adam Hilger, Bristol.

Maestri, J.J.V., Landau, R.H., and Páez, M.J. (2000) Two-particle Schrödinger equation animations of wave packet-wave packet scattering. *Am. J. Phys.*, **68**, 1113; http://physics.oregonstate.edu/~rubin/nacphy/ComPhys/PACKETS/.

Mallat, P.G. (1982) A theory for multiresolution signal decomposition: The wavelet representation. *IEEE Trans. Pattern Anal. Mach. Intell.*, **11** (7), 674.

Mandelbrot, B. (1967) How long is the coast of Britain? *Science*, **156**, 638.

Mandelbrot, B. (1982) *The Fractal Geometry of Nature*, Freeman, San Francisco.

Manneville, P. (1990) *Dissipative Structures and Weak Turbulence*, Academic Press, San Diego.

Mannheim, P.D. (1983) The physics behind path integrals in quantum mechanics. *Am. J. Phys.*, **51**, 328.

Marion, J.B. and Thornton, S.T. (2003) *Classical Dynamics of Particles and Systems*, 5th edn, Harcourt Brace Jovanovich, Orlando.

Mathews, J. (2002) *Numerical Methods for Mathematics, Science and Engineering*, Prentice-Hall, Upper Saddle River.

Metropolis, M., Rosenbluth, A.W., Rosenbluth, M.N., Teller, A.H., and Teller, E. (1953) *J. Chem. Phys.*, **21**, 1087.

Moon, F.C. and Li, G.-X. (1985) *Phys. Rev. Lett.*, **55**, 1439.

Morse, P.M. and Feshbach, H. (1953) *Methods of Theoretical Physics*, McGraw-Hill, New York.

Motter, A. and Campbell, D. (2013) Chaos at fifty. *Phys. Today*, **66** (5), 27.

Nelson, M., Humphrey, W., Gursoy, A., Dalke, A., Kalé, L., Skeel, R.D., and Schulten, K. (1996) NAMD – Scalable Molecular Dynamics. *J. Supercomput. Apps. High Perform. Comput.*, **10**, 251–268, www.ks.uiuc.edu/Research/namd (accessed 22 March 2015).

Nesvizhevsky, V.V., Borner, H.G., Petukhov, A.K., Abele, H., Baessler, S., Ruess, F.J., Stoferle, T., Westphal, A., Gagarski, A.M., Petrov, G.A., and Strelkov, A.V. (2002) Quantum states of neutrons in the Earth's gravitational field. *Nature*, **415**, 297.

NIST Digital Library of Mathematical Functions (2014) dlmf.nist.gov/ (accessed 22 March 2015).

Numerical Python (2013) NumPy numpy.scipy.org (accessed 22 March 2015).

NumPy Tutorial, Tentative (2015) Tentative NumPy Tutorial wiki.scipy.org/Tentative_NumPy_Tutorial (accessed 22 March 2015).

Oliphant, T.E. (2006) *Guide to NumPy*, csc.ucdavis.edu/~chaos/courses/nlp/Software/NumPyBook.pdf (accessed 22 March 2015).

Ott, E. (2002) *Chaos in Dynamical Systems*, Cambridge University Press, Cambridge.

Otto A. (2011) Numerical Simulations of Fluids and Plasmas, how.gi.alaska.edu/ao/sim (accessed 22 March 2015).

Pancake, C.M. (1996) Is parallelism for you?, *Comput. Sci. Eng.*, **3**, 18.

Peitgen, H.-O., Jürgens, H., and Saupe, D. (1992) *Chaos and Fractals*, Springer, New York.

Penna, T.J.P. (1994) *Comput. Phys.*, **9**, 341.

Perez, F., Granger, B.E. and Hunter, J.D. (2010) Python: An Ecosystem for Scientifc Computing. *Comput. Sci. Eng.*, **13** (2), www.computer.org/web/computingnow/cise (accessed 22 March 2015).

Perlin, K. (1985) An Image Synthesizer, *Computer Graphics* (Proceedings of ACM SIGGRAPH 85) **24**, 3.

Phatak, S.C. and Rao, S.S. (1995) Logistic map: A possible random-number generator. *Phys. Rev. E*, **51**, 3670.

Plischke, M. and Bergersen, B. (1994) *Equilibrium Statistical Physics*, 2nd edn, World Scientific, Singapore.

Polikar, R. (2001) The Wavelet Tutorial, users.rowan.edu/~polikar/WAVELETS/WTtutorial.html (accessed 22 March 2015).

Polycarpou, A.C. (2006) *Introduction to the Finite Element Method in Electromagnetics*, Morgan and Claypool, San Rafael.

Potvin, J. (1993) *Comput. Phys.*, **7**, 149.

(2013) Pov-Ray, Persistence of Vision Raytracer, www.povray.org (accessed 22 March 2015).

Press, W.H., Flannery, B.P., Teukolsky, S.A., and Vetterling, W.T. (1994) *Numerical Recipes*, Cambridge University Press, Cambridge.

Python (2014) Python for Programmers, https://wiki.python.org/moin/BeginnersGuide/Programmers (accessed 22 March 2015).

LearnPython.org (2014) Interactive Python Tutorial, http://www.learnpython.org/ (accessed 22 March 2015).

(2014) The Python Tutorial, docs.python.org/2/tutorial/ (accessed 22 March 2015).

(2014) Python Index of Packages, pypi.python.org/pypi (accessed 22 March 2015).

(2014) Python Documentation, www.python.org/doc (accessed 22 March 2015).

Quinn, M.J. (2004) *Parallel Programming in C with MPI and OpenMP*, McGraw-Hill, New York.

Ramasubramanian, K. and Sriram, M.S. (2000) A comparative study of computation of Lyapunov spectra with different algorithms. *Physica D*, **139**, 72.

Rapaport, D.C. (1995) *The Art of Molecular Dynamics Simulation*, Cambridge University Press, Cambridge.

Rasband, S.N. (1990) *Chaotic Dynamics of Nonlinear Systems*, John Wiley & Sons, New York.

Rawitscher, G., Koltracht, I., Dai, H., and Ribetti, C. (1996) *Comput. Phys.*, **10**, 335.

Reddy, J.N. (1993) *An Introduction to the Finite Element Method*, 2nd edn, McGraw-Hill, New York.

Refson, K. (2000) Moldy, A General-Purpose Molecular Dynamics Simulation Program, cc-ipcp.icp.ac.ru/Moldy_2_16.html (accessed 22 March 2015).

Reynolds, O. (1883) *Proc. R. Soc. Lond.*, **35**, 84.

Richardson. L.F. (1961) Problem of contiguity: an appendix of statistics of deadly quarrels. *General Syst. Yearbook*, **6**, 139.

Rowe, A.C.H. and Abbott, P.C. (1995) Daubechies wavelets and mathematica. *Comput. Phys.*, **9**, 635.

Russell, J.S. (1844) *Report of the 14th Meeting of the British Association for the Advancement of Science*, John Murray, London.

Sander, E., Sander, L.M., and Ziff, R.M. (1994) *Comput. Phys.*, **8**, 420.

Sanders, J. and Kandrot, E. (2011) *Cuda by Example*, Addison Wesley, Upper Saddle River.

Satoh, A. (2011) *Introduction to Practice of Molecular Simulation*, Elsevier, Amsterdam.

Scheck, F. (1994) *Mechanics, from Newton's Laws to Deterministic Chaos*, 2nd edn, Springer, New York.

Shannon, C.E. (1948) A mathematical theory of communication. *Bell Syst. Tech. J.*, **27**, 379.

(2014) SciPy, a Python-based ecosystem, www.scipy.org (accessed 22 March 2015).

Shaw C.T. (1992) *Using Computational Fluid Dynamics*, Prentice-Hall, Englewood Cliffs, NJ.

Singh, P.P. and Thompson, W.J. (1993) *Comput. Phys.*, **7**, 388.

Sipper, M. (1997) *Evolution of Parallel Cellular Machines*, Springer, Heidelberg, cell-auto.com (accessed 22 March 2015).

Smith, D.N. (1991) *Concepts of Object-Oriented Programming*, McGraw-Hill, New York.

Smith, L.I. (2002) A Tutorial on Principal Components Analysis, www.cs.otago.ac.nz/cosc453/student_tutorials/principal_components.pdf (accessed 22 March 2015).

Smith, S.W. (1999) *The Scientist and Engineer's Guide to Digital Signal Processing*, California Technical Publishing, San Diego.

Stetz, A., Carroll, J., Chirapatpimol, N., Dixit, M., Igo, G., Nasser, M., Ortendahl, D., and Perez-Mendez, V. (1973) *Determination of the Axial Vector Form Factor in the Radiative Decay of the Pion*, LBL 1707.

Sullivan, D. (2000) *Electromagnetic Simulations Using the FDTD Methods*, IEEE Press, New York.

Tabor, M. (1989) *Chaos and Integrability in Nonlinear Dynamics*, John Wiley & Sons, New York.

Taflove, A. and Hagness, S. (2000) *Computational Electrodynamics: The Finite Difference Time Domain Method*, 2nd edn, Artech House, Boston.

Tait, R.N., Smy, T., and Brett, M.J. (1990) *Thin Solid Films*, **187**, 375.

Thijssen J.M. (1999) *Computational Physics*, Cambridge University Press, Cambridge.

Thompson, W.J. (1992) *Computing for Scientists and Engineers*, John Wiley & Sons, New York.

Tickner, J. (2004) Simulating nuclear particle transport in stochastic media using Perlin noise functions. *Nucl. Instrum. Methods B*, **203**, 124.

Vallée, O. (2000) Comment on a quantum bouncing ball. *Am. J. Phys.*, **68**, 672.

van de Velde, E.F. (1994) *Concurrent Scientific Computing*, Springer, New York.

van den Berg, J.C. (ed.) (1999) *Wavelets in Physics*, Cambridge University Press, Cambridge.

Vano, J.A., Wildenberg, J.C., Anderson, M.B., Noel, J.K., and Sprott, J.C. (2006) Chaos in low-dimensional Lotka–Volterra models of competition. *Nonlinearity*, **19**, 2391–2404.

Visscher, P.B. (1991) *Comput. Phys.*, **5**, 596.

Vold, M.J. (1959) *J. Colloid Sci.*, **14**, 168.

Volterra, V. (1926) Variazioni e fluttuazioni del numero d'individui in specie animali conviventi. *Mem. R. Accad. Naz. Lincei. Ser. VI*, **2**.

Warburton, R.D.H. and Wang, J. (2004) Analysis of asymptotic projectile motion with air resistance using the Lambert W function. *Am. J. Phys.*, **72**, 1404.

Ward, D.W. and Nelson, K.A. (2005) Finite difference time domain, FDTD, simulations of electromagnetic wave propagation using a spreadsheet. *Comput. Appl. Eng. Educat.*, **13** (3), 213–221.

Whineray, J. (1992) An energy representation approach to the quantum bouncer. *Am. J. Phys.*, **60**, 948.

(2014) *Principal component analysis*, en.wikipedia.org/wiki/Principal_component_analysis (accessed 22 March 2015).

Williams, G.P. (1997) *Chaos Theory Tamed*, Joseph Henry Press, Washington.

Witten, T.A. and Sander, L.M. (1981) *Phys. Rev. Lett.*, **47**, 1400; Witten, T.A. and Sander, L.M. (1983) *Phys. Rev. B*, **27**, 5686.

Wolf, A., Swift, J.B., Swinney, H.L., and Vastano, J.A. (1985) Determining Lyapunov exponents from a time series. *Physica D*, **16**, 285.

Wolfram S. (1983) Statistical mechanics of cellular automata. *Rev. Mod. Phys.*, **55**, 601.

Yang, C.N. (1952) The spontaneous magnetization of a two-dimensional Ising model. *Phys. Rev.*, **85**, 809.

Yee, K. (1966) *IEEE Trans. Ant. Propagat.*, **AP-14**, 302.

Yue, K., Fiebig, K.M., Thomas, P.D., Chan, H.S., Shakhnovich, E.I., and Dill, A. (1995) *Proc. Natl. Acad. Sci. USA*, **92**, 325.

Zabusky, N.J. and Kruskal, M.D. (1965) *Phys. Rev. Lett.*, **15**, 240.

Zeller, C. (2008) High Performance Computing with CUDA, www.nvidia.com/object/sc10_cuda_tutorial.htmlP (accessed 22 March 2015).

Index

CPSIA information can be obtained
at www.ICGtesting.com
Printed in the USA
LVHW050735180223
739615LV00021B/104

9 783527 413157